P9-AQE-220

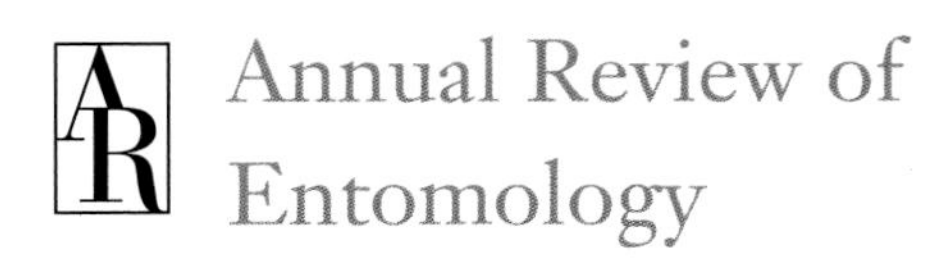
Annual Review of
Entomology

Annual Review of Entomology

Volume 56, 2011

May R. Berenbaum, *Editor*
University of Illinois, Urbana-Champaign

Ring T. Cardé, *Associate Editor*
University of California, Riverside

Gene E. Robinson, *Associate Editor*
University of Illinois, Urbana-Champaign

www.annualreviews.org • science@annualreviews.org • 650-493-4400

Annual Reviews
4139 El Camino Way • P.O. Box 10139 • Palo Alto, California 94303-0139

Annual Reviews
Palo Alto, California, USA

International Standard Serial Number: 0066-4170
International Standard Book Number: 978-0-8243-0156-9
Library of Congress Catalog Card Number: A56-5750

∞ The paper used in this publication meets the minimum requirements of American National Standards for Information Sciences—Permanence of Paper for Printed Library Materials, ANSI Z39.48-1992.

TYPESET BY APTARA
PRINTED AND BOUND BY FRIESENS CORPORATION, ALTONA, MANITOBA, CANADA

Preface

Every scientist would like to publish a paradigm-shattering paper. What we scientists do most of the time is much more modest but no less noble. We publish research papers that incrementally extend, brick by brick, the edifice that is the sum total of scientific knowledge, an edifice built over centuries and millennia by all our colleagues who came before. Although this shared human enterprise has been punctuated by occasional setbacks (think Library of Alexandria), our current "global information database" is the product of the most productive period of uninterrupted progress in history, with a growth rate that is arguably exponential. It's an exhilarating time to be doing science, but it's also a daunting one. As scientists, it is not only our desire but also our duty to keep informed of progress within our fields of concentration and in closely related fields. On top of that, most of us are interested in other, more distant fields as well. Even as the information content of each field increases, the number of fields also increases. How do we keep up with the unrelenting torrent of new information?

Information overload is a general problem for society that extends well beyond science. Very clever people continue to invent very clever systems for organizing and interpreting information, spanning the spectrum from "unfiltered" to "authoritative." At the unfiltered end of the spectrum we have the example of Google, which attempts to find everything and organizes it only by number of hits. (A search on "Formicidae" in Google Web nets 2,320,000 results; Google Scholar nets 45,300 results.) Somewhere in the middle of the spectrum we have Wikipedia, which depends on self-nominated, self-assembling groups of contributors to organize information. (Searching for "Formicidae" in Wikipedia shunts you to the page for "Ants," which, despite containing a number of errors, turns out to be pretty good.) The authoritative end of the information-organizing spectrum is the most challenging because it depends on senior-level experts with exhaustive knowledge of their fields. An authoritative review must simultaneously provide an entrée into a particular field's primary literature, summarize current knowledge in that field, and judiciously interpret that knowledge. Frequently, as information in a particular field continues to expand, a new review must cite and build upon past reviews. It is this end of the information-organizing spectrum upon which we scientists depend in order to stay informed. And it is at this end of the spectrum that we encounter the Annual Review model, presciently created in 1932 and eighty years later proving eminently adaptable to the brave new world of electronic information, e.g., with online access and online supplementary materials such as video and audio clips. As someone who has on more than one occasion been asked by a journal editor to shorten my bibliography, I am especially pleased to point out that Annual Review bibliographies are luxuriously exhaustive.

Many diverse subjects are covered by the forty journals in the Annual Review series. Why, specifically, do we need an *Annual Review of Entomology*? As pointed out by Gene Robinson in last year's preface, we are in the midst of a trend away from "vertical," taxon-oriented categories in biology such as botany, zoology, and entomology. Instead, courses, majors, departments, and journals are increasingly arranged around "horizontal" categories such as ecology, evolutionary biology, behavior, and physiology. Correlated with this trend, departments of entomology have been dissolved and their faculties divided and "reintegrated" into departments of biology, ecology and evolutionary biology, forestry, horticulture, neurobiology, biochemistry, etc. So why an *Annual Review of Entomology*? What is the rationale behind publishing a yearly collection of entomocentric reviews? The answer is that there are very good reasons that entomology has persisted as a distinct discipline for over 300 years. First, because insects comprise over 50% of all life on Earth (and arthropods over 60%), it could be argued that the subject of entomology is, if anything, pretty darned inclusive. Second, a quick look at their Web site indicates that the *BIOSIS* "Abstracts of Entomology" adds an average of 20,000 references per year, and, because they don't index all the entomology journals, this is certainly an underestimate of the rate at which new entomological publications are being generated. Clearly, entomology as a subject has never been healthier. Slicing through the vast and ever-expanding body of scientific knowledge in new, nontaxonomic ways no doubt produces important insights (eusocial insects, meet naked mole rats). But even as we experiment with novel ways of organizing information, we should remember that the vertical, taxon-oriented approach has not only paid dividends for centuries but also permits us to compare biological phenomena against the very real scaffold of phylogeny (genomic data, anyone?). In short, the discipline of entomology isn't going anywhere and neither is the *Annual Review of Entomology*. (And it might be wise to reconsider those scattered departments of entomology, too.)

Because I'm proud to be an entomologist and because I'm a disciple of expertise-driven, authoritative reviews, I am honored to be part of the *Annual Review of Entomology* editorial committee. A large part of the editorial process involves carefully reading successive drafts of review articles covering entomological subjects well outside of my usual areas of concentration. Reading these reviews, which are almost always well-written and deeply knowledgeable, has inspired in me an even greater enthusiasm for entomology. Likewise, the care that my fellow committee members employ in reviewing proposals and manuscripts sets an inspiringly high standard that I endeavor to emulate. It's my pleasure, therefore, to welcome you to Volume 56 of the *Annual Review of Entomology*, wherein you will discover reliable, up-to-date surveys of diverse entomological subdisciplines. These reviews summarize current knowledge, place that knowledge into a larger context, and direct you to the primary literature. With the bibliographies in hand, you can use the Web (recent publications) or your library (older publications) to further explore the brick-by-brick contributions of our many past and present entomological colleagues . . . and to kick through the rubble of the occasional shattered paradigm as well.

Ted Schultz
For the Editorial Committee

Annual Review of Entomology

Volume 56, 2011

Contents

Indexes

Errata

An online log of corrections to *Annual Review of Entomology* articles may be found at http://ento.annualreviews.org/errata.shtml

Related Articles

From the ***Annual Review of Ecology, Evolution, and Systematics***, Volume 41 (2010)

From the ***Annual Review of Genetics***, Volume 44 (2010)

From the ***Annual Review of Microbiology***, Volume 64 (2010)

Structure, Function, and Evolution of Linear Replicons in *Borrelia*
George Chaconas and Kerri Kobryn

From the ***Annual Review of Phytopathology***, Volume 48 (2010)

Go Where the Science Leads You
Richard S. Hussey

Current Epidemiological Understanding of Citrus Huanglongbing
Tim R. Gottwald

Principles of Predicting Plant Virus Disease Epidemics
Roger A.C. Jones, Moin U. Salam, Timothy J. Maling, Arthur J. Diggle, and Deborah J. Thackray

Potyviruses and the Digital Revolution
Adrian Gibbs and Kazusato Ohshima

Managing Nematodes Without Methyl Bromide
Inga A. Zasada, John M. Halbrendt, Nancy Kokalis-Burelle, James LaMondia, Michael V. McKenry, and Joe W. Noling

Ecology of Plant and Free-Living Nematodes in Natural and Agricultural Soil
Deborah A. Neher

Annual Reviews is a nonprofit scientific publisher established to promote the advancement of the sciences. Beginning in 1932 with the *Annual Review of Biochemistry*, the Company has pursued as its principal function the publication of high-quality, reasonably priced *Annual Review* volumes. The volumes are organized by Editors and Editorial Committees who invite qualified authors to contribute critical articles reviewing significant developments within each major discipline. The Editor-in-Chief invites those interested in serving as future Editorial Committee members to communicate directly with him. Annual Reviews is administered by a Board of Directors, whose members serve without compensation.

Annual Reviews of

Analytical Chemistry
Anthropology
Astronomy and Astrophysics
Biochemistry
Biomedical Engineering
Biophysics
Cell and Developmental Biology
Chemical and Biomolecular Engineering
Clinical Psychology
Condensed Matter Physics
Earth and Planetary Sciences
Ecology, Evolution, and Systematics
Economics
Entomology
Environment and Resources
Financial Economics
Fluid Mechanics
Food Science and Technology
Genetics
Genomics and Human Genetics
Immunology
Law and Social Science
Marine Science
Materials Research
Medicine
Microbiology
Neuroscience
Nuclear and Particle Science
Nutrition
Pathology: Mechanisms of Disease
Pharmacology and Toxicology
Physical Chemistry
Physiology
Phytopathology
Plant Biology
Political Science
Psychology
Public Health
Resource Economics
Sociology

SPECIAL PUBLICATIONS
Excitement and Fascination of Science, Vols. 1, 2, 3, and 4

Bemisia tabaci: A Statement of Species Status

Paul J. De Barro,[1] Shu-Sheng Liu,[2] Laura M. Boykin,[3] and Adam B. Dinsdale[1,4]

[1]CSIRO Entomology, Indooroopilly, QLD 4068, Australia; email; paul.debarro@csiro.au

[2]Institute of Insect Sciences, Zhejiang University, Hangzhou 310029, China; email: shshliu@zju.edu.cn

[3]BioProtection Research Centre, Lincoln University, Lincoln 7647, Christchurch, New Zealand; email: laura.boykin@lincoln.ac.nz

[4]School of Biological Sciences, University of Queensland, St Lucia, QLD 4072, Australia; email: abdinsdale@gmail.com

Annu. Rev. Entomol. 2011. 56:1–19

First published online as a Review in Advance on July 23, 2010

The *Annual Review of Entomology* is online at ento.annualreviews.org

This article's doi: 10.1146/annurev-ento-112408-085504

0066-4170/11/0107-0001$20.00

Key Words

whitefly, biotype, host race, mitochondrial cytochrome oxidase 1, phylogeography, alien species

Abstract

Bemisia tabaci has long been considered a complex species. It rose to global prominence in the 1980s owing to the global invasion by the commonly named B biotype. Since then, the concomitant eruption of a group of plant viruses known as begomoviruses has created considerable management problems in many countries. However, an enduring set of questions remains: Is *B. tabaci* a complex species or a species complex, what are *Bemisia* biotypes, and how did all the genetic variability arise? This review considers these issues and concludes that there is now sufficient evidence to state that *B. tabaci* is not made up of biotypes and that the use of biotype in this context is erroneous and misleading. Instead, *B. tabaci* is a complex of 11 well-defined high-level groups containing at least 24 morphologically indistinguishable species.

INTRODUCTION

What Is *Bemisia tabaci* and Where Is It Found?

mtCO1: mitochondrial cytochrome oxidase 1

Bemisia tabaci (Gennadius) (Insecta: Hemiptera: Homoptera: Sternorrhyncha: Aleyrodoidea: Aleyrodidae: Aleyrodinae) is a phloem-feeding insect that lives predominantly on herbaceous species. It is a considerable pest of ornamental, vegetable, grain legume, and cotton production, causing damage directly through feeding and indirectly through the transmission of plant pathogenic viruses, primarily begomoviruses (77). It has a global distribution (18, 60). As *B. tabaci* has been the subject of several reviews (24, 44, 66, 76, 105), we focus on key questions not already addressed or avoided.

B. tabaci rose to international prominence in the mid- to late 1970s in Sudan and again in the 1980s in the southwestern United States and since then has risen in status to one of the most globally damaging pests of open field and protected cropping (Global Invasive Species Database, **http://www.issg.org/database**). Although there is no doubt regarding its impact, whether *B. tabaci* is a complex species or a species complex has been a source of debate (5, 36, 113) since the raising of *B. tabaci* biotype B to separate species status and its redescription as *B. argentifolii* Bellows & Perring (8, 111). In an attempt to draw a conclusion, the following question was posed, "Is devastating whitefly invader really a new species?" (4), but by the article's end the answer remained elusive. A 1995 review of the subject (24) again came to no conclusion. There the debate rested until an analysis of the available data concluded that the lack of biological data and a clear picture of the phylogenetic structure of *B. tabaci* meant that there was insufficient data to support species status (54). Since then, a series of studies, both biological and molecular, have been undertaken and it is now time to consider whether these studies enable us to make a more definitive statement on the species question.

Figure 1 shows a summary of the evolutionary reconstruction using a Bayesian analysis of mitochondrial cytochrome oxidase 1 (mtCO1) (60). This is the best picture we have of the genetic structure of *B. tabaci* together with the assignment of biotypes to particular genetic groups. The nomenclature used here for the 24 low-level groups is applied throughout the rest of the review, unless otherwise indicated, and where a biotype name is mentioned the genetic group to which it belongs is in parentheses.

History of Pest Status and Global Invasion

The history of *B. tabaci* as a pest stretches back to 1889 (24, 66, 88). However, much of the current attention on *B. tabaci* centers on a series of major invasion events, and we focus on these as it is the mechanisms and consequences associated with these invasions that provide some of the most compelling biological data that support the species argument.

The first major global invasion event that we know of is that of the B biotype (Middle East–Asia Minor 1), which commenced sometime in the late 1980s principally via the trade in ornamentals (19, 38), from its origins in the Middle East–Asia Minor region (Iran, Israel, Jordan, Kuwait, Pakistan, Saudi Arabia, Syria, United Arab Republic, Yemen) to at

Figure 1

The following is a summary of the evolutionary relationships of the 11 high-level (*blue boxes*) and 24 low-level (*black boxes*) groups based on a Bayesian analysis of 201 unique mtCO1 haplotypes (60) using the HKY + I + G model. Posterior probabilities for the branches are given. Thirty-three biotypes (*in parentheses*) have been assigned across the various genetic groups (109). The biotypes not assigned are the BR, E, I, Cassava, and Okra biotypes. The BR and I biotypes have no mtCO1 sequence but likely belong to either the Asia I or the Asia II high-level group. Whereas the E, Cassava, and Okra biotypes belong to the sub-Saharan Africa high-level group, placement in a low-level group is not possible due to the lack of suitable mtCO1 sequences. A further point to recognize is the location of *Bemisia atriplex*, which sits within the *B. tabaci* ingroup; however, this may be an artefact based on the limited number of outgroups that will influence topology but not grouping. The placement of ZHJ1 in Reference 60 is incorrect; it belongs in AsiaII_3.

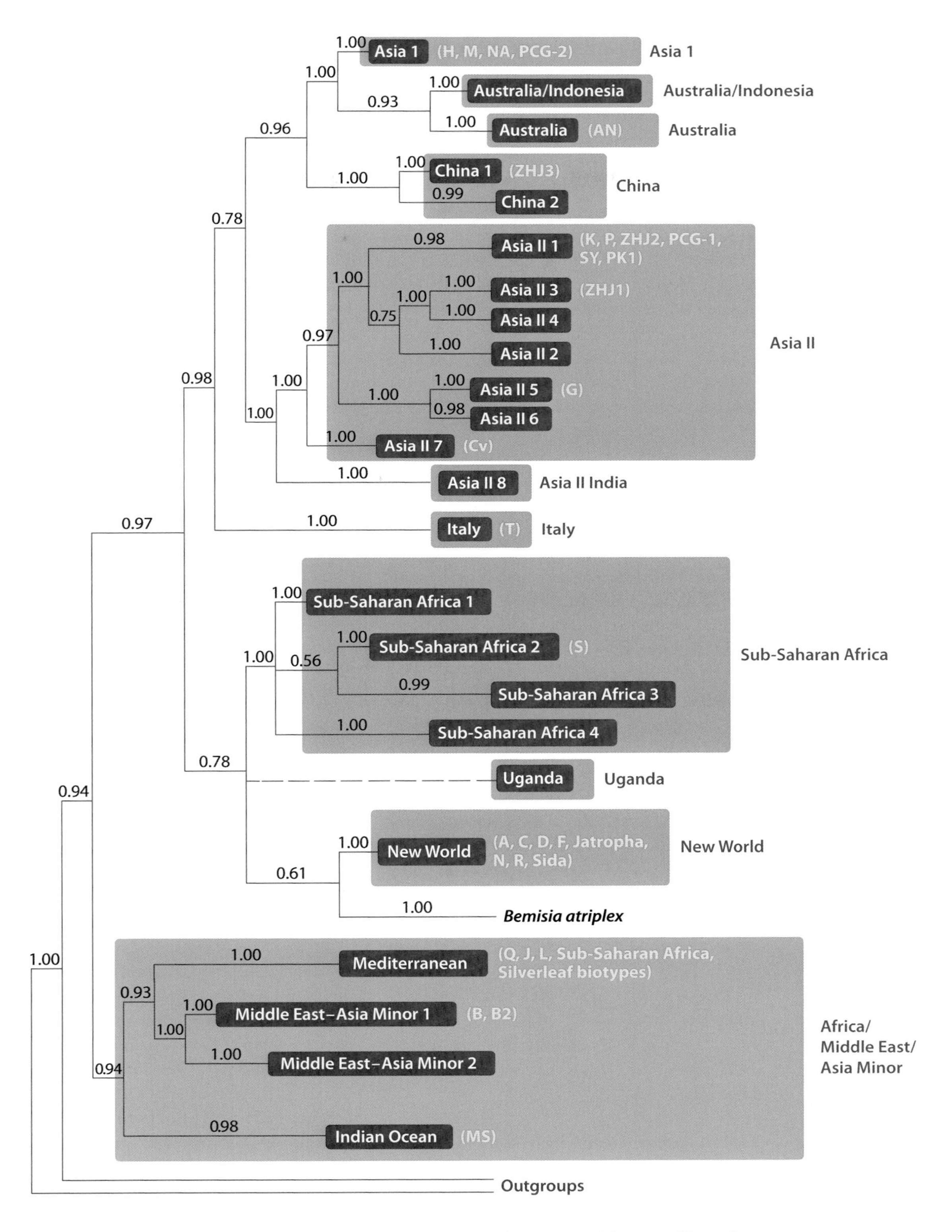

Asia 1 (H, M, NA, PCG-2)
Asia 1
Australia/Indonesia
Australia/Indonesia
Australia (AN)
Australia
China 1 (ZHJ3)
China 2
China
Asia II 1 (K, P, ZHJ2, PCG-1, SY, PK1)
Asia II 3 (ZHJ1)
Asia II 4
Asia II 2
Asia II 5 (G)
Asia II 6
Asia II 7 (Cv)
Asia II
Asia II 8
Asia II India
Italy (T)
Italy
Sub-Saharan Africa 1
Sub-Saharan Africa 2 (S)
Sub-Saharan Africa 3
Sub-Saharan Africa 4
Sub-Saharan Africa
Uganda
Uganda
New World (A, C, D, F, Jatropha, N, R, Sida)
New World
Bemisia atriplex
Mediterranean (Q, J, L, Sub-Saharan Africa, Silverleaf biotypes)
Middle East–Asia Minor 1 (B, B2)
Middle East–Asia Minor 2
Indian Ocean (MS)
Africa/ Middle East/ Asia Minor
Outgroups

Sympatric: species whose habitats overlap

Allopatric: species whose habitats are separated and not adjoining

Parapatric: species whose habitats are separated, but adjoining

RAPD-PCR: randomly amplified polymorphic DNA–polymerase chain reaction

least 54 countries (American Samoa, Antigua & Barbados, Argentina, Australia, Austria, Belize, Brazil, Canada, China, Colombia, Cook Islands, Costa Rica, Cyprus, Denmark, Dominican Republic, Egypt, Fiji, France, French Polynesia, Germany, Greece, Grenada, Guadeloupe, Guam, Guatemala, Honduras, India, Italy, Japan, Martinique, Marshall Islands, Mauritius, Mayotte, Mexico, New Caledonia, New Zealand, the Netherlands, Niue, North Mariana Islands, Norway, Panama, Poland, Puerto Rico, Reunion, Saint Kitts-Nevis, South Africa, South Korea, Spain, Taiwan, Tonga, Trinidad-Tobago, Tunisia, United States, Venezuela). In recent years this has been followed by the global spread of Q (Mediterranean), which has spread from its origin in the countries bordering the Mediterranean Basin (Algeria, Crete, Croatia, Egypt, France, Greece, Israel, Italy, Morocco, Portugal, Spain, Sudan, Syria, Turkey) to at least 10 countries (Canada, China, Guatemala, Japan, Mexico, the Netherlands, New Zealand, South Korea, Uruguay, United States) (country records obtained from GenBank) [the recent additions of new mtCO1 sequences into GenBank indicate that B and Q are linked to East and West Africa, with records of B from Uganda and of Q from Burkina Faso, Cameroon, and Uganda that are genetically within the Middle East Asia Minor 1 and Mediterranean groups (**Figure 1**), respectively]. The other significant invasion event, although disputed, is associated with the severe African cassava mosaic virus epidemic in Uganda and involves the possible displacement of one sub-Saharan African group with another (81, 132).

HOST RACES AND BIOTYPES

A key step in being able to define species requires a capacity to identify and distinguish the set of characteristics that provide a unique description.

Early Observations

The early observations are covered extensively in a previous review (24) and are not broadly discussed here. In summary, the existence of host-plant-related races, or biotypes, was first proposed on the basis of observations of the narrowly specific Jatropha and the more polyphagous Sida races in Puerto Rico (11). Similar observations were made in Ivory Coast of a biotype largely confined to cassava and a polyphagous biotype that did not include cassava as a host (26). Furthermore, the variability was linked to the capacity to transmit begomoviruses (12), with the polyphagous Sida race transmitting several viruses whereas the monophagous Jatropha race transmitted just one. Similar findings in relation to host race were made in Brazil (45). It is highly likely that these populations with different host ranges reflect underlying genetic differences and so our earliest concept of genetic variability between different groups of *B. tabaci* was based on biological differences in the form of host utilization. A critically important point to recognize here is that each of these distinctions was made between sympatric populations and allopatric or parapatric populations were not considered.

History of Biotype Determination

The concept of biotype in *B. tabaci* came to prominence with the invasion of the southern United States by a *B. tabaci* that behaved quite differently from the indigenous population. The invader had a different esterase profile and host range from the indigenous population and, based on the earlier observations of differential host utilization (11), was referred to as the B biotype and the indigenous population as the A biotype (New World) (7, 22, 46). Allozymes (110) and RAPD-PCR (randomly amplified polymorphic DNA–polymerase chain reaction) (65, 110) showed further consistent differences between A and B biotypes. Furthermore, they were reproductively incompatible (47, 110). But unlike the earlier applications of the term host race or biotype, the term was now being applied to two groups that had, until the very recent human intervention, occupied allopatric distributions (18, 60). The application of biotype was understandable given the level of knowledge at the time. Since the determination of A and B, there has been an explosion in biotype designation, with C through T being added to the list (**Figure 1**) (109). So the question remains: Is there a clear set of biological characteristics that define each biotype?

The most comprehensive comparison we can find compares 11 biotypes (A, B, B2, D, E, G, H, K, J, L, M) against a range of biological characters including capacity to transmit begomoviruses, capacity to induce silverleafing in squash and yellow vein in honeysuckle and nightshade, host range, and capacity to produce female offspring following interbiotype mating trials (6). We now know through the comparison of mtCO1 that these biotypes are mapped to the Asia I (H, M), Asia II 5 (G), Asia II 1 (K), New World (A, D), Mediterranean (J, L), Middle East–Asia Minor 1 (B, B2), and sub-Saharan Africa (E, high-level group only as available mtCO1 sequences are too short to allow accurate placement) genetic groups (18, 60) (**Figure 1**). The results were mostly equivocal. Of the 11 biotypes tested, only B and B2 induced physiological changes in host plants. In the case of begomovirus transmission there was some variation, but for the most part the differences were uninformative. Similarly, differences in host plant adaptation were not significant, although B showed higher levels of survival than the non-B biotypes (6). Variation in host range is an often repeated argument in regards to differences between biotypes (21, 24, 109); however, this is largely assumed as experimental studies comparing performance across different hosts are few and restricted not only in the numbers of host plant species included, but also primarily to comparisons between B and other non-B biotypes (6, 10, 46, 57, 145).

The only other clear biological difference between A and B (6) was the capacity to induce a number of physiological changes in a range of host plants (49) that were first associated with *B. tabaci* in the United States (98, 144). The capacity of B to induce squash silverleafing became one of the means by which it was characterized (6, 46, 111). However, this capacity has since been observed for the Middle East–Asia Minor 1 (B2) (6), Indian Ocean (MS) (58), and Mediterranean (J) groups (55, 131). All three belong to the cluster of *B. tabaci* that originated in the region encompassed by Africa, the Mediterranean Basin, the Middle East, and Asia Minor. Note that the Mediterranean group contains two distinct elements: a nonsilverleafing subgroup that contains the invasive Q and the silverleafing subgroup from Ghana, Ivory Coast, and Nigeria (J) (60, 131). This indicates that silverleafing has an evolutionary relationship in which the ancestral state was nonsilverleafing and one major clade then evolved the capacity before losing it again (18, 60) (**Figure 1**).

Other traits used to characterize biotypes include the capacity to disperse widely and insecticide resistance (24). In the case of capacity to disperse, this trait (13–15, 28) has been studied only for B. In the case of insecticide resistance, this trait is not unique to B, having been detected in individuals belonging to the Asia I, Mediterranean, and New World genetic groups (2, 30, 32–34, 62, 73). Furthermore, it has been argued that although B contains both susceptible and resistant genotypes (30, 33, 35), it lacks the variability found in non-B populations (2). However, our current knowledge indicates that most of the collections used to examine resistance in B were from invading populations and not indigenous populations and as such may represent only a portion of the home range genetic variability. Given that the global spread of B is firmly linked to trade in ornamentals, the widespread use of insecticides in this industry is likely to have eliminated susceptible genotypes. This is further supported by observations that where susceptible B has been recovered, the collections were made in its geographic origin (2, 30, 33).

Differences in biological traits between groups are likely to exist, but what is lacking are quantified measures of separation that can be applied across the whole of *B. tabaci*. Instead, capacity to separate across the whole of *B. tabaci* relies on estimates of the degree of genetic relatedness.

MORPHOLOGICAL SPECIES, SYNONYMIZATION, AND LACK OF VARIABILITY

SCAR: sequence characterized amplified region

CAPS: cleavage amplified polymorphic sequence

ribosomal ITS1: ribosomal intergenic transcribed spacer 1

Whitefly taxonomy relies primarily on morphological characters of the fourth instar (99). *B. tabaci* was first described in 1889, as *Aleurodes tabaci* (Gennadius, 1889), a pest of tobacco in Greece, and was subsequently redescribed under 22 additional names (101). Given that juveniles are unlikely to be under strong sexual selection, the characters present in the fourth instar are more likely to be associated with avoidance of natural enemies, desiccation, exposure to UV, and host plant characteristics than with mate finding and as such are probably convergent. This high level of morphological plasticity has no doubt contributed to the serial redescriptions. This situation may well have been exacerbated by geographic isolation between the different taxonomists of the time, resulting in a lack of information exchange and the consequential assignment of multiple names to a globally distributed insect. The recognition of the confusion caused by the redescriptions led to the synonymization under *B. tabaci* (Gennadius) (99, 123).

The invasion of the United States by biotype B led to renewed interest in the morphometrics of *B. tabaci*. An initial examination of morphological characters did not allow A to be distinguished from B (67), but then additional characters were discovered, such as absence of the fourth anterior margin setal pair, width of the thoracic tracheal folds, and width of wax extrusions from the folds, that enabled B to be separated from A (8). However, that study did not include *B. tabaci* from other parts of the world and as such failed to consider the full range of available variability. When *B. tabaci* from other parts were included (120), the subsequent comparison across 5 of the 11 major groups (Asia I, Asia II, Africa/Middle East/Asia Minor, New World, sub-Saharan Africa) (**Figure 1**) showed that the morphological characters previously used to separate A from B displayed considerable within- and between-group variability and were not useful for the reliable separation of any of the groups. The use of adult characters has similarly failed (35).

GLOBAL PHYLOGEOGRAPHIC GENETIC STRUCTURE

Allozymes and RAPD-PCR

The use of allozymes to separate different *B. tabaci* was first shown in 1989 (141), and in 1991 a comparison of individuals from Colombia, Israel, Kenya, and the United States led to the first suggestion of the existence of geographical races (139). Since then, numerous studies using a range of enzymes, but primarily esterases (6, 20, 23–26, 29, 31, 43, 46, 47, 59, 61, 70, 81, 86, 90, 110, 124, 127, 128, 138–140), to identify differences between populations of *B. tabaci* have been published.

RAPD-PCR was first applied to *B. tabaci* in 1993 (65). This has since been followed by numerous studies (1, 16, 51, 58, 70, 72, 82, 85, 94, 102, 116, 117, 125, 127, 128, 135, 149). The use of RAPDs has since evolved into the use of SCAR (sequence characterized amplified region) and CAPS (cleavage amplified polymorphic sequence) (16, 42, 78, 79, 124, 146) to produce specific markers.

Mitochondrial CO1, 16s, Nuclear Ribosomal ITS1, AFLPs, and Microsatellites

Although allozymes and RAPDs provided the first real insights into genetic variability within *B. tabaci*, it was the use of mitochondrial 16s and cytochrome oxidase 1 (mtCO1) and the nuclear ribosomal intergenic transcribed spacer 1 (ITS1) that uncovered the phylogenetic structure. Mitochondrial CO1 and 16s were first used to reconstruct a phylogeography of 10 collections from the New World, India, the Middle East, and North Africa, leading to the conclusion that the B biotype was introduced into the New World from desert/sahel-like regions in or around Israel and Yemen (64). Shortly after, a second study using nuclear ribosomal ITS1 (52) to analyze 31 collections from

Australia, Africa, Asia, the Mediterranean, North America, and the Middle East revealed strong global geographic patterns in the distributions of the different genetic groups. Since then, a number of further studies using either AFLPs, mtCO1, 16s, or ITS1 have reaffirmed these results (1, 9, 18, 25, 37, 41, 54, 55, 58, 60, 74, 75, 80, 82, 83, 95, 96, 109, 117, 118, 131–133, 136, 142), with specific primers (126) and restriction fragment length polymorphisms (91) identified to distinguish the invasive B and Q from other groups. Only one study has compared both mtCO1 and ITS1 and showed that they produced the same genetic groupings (54). Finally, microsatellites (48, 50, 56, 58, 129) have also begun to reveal not only patterns of relatedness but also directions of gene flow.

Of these studies, a subset has analyzed phylogenetic structure and has led to the development of a broad geographical understanding of the distribution of the different groups making up *B. tabaci* (1, 9, 25, 52, 54, 55, 58, 64, 83, 97, 118, 136). All the studies from this subset have varied in regard to the types of analyses applied and their inherent constraints, the degree to which phylogenetic difference and structure have been interpreted, the number of taxa included, and the breadth of the genetic variation sampled. This variability has limited our overall interpretation of the global relationships of *B. tabaci*. Despite the limitations, it is clear that several major geographic groups exist, namely sub-Saharan Africa, North Africa/Mediterranean Basin/Middle East/Asia Minor, Asia, Australia, and the Americas.

The variation in structure observed across the studies might be explained through limited taxon sampling, which adds greatly to the phylogenetic estimation error (115, 150) and as such has meant that the global relationships of *B. tabaci* have been difficult to assess. Two studies have directly addressed this lack of resolution (18, 60). The first did so by considerably increasing the size of the mtCO1 dataset analyzed to 366 sequences and in so doing succeeded in decreasing the error (18). However, this study did not take into account the effects duplicate haplotypes, indels, unknown bases, and pseudogenes might have had on the analysis. These deficits were addressed in a subsequent analysis that used 201 unique haplotypes (60). Through comparisons of a large number of individual sequences of *B. tabaci* populations, global relationships were resolved with high levels of branch support provided by Bayesian analyses in which previously predicted relationships using neighbor joining, maximum parsimony, and maximum likelihood were unresolved. The first Bayesian approach provided resolution of 12 major clades (18), whereas the latter analysis yielded 11 major clades (60). In this latter study, the elimination of problematic sequences (see above), the inclusion of sequences representing the full genetic diversity of the species complex, and the application of appropriate phylogenetic methodologies resolved the relationships between the different clades, an outcome lacking in many of the previous studies. This study (60) went on to show that the 11 high-level groups could be further resolved to 24 low-level groups. This study is particularly important because it produced for the first time quantifiable bounds with which to subdivide *B. tabaci*. It did so by generating a frequency distribution based on all pairwise comparisons of genetic distance and as a result identified two clear breaks in the distribution, one at 11% and the other at 3.5%. A consensus sequence for each group was then created that allowed each sequence to be compared against the consensus sequence to determine whether the subsequent assignment matched the placement made using the Bayesian analysis. Placement at both 11% and 3.5% showed a high degree of rigor, with the probability of sequence assignment being 100% when pairwise genetic distances were compared. In other words, if an unknown sequence compared against the consensus sequence lies within these limits, then that is the identity of the unknown.

What all these protein and DNA markers show is that there is a remarkable level of genetic variability at multiple loci within *B. tabaci*; however, what none of the studies provides is a set of bounds that enables us to clearly define any biotype. This lack of bounds has caused

Microsatellites: a microsatellite consists of a specific sequence of DNA bases or nucleotides that contains mono, di, tri, etc. tandem repeats

Haplotypes: a set of closely linked genetic markers present on one chromosome that tend to be inherited together

Indels: DNA mutations involving one or more base pairs

Pseudogenes: an inactive gene derived from an ancestral active gene such as mtCO1 that has become incorporated into the nuclear genome

Pairwise genetic distance: a measure of the similarity between two DNA sequences

the proliferation of biotypes; at this stage the combined use of biological data on host range, host utilization, capacity to induce physiological changes in some hosts, insecticide resistance, capacity to disperse widely, capacity to transmit begomoviruses, esterase and RAPD profiles, as well as sequence variation data has so far seen the identification of at least 36 biotypes: A, AN, B, B2, BR, C, Cassava, Cv, D, E, F, G (India), G (Guatemala), H, I, J, Jatropha, K, L, M, N, NA, Okra, P, PCG-1, PCG-2, PK1, Q, R, S, Sida, SY, T, ZHJ1, ZHJ2, and ZHJ3 (22, 109, 116, 127, 128, 145). It is certain and well known that there is genetic structure that enables us to break up *B. tabaci* into a series of well-defined subgroups, and it is equally certain that there are not 36 biotypes. Moreover, because the descriptions of the vast majority of the biotypes are based primarily on genetic markers, whether protein or DNA, and not biological data, the term biotype in the context of *B. tabaci* is a surrogate term for a genetic group. Therefore, it is time to put the biotype terminology first promulgated by Bird in 1957 aside and instead focus on genetic structure.

POSSIBLE MECHANISMS DRIVING PHYLOGENETIC STRUCTURE

The origin of *B. tabaci* has been open to debate; Asia (99, 100), India (63), Africa (71), and the New World (67) have all been proposed. There have been numerous evolutionary reconstructions of *B. tabaci* phylogenies using primarily mtCO1 and ITS1; however, as we have shown many of the reconstructions are flawed due to phylogenetic estimation error caused by small (9, 41, 55, 75, 79, 82, 118, 131, 132, 136, 143), often highly biased sample sets (9, 41, 55, 75, 79, 118, 136). Further, none of the estimations considered levels of divergence. As a result, groups that had considerable differences in levels of sequence divergence were created (9, 18, 55, 82, 131, 132, 142) and potential groups were ignored even if they were quite divergent (9, 18, 55, 82, 131, 132, 142). The most recent analysis that addresses this inconsistency is summarized in **Figure 1**. *B. tabaci* is composed of several major clusters and across these there is no obvious ancestral hierarchy. The Africa–Middle East–Asia Minor cluster splits from a very large cluster containing the Asian, sub-Saharan African, and New World *B. tabaci*. This latter cluster then splits into two large clusters, one composed of Asian, Australian, and Italian (Sicily) individuals and the other composed of the sub-Saharan African and New World *B. tabaci*. The individuals from sub-Saharan Africa show mtCO1 sequence divergences of between 20% and 24% from all other *B. tabaci*, which represent the greatest degree of overall divergence of any of the *B. tabaci* groups. This suggests that sub-Saharan Africa may be the origin of *B. tabaci*, but the anomalous position of the New World group suggests more work is needed to resolve this. The placement of *B. atriplex* within the *B. tabaci* ingroup (60) and the overall effect outgroup choice has on tree topology indicate that more work is needed to determine the evolutionary (topology) relationships between groups. One approach would be to employ Bayesian techniques for rooting phylogenetic trees to determine the root of the *B. tabaci* phylogeny (17). That said, if we use a ballpark estimate of 5% divergence per million years (based on mtCO1 molecular clocks), then the origin for *B. tabaci* stretches back approximately 4 million years. How this complex structure evolved is unclear, but there are several possible explanations.

Host Plant Utilization as a Driver for Genetic Structure

As we have seen, the early observations on differences in host associations indicate the presence of sympatric populations displaying clear differences in host utilization. However, few studies on *B. tabaci* really consider the issue of host range, so the available data is scant. With the exceptions that have already been discussed, *B. tabaci* is more commonly observed to be polyphagous, so monophagy appears to be the exception. The lack of research focusing directly on the role host plant utilization may play in determining population genetic structure may help explain why the Jatropha race (11) and the cassava race (1, 26) are not supported by more examples.

The argument that host plants may be driving genetic differentiation is perhaps best considered by an examination of the sub-Saharan Africa cassava race (1, 9, 26). Cassava originated in the New World, but the whiteflies that make up this group are not derived from the New World and instead are unique to this part of Africa (18, 52, 54, 60, 64). Further, the association of these whiteflies with cassava could have arisen only in the past few hundred years, as cassava was introduced during the late sixteenth century from the New World and for all but the past 100 years has been restricted to a small part of West Africa. If one considers the levels of mtCO1 sequence divergence for the cassava and noncassava populations found in this region, one finds levels of divergence as high as 16.5% (9), suggesting the two diverged (based on 5% per Mya) around 3 Mya. Furthermore, there was considerable evidence for allopatric genetic structure, with the cassava-feeding *B. tabaci* from West Africa differing from those from southern Africa by at least 2%, and an Ivory Coast individual differing from the West African clade by approximately 6% (9). Overall, five distinct cassava-feeding genetic groups of sub-Saharan *B. tabaci* were identified, all separated by levels of divergence between 2% and 4%, differences that suggest separations of several 100,000 years (9, 60). Given that cassava was first introduced 400 years ago, the levels of divergence are simply too great to have been driven by this host, and therefore other explanations should be sought to explain the shifts from indigenous hosts to cassava in five genetically distinct populations in at least three separate geographical areas.

Asymmetric Mating Interference

Since 1994, several studies have observed mating interactions between individuals belonging to different genetic groups. These studies have shown that reproductive interference influences the relative availability of male and female whiteflies and can have profound ecological impacts on whitefly abundance (87). Males belonging to Middle East–Asia Minor 1 (B) have been found to interfere with the courtship of the New World group (A) (112, 114) and Mediterranean group (Q) (**Figure 1**) (92, 107, 108). As whiteflies are haplodiploid, a key feature of the interaction involved the initial shift in progeny sex ratio from a female to male bias (53, 87, 99). However, unilaterally increasing copulations by B females, but not Australia 1 (AN) or Asia II 3 (ZHJ1) females, elevated the proportion of female progeny in B to levels above that seen in precompetition populations (**Figure 1**) (87, 89, 147). Moreover, invader males interrupted copulation by indigenous individuals more frequently than did indigenous males that interrupted copulation by invader individuals (87). This is further evidence of species-level divergence and gives an insight into the mechanism that could be driving both species dominance and speciation.

Bacterial Symbionts: *Wolbachia* and Other Reproduction-Altering Bacteria

B. tabaci contains the primary symbiont *Porteria* as well as six secondary symbionts belonging to the genera *Arsenophonus*, *Cardinium*, *Fritschea*, *Hamiltonella*, *Rickettsia*, and *Wolbachia* (39, 68, 69, 121), but infections were not observed in all individuals belonging to a population. *Wolbachia* have been found in individuals belonging to the Asia I, Asia II, Australia, Africa/Middle East/Asia Minor, New World, and sub-Saharan African high-level genetic groups (**Figure 1**) (40, 50, 84, 103, 121, 148), but again infections were not observed in all individuals belonging to a population. Of the symbionts, *Wolbachia* (134) and *Cardinium* (137) alter reproduction processes in many arthropods. To date, although there is some evidence of association with some genetic groups, the data are too preliminary to determine whether symbionts are playing a role in driving patterns of genetic variation.

Cuticular Hydrocarbons

Insect cuticular hydrocarbons act as contact signals for species, kin, and mate recognition in numerous insect species (130). The role for cuticular hydrocarbons in mate recognition in *B. tabaci* is purely speculative, but published observations on other insects (59, 87, 114) indicate that the point of recognition occurs during the courtship phase after male-female contact has been made, but before copulation.

CAPACITY TO INTERBREED

Since 1993, a series of studies (6, 31, 47, 53, 94, 96, 104, 106, 119, 143, 147) have shown that members of the 24 low-level groups (**Figure 1**) are for the most part reproductively incompatible (143). One study has suggested some small capacity to produce F1 females in crosses between A and B, but these females always had the same profiles as the female parent, suggesting possible contamination rather than hybridization (47), a point that is further supported by the failure to observe copulation (114). Mating studies between Australia 1 (AN) and Middle East–Asia Minor 1 (B) (53) showed that F1 females could occasionally be produced but were always sterile. The only studies to show some capacity to copulate and produce viable F1 progeny come from laboratory experiments involving the individuals from Middle East–Asia Minor 1 (B), Mediterranean (Q, L, and the silverleafing population from okra in Ghana), sub-Saharan Africa 1, and Indian Ocean (MS) (31, 56, 96, 102, 106, 119, 143), as well as Middle East–Asia Minor 1 × Indian Ocean in the field (56).

A key issue in the interpretation of mating studies is what defines success. In one study, reciprocal crossing experiments between Asia II 1 (ZHJ2) and Asia II 3 (ZHJ1) produced no female progeny in the ZHJ2 female × ZHJ1 male cross. Furthermore, of the 45 replicates of the ZHJ1 female × ZHJ2 male cross, only 4 produced female progeny and in each case the percentage was 0.7%, well below the >50% observed in the putative intraspecies crosses (143). However, another study concluded there was full compatibility between groups in which the mtCO1 diverges by 7%–8% (96). However, this conclusion is questionable, as both F1 and F2 generations show a marked shift in sex ratio from strong female bias to strong male bias. In addition, one-half to one-third of the replicate crosses produced no female progeny, whereas all replicates for the putative same species crosses produced female progeny (96). These results are more consistent with partial mating compatibility, which is unsurprising, and there are many examples in the literature of incomplete reproductive isolation between closely related species (93). In addition to mating studies, two microsatellite studies (50, 56) have shown both private loci and alleles, adding further support for the lack of gene flow. Although mating trials involving all possible combinations have not been undertaken, there is a clear pattern emerging. Individuals belonging to different genetic groups are reproductively isolated or they demonstrate significantly lower degrees of compatibilities than one would expect with full compatibility (143).

SPECIES OR HOST RACES? THE MYTHOLOGY OF BIOTYPES: USE AND ABUSE

Over the past 18 years there has been a proliferation in biotypes as a means of describing the inherent genetic variability within *B. tabaci*. The use of the term biotype has passed into the vernacular within the *B. tabaci* lexicon, but we now must reassess its validity given our current knowledge. **Figure 1** shows the relationship between biotypes and genetic groups identified using 11% and 3.5% sequence divergence limits, and what becomes obvious is that several groups are associated with two or more biotypes and others with none. As we have already discussed there is no unambiguous support for most of these biotypes. The one exception is the Mediterranean group, which has two subgroups, one containing Q and L, which are unable to induce silverleafing, and the other containing J, which does induce this symptom.

This leaves us with two choices: Either persist with the seemingly arbitrary unquantified use of biotypes or shift to the largely unambiguous, quantified structure using consensus sequences and the 3.5% sequence divergence limit.

CONCLUSION

Does this level of divergence warrant species-level separation? At present, the ultimate test for specific identity, namely the capacity to interbreed, has limited biological data available to support it. Moreover, because most of the groups are allopatric, interbreeding can be tested only in the laboratory or when one group invades a region occupied by another. However, existing evidence shows that matings between members of different groups either are unable to produce female progeny or produce proportions of female progeny that are well below those expected from within-group matings (6, 31, 47, 53, 86, 87, 94, 96, 102, 119, 143, 147). Further, hybrids are observed primarily in laboratory mating studies rather than in the field, and hybrid females are either sterile or far less fecund than their parents. Absence of biological data to support species designation is not unusual in such studies, as most species designations are based on morphological characters with little or no underlying knowledge of biology or ecology (99a).

In our case, given the limited biological data available, we rely on genetic data. The groups identified using the 3.5% divergence limits observed for mtCO1 are comparable to those observed using ribosomal ITS1 region, indicating that the differences identified by mtCO1 are generally applicable (18, 52, 54). Although species delineation using mtCO1 has been controversial (122), there is an increasing body of data (3, 27) that supports such species-level separation at 2–3% in vertebrates and invertebrates. While these data so far reflect only a limited range of taxa, the evidence is compelling and is consistent with the observed 3.5% break in divergence distributions (60) in *B. tabaci*. A key point when considering the observed breaks in the divergence frequency distributions is that they are representative of individuals belonging to clusters that are not uniformly dispersed across genetic space. Whereas the 11% break is not associated with patterns of reproductive isolation and so may represent a distant evolutionary event, the 3.5% break and resulting groupings suggest the lack of between-group gene flow. This finding is supported by the available data on reproductive isolation (143) and provides further evidence of distinct species. On this basis it is time to move from the recognition of *B. tabaci* as a conglomerate of biotypes and accept it is a complex of 24 morphologically indistinguishable species. These species can be clearly defined by comparisons against consensus sequences and delimited by 3.5% mtCO1 sequence pairwise genetic distance divergence.

SUMMARY POINTS

1. There is no definitive set of biological data that can quantitatively distinguish any *B. tabaci*. Differences in host association and capacity to disperse, to transmit begomoviruses, to develop resistance to insecticides, and to induce physiological changes have all been used to distinguish different groups of *B. tabaci*. These characters either are shared between genetic groups or show within-group variability.
2. Considerable genetic differences have been shown by using a wide array of methodologies, protein allozymes electrophoresis, RAPD-PCR, AFLP, mtCO1, and 16s, ribosomal ITS1, microsatellites, CAPS, and SCAR. mtCO1 genetic distances show two distinct breaks in the frequency distribution, one at 11% and the other at 3.5%. These breaks identify 11 groups at 11% divergence limits and 24 groups at 3.5%. Available biological data indicate at 3.5% the groups are reproductively isolated or nearly so. At 3.5%

the groups are not uniformly dispersed across genetic space and this suggests the lack of between-group gene flow. Consensus sequences for each of the groups have been determined, which allow unknown sequences to be assigned unambiguously.

3. Asymmetrical mating interference provides a clear mechanism that could contribute to the formation of the global genetic structure, but there is no evidence that *Wolbachia* or other bacterial symbionts are currently contributing to the genetic structure. However, the data are too limited to be conclusive.
4. *B. tabaci* is a complex of at least 24 morphologically indistinguishable species. These species can be clearly defined by comparison against consensus sequences and delimited by 3.5% mtCO1 sequence pairwise genetic distance divergence.

FUTURE ISSUES

1. The mechanism behind species recognition in *B. tabaci* is not known. We know recognition is achieved before attempts to copulate, but after body-to-body contact is made. This suggests that cuticle surface chemistry may be involved.
2. The identification of many different species within the *B. tabaci* complex using primarily mtCO1 sequence data and limited information on mating biology and behavior is bound to raise concern. This paper seeks to set a measurable benchmark and, in so doing, challenge the research community to further explore the question of species status. A focus on the evolution of mitochondrial DNA in *Bemisia* may yield further clues to its evolution.
3. Nuclear genes, where examined, support the mtCO1 phylogeny; however, further attention is needed on the evolution of nuclear genes as well as those from the mitochondrion. Sequencing of the *B. tabaci* genome would make considerable contribution to this.
4. The question of what makes some *B. tabaci* invasive remains. Some mechanisms have been identified and others proposed. The role of insecticide resistance and access to an invasion pathway bear further exploration.
5. A robust evolutionary frame provides a good basis for questions such as the evolution of host range and insecticide resistance. The topology of the *B. tabaci* phylogeny would appear to be influenced by the number and type of outgroup. Inclusion of more outgroups, particularly those belonging to the former *Lipaleyrodes* genus, and key basal ingroups may well help to stabilize topology and give further clues to the evolution of this species complex.

DISCLOSURE STATEMENT

The authors are not aware of any affiliations, memberships, funding, or financial holdings that might be perceived as affecting the objectivity of this review.

ACKNOWLEDGMENTS

We thank Steve Castle, Don Frohlich, and Dan Gerling for their comments on the manuscript. This work was funded partially by the Grains and Cotton Research and Development

Corporations and Horticulture Australia, and the National Basic Research Program of China (Project 2009CB119200 to S.S.L.).

LITERATURE CITED

1. Abdullahi I, Winter S, Atiri GI, Thottappilly G. 2003. Molecular characterization of whitefly, *Bemisia tabaci* (Hemiptera: Aleyrodidae) populations infesting cassava. *Bull. Entomol. Res.* 93:97–106
2. Anthony NM, Brown JK, Markham PG, ffrench-Constant RH. 1995. Molecular analysis of cyclodiene resistance-associated mutations among populations of the sweetpotato whitefly *Bemisia tabaci*. *Pestic. Biochem. Physiol.* 51:220–28
3. Barat M, Tarayre M, Atlan A. 2008. Genetic divergence and ecological specialisation of seed weevils (*Exapion* spp.) on gorses (*Ulex* spp.). *Ecol. Entomol.* 33:328–36
4. Barinaga M. 1993. Is devastating whitefly invader really a new species? *Science* 259:30
5. Bartlett AC, Gawel NJ. 1993. Determining whitefly species. *Science* 261:1333–34
6. **Bedford ID, Briddon RW, Brown JK, Rosell RC, Markham PG. 1994. Geminivirus transmission and biological characterization of *Bemisia tabaci* (Gennadius) biotypes from different geographic regions. *Ann. Appl. Biol.* 125:311–25**

   6. The first paper to attempt to systematically compare a number of different genetic groups of *B. tabaci*.
7. Bedford ID, Briddon RW, Markham PG, Brown JK, Rosell RC. 1993. A new species of *Bemisia* or biotype of *Bemisia tabaci* (Genn.) as a future pest of European agriculture. *Plant Health Eur. Single Market BCPC Monogr.* 54:381–86
8. **Bellows TS Jr, Perring TM, Gill RJ, Headrick DH. 1994. Description of a species of *Bemisia* (Homoptera: Aleyrodidae). *Ann. Entomol. Soc. Am.* 87:195–206**

   8. Initiated the species debate by raising the commonly named B biotype to species status.
9. Berry SD, Fondong VN, Rey C, Rogan D, Fauquet CM, Brown JK. 2004. Molecular evidence for five distinct *Bemisia tabaci* (Homoptera: Aleyrodidae) geographic haplotypes associated with cassava plants in sub-Saharan Africa. *Ann. Entomol. Soc. Am.* 97:852–59
10. Bethke JA, Paine TD, Nuessly GS. 1991. Comparative biology, morphometrics, and development of two populations of *Bemisia tabaci* (Homoptera: Aleyrodidae) on cotton and poinsettia. *Ann. Entomol. Soc. Am.* 84:407–11
11. **Bird J. 1957. A whitefly-transmitted mosaic of *Jatropha gossypifolia*. *Tech. Pap. Agric. Exp. Stn. Univ. P. R.* 22:1–35**

    11. The earliest paper to provide evidence of significant underlying genetic structure within *B. tabaci*.
12. Bird J, Maramorosch K. 1978. Viruses and virus diseases associated with whiteflies. *Adv. Virus Res.* 22:55–110
13. Blackmer JL, Byrne DN. 1993. Environmental and physiological factors influencing phototactic flight of *Bemisia tabaci*. *Physiol. Entomol.* 18:336–42
14. Blackmer JL, Byrne DN. 1993. Flight behavior of *Bemisia tabaci* in a vertical flight chamber: effect of time of day, sex, age and host quality. *Physiol. Entomol.* 18:223–32
15. Blackmer JL, Byrne DN, Tu Z. 1995. Behavioral, morphological, and physiological traits associated with migratory *Bemisia tabaci* (Homoptera: Aleyrodidae). *J. Insect Behav.* 8:251–67
16. Boukhatem N, Jdaini S, Mukovski Y, Jacquemin JM, Bouali A. 2007. Identification of *Bemisia tabaci* (Gennadius) (Homoptera: Aleyrodidae) based on RAPD and design of two SCAR markers. *J. Biol. Res.* 8:167–76
17. Boykin LM, Kubatko LS, Lowrey TK. 2010. Comparison of rooting phylogenetic trees: a case study using Orcuttieae (Poaceae: Chloridoideae). *Mol. Phylogenet. Evol.* 54:687–700
18. **Boykin LM, Shatters RG Jr, Rosell RC, McKenzie CL, Bagnall RA, et al. 2007. Global relationships of *Bemisia tabaci* (Hemiptera: Aleyrodidae) revealed using Bayesian analysis of mitochondrial COI DNA sequences. *Mol. Phylogenet. Evol.* 44:1306–19**

    18. The first study to create a comprehensive phylogeny of *B. tabaci* using mtCO1 and thus overcome many of the problems associated with previous analyses.
19. Broadbent AB, Foottit RG, Murphy GD. 1989. Sweetpotato whitefly *Bemisia tabaci* (Gennadius) (Homoptera: Aleyrodidae), a potential insect pest in Canada. *Can. Entomol.* 121:1027–28
20. Brown JK, Bird J. 1995. Variability within the *Bemisia tabaci* species complex and its relation to new epidemics caused by geminiviruses. *Ceiba* 36:73–80
21. Brown JK, Bird J, Frohlich D, Rosell RC, Bedford ID, Markham PG. 1996. The relevance of variability within the *Bemisia tabaci* species complex to epidemics caused by subgroup III geminiviruses. See Ref. 66, pp. 77–89

22. Brown JK, Coats SA, Bedford ID, Markham PG, Bird J. 1992. Biotypic characterization of *Bemisia tabaci* populations based on esterase profiles, DNA fingerprinting, virus transmission, and bioassay to key host plant species. *Phytopathology* 82:1104
23. Brown JK, Coats SA, Bedford ID, Markham PG, Bird J, Frohlich DR. 1995. Characterization and distribution of esterase electromorphs in the whitefly, *Bemisia tabaci* (Genn.) (Homoptera: Aleyrodidae). *Biochem. Genet.* 33:205–14
24. Brown JK, Frohlich DR, Rosell RC. 1995. The sweetpotato or silverleaf whiteflies: biotypes of *Bemisia tabaci* or a species complex? *Annu. Rev. Entomol.* 40:511–34
25. Brown JK, Perring TM, Cooper AD, Bedford ID, Markham PG. 2000. Genetic analysis of *Bemisia* (Hemiptera: Aleyrodidae) populations by isoelectric focusing electrophoresis. *Biochem. Genet.* 38:13–25
26. Burban C, Fishpool LDC, Fauquet C, Fargette D, Thouvenel JC. 1992. Host-associated biotypes within West African populations of the whitefly *Bemisia tabaci* (Genn.), (Hom., Aleyrodidae). *J. Appl. Entomol.* 113:416–23
27. Burns J, Janzen D, Hajibabaei M, Hallwachs W, Hebert P. 2008. DNA barcodes and cryptic species of skipper butterflies in the genus Perichares in Area de Conservacion Guanacaste, Costa Rica. *Proc. Natl. Acad. Sci. USA* 105:6350–55
28. Byrne DN. 1999. Migration and dispersal by the sweet potato whitefly, *Bemisia tabaci*. *Agric. For. Meteorol.* 97:309–16
29. Byrne FJ, Bedford ID, Devonshire AL, Markham PG. 1995. Esterase variation and squash induction in B-type *Bemisia tabaci* (Homoptera: Aleyrodidae). *Bull. Entomol. Res.* 85:175–79
30. Byrne FJ, Cahill M, Denholm I, Devonshire AL. 1994. A biochemical and toxicological study of the role of insensitive acetylcholinesterase in organophosphorus resistant *Bemisia tabaci* (Homoptera: Aleyrodidae) from Israel. *Bull. Entomol. Res.* 84:179–84
31. Byrne FJ, Cahill M, Denholm I, Devonshire AL. 1995. Biochemical identification of interbreeding between B-type and non B-type strains of tobacco whitefly *Bemisia tabaci*. *Biochem. Genet.* 33:13–23
32. Cahill M, Byrne FJ, Denholm I, Devonshire AL, Gorman KJ. 1994. Insecticide resistance in *Bemisia tabaci*. *Pestic. Sci.* 42:137–39
33. Cahill M, Byrne FJ, Gorman K, Denholm I, Devonshire AL. 1995. Pyrethroid and organophosphate resistance in the tobacco whitefly *Bemisia tabaci* (Homoptera: Aleyrodidae). *Bull. Entomol. Res.* 85:181–87
34. Cahill M, Denholm I, Ross G, Gorman K, Johnston D. 1996. Relationship between bioassay data and the simulated field performance of insecticides against susceptible and resistant adult *Bemisia tabaci* (Homoptera: Aleyrodidae). *Bull. Entomol. Res.* 86:109–16
35. Calvert LA, Cuervo M, Arroyave JA, Constantino LM, Bellotti A, Frohlich D. 2001. Morphological and mitochondrial DNA marker analyses of whiteflies (Homoptera: Aleyrodidae) colonizing cassava and beans in Colombia. *Ann. Entomol. Soc. Am.* 94:512–19
36. Campbell BC, Duffus JE, Baumann P. 1993. Determining whitefly species. *Science* 261:1333–35
37. Cervera MT, Cabezas JA, Simon B, Martinez-Zapater JM, Beitia F, Cenis JL. 2000. Genetic relationships among biotypes of *Bemisia tabaci* (Hemiptera: Aleyrodidae) based on AFLP analysis. *Bull. Entomol. Res.* 90:391–96
38. Cheek S, Macdonald O. 1994. Statutory controls to prevent the establishment of *Bemisia tabaci* in the United Kingdom. *Pestic. Sci.* 42:135–42
39. Chiel E, Gottlieb Y, Zchori-Fein E, Mozes-Daube N, Katzir N, et al. 2007. Biotype-dependent secondary symbiont communities in sympatric populations of *Bemisia tabaci*. *Bull. Entomol. Res.* 97:407–13
40. Chu D, Liu GX, Tao YL, Zhang YJ. 2006. Diversity of endosymbionts in *Bemisia tabaci* (Gennadius) complex and their biological implications. *Acta Entomol. Sin.* 49:687–94
41. Chu D, Wan FH, Tao YL, Liu GX, Fan ZX, Bi YP. 2008. Genetic differentiation of *Bemisia tabaci* (Gennadius) (Hemiptera: Aleyrodidae) biotype Q based on mitochondrial DNA markers. *Insect Sci.* 15:115–23
42. Chu D, Zhang YJ, Cong B, Xu BY, Wu QJ. 2004. Developing sequence characterized amplified regions (SCARs) to identify *Bemisia tabaci* and *Trialeurodes vaporariorum*. *Plant Prot.* 30:27–30
43. Coats SA, Brown JK, Hendrix DL. 1994. Biochemical characterization of biotype-specific esterases in the whitefly, *Bemisia tabaci* Genn (Homoptera: Aleyrodidae). *Insect Biochem. Mol. Biol.* 24:723–28

44. Cock MJW, ed. 1986. Bemisia tabaci, *A Literature Survey on the Cotton Whitefly with an Annotated Bibliography*. Food and Agriculture Organization of the U.N., CAB Int. Inst. Biol. Control. 121 pp.
45. Costa AS, Russell LM. 1975. Failure of *Bemisia tabaci* to breed on cassava plants in Brazil (Homoptera: Aleyrodidae). *Cienc. Cult.* 27:388–90
46. Costa HS, Brown JK. 1991. Variation in biological characteristics and esterase patterns among populations of *Bemisia tabaci*, and the association of one population with silverleaf symptom induction. *Entomol. Exp. Appl.* 61:211–19
47. Costa HS, Brown JK, Sivasupramaniam S, Bird J. 1993. Regional distribution, insecticide resistance, and reciprocal crosses between the A-biotype and B-biotype of *Bemisia tabaci*. *Insect Sci. Appl.* 14:255–66
48. Dalmon A, Halkett F, Granier M, Delatte H, Peterschmitt M. 2008. Genetic structure of the invasive pest *Bemisia tabaci*: evidence of limited but persistent genetic differentiation in glasshouse populations. *Heredity* 100:316–25
49. De Barro P, Khan S. 2007. Adult *Bemisia tabaci* biotype B can induce silverleafing in squash. *Bull. Entomol. Res.* 97:433–36
50. De Barro PJ. 2005. Genetic structure of the whitefly *Bemisia tabaci* in the Asia-Pacific region revealed using microsatellite markers. *Mol. Ecol.* 14:3695–718
51. De Barro PJ, Driver F. 1997. Use of RAPD PCR to distinguish the B biotype from other biotypes of *Bemisia tabaci* (Gennadius) (Hemiptera: Aleyrodidae). *Aust. J. Entomol.* 36:149–52
52. De Barro PJ, Driver F, Trueman JWH, Curran J. 2000. Phylogenetic relationships of world populations of *Bemisia tabaci* (Gennadius) using ribosomal ITS1. *Mol. Phylogenet. Evol.* 16:29–36
53. De Barro PJ, Hart PJ. 2000. Mating interactions between two biotypes of the whitefly, *Bemisia tabaci* (Hemiptera: Aleyrodidae) in Australia. *Bull. Entomol. Res.* 90:103–12
54. De Barro PJ, Trueman JWH, Frohlich DR. 2005. *Bemisia argentifolii* is a race of *B. tabaci* (Hemiptera: Aleyrodidae): the molecular genetic differentiation of *B. tabaci* populations around the world. *Bull. Entomol. Res.* 95:193–203
55. de la Rua P, Simon B, Cifuentes D, Martinez-Mora C, Cenis JL. 2006. New insights into the mitochondrial phylogeny of the whitefly *Bemisia tabaci* (Hemiptera: Aleyrodidae) in the Mediterranean Basin. *J. Zool. Syst. Evol. Res.* 44:25–33
56. Delatte H, David P, Granier M, Lett JM, Goldbach R, et al. 2006. Microsatellites reveal extensive geographical, ecological and genetic contacts between invasive and indigenous whitefly biotypes in an insular environment. *Genet. Res.* 87:109–24
57. Delatte H, Duyck PF, Triboire A, David P, Becker N, et al. 2009. Differential invasion success among biotypes: case of *Bemisia tabaci*. *Biol. Invas.* 11:1059–70
58. Delatte H, Reynaud B, Granier M, Thornary L, Lett JM, et al. 2005. A new silverleaf-inducing biotype Ms of *Bemisia tabaci* (Hemiptera: Aleyrodidae) indigenous to the islands of the south-west Indian Ocean. *Bull. Entomol. Res.* 95:29–35
59. Demichelis S, Bosco D, Manino A, Marian D, Caciagli P. 2000. Distribution of *Bemisia tabaci* (Hemiptera: Aleyrodidae) biotypes in Italy. *Can. Entomol.* 132:519–27
60. Dinsdale A, Cook L, Riginos C, Buckley YM, De Barro P. 2010. Refined global analysis of *Bemisia tabaci* (Hemiptera: Sternorrhyncha: Aleyrodoidea: Aleyrodidae) mitochondrial cytochrome oxidase 1 to identify species level genetic boundaries. *Ann. Entomol. Soc. Am.* 103:196–208

60. Quantifiable bounds to delineate the genetic structure of *B. tabaci* were produced for the first time, enabling unambiguous assignment to species group.

61. Dittrich V, Ernst GH, Ruesch O, Uk S. 1990. Resistance mechanisms in sweetpotato whitefly (Homoptera: Aleyrodidae) populations from Sudan, Turkey, Guatemala, and Nicaragua. *J. Econ. Entomol.* 83:1665–70
62. Elbert A, Nauen R. 2000. Resistance of *Bemisia tabaci* (Homoptera: Aleyrodidae) to insecticides in southern Spain with special reference to neonicotinoids. *Pest Manag. Sci.* 56:60–64
63. Fishpool LDC, Burban C. 1994. *Bemisia tabaci*: the whitefly vector of African cassava mosaic geminivirus. *Trop. Sci.* 34:55–72
64. Frohlich DR, Torres-Jerez I, Bedford ID, Markham PG, Brown JK. 1999. A phylogeographical analysis of the *Bemisia tabaci* species complex based on mitochondrial DNA markers. *Mol. Ecol.* 8:1683–91

64. The first phylogeographic study of *B. tabaci*.

65. Gawel NJ, Bartlett AC. 1993. Characterization of differences between whiteflies using RAPD-PCR. *Insect Mol. Biol.* 2:33–38

66. Gerling D, Mayer RT, eds. 1996. Bemisia*: 1995. Taxonomy, Biology, Damage, Control and Management*. Andover, UK: Intercept
67. Gill RJ. 1992. A review of the sweetpotato whitefly in Southern California. *Pan-Pac. Entomol.* 68:144–52
68. Gottlieb Y, Ghanim M, Chiel E, Gerling D, Portnoy V, et al. 2006. Identification and localization of a *Rickettsia* sp. in *Bemisia tabaci* (Homoptera: Aleyrodidae). *Appl. Environ. Microbiol.* 72:3646–52
69. Gottlieb Y, Ghanim M, Gueguen G, Kontsedalov S, Vavre F, et al. 2008. Inherited intracellular ecosystem: Symbiotic bacteria share bacteriocytes in whiteflies. *FASEB J.* 22:2591–99
70. Guirao P, Beitia F, Cenis JL. 1997. Biotype determination of Spanish populations of *Bemisia tabaci* (Hemiptera: Aleyrodidae). *Bull. Entomol. Res.* 87:587–93
71. Hidalgo SO, Leon QG, Lindo EO, Vaughan RM. 1975. *Informe de la mission de studio de al mosca blanc*. Banco Nacionale de Nicaragua, Comision Nacionale del Algodon and Misisterio de Agricultura y Ganaderia. 6 pp.
72. Homam BH, Shedeed MI, Mohamed MA. 2005. Evidence for presence many biotypes of *Bemisia tabaci* (Genn.) within Egypt. *Egypt J. Agric. Res.* 83:141–50
73. Horowitz AR, Kontsedalov S, Khasdan V, Ishaaya I. 2005. Biotypes B and Q of *Bemisia tabaci* and their relevance to neonicotinoid and pyriproxyfen resistance. *Arch. Insect Biochem. Physiol.* 58:216–25
74. Hsieh CH, Wang CH, Ko CC. 2005. Identification of biotypes of *Bemisia tabaci* (Hemiptera: Aleyrodidae) in Taiwan based on mitochondrial 16S rDNA sequences. *Formos. Entomol.* 25:255–67
75. Hsieh CH, Wang CH, Ko CC. 2006. Analysis of *Bemisia tabaci* (Hemiptera: Aleyrodidae) species complex and distribution in Eastern Asia based on mitochondrial DNA markers. *Ann. Entomol. Soc. Am.* 99:768–75
76. Inbar M, Gerling D. 2008. Plant-mediated interactions between whiteflies, herbivores, and natural enemies. *Annu. Rev. Entomol.* 53:431–48
77. Jones DR. 2003. Plant viruses transmitted by whiteflies. *Eur. J. Plant Pathol.* 109:195–219
78. Khasdan V, Levin I, Rosner A, Morin S, Kontsedalov S, et al. 2005. DNA markers for identifying biotypes B and Q of *Bemisia tabaci* (Hemiptera: Aleyrodidae) and studying population dynamics. *Bull. Entomol. Res.* 95:605–13
79. Ko CC, Hung YC, Wang CH. 2007. Sequence characterized amplified region markers for identifying biotypes of *Bemisia tabaci* (Hem., Aleyrodidae). *J. Appl. Entomol.* 131:542–47
80. Legg JP, French R, Rogan D, Okao-Okuja G, Brown JK. 2002. A distinct *Bemisia tabaci* (Gennadius) (Hemiptera: Sternorrhyncha: Aleyrodidae) genotype cluster is associated with the epidemic of severe cassava mosaic virus disease in Uganda. *Mol. Ecol.* 11:1219–29
81. Legg JP, Gibson RW, Otim-Nape GW. 1994. Genetic polymorphism among Ugandan populations of *Bemisia tabaci* (Gennadius) (Homoptera: Aleyrodidae), vector of African cassava mosaic geminivirus. *Trop. Sci.* 34:73–81
82. Li ZX. 2006. Molecular phylogenetic analysis reveals at least five genetic races of *Bemisia tabaci* in China. *Phytoparasitica* 34:431–40
83. Li ZX, Hu DX. 2005. Rapid identification of B biotype of *Bemisia tabaci* (Homoptera: Aleyrodidae) based on analysis of internally transcribed spacer 1 sequence. *Insect Sci.* 12:421–27
84. Li ZX, Lin HZ, Guo XP. 2007. Prevalence of *Wolbachia* infection in *Bemisia tabaci*. *Curr. Microbiol.* 54:467–71
85. Lima LHC, Navia D, Inglis PW, Oliveira MRVd. 2000. Survey of *Bemisia tabaci* (Gennadius) (Hemiptera: Aleyrodidae) biotypes in Brazil using RAPD markers. *Genet. Mol. Biol.* 23:781–85
86. Liu HY, Cohen S, Duffus JE. 1992. The use of isozyme patterns to distinguish sweetpotato whitefly (*Bemisia tabaci*) biotypes. *Phytoparasitica* 20:187–94
87. Liu SS, De Barro PJ, Xu J, Luan JB, Zang LS, et al. 2007. Asymmetric mating interactions drive widespread invasion and displacement in a whitefly. *Science* 318:1769–72

87. Identified a key mechanism that enables the invasive B biotype to displace competitors.

88. Lopez-Avila A, Cock MJW. 1986. Economic damage. See Ref. 44, pp. 51–53
89. Luan JB, Ruan YM, Zhang L, Liu SS. 2008. Pre-copulation intervals, copulation frequencies, and initial progeny sex ratios in two biotypes of whitefly, *Bemisia tabaci*. *Entomol. Exp. Appl.* 129:316–24
90. Ma DY, Gorman K, Devine G, Luo WC, Denholm I. 2007. The biotype and insecticide-resistance status of whiteflies, *Bemisia tabaci* (Hemiptera: Aleyrodidae), invading cropping systems in Xinjiang Uygur Autonomous Region, northwestern China. *Crop. Prot.* 26:612–17

91. Ma DY, Li XC, Dennehy TJ, Lei CL, Wang M, et al. 2009. Utility of mtCO1 polymerase chain reaction-restriction fragment length polymorphism in differentiating between Q and B whitefly *Bemisia tabaci* biotypes. *Insect Sci. Appl.* 16:107–14
92. Mabbett T. 2004. Mating interactions of *Bemisia tabaci* biotypes in Cyprus. *Resist. Pest Manag. Newsl.* 13:3–4
93. Mallet J. 2005. Hybridization as an invasion of the genome. *Trends Ecol. Evol.* 20:229–37
94. Maruthi MN, Colvin J, Seal S. 2001. Mating compatibility, life-history traits, and RAPD-PCR variation in *Bemisia tabaci* associated with the cassava mosaic disease pandemic in East Africa. *Entomol. Exp. Appl.* 99:13–23
95. Maruthi MN, Colvin J, Seal S, Gibson G, Cooper J. 2002. Co-adaptation between cassava mosaic geminiviruses and their local vector populations. *Virus Res.* 86:71–85
96. Maruthi MN, Colvin J, Thwaites RM, Banks GK, Gibson G, Seal SE. 2004. Reproductive incompatibility and cytochrome oxidase I gene sequence variability among host-adapted and geographically separate *Bemisia tabaci* populations (Hemiptera: Aleyrodidae). *Syst. Entomol.* 29:560–68
97. Maruthi MN, Seal S, Colvin J, Briddon RW, Bull SE. 2004. East African cassava mosaic Zanzibar virus–a recombinant begomovirus species with a mild phenotype. *Arch. Virol.* 149:2365–77
98. Maynard DN, Cantliffe DJ. 1989. *Squash silverleaf and tomato ripening: new vegetable disorders in Florida.* Univ. Florida, Vegetable Crop Fact Sheet VC-37
99. Mound LA. 1963. Host-correlated variation in *Bemisia tabaci* (Gennadius) (Homoptera: Aleyrodidae). *Proc. R. Entomol. Soc. Lond. A* 38:171–80

99. The first study to identify the host-plant-related morphological plasticity of *B. tabaci.*

99a. Miller SE. 2007. DNA barcoding and the renaissance of taxonomy. *Proc. Natl. Acad. Sci. USA* 104:4775–76
100. Mound LA. 1983. Biology and identity of whitefly vectors of plant pathogens. In *Plant Virus Epidemiology. The Spread and Control of Insect-borne Viruses*, ed. RT Plumb, JM Thresh, pp. 305–13. Oxford: Blackwell Publ.
101. Mound LA, Halsey SH. 1978. *Whitefly of the World. A Systematic Catalogue of the Aleyrodidae (Homoptera) with Host Plant and Natural Enemy Data*. Chichester, UK: Wiley. 340 pp.
102. Moya A, Guirao P, Cifuentes D, Beitia F, Cenis JL. 2001. Genetic diversity of Iberian populations of *Bemisia tabaci* (Hemiptera: Aleyrodidae) based on random amplified polymorphic DNA-polymerase chain reaction. *Mol. Ecol.* 10:891–97
103. Nirgianaki A, Banks GK, Frohlich DR, Veneti Z, Braig HR, et al. 2003. *Wolbachia* infections of the whitefly *Bemisia tabaci*. *Curr. Microbiol.* 47:93–101
104. Ochando MD, Callejas C. 2003. Molecular identification of insect quarantine pests by RAPD-PCR. *Bull. OILB/SROP* 26:181–86
105. Oliveira MRV, Henneberry TJ, Anderson P. 2001. History, current status, and collaborative research projects for *Bemisia tabaci*. *Crop. Prot.* 20:709–23
106. Omondi BA, Sseruwagi P, Obeng-Ofori D, Danquah EY, Kyerematen RA. 2005. Mating interactions between okra and cassava biotypes of *Bemisia tabaci* (Homoptera: Aleyrodidae) on eggplant. *Int. J. Trop. Insect Sci.* 25:159–67
107. Pascual S. 2006. Mechanisms in competition, under laboratory conditions, between Spanish biotypes B and Q of *Bemisia tabaci* (Gennadius). *Span. J. Agric. Res.* 4:351–54
108. Pascual S, Callejas C. 2004. Intra- and interspecific competition between biotypes B and Q of *Bemisia tabaci* (Hemiptera: Aleyrodidae) from Spain. *Bull. Entomol. Res.* 94:369–75
109. Perring TM. 2001. The *Bemisia tabaci* species complex. *Crop. Prot.* 20:725–37
110. Perring TM, Cooper A, Kazmer DJ. 1992. Identification of the poinsettia strain of *Bemisia tabaci* (Homoptera: Aleyrodidae) on broccoli by electrophoresis. *J. Econ. Entomol.* 85:1278–84
111. Perring TM, Cooper AD, Rodriguez RJ, Farrar CA, Bellows TS Jr. 1993. Identification of a whitefly species by genomic and behavioral studies. *Science* 259:74–77
112. Perring TM, Farrar CA, Cooper AD. 1994. Mating behavior and competitive displacement in whiteflies. In *Supplement to the Five-Year National Research and Action Plan*, ed. TJ Henneberry, NC Toscano, RM Faust, JR Coppedge, p. 25. USDA:ARS-125
113. Perring TM, Farrar CA, Cooper AD, Bellows TS Jr. 1993. Determining whitefly species. *Science* 261:1334–35

114. Perring TM, Symmes EJ. 2006. Courtship behavior of *Bemisia argentifolii* (Hemiptera: Aleyrodidae) and whitefly mate recognition. *Ann. Entomol. Soc. Am.* 99:598–606
115. Pollock DD, Zwickl DJ, McGuire JA, Hillis DM. 2002. Increased taxon sampling is advantageous for phylogenetic inference. *Syst. Biol.* 51:664–71
116. Qiu BL, Ren SX, Mandour NS, Wen SY. 2006. Population differentiation of *Bemisia tabaci* (Gennadius) (Hemiptera: Aleyrodidae) by DNA polymorphism in China. *J. Entomol. Res.* 30:1–6
117. Rabello AR, Queiroz PR, Simoes KCC, Hiragi CO, Lima LHC, et al. 2008. Diversity analysis of *Bemisia tabaci* biotypes: RAPD, PCR-RFLP and sequencing of the ITS1 rDNA region. *Genet. Mol. Biol.* 31:585–90
118. Rekha AR, Maruthi MN, Muniyappa V, Colvin J. 2005. Occurrence of three genotypic clusters of *Bemisia tabaci* and the rapid spread of the B biotype in south India. *Entomol. Exp. Appl.* 117:221–33
119. Ronda MA, Adan A, Cifuentes D, Cenis JL, Beitia F. 1999. *Laboratory evidence of interbreeding between biotypes of* Bemisia tabaci *(Homoptera, Aleyrodidae) present in Spain*. Presented at V11th Int. Plant Virus Epidemiol. Symp.-Plant Virus Epidemiol.: Curr. Status Future Prospects, Aguadulce, Spain
120. Rosell RC, Bedford ID, Frohlich DR, Gill RJ, Brown JK, Markham PG. 1997. Analysis of morphological variation in distinct populations of *Bemisia tabaci* (Homoptera: Aleyrodidae). *Ann. Entomol. Soc. Am.* 90:575–89

120. Demonstrated that the morphological characters used to separate *B. tabaci* from *B. argentifolii* were themselves variable and could not reliably separate any *B. tabaci*.

121. Ruan YM, Liu SS. 2005. Detection and phylogenetic analysis of prokaryotic endosymbionts in *Bemisia tabaci*. *Acta Entomol. Sin.* 48:859–65
122. Rubinoff D, Cameron A, Will K. 2006. A genomic perspective on the shortcomings of mitochondrial DNA for "barcoding" identification. *J. Hered.* 97:581–94
123. Russell LM. 1958. Synonyms of *Bemisia tabaci* (Gennadius) (Homoptera, Aleyrodidae). *Bull. Brooklyn Entomol. Soc.* 52:122–23
124. Shankarappa KS, Rangaswamy KT, Narayana DSA, Rekha AR, Raghavendra N, et al. 2007. Development of silverleaf assay, protein and nucleic acid-based diagnostic techniques for the quick and reliable detection and monitoring of biotype B of the whitefly, *Bemisia tabaci* (Gennadius). *Bull. Entomol. Res.* 97:503–13
125. Sharaf N, Hasan H. 2003. The identification of two biotypes of *Bemisia tabaci* in Jordan. *Agric. Sci.* 30:101–8
126. Shatters RG Jr, Powell CA, Boykin LM, He LS, McKenzie CL. 2009. Improved DNA barcoding method for *Bemisia tabaci* and related Aleyrodidae: development of universal and *Bemisia tabaci* biotype-specific mitochondrial cytochrome c oxidase chain reaction primers. *J. Econ. Entomol.* 102:750–58
127. Simon B, Cenis JL, Beitia F, Saif K, Moreno IM, et al. 2003. Genetic structure of field populations of begomoviruses and of their vector *Bemisia tabaci* in Pakistan. *Phytopathology* 93:1422–29
128. Simon B, Cenis JL, Demichelis S, Rapisarda C, Caciagli P, Bosco D. 2003. Survey of *Bemisia tabaci* (Hemiptera: Aleyrodidae) biotypes in Italy with the description of a new biotype (T) from *Euphorbia characias*. *Bull. Entomol. Res.* 93:259–64
129. Simón B, Cenis JL, Rua PDL. 2007. Distribution patterns of the Q and B biotypes of *Bemisia tabaci* in the Mediterranean Basin based on microsatellite variation. *Entomol. Exp. Appl.* 124:327–36
130. South A, Levan K, Leombruni L, Orians CM, Lewis SM. 2008. Examining the role of cuticular hydrocarbons in firefly species recognition. *Ethology* 114:916–24
131. Sseruwagi P, Legg JP, Maruthi MN, Colvin J, Rey MEC, Brown JK. 2005. Genetic diversity of *Bemisia tabaci* (Gennadius) (Hemiptera: Aleyrodidae) populations and presence of the B biotype and a non-B biotype that can induce silverleaf symptoms in squash, in Uganda. *Ann. Appl. Biol.* 147:253–65
132. Sseruwagi P, Maruthi MN, Colvin J, Rey MEC, Brown JK, Legg JP. 2006. Colonization of noncassava plant species by cassava whiteflies (*Bemisia tabaci*) in Uganda. *Entomol. Exp. Appl.* 119:145–53
133. Sseruwagi P, Sserubombwe WS, Legg JP, Ndunguru J, Thresh JM. 2004. Methods of surveying the incidence and severity of cassava mosaic disease and whitefly vector populations on cassava in Africa: a review. *Virus Res.* 100:129–42
134. Stouthamer R, Breeuwer JAJ, Hurst GDD. 1999. *Wolbachia pipientis*: microbial manipulator of arthropod reproduction. *Annu. Rev. Microbiol.* 53:71–102
135. Tahiri A, Sekkat A, Bennani A, Granier M, Delvare G, Peterschmitt M. 2006. Distribution of tomato-infecting begomoviruses and *Bemisia tabaci* biotypes in Morocco. *Ann. Appl. Biol.* 149:175–86

136. Viscarret MM, Torres-Jerez I, Agostini de Manero E, Lopez SN, Botto EE, Brown JK. 2003. Mitochondrial DNA evidence for a distinct New World group of *Bemisia tabaci* (Gennadius) (Hemiptera: Aleyrodidae) indigenous to Argentina and Bolivia, and presence of the Old World B biotype in Argentina. *Ann. Entomol. Soc. Am.* 96:65–72
137. Weeks AR, Velten R, Stouthamer R. 2003. Incidence of a new sex-ratio-distorting endosymbiotic bacterium among arthropods. *Proc. R. Soc. Biol. Sci. Ser. B* 270:1857–65
138. Wool D, Calvert L, Constantino LM, Bellotti AC, Gerling D. 1994. Differentiation of *Bemisia tabaci* (Genn.) (Hom., Aleyrodidae) populations in Colombia. *J. Appl. Entomol.* 117:122–34
139. Wool D, Gerling D, Bellotti A, Morales F, Nolt B. 1991. Spatial and temporal genetic variation in populations of the whitefly *Bemisia tabaci* (Genn.) in Israel and Colombia: an interim report. *Insect Sci. Appl.* 12:225–30
140. Wool D, Gerling D, Bellotti AC, Morales FJ. 1993. Esterase electrophoretic variation in *Bemisia tabaci* (Genn.) (Hom., Aleyrodidae) among host plants and localities in Israel. *J. Appl. Entomol.* 115:185–96
141. Wool D, Gerling D, Nolt BL, Constantino LM, Bellotti AC, Morales FJ. 1989. The use of electrophoresis for identification of adult whiteflies (Homoptera, Aleyrodidae) in Israel and Colombia. *J. Appl. Entomol.* 107:344–50
142. Wu XX, Li ZX, Hu DX, Shen ZR. 2003. Identification of Chinese populations of *Bemisia tabaci* (Gennadius) by analyzing ITS1 sequence. *Prog. Nat. Sci.* 13:276–81
143. Xu J, Liu SS, De Barro P. 2010. Reproductive incompatibility among genetic groups of *Bemisia tabaci* supports the proposition that the whitefly is a cryptic species complex. *Bull. Entomol. Res.* 100:359–66
144. Yokomi RK, Hoelmer KA, Osborne LS. 1990. Relationships between the sweetpotato whitefly and the squash silverleaf disorder. *Phytopathology* 80:895–900
145. Zang LS, Chen WQ, Liu SS. 2006. Comparison of performance on different host plants between the B biotype and a non-B biotype of *Bemisia tabaci* from Zhejiang, China. *Entomol. Exp. Appl.* 121:221–27
146. Zang LS, Jiang T, Xu J, Liu SS, Zhang YJ. 2006. SCAR molecular markers of the B biotype and two non-B populations of the whitefly, *Bemisia tabaci* (Hemiptera: Aleyrodidae). *J. Agric. Biotechnol.* 14:208–12
147. Zang LS, Liu SS. 2007. A comparative study on mating behavior between the B biotype and a non-B biotype of *Bemisia tabaci* (Hemiptera: Aleyrodidae) from Zhejiang, China. *J. Insect Behav.* 20:157–71
148. Zchori-Fein E, Brown JK. 2002. Diversity of prokaryotes associated with *Bemisia tabaci* (Gennadius) (Hemiptera: Aleyrodidae). *Ann. Entomol. Soc. Am.* 95:711–18
149. Zhang LP, Zhang YJ, Zhang WJ, Wu QJ, Xu BY, Chu D. 2005. Analysis of genetic diversity among different geographical populations and determination of biotypes of *Bemisia tabaci* in China. *J. Appl. Entomol.* 129:121–28
150. Zwickl DJ, Hillis DM. 2002. Increased taxon sampling greatly reduces phylogenetic error. *Syst. Biol.* 51:588–98

Insect Seminal Fluid Proteins: Identification and Function

Frank W. Avila, Laura K. Sirot, Brooke A. LaFlamme, C. Dustin Rubinstein, and Mariana F. Wolfner

Department of Molecular Biology and Genetics, Cornell University, Ithaca, New York 14853; email: fwa5@cornell.edu, ls286@cornell.edu, bal44@cornell.edu, cdr25@cornell.edu, mfw5@cornell.edu

Annu. Rev. Entomol. 2011. 56:21–40

First published online as a Review in Advance on September 24, 2010

The *Annual Review of Entomology* is online at ento.annualreviews.org

This article's doi: 10.1146/annurev-ento-120709-144823

0066-4170/11/0107-0021$20.00

Key Words

egg production, mating receptivity, sperm storage, mating plug, sperm competition, feeding, reproduction, Acp

Abstract

Seminal fluid proteins (SFPs) produced in reproductive tract tissues of male insects and transferred to females during mating induce numerous physiological and behavioral postmating changes in females. These changes include decreasing receptivity to remating; affecting sperm storage parameters; increasing egg production; and modulating sperm competition, feeding behaviors, and mating plug formation. In addition, SFPs also have antimicrobial functions and induce expression of antimicrobial peptides in at least some insects. Here, we review recent identification of insect SFPs and discuss the multiple roles these proteins play in the postmating processes of female insects.

INTRODUCTION

Seminal fluid protein (SFP): protein expressed from tissues of the male reproductive tract and likely transferred to females during mating

RT: reproductive tract

AG: accessory gland

In many insect species, mating initiates a behavioral and physiological switch in females, triggering responses in several processes related to fertility. Receipt of seminal fluid—a mixture of proteins and other molecules—by the female is a major component of this switch. Insect seminal fluid proteins (SFPs) are the products of male reproductive tract (RT) secretory tissues—accessory glands (AGs), seminal vesicles, ejaculatory duct, ejaculatory bulb, and testes. SFPs are transferred to females with sperm during mating. They are major effectors of a wide range of female postmating responses, including changing female likelihood of remating, increasing ovulation and egg-laying rates, changing female flight and feeding behavior, inducing antimicrobial activities, and modulating sperm storage parameters. Absence of SFPs from the ejaculate adversely affects the reproductive success of both sexes. SFPs identified to date represent numerous protein classes, including proteases/protease inhibitors, lectins, prohormones, peptides, and protective proteins such as antioxidants; these protein classes are present in the ejaculate of organisms from arthropods to mammals (121). While nonprotein molecules are also present in seminal fluid (e.g., prostaglandins in crickets: 86; steroid hormones in mosquitoes: 124), research on the effects of seminal fluid receipt has focused largely on the action of SFPs. Although the focus of this review is on insect SFPs, progress on the identification and function of SFPs in tick species is also included.

The past few years have witnessed an explosion in the identification and functional analyses of SFPs in insects due to new proteomic and RNA interference technologies. Because earlier results in this field were reviewed by Gillott (56), Chen (29), and Leopold (84), we focus here primarily on recent developments, referring readers to those comprehensive reviews for details on earlier studies.

Dissection of the nature and function of insect SFPs has relevance beyond understanding insect reproductive molecules and their actions. SFPs provide intriguing targets for the control of disease vectors and agricultural pests. As we discuss below, SFPs alter reproductive and/or feeding behaviors in a number of arthropods, including insects that cause economic damage or spread disease. In many of these species, there are no approved and effective methods to control the damage they cause. For example, vaccines for tick-borne pathogens have been developed for a limited number of tick antigens (35) and the principal vectors of malaria and dengue fever—the mosquitoes *Aedes aegypti* and *Anopheles gambiae*, respectively—have been notoriously difficult to control. Thus, the best current method of limiting these diseases is to control the spread of their insect vectors. In another area, increased resistance to pesticides has made population control by conventional means difficult for pests such as the cotton bollworm/corn earworm, *Helicoverpa armigera* (61); the bed bug, *Cimex lectularius* (80, 143); and ticks (57). As more is learned about the reproductive biology of specific arthropods, their SFPs may provide tools or targets for the control of disease vectors and agricultural pests.

The study of SFPs also provides insight into the evolutionary patterns of reproductive traits. Although the functional classes of SFPs are conserved, a significant fraction of individual SFPs show signs of unusual, often rapid, evolution at the primary sequence level. The forces driving this pattern are not understood, and the study of SFPs may allow for their identification and dissection. Comparative studies of SFPs, individually and in aggregate, are important because (*a*) lineage-specific SFPs may be involved in the reproductive isolation between species; (*b*) highly conserved SFPs or SFP classes may be essential for reproduction; and (*c*) SFP divergence between closely related species may illuminate selective pressures underlying SFP evolution. Recent reviews have focused on the evolutionary dynamics of SFPs (49, 53, 120, 148); therefore, we refer the reader to those reviews and focus here on the nature and function of SFPs.

IDENTIFICATION OF SEMINAL FLUID PROTEINS

The identification of proteins produced in secretory tissues of the male RT and demonstrated to be, or likely to be, transferred to females during mating is the primary step in SFP identification. Transcriptomic [expressed sequence tag (EST), microarray; 5, 16, 20, 30, 34, 41, 112, 130, 151, 158, 161, 165] and proteomic (6, 10, 32, 48, 50, 75, 136, 140, 149, 153, 160, 162) methods have given a global view of proteins produced in arthropod male RT glands and, in some cases, of proteins transferred to females during mating (**Table 1**). For the purposes of this review, proteins within the seminal fluid and transferred to females are called SFPs. Proteins synthesized in male AGs are called Acps.

Accessory gland protein (Acp): protein made by, and expected to be secreted from, the accessory gland of male insect reproductive tracts

Most of the Acp/SFP identification studies in **Table 1** examined RNA or proteins found in tissues of the male RT, but did not demonstrate SFP transfer during mating. A novel proteomic method directly identified 146 *Drosophila melanogaster* SFPs (39 previously unannotated), 126 *D. simulans* SFPs, and 116 *D. yakuba* SFPs that are transferred to

Table 1 Recently identified SFPs and RT-expressed genes in insects

Order	Family	Genus sp.	Method (Ref.)	No. identified (Ref.)	
				In RT	Transferred
Diptera	Drosophilidae	*Drosophila simulans*	EST screen (151), proteomic (48)	57[a] (151)	3 (48)
		Drosophila melanogaster	Cumulative review (130), EST screen (151), peptide purification (91, 137), proteomic (48, 104, 153, 160), microarray (30, 130)	112[a] (130), 46 (30, 130), 440 (153), 13[a] (160), 34[a] (104, 151)	138 (50), 8 (48), 14 (130), 3 (91, 137, 169)
		Drosophila yakuba	EST screen (16), proteomic (48, 50)	119 (16)	107 (50), 8 (48)
		Drosophila erecta	EST screen (16)	114 (16)	—
		Drosophila mojavensis	EST screen (158), proteomic (75)	54 (158), 786 (75)	—
	Tephritidae	*Ceratitis capitata*	EST screen (34)	13 (34)	—
	Culicidae	*Aedes aegypti*	Bioinf. candidates/RT-PCR (149), proteomic (147)	63[a] (149)	56 (147)
		Anopheles gambiae	Bioinf. candidates/RT-PCR (41), proteomic (140)	46 (41), 25 (140)	15 (140)
Lepidoptera	Nymphalidae	*Heliconius erato*	EST screen/bioinf. filter/RT-PCR (161), proteomic (162)	371 (161)	25 (162)
		Heliconius melpomene	EST screen/bioinf. filter/RT-PCR (161), proteomic (162)	340 (161)	10 (162)
Hymenoptera	Apidae	*Apis mellifera*	Proteomic (10, 32)	69 (32)	33 (32), 57 (10)
Coleoptera	Tenebrionidae	*Tribolium castaneum*	Microarray (112)	112 (112)	—
Orthoptera	Gryllidae	*Allonebrius*	EST screen (20)	183 (20)	—
		Gryllus firmus	EST screen (5, 20)	247 (5, 20)	—
		Gryllus pennsylvanicus	EST screen (5), proteomic (6)	277 (5)	22 (6)
Arachnida: Ixodida	Ixodidae	*Amblyomma hebraeum*	EST screen (165)	35 (165)	—

[a]Candidate SFPs were based on criteria of accessory gland–specific or –enriched expression and/or presence of a predicted secretion signal sequence.
Abbreviations: Acp, accessory gland protein; EST, expressed sequence tag; RT, reproductive tract; RT-PCR, reverse transcriptase–polymerase chain reaction; SFP, seminal fluid protein.

females during mating (48, 50). Findlay et al. (50) fed females a diet enriched in ^{15}N so that the females produced isotopically heavy proteins. After these females were mated to unlabeled males, only proteins transferred from males were detected when the female RTs were analyzed by mass spectrometry. This method was subsequently adapted to identify transferred SFPs in *Aedes aegypti* (147).

SFPs identified in these studies (**Table 1**) include peptides and prohormones and protein classes predicted to play roles in numerous functions including sperm binding (lectins and cysteine-rich secretory proteins), proteolysis, lipases, and immunity-related functions. These protein classes are seen in the ejaculates of several insect species, providing evidence that, although the primary sequence of some SFPs evolves rapidly, the protein classes represented in seminal fluid are constrained (104). Further, examining the seminal fluid of the extensively studied *Drosophila* species reveals rapid gain/loss of Acp genes to be a common feature of *Drosophila* seminal fluid evolution (50, 104), suggesting that Acp genes evolve de novo, perhaps from noncoding DNA (15, 159). However, even as much of the knowledge of insect SFPs has been obtained via studies in *Drosophila* species, a recent proteomic study showed that *Apis mellifera* (honey bee) SFPs shared more sequence similarities with human SFPs than with *D. melanogaster* SFPs (10). Therefore, future studies of SFPs across representative taxonomic groups should elucidate the fascinating evolutionary history of these proteins.

FUNCTION OF IDENTIFIED SEMINAL FLUID PROTEINS

Historically, insect SFP function has been analyzed by several approaches, including the injection of a purified SFP or protein fractions into virgin females, biochemical analysis, removal of putative SFPs by RNAi or mutation in *Drosophila*, and ectopic expression of SFPs in unmated *Drosophila* females. Moreover, the increasing availability of genomic and predicted protein annotations has made functional prediction of SFPs by sequence comparison much easier. For example, comparative structural modeling suggested the structure/function of 28 predicted Acps of *D. melanogaster* males (105). FlyBase annotations to *D. melanogaster* genes were used to predict the functional classes of 240 candidate SFPs in *D. mojavensis* (75). Cross-species comparisons to *D. simulans* and *D. yakuba* led to the identification of 19 *D. melanogaster* proteins previously unreported as SFPs (50). The newly identified putative SFPs of these species fall into the same categories previously identified in *D. melanogaster* (105).

Aside from putative function based on sequence analysis, direct assessment of specific SFP tissue targets within the mated female may hint at those SFP functions (e.g., localization to the sperm storage organs may suggest a role in sperm storage or maintenance, as seen for a network of *D. melanogaster* Acps; 132). Thirteen *D. melanogaster* Acps target multiple tissues within the mated female RT, each having a unique targeting pattern (133). In addition, a subset of Acps leave the female RT and enter the hemolymph (90, 99, 119, 133), potentially reaching nervous and/or endocrine system targets.

More direct methods have also identified the roles of specific SFPs in processes such as gene regulation, behavior, and physiological processes such as sperm storage. These results are discussed below.

Transcriptome Changes

Changes in female gene expression postmating have been examined in *D. melanogaster* and, to a lesser extent, in *An. gambiae* and *Apis mellifera*. In *D. melanogaster*, the effect of SFPs and sperm on transcriptional change in mated females has been dissected by microarray analyses. Levels of over 1,700 transcripts are altered at 1–3 h postmating in females (82, 95). The mating-dependent genes have predicted functions in a multitude of biological processes, including metabolism, immune defense, and protein modification. However, only a handful of the mating-responsive changes in RNA levels are greater than twofold, consistent with the hypothesis that sexually mature females are poised to respond to mating (64, 95). By 6 h postmating, larger-magnitude changes in RNA levels are observed in a smaller number of genes (94). After a second mating, the expression of immunity-related genes is more pronounced (67), suggesting that previously mated females have sufficiently upregulated metabolic and/or structural genes required for postmating processes (e.g., ovulation and egg laying) to continue.

In the lower RT (the lower common oviduct, seminal receptacle, female AGs, spermathecae, and anterior uterus), the levels of over 500 transcripts are changed postmating (92). A distinct shift—from gene silence to activation—is observed soon after the onset of mating (92). A dramatic peak in differential gene expression is seen 6 h postmating (92), consistent with the whole-body transcriptome results described above (94). In the oviduct, mating induces an upregulation of immune-related transcripts and increases levels of RNA for cytoskeleton-related proteins (72). Some oviduct mating-responsive genes respond only to the first mating, and others to both the first and second matings (73). The female RT transcriptome suggests that the structural changes occurring after the first mating (presumably due to mating-responsive gene expression) are sufficient for continued postmating processes.

Because some of the mating-dependent gene expression change is due to SFP receipt (95), transcriptome change in mates of males lacking specific SFPs was investigated. The ovulation-inducing SFP ovulin and the sperm storage protein Acp36DE do not contribute extensively to female transcriptome change at 1–3 h postmating (94). However, two other SFPs, Acp29AB and Acp62F, substantially affect the female transcriptome (94). Surprisingly, Acp29AB and Acp62F contribute to the upregulation of genes involved in egg production and muscle development, even though analyses of mates of Acp29AB- or Acp62F-null males do not detect ovulation or egg-laying defects (102, 169).

Sex peptide (SP), a 36-amino-acid Acp with roles in egg production, receptivity, feeding, and sleep behaviors in mated females (25, 69, 85), affects expression of 52 genes in the head and abdomen of mated females. The majority of these RNAs changed only two- to threefold (39). In the head, SP regulated RNA levels of genes for proteins involved in metabolism, proteolysis, signal transduction, and transcription. In abdomens, the SP upregulated antimicrobial peptide genes via the Toll and IMD pathways (115); a C-terminal motif of SP is responsible for mediating this effect (39). Despite the induction of antimicrobial peptide genes by mating, hemolymph challenge did not detect an immune response in mated females (47, 167).

SP: sex peptide

RNA levels for 141 genes in *An. gambiae* females experience changes at 2, 6, and 24 h postmating (141). The number of genes with changes in expression levels twofold or greater increased with time since mating. Changes in transcript levels of many of these genes persist for at least 4 days postmating. Mating-responsive expression changes were examined specifically in the head, the gut, the ovaries, the lower RT (tissues below the ovaries), and the two major organs of the lower RT (the atrium, where the ejaculate is received, and the spermatheca). In both the gut and the lower RT, the expression levels of several genes change postmating. Many of these are predicted proteolysis regulators. In the spermatheca, a predicted vitellogenin was also highly upregulated postmating. As with *D. melanogaster*, Rogers et al. (141) conclude that the female atrium is poised to respond to mating. However, they propose that the spermatheca may rely on signals received during mating to regulate genes involved in sperm storage and maintenance.

Large-scale transcriptional changes occur in the ovaries and brains of *Apis mellifera* queens postmating (78, 79). In the ovaries, 366 transcripts are differentially expressed postmating, with the regulated genes largely involved in cell division, gametogenesis, reproduction, and oogenesis (78). The RNA levels of 971 genes are differentially expressed in female *A. mellifera* brains postmating, with an overrepresentation of genes involved in protein folding, protein catabolism, and the stress response (78). The types of genes regulated by mating in *A. mellifera* overlap with those seen in the previously mentioned *D. melanogaster* studies (e.g., genes involved in the immune response), suggesting that the postmating transcriptional response may be conserved across species (79). In addition, insemination quantity affects gene expression in the brain (138), suggesting that ejaculate volume and, possibly, quantity of specific SFPs received may act as cues for this processes in mated *A. mellifera* females.

Antimicrobial Functions of SFPs

Aside from roles in mediating the upregulation of antimicrobial genes in mated females (95, 115), some SFPs have an intrinsic antimicrobial function. Three *D. melanogaster* SFPs (from the AG and ejaculatory duct) have antimicrobial activity on *Escherichia coli* growth in vitro (88). An additional three *D. melanogaster* SFPs have antimicrobial activity in vivo—upon ectopic expression, they reduce bacterial loads in females infected with *Serratia marcescens* (103); the relationship of these three genes to the three identified biochemically is unknown. Although analogous antimicrobial activities have not been detected in the seminal fluid of the bed bug, *C. lectularius*, its seminal fluid does contain bacteriolytic activity, specifically, a lysozyme-like immune activity capable of degrading bacteria (111). These findings suggest that SFPs might play a protective role within the RTs of mated females, possibly aiding females' ability to clear microbes introduced during mating.

Structural and Conformational Changes of the Female RT

The receipt of seminal fluid induces physiological and structural changes of female RTs. In the oviduct of *D. melanogaster* females, tissue-wide postmating changes include the differentiation of cellular junctions, remodeling of the extracellular matrix, increased myofibril formation, and increased innervation of this tissue (72). Postmating increases in neural activity to the oviduct occur in the form of vesicle release from RT nerve termini and are modulated distinctly by mating, Acp receipt, and sperm receipt. Mating and/or the receipt of Acps or sperm has differing effects on vesicle release in different regions of the RT at different times postmating, inferred from the intensity of labeled vesicles (64). Immediate postmating change in neural vesicle release occurs in the lower common oviduct, seminal receptacle, and uterus. At 3 h postmating—when females are ovulating at high rates and egg production has reached maximal levels (18)—vesicle release is inhibited in the common oviduct and lateral oviducts, with Acps modulating changes in nerve termini innervating the seminal receptacle (64).

SFP receipt also affects the lower RT, inducing a series of conformational changes in the uteri of mated *D. melanogaster* females, initiating during the first moments of copulation and continuing after mating has ended (1). At least part of this process aids in the storage of sperm, allowing them to access the storage organs. Acps, and not sperm, are the ejaculatory components required to trigger these changes (1). The Acp(s) that initiates this process is unknown. However, Acp36DE is essential for the progression of these changes (8). Incomplete progression of these changes in the absence of Acp36DE leaves sperm lagging in the mid-uterus instead of forming a dense mass adjacent to the sperm storage organ entrances (8). This finding, coupled with the abnormally low numbers of sperm stored in Acp36DE-null mates (19, 108), suggests that the postmating uterine conformation changes aid sperm movement, en masse, toward storage.

An. gambiae female RTs undergo structural changes upon mating (141). In virgin females, the apical cytoplasm of atrium cells has extensive, smooth endoplasmic reticulum surrounded by high numbers of mitochondria. The basal poles of the cells have a high density of rough endoplasmic reticulum. In mated females, both the smooth and the rough endoplasmic reticula mostly disappear, and the mitochondria become distributed throughout the cells. Rogers et al. (141) propose that these structural changes may result in a barrier to remating.

Sperm Maintenance in, and Release from, Storage

In addition to roles of SFPs in sperm storage, SFPs are involved in the maintenance of sperm viability in, and their release from, storage (e.g., 170). Seminal fluid secretions from male AGs of both the honey bee, *A. mellifera*, and the leafcutter ant, *Atta colombica*, promote sperm viability (37, 38). However, AG secretions from one male do not positively affect the viability of another male's sperm (36). In ants and bees, the effects of AG secretions on sperm survival differ between monandrous and polyandrous species (36). AG secretions from monandrous species promote sperm survival, even when the seminal fluid and sperm are from different males. AG secretions from polyandrous species, however, are detrimental to sperm survival—even to sperm of related males—suggesting a sensitive recognition system exists during sperm competition (36). The negative effects of AG fluid on sperm survival are mitigated by spermathecal secretions in the leafcutter ant, suggesting that females of this species control ejaculate competition once sperm are stored (36). Similarly, *D. melanogaster* seminal fluid has a protective function, improving the survival of even rival sperm (66).

In *D. melanogaster*, Acps are necessary for the efficient utilization of stored sperm, as the few sperm stored in the absence of Acps are not used to fertilize eggs (170). Utilization of sperm involves their retention in, and release from, storage (131). The removal of five Acps, individually, from the male ejaculate led to sperm retention in both storage organs after mating (7, 131). Four of these proteins (CG9997, CG1652, CG1656, CG17575—a serine protease, two C-type lectins, and a cysteine-rich secretory protein, respectively) are required for the localization of the fifth protein, SP, to sperm (132), acting in a functional pathway that targets SP to the storage organs in mated females (132). SP, responsible for eliciting numerous postmating responses, is unique in exerting its effects on mated females for several days. The long-term effects of SP result from its physically binding sperm, maintaining its presence in the female RT as long as sperm remain in storage (114). Sperm binding is a function of the N terminus of the peptide; the C terminus of SP—which contains the receptivity-modulating activity—is gradually cleaved from sperm tails (114). These findings suggest that the phenotypes (in sperm storage but also in receptivity and egg laying) elicited by the absence of CG9997, CG1652, CG1656, and CG17575 from the ejaculate may be attributable to the inability of SP to localize to sperm.

SFPs do not affect sperm release. Acp29AB, a predicted lectin, is needed for sperm to be retained within the sperm storage organs. Sperm from males homozygous for a Acp29AB loss-of-function mutation enter into but are not well maintained within storage, consequently faring poorly in a sperm-competitive environment (169). The latter result is likely due to the reduced numbers of stored sperm—a phenotype analogous to that seen in mates of Acp36DE-null males (28).

Receptivity to Remating

Decreased sexual receptivity of mated females occurs in a wide range of insects, and it has been suggested that inducing this change in females is of benefit to males by decreasing the likelihood of sperm competition. In *D. melanogaster*, the receipt of SFPs changes female behavior—mated females actively reject courting males. The SP plays a central role in inducing this change in female receptivity (26, 60, 85, 132, 173, 174). The four Acps required for SP sperm localization also influence receptivity (131). How SP accomplishes its regulation of female receptivity is not known, but its action requires a G-coupled-protein receptor (174) and specific neurons (60, 173) in females.

C. capitata females are less receptive to male courtship and are less likely to mate for several days after a single mating than are virgin females (70, 97). Sperm storage may play a role in female receptivity in this species, as females who store less sperm postmating are more likely to remate sooner (97, 100). In addition, *C. capitata* females switch from a male pheromone odor preference to a host plant odor preference postmating (70), a switch possibly mediated by factors in male seminal fluid.

When mated to irradiated sterile males (and thus subsequently storing little or no sperm), Queensland fruit fly females, *Bactrocera tryoni*, show no difference in sexual receptivity when compared with mates of nonirradiated males (58), suggesting that products of the seminal fluid, not sperm, are responsible for the reduced postmating receptivity observed. In support of this hypothesis, virgin *B. tryoni* females injected with male RT extracts experienced diminished sexual receptivity and shorter copulation times when subsequently mated, similar to behaviors seen in previously mated females (126). That *B. tryoni* male AG size decreases after mating suggests that this tissue is a major site of SFP synthesis (127).

In *An. gambiae*, male reproductive gland proteins also mediate female likelihood of remating (23, 24). This conclusion was initially suggested by studies involving females mated to hybrid males with reduced AGs (23, 24) and subsequently verified by injections of male AG homogenates into virgin females (145).

In several moth species, sexually receptive females produce sex pheromones to attract mates. Sex pheromone production declines substantially after mating. Calling behaviors cease and oviposition behaviors initiate (13). In several of these species, pheromone production is under neuroendocrine control, resulting from the release of pheromone biosynthesis activating neuropeptide (PBAN) into the female hemolymph (128). Reduction of female pheromone levels postmating is a consequence of PBAN reduction in the female hemolymph (106). Synthetic *D. melanogaster* SP and the pheromone suppression peptide *Hez*PSP—from *Helicoverpa zea* AGs (76, 129)—suppress pheromone production after injection into unmated *H. armigera* females. This effect occurs in a dose-dependent fashion (44, 45). In addition, antibodies raised against the *D. melanogaster* SP detect signals from *H. armigera* male RTs (107).

There is growing evidence that SFPs have important effects on female postmating behavior in lady beetles (117, 118), seed beetles (101, 144, 172), and ground beetles (152). Injection of testis extracts reduces the probability of mating at 3 h and 2 days postinjection, whereas injections of AG extracts reduce the probability of mating only at 2 days postinjection. Further, injection of a small-molecular-weight (<3 kDa) fraction of male RTs results in a short-term decrease in the probability of mating (1 and 3 h postinjection), whereas injection of a higher-molecular-weight (>14 kDa) fraction results in longer-term inhibition of mating (2 and 4 days after injection; 171). In the ground beetle *Leptocarabus procerulus*, injections of either testes or AG homogenates into virgin females independently decrease the probability of mating (152).

Egg Production

A frequent effect of seminal fluid receipt is an increase in egg production, ovulation, and/or egg-laying rates in female insects. Transferring SFPs that upregulate these processes can benefit males, ensuring their sperm fertilize the maximum number of eggs before the female remates, and can also benefit females, allowing increased egg production only when sperm are present to fertilize those eggs. *D. melanogaster* SP stimulates egg laying in mated females (56, 130), and the long-term persistence of this activity requires the four Acps that localize SP to sperm (114, 131, 132).

The prohormone-like SFP ovulin (Acp26Aa) stimulates ovulation (62). Its mechanism is unknown but could involve two nonmutually exclusive mechanisms: directly via ovulin interaction with neuromuscular targets along the lateral oviducts of the female RT, or indirectly by affecting the activity of the neuroendocrine system (62, 64). Ovulin is proteolytically cleaved in the female RT in a step-wise manner (reviewed in Reference 148), a process dependent on at least one other Acp, CG11864 (134), a predicted astacin family metalloprotease that is itself cleaved in the male RT during transfer to females (134). Ectopic expression of full-length ovulin, or either of ovulin's two C-terminal cleavage products, is sufficient to stimulate ovulation in unmated females (63), suggesting that ovulin cleavage may increase its activity by generating more bioactive components of the protein.

Little is known about SFP-mediated effects on ovulation in other insect species. In *Apis mellifera*, mating stimulates vitellogenesis and oocyte maturation in females (79, 154). In *Ae. aegypti*, male reproductive gland proteins modulate an increase in oviposition (reviewed in References 31, 56, 77, and 149). In *Anopheles* sp., there is indirect evidence that SFPs regulate female fecundity (43, 68): Males have angiotensin converting enzyme (ACE) activity in their reproductive glands and females mated to males fed ACE inhibitors lay fewer eggs than females mated to control males. In *H. armigera*, crude extracts of the male AGs stimulate egg maturation and oviposition when injected into virgin females, similar to effects seen after mating (71). The receipt of the male ejaculate increases *C. lectularius* female reproductive rates in terms of lifetime egg production, and females receiving more ejaculate enter reproductive senescence later than females who receive less ejaculate (135). These findings suggest that ejaculate components may compensate for the costs of elevated reproductive rates by delaying reproductive senescence in this species. Ejaculate volume affects seed beetle fecundity, as females receiving smaller ejaculates have lower fecundity than females receiving larger ejaculates, although this effect is not seen in all beetle species (101, 144). In the twospotted lady beetle, *Adalia bipunctata*, females ingest SFPs in the male spermatophore. Females prevented from consuming spermatophores have a longer latency to oviposition as well as a lower duration of resistance to remating than do control females (117).

Mating Plug Formation

In several insect species, a mating plug is formed within the female RT during and/or after mating. Mating plugs often contain SFPs, and their formation is dependent on receipt of SFPs. Mating plugs have a wide range of functions, such as in sperm competition (17, 42, 146); formation of a physical barrier to remating, as in butterflies (110); and switching off female receptivity entirely, as in the bumble bee, *Bombus terrestris* (11).

In *D. melanogaster*, a mating plug is formed shortly after mating begins. This structure has two major regions: a posterior region composed of ejaculatory bulb proteins (PEB-me, PEBII, and PEBIII; 22, 91) and an anterior region composed of Acps (91). Evidence for a role of the mating plug, and the SFPs within it, in reducing female receptivity has been shown in *D. melanogaster*: Mates to PEBII knockdown males (who form smaller mating plugs) are more receptive to remating than controls in the short term (4 h; 22). These results suggest that the mating plug mediates a short-term decline in receptivity before the long-term effects of other SFPs set in. A similar effect is seen in *D. hibisci*, in which the mating plug inhibits courtship by subsequent males and reduces female receptivity (123). In *D. hibisci*, the mating plug is also suggested to facilitate sperm storage by preventing the back flow of sperm away from the storage organs (122).

In *An. gambiae*, the mating plug is necessary for proper sperm storage but does not prevent remating by the female (140). Further, a male AG-specific transglutaminase is necessary to form the mating plug (140). Although transglutaminases are made in other mosquito species, male AG-specific transglutaminases are only found in mosquitoes that form mating plugs (140).

In *D. mojavensis* and related species, females experience an insemination reaction mass (113). Although not a mating plug per se, it fills the entire uterus and persists for hours, absorbing nutrients from the male ejaculate that are incorporated into female somatic tissue (93).

Proteins with sequence similarity to larval clotting factors in *D. melanogaster* (75) found in the AGs of *D. mojavensis*, along with proteins with fibrinogen domains found in *D. mayaguana* (3) and *D. mojavensis* (75) AGs, are good candidates for proteins involved in forming the clot-like insemination reaction mass.

Longevity

The longevity of mated females is decreased in some (e.g., *D. melanogaster*: 166) but not all (e.g., cricket: 157) insects. In *D. melanogaster*, Acps mediate at least part of this longevity reduction (27) for reasons that are as yet unknown, and SP plays a major role in Acp-mediated decrease in longevity (166). In addition, SP and three other Acps (the protease inhibitors Acp62F and CG8137, and the peptide CG10433) are toxic to *D. melanogaster* upon ectopic expression (89, 103), possibly reflecting the negative effect of their action under normal mating conditions. However, the mechanism(s) by which these Acps decrease longevity is unknown, and as ectopic expression produces protein levels higher than normally encountered during mating, the toxicity observed may not reflect the true effects of these Acps, suggesting that the longevity effects associated with mating may be an indirect effect of SFP receipt (102).

Feeding

SFPs affect the feeding behavior of some female arthropods. *D. melanogaster* SP increases female feeding postmating (25). This behavioral change is substantially reduced in eggless females and increased in virgin females with experimentally elevated rates of egg production, suggesting that increased feeding is tied to the postmating increases in ovulation and/or oviposition (14). However, egg-less *D. melanogaster* females continue to show mating-dependent decreases in life span similar to that of fertile, wild-type females, suggesting that the decreased longevity observed in mated *D. melanogaster* females is not attributable to overfeeding or to the energetic costs of egg production (14).

In female ticks, the feeding cycle consists of a preparatory phase, a slow feeding phase, and a rapid feeding phase (12). After completion of this cycle, females will have increased in weight almost 100-fold—an engorgement process that lasts ~6–10 days and completes before females lay an egg batch (12). The transition weight between the slow and rapid feeding phases is termed the critical weight. Most virgin females do not feed past the critical weight (54, 59). Initiation of the rapid feeding phase is dependent on the receipt of a testis/vas deferens–derived engorgement factor called voraxin (74, 87, 164). Voraxin consists of two components (voraxin α and β) and is sufficient to stimulate engorgement of feeding when injected into virgin females (164). In addition, female feeding to engorgement was reduced by 74% when reared on rabbits immunized with recombinant voraxin (164). Paradoxically, RNAi knockdown of voraxin had no effect on female engorgement after mating with knockdown males (150), and experiments with the American dog tick, *Dermacentor variabilis*, found that silencing engorgement factor α and β homologs via RNAi failed to reduce engorgement (40). Thus, the feeding role of these proteins has yet to be fully ascertained.

Activity Levels

The increase in *D. melanogaster* female feeding observed postmating coincides with a decrease in female siesta sleep (a quiescent sleep-like state) postmating (69). This effect is mediated by receipt of SP, which decreases siesta sleep by 70% (69), consequently increasing foraging and egg-laying activity of mated females. In conjunction with the negative impact of SP on female life span (166), the effect of SP on female longevity may be the result of increases in stress due to sleep deprivation and to increased locomotor activity (69).

Flight behavior is altered postmating in *A. mellifera* queens. At ~1–2 weeks of age, queens mate multiply during mating flights—inseminated by an average of 12 males (155).

Mating makes queens less likely to attempt flight again (79). In addition, insemination by single versus multiple drones affects several behaviors, including flight behavior (79), suggesting that queens might use ejaculate volume (and possibly contents) as a cue for flight attempts.

FEMALE EFFECTORS OF SEMINAL FLUID PROTEINS

Little is known about the female molecules that interact with SFPs and are subsequently responsible for inducing the myriad postmating changes observed in insects and other arthropods. A notable exception is the receptor for the *D. melanogaster* SP, the sex peptide receptor (SPR). SPR, identified in an extensive RNAi screen, is a G-protein-coupled receptor that acts through a cAMP-dependent pathway (174). The ejaculatory duct peptide DUP99B, which has a C terminus similar to that of SP, also interacts with SPR (137, 174). SPR expression in neurons that express the sex-specific *fruitless* transcript is necessary and sufficient to reestablish the receptivity and egg-laying effects of SPR (174). Further, SPR expression is necessary in sensory neurons innervating the female RT that express the *pickpocket* marker, possibly reducing the output of these neurons to the central nervous system (60, 173). The ability of the *D. melanogaster* SP to interact in vitro with *Ae. aegypti* and *Bombyx mori* SPR orthologs (174) suggests that SFPs analogous to the SP are present in the seminal fluid of these, and potentially other, insects. This interpretation is consistent with the ability of *D. melanogaster* SP to induce postmating responses when injected into unmated *H. armigera* females (45, 46, 107).

Ultimately, SFPs must interact in the context of the female RT. Thus, progress in understanding the signaling mechanisms involved in insect reproductive processes may illuminate mechanisms of SFP modulation in female physiology and behavior. Recent reviews have addressed neuropeptide control of insect hormones and sexual receptivity (55) in the reproductive physiology of the locust *Locusta migratoria* (e.g., 81). SFPs may act upstream of traditional neural signaling systems. For example, ovulation and subsequent egg laying are presumably mediated by contraction of the female RT. The biogenic amine octopamine (OA) is an important regulator of ovulation-related contractions of the female RT in *Locusta* (109), *Stomoxys* (33), and *Drosophila* (96, 98, 139). Further, RT extracts from male *Stomoxys* induce changes in muscle contraction in female RTs (33). The *Drosophila* receptor, OAMB, critical for the ovulation effect, is selectively expressed in oviductal epithelium (83). OA mediates ovary muscle contraction and oviduct muscle relaxation. These opposing effects may serve to expel the egg from the ovary while facilitating entry into the common oviduct (81, 96, 156). Perhaps signaling systems, such as OA, or their proximate downstream targets may serve as substrates that are modulated by SFPs.

SPR: sex peptide receptor

SOCIAL BEHAVIOR EFFECTS

The amount of SFPs transferred may depend, in part, on the mating status of males and females and their social environment before or during mating. Mating status of both sexes can affect the magnitude of female postmating responses (51, 65, 116). In a number of insects, females mated to recently mated males show less pronounced postmating changes in receptivity and egg production than do females mated to virgin males (e.g., *D. melanogaster*: 65; *Anastrepha obliqua*: 116). In some insects, mates of nutritionally stressed males have less pronounced postmating changes in receptivity to remating than mates of control males (4). These studies suggest that males are limited in the amount of SFPs they can produce and/or store at a given time and that SFP production may be resource limited.

Given this potential limitation and the importance of SFPs in determining male reproductive success (e.g., via effects on sperm storage, egg production, and remating), selective pressures should exist for males to allocate the ejaculate in a manner that maximizes their reproductive success. One way this could be accomplished would be to allocate more SFPs to females mated under conditions of higher sperm competition risk (elevated either because the female has previously mated or because other males are in the vicinity of the mating pair). There is support for such strategic allocation of sperm in a number of insect species, including beetles, crickets, and medflies (reviewed in Reference 163). Recent evidence has demonstrated strategic allocation of SFPs as well (reviewed in Reference 148). Briefly, *D. melanogaster* males transfer more sex peptide when they are exposed to another male before and during mating than when they are alone with the female before and during mating (168). Other evidence is consistent with the hypothesis that strategic SFP allocation increases male reproductive success. For example, the mates of males exposed to other males before mating have longer latencies to remating and higher fecundity than do mates of males not exposed to other males before mating (21). Thus, *D. melanogaster* males are able to adjust their ejaculate composition in response to risk of sperm competition, an adjustment that appears to increase male reproductive success. Future research in this area should test for strategic SFP allocation in other insect species.

CONCLUSIONS

SFPs have roles in modulating many female behavioral and physiological processes across a wide range of insect species. The recent rapid pace of technological advances in transcript and protein identification has resulted in greatly increased knowledge of suites of SFPs in a number of insect species, and the roles of individual SFPs in female postmating responses are being elucidated. However, several questions still need to be addressed.

First, how do male SFPs interact with each other and with female molecules to effect the changes observed in mated females? Downstream female effectors with or through which SFPs exert their functions remain unknown, with the exception of *D. melanogaster* SP and its receptor SPR. Proteins secreted from female RTs, including the sperm storage organs, offer an exciting list of candidates to test for roles in mediating SFP responses (2, 9, 125). That hundreds of SFPs are transferred to females suggest that many molecular pathways may be involved in female postmating responses. Interactions affect protein localization (e.g., SP to sperm; 132) and proteolytic cascades (134) in *D. melanogaster*, but much needs to be done to characterize these and other pathways.

Second, what extrinsic factors affect the production and transfer of SFPs and the magnitude of their effects on mated females? Few studies have investigated this question, but those that have suggest that effects of SFPs on female postmating response are influenced by a number of factors. For example, adult female nutrition alters the magnitude of the effects of SP on different phenotypic traits, which show that the responses to mating in general, and SFPs in particular, can vary under different environmental conditions (52, 142). Furthermore, males transfer different amounts of SFPs in different contexts, such as a competitive environment (168). These effects observed in the laboratory suggest that modulation of SFP action and allocation in natural settings will be important to consider for fundamental reasons and also in insect pest control.

Third, how do different proteins come to play similar roles in different species, and what forces lead to the rapid SFP sequence evolution? Conservation of the effects of SFPs on females and of SFP classes across the seminal fluid of a wide taxonomic range indicates fundamentally conserved roles for SFPs. Yet, individual SFPs tend to evolve rapidly and are not well conserved even across closely related species. Studies of SFP evolutionary dynamics will provide insight not only into SFP evolution but also into the mechanisms by which sexual selection shapes the structure and function of molecules. These are only a small sampling of the fascinating questions that await answers.

SUMMARY POINTS

1. SFPs have roles in modulating female behavioral and physiological processes in numerous insect species. SFPs are being identified in an increasing number of insects.
2. SFPs and proteins of Diptera, Lepidoptera, Hymenoptera, Coleoptera, Orthoptera, Hemiptera, and Ixodida species are described.
3. Mating and SFPs mediate female postmating responses in processes such as transcriptional and RT structural changes, upregulation of antimicrobial peptide genes, altered receptivity to remating, sperm storage, mating plug formation, postmating feeding, and female activity levels.

FUTURE ISSUES

1. How do male SFPs interact with each other and with female molecules to effect the changes observed in mated females?
2. How do extrinsic factors (e.g., nutrition and differing social conditions) affect the production and transfer of SFPs and the magnitude of their effects in mated females?
3. How do SFPs regulate similar reproductive processes across numerous species in the face of selective pressures and rapid evolution? The forces that drive these changes ought to be examined.

DISCLOSURE STATEMENT

The authors are not aware of any affiliations, memberships, funding, or financial holdings that might be perceived as affecting the objectivity of this review.

ACKNOWLEDGMENTS

We would like to thank our colleagues in the reproductive biology and SFP fields for useful discussion, Jessica Sitnik and Erin Kelleher for helpful comments on the manuscript, and NIH grant RO1-HD038921 (to MFW) for support. We apologize to colleagues whose work could not be included or receive detailed description because of length restrictions. Brooke A. LaFlamme and C. Dustin Rubinstein contributed equally to this review.

LITERATURE CITED

1. Adams EM, Wolfner MF. 2007. Seminal proteins but not sperm induce morphological changes in the *Drosophila melanogaster* female reproductive tract during sperm storage. *J. Insect Physiol.* 53:319–31
2. Allen AK, Spradling AC. 2008. The Sf1-related nuclear hormone receptor Hr39 regulates *Drosophila* female reproductive tract development and function. *Development* 135:311–21
3. Almeida FC, Desalle R. 2009. Orthology, function and evolution of accessory gland proteins in the *Drosophila repleta* group. *Genetics* 181:235–45
4. Aluja M, Rull J, Sivinski J, Trujillo G, Perez-Staples D. 2009. Male and female condition influence mating performance and sexual receptivity in two tropical fruit flies (Diptera: Tephritidae) with contrasting life histories. *J. Insect Physiol.* 55:1091–98

5. Andres JA, Maroja LS, Bogdanowicz SM, Swanson WJ, Harrison RG. 2006. Molecular evolution of seminal proteins in field crickets. *Mol. Biol. Evol.* 23:1574–84
6. Andres JA, Maroja LS, Harrison RG. 2008. Searching for candidate speciation genes using a proteomic approach: seminal proteins in field crickets. *P. R. Soc. B* 275:1975–83
7. Avila FW, Ravi Ram K, Bloch Qazi M, Wolfner M. 2010. Sex peptide is required for the efficient release of sperm stored in mated *Drosophila* females. *Genetics*. Epub ahead of print
8. Avila FW, Wolfner MF. 2009. Acp36DE is required for uterine conformational changes in mated *Drosophila* females. *Proc. Natl. Acad. Sci. USA* 106:15796–800
9. Baer B, Eubel H, Taylor NL, O'Toole N, Millar AH. 2009. Insights into female sperm storage from the spermathecal fluid proteome of the honeybee *Apis mellifera*. *Genome Biol.* 10:R67
10. Baer B, Heazlewood JL, Taylor NL, Eubel H, Millar AH. 2009. The seminal fluid proteome of the honeybee *Apis mellifera*. *Proteomics* 9:2085–97
11. Baer B, Morgan ED, Schmid-Hempel P. 2001. A nonspecific fatty acid within the bumblebee mating plug prevents females from remating. *Proc. Natl. Acad. Sci. USA* 98:3926–28
12. Balashov YS. 1972. Bloodsucking ticks (Ixodoidea)—vectors of disease of man and animals. *Misc. Publ. Entomol. Soc. Am.* 8:161–376
13. Bali G, Raina AK, Kingan TG, Lopez JD. 1996. Ovipositional behavior of newly colonized corn earworm (Lepidoptera: Noctuidae) females and evidence for an oviposition stimulating factor of male origin. *Ann. Entomol. Soc. Am.* 89:475–80
14. Barnes AI, Wigby S, Boone JM, Partridge L, Chapman T. 2008. Feeding, fecundity and lifespan in female *Drosophila melanogaster*. *Proc. Biol. Sci.* 275:1675–83
15. Begun DJ, Lindfors HA. 2005. Rapid evolution of genomic Acp complement in the *melanogaster* subgroup of *Drosophila*. *Mol. Biol. Evol.* 22:2010–21
16. Begun DJ, Lindfors HA, Thompson ME, Holloway AK. 2006. Recently evolved genes identified from *Drosophila yakuba* and *D. erecta* accessory gland expressed sequence tags. *Genetics* 172:1675–81
17. Birkhead T, Moller A. 1993. Female control of paternity. *Trends Ecol. Evol.* 8:100–4
18. Bloch Qazi MC, Heifetz Y, Wolfner MF. 2003. The developments between gametogenesis and fertilization: ovulation and female sperm storage in *Drosophila melanogaster*. *Dev. Biol.* 256:195–211
19. Bloch Qazi MC, Wolfner MF. 2003. An early role for the *Drosophila melanogaster* male seminal protein Acp36DE in female sperm storage. *J. Exp. Biol.* 206:3521–28
20. Braswell WE, Andres JA, Maroja LS, Harrison RG, Howard DJ, Swanson WJ. 2006. Identification and comparative analysis of accessory gland proteins in Orthoptera. *Genome* 49:1069–80
21. Bretman A, Fricke C, Chapman T. 2009. Plastic responses of male *Drosophila melanogaster* to the level of sperm competition increase male reproductive fitness. *Proc. Biol. Sci.* 276:1705–11
22. Bretman A, Lawniczak MK, Boone J, Chapman T. 2010. A mating plug protein reduces early female remating in *Drosophila melanogaster*. *J. Insect Physiol.* 56:107–13
23. Bryan JH. 1968. Results of consecutive matings of female *Anopheles gambiae* species b with fertile and sterile males. *Nature* 218:489
24. Bryan JH. 1972. Further studies on consecutive matings in *Anopheles gambiae* complex. *Nature* 239:519–20
25. Carvalho GB, Kapahi P, Anderson DJ, Benzer S. 2006. Allocrine modulation of feeding behavior by the Sex Peptide of *Drosophila*. *Curr. Biol.* 16:692–96
26. Chapman T, Bangham J, Vinti G, Seifried B, Lung O, et al. 2003. The sex peptide of *Drosophila melanogaster*: female post-mating responses analyzed by using RNA interference. *Proc. Natl. Acad. Sci. USA* 100:9923–28
27. Chapman T, Liddle LF, Kalb JM, Wolfner MF, Partridge L. 1995. Cost of mating in *Drosophila melanogaster* females is mediated by male accessory-gland products. *Nature* 373:241–44
28. Chapman T, Neubaum DM, Wolfner MF, Partridge L. 2000. The role of male accessory gland protein Acp36DE in sperm competition in *Drosophila melanogaster*. *Proc. Biol. Sci.* 267:1097–105
29. Chen PS. 1984. The functional morphology and biochemistry of insect male accessory-glands and their secretions. *Annu. Rev. Entomol.* 29:233–55
30. Chintapalli VR, Wang J, Dow JAT. 2007. Using FlyAtlas to identify better *Drosophila melanogaster* models of human disease. *Nat. Genet.* 39:715–20

31. Clements AN. 2000. *The Biology of Mosquitoes*. London: Chapman & Hall
32. Collins AM, Caperna TJ, Williams V, Garrett WM, Evans JD. 2006. Proteomic analyses of male contributions to honey bee sperm storage and mating. *Insect Mol. Biol.* 15:541–49
33. Cook BJ, Wagner RM. 1992. Some pharmacological properties of the oviduct muscularis of the stable fly *Stomoxys calcitrans*. *Comp. Biochem. Physiol. C Comp. Pharmacol. Toxicol.* 102:273–80
34. Davies SJ, Chapman T. 2006. Identification of genes expressed in the accessory glands of male Mediterranean fruit flies (*Ceratitis capitata*). *Insect Biochem. Mol. Biol.* 36:846–56
35. de la Fuente J, Kocan KM, Blouin EF. 2007. Tick vaccines and the transmission of tick-borne pathogens. *Vet. Res. Commun.* 31:85–90
36. den Boer SP, Baer B, Boomsma JJ. 2010. Seminal fluid mediates ejaculate competition in social insects. *Science* 327:1506–9
37. den Boer SP, Baer B, Dreier S, Aron S, Nash DR, Boomsma JJ. 2009. Prudent sperm use by leaf-cutter ant queens. *Proc. Biol. Sci.* 276:3945–53
38. den Boer SP, Boomsma JJ, Baer B. 2009. Honey bee males and queens use glandular secretions to enhance sperm viability before and after storage. *J. Insect Physiol.* 55:538–43
39. Domanitskaya EV, Liu H, Chen S, Kubli E. 2007. The hydroxyproline motif of male sex peptide elicits the innate immune response in *Drosophila* females. *FEBS J.* 274:5659–68
40. Donohue KV, Khalil SMS, Ross E, Mitchell RD, Roe RM, Sonenshine DE. 2009. Male engorgement factor: role in stimulating engorgement to repletion in the ixodid tick, *Dermacentor variabilis*. *J. Insect Physiol.* 55:909–18
41. Dottorini T, Nicolaides L, Ranson H, Rogers DW, Crisanti A, Catteruccia F. 2007. A genome-wide analysis in *Anopheles gambiae* mosquitoes reveals 46 male accessory gland genes, possible modulators of female behavior. *Proc. Natl. Acad. Sci. USA* 104:16215–20
42. Eberhard WG. 1996. *Female Control: Sexual Selection by Cryptic Female Choice*. Princeton, NJ: Princeton Univ. Press. 472 pp.
43. Ekbote U, Looker M, Isaac RE. 2003. ACE inhibitors reduce fecundity in the mosquito, *Anopheles stephensi*. *Comp. Biochem. Physiol. B Biochem. Mol. Biol.* 134:593–98
44. Eliyahu D, Nagalakshmi V, Applebaum SW, Kubli E, Choffat Y, Rafaeli A. 2003. Inhibition of pheromone biosynthesis in *Helicoverpa armigera* by pheromonostatic peptides. *J. Insect Physiol.* 49:569–74
45. Fan Y, Rafaeli A, Gileadi C, Kubli E, Applebaum SW. 1999. *Drosophila melanogaster* sex peptide stimulates juvenile hormone synthesis and depresses sex pheromone production in *Helicoverpa armigera*. *J. Insect Physiol.* 45:127–33
46. Fan Y, Rafaeli A, Moshitzky P, Kubli E, Choffat Y, Applebaum SW. 2000. Common functional elements of *Drosophila melanogaster* seminal peptides involved in reproduction of *Drosophila melanogaster* and *Helicoverpa armigera* females. *Insect Biochem. Mol. Biol.* 30:805–12
47. Fedorka KM, Linder JE, Winterhalter W, Promislow D. 2007. Post-mating disparity between potential and realized immune response in *Drosophila melanogaster*. *Proc. Biol. Sci.* 274:1211–17
48. Findlay GD, MacCoss MJ, Swanson WJ. 2009. Proteomic discovery of previously unannotated, rapidly evolving seminal fluid genes in *Drosophila*. *Genome Res.* 19:886–96
49. Findlay GD, Swanson WJ. 2010. Proteomics enhances evolutionary and functional analysis of reproductive proteins. *Bioessays* 32:26–36
50. Findlay GD, Yi X, Maccoss MJ, Swanson WJ. 2008. Proteomics reveals novel *Drosophila* seminal fluid proteins transferred at mating. *PLoS Biol.* 6:e178
51. Friberg U. 2006. Male perception of female mating status: its effect on copulation duration, sperm defence and female fitness. *Anim. Behav.* 72:1259–68
52. Fricke C, Bretman A, Chapman T. 2010. Female nutritional status determines the magnitude and sign of responses to a male ejaculate signal in *Drosophila melanogaster*. *J. Evol. Biol.* 23:157–65
53. Fricke C, Perry J, Chapman T, Rowe L. 2009. The conditional economics of sexual conflict. *Biol. Lett.* 5:671–74
54. Friesen KJ, Kaufman WR. 2009. Salivary gland degeneration and vitellogenesis in the ixodid tick *Amblyomma hebraeum*: Surpassing a critical weight is the prerequisite and detachment from the host is the trigger. *J. Insect Physiol.* 55:936–42

55. Gäde G, Hoffmann K. 2005. Neuropeptides regulating development and reproduction in insects. *Physiol. Entomol.* 30:103–21
56. Gillott C. 2003. Male accessory gland secretions: modulators of female reproductive physiology and behavior. *Annu. Rev. Entomol.* 48:163–84
57. Graf JF, Gogolewski R, Leach-Bing N, Sabatini GA, Molento MB, et al. 2004. Tick control: an industry point of view. *Parasitology* 129(Suppl.):S427–42
58. Harmer AM, Radhakrishnan P, Taylor PW. 2006. Remating inhibition in female Queensland fruit flies: effects and correlates of sperm storage. *J. Insect Physiol.* 52:179–86
59. Harris RA, Kaufman WR. 1984. Neural involvement in the control of salivary-gland degeneration in the ixodid tick *Amblyomma hebraeum*. *J. Exp. Biol.* 109:281–90
60. Hasemeyer M, Yapici N, Heberlein U, Dickson BJ. 2009. Sensory neurons in the *Drosophila* genital tract regulate female reproductive behavior. *Neuron* 61:511–18
61. Heckel DG, Gahan LJ, Baxter SW, Zhao JZ, Shelton AM, et al. 2007. The diversity of Bt resistance genes in species of Lepidoptera. *J. Invertebr. Pathol.* 95:192–97
62. Heifetz Y, Lung O, Frongillo EA Jr, Wolfner MF. 2000. The *Drosophila* seminal fluid protein Acp26Aa stimulates release of oocytes by the ovary. *Curr. Biol.* 10:99–102
63. Heifetz Y, Vandenberg LN, Cohn HI, Wolfner MF. 2005. Two cleavage products of the *Drosophila* accessory gland protein ovulin can independently induce ovulation. *Proc. Natl. Acad. Sci. USA* 102:743–48
64. Heifetz Y, Wolfner MF. 2004. Mating, seminal fluid components, and sperm cause changes in vesicle release in the *Drosophila* female reproductive tract. *Proc. Natl. Acad. Sci. USA* 101:6261–66
65. Hihara F. 1981. Effects of the male accessory gland secretion on oviposition and remating in females of *Drosophila melanogaster*. *Zool. Mag.* 90:303–16
66. Holman L. 2009. *Drosophila melanogaster* seminal fluid can protect the sperm of other males. *Funct. Ecol.* 23:180–86
67. Innocenti P, Morrow EH. 2009. Immunogenic males: a genome-wide analysis of reproduction and the cost of mating in *Drosophila melanogaster* females. *J. Evol. Biol.* 22:964–73
68. Isaac RE, Lamango NS, Ekbote U, Taylor CA, Hurst D, et al. 2007. Angiotensin-converting enzyme as a target for the development of novel insect growth regulators. *Peptides* 28:153–62
69. Isaac RE, Li C, Leedale AE, Shirras AD. 2010. *Drosophila* male sex peptide inhibits siesta sleep and promotes locomotor activity in the post-mated female. *Proc. Biol. Sci.* 277:65–70
70. Jang EB. 1995. Effects of mating and accessory-gland injections on olfactory-mediated behavior in the female mediterranean fruit-fly, *Ceratitis capitata*. *J. Insect Physiol.* 41:705–10
71. Jin ZY, Gong H. 2001. Male accessory gland derived factors can stimulate oogenesis and enhance oviposition in *Helicoverpa armigera* (Lepidoptera: Noctuidae). *Arch. Insect Biochem.* 46:175–85
72. Kapelnikov A, Rivlin PK, Hoy RR, Heifetz Y. 2008. Tissue remodeling: a mating-induced differentiation program for the *Drosophila* oviduct. *BMC Dev. Biol.* 8:114
73. Kapelnikov A, Zelinger E, Gottlieb Y, Rhrissorrakrai K, Gunsalus KC, Heifetz Y. 2008. Mating induces an immune response and developmental switch in the *Drosophila* oviduct. *Proc. Natl. Acad. Sci. USA* 105:13912–17
74. Kaufman WR, Lomas LO. 1996. "Male factors" in ticks: their role in feeding and egg development. *Invertebr. Reprod. Dev.* 30:191–98
75. Kelleher ES, Watts TD, LaFlamme BA, Haynes PA, Markow TA. 2009. Proteomic analysis of *Drosophila mojavensis* male accessory glands suggests novel classes of seminal fluid proteins. *Insect Biochem. Mol. Biol.* 39:366–71
76. Kingan TG, Thomas-Laemont PA, Raina AK. 1993. Male accessory gland factors elicit change from 'virgin' to 'mated' behaviour in the female corn earworm moth *Helicoverpa zea*. *J. Exp. Biol.* 183:61–76
77. Klowden MJ. 1999. The check is in the male: Male mosquitoes affect female physiology and behavior. *J. Am. Mosquito Control Assoc.* 15:213–20
78. Kocher SD, Richard FJ, Tarpy DR, Grozinger CM. 2008. Genomic analysis of post-mating changes in the honey bee queen (*Apis mellifera*). *BMC Genomics* 9:232
79. Kocher SD, Tarpy DR, Grozinger CM. 2009. The effects of mating and instrumental insemination on queen honey bee flight behaviour and gene expression. *Insect Mol. Biol.* 19:153–62

80. Kolb A, Needham GR, Neyman KM, High WA. 2009. Bedbugs. *Dermatol. Ther.* 22:347–52
81. Lange AB. 2009. Neural mechanisms coordinating the female reproductive system in the locust. *Front. Biosci.* 14:4401–15
82. Lawniczak MK, Begun DJ. 2004. A genome-wide analysis of courting and mating responses in *Drosophila melanogaster* females. *Genome* 47:900–10
83. Lee H-G, Seong C-S, Kim Y-C, Davis RL, Han K-A. 2003. Octopamine receptor OAMB is required for ovulation in *Drosophila melanogaster*. *Dev. Biol.* 264:179–90
84. Leopold RA. 1976. Role of male accessory glands in insect reproduction. *Annu. Rev. Entomol.* 21:199–221
85. Liu H, Kubli E. 2003. Sex-peptide is the molecular basis of the sperm effect in *Drosophila melanogaster*. *Proc. Natl. Acad. Sci. USA* 100:9929–33
86. Loher W, Ganjian I, Kubo I, Stanleysamuelson D, Tobe SS. 1981. Prostaglandins—their role in egg-laying of the cricket *Teleogryllus commodus*. *Proc. Natl. Acad. Sci. USA* 78:7835–38
87. Lomas LO, Kaufman WR. 1992. The influence of a factor from the male genital-tract on salivary-gland degeneration in the female ixodid tick, *Amblyomma hebraeum*. *J. Insect Physiol.* 38:595–601
88. Lung O, Kuo L, Wolfner MF. 2001. *Drosophila* males transfer antibacterial proteins from their accessory gland and ejaculatory duct to their mates. *J. Insect Physiol.* 47:617–22
89. Lung O, Tram U, Finnerty C, Eipper-Mains M, Kalb JM, Wolfner MF. 2002. The *Drosophila melanogaster* seminal fluid protein Acp62F is a protease inhibitor that is toxic upon ectopic expression. *Genetics* 160:211–24
90. Lung O, Wolfner MF. 1999. *Drosophila* seminal fluid proteins enter the circulatory system of the mated female fly by crossing the posterior vaginal wall. *Insect Biochem. Mol. Biol.* 29:1043–52
91. Lung O, Wolfner MF. 2001. Identification and characterization of the major *Drosophila melanogaster* mating plug protein. *Insect Biochem. Mol. Biol.* 31:543–51
92. Mack PD, Kapelnikov A, Heifetz Y, Bender M. 2006. Mating-responsive genes in reproductive tissues of female *Drosophila melanogaster*. *Proc. Natl. Acad. Sci. USA* 103:10358–63
93. Markow TA, Ankney PF. 1984. *Drosophila* males contribute to oogenesis in a multiple mating species. *Science* 224:302–3
94. McGraw LA, Clark AG, Wolfner MF. 2008. Post-mating gene expression profiles of female *Drosophila melanogaster* in response to time and to four male accessory gland proteins. *Genetics* 179:1395–408
95. McGraw LA, Gibson G, Clark AG, Wolfner MF. 2004. Genes regulated by mating, sperm, or seminal proteins in mated female *Drosophila melanogaster*. *Curr. Biol.* 14:1509–14
96. Middleton CA, Nongthomba U, Parry K, Sweeney ST, Sparrow JC, Elliott CJH. 2006. Neuromuscular organization and aminergic modulation of contractions in the *Drosophila* ovary. *BMC Biol.* 4:17
97. Miyatake T, Chapman T, Partridge L. 1999. Mating-induced inhibition of remating in female Mediterranean fruit flies *Ceratitis capitata*. *J. Insect Physiol.* 45:1021–28
98. Monastirioti M. 2003. Distinct octopamine cell population residing in the CNS abdominal ganglion controls ovulation in *Drosophila melanogaster*. *Dev. Biol.* 264:38–49
99. Monsma SA, Harada HA, Wolfner MF. 1990. Synthesis of two *Drosophila* male accessory gland proteins and their fate after transfer to the female during mating. *Dev. Biol.* 142:465–75
100. Mossinson S, Yuval B. 2003. Regulation of sexual receptivity of female Mediterranean fruit flies: old hypotheses revisited and a new synthesis proposed. *J. Insect Physiol.* 49:561–67
101. Moya-Larano J, Fox CW. 2006. Ejaculate size, second male size, and moderate polyandry increase female fecundity in a seed beetle. *Behav. Ecol.* 17:940–46
102. Mueller JL, Linklater JR, Ravi Ram K, Chapman T, Wolfner MF. 2008. Targeted gene deletion and phenotypic analysis of the *Drosophila melanogaster* seminal fluid protease inhibitor Acp62F. *Genetics* 178:1605–14
103. Mueller JL, Page JL, Wolfner MF. 2007. An ectopic expression screen reveals the protective and toxic effects of *Drosophila* seminal fluid proteins. *Genetics* 175:777–83
104. Mueller JL, Ravi Ram K, McGraw LA, Bloch Qazi MC, Siggia ED, et al. 2005. Cross-species comparison of *Drosophila* male accessory gland protein genes. *Genetics* 171:131–43
105. Mueller JL, Ripoll DR, Aquadro CF, Wolfner MF. 2004. Comparative structural modeling and inference of conserved protein classes in *Drosophila* seminal fluid. *Proc. Natl. Acad. Sci. USA* 101:13542–47

106. Nagalakshmi VK, Applebaum SW, Azrielli A, Rafaeli A. 2007. Female sex pheromone suppression and the fate of sex-peptide-like peptides in mated moths of *Helicoverpa armigera*. *Arch. Insect Biochem. Physiol.* 64:142–55
107. Nagalakshmi VK, Applebaum SW, Kubli E, Choffat Y, Rafaeli A. 2004. The presence of *Drosophila melanogaster* sex peptide-like immunoreactivity in the accessory glands of male *Helicoverpa armigera*. *J. Insect Physiol.* 50:241–48
108. Neubaum DM, Wolfner MF. 1999. Mated *Drosophila melanogaster* females require a seminal fluid protein, Acp36DE, to store sperm efficiently. *Genetics* 153:845–57
109. Orchard I, Lange AB. 1985. Evidence for octopaminergic modulation of an insect visceral muscle. *J. Neurobiol.* 16:171–81
110. Orr AG, Rutowski R. 1991. The function of the sphragis in *Cressida cressida* (Fab.) (Lepidoptera, Papilionidae): a visual deterrent to copulation attempts. *J. Nat. Hist.* 25:703–10
111. Otti O, Naylor RA, Siva-Jothy MT, Reinhardt K. 2009. Bacteriolytic activity in the ejaculate of an insect. *Am. Nat.* 174:292–95
112. Parthasarathy R, Tan A, Sun Z, Chen Z, Rankin M, Palli SR. 2009. Juvenile hormone regulation of male accessory gland activity in the red flour beetle, *Tribolium castaneum*. *Mech. Dev.* 126:563–79
113. Patterson JT. 1946. A new type of isolating mechanism in *Drosophila*. *Proc. Natl. Acad. Sci. USA* 32:202–8
114. Peng J, Chen S, Busser S, Liu H, Honegger T, Kubli E. 2005. Gradual release of sperm bound sex-peptide controls female postmating behavior in *Drosophila*. *Curr. Biol.* 15:207–13
115. Peng J, Zipperlen P, Kubli E. 2005. *Drosophila* sex-peptide stimulates female innate immune system after mating via the Toll and Imd pathways. *Curr. Biol.* 15:1690–94
116. Perez-Staples D, Aluja M, Macias-Ordonez R, Sivinski J. 2008. Reproductive trade-offs from mating with a successful male: the case of the tephritid fly *Anastrepha obliqua*. *Behav. Ecol. Sociobiol.* 62:1333–40
117. Perry JC, Rowe L. 2008. Ingested spermatophores accelerate reproduction and increase mating resistance but are not a source of sexual conflict. *Anim. Behav.* 76:993–1000
118. Perry JC, Rowe L. 2008. Neither mating rate nor spermatophore feeding influences longevity in a ladybird beetle. *Ethology* 114:504–11
119. Pilpel N, Nezer I, Applebaum SW, Heifetz Y. 2008. Mating increases trypsin in female *Drosophila* hemolymph. *Insect Biochem. Mol. Biol.* 38:320–30
120. Pizzari T, Snook RR. 2003. Perspective: sexual conflict and sexual selection: chasing away paradigm shifts. *Evolution* 57:1223–36
121. Poiani A. 2006. Complexity of seminal fluid: a review. *Behav. Ecol. Sociobiol.* 60:289–310
122. Polak M, Starmer WT, Barker JSF. 1998. A mating plug and male mate choice in *Drosophila hibisci* Bock. *Anim. Behav.* 56:919–26
123. Polak M, Wolf LL, Starmer WT, Barker JSF. 2001. Function of the mating plug in *Drosophila hibisci* Bock. *Behav. Ecol. Sociobiol.* 49:196–205
124. Pondeville E, Maria A, Jacques JC, Bourgouin C, Dauphin-Villemant C. 2008. *Anopheles gambiae* males produce and transfer the vitellogenic steroid hormone 20-hydroxyecdysone to females during mating. *Proc. Natl. Acad. Sci. USA* 105:19631–36
125. Prokupek AM, Kachman SD, Ladunga I, Harshman LG. 2009. Transcriptional profiling of the sperm storage organs of *Drosophila melanogaster*. *Insect Mol. Biol.* 18:465–75
126. Radhakrishnan P, Taylor PW. 2007. Seminal fluids mediate sexual inhibition and short copula duration in mated female Queensland fruit flies. *J. Insect Physiol.* 53:741–45
127. Radhakrishnan P, Taylor PW. 2008. Ability of male Queensland fruit flies to inhibit receptivity in multiple mates, and the associated recovery of accessory glands. *J. Insect Physiol.* 54:421–28
128. Rafaeli A. 2002. Neuroendocrine control of pheromone biosynthesis in moths. *Int. Rev. Cytol.* 213:49–91
129. Raina AK, Kingan TG, Giebultowicz JM. 1994. Mating-induced loss of sex pheromone and sexual receptivity in insects with emphasis on *Helicoverpa zea* and *Lymantria dispar*. *Arch. Insect Biochem.* 25:317–27
130. Ram KR, Wolfner MF. 2007. Seminal influences: *Drosophila* Acps and the molecular interplay between males and females during reproduction. *Integr. Comp. Biol.* 47:427–45
131. Ram KR, Wolfner MF. 2007. Sustained post-mating response in *Drosophila melanogaster* requires multiple seminal fluid proteins. *PLoS Genet.* 3:e238

132. Ram KR, Wolfner MF. 2009. A network of interactions among seminal proteins underlies the long-term postmating response in *Drosophila*. *Proc. Natl. Acad. Sci. USA* 106:15384–89
133. Ravi Ram K, Ji S, Wolfner MF. 2005. Fates and targets of male accessory gland proteins in mated female *Drosophila melanogaster*. *Insect Biochem. Mol. Biol.* 35:1059–71
134. Ravi Ram K, Sirot LK, Wolfner MF. 2006. Predicted seminal astacin-like protease is required for processing of reproductive proteins in *Drosophila melanogaster*. *Proc. Natl. Acad. Sci. USA* 103:18674–79
135. Reinhardt K, Naylor RA, Siva-Jothy MT. 2009. Ejaculate components delay reproductive senescence while elevating female reproductive rate in an insect. *Proc. Natl. Acad. Sci. USA* 106:21743–47
136. Reinhardt K, Wong CH, Georgiou AS. 2009. Detection of seminal fluid proteins in the bed bug, *Cimex lectularius*, using two-dimensional gel electrophoresis and mass spectrometry. *Parasitology* 136:283–92
137. Rexhepaj A, Liu HF, Peng J, Choffat Y, Kubli E. 2003. The sex-peptide DUP99B is expressed in the male ejaculatory duct and in the cardia of both sexes. *Eur. J. Biochem.* 270:4306–14
138. Richard FJ, Tarpy DR, Grozinger CM. 2007. Effects of insemination quantity on honey bee queen physiology. *PLoS One* 2:e980
139. Rodríguez-Valentín R, López-González I, Jorquera R, Labarca P, Zurita M, Reynaud E. 2006. Oviduct contraction in *Drosophila* is modulated by a neural network that is both, octopaminergic and glutamatergic. *J. Cell. Physiol.* 209:183–98
140. Rogers DW, Baldini F, Battaglia F, Panico M, Dell A, et al. 2009. Transglutaminase-mediated semen coagulation controls sperm storage in the malaria mosquito. *PLoS Biol.* 7:e1000272
141. Rogers DW, Whitten MM, Thailayil J, Soichot J, Levashina EA, Catteruccia F. 2008. Molecular and cellular components of the mating machinery in *Anopheles gambiae* females. *Proc. Natl. Acad. Sci. USA* 105:19390–95
142. Rogina B. 2009. The effect of sex peptide and calorie intake on fecundity in female *Drosophila melanogaster*. *ScientificWorldJournal* 9:1178–89
143. Romero A, Potter MF, Potter DA, Haynes KF. 2007. Insecticide resistance in the bed bug: a factor in the pest's sudden resurgence? *J. Med. Entomol.* 44:175–78
144. Ronn J, Katvala M, Arnqvist G. 2006. The costs of mating and egg production in *Callosobruchus* seed beetles. *Anim. Behav.* 72:335–42
145. Shutt B, Stables L, Aboagye-Antwi F, Moran J, Tripet F. 2010. Male accessory gland proteins induce female monogamy in anopheline mosquitoes. *Med. Vet. Entomol.* 24:91–94
146. Simmons LW. 2001. *Sperm Competition and its Evolutionary Consequences in the Insects*. Princeton, NJ: Princeton Univ. Press. 448 pp.
147. Sirot LK, Hardstone MC, Helinski MEH, Marinotti O, Kimura M, et al. 2010. Towards an ejaculatome of the dengue vector mosquito: protein identification and potential functions. *PLoS Negl. Trop. Dis.* Under review
148. Sirot LK, LaFlamme BA, Sitnik JL, Rubinstein CD, Avila FW, et al. 2009. Molecular social interactions: *Drosophila melanogaster* seminal fluid proteins as a case study. *Adv. Genet.* 68:23–56
149. Sirot LK, Poulson RL, McKenna MC, Girnary H, Wolfner MF, Harrington LC. 2008. Identity and transfer of male reproductive gland proteins of the dengue vector mosquito, *Aedes aegypti*: potential tools for control of female feeding and reproduction. *Insect Biochem. Mol. Biol.* 38:176–89
150. Smith A, Guo X, de la Fuente J, Naranjo V, Kocan KM, Kaufman WR. 2009. The impact of RNA interference of the subolesin and voraxin genes in male *Amblyomma hebraeum* (Acari: Ixodidae) on female engorgement and oviposition. *Exp. Appl. Acarol.* 47:71–86
151. Swanson WJ, Clark AG, Waldrip-Dail HM, Wolfner MF, Aquadro CF. 2001. Evolutionary EST analysis identifies rapidly evolving male reproductive proteins in *Drosophila*. *Proc. Natl. Acad. Sci. USA* 98:7375–79
152. Takami Y, Sasabe M, Nagata N, Sota T. 2008. Dual function of seminal substances for mate guarding in a ground beetle. *Behav. Ecol.* 19:1173–78
153. Takemori N, Yamamoto MT. 2009. Proteome mapping of the *Drosophila melanogaster* male reproductive system. *Proteomics* 9:2484–93
154. Tanaka ED, Hartfelder K. 2004. The initial stages of oogenesis and their relation to differential fertility in the honey bee (*Apis mellifera*) castes. *Arthropod Struct. Dev.* 33:431–42
155. Tarpy DR, Page RE. 2001. The curious promiscuity of queen honey bees (*Apis mellifera*): evolutionary and behavioral mechanisms. *Ann. Zool. Fenn.* 38:255–65

156. Thomas A. 1979. Nervous control of egg progression into the common oviduct and genital chamber of the stick-insect *Carausius morosus*. *J. Insect Physiol.* 25:811–21
157. Wagner WE Jr, Kelley RJ, Tucker KR, Harper CJ. 2001. Females receive a life-span benefit from male ejaculates in a field cricket. *Evolution* 55:994–1001
158. Wagstaff BJ, Begun DJ. 2005. Molecular population genetics of accessory gland protein genes and testis-expressed genes in *Drosophila mojavensis* and *D. arizonae*. *Genetics* 171:1083–101
159. Wagstaff BJ, Begun DJ. 2007. Adaptive evolution of recently duplicated accessory gland protein genes in desert *Drosophila*. *Genetics* 177:1023–30
160. Walker MJ, Rylett CM, Keen JN, Audsley N, Sajid M, et al. 2006. Proteomic identification of *Drosophila melanogaster* male accessory gland proteins, including a pro-cathepsin and a soluble gamma-glutamyl transpeptidase. *Proteome Sci.* 4:9
161. Walters JR, Harrison RG. 2008. EST analysis of male accessory glands from *Heliconius* butterflies with divergent mating systems. *BMC Genomics* 9:592
162. Walters JR, Harrison RG. 2010. Combined EST and proteomic analysis identifies rapidly evolving seminal fluid proteins in *Heliconius* butterflies. *Mol. Biol. Evol.* 27:2000–13
163. Wedell N, Gage MJG, Parker GA. 2002. Sperm competition, male prudence and sperm-limited females. *Trends Ecol. Evol.* 17:313–20
164. Weiss BL, Kaufman WR. 2004. Two feeding-induced proteins from the male gonad trigger engorgement of the female tick *Amblyomma hebraeum*. *Proc. Natl. Acad. Sci. USA* 101:5874–79
165. Weiss BL, Stepczynski JM, Wong P, Kaufman WR. 2002. Identification and characterization of genes differentially expressed in the testis/vas deferens of the fed male tick, *Amblyomma hebraeum*. *Insect Biochem. Mol. Biol.* 32:785–93
166. Wigby S, Chapman T. 2005. Sex peptide causes mating costs in female *Drosophila melanogaster*. *Curr. Biol.* 15:316–21
167. Wigby S, Domanitskaya EV, Choffat Y, Kubli E, Chapman T. 2008. The effect of mating on immunity can be masked by experimental piercing in female *Drosophila melanogaster*. *J. Insect Physiol.* 54:414–20
168. Wigby S, Sirot LK, Linklater JR, Buehner N, Calboli FC, et al. 2009. Seminal fluid protein allocation and male reproductive success. *Curr. Biol.* 19:751–57
169. Wong A, Albright SN, Giebel JD, Ram KR, Ji S, et al. 2008. A role for Acp29AB, a predicted seminal fluid lectin, in female sperm storage in *Drosophila melanogaster*. *Genetics* 180:921–31
170. Xue L, Noll M. 2000. *Drosophila* female sexual behavior induced by sterile males showing copulation complementation. *Proc. Natl. Acad. Sci. USA* 97:3272–75
171. Yamane T, Kimura Y, Katsuhara M, Miyatake T. 2008. Female mating receptivity inhibited by injection of male-derived extracts in *Callosobruchus chinensis*. *J. Insect Physiol.* 54:501–7
172. Yamane T, Miyatake T, Kimura Y. 2008. Female mating receptivity after injection of male-derived extracts in *Callosobruchus maculatus*. *J. Insect Physiol.* 54:1522–27
173. Yang CH, Rumpf S, Xiang Y, Gordon MD, Song W, et al. 2009. Control of the postmating behavioral switch in *Drosophila* females by internal sensory neurons. *Neuron* 61:519–26
174. Yapici N, Kim YJ, Ribeiro C, Dickson BJ. 2008. A receptor that mediates the post-mating switch in *Drosophila* reproductive behaviour. *Nature* 451:33–37

Using Geographic Information Systems and Decision Support Systems for the Prediction, Prevention, and Control of Vector-Borne Diseases

Lars Eisen[1] and Rebecca J. Eisen[2]

[1]Department of Microbiology, Immunology, and Pathology, Colorado State University, Fort Collins, Colorado 80523; email: lars.eisen@colostate.edu

[2]Division of Vector-Borne Infectious Diseases, Coordinating Center for Infectious Diseases, Centers for Disease Control and Prevention, Fort Collins, Colorado 80522; email: dyn2@cdc.gov

Annu. Rev. Entomol. 2011. 56:41–61

First published online as a Review in Advance on September 24, 2010

The *Annual Review of Entomology* is online at ento.annualreviews.org

This article's doi: 10.1146/annurev-ento-120709-144847

0066-4170/11/0107-0041$20.00

Key Words

GIS, modeling, risk map, Lyme borreliosis, plague, malaria, dengue, West Nile virus disease

Abstract

Emerging and resurging vector-borne diseases cause significant morbidity and mortality, especially in the developing world. We focus on how advances in mapping, Geographic Information System, and Decision Support System technologies, and progress in spatial and space-time modeling, can be harnessed to prevent and control these diseases. Major themes, which are addressed using examples from tick-borne Lyme borreliosis; flea-borne plague; and mosquito-borne dengue, malaria, and West Nile virus disease, include (*a*) selection of spatial and space-time modeling techniques, (*b*) importance of using high-quality and biologically or epidemiologically relevant data, (*c*) incorporation of new technologies into operational vector and disease control programs, (*d*) transfer of map-based information to stakeholders, and (*e*) adaptation of technology solutions for use in resource-poor environments. We see great potential for the use of new technologies and approaches to more effectively target limited surveillance, prevention, and control resources and to reduce vector-borne and other infectious diseases.

Vector-borne disease (VBD): disease caused by a pathogen transmitted by an arthropod vector

WNV: West Nile virus

Geographic Information System (GIS): software with database, mapping, and data analysis capacity

Remote sensing (RS): gathering of data about the physical world by detecting and measuring signals composed of radiation, particles, and fields emanating from objects located beyond the immediate vicinity of the sensor device

Decision Support System (DSS): an interactive system that aids the process of gathering, storing, and analyzing data; presenting data outputs; gaining new insights; generating alternatives; and making decisions

Spatial risk model: a statistical model used, in the case of VBDs, to estimate or predict vector presence or abundance, or VBD case presence or incidence, within a particular geographical area

INTRODUCTION

Emerging and resurging vector-borne diseases (VBDs) cause significant morbidity and mortality, especially in the developing world (42). VBDs account for 7 of 10 neglected infectious diseases that disproportionately affect poor and marginalized populations and therefore have been targeted for control programs by the World Health Organization (WHO) Special Program for Research and Training in Tropical Diseases (**http://apps.who.int/tdr/**). These diseases include, among others, malaria, with an estimated 247 million cases and nearly a million deaths in 2006, and dengue, with up to 50 million dengue infections and 500,000 cases of severe dengue hemorrhagic fever estimated to occur each year (119–120). Furthermore, new VBDs have emerged and have become established in developed regions of the world. Examples of VBDs that now are a fact of life in such areas, and highly unlikely to be eliminated, include West Nile virus (WNV) disease in North America and Lyme borreliosis in Asia, Europe, and North America (62, 100).

Technological advances over the last decades with relevance to VBDs include the emergence of molecular techniques for vector species identification and pathogen detection and identification, and a rapid evolution in hardware and software options to support data collection, management, and analysis. These advances are now dramatically changing our capacity to predict, prevent, and control VBDs. In this review, we focus on how advances in mapping, Geographic Information System (GIS), Remote Sensing (RS), and Decision Support System (DSS) technologies, and progress in the fields of spatial and space-time modeling, can be harnessed to reduce the burden that VBDs inflict on humans. Major themes to be explored include (*a*) selection of appropriate spatial and space-time modeling techniques, (*b*) the importance of using high-quality and biologically and/or epidemiologically relevant data, (*c*) incorporation of new technologies and approaches into operational vector and disease control programs, (*d*) transfer of map-based information to the stakeholder community, and (*e*) adaptation of technology solutions for use in resource-poor environments. These will be addressed using examples from a broad range of VBDs including tick-borne Lyme borreliosis, flea-borne plague, and mosquito-borne dengue, malaria, and WNV disease (**Table 1**). It should be noted that the literature is too extensive for exhaustive reviews of all related published papers; therefore, we present only selected, representative publications as examples.

SPATIAL AND SPACE-TIME RISK MODELS AND THEIR APPLICATIONS AS PUBLIC HEALTH TOOLS

Statistical modeling techniques are commonly incorporated into a GIS framework to (*a*) identify spatial and space-time patterns of vectors and VBD cases, (*b*) improve our understanding of how environmental factors affect arthropod vectors and influence transmission of vector-borne pathogens, and (*c*) predict future changes in spatial risk of exposure to vectors and vector-borne pathogens in response to shifting land use or climatic patterns. The ultimate goal of these activities is to reduce disease burdens by generating information that empowers the public to take protective action and helps public health agencies to allocate limited prevention, surveillance, and control resources to best effect.

Introduction to Spatial Risk Models

Spatial risk models are defined, in the context of this review, as GIS-based statistical models used to estimate vector presence or abundance, or VBD case presence or incidence, within a particular geographical area. Model outputs typically are displayed in map format (**Figure 1**). Basic spatial modeling approaches include (*a*) interpolation based on spatial dependence in vector or VBD data and (*b*) extrapolation based on associations between vector or VBD data and environmental or socioeconomic predictor variables. Importantly, this allows for

Table 1 Characteristics of vector-borne diseases used as examples in the review

Disease	Causative agent(s)	Primary vectors	Primary vertebrate reservoirs	Primary current disease foci
Mosquito-borne viral disease				
Dengue	Dengue virus	*Aedes aegypti*, *Ae. albopictus*	Humans	Subtropics and tropics, especially in Asia and the Americas[a]
West Nile virus disease	West Nile virus	*Culex* spp.	Birds	North America[a]
Tick-borne bacterial disease				
Lyme borreliosis	*Borrelia burgdorferi* sensu lato	*Ixodes* spp.	Rodents	Temperate areas in Asia, Europe, and North America
Flea-borne bacterial disease				
Plague	*Yersinia pestis*	Various flea species	Rodents	Africa
Mosquito-borne parasitic disease				
Malaria	*Plasmodium* spp.	*Anopheles* spp.	Humans	Subtropics and tropics in Africa, the Americas, and Asia

[a]Disease burden in Africa is poorly understood.

development of continuous risk surfaces that include locations where surveillance data are lacking and the level of risk therefore is not known prior to the modeling exercise. Coupled with demographic data, such risk surfaces can be used to assess the number of individuals or the proportion of a human population that is potentially at risk for exposure to vectors and vector-borne pathogens. Another benefit is the potential to reveal spatial heterogeneity in risk patterns at fine scales relevant for practical prevention and control activities.

Good models should have high accuracy in explaining the data used to build the model and in predicting new, independent data. They also should be parsimonious, based on the fewest possible predictor variables, and only include predictor variables that can be reasoned to be of biological or epidemiologic relevance. Numerous reviews have compared the pros and cons of the various techniques employed in spatial modeling, and the statistical methods have been described in detail elsewhere (34–35, 44, 52, 58, 73, 82, 84, 89, 95, 115). Here, we introduce the most commonly used techniques and summarize findings of studies that utilized them to develop spatial risk models of vectors or VBDs.

Spatial Risk Interpolation Models

Spatial dependence for vector abundance or VBD case count or incidence is frequently observed at fine spatial scales (58). For example, areas with high vector abundance or high disease incidence often border on other areas with high vector abundance or high disease incidence, and the similarity in the response variable decreases with increasing distance. In such instances, kriging or other types of interpolation models are used to produce smooth interpolated maps of the response variable (12, 20, 22, 32, 41, 58, 64, 80, 85, 99). High-end GIS software packages, such as ArcGIS® have extensive capacity for interpolation modeling.

Although interpolation methods are useful for transforming point-based data into smooth risk surfaces that can be used to infer risk in areas that were not sampled, they are most useful at fine spatial scales and often are unreliable beyond the geographical area within which point data were gathered (20, 58). Because of this limitation, other techniques that depend on identifying environmental predictors of vector abundance or VBD incidence are used to extrapolate risk surfaces beyond the local areas where vector or VBD data were collected. In

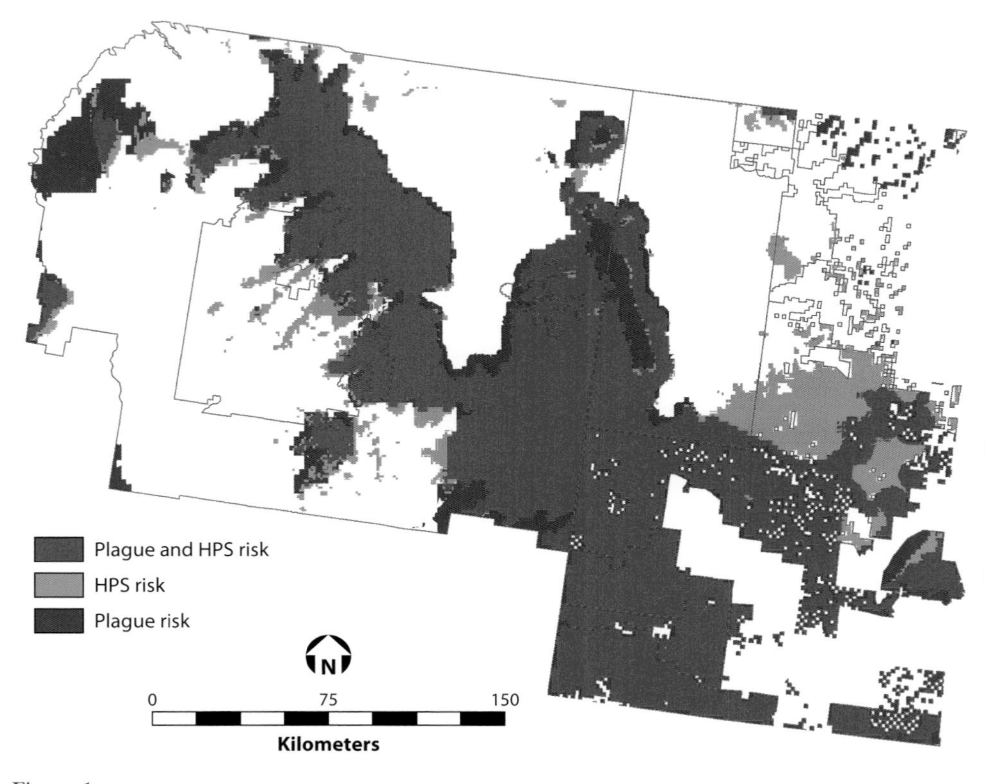

Figure 1

Map of areas predicted to pose elevated risk of plague, hantavirus pulmonary syndrome (HPS), or both on tribally or privately owned land in northeastern Arizona and northwestern New Mexico. The spatial risk models were developed using logistic regression modeling. This figure was previously published in the *American Journal of Tropical Medicine and Hygiene* (30).

instances where spatial dependence is very strong, interpolation may be incorporated into the alternative modeling approaches (12, 19–20, 23, 60).

Spatial Risk Extrapolation: Model Development Based on Generalized Linear Models

Spatial patterns of vector abundance or VBD incidence often can be predicted on the basis of their associations with environmental or socioeconomic variables, such as land use, soil type, temperature-related factors, or rainfall. In this scenario, the GIS software is first used to extract spatially explicit data for environmental factors of interest for the point locations (e.g., mosquito trap locations) or geographical areas (e.g., boundary units for which VBD incidence was calculated) where data were collected. Thereafter, a predictive model is developed in a statistical software package and the model equation is then applied in the GIS, for example, using the Raster Calculator in ArcGIS, to extrapolate a surface for the risk measure of interest. This basic approach has multiple benefits including (*a*) the potential for identifying environmental or socioeconomic predictors for risk of exposure to vectors or vector-borne pathogens, (*b*) development of continuous spatial surfaces that present estimates of risk for exposure to vectors or vector-borne pathogens and can be delivered to stakeholders in readily understandable map format, and (*c*) use of map outputs to guide decisions regarding allocation of surveillance, prevention, and control resources.

Commonly used techniques for the model development step include linear or Poisson regression models for continuous response variables, binomial logistic regression models for binary response variables, and multinomial or ordinal logistic regression models for multicategorical response variables. There also is increasing use of other techniques such as generalized additive models, which extend the generalized linear models to include nonparametric fits, and Bayesian approaches, which include a more rigorous accounting of uncertainty compared with models based on frequency probability (19, 51, 97, 98). Generalized linear models also may include terms that account for spatial dependence in the response variable. Below, we provide examples of the use of generalized linear models to identify factors predictive of elevated risk for exposure to vectors or vector-borne pathogens, and explore some issues relating to extrapolation of spatial risk surfaces from these models.

Examples of Identification of Risk Factors for Exposure to Vectors or Vector-Borne Pathogens Based on Use of Generalized Linear Models

Generalized linear models are effective tools to identify factors that are associated with elevated vector abundance or VBD incidence. Logistic regression modeling based on epidemiologic data was used to identify environmental predictors of elevated risk for human plague in the southwestern United States and Uganda; this consistently identified elevation as a primary risk predictor together with vegetation type, moisture, and temperature (29–31, 33, 124). Furthermore, use of linear, Poisson, and logistic regression modeling has revealed that high abundances of *Ixodes pacificus* and *I. scapularis*, which are key tick vectors of Lyme borreliosis spirochetes in North America, can be predicted by GIS- or RS-based environmental factors related to elevation, slope of the landscape, vegetation type, soil type, temperature, and moisture (10, 12, 20, 28, 38, 43).

The introduction of WNV into North America in 1999 spurred a series of similar modeling exercises for abundance of *Culex* WNV vectors and incidence of human WNV disease, which revealed that environmental predictive factors for elevated mosquito vector abundance or WNV disease incidence among the human population include availability of water sources, elevation, vegetation type, and temperature-related factors (11, 21, 121, 123). Furthermore, a detailed study from the greater Chicago area revealed that in urban settings factors predicting high WNV disease incidence among the human population can include environmental factors (presence of vegetation) as well as socioeconomic factors (income, age, housing age) and presence or absence of mosquito control activities (93).

Generalized linear models have been used extensively to identify environmental factors predictive of elevated abundances of anopheline malaria vectors and malaria prevalence in humans in sub-Saharan Africa (3, 7, 19, 46, 53, 60, 98, 99, 109, 126), and a few examples of model results are given below. Logistic regression modeling was used to predict the spatial abundance pattern of two malaria vectors (*Anopheles gambiae* sensu stricto and *Anopheles arabiensis*) in Mali based on vegetation indices, soil features, distance to water, temperature, and rainfall (98). In the same country, a logistic regression modeling approach was used to identify the normalized difference vegetation index (NDVI), distance to water, temperature, and rainfall as predictors of malaria prevalence in children (60). The model predictions in the latter study were further refined by incorporating a spatial interpolation (kriging) of the model residuals. Studies from other African countries have produced similar results; prevalence of malaria infection in humans was associated with rainfall, temperature, and elevation in Botswana (19), and elevation alone predicted 73% of households where an occupant had splenomegaly associated with a malaria infection (an indicator of repeated attacks and prolonged exposure to malaria parasites) in the Usambara Mountains of Tanzania (3).

The models described above for anopheline vectors and malaria provide good examples of instances in which the environmental predictors make biological sense based on our understanding of the vectors. Water sources, ranging from rice fields to cattle hoof prints, provide development sites for the immature life stages of the anopheline mosquito vectors, and temperature, which decreases with increasing elevation, affects the development time for immatures, the length of the gonotrophic cycle in the female mosquito, and the extrinsic incubation period for the malaria parasite within the vector (46, 66). This type of clear linkage between readily understandable predictor variables with obvious biological relevance and entomological or epidemiologic model outcomes is an important factor for decisions to use model results to guide operational vector and disease control program activities.

There also is a growing literature for use of generalized linear models to identify factors that can predict elevated spatial risk of exposure to dengue virus or to the primary vector, *Aedes aegypti*. Because this endophilic and endophagic mosquito is closely associated with human habitation and utilizes containers in the peridomestic environment as larval development sites, fine-scale modeling efforts in urban areas to identify predictors of elevated vector abundance or dengue infection rate need to incorporate factors relating to socioeconomic conditions such as presence of piped water, housing characteristics, and income (8, 75, 91, 107). In more rural settings in Thailand, land cover around the home was a useful predictive factor for dengue virus exposure, perhaps related to the presence of another dengue virus vector, *Ae. albopictus* (110, 112). There also have been some models developed for larger scales. Dengue risk in Taiwan was found to increase with average annual temperatures above 18°C and the degree of urbanization (125), and a study from Puerto Rico showed that the relative strength of temperature versus precipitation as predictors for dengue virus transmission varied between different parts of the island (55). One complicating factor for development of spatial risk models based on epidemiologic dengue data is that the disease can be caused by four different serotypes of dengue virus and that infection with one serotype does not provide long-term cross-protection against the other serotypes. This results in a dynamic situation with potential for rapid fluctuations in serotype-specific levels of susceptibility among the human population, which can confound spatial modeling efforts.

Spatial Risk Extrapolation: Production of Risk Surfaces Based on Use of Generalized Linear Models

Extrapolation of continuous spatial surfaces estimating the level of risk for exposure to vectors or vector-borne pathogens, based on application of a model equation in a GIS, was incorporated into many of the studies described above (see example for plague in the southwestern United States in **Figure 1**). One important but poorly studied aspect of this activity is the selection of the geographical area over which the risk model can reliably be extrapolated. A commonsense approach is to restrict the extrapolation of a model to geographical areas that fall within the data range of the ecological or climatic predictor variables used to develop the model. This was illustrated in a study on the WNV vector *Culex tarsalis* in Colorado, where a mosquito abundance model was developed on the basis of an association with cooling degree days (CDD) for the eastern slope of the Rocky Mountains (121). The model then was scaled up to different parts of the state of Colorado, and correlations between the extent of area with high predicted *Cx. tarsalis* abundance and WNV disease incidence in humans were examined at the census tract scale. This revealed a positive correlation between model-predicted areas with high vector abundance and WNV disease incidence in the western, mountainous part of the state, which has a CDD range similar to the model development area and to which extrapolation thus is appropriate. In contrast, a negative correlation was observed in the eastern Colorado plains, where the CDD range is distinctly different from the model development area; extrapolation therefore should not be assumed to produce reliable results.

When model extrapolation is restricted to areas with ecological and climatic characteristics similar to those of the model development area, this approach can be a powerful tool to gain insights into levels of risk within areas where surveillance data are lacking or unreliable. As an example, in the most recent revision of WHO's International Health Regulations (117), only the most severe form of plague (pneumonic plague) is an internationally notifiable disease whereas other manifestations are notifiable only from nonendemic localities. Compliance with these regulations thus requires a clear understanding of the locations of plague foci, and WHO recommends use of GIS technology and modeling to refine our knowledge of these risk areas and to cost-effectively target surveillance resources (118). In the West Nile region of Uganda, a linear regression modeling approach revealed that incidence of plague cases was higher above 1,300 m and in parishes with higher surface temperature and greater land cover variety, and with certain remotely sensed indicators of soil and vegetation type (124). This spatial risk model, and an additional model for the same area that focuses on the finer village and subvillage scale and therefore is better suited for allocation of scarce plague control resources (31), has great potential for extrapolation from Uganda's West Nile region into neighboring areas of the Democratic Republic of Congo with similar ecological, socioeconomic, and climatic characteristics but where plague surveillance activities are lacking (**Figure 2**).

Spatial Risk Assessments Based on Machine Learning Algorithms or Dynamic Simulation Models

Low confidence in data related to absence of vectors or VBD cases is often an obstacle to the use of generalized linear models. In this case, presence-only machine learning (rule-based) algorithms or dynamic simulation models serve as alternatives (84, 101, 103). Examples include various ecological niche modeling algorithms such as genetic algorithm for rule-set prediction (GARP) (101) and MAXENT, a machine learning algorithm based on maximum entropy (84), and simulation models such as CLIMEX (**http://www.climatemodel.com/index.html**). Ecological niche models have, for example, been used to model distributions of anopheline malaria vectors at continental scales using archived species occurrence data (74, 81). Likewise, risk of human plague in the United States and in Africa has been modeled using GARP (77–79) and MAXENT (50). In instances where model outputs of generalized linear models and GARP could be directly compared, the former typically provided a more restricted area that was expected to pose elevated risk of plague (29, 33, 77, 79, 124). The more restricted risk surfaces that resulted from the generalized linear models are well suited for prevention and control resource allocation, whereas the broader ecological niche that is captured by GARP may be useful for identifying new areas where vectors or VBDs are likely to emerge.

One stated benefit of the dynamic simulation model CLIMEX, which has been used to model areas in Australia at risk for malaria under different climate change scenarios (102), is that absence data in areas estimated as suitable from presence data are attributed as unknowns, which prevents the model from restricting its parameter values to simulating only the presence records and stimulates a search for other explanations (103). This type of model is particularly useful for anticipating future expansions of vector or pathogen distributions, as opposed to current distributions that are often accurately portrayed using generalized linear models (103). Another recently emerged simulation model is the stochastic and spatially explicit Skeeter Buster, which focuses on container-breeding dengue virus mosquito vectors and operates at the spatial scale of individual

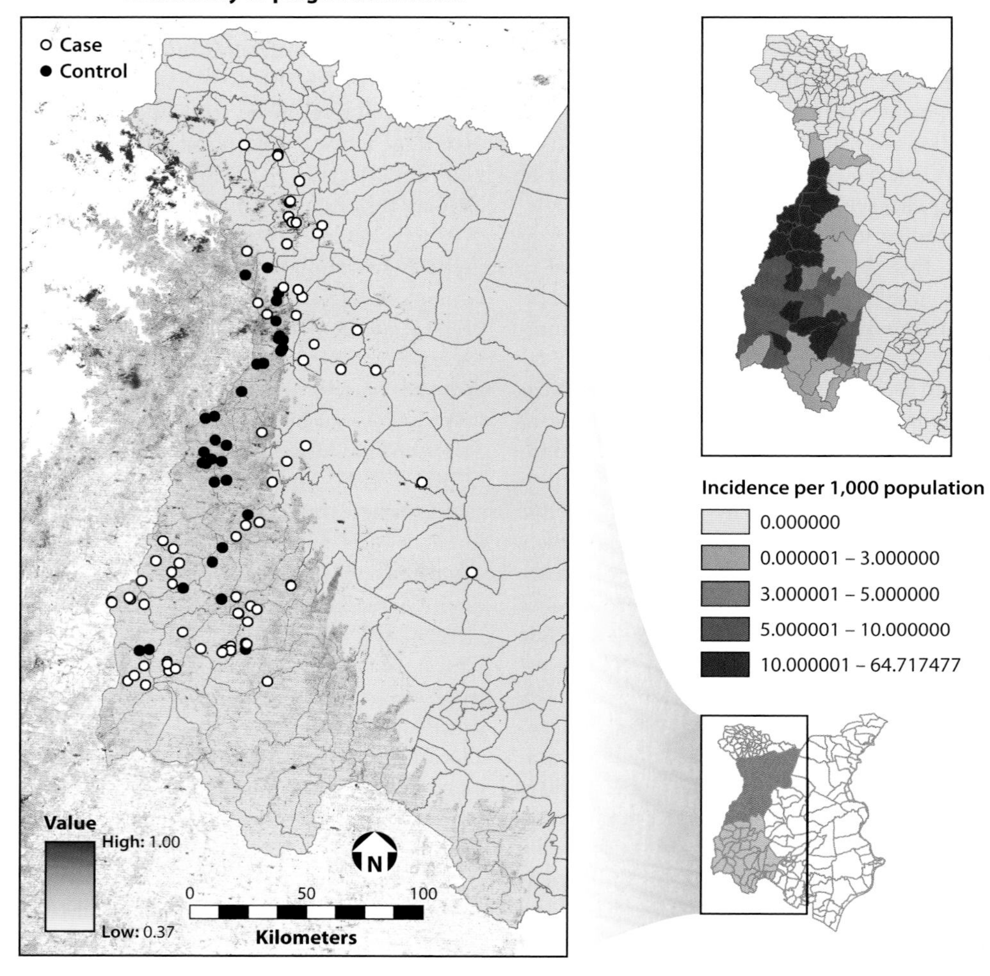

Figure 2

Predicted distribution of areas where humans are at elevated risk for exposure to *Yersinia pestis* within the area of interest in the West Nile region of Uganda and in neighboring regions of the Democratic Republic of Congo. Shaded areas represent pixels classified as elevated risk; the color gradient indicates the probability of case occurrence within areas of elevated risk based on dichotomization at a probability value of 0.37. Input variables used in the predictive model include positive associations with elevation above 1,300 m, brightness, and Landsat ETM+ band 3, and a negative association with Landsat ETM+ band 6. Insets show the area of interest and parish level incidence for Ugandan parishes within the area of interest. This figure was previously published in the *American Journal of Tropical Medicine and Hygiene* (31).

water-filled containers and individual premises (70).

Space-Time Risk Models

Space-time risk models are used to explore spatial clustering of vectors or VBD cases and are useful for identifying changing risk patterns. This is especially relevant for VBDs such as dengue, which is characterized by outbreaks with rapidly building dengue case numbers and explosive spread within affected areas. Space-time modeling may aid in identifying underlying factors that regulate the spread of vectors or the occurrence of VBD cases and may be more sensitive than purely spatial or temporal models in detecting local outbreaks (40, 63). Therefore, outputs of these models may be used as early-warning systems and could guide vector control or surveillance activities. Commonly used methods include space-time permutation scan statistics (e.g., SaTScan) (63, 113, 116), Knox tests (61), generalized additive mixed models (69), and Bayesian hierarchical regression models (39, 65).

For dengue, space-time modeling consistently demonstrates that dengue cases are clustered in space and time (4, 13, 57, 71, 76, 92). Part of the challenge to ultimately forecast space-time patterns of a dengue outbreak, and rapidly implement vector control response to curb the outbreak, is that the virus spreads in two different clustering diffusion patterns: (*a*) a predictable contiguous pattern of dengue diffusion over short distances from an index case that probably is driven by a combination of movement by infected mosquito vectors and short-range movement of infectious humans, and (*b*) an unpredictable relocation pattern of dengue diffusion with cases jumping to new, unconnected areas and initiating new local dengue foci, most likely resulting from longer-range movements of infectious humans (57, 96).

For malaria, space-time permutation scan statistics were incorporated into a malaria early-warning system to detect local malaria clusters and to target vector control and health education activities in South Africa's Mpumalanga Province (16). They were also used to detect temporally stable malaria clusters in Ethiopia (83). In Kenya, regression analyses were used to identify meteorological factors and remotely sensed vegetation indices that were predictive of mean monthly percentages of total annual malaria admissions; model thresholds were then extrapolated spatially in a clinical disease seasonality map to define the number of months per location with expected malaria admissions (48). Similarly, a malaria seasonality risk map for Zimbabwe was created using a space-time regression model within a Bayesian framework (68).

Space-time models also have been applied to Lyme borreliosis and WNV disease in the United States to gain a better understanding of how the geographic coverages of these diseases may change over time. Waller et al. (114) used hierarchical linear models to describe an increasing county-level Lyme borreliosis incidence pattern for the central northeastern Atlantic coast with branching to the north and west and stable or slightly decreasing incidence in counties located in western New York State. Within the state of New York, Chen et al. (15) implemented a Bayesian hierarchical regression approach to reveal a northerly progression from 1990 to 2000 in the location of counties reporting the highest Lyme borreliosis incidence. The emergence of WNV in the United States prompted the development of the Dynamic Continuous-Area Space-Time (DYCAST) system, which uses a localized Knox test to capture the space-time interaction of WNV surveillance data such as dead birds (108).

Space-time risk model: a model that incorporates data with both spatial and temporal characteristics, is used to explore spatial clustering over time, and is useful for identifying changing risk patterns

THE IMPORTANCE OF HIGH-QUALITY AND BIOLOGICALLY OR EPIDEMIOLOGICALLY RELEVANT DATA

Perhaps the most important thing to remember when embarking on a modeling project is that models are only as good as the data upon which they are based. Furthermore, the data requirements for a meaningful model with public health utility are dependent on first determining the intent of the modeling exercise, e.g., hypothesis testing, identifying knowledge gaps, providing direction for surveillance and control efforts, or evaluating effectiveness of an intervention (59). For example, modeling the probability of plague case occurrence at regional or even continental scales may be useful to broadly determine the need for surveillance activities (77–79), but finer-resolution models are required for targeting prevention and control resources implemented at local spatial scales (29–30, 33, 124).

Risk of human exposure to vector-borne pathogens may be assessed on the basis of modeling the spatial distribution of human disease cases using epidemiologic data, or on the basis of vector data. The advantages and disadvantages of using vector data versus epidemiologic data in spatial risk modeling were reviewed recently (27) and are summarized briefly here. Benefits of training spatial risk models on epidemiologic data include (*a*) that a human VBD case is unequivocally linked with exposure to the disease agent and (*b*) there is potential for identifying socioeconomic factors associated with exposure risk. However, there are numerous drawbacks to using epidemiologic data, including (*a*) not all VBDs are notifiable, (*b*) case definitions and reporting practices change over space and time, (*c*) socioeconomic factors influence care-seeking behavior, and (*d*) for some VBDs, such as dengue and WNV disease, asymptomatic infections are common. Furthermore, location of pathogen exposure is often not investigated or reported and instead the location of residence is used as a surrogate, often without a detailed understanding of the spatial dimension of pathogen transmission to humans. To account for patient privacy, outputs from models based on epidemiologic data often are displayed at coarse spatial scales (e.g., county of residence in the United States) and fail to account for fine-scale variability in risk patterns (25). Finally, spatial models based on epidemiologic data cannot be used to assess risk of exposure on publicly owned lands where humans do not reside but where visitors may be at risk for recreational exposure to vector-borne pathogens.

Modeling spatial risk on the basis of vector data can be advantageous because (*a*) it allows for fine-scale spatial precision in the vector collection location, (*b*) many vectors transmit multiple pathogens and therefore assessing abundance of a single vector may be informative for risk assessments of several VBDs, (*c*) many VBDs are not notifiable and therefore epidemiologic data are lacking, and (*d*) models can be developed for public or private land because the vector abundance estimate is independent of a human population base. Limitations of basing spatial risk models on vector data include (*a*) the cost and effort associated with field collections and pathogen detection and (*b*) the lack of a direct correlation between abundance of vectors or infected vectors and human disease (prevention activities such as use of repellents may limit contact between humans and vectors even in areas with high vector abundance).

The quality of data for predictor variables is equally important. As noted previously, risk models for vectors or VBDs have been based on associations with socioeconomic conditions or environmental factors such as elevation, soil type, vegetation type, land cover, and climatic or meteorological variables. In the United States, GIS-based data layers containing a wide range of socioeconomic variables can be accessed from the U.S. Census Bureau and fine-resolution GIS-based data layers for elevation, soil type, hydrological features, vegetation type, and land use are available from sources such as the U.S. Geological Survey and state Gap Analysis Program projects. Furthermore, GIS-based climatic and meteorological data layers are readily available from Oregon State University's PRISM climate group at spatial resolutions in the 2- to 4-km range and are commonly reported at monthly intervals. Although such classified data layers are often lacking from developing countries, RS data for vegetation indices (NDVI, greenness), soil and water characteristics (brightness, wetness), and some meteorological variables can be derived using images from satellites such as Landsat (15- to 60-m resolution, covers entire earth every 16 days) and MODIS (moderate-resolution imaging spectroradiometer; 250- to 1,000-m resolution, covers entire earth every 1 to 2 days). Use of RS data for modeling risk of exposure to vectors and VBDs was reviewed previously (90). Finally, when compiling the predictor variable data layers, they should represent epidemiologically relevant spatial and temporal scales in order to produce model outputs of direct use in prevention and control efforts (54, 94).

Potential stumbling blocks to consider for use of GIS and RS data in risk modeling for vectors and VBDs include lack of fine-scale GIS-based data for zoonotic pathogen reservoirs (e.g., for modeling risk of exposure to the etiologic agents of Lyme borreliosis, plague, and WNV disease in the United States), the challenge of acquiring cloud-free satellite imagery in the tropics, and the need for an enhanced meteorological observation network in developing countries to develop better GIS-based climate data.

OPERATIONAL USE OF GEOGRAPHICAL INFORMATION SYSTEM TECHNOLOGY IN VECTOR AND DISEASE CONTROL PROGRAMS

Numerous reviews have broadly addressed the use of GIS/RS and spatial and space-time modeling approaches in the field of VBDs (5, 27, 58, 90). However, the critically important issue of the potential for such technologies and methodologies to be used for operational surveillance and control of VBDs has not received the attention it deserves, especially for neglected tropical diseases. Eisen & Lozano-Fuentes (26) recently reviewed this topic for dengue and concluded that there is tremendous potential for moving mapping and modeling approaches from the research arena to practical applications that can enhance operational vector and dengue control. Similar conclusions were drawn for malaria in a review by Saxena et al. (95). Below, we provide some examples of GIS technology used in operational vector and disease control.

In South Africa, GIS technology was incorporated into operational malaria control programs in the 1990s and is now used for a variety of purposes including malaria case mapping and monitoring of vector control coverage (9, 72). Australia and Singapore are making extensive use of GIS in their operational vector and dengue control programs. For example, Queensland Health in Australia employs GIS in ongoing mapping of dengue case locations in relation to spatial coverage of implemented vector control to help determine if response activities have adequate spatial coverage (86–87). Singapore's National Environment Agency uses GIS in a wide range of operational vector and dengue control activities including tracking of dengue case locations, vector mosquito surveillance, and monitoring of vector control coverage (1, 105–106). Nicaragua's Ministry of Health also is starting to use GIS and mapping software in vector and dengue control activities such as dengue case mapping and mosquito vector surveillance (2, 14).

Mapping software: software with mapping and editing capability but no or minimal data analysis capability

In the United States, the emergence of WNV in 1999 resulted in a new national GIS-based surveillance system for arboviral diseases: ArboNET. This system compiles surveillance data for a wide range of arboviral diseases (WNV disease, St. Louis encephalitis, eastern equine encephalitis, western equine encephalitis, La Crosse encephalitis, and Powassan encephalitis) and includes data for disease in humans and domestic animals as well as surveillance data for infection in vertebrates (e.g., sentinel chickens or wild birds) and vectors (e.g., testing of *Culex* mosquitoes for WNV). Map-based outputs from ArboNET are made available online by the Centers for Disease Control and Prevention (**http://www.cdc.gov/ncidod/dvbid/westnile/surv&control.htm**) and the U.S. Geological Survey (**http://diseasemaps.usgs.gov/**).

To achieve increased use of mapping and GIS technologies in operational vector and disease control, it is critically important for control programs to share their experiences with GIS and other emerging technologies through publications and other information delivery mechanisms. Although it may be difficult to assign time to such undertakings in the midst of the day-to-day control activities, it must be stressed that operational vector and disease control programs around the world have gained invaluable experiences regarding the benefits and drawbacks of using GIS and other emerging technologies that need to be shared with their counterparts in other disease-endemic areas.

TRANSFER OF MAP-BASED INFORMATION FOR VECTOR-BORNE DISEASES TO THE STAKEHOLDER COMMUNITY

Basic options to present information for spatial risk of VBDs in map formats to the stakeholder community include point locations for disease cases, aggregation of disease case counts or disease incidence to administrative boundary units, or smoothing. A map showing individual case point locations is obviously the most precise way to present spatial disease data. However, this has distinct disadvantages including (*a*) the possibility that the address of residence is not the site of pathogen exposure, (*b*) a lack of accounting for population size, and (*c*) in some countries, including the United States, strict regulations to guide the use of patient health information. To avoid privacy issues, it is common practice to aggregate disease case counts or disease incidence to administrative boundaries in the display of spatial risk patterns.

GIS software and new mapping software, such as Google Earth™ (**http://earth.google.com/**), now provide capacity to generate risk maps in a variety of formats including overlays on satellite imagery and dynamic illustrations of space-time patterns that can be played as movie clips (6, 26, 67). This is accompanied by explosive development in the field of Web-based information delivery, which now provides an effective medium to distribute risk maps to a wide range of stakeholders including the medical community, vector control practitioners, policy makers, and the public at large (36, 37). The Malaria Atlas Project (**http://www.map.ox.ac.uk/**) is one example of effective online map delivery, including maps of the spatial limits of *Plasmodium falciparum* transmission (45, 47). Another is the MARA/ARMA project with online delivery of a wide range of malaria-related maps (**http://www.mara.org.za/**).

As illustrated in **Figure 3** for WNV disease in Colorado, maps showing case counts and disease incidence for different spatial boundary units can be used as tools to target limited prevention, surveillance, and control resources to high-risk areas for WNV exposure, and to inform the public about local risk levels (122). For example, a mosquito control program aiming to implement control activities to suppress vector mosquitoes and reduce the burden of WNV disease likely will be most interested in finding out where high numbers of WNV disease cases occur in order to focus expensive prevention efforts. On the other hand, a member of the public seeking information to help determine his/her personal risk of exposure to WNV, and the need for use of personal protective measures such as repellents, will be more interested in spatial risk estimates based on WNV disease incidence (which account for population size).

However, with this new technological capacity to present spatial risk patterns comes a series of questions regarding how it should be used responsibly in public health (54, 94). One key question relates to the spatial scale at which risk maps based on epidemiologic data should be presented to best balance the needs to provide spatially explicit and accurate risk information while protecting patient privacy. Using finer spatial scales (and smaller population bases) can result in more informative maps but also may result in analysis artifacts and misleading risk maps. Further studies are urgently needed for important VBDs to determine the benefits and drawbacks of presenting risk maps at different spatial scales. In addition, we need to gain a better understanding of what type of information different types of stakeholders feel

→

Figure 3

West Nile virus disease case counts and incidences per 100,000 person-years in Colorado assorted by county, census tract, and zip code based on combined data for 2003 and 2007 and classified as no cases reported, or by quartile for the spatial boundary units reporting cases. This figure was previously published in the *American Journal of Tropical Medicine and Hygiene* (122).

Total cases

Incidence per 100,000 person-years

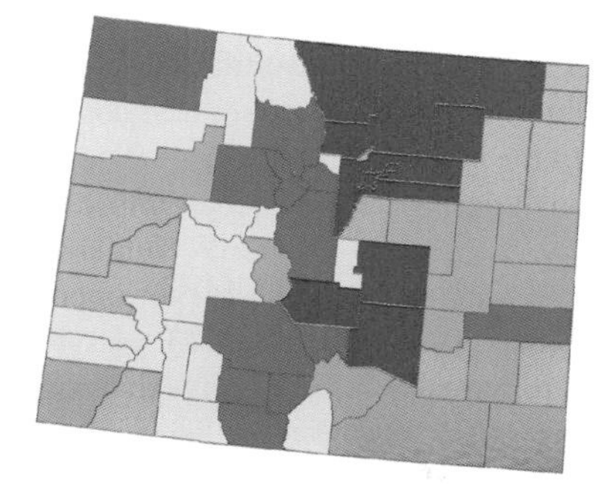

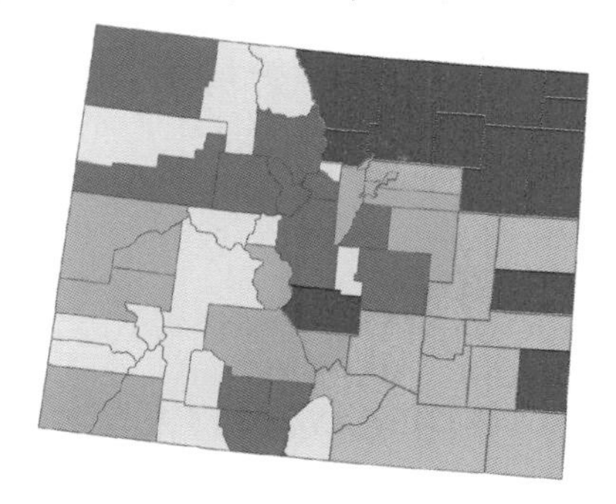

County scale

Census tract scale

Zip code scale

Quartile:
(Range of cases for county, census tract, and zip code)

- **1st** [(1–3), (1), (1–2)]
- **2nd** [(4–11), (2–3), (3–6)]
- **3rd** [(12–67), (4–7), (7–15)]
- **4th** [(>67), (>7), (>15)]
- No cases reported

Quartile:
(Range of incidence for county, census tract, and zip code)

- **1st** [(1–11.7), (0–15.6), (2.8–14.1)]
- **2nd** [(11.8–26.9), (15.7–31.4), (14.2–39.2)]
- **3rd** [(27–88.4), (31.5–70.9), (39.3–106.4)]
- **4th** [(>88.4), (>70.9), (>106.4)]
- Incidence of 0

that they require, and to determine optimal map and text formats to ensure that the message we aim to transmit in fact is clear to the user (6). Finally, there is a critical need to determine if presenting risk maps to the public has any impact, positive or negative, on preventive behaviors relating to VBDs.

TECHNOLOGIES FOR COLLECTION, MANAGEMENT, ANALYSIS, AND DISPLAY OF VECTOR AND DISEASE DATA

Management and analysis of data can be a complex undertaking and is greatly facilitated by use of database and statistical analysis software. A basic practical example involves entering and manipulating data in a Microsoft® Excel spreadsheet and then exporting the data to a statistical software package for analysis. This, however, becomes challenging when large amounts and diverse types of data are handled. We then instead can make use of a relational database software package, e.g., Microsoft® Access, Microsoft® SQL Server, or an open source alternative such as PostgreSQL (**http://www.postgresql.org/**). The primary benefits of relational databases are that they can be customized with regards to data entry and that the relationships defined within the database allows for effective querying and extraction of data. GIS software typically includes a relational database management system to facilitate entry, storage, extraction, and visualization of data.

With the current expansion in mobile data capturing technology, we now also have the opportunity to move the stage of electronic data capture all the way down to the initial data capturing session in the field (2, 56, 111). This can be accomplished using a laptop or netbook computer, a personal digital assistant (PDA), or even a smart phone. The basic workflow involves capture of data on the mobile electronic device followed by download into a central database by means of direct connection or transmission of data over the Internet and/or cell phone network. In an ideal scenario, the data-capturing device also has capacity to act as a Global Positioning System receiver and thus generate data for the location at which the data were entered. Mobile data-capturing technology is a field where we expect to see tremendous advances in the coming decade.

Data analysis capacity also has improved dramatically in recent decades. Numerous software packages are now available to support statistical analyses, to process GIS and RS data, and to visualize maps. An exhaustive review of such software is beyond the scope of this paper, but there are some developments worth noting. First, at-cost software packages are rapidly being complemented by freely available alternatives, many of which also are open source with access to the source code. Examples of software that can be downloaded and used at no cost include the statistical package R (**http://www.r-project.org/**), the previously mentioned relational database software PostgreSQL and the mapping software Google Earth™, and the space-time pattern analysis software SaTScan™ (**http://www.satscan.org/**). Second, the availability of environmental GIS and RS data and demographic and socioeconomic GIS data is rapidly increasing and there is a positive trend toward such data being made freely available for download. Third, there is a drive toward developing DSS software packages for operational control of VBDs, including neglected tropical diseases such as dengue, malaria, and human African trypanosomiasis, that combine user-friendly data entry with data analysis and visualization capacity (17–18, 24, 49, 67, 88, 104).

ADAPTATION OF TECHNOLOGY SOLUTIONS FOR USE IN RESOURCE-POOR ENVIRONMENTS: DECISION SUPPORT SYSTEMS

Adaptation of technology solutions for use in resource-constrained disease-endemic environments that experience the most severe VBD burdens must be made part of the new frontier in VBD research. For example, use of cell phones for rapid and inexpensive information transfer has great potential for implementation in malaria- and dengue-endemic areas with poor Internet access but well-developed cell phone networks (56). Development of DSS software packages for malaria and dengue that are composed entirely of software components that can be distributed to users without licensing costs, which ensures that they can be implemented in resource-poor environments, is another positive example (17, 24, 49). These systems incorporate a wide range of data from entomological surveillance, disease case surveillance, vector and disease control intervention monitoring, and stock control. They also include a GIS backbone and reporting tools that allow the user to produce a wide range of outputs including tables, graphs, and maps. Key benefits of implementing a DSS include (*a*) improved capacity for electronic data storage; (*b*) compilation of a wide range of data in a single system, which allows the user to produce outputs combining different types of data such as coverage of vector control in relation to disease case incidence; and (*c*) improved capacity for monitoring and evaluation of control program performance.

SUMMARY POINTS

1. Advances in mapping, GIS, and DSS technologies, and progress in the fields of spatial and space-time modeling, provide new opportunities to prevent and control emerging and resurging VBDs.
2. Benefits of spatial and space-time risk modeling include identification of risk patterns for exposure to vectors and vector-borne pathogens, and an improved understanding of how socioeconomic and environmental factors affect the vectors and influence transmission of their associated pathogens.
3. Perhaps the most important thing to remember when embarking on a mapping or modeling project is that map or model outputs are only as good as the data on which they are based.
4. GIS-based spatial and space-time risk modeling have proven effective tools to develop risk surfaces (maps) to inform policy makers, control programs, and the public.
5. There needs to be a stronger emphasis on moving GIS technology and modeling approaches from the research arena into operational vector and disease control programs.
6. Clear linkage between readily understandable predictor variables with obvious biological relevance and entomological or epidemiologic model outcomes is an important factor for decisions to use model results to guide operational vector and disease control program activities.
7. We need to determine what type of map-based information different stakeholders require in order to make practical use of the maps, and to determine if presenting risk maps to the public has any impact, positive or negative, on preventive behaviors.
8. Adaptation of technology solutions for use in resource-constrained environments that experience the most severe disease burdens must be made part of the new frontier in VBD research.

FUTURE ISSUES

1. In addition to the strong current focus on improving statistical modeling techniques, there is a need to allocate resources for development of the high-quality data sets for vector and epidemiologic data without which development of high-quality models and risk maps is impossible. This includes determination of probable pathogen exposure sites for VBD patients to complement information for residence location.
2. Concerted efforts are needed to ensure ready and inexpensive access for the academic and public health communities in developing countries to both high-end GIS software and high-quality GIS/RS-based data for socioeconomic and environmental factors as well as administrative boundaries and natural features.
3. The research community is very adept at using GIS/RS-based data to develop predictive models for spatial or space-time patterns of VBDs and to display these as risk maps. There is, however, a disconnect between this model and risk map production process and the practical use of the models and risk maps for prevention and control purposes. To bridge this gap, studies are urgently needed to determine how stakeholders make use of model findings and map-based risk information.

DISCLOSURE STATEMENT

The authors are not aware of any affiliations, memberships, funding, or financial holdings that might be perceived as affecting the objectivity of this review.

ACKNOWLEDGMENTS

This review was developed, in part, based on funding from the Innovative Vector Control Consortium and the National Institutes of Allergy and Infectious Diseases (contract N01-AI-25489).

LITERATURE CITED

1. Ai-leen GT, Song RJ. 2000. The use of GIS in ovitrap monitoring for dengue control in Singapore. *Dengue Bull.* 24:110–16
2. Aviles W, Ortega O, Kuan G, Coloma J, Harris E. 2008. Quantitative assessment of the benefits of specific information technologies applied to clinical studies in developing countries. *Am. J. Trop. Med. Hyg.* 78:311–15
3. Balls MJ, Bodker R, Thomas CJ, Kisinza W, Msangeni HA, Lindsay SW. 2004. Effect of topography on the risk of malaria infection in the Usambara Mountains, Tanzania. *Trans. R. Soc. Trop. Med. Hyg.* 98:400–8
4. Barreto FR, Teixeira MG, Costa MDCN, Carvalho MS, Barreto ML. 2008. Spread pattern of the first dengue epidemic in the city of Salvador, Brazil. *BMC Publ. Health* 8:51
5. Beck LR, Lobitz BM, Wood BL. 2000. Remote sensing and human health: New sensors and new opportunities. *Emerg. Infect. Dis.* 6:217–27
6. Bell BS, Hoskins RE, Pickle LW, Wartenberg D. 2006. Current practices in spatial analysis of cancer data: mapping health statistics to inform policymakers and the public. *Int. J. Health Geogr.* 5:49
7. Bogh C, Lindsay SW, Clarke SE, Dean A, Jawara M, et al. 2007. High spatial resolution mapping of malaria transmission risk in the Gambia, West Africa, using LANDSAT TM satellite imagery. *Am. J. Trop. Med. Hyg.* 76:875–81

8. Bohra A, Andrianasolo H. 2001. Application of GIS in modeling of dengue risk based on sociocultural data: case of Jalore, Rajasthan, India. *Dengue Bull.* 25:92–102
9. Booman M, Durrheim DN, La Grange K, Martin C, Mabuza AM, et al. 2000. Using a geographical information system to plan a malaria control program in South Africa. *Bull. WHO* 78:1438–44
10. Brownstein JS, Holford TR, Fish D. 2003. A climate-based model predicts the spatial distribution of the Lyme disease vector *Ixodes scapularis* in the United States. *Environ. Health Perspect.* 111:1152–57
11. Brownstein JS, Rosen H, Purdy D, Miller JR, Merlino M, et al. 2002. Spatial analysis of West Nile virus: rapid risk assessment of an introduced vector-borne zoonosis. *Vector Borne Zoonotic Dis.* 2:157–64
12. Bunnell JE, Price SD, Das A, Shields TM, Glass GE. 2003. Geographic information systems and spatial analysis of adult *Ixodes scapularis* (Acari: Ixodidae) in the Middle Atlantic region of the USA. *J. Med. Entomol.* 40:570–76
13. Chadee DD, Williams FLR, Kitron UD. 2005. Impact of vector control on a dengue fever outbreak in Trinidad, West Indies, in 1998. *Trop. Med. Int. Health* 10:748–54
14. Chang A, Parrales M, Jimenez J, Sobieszczyk M, Hammer S, et al. 2009. Combining Google Earth and GIS mapping technologies in a dengue surveillance system for developing countries. *Int. J. Health Geogr.* 8:49
15. Chen H, Stratton HH, Caraco TB, White DJ. 2006. Spatiotemporal Bayesian analysis of Lyme disease in New York state, 1990–2000. *J. Med. Entomol.* 43:777–84
16. Coleman M, Coleman M, Mabuza AM, Kok G, Coetzee M, Durrheim DN. 2009. Using the SaTScan method to detect local malaria clusters for guiding malaria control programmes. *Malar. J.* 8:68
17. Coleman M, Hemingway J. 2008. The implications of entomological monitoring and evaluation for arthropod vector-borne disease control programs. In *Vector-Borne Disease – Understanding the Environmental, Human Health, & Ecological Connections. Institute of Medicine Workshop Summary*, ed. SM Lemon, PF Sparling, MA Hamburg, DA Relman, ER Choffnes, A Mack, pp. 178–89. Washington, DC: Natl. Acad. Press
18. Coleman M, Sharp B, Seocharan I, Hemingway J. 2006. Developing an evidence-based decision support system for rational insecticide choice in the control of African malaria vectors. *J. Med. Entomol.* 43:663–68
19. Craig MH, Sharp BL, Mabaso ML, Kleinschmidt I. 2007. Developing a spatial-statistical model and map of historical malaria prevalence in Botswana using a staged variable selection procedure. *Int. J. Health Geogr.* 6:44
20. Das A, Lele SR, Glass GE, Shields T, Patz J. 2002. Modelling a discrete spatial response using generalized linear mixed models: application to Lyme disease vectors. *Int. J. Geogr. Inf. Sci.* 16:151–66
21. Diuk-Wasser MA, Brown HE, Andreadis TG, Fish D. 2006. Modeling the spatial distribution of mosquito vectors for West Nile virus in Connecticut, USA. *Vector-Borne Zoonotic Dis.* 6:283–95
22. Diuk-Wasser MA, Gatewood AG, Cortinas MR, Yaremych-Hamer S, Tsao J, et al. 2006. Spatiotemporal patterns of host-seeking *Ixodes scapularis* nymphs (Acari: Ixodidae) in the United States. *J. Med. Entomol.* 43:166–76
23. Diuk-Wasser MA, Vourc'h G, Cislo P, Gatewood Hoen AG, Melton F, et al. 2010. Field and climate-based model for predicting the density of host-seeking nymphal *Ixodes scapularis*, an important vector of tick-borne disease agents in the eastern United States. *Global Ecol. Biogeogr.* 19:504–14
24. Eisen L, Beaty BJ. 2008. Innovative decision support and vector control approaches to control dengue. In *Vector-Borne Disease – Understanding the Environmental, Human Health, & Ecological Connections. Institute of Medicine Workshop Summary*, ed. SM Lemon, PF Sparling, MA Hamburg, DA Relman, ER Choffnes, A Mack, pp. 150–61. Washington, DC: Natl. Acad. Press
25. Eisen L, Eisen RJ. 2007. Need for improved methods to collect and present spatial epidemiologic data for vectorborne diseases. *Emerg. Infect. Dis.* 13:1816–20
26. Eisen L, Lozano-Fuentes S. 2009. Use of mapping and spatial and space-time modeling approaches in operational control of *Aedes aegypti* and dengue. *PLoS Negl. Trop. Dis.* 3:e411
27. Eisen RJ, Eisen L. 2008. Spatial modeling of human risk of exposure to vector-borne pathogens based on epidemiological versus arthropod vector data. *J. Med. Entomol.* 45:181–92
28. Eisen RJ, Eisen L, Lane RS. 2006. Predicting density of *Ixodes pacificus* nymphs in dense woodlands in Mendocino County, California, based on geographic information systems and remote sensing versus field-derived data. *Am. J. Trop. Med. Hyg.* 74:632–40

29. Eisen RJ, Enscore RE, Biggerstaff BJ, Reynolds PJ, Ettestad P, et al. 2007. Human plague in the southwestern United States, 1957–2004: spatial models of elevated risk of human exposure to *Yersinia pestis*. *J. Med. Entomol.* 44:530–37
30. Eisen RJ, Glass GE, Eisen L, Cheek J, Enscore RE, et al. 2007. A spatial model of shared risk for plague and hantavirus pulmonary syndrome in the southwestern United States. *Am. J. Trop. Med. Hyg.* 77:999–1004
31. Eisen RJ, Griffith KS, Borchert JN, MacMillan K, Apangu T, et al. 2010. Assessing human risk of exposure to plague bacteria in northwestern Uganda based on remotely sensed predictors. *Am. J. Trop. Med. Hyg.* 82:904–11
32. Eisen RJ, Lane RS, Fritz CL, Eisen L. 2006. Spatial patterns of Lyme disease risk in California based on disease incidence data and modeling of vector-tick exposure. *Am. J. Trop. Med. Hyg.* 75:669–76
33. Eisen RJ, Reynolds PJ, Ettestad P, Brown T, Enscore RE, et al. 2007. Residence-linked human plague in New Mexico: a habitat-suitability model. *Am. J. Trop. Med. Hyg.* 77:121–25
34. Elith J, Graham CH, Anderson RP, Dudik M, Ferrier S, et al. 2006. Novel methods improve prediction of species' distributions from occurrence data. *Ecography* 29:129–51
35. Fielding AH, Bell JF. 1997. A review of methods for the assessment of prediction errors in conservation presence/absence models. *Environ. Conserv.* 24:38–49
36. Gao S, Mioc D, Anton F, Yi X, Coleman D. 2008. Online GIS services for mapping and sharing disease information. *Int. J. Health Geogr.* 7:8
37. Gao S, Mioc D, Yi X, Anton F, Oldfield E, Coleman D. 2009. Towards Web-based representation and processing of health information. *Int. J. Health Geogr.* 8:3
38. Gatewood AG, Liebman KA, Vourc'h G, Bunikis J, Hamer SA, et al. 2009. Climate and tick seasonality are predictors of *Borrelia burgdorferi* genotype distribution. *Appl. Environ. Microbiol.* 75:2476–83
39. Gelman A, Carlin JB, Stern HS, Rubin DB. 2004. *Bayesian Data Analysis*. Boca Raton, FL: Chapman & Hall/CRC. 668 pp.
40. Gething PW, Atkinson PM, Noor AM, Gikandi PW, Hay SI, Nixon MS. 2007. A local space–time kriging approach applied to a national outpatient malaria data set. *Comput. Geosci.* 33:1337–50
41. Getis A, Morrison AC, Gray K, Scott TW. 2003. Characteristics of the spatial pattern of the dengue vector, *Aedes aegypti*, in Iquitos, Peru. *Am. J. Trop. Med. Hyg.* 69:494–505
42. Gratz NG. 1999. Emerging and resurging vector-borne diseases. *Annu. Rev. Entomol.* 44:51–75
43. Guerra M, Walker E, Jones C, Paskewitz S, Cortinas MR, et al. 2002. Predicting the risk of Lyme disease: habitat suitability for *Ixodes scapularis* in the north central United States. *Emerg. Infect. Dis.* 8:289–97
44. Guisan A, Edwards TC, Hastie T. 2002. Generalized linear and generalized additive models in studies of species distributions: setting the scene. *Ecol. Model.* 157:89–100
45. Hay SI, Guerra CA, Gething PW, Patil AP, Tatem AJ, et al. 2009. A world malaria map: *Plasmodium falciparum* endemicity in 2007. *PLoS Med.* 6:e1000048
46. Hay SI, Omumbo JA, Craig MH, Snow RW. 2000. Earth observation, geographic information systems and *Plasmodium falciparum* malaria in sub-Saharan Africa. *Adv. Parasitol.* 47:173–215
47. Hay SI, Snow RW. 2006. The Malaria Atlas Project: developing global maps of malaria risk. *PLoS Med.* 3:e473
48. Hay SI, Snow RW, Rogers DJ. 1998. Predicting malaria seasons in Kenya using multitemporal meteorological satellite sensor data. *Trans. R. Soc. Trop. Med. Hyg.* 92:12–20
49. Hemingway J, Beaty BJ, Rowland M, Scott TW, Sharp BL. 2006. The Innovative Vector Control Consortium: improved control of mosquito-borne diseases. *Parasitol. Today* 22:308–12
50. Holt AC, Salkeld DJ, Fritz CL, Tucker JR, Gong P. 2009. Spatial analysis of plague in California: niche modeling predictions of the current distribution and potential response to climate change. *Int. J. Health Geogr.* 8:38
51. Honório NA, Nogueira RMR, Codeço CT, Carvalho MS, Cruz OG, et al. 2009. Spatial evaluation and modeling of dengue seroprevalence and vector density in Rio de Janeiro, Brazil. *PLoS Negl. Trop. Dis.* 3:e545
52. Hosmer DW, Lemeshow S. 2000. *Applied Logistic Regression*. New York: Wiley. 375 pp.

53. Jacob BG, Arheart KL, Griffith DA, Mbogo CM, Githeko AK, et al. 2005. Evaluation of environmental data for identification of *Anopheles* (Diptera: Culicidae) aquatic larval habitats in Kisumu and Malindi, Kenya. *J. Med. Entomol.* 42:751–55
54. Jacquez G. 2004. Current practices in the spatial analysis of cancer: flies in the ointment. *Int. J. Health Geogr.* 3:22
55. Johansson MA, Dominici F, Glass GE. 2008. Local and global effects of climate on dengue transmission in Puerto Rico. *PLoS Negl. Trop. Dis.* 3:e382
56. Johnson PR, Blazes DL. 2007. Using cell phone technology for infectious disease surveillance in low-resource environments: a case study from Peru. In *Global Infectious Disease Surveillance and Detection: Assessing the Challenges – Finding Solutions. Institute of Medicine Workshop Summary*, ed. SM Lemon, PF Sparling, MA Hamburg, DA Relman, ER Choffnes, A Mack, pp. 136–57. Washington, DC: Natl. Acad. Press
57. Kan C-C, Lee P-F, Wen T-H, Chao D-Y, Wu M-H, et al. 2008. Two clustering diffusion patterns identified from the 2001–2003 dengue epidemic, Kaohsiung, Taiwan. *Am. J. Trop. Med. Hyg.* 79:344–52
58. Kitron U. 1998. Landscape ecology and epidemiology of vector-borne diseases: tools for spatial analysis. *J. Med. Entomol.* 35:435–45
59. Kitron UD. 2000. Risk maps: transmission and burden of vector-borne diseases. *Parasitol. Today* 16:324–25
60. Kleinschmidt I, Bagayoko M, Clarke GPY, Craig M, Le Sueur D. 2000. A spatial statistical approach to malaria mapping. *Int. J. Epidemiol.* 29:355–61
61. Knox GE. 1964. The detection of space-time iterations. *J. R. Stat. Soc.* 13:25–29
62. Kramer LD, Styer LM, Ebel GD. 2008. A global perspective on the epidemiology of West Nile virus. *Annu. Rev. Entomol.* 53:61–81
63. Kulldorff M, Heffernan R, Hartman J, Assuncao R, Mostashari F. 2005. A space-time permutation scan statistic for disease outbreak detection. *PLoS Med.* 2:e59
64. Lagrotta MT, Silva Wda C, Souza-Santos R. 2008. Identification of key areas for *Aedes aegypti* control through geoprocessing in Nova Iguacu, Rio de Janeiro State, Brazil. *Cad. Saude Publ.* 24:70–80
65. Lawson AB. 2006. Disease cluster detection: a critique and a Bayesian proposal. *Stat. Med.* 25:897–916
66. Lindsay SW, Birley MH. 1996. Climate change and malaria transmission. *Ann. Trop. Med. Parasitol.* 90:573–88
67. Lozano-Fuentes S, Elizondo-Quiroga D, Farfan-Ale JA, Loroño-Pino MA, Garcia-Rejon J, et al. 2008. Use of Google Earth to strengthen public health capacity and facilitate management of vector-borne diseases in resource-poor environments. *Bull. WHO* 86:718–25
68. Mabaso ML, Craig M, Vounatsou P, Smith T. 2005. Towards empirical description of malaria seasonality in southern Africa: the example of Zimbabwe. *Trop. Med. Int. Health* 10:909–18
69. MacNab YC, Dean CB. 2002. Spatio-temporal modelling of rates for the construction of disease maps. *Stat. Med.* 21:347–58
70. Magori K, Legros M, Puente ME, Focks DA, Scott TW, et al. 2009. Skeeter Buster: a stochastic, spatially explicit modeling tool for studying *Aedes aegypti* population replacement and population suppression strategies. *PLoS Negl. Trop. Dis.* 3:e508
71. Mammen MP, Pimgate C, Koenraadt CJM, Rothman AL, Aldstadt J, et al. 2008. Spatial and temporal clustering of dengue virus transmission in Thai villages. *PLoS Med.* 5:e205
72. Martin C, Curtis B, Fraser C, Sharp B. 2002. The use of a GIS-based malaria information system for malaria research and control in South Africa. *Health Place* 8:227–36
73. McCullagh P, Nelder JA. 1989. *Generalized Linear Models*. Boca Raton, FL: Chapman & Hall/CRC. 511 pp.
74. Moffett A, Shackelford N, Sarkar S. 2007. Malaria in Africa: vector species' niche models and relative risk maps. *PLoS One* 2:e824
75. Mondini A, Chiaravalloti-Neto F. 2008. Spatial correlation of incidence of dengue with socioeconomic, demographic and environmental variables in a Brazilian city. *Sci. Total Environ.* 393:241–48
76. Morrison AC, Getis A, Santiago M, Rigau-Perez JG, Reiter P. 1998. Exploratory space-time analysis of reported dengue cases during an outbreak in Florida, Puerto Rico, 1991–1992. *Am. J. Trop. Med. Hyg.* 58:287–98

77. Nakazawa Y, Williams R, Peterson AT, Mead P, Staples E, Gage KL. 2007. Climate change effects on plague and tularemia in the United States. *Vector-Borne Zoonotic Dis.* 7:529–40
78. Neerinckx S, Peterson AT, Gulinck H, Deckers J, Kimaro D, Leirs H. 2010. Predicting potential risk areas of human plague for the western Usambara Mountains, Lushoto district, Tanzania. *Am. J. Trop. Med. Hyg.* 82:492–500
79. Neerinckx SB, Peterson AT, Gulinck H, Deckers J, Leirs H. 2008. Geographic distribution and ecological niche of plague in sub-Saharan Africa. *Int. J. Health Geogr.* 7:54
80. Nicholson MC, Mather TN. 1996. Methods for evaluating Lyme disease risks using geographic information systems and geospatial analysis. *J. Med. Entomol.* 33:711–20
81. Peterson AT. 2009. Shifting suitability for malaria vectors across Africa with warming climates. *BMC Infect. Dis.* 9:59
82. Peterson AT, Papes M, Eaton M. 2007. Transferability and model evaluation in ecological niche modeling: a comparison of GARP and Maxent. *Ecography* 30:550–60
83. Peterson I, Borrell LN, El-Sadr W, Teklehaimanot A. 2009. A temporal-spatial analysis of malaria transmission in Adama, Ethiopia. *Am. J. Trop. Med. Hyg.* 81:944–49
84. Phillips SJ, Anderson RP, Schapire RE. 2006. Maximum entropy modeling of species geographic distributions. *Ecol. Model.* 190:231–59
85. Ribeiro JMC, Seulu F, Abose T, Kidane G, Teklehaimanot A. 1996. Temporal and spatial distribution of anopheline mosquitoes in an Ethiopian village: implications for malaria control strategies. *Bull. WHO* 74:299–305
86. Ritchie SA, Hanna JN, Hills SL, Piispanen JP, McBride WJH, et al. 2002. Dengue control in North Queensland, Australia: case recognition and selective indoor residual spraying. *Dengue Bull.* 26:7–13
87. Ritchie SA, Hart A, Long S, Montgomery B, Walsh I, Foley P. 2001. Update on dengue in North Queensland. *Arbovirus Res. Aust.* 8:294–99
88. Robinson TP, Harris RS, Hopkins JS, Williams BG. 2002. An example of decision support for trypanosomiasis control using a geographical information system in eastern Zambia. *Int. J. Geogr. Inf. Sci.* 16:345–60
89. Rogers DJ. 2006. Models for vectors and vector-borne diseases. *Adv. Parasitol.* 62:1–35
90. Rogers DJ, Randolph SE. 2003. Studying the global distribution of infectious diseases using GIS and RS. *Nat. Rev. Microbiol.* 1:231–37
91. Rosa-Freitas MG, Tsouris P, Sibajev A, de Souza Weimann ET, Marques AU, et al. 2003. Exploratory temporal and spatial distribution analysis of dengue notifications in Boa Vista, Roraima, Brazilian Amazon, 1999–2001. *Dengue Bull.* 27:63–80
92. Rotela C, Fouque F, Lamfri M, Sabatier P, Introini V, et al. 2007. Space-time analysis of the dengue spreading dynamics in the 2004 Tartagal outbreak, Northern Argentina. *Acta Trop.* 103:1–13
93. Ruiz M, Tedesco C, McTighe T, Austin C, Kitron U. 2004. Environmental and social determinants of human risk during a West Nile virus outbreak in the greater Chicago area, 2002. *Int. J. Health Geogr.* 3:8
94. Rytkönen MJP. 2004. Not all maps are equal: GIS and spatial analysis in epidemiology. *Int. J. Circumpolar Health* 63:9–24
95. Saxena R, Nagpal BN, Srivastava A, Gupta SK, Dash AP. 2009. Application of spatial technology in malaria research and control: some new insights. *Indian J. Med. Res.* 130:125–32
96. Scott TW, Morrison AC. 2010. Vector dynamics and transmission of dengue virus: implications for dengue surveillance and prevention strategies. *Curr. Top. Microbiol. Immunol.* 338:115–28
97. Siqueira JB, Maciel IJ, Barcellos C, Souza WV, Carvalho MS, et al. 2008. Spatial point analysis based on dengue surveys at household level in central Brazil. *BMC Public Health* 8:361
98. Sogoba N, Vounatsou P, Bagayoko MM, Doumbia S, Dolo G, et al. 2007. The spatial distribution of *Anopheles gambiae* sensu stricto and *An. arabiensis* (Diptera: Culicidae) in Mali. *Geospat. Health* 1:213–22
99. Sogoba N, Vounatsou P, Bagayoko MM, Doumbia S, Dolo G, et al. 2008. Spatial distribution of the chromosomal forms of *Anopheles gambiae* in Mali. *Malar. J.* 7:205
100. Sonenshine DE. 1993. *Biology of Ticks. Volume 2*. Oxford: Oxford Univ. Press. 465 pp.
101. Stockwell D, Peters D. 1999. The GARP modelling system: problems and solutions to automated spatial prediction. *Int. J. Geogr. Inf. Sci.* 13:143–58

102. Sutherst RW. 2004. Global change and human vulnerability to vector-borne diseases. *Clin. Microbiol. Rev.* 17:136–73
103. Sutherst RW, Bourne AS. 2009. Modelling nonequilibrium distributions of invasive species: a tale of two modelling paradigms. *Biol. Inv.* 11:1231–37
104. Symeonakis E, Robinson T, Drake N. 2007. GIS and multiple-criteria evaluation for the optimisation of tsetse fly eradication programmes. *Environ. Mon. Assess.* 124:89–103
105. Tang CS, Pang FY, Ng LC, Appoo SS. 2006. Surveillance and control of dengue vectors in Singapore. *Epidemiol. News Bull.* 32:1–9
106. Teng TB. 2001. New initiatives in dengue control in Singapore. *Dengue Bull.* 25:1–6
107. Thammapalo S, Chongsuvivatwong V, Geater A, Dueravee M. 2008. Environmental factors and incidence of dengue fever and dengue haemorrhagic fever in an urban area, Southern Thailand. *Epidemiol. Infect.* 136:135–43
108. Theophilides CN, Ahearn SC, Grady S, Merlino M. 2003. Identifying West Nile virus risk areas: the dynamic continuous-area space-time system. *Am. J. Epidemiol.* 157:843–54
109. Thomson MC, Connor SJ, D'Alessandro U, Rowlingson B, Diggle P, et al. 1999. Predicting malaria infection in Gambian children from satellite data and bed net use surveys: the importance of spatial correlation in the interpretation of results. *Am. J. Trop. Med. Hyg.* 61:2–8
110. Van Benthem BHB, Vanwambeke SO, Khantikul N, Burghoorn-Maas C, Panart K, et al. 2005. Spatial patterns of and risk factors for seropositivity for dengue infection. *Am. J. Trop. Med. Hyg.* 72:201–8
111. Vanden Eng JL, Wolkon A, Frolov AS, Terlouw DJ, Eliades MJ, et al. 2007. Use of handheld computers with Global Positioning Systems for probability sampling and data entry in household surveys. *Am. J. Trop. Med. Hyg.* 77:393–99
112. Vanwambeke SO, Somboon P, Harbach RE, Isenstadt M, Lambin EF, et al. 2007. Landscape and land cover factors influence the presence of *Aedes* and *Anopheles* larvae. *J. Med. Entomol.* 44:133–44
113. Wallenstein S. 1980. A test for detection of clustering over time. *Am. J. Epidemiol.* 111:367–72
114. Waller LA, Goodwin BJ, Wilson ML, Ostfeld RS, Marshall SL, Hayes EB. 2007. Spatio-temporal patterns in county-level incidence and reporting of Lyme disease in the northeastern United States, 1990–2000. *Environ. Ecol. Stat.* 14:83–100
115. Waller LA, Gotway CA. 2004. *Applied Spatial Statistics for Public Health Data*. New York: Wiley. 520 pp.
116. Weinstock MA. 1981. A generalised scan statistic test for the detection of clusters. *Int. J. Epidemiol.* 10:289–93
117. WHO. 2005. *International Health Regulations*. Geneva, Switzerland. 2nd ed.
118. WHO. 2006. International meeting on preventing and controlling plague: The old calamity still has a future. *Wkly. Epidemiol. Rec.* 28:279–84
119. WHO. 2007. *Scientific Working Group Report on dengue*. Geneva, Switzerland
120. WHO. 2008. *World malaria report 2008*. Geneva, Switzerland
121. Winters AM, Bolling BG, Beaty BJ, Blair CD, Eisen RJ, et al. 2008. Combining mosquito vector and human disease data for improved assessment of spatial West Nile virus disease risk. *Am. J. Trop. Med. Hyg.* 78:654–65
122. Winters AM, Eisen RJ, Delorey MJ, Fischer M, Nasci RS, et al. 2010. Spatial risk assessments based on vector-borne disease epidemiologic data: importance of scale for West Nile virus disease in Colorado. *Am. J. Trop. Med. Hyg.* 82:945–53
123. Winters AM, Eisen RJ, Lozano-Fuentes S, Moore CG, Pape WJ, Eisen L. 2008. Predictive spatial models for risk of West Nile virus exposure in eastern and western Colorado. *Am. J. Trop. Med. Hyg.* 79:581–90
124. Winters AM, Staples JE, Ogen-Odoi A, Mead PS, Griffith K, et al. 2009. Spatial risk models for human plague in the West Nile region of Uganda. *Am. J. Trop. Med. Hyg.* 80:1014–22
125. Wu PC, Lay JG, Guo HR, Lin CY, Lung SC, Su HJ. 2009. Higher temperature and urbanization affect the spatial patterns of dengue fever transmission in subtropical Taiwan. *Sci. Total Environ.* 407:2224–33
126. Zhou G, Munga S, Minakawa N, Githeko AK, Yan G. 2007. Spatial relationship between adult malaria vector abundance and environmental factors in western Kenya highlands. *Am. J. Trop. Med. Hyg.* 77:29–35

Salivary Gland Hypertrophy Viruses: A Novel Group of Insect Pathogenic Viruses

Verena-Ulrike Lietze,[1] Adly M.M. Abd-Alla,[2] Marc J.B. Vreysen,[2] Christopher J. Geden,[3] and Drion G. Boucias[1]

[1]Entomology and Nematology Department, University of Florida, Gainesville, Florida 32611; email: vlietze@ufl.edu, pathos@ufl.edu

[2]Insect Pest Control Laboratory, Joint FAO/IAEA Program of Nuclear Techniques in Food and Agriculture, A-1400 Vienna, Austria; email: A.M.M.Abd-Alla@iaea.org, m.vreysen@iaea.org

[3]Center for Medical, Agricultural and Veterinary Entomology, USDA, ARS, Gainesville, Florida 32608; email: chris.geden@ars.usda.gov

Annu. Rev. Entomol. 2011. 56:63–80

First published online as a Review in Advance on July 27, 2010

The *Annual Review of Entomology* is online at ento.annualreviews.org

This article's doi:
10.1146/annurev-ento-120709-144841

0066-4170/11/0107-0063$20.00

Key Words

insect DNA virus, *Hytrosaviridae*, Diptera, pathology, sterilizing effect, sterile insect technique

Abstract

Salivary gland hypertrophy viruses (SGHVs) are a unique, unclassified group of entomopathogenic, double-stranded DNA viruses that have been reported from three genera of Diptera. These viruses replicate in nuclei of salivary gland cells in adult flies, inducing gland enlargement with little obvious external disease symptoms. Viral infection inhibits reproduction by suppressing vitellogenesis, causing testicular aberrations, and/or disrupting mating behavior. Historical and present research findings support a recent proposal of a new virus family, the *Hytrosaviridae*. This review describes the discovery and prevalence of different SGHVs, summarizes their biochemical characterization and taxonomy, compares morphological and histopathological properties, and details transmission routes and the influence of infection on host biology and reproduction. In addition, the potential use of SGHVs as sterilizing agents for house fly control and the deleterious impact of SGHVs on colonized tsetse flies reared for sterile insect technique are discussed.

SGH: salivary gland hypertrophy

SGHV: salivary gland hypertrophy virus

DISCOVERYS AND SIGNIFICANCE OF SGHV

The few described viruses associated with symptoms of salivary gland hypertrophy (SGH) in adult dipteran insects were discovered in the early 1970s. The need to dissect insects to detect hypertrophied glands may explain, in part, the limited number of known insect species harboring the salivary gland hypertrophy viruses (SGHVs). To date, there is only one report of SGHV infection in populations of the adult narcissus bulb fly, *Merodon equestris* (Diptera: Syrphidae) (8, 56). A survey conducted in southern France in the early 1970s revealed high incidences of SGH in flies from two varieties of this insect species. SGH was recorded in 31% and 54% of adult *M. equestris* var. *nobilis* and *M. equestris* var. *transversalis*, respectively; this symptom was accompanied by atrophied gonads in both genders (56). Long, rod-shaped virus particles isolated from the hypertrophied salivary glands showed ultrastructural similarities to certain baculoviruses and were speculated to cause SGH (8).

The first description of SGHV in tsetse flies (Diptera: Glossinidae) also dates to the 1970s, when Jenni & Steiger (37) published results from their ultrastructural examination of trypanosome development in the salivary glands of adult *Glossina morsitans centralis* collected from Singida, United Republic of Tanzania. On the basis of morphological characteristics of the detected virus particles, the authors suggested a resemblance to arboviruses (37), an incorrect, misleading association. Several years later, virus particles were discovered in nuclei, cytoplasm, intercellular spaces, and lumina of enlarged salivary glands of adult *G. pallidipes* collected from Kibwezi Forest, Kenya (34). In these collections, the percentage of flies with SGH symptoms was low and varied between 1% and 2%. A significant proportion of the symptomatic *G. pallidipes* displayed atrophied gonads in both genders, indicating a sterilizing effect of SGHV infection on the tsetse host (34). SGHV infections recently have been linked to the collapse of valuable colonies of *G. pallidipes* at the Insect Pest Control Laboratory (former Entomology Unit) of the FAO/IAEA Joint Program in Seibersdorf, Austria (1). Following these discoveries, SGHVs from eight *Glossina* species collected from seven different African countries have been described (**Table 1** and references therein).

In the 1990s, Coler et al. (15) discovered a third dipteran genus harboring SGHV. In dissections of adult house flies, *Musca domestica* (Diptera: Muscidae), sampled from populations in Florida for a survey of parasitic nematodes, several flies contained grossly enlarged, discolored, whitish-blue salivary glands. Histological and biochemical examination revealed that a rod-shaped, double-stranded DNA (dsDNA) virus was associated with these symptomatic glands, and the authors pointed out the striking similarity between this virus and the above described SGHVs. Most females (95%) displaying SGH had undeveloped ovaries, again demonstrating a sterilizing effect of viral infection (15).

Over the past four decades, several research groups in Europe, Africa, and the United States have investigated the SGHVs infecting tsetse flies and house flies to identify ultrastructural, biochemical, and molecular characteristics, pathology, transmission, field incidence, and geographical distribution of these viruses. The results are summarized and discussed in this review.

CHARACTERIZATION AND TAXONOMY OF SGHV

Structure and Composition

Transmission electron microscopy (TEM) examination of ultrathin sections of hypertrophied salivary glands from different host flies revealed the presence of numerous rod-shaped virus particles in the nucleus, cytoplasm, intercellular spaces, and gland lumen (**Figure 1*a***) (34). Nucleocapsids assemble in the nuclei and aggregate in spaces adjacent to the virogenic stroma, the presumed site of DNA replication. The size of the nucleocapsids varies among the different SGHVs; the *Musca domestica* SGHV

Table 1 Reported host species of SGHVs and prevalence of SGH symptoms in host populations

Species	Location	Prevalence (%)[a]	Reference
Glossina austeni Newstead	Tanzania (Amani)	1.6 (432)	13[b,c]
G. morsitans Westwood	Tanzania (Singida, Kondoa)	0.1 (8,916)	13[b,d]
G. m. centralis Machado	Tanzania (Singida)	No data	36,[e] 37
G. m. morsitans Westwood	Zimbabwe (Zambezi valley)	0.5 (1,162)	18
G. m. morsitans Westwood	Lab colony, Kenya (Nairobi)	No data	50
G. nigrofusca nigrofusca Newstead	Ivory Coast (Vavoua)	0.7 (143)	28
G. pallicera pallicera Bigot	Ivory Coast (Vavoua)	2.1 (287)	28
G. pallidipes Austen	Kenya (East coast, various sites)	2.7 (17,180)	62
G. pallidipes Austen	Kenya (Kiboko)	0.4 (23,960)	59
G. pallidipes Austen	Kenya (Kibwezi forest)	1.0 (5,361) in 1975/1976; 1.2 (491) in 1980	34, 67
G. pallidipes Austen	Kenya (Lambwe Valley Game Reserve)	1.6 (929)	40, 41, 51, 67
G. pallidipes Austen	Kenya (Meru National Park)	0.9 (439)	67
G. pallidipes Austen	Kenya (Mombasa)	1.8 (18,410)	64
G. pallidipes Austen	Kenya (Shimba Hills Game Reserve)	5.4 (204)	67
G. pallidipes Austen	Kenya (Sindo)	1.1 (8,403)	59
G. pallidipes Austen	Kenya (South coast, primary forest)	7.0 (1,213)	63
G. pallidipes Austen	Kenya (South coast, secondary forest)	3.8 (662)	63
G. pallidipes Austen	Kenya (South coast, shrub area)	3.2 (464)	63
G. pallidipes Austen	Kenya (South coast, fallow land)	1.8 (329)	63
G. pallidipes Austen	South Africa (Zululand)	2.9 (1,129)	90, 91[b]
G. pallidipes Austen	Tanzania (Amani)	1.7 (376)	13[b]
G. pallidipes Austen	Zimbabwe (Zambezi valley)	2.0 (886)	18
G. pallidipes Austen	Laboratory colony, Austria (Seibersdorf); origin: Ethiopia (Southern Rift Valley)	>85.0 (unknown)	1
G. pallidipes Austen	Laboratory colony, Austria (Seibersdorf); origin: Uganda (Tororo)	3.8 (2,011)	1, 4
G. palpalis palpalis Rob. Desv.	Ivory Coast (Vavoua)	0.3 (1,351)	28
Musca domestica L.	California	2.0 (100)	71
Musca domestica L.	Denmark (Havbro)	1.0 (100)	71
Musca domestica L.	Denmark (Morum)	1.0 (200)	71
Musca domestica L.	Denmark (Slangerup)	1.0 (200)	71
Musca domestica L.	Denmark (Tønder)	4.7 (169)	71
Musca domestica L.	Florida (various sites)	6.3 (11,110) in 1991 0.5–10.0 (28,800) in 2005/2006	15, 24, 71
Musca domestica L.	Kansas	0.7 (155)	71
Musca domestica L.	New Zealand	No data	71
Musca domestica L.	Thailand	11.1 (45) in 2008; 1.8 (110) in 2009	71
Musca domestica L.	Virgin Islands	1.2 (276)	71
Merodon equestris F.	France	30.8 (39) and 54.1 (37)	8, 56

[a]Numbers in parentheses indicate total numbers of flies dissected.
[b]Correlation with virus infection not established in these articles.
[c]Statements in the article are contradictory: Text states no SGH in this species, whereas table indicates 7 of 432 flies with SGH.
[d]Tables in the article are contradictory: One table indicates no SGH in this species, another table and the text state 7 of 8,916 flies with SGH.
[e]No salivary gland enlargement described in this study.

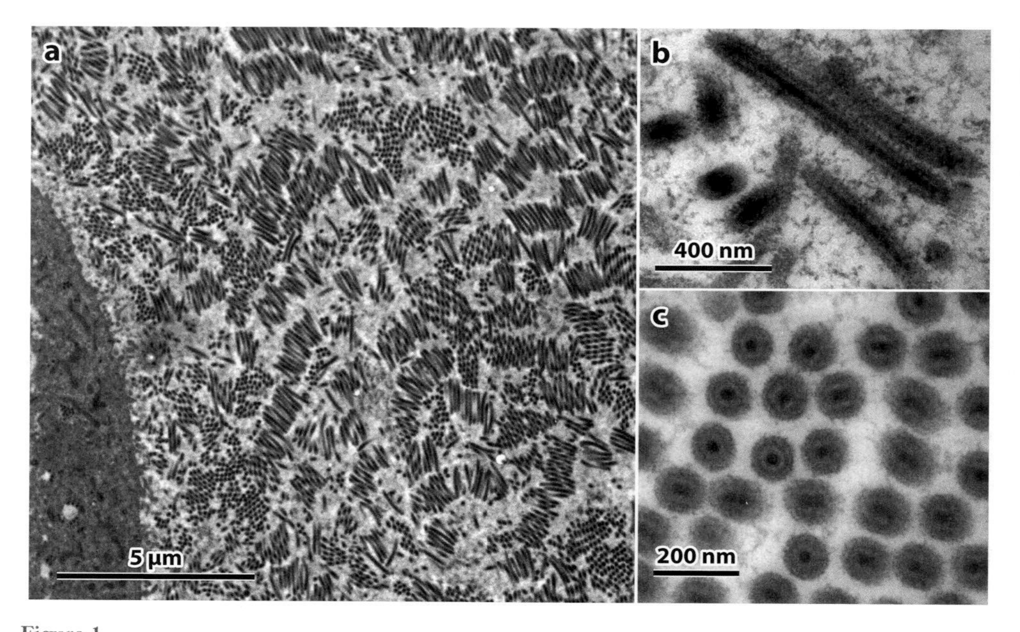

Figure 1

Transmission electron micrographs of *Glossina pallidipes* salivary gland hypertrophy virus (GpSGHV). (*a*) Luminal region of tsetse fly salivary gland displaying the numerous enveloped GpSGHV virions. High magnification images of the enveloped GpSGHV in (*b*) longitudinal section and (*c*) cross section demonstrate the complex structure of the particles.

(MdSGHV) and the *Merodon equestris* SGHV (MeSGHV) measure ~500–600 nm in length by 50–60 nm in diameter, whereas the nucleocapsids associated with tsetse fly SGHVs are significantly longer and measure 800–1200 nm in length by 50–60 nm in diameter (**Figure 1*b***) (8, 24, 34). Negative staining revealed that the nucleocapsids comprise structural units arranged as a series of stacked rings 7 to 10 nm wide (18, 64). The nucleocapsids exit the nuclei via nuclear pores, associate with the Golgi apparatus, and acquire their envelope in situ in the cytoplasm. Enveloped virus particles, measuring 70–80 nm in diameter, consist of an inner membrane that encloses the nucleocapsid and an outer membrane separated from the inner membrane by a narrow space (**Figure 1*c***) (34). Both the *Glossina pallidipes* SGHV (GpSGHV) and the MdSGHV band at a density of 1.153 g cm^{-3} when subjected to 10–60% Nycodenz® gradient centrifugation (22).

Biochemical analysis demonstrated that the SGHVs contain a complex array of major and minor structural proteins. SDS-PAGE analysis of the Ugandan GpSGHV isolate revealed more than 35 protein bands ranging in size from 10 to 220 kDa (3). At least six bands are larger than 100 kDa, with the major bands having a molecular mass of 39 and 40 kDa. A similar analysis conducted on Nycodenz-purified MdSGHV revealed a complex of major and minor bands that range from 10 to 200 kDa (15, 22). SDS-PAGE and nanocapillary liquid chromatography tandem mass spectrometry (GeLC-MS/MS) analysis of MdSGHV peptides separated on SDS gels identified unique peptide fragments that were encoded on 29 open reading frames (ORFs) (22). Sixteen of the MdSGHV structural ORFs have homologs detected in the GpSGHV genome, whereas only four ORFs are homologous to non-SGHV genes (21, 22). No protein inclusions or viral occlusions have been detected in SGHV-infected cells.

Genome Organization

Early work demonstrated that the MdSGHV contained a relatively large (>100 kbp) single dsDNA molecule (15). The detection of two bands (supercoiled and relaxed forms) after agarose electrophoresis of purified DNA and the lack of end-labeling of undigested DNA indicated SGHVs possessed circular genomes. The genomes of both the GpSGHV (NC_010356.1) and the MdSGHV (NC_010671) have been fully sequenced by a combination of conventional sequencing of viral clones and pyrosequencing (3, 22). The longer GpSGHV encapsidates a 190,032-bp genome (28% G + C), whereas the smaller MdSGHV virion encapsidates a 124,279-bp genome (44% G + C). Both genomes were derived from wild-type virus isolates, and attempts to replicate SGHVs in insect cell cultures have failed, precluding access to clonal preparations. Analysis of sequence data demonstrated the presence of polymorphic sites involving single-base substitutions located randomly throughout the genome (22).

In silico analysis demonstrated that the GpSGHV genome encodes for 322 potential ORFs for proteins composed of at least 50 amino acids with a methionine start codon (3). Of these, only 160 ORFs possess either no or minimal overlap with adjacent ORFs (3). The predicted 160 ORFs, representing 86% of the genome, are distributed evenly on both strands (51% forward and 49% reverse) with many arranged in unidirectional gene clusters. A total of 108 putative MdSGHV ORFs were identified in silico (22). The transcriptional orientation of the predicted ORFs was slightly different, with 53 ORFs (49%) in the clockwise direction and 55 (51%) in the opposite direction. Similar to the GpSGHV, several clusters of MdSGHV ORFs were transcribed in one direction. The majority of ORFs identified on the two SGHV genomes have no detectable homologs when subjected to BLAST analysis (3, 21, 22). For example, only 30 and 47 of the 108 and 160 putative ORFs detected in the MdSGHV and GpSGHV, respectively, could be assigned to any homolog. The identified homologs encode for structural proteins, proteins involved in DNA replication (e.g., DNA polymerase, helicase), and protein-modifying enzymes (e.g., protein kinase). A total of 101 MdSGHV ORFs have been validated using rapid amplification of cDNA 3′ ends and reverse transcriptase PCR (73). Most are transcribed as individual transcripts, whereas 34 ORFs are transcribed in tandem with adjacent ORFs (73); similar events have occurred in other insect dsDNA viruses (20, 30, 69). Analysis of the MdSGHV 3′-untranslated regions (3′-UTRs) revealed extensive heterogeneity in both the polyadenylation signals and cleavage sites present on the MdSGHV ORFs (73). Convergent, unidirectional, and divergent overlap found in the 3′-UTRs of 34 transcript pairs suggests *cis*-encoded natural antisense viral transcription (82). Promoter analysis revealed that the 5′ upstream regions in both the MdSGHV and the GpSGHV are highly enriched with a TAAG motif, which is identical to the canonical baculovirus late transcription initiation sequence (3, 22). In addition to the ORFs and their associated UTRs, a series of direct repeats (drs) is distributed throughout the GpSGHV (14 drs) and MdSGHV (18 drs) genomes.

Classification of *Hytrosaviridae*

When discovered, the SGHVs were tentatively associated with arboviruses and the nonoccluded baculoviruses (formerly subgroup C baculoviruses, presently nudiviruses) (34, 37). At that time, available morphological (enveloped, rod-shaped), molecular (large circular dsDNA genome), and biological (oral infectivity, nuclear replication) information justified placement of the SGHVs with the nudiviruses. However, as sequence information became available (2, 24), the relationships between the SGHVs and other dsDNA insect viruses

Hypertrophy: enlargement or overgrowth of an organ or part of the body due to the increased size of the constituent cells

became less defined. Gene parity plot analyses demonstrated colinear regions between the MdSGHV and the GpSGHV but failed to display any linear correspondence between the SGHVs and the sequenced nudiviruses GbNV and HzNV-1 (21). Furthermore, syntenic map analysis displayed comparable colinearity between the MdSGHV and GpSGHV genomes, which was not found when either genome was compared to GbNV or HzNV-1. Finally, phylogenetic analysis of selected genes (e.g., DNA polymerase, per os infectivity factors) failed to show any association with homologs from other dsDNA insect viruses (21). These genetic differences, in combination with the unique (patho)biological properties displayed by SGHVs, justify the proposal of a new virus family named *Hytrosaviridae* (2). The SGHVs infect and replicate in adult flies and cause distinct SGH symptoms that result in insect sterility. Although the GpSGHV and MdSGHV share general relatedness in some characters, they possess many unique properties suggesting separate genera, the *Glossinavirus* and *Muscavirus*, respectively.

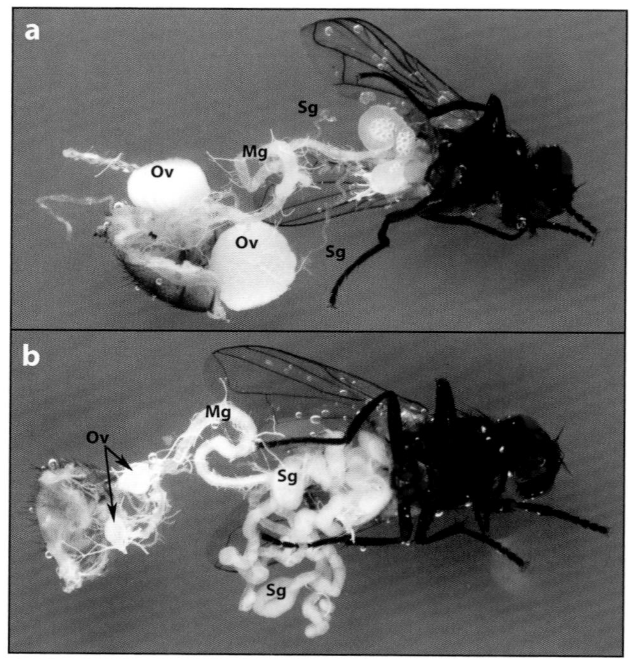

Figure 2

Musca domestica females with (*a*) healthy and (*b*) hypertrophied salivary glands showing the lack of ovarian development in the SGHV-infected fly (*b*). Abbreviations: Mg, midgut; Ov, ovary; Sg, salivary gland.

PATHOLOGICAL EFFECTS OF SGHV

External Morphology of the Infected Host

Although SGHV infection enlarges salivary glands, which eventually fill most of the abdominal cavity, flies with SGH cannot be distinguished easily from normal flies by external visual examination (62). However, in teneral tsetse flies, the enlarged, bluish-white salivary glands may appear as a pale outline through the abdominal integument and form irregular ridges on the soft cuticle. In addition, a bloated abdomen in *G. pallidipes* males that contains traces of blood two days after feeding suggests the presence of SGH (4). In older flies, pigmentation, thickening of the cuticle, or enlargement of the abdomen in gravid females precludes detection of SGH.

Impact of SGHV Infection on Salivary Glands

Morphological characteristics of SGH were described first by Whitnall in 1932: Enlarged salivary glands of *G. pallidipes* were swollen to almost four times their normal thickness (90, 91). During dissections of the various hosts, hypertrophied glands are distinguishable from normal, transparent salivary glands due to their large size and chalky-white or bluish appearance (**Figure 2**) (1, 8, 15, 28, 54). Typically, the paired salivary glands are equally affected, and hypertrophy is uniform over the entire length of the distal part (15, 62). Whereas the diameter of the distal part of a normal salivary gland is about 100 μm with almost no variation throughout the life of the tsetse fly, the diameter of hypertrophied salivary glands increases to up to 1.1 mm and has been used to categorize glands with SGH symptoms into nine size classes (62). In the smallest size category, i.e., at an early stage of disease, discoloration without enlargement is observed because epithelial cells expand toward the gland lumina, resulting in a constricted lumen. In hypertrophied glands of category 9, the entire abdominal hemocoel is filled with grossly enlarged and highly coiled salivary glands. Late-stage hypertrophied glands are about twice as long as normal salivary glands (62).

Ultrastructural studies have shown that in tsetse flies and in the narcissus bulb fly, gland enlargement is caused by cellular proliferation of the glandular epithelial cells and hypertrophy of their nuclei and cytoplasm, resulting in an abnormal multilayered epithelium and a reduced gland lumen (8, 34, 50, 67). In contrast, SGH in house flies appears to be caused by nuclear and cellular hypertrophy without any increase of cell numbers. In all known SGHV hosts, diseased salivary glands develop heavily vacuolated cytoplasm, cell membranes separate from adjacent epithelial cells, and numerous rod-shaped virus particles are observed in the nuclei, cytoplasm, intercellular spaces, and gland lumina (8, 24, 34, 67). The presence of both nucleocapsids and enveloped virions in the cytoplasm of hypertrophied salivary gland cells indicates that the SGHVs assemble their envelope in the cytoplasm (18, 24).

It is not known whether the enzyme composition and/or functions of hypertrophied salivary glands are negatively affected by the disease. The observed histopathological aberrations suggest that feeding and digestion may be impaired in diseased flies, and several studies have examined the impact of SGHV infection on fitness and on survival rates (see below). The long survival time of infected flies indicates that salivary gland function is maintained.

Tissue Tropism and Impact of the Virus on Other Tissues of Infected Flies

The primary tissue infected with SGHV is the salivary gland, but the virus is also present in other tissues of infected flies as demonstrated by TEM (40, 41, 49, 55, 76), diagnostic PCR (1, 4, 54), and infection bioassays (54, 55). In TEM studies, virus particles have been observed in the crop and midgut lumen (55, 77), in intercellular spaces of muscle tissue (55), in nuclei and cytoplasm of milk gland cells (76), in germ cell nuclei and nurse cell and oocyte cytoplasm of ovarioles (40, 41), and in the lumen and epithelial cells of male accessory reproductive glands (ARGs) (49). Detection of both virogenic stroma and associated nucleocapsids within nuclei is proof of viral replication in ovarian germ cells, follicles, and milk gland cells (40, 76). In several reports, virions were not detected in testes, male ARGs, flight muscle, fat body, and spermathecae of SGH-positive flies (67, 75). Detection of viral DNA by PCR in excised legs indicated the presence of hemolymph-borne virus (1). Bioassays confirmed that virions detected by PCR in crops and ovaries of viremic flies were infectious and induced SGH when delivered orally or through intrahemocoelic injection (54, 55).

The most obvious impact of SGHV infection on nonsalivary gland tissues is the abnormal development of gonads (40, 54, 56). Ovaries of SGHV-infected house flies remain at the previtellogenic stage (**Figure 2**), and the expression of the female-specific hexamerin and yolk protein genes by fat body cells is downregulated (54). Infected tsetse females display irregular ovariolar development (34) or severe necrosis and degeneration of the germaria (40). Examination of testes from viremic tsetse flies indicated a complete arrest of spermatogenesis, with follicles containing highly vacuolated, degenerate spermatogenic cells (40); these males did not produce spermatophores (75). Although virus particles may not be detected in male ARGs, these tissues can be affected by SGHV infection of the insect, showing reduction in size, disintegration of epithelial cell organelles, and detachment of adjacent muscle cells from the basal lamina (75). In some cases, viremic flies show midgut hypertrophy (62) and midgut epithelial cell necrosis, degeneration, and lysis in the anterior, secretory, and posterior parts (77). Presumably, secretory and absorptive functions of the midgut are impaired, which, in concert with dysfunctional salivary glands, decreases nutrient assimilation and leads to starvation (77). Virus particles also were detected in the milk glands of female tsetse flies (*G. morsitans centralis*) with SGH (76). Viral replication and severe necrotic lesions in the secretory cell layers of these glands suppress milk synthesis, which hinders F_1 larval development and decreases viability (76).

Impact of Viral Infection on Host Fitness and Behavior

Susceptibility/resistance to infection. Typically, injection with viral preparations results in 100% symptomatic infection within several days after injection (42, 54, 71). While tsetse flies may be injected as larvae to produce infected adults (42, 75), only adults have been used for injection experiments with house flies (54, 71). Per os infectivity of the MdSGHV to newly emerged house flies varies between 2% and 83% (24, 55, 71); the origin of viral inoculum (salivary glands, crops, saliva, feces) as well as differences in viral titers between inocula may explain this variation. Significantly, adult house flies develop resistance to oral infection within hours after eclosion; *per os* treatment of 24-h-old and 2-h-old flies with the same MdSGHV preparation yields, on average, sixfold-lower infection rates in the older flies (71). The factors responsible for age-related resistance to MdSGHV infection, although unknown, could be related to the maturation of the peritrophic matrix.

Mortality. Laboratory tsetse and house flies with SGH can survive for at least two weeks, often longer (15, 35, 54, 77). The few studies examining survival rates have reported either cumulative mortality or mean life span. The detection of SGH in field-collected tsetse flies that harbored mature trypanosomes, indicating that these flies were at least three weeks old, suggests that viremic flies survive for several weeks (91). Alternatively, adult field flies may acquire the virus or develop SGH after a period of chronic asymptomatic infection later in their life. Field data have shown that a high proportion of *G. pallidipes* with SGH (63%) are young flies (nulliparous females and teneral males), whereas in the asymptomatic group only 14% are young flies, suggesting either high mortality of infected flies (35) or reduced susceptibility as flies age. Although some data show that infected flies survive as long as healthy flies (15), most reports demonstrate that SGHV infection reduces the life span of the insect. Laboratory healthy and viremic *G. pallidipes* females, for example, have a maximum life span of 112–161 days and 56–58 days, respectively. Similarly, the life span of *G. m. centralis* is significantly reduced in viremic flies (63 and 29 days in heavily infected females and males, respectively) compared with asymptomatic flies (106 and 92 days in females and males, respectively), and the reduction in life span is positively correlated to the severity of infection (77). Cumulative mortality of viremic female *M. domestica* significantly increases to 69% after 16 days compared with 24% mortality of healthy females (54).

Flight and feeding. In the field, tsetse flies with SGH show normal flight behavior, and their blood-filled gut indicates they are capable of securing a blood meal (62). In the laboratory, *G. pallidipes* with SGH feed normally on rabbits (62) or on a membrane-feeding system (5). In contrast, in *G. m. centralis*, SGH impairs the ability to feed: Symptomatic flies probe more often, take more time, and imbibe less blood during feeding than nonsymptomatic flies do (77). These smaller blood meals could be attributed to reduced saliva secretion hampering blood uptake and/or enlarged salivary glands occupying the entire abdominal space and limiting blood ingestion. In addition, *G. m. centralis* with SGH have difficulty digesting blood, as indicated by the presence of blood clots in the crop (77). Blood clotting might be due to insufficient release of saliva in the hypertrophied salivary glands, leading to incomplete blood anticoagulation. The difficulty in feeding on and digesting blood, in combination with the subsequent rupture of the crop and/or midgut, may explain the high mortality observed in viremic *G. m. centralis*. Multiple attempts to obtain a blood meal from a host by infected tsetse flies may induce host defensive behaviors, leading to interrupted or aborted feeding. House flies with SGH are capable of ingesting and digesting protein-containing food, although proteolytic activity in the midgut is reduced compared with healthy flies (54).

Reproduction. The effect of SGHV infection on the reproductive potential of infected flies appears different within and between various tsetse fly species and house flies. Female *G. pallidipes* with SGH, for example, mate with normal males and produce offspring, whereas the viremic males are mostly unable to inseminate female flies (34). However, significant proportions of both female (45%) and male (71%) *G. pallidipes* with SGH show abnormal gonad development, indicating that SGHV-infection sterilizes both genders (34). In *G. m. morsitans* and *G. m. centralis*, insemination by males with SGH is also impaired (39). Mating behavior of tsetse flies with SGH appears to be normal. Both mating duration and time to reach the jerking phase before separation are similar for healthy and viremic *G. m. morsitans* (39). Although no sperm are transferred to the female spermathecae by male *G. m. morsitans* with SGH, a portion of the females becomes refractory to remating, as is the case when the first mating partner is a normal male (39). Jura & Davies-Cole (39) speculated that males sterilized by SGHV infection retain a competitive mating efficiency and may be useful in a sterile male release program. However, choice assays or field trials have not been conducted to verify this claim. The fecundity of females with SGH is significantly reduced in both tsetse and house flies. Viremic *G. m. centralis* females have longer pregnancy cycles and produce pupae with lower weights than do healthy females (74). In *M. domestica*, SGHV infection completely inhibits vitellogenesis (54). Only females that acquire the virus after completion of the first gonadotropic cycle are able to deposit one fertilized batch of eggs. However, with the onset of SGH symptoms, virgin, egg-containing females become unresponsive to mating attempts and do not copulate (54). Male house flies with SGH are able to copulate and deliver viable sperm, but with progressing infection, their avidity is reduced (54). While sterile insects could be expected to have an extended life span, increased mortality of flies with SGH can be partially explained by impaired digestion.

SGH and Infection with Other Microorganisms

SGH was first observed during prevalence studies of trypanosomes in tsetse flies. Prevalence of SGH positively correlated with that of *Trypanosoma* spp. in *G. pallidipes* but not in *G. morsitans*, *G. p. palpalis*, *G. p. pallicera*, and *G. n. nigrofusa* (13, 28, 67, 91). Jaenson (34) therefore cautioned that the impact of viremia on the vector potential of *G. pallidipes* would need assessment before considering the use of SGHVs as biological control agents against tsetse flies. However, later surveys found no relationship between SGH and trypanosome infection (18, 59). The impact of SGHV infection on trypanosome transmission is unclear (67). Histopathological examination of tsetse salivary glands indicated that SGHV-trypanosome mixed infections caused severe cellular disintegration (cytoplasmic vacuolation, lacerated basal plasma membranes, numerous lysosomes, and residual bodies) and showed significant degeneration of trypanosomes in these cells (51). The low incidence of flies with dual infections (0.02%) suggests that these flies suffer high mortality (51).

Co-infection of SGH-infected tsetse flies with rickettsia-like organisms (RLOs) was observed in two studies in the late 1980s (18, 50). Quantitative data are available from a small sample size ($n = 12$) and demonstrate a high incidence (83%) of RLO infection in *G. pallidipes* with SGH, suggesting that the presence of RLOs may increase susceptibility of salivary glands to viral infection or vice versa (18). It is unknown if RLOs induce additional pathological effects in the enlarged, SGHV-infected salivary glands, nor is there any proven role of RLOs in predisposing tsetse flies to trypanosome infections. Based on current knowledge, these reported RLOs may represent bacterial symbionts associated with tsetse flies (7).

TRANSMISSION OF SGHV

Transmission of SGHVs is dictated by the biology of the host. Viral transmission in field and laboratory populations of the viviparous, hematophagous tsetse has been the subject of many studies (4, 34, 35, 39–41, 74–77), and two main routes have been suggested: (*a*) vertical transmission from mother to offspring and (*b*) horizontal transmission by oral infection through the gut. The presence of SGH in teneral (no blood meal yet taken) progeny of viremic females mated with normal males supports the hypothesis that the virus could be transmitted from mother to offspring (34), possibly by the transovarial/transovum route (62). The high SGHV genome copy numbers in pupae produced by females with SGH, compared to the low numbers in pupae produced by females with normal salivary glands, likewise support this hypothesis (4). The presence of virus particles within germarial cells, nurse cells, and oocytes of ovaries is an additional indicator of transovarial transmission (41). Virus particles present in milk glands suggest transmission from mother to offspring through oral ingestion of the milk by the larva developing in the uterus (74). Effective SGHV transmission from females with SGH to their progeny is hindered by their reduced fecundity (74), which is likely caused by malnutrition due to impaired feeding (77) and by necrotic lesions in ovaries (40, 41) and milk glands (76). In females that do produce offspring, vertical virus transmission to the F_1 is not an absolute outcome. For instance, only 21% and 48% of progeny produced by SGH-positive *G. m. centralis* and *G. m. morsitans*, respectively, developed SGH (74). A low transmission rate can explain the low prevalence of SGH in the field, but it is not clear which other mechanism exists that enables the virus to maintain itself in nature. Hypertrophied salivary glands of category 1 (0.1-mm diameter of distal part indicating an early stage of SGH) were found in very old flies, explained by latent infections originating from low doses of virus transmitted from mother to offspring or by adult tsetse flies acquiring the virus infection from the environment (62).

In a laboratory colony of *G. pallidipes* that was maintained using the membrane-feeding technique, a very high rate of SGH (85%) was due to horizontal transmission during membrane feeding (1). Jaenson (34) speculated on the possibility of horizontal transmission of SGHV, and Odindo et al. (65) demonstrated horizontal transmission by feeding a suspension of hypertrophied salivary glands to *G. pallidipes* teneral flies, of which 31% developed SGH. Moreover, one viremic fly can deposit 10^7 viral genome copies into the blood under the membrane during feeding, and the observation that these secreted viruses can initiate an infection in healthy tsetse flies confirms the horizontal transmission route (5).

Horizontal transmission is the major transmission route of MdSGHV in the gregarious house fly. During a few seconds of feeding, infected flies deposit an average 10^6 viral genome copies onto a solid food substrate. The virus released in the salivary secretions is highly infectious to newly emerged conspecifics, causing 66% infection rates in the challenged flies (55). There is evidence that the virus is acquired only during the adult stage; flies exposed as larvae to oral treatments of virus preparations do not develop SGH (24), whereas similar treatments of newly emerged adults result in average infection rates varying between 30% and 83% at 6–7 days posttreatment (55, 71). In addition, larvae collected from field sites with high incidence of SGH in adults (max. 34%) and reared to adulthood in the laboratory do not display SGH as seven-day-old adults (24). Mating experiments have demonstrated that MdSGHV is transmitted neither sexually between healthy and infected mating partners nor vertically to progeny (24, 54).

GEOGRAPHICAL DISTRIBUTION AND PREVALENCE OF SGHV

It is apparent that the narrow host range of SGHVs dictates their geographical distribution (**Table 1**). In the only sampled field populations of narcissus bulb flies in southern France, the prevalence of viral infection was high (>31%) (56). At present, no assumptions can be made about the geographical distribution of MeSGHV. Distribution of the tsetse SGHVs is restricted to the African continent and tied to the distribution patterns of the hosts. In *Glossina* spp., field incidence of SGH symptoms varies from as little as 0.08% (13) to as much as 15.6% (62). In contrast, the MdSGHV infecting house flies has a global distribution with average prevalence rates at individual sites ranging from 0.5% to 10% (**Table 1**).

Factors Influencing SGHV Dynamics in Natural Populations

Although flies with SGH can be found in most surveyed areas, the number of symptomatic flies varies by location and even by trap site (24, 62, 63, 67, 71). In addition, seasonal fluctuations in SGH prevalence were observed within trap sites (24, 62). There is evidence that both the ecosystem and fly density affect the prevalence of SGH. As expected with an orally transmitted disease, house fly density correlates positively with SGH incidence (24). On the other hand, prevalence of vertically transmitted tsetse SGHV is inversely correlated with host density (62). This inverse relationship suggests that the virus may be one of the population-regulating factors in the viremic populations. Only one study has attempted to identify climatic factors that impact field incidence of SGH (63). In a forest ecosystem in Kenya, prevalence of SGH was positively correlated with vegetation density and rainfall (humidity) and negatively correlated with temperature and age structure of the tsetse population (63). SGH is found in all age categories of the adult host (62). Although the diameter of hypertrophied salivary glands increases as flies age, glands with a high degree of hypertrophy (0.8-mm diameter) can be found in newly emerged flies and glands with a low degree of hypertrophy (0.1-mm diameter) can be found in aged flies (34, 62). These variations in disease expression could be due to different incubation periods of the vertically or horizontally transmitted virus, which may exist in an asymptomatic stage before unknown factors trigger expression of SGH. Alternatively, the host may have acquired the virus during an earlier life stage (likely to occur in tsetse flies but not in house flies) or as an older adult.

In several host populations, the incidence of SGH is higher in males than in females. In the narcissus bulb fly, for example, symptomatic SGH was recorded in 88% of males and 16% of females of *M. equestris* var. *nobilis* and in 93% of males and 30% of females of *M. equestris* var. *transversalis*, accounting for 5.5-fold- and 3.1-fold-higher incidence, respectively, in males than in females (56). In house fly and tsetse field populations, incidence of SGH may be up to 2-fold and 4.6-fold higher, respectively, in males than in females (24, 28). Similar observations were made in laboratory colonies of *G. pallidipes* (1).

Prevalence of SGH in Laboratory Tsetse Fly Colonies

Whereas the prevalence of SGH in wild tsetse flies is low (0.2–5.4%), with the exception of the high rate (15.6%) in Kenya as reported by Odindo (62), the prevalence of SGH in laboratory tsetse colonies varies greatly with colony and over time. The SGH prevalence in a *G. pallidipes* colony that originated from Tororo, Uganda, and was maintained at the Insect Pest Control Laboratory of the FAO/IAEA in Seibersdorf, Austria, varied between 3.8% in 2007 and 10% in 2008 and 2009, a prevalence low enough to enable maintenance of the colony. However, a *G. pallidipes* colony originating from the Southern Rift Valley in Ethiopia and established at IAEA, Austria, became extinct in 2002 as a result of high SGH prevalence (85%) (1). In a *G. pallidipes* colony of the same origin, maintained at Kaliti, Addis

Ababa, Ethiopia, SGH rates up to 45% were observed in 2008 (5).

IPM: integrated pest management

SIT: sterile insect technique

The majority of prevalence studies of SGH in field populations relied on visual detection of SGH in dissected flies. Hence, reports only reflect prevalence of the SGH symptom and not general SGHV infection. Recent findings have shown that the GpSGHV can exist in an asymptomatic stage. In laboratory colonies of *G. pallidipes*, symptomatic SGH can be as low as 3.8%, but PCR-based assays detected the virus in almost all tested flies (1). A search (TBLASTX) for putative GpSGHV mRNAs in the expressed sequence tag database of another tsetse fly species, *G. morsitans morsitans* (**http://old.genedb.org/genedb/glossina/blast.jsp**), revealed significant identities (64%–100%) to 12 GpSGHV ORFs, suggesting that these expressed sequence tags were derived from cDNA synthesized from mRNA from tsetse flies asymptomatically infected with GpSGHV. It can be anticipated that SGHV prevalence in field populations is significantly higher than the reported prevalence of SGH.

SIGNIFICANCE OF SGHV IN TSETSE AND HOUSE FLY HOSTS

Glossina spp.

The obligatory blood-feeding tsetse flies (Diptera: Glossinidae) are solely responsible for the cyclical transmission of *Trypanosoma* parasites, the causative agents of human African trypanosomosis (HAT) (or sleeping sickness) in humans and African animal trypanosomosis (AAT) (or nagana) in livestock (52). An estimated 60 million people and 45–50 million cattle live under the constant risk of contracting the disease (14, 79). AAT is considered the single greatest constraint to improved livestock production in sub-Saharan Africa, with estimated direct annual cattle production losses of USD $600 million to $1.2 billion (33) and an annual lost potential in livestock and crop production of USD $4.75 billion (12).

Although AAT and HAT are contained mostly through curative and prophylactic treatment with trypanocidal drugs, the sustainable removal of the vector theoretically remains the most desirable strategy to contain the two diseases (38, 52). A variety of efficient tsetse control tactics is available, which can be combined in an integrated pest management (IPM) approach: a strategy that is derived from the principle that favorable aspects of different control methods complement each other, making the limitations of each method less important (11, 19). Environmentally acceptable tsetse control tactics include stationary bait techniques (29), the live bait technique (10), the sequential aerosol technique (45), and the sterile insect technique (SIT) (66, 70, 87). Applying the control effort on an area-wide (AW) basis, i.e., against an entire tsetse population within a circumscribed area (46, 48, 85), has resulted in more sustainable control (17, 45, 80, 87) compared with localized IPM where the control effort was directed against only parts of the tsetse population (9).

In 1996, the government of Ethiopia embarked on such an AW-IPM program to remove *G. pallidipes* Austen from 25,000 km^2 in the Southern Rift Valley. After the collection of the entomological baseline data (86), it was decided that eradication should be the strategy, with SIT as the final eradication component. Colonies of the local *G. pallidipes* strain were established in Addis Ababa and in Seibersdorf, but serious difficulties were experienced during the adaptation process, with up to 85% of the colonized flies showing SGH (5). Such high prevalence rates have drastic consequences for the mass rearing and the efficiency of SIT.

SIT requires rearing large numbers of target species individuals, which can then be sterilized using ionizing radiation and released sequentially over the target area (47). Rearing tsetse flies is challenging in view of their low reproductive capacity and the lack of an artificial diet. A high prevalence of SGH in a fly colony entails a high proportion of sterile males and of females with reduced fecundity. Above a certain threshold, such a colony cannot sustain itself and will decline and eventually collapse.

For SIT to be successful, the released sterile male flies should be able to compete with the wild target population (68, 84). Male *G. pallidipes* with SGH symptoms show testicular degeneration and have lost most of their potential to transfer viable sperm. Even if these male flies could track and mate with virgin female flies in nature, the absence of sperm transfer would not contribute to the induction of sterility in the native population. From laboratory studies in small cages, it is known that a proportion of female *G. austeni* and *G. tachinoides* Westwood will accept multiple matings (83), and assuming the same is true for *G. pallidipes*, any subsequent mating of such a female with a normal, wild male will result in a fully fertile female fly.

Strategies to manage the virus in these *G. pallidipes* colonies will therefore be required to produce high-quality, sterile male flies. Promising results have been obtained to reduce the virus load by using new, clean blood for each cage of flies for each feeding opportunity, rather than using the same blood for several successive feeds as is normally practiced for tsetse colonies (5). The high cost of the clean feeding method probably prohibits its use for large-scale rearing, but the clean feeding system could be used to establish a seed colony with low virus load. This colony could then be transferred to a normal feeding system combined with various virus management strategies that reduce virus replication [i.e., antiviral drugs and/or RNA interference (RNAi) for virus-specific genes] or that block horizontal transmission by neutralizing the virus infection in the blood using specific virus antibodies.

Musca domestica

The house fly, a global pest of agricultural and public health importance, has been known since antiquity (89). The ability of the fly to exploit a vast range of patchily distributed and ephemeral organic larval substrates has enabled it to plague virtually any area where humans and their animals congregate. Adult flies pose nuisance problems to farmworkers and to neighboring residents, but the habit of adult flies to defecate and regurgitate on animal and human food led to the early recognition of their role as vectors of human and animal pathogens, especially those responsible for enteric diseases (32).

RNAi: RNA interference

Because adult house flies can consume only liquids, they must regurgitate fluids from the alimentary system onto solid food in order to consume it in liquid form. This behavior is an important element in the movement and transmission of SGHV as well as human pathogens. Indeed, recent concerns about food-borne human illnesses have led to renewed documentation of the role of house flies in spreading disease-causing organisms, especially *Escherichia coli*, *Shigella* spp., and *Salmonella* spp. (6, 31, 57, 61). Pathogen-carrying flies are commonly found around human and animal waste and landfills, from which they disperse to areas of human habitation and activity (58, 81).

Conventional management of house flies has relied on the use of residual insecticides applied to fly resting sites, pyrethrin space sprays, and sugar baits containing toxicants. The rapidity with which house flies develop high levels of resistance to residual insecticides is legendary and has made it exceedingly difficult to control flies in areas with long histories of the use of common toxicants such as permethrin and cyfluthrin (27, 78). Cross-resistance and the high innate tolerance of flies have led to surprisingly high levels of resistance to novel insecticides such as imidacloprid within a few years of their introduction, even when these toxicants are deployed as baits (44).

Other methods of fly management include cultural control, especially removal of manure and other breeding habitats, the use of various types of traps, and biological control. Most of the biological control efforts have targeted the immature stages of the fly, with main emphases on egg predators and pupal parasitoids (23, 72). In contrast, biological control options for adult house flies have received relatively little attention. The entomopathogenic fungus *Entomophthora muscae* often produces spectacular epizootics in house fly populations (60, 88), but attempts to manipulate this pathogen have been limited by the need for high fly populations to sustain epizootics (25) and the ability of the flies to mitigate the effects of infection by resting in warm areas to raise their body temperature (43). Similarly, adult house flies are susceptible to *Beauveria bassiana* and *Metarhizium anisopliae* (16, 26, 53), but attempts to use these pathogens in the field have met with mixed results. MdSGHV is a particularly attractive candidate for fly biocontrol because it is already compatible with the ecology and behavior of the fly, and its ovary-suppressing effect is unique among natural enemies of the fly.

SUMMARY POINTS

1. SGHVs, unlike many other insect viruses, have unique pathological properties: They infect the adult stage, cause a chronic infection that produces little if any external symptoms, and at the cellular level cause a unique pathology in the salivary gland.
2. This virus group is novel and has been proposed to constitute a new virus family, the *Hytrosaviridae*.
3. The discovery of *Hytrosaviridae* is hindered due to the absence of obvious external symptoms and/or acute mortality associated with infection.
4. The ability of MdSGHV to downregulate vitellogenesis and disrupt mating behavior in the infected host provides for potential manipulation as a natural sterilizing control agent.

DISCLOSURE STATEMENT

The authors are not aware of any affiliations, memberships, funding, or financial holdings that might be perceived as affecting the objectivity of this review.

ACKNOWLEDGMENTS

We thank Lyle Buss and Jane Medley for assistance during figure production.

LITERATURE CITED

1. Abd-Alla A, Bossin H, Cousserans F, Parker A, Bergoin M, Robinson A. 2007. Development of a non-destructive PCR method for detection of the salivary gland hypertrophy virus (SGHV) in tsetse flies. *J. Virol. Methods* 139:143–49
2. **Abd-Alla A, Vlak JM, Bergoin M, Maruniak JE, Parker A, et al. 2009. *Hytrosaviridae*: a proposal for classification and nomenclature of a new insect virus family. *Arch. Virol.* 154:909–18**

2. Proposal of a new virus family.

3. **Abd-Alla AMM, Cousserans F, Parker AG, Jehle JA, Parker NJ, et al. 2008. Genome analysis of a *Glossina pallidipes* salivary gland hypertrophy virus (GpSGHV) reveals a novel large double-stranded circular DNA virus. *J. Virol.* 82:4595–611**

3. GpSGHV genome sequenced.

4. Abd-Alla AMM, Cousserans F, Parker AG, Jridi C, Bergoin M, Robinson AS. 2009. Quantitative PCR analysis of the salivary gland hypertrophy virus (GpSGHV) in a laboratory colony of *Glossina pallidipes*. *Virus Res.* 139:48–53
5. Abd-Alla AMM, Kariithi H, Parker AG, Robinson AS, Kiflom M. 2010. Dynamics of the salivary gland hypertrophy virus in laboratory colonies of *Glossina pallidipes* (Diptera: Glossinidae). *Virus Res.* 150:103–10
6. Ahmad A, Nagaraja TG, Zurek L. 2007. Transmission of *Escherichia coli* O157:H7 to cattle by house flies. *Prev. Vet. Med.* 80:74–81

7. Aksoy S. 2003. Control of tsetse flies and trypanosomes using molecular genetics. *Vet. Parasitol.* 115:125–45
8. Amargier A, Lyon JP, Vago C, Meynadier G, Veyrunes JC. 1979. Mise en evidence et purification d'un virus dans la proliferation monstreuse glandulaire d'insectes. Etude sur *Merodon equestris* (Diptera, Syrphidae). Note. *C. R. Seanc. Acad. Sci. Ser. D Sci. Nat.* 289:481–84
9. Barrett K, Okali C. 1998. Partnership for tsetse control-community participation and other options. *WAR/RMZ* 1:39–46
10. Bauer B, Kabore I, Liebisch A, Meyer F, Petrich-Bauer J. 1992. Simultaneous control of ticks and tsetse flies in Satiri, Burkina Faso, by the use of flumethrin pour on for cattle. *Trop. Med. Parasitol.* 43:41–46
11. Brader L. 1979. Integrated pest control in the developing world. *Annu. Rev. Entomol.* 24:225–54
12. Budd LT. 1999. *DFID-Funded Tsetse and Trypanosomosis Research and Development Since 1980.* Vol. 2. *Economic Analysis*. London: DIFD. 123 pp.
13. Burtt E. 1945. Hypertrophied salivary glands in *Glossina*: evidence that *G. pallidipes* with this abnormality is peculiarly suited to trypanosome infection. *Ann. Trop. Med. Parasitol.* 39:11–13
14. Cattand P, Jannin J, Lucas P. 2001. Sleeping sickness surveillance: an essential step towards elimination. *Trop. Med. Int. Health* 6:348–61
15. **Coler RR, Boucias DG, Frank JH, Maruniak JE, Garcia-Canedo A, Pendland JC. 1993. Characterization and description of a virus causing salivary gland hyperplasia in the housefly, *Musca domestica*. *Med. Vet. Entomol.* 7:275–82**

15. First published record of SGH in *M. domestica*.

16. Darwish E, Zayed A. 2002. Pathogenicity of two entomopathogenic hyphomycetes, *Beauveria bassiana* and *Metarhizium anisopliae*, to the housefly *Musca domestica* L. *J. Egypt. Soc. Parasitol.* 32:785–96
17. du Toit R. 1954. Trypanosomiasis in Zululand and the control of tsetse flies by chemical means. *Onderstepoort J. Vet. Res.* 26:317–85
18. Ellis DS, Maudlin I. 1987. Salivary gland hyperplasia in wild caught tsetse from Zimbabwe. *Entomol. Exp. Appl.* 45:167–73
19. FAO/IAEA/USDA. 2003. *Manual for Product Quality Control and Shipping Procedures for Sterile Mass-Reared Tephritid Fruit Flies*. Vienna: IAEA. 84 pp.
20. Friesen PD, Miller LK. 1985. Temporal regulation of baculovirus RNA: overlapping early and late transcripts. *J. Virol.* 54:392–400
21. Garcia-Maruniak A, Abd-Alla A, Salem TZ, Parker A, Lietze V-U, et al. 2009. Two viruses that cause salivary gland hypertrophy in *Glossina pallidipes* and *Musca domestica* are related and form a distinct phylogenetic clade. *J. Gen. Virol.* 90:334–46
22. **Garcia-Maruniak A, Maruniak JE, Farmerie W, Boucias DG. 2008. Sequence analysis of a nonclassified, nonoccluded DNA virus that causes salivary gland hypertrophy of *Musca domestica*, MdSGHV. *Virology* 377:184–96**

22. MdSGHV genome sequenced.

23. Geden CJ, Hogsette JA. 2006. Suppression of house flies (Diptera: Muscidae) in Florida poultry houses by sustained releases of *Muscidifurax raptorellus* and *Spalangia cameroni* (Hymenoptera: Pteromalidae). *Environ. Entomol.* 35:75–82
24. Geden CJ, Lietze V-U, Boucias D. 2008. Seasonal prevalence and transmission of salivary gland hypertrophy virus of house flies (Diptera: Muscidae). *J. Med. Entomol.* 45:42–51
25. Geden CJ, Steinkraus DC, Rutz DA. 1993. Evaluation of two methods for release of *Entomophthora muscae* (Entomophthorales: Entomophthoraceae) to infect house flies (Diptera: Muscidae) on dairy farms. *Environ. Entomol.* 20:1201–8
26. Geden CJ, Steinkraus DC, Rutz DA. 1995. Virulence of different isolates and formulations of *Beauveria bassiana* for house flies and the parasitoid *Muscidifurax raptor*. *Biol. Control* 5:615–21
27. Georghiou GP, Mellon R. 1983. Pesticide resistance in time and space. In *Pest Resistance to Pesticides*, ed. GP Georghiou, T Saito, pp. 1–46. New York: Plenum
28. Gouteux JP. 1987. Prevalence of enlarged salivary glands in *Glossina palpalis*, *G. pallicera*, and *G. nigrofusca* (Diptera: Glossinidae) from the Vavoua area, Ivory Coast. *J. Med. Entomol.* 24:268
29. Green CH. 1994. Bait methods for tsetse fly control. *Adv. Parasitol.* 34:229–91
30. Gross CH, Rohrmann GF. 1993. Analysis of the role of 5′ promoter elements and 3′ flanking sequences on the expression of a baculovirus polyhedron envelope protein gene. *Virology* 192:273–81

31. Holt PS, Geden CJ, Moore RW, Gast RK. 2007. Isolation of *Salmonella enterica* serovar *enteriditis* from houseflies (Musca domestica) found in rooms containing *Salmonella* serovar *enteriditis*-challenged hens. *Appl. Environ. Microbiol.* 73:6030–35
32. Howard LO. 1911. *The House Fly—Disease Carrier*. New York: Stokes. 312 pp.
33. Hursey BS, Slingenbergh J. 1995. The tsetse fly and its effects on agriculture in sub-Saharan Africa. *World Anim. Rev.* 84/85:67–73
34. Jaenson TGT. 1978. Virus-like rods associated with salivary gland hyperplasia in tsetse, *Glossina pallidipes*. *Trans. R. Soc. Trop. Med. Hyg.* 72:234–38
35. Jaenson TGT. 1986. Sex-ratio distortion and reduced life-span of *Glossina pallidipes* infected with the virus causing salivary gland hyperplasia. *Entomol. Exp. Appl.* 41:265–71
36. Jenni L. 1973. Virus-like particles in a strain of *G. morsitans centralis*, Machado 1970. *Trans. R. Soc. Trop. Med. Hyg.* 67:295
37. Jenni L, Steiger R. 1974. Viruslike particles in the tsetse fly, *Glossina morsitans* sspp. Preliminary results. *Rev. Suisse Zool.* 81:663–66
38. Jordan AM. 1986. *Trypanosomiasis Control and African Rural Development*. London: Longman. 357 pp.
39. Jura WGZO, Davies-Cole JOA. 1992. Some aspects of mating behavior of *Glossina morsitans morsitans* males infected with a DNA virus. *Biol. Control* 2:188–92
40. Jura WGZO, Odhiambo TR, Otieno LH, Tabu NO. 1988. Gonadal lesions in virus-infected male and female tsetse, *Glossina pallidipes* (Diptera: Glossinidae). *J. Invertebr. Pathol.* 52:1–8
41. Jura WGZO, Otieno LH, Chimtawi M. 1989. Ultrastructural evidence for trans-ovum transmission of the DNA virus of tsetse *Glossina pallidipes* (Diptera: Glossinidae). *Curr. Microbiol.* 18:1–4
42. Jura WGZO, Zdarek J, Otieno LH. 1993. A simple method for artificial infection of tsetse, *Glossina morsitans morsitans* larvae with the DNA virus of *G. pallidipes*. *Insect Sci. Appl.* 14:383–87
43. Kalsbeek V, Mullens BA, Jespersen JB. 2001. Field studies of *Entomophthora* (Zygomycetes: Entomophthorales)—induced behavioral fever in *Musca domestica* (Diptera: Muscidae) in Denmark. *Biol. Control* 21:264–73
44. Kaufman PE, Nunez S, Mann RS, Geden CJ, Scharf ME. 2010. Nicotinoid and pyrethroid insecticide resistance in house flies (Diptera: Muscidae) collected from Florida dairies. *Pest Manage. Sci.* 66:290–94
45. Kgori PM, Modo S, Torr SJ. 2006. The use of aerial spraying to eliminate tsetse from the Okavango Delta of Botswana. *Acta Trop.* 99:184–99
46. Klassen W, Curtis CF. 2005. History of the sterile insect technique. In *Sterile Insect Technique: Principles and Practice in Area-Wide Integrated Pest Management*, ed. VA Dyck, J Hendrichs, AS Robinson, pp. 3–36. Dordrecht, The Neth.: Springer
47. Knipling EF. 1955. Possibilities of insect control or eradication through the use of sexually sterile males. *J. Econ. Entomol.* 48:459–62
48. Knipling EF. 1972. Sterilization and other genetic techniques. In *Proc. Symp. Pest Control: Strategies Future*, pp. 272–87. Washington, DC: Natl. Acad. Sci.
49. Kokwaro ED. 2006. Virus particles in male accessory reproductive glands of tsetse, *Glossina morsitans morsitans* (Diptera: Glossinidae) and associated tissue changes. *Int. J. Trop. Inst. Sci.* 26:266–72
50. Kokwaro ED, Nyindo M, Chimtawi M. 1990. Ultrastructural changes in salivary glands of tsetse, *Glossina morsitans morsitans*, infected with virus and rickettsia-like organisms. *J. Invertebr. Pathol.* 56:337–46
51. Kokwaro ED, Otieno LH, Chimtawi M. 1991. Salivary glands of the tsetse *Glossina pallidipes* Austen infected with *Trypanosoma brucei* and virus particles: ultrastructural study. *Insect Sci. Appl.* 12:661–69
52. Leak SGA. 1998. *Tsetse Biology and Ecology: Their Role in the Epidemiology and Control of Trypanosomasiasis*. Wallingford, UK: CABI Publ. 570 pp.
53. Lecuona RE, Turica M, Tarocco F, Crespo DC. 2005. Microbial control of *Musca domestica* (Diptera: Muscidae) with selected strains of *Beauveria bassiana*. *J. Med. Entomol.* 42:332–36
54. Lietze V-U, Geden CJ, Blackburn P, Boucias DG. 2007. Effects of salivary gland hypertrophy virus on the reproductive behavior of the house fly, *Musca domestica*. *Appl. Environ. Microbiol.* 73:6811–18
55. Lietze V-U, Sims KR, Salem TZ, Geden CJ, Boucias DG. 2009. Transmission of MdSGHV among adult house flies, *Musca domestica* (Diptera: Muscidae), occurs via salivary secretions and excreta. *J. Invertebr. Pathol.* 101:49–55

34. First identification of virus particles unambiguously associated with SGH symptoms.

56. **Lyon JP. 1973. La mouche des narcisses (*Merodon equestris* F., Diptere Syrphidae). I. Identification de l'insecte et de ses degats et biologie dans le sud-est de la France. *Rev. Zool. Agric. Pathol. Veg.* 72:65–92**

56. **First published record of SGH in *M. equestris*.**

57. Macovei L, Miles B, Zurek L. 2008. The potential of house flies to contaminate ready-to-eat food with antibiotic resistant enterococci. *J. Food Prot.* 71:432–39
58. Mian LS, Maag H, Tacal JV. 2002. Isolation of *Salmonella* from muscoid flies at commercial animal establishments in San Bernardino County, California. *J. Vector Ecol.* 27:82–85
59. Minter-Goedbloed E, Minter DM. 1989. Salivary gland hyperplasia and trypanosome infection of *Glossina* in two areas of Kenya. *Trans. R. Soc. Trop. Med. Hyg.* 83:640–41
60. Mullens BA, Rodriguez JL, Meyer JA. 1987. An epizootiological study of *Entomophthora muscae* in muscoid fly populations on Southern California poultry facilities, with emphasis on *Musca domestica*. *Hilgardia* 55:1–41
61. Nayduch D, Stutzenberger F. 2001. The housefly (*Musca domestica*) as a vector for emerging bacterial enteropathogens. *Rec. Res. Dev. Microbiol.* 5:205–9
62. Odindo MO. 1982. Incidence of salivary gland hypertrophy in field populations of the tsetse *Glossina pallidipes* on the south Kenyan coast. *Insect Sci. Appl.* 3:59–64
63. Odindo MO, Amutalla PA. 1986. Distribution pattern of the virus of *Glossina pallidipes* Austen in a forest ecosystem. *Insect Sci. Appl.* 7:79–84
64. Odindo MO, Payne CC, Crook NE, Jarrett P. 1986. Properties of a novel DNA virus from the tsetse fly, *Glossina pallidipes*. *J. Gen. Virol.* 67:527–36
65. Odindo MO, Sabwa DM, Amutalla PA, Otieno WA. 1981. Preliminary tests on the transmission of virus-like particles to the tsetse *Glossina pallidipes*. *Insect Sci. Appl.* 2:219–21
66. Oladunmade MA, Feldmann U, Takken W, Tenabe SO, Hamann HJ, et al. 1990. Eradication of *Glossina palpalis palpalis* (Robineau-Desvoidy) (*Diptera: Glossinidae*) from agropastoral land in central Nigeria by means of the sterile insect technique, pp. 5–23. Presented at *Sterile Insect Techn. Tsetse Control Erad.* (*Proc. Final Res. Coord. Meet.*), *Vom, Nigeria, 6–10 June, 1988*. Vienna: IAEA/RC/319.3/1
67. Otieno LH, Kokwaro ED, Chimtawi M, Onyango P. 1980. Prevalence of enlarged salivary glands in wild populations of *Glossina pallidipes* in Kenya, with a note on the ultrastructure of the affected organ. *J. Invertebr. Pathol.* 36:113–18
68. Parker AG. 2005. Mass-rearing for sterile insect release. In *Sterile Insect Technique. Principles and Practice in Area-Wide Integrated Pest Management*, ed. VA Dyck, J Hendrichs, AS Robinson, pp. 209–32. Dordrecht, The Neth.: Springer
69. Passarelli AL, Guarino LA. 2007. Baculovirus late and very late gene regulation. *Curr. Drug Targets* 8:1103–15
70. Politzar H, Cuisance D. 1982. SIT in the control and eradication of *Glossina palpalis gambiensis*. *Rep. IAEA-SM-255/4*
71. Prompiboon P, Lietze V-U, Denton JSS, Geden CJ, Steenberg T, Boucias DG. 2010. *Musca domestica* salivary gland hypertrophy virus: a globally distributed insect virus that infects and sterilizes female houseflies. *Appl. Environ. Microbiol.* 76:994–98
72. Rutz DA, Patterson RS. 1990. *Biocontrol of Arthropods Affecting Livestock and Poultry*. Boulder, CO: Westview
73. Salem TZ, Garcia-Maruniak A, Lietze V-U, Maruniak JE, Boucias DG. 2009. Analysis of transcripts from predicted open reading frames of the *Musca domestica* salivary gland hypertrophy virus. *J. Gen. Virol.* 90:1270–80
74. Sang RC, Jura WGZO, Otieno LH, Mwangi RW. 1998. The effects of a DNA virus infection on the reproductive potential of female tsetse flies, *Glossina morsitans centralis* and *Glossina morsitans morsitans* (Diptera: Glossinidae). *Mem. Inst. Oswaldo Cruz* 93:861–64
75. Sang RC, Jura WGZO, Otieno LH, Mwangi RW, Ogaja P. 1999. The effects of a tsetse DNA virus infection on the functions of the male accessory reproductive gland in the host fly *Glossina morsitans centralis* (Diptera: Glossinidae). *Curr. Microbiol.* 38:349–54
76. Sang RC, Jura WGZO, Otieno LH, Ogaja P. 1996. Ultrastructural changes in the milk gland of tsetse *Glossina morsitans centralis* (Diptera: Glossinidae) female infected by a DNA virus. *J. Invertebr. Pathol.* 68:253–59

77. Sang RC, Jura WGZO, Otieno LH, Tukei PM, Mwangi RW. 1997. Effects of tsetse DNA virus infection on the survival of a host fly, *Glossina morsitans centralis* (Diptera: Glossinidae). *J. Invertebr. Pathol.* 69:253–60
78. Scott JG, Alefantis TG, Kaufman PE, Rutz DA. 2000. Insecticide resistance in house flies from caged-layer poultry facilities. *Pest Manag. Sci.* 56:147–53
79. Shaw A, Torr S, Waiswa C, Robinson T. 2007. *Comparable costings of alternatives for dealing with tsetse: estimates for Uganda*. PPLPI Work. Pap., *FAO* 40:vii–59
80. Spielberger U, Naisa BK, Abdurrahim U. 1977. Tsetse (Diptera: Glossinidae) eradication by aerial (helicopter) spraying of persistent insecticides in Nigeria. *Bull. Entomol. Res.* 67:589–98
81. Sulaiman S, Othman MZ, Aziz AH. 2000. Isolations of enteric pathogens from synanthropic flies trapped in downtown Kuala Lumpur. *J. Vector Ecol.* 25:90–93
82. Sun M, Hurst LD, Carmichael GG, Chen JJ. 2005. Evidence for a preferential targeting of 3′-UTRs by *cis*-encoded natural antisense transcripts. *Nucleic Acids Res.* 33:5533–43
83. Vreysen MJB. 1995. *Radiation induced sterility to control tsetse flies. The effect of ionising radiation and hybridisation on tsetse biology and the use of the sterile insect technique in integrated tsetse control.* PhD thesis. Wageningen Agric. Univ., Wageningen, The Netherlands. 282 pp.
84. Vreysen MJB. 2005. Monitoring sterile and wild insects in area-wide integrated pest management programmes. In *Sterile Insect Technique. Principles and Practice in Area-Wide Integrated Pest Management*, ed. VA Dyck, J Hendrichs, AS Robinson, pp. 325–62. Dordrecht, The Neth.: Springer
85. Vreysen MJB, Gerardo-Abaya J, Cayol JP. 2007. Lessons from area-wide integrated pest management (AW-IPM) programmes with an SIT component: an FAO/IAEA perspective. In *Area-Wide Control of Insect Pests from Research to Field Implementation*, ed. MJB Vreysen, AS Robinson, J Hendrichs, pp. 723–44. Dordrecht, The Neth.: Springer
86. Vreysen MJB, Mebrate A, Menjeta M, Bancha B, Woldeyes G, et al. 1999. The distribution and relative abundance of tsetse flies in the Southern Rift Valley of Ethiopia: Preliminary survey results. *Proc. 25th Meet. Int. Sci. Counc. Trypanosomiasis Res. Control, Mombasa, Kenya, 27 Sept. – 1 Oct.*
87. Vreysen MJB, Saleh KM, Ali MY, Abdulla AM, Zhu ZR, et al. 2000. *Glossina austeni* (Diptera: Glossinidae) eradicated on the Island of Unguja, Zanzibar, using the sterile insect technique. *J. Econ. Entomol.* 93:123–35
88. Watson DW, Petersen JJ. 1993. Seasonal activity of *Entomophthora muscae* (Zygomycetes: Entomophthorales) in *Musca domestica* L. (Diptera: Muscidae) with reference to temperature and relative humidity. *Biol. Control* 3:182–90
89. West LS. 1951. *The House Fly*. Ithaca, NY: Comstock Publ. 584 pp.
90. **Whitnall ABM. 1932. The trypanosome infections of *Glossina pallidipes* in the Umfolosi Game Reserve, Zululand (preliminary report). *Rep. Dir. Vet. Serv. S. Afr.* 18:21–30**
91. Whitnall AMB. 1934. The trypanosome infections of *Glossina pallidipes* in the Umfolosi game reserve, Zuhuland. *Onderstepoort J. Vet. Sci. Anim. Ind.* 11:7–21

90. First published record of SGH in *Glossina* spp.

Insect-Resistant Genetically Modified Rice in China: From Research to Commercialization

Mao Chen,[1,2] Anthony Shelton,[1] and Gong-yin Ye[2]

[1]Department of Entomology, Cornell University/NYSAES, Geneva, New York 14456; email: mc447@cornell.edu, ams5@cornell.edu

[2]State Key Laboratory of Rice Biology, Ministry of Agriculture Key Laboratory of Molecular Biology of Crop Pathogens and Insects, Institute of Insect Sciences, Zhejiang University, Hangzhou, Zhejiang 310029, China; email: chu@zju.edu.cn

Annu. Rev. Entomol. 2011. 56:81–101

First published online as a Review in Advance on September 24, 2010

The *Annual Review of Entomology* is online at ento.annualreviews.org

This article's doi:
10.1146/annurev-ento-120709-144810

0066-4170/11/0107-0081$20.00

Key Words

agricultural biotechnology, *Bacillus thuringiensis*, field testing, risk assessments

Abstract

From the first insect-resistant genetically modified (IRGM) rice transformation in 1989 in China to October 2009 when the Chinese Ministry of Agriculture issued biosafety certificates for commercial production of two *cry1Ab/Ac Bacillus thuringiensis* (*Bt*) lines, China made a great leap forward from IRGM rice basic research to potential commercialization of the world's first IRGM rice. Research has been conducted on developing IRGM rice, assessing its environmental and food safety impacts, and evaluating its socioeconomic consequences. Laboratory and field tests have confirmed that these two *Bt* rice lines can provide effective and economic control of the lepidopteran complex on rice with less risk to the environment than present practices. Commercializing these *Bt* plants, while developing other GM plants that address the broader complex of insects and other pests, will need to be done within a comprehensive integrated pest management program to ensure the food security of China and the world.

INTRODUCTION

Rice (*Oryza sativa*) is the most widely consumed food crop and was grown on over 159 million ha worldwide in 2008, with over 18% grown in China (32). As one of the centers of origin, China has been cultivating rice for over 7,000 years (126). Rice is a staple food for over 1 billion people in China in addition to 2 billion people in other countries (39). Although different synthetic insecticides are applied frequently in order to control insect pests of rice, tremendous economic and environmental losses still occur regularly. For instance, rice stem borers, a major group of lepidopteran pests of rice, cause annual losses of 11.5 billion yuan (US$1.69 billion) (79, 80). In addition, another major group of insect pests, planthoppers, has caused large annual yield losses across the country since the 1970s (24, 25).

Cry proteins: insecticidal proteins produced by *B. thuringiensis* during sporulation phase as parasporal crystals

IRGM: insect-resistant genetically modified

***indica* rice:** *O. sativa* subspecies, characterized by long grains, less stickiness, and a higher photosynthetic rate

***japonica* rice:** *O. sativa* subspecies, characterized by short grains and stickiness

Genes from the bacterium *Bacillus thuringiensis* (*Bt*) that code for insecticidal Crystal (Cry) proteins were engineered into plants in the mid-1980s to develop the first insect-resistant genetically modified (IRGM) plants (89). Soon after, Chinese scientists began to use genetic engineering techniques to develop new control tactics for insect pests of rice. In 1989, scientists from the Chinese Academy of Agricultural Sciences (CAAS), by means of polyethyl glycol, generated the first IRGM rice plant with a *Bt* delta-endotoxin gene under control of the *CaMV 35S* promoter (108). After 20 years of extensive laboratory and field studies, on October 22, 2009, China's Ministry of Agriculture issued its first two biosafety certificates for commercial production of two *Bt* rice lines (*cry1Ab/Ac* Huahui No. 1 and *cry1Ab/Ac Bt* Shanyou 63) for Hubei Province (**http://www.stee.agri.gov.cn/biosafety/spxx/t20091022_819217.htm**). By this action, China had not only made great strides from basic research to commercialization of IRGM rice, but likely provided the impetus for the development of other IRGM food crops worldwide, thus moving toward the goal of fighting global poverty and food scarcity (45).

In this review, we examine the history and importance of rice production in China, IRGM rice research, the development of regulations for IRGM crops as they relate to food safety and the environment, and the socioeconomic impact of IRGM rice. Knowledge gaps and future directions for China's IRGM rice research are also discussed.

RICE PRODUCTION IN CHINA

China is the largest rice producer and consumer in the world and has a long history of rice cultivation in many geographic regions of the country. According to records of pollen, algal, and fungal spores and microcharcoal data from sediments dating back 7,700 years, local communities in the lower Yangtze region of China, a center of rice domestication, cultivated rice in lowland swamps after using fire to clear alder-dominated wetland scrub and dirt banks to control brackish water flooding (126). Today, China's first priority is to feed its population of 1.3 billion. Among all the agricultural crops (e.g., other grain crops, fruit and vegetable crops, oil crops, fiber crops, sugar, and tobacco) planted in China, the largest share of land is devoted mostly to rice production (ca. 20%), surpassing that devoted to corn (ca. 18%) and cotton (ca. 3–4%) (66).

Traditionally, there are six rice-growing regions in China (**Figure 1**). Among the six regions, the south China region (region I), central China region (region II), and southwest China region (region III) have the best climatic conditions and are the major rice-growing areas. Both *indica* and *japonica* rice subspecies are grown in regions I, II, and III, with approximately two planting seasons a year. However, there is only one season for *japonica* rice in region IV (north China), V (northeast China), and VI (northwest China) (**Figure 1**) (34).

It has been estimated that rice yield needs to reach 7.85×10^3 kg ha^{-1} by 2030 to feed China's anticipated population of 1.6 billion (26). To do this, China needs 0.2 billion kilograms of rice per year, which is equal to the present total rice production worldwide. Because the yield reduction caused by insect pests is a threat to food security and because the

Figure 1

The six rice-growing regions in China.

present heavy reliance on traditional insecticides is recognized as a problem for food and environmental safety, there is increased interest in using other technologies such as genetic engineering (115) and integrated pest management (IPM) (103). The Green Super Rice (GRS) concept integrates traditional, transgenic, and marker-assisted breeding strategies

IPM: integrated pest management

GRS: green super rice

to battle rice yield constraints (e.g., insect pests) and improve rice quality (115). From a global standpoint, it should be recognized that if China imports rice to feed its growing population, food scarcity will increase in other parts of the world and may cause a crisis in the price of rice (117). It is important to the world that China is able to meet its own food needs; rice production and the ability to manage rice pests will be key to meeting this goal.

Rice Insect Pest Complex and Damage

Across the rice-growing regions in China, there are more than 200 species of insect pests that damage different life stages and different parts of rice (23, 52). In general, insect pests of rice are divided into two groups: chewing insects (e.g., rice stem borers, leaffolders, rice water weevils) and sucking insects (e.g., planthoppers and leafhoppers). There are five primary insect pests of rice in China, of which three are lepidopterans and two hemipterans (**Table 1**). Lepidopteran stem borers are chronic pests in rice ecosystems. The earliest documented stem borer infestation on rice in China was during the Song Dynasty (960–1279 AD) (80). Prior to the 1950s, the Asiatic stem borer, *Chilo suppressalis* (Crambidae), was the most dominant stem borer species throughout China. However, during the 1950s–1970s the yellow stem borer, *Scirpophaga incertulas* (Crambidae), became a more important pest (23, 80). Rice stem borers generally caused negligible yield losses to rice production in China in the 1970s. However, since 1993, stem borer infestations have

Table 1 Primary and secondary insect pests of rice in China

Order	Primary insect pests	Secondary insect pests
Lepidoptera	*Scirpophaga incertulas* (Walker) *Chilo suppressalis* (Walker) *Cnaphalocrocis medinalis* (Guenée)	*Chrysaspidia festucae* (Graeser) *Leucania loreyi* (Duponchel) *Leucania separata* (Walker) *Parnara guttata* (Bremer et Gray) *Naranga aenescens* (Moore) *Sesamia inferens* (Walker) *Spodoptera mauritia* (Boisduval) *Pelopidas mathias* (Fabricius)
Hemiptera	*Nilaparvata lugens* (Stål) *Sogatella furcifera* (Harváth)	*Thaia rubiginosa* (Kuoh) *Nephotettix cincticeps* (Uhler) *Nephotettix virescens* (Distant) *Recilia dorsalis* (Motschulsky) *Laodelphax striatellus* (Fallén) *Macrosiphum avenae* (Fabricius) *Niphe elongata* (Dallas) *Leptocorisa acuta* (Thunberg)
Diptera	–	*Orseolia oryzae* (Wood-Mason) *Ephydra macellaria* (Egger) *Hydrellia sinica* (Fan et Xia) *Chlorops oryzae* (Matsumura)
Coleoptera	–	*Oulema oryzae* (Kuwayama) *Donacia provosti* (Fairmaire) *Echinocnemus squameus* (Billberg)
Orthoptera	–	*Oxya chinensis* (Thunberg)
Thysanoptera	–	*Frankliniella intonsa* (Trybom) *Stenchaetothrips biformis* (Bagnall) *Haplothrips aculeatus* (Fabricius)

been severe every year, with a disastrous outbreak in 1996 (80), likely owing to extensive use of synthetic insecticides, insecticide resistance, global warming, changing farming practices (no tillage or reduced tillage), expanded rice production in some areas, and coexistence of early-, mid-, and late-rice varieties (60). Many other foliage-feeding lepidopteran species also occur on rice, the most important of which is the rice leaffolder, *Cnaphalocrocis medinalis* (Pyralidae), which occurs throughout the country.

Yield losses caused by stem borers have been severe in the last decades. Sheng et al. (79, 80) estimated a yearly 3.1% loss nationally (approximately 6.3 billion kilograms of rice), equal to 6.5 billion Chinese yuan (US$956 million), in addition to 5 billion yuan (US$735 million) direct control cost (e.g., insecticides and labor fees). Although yield losses due to leaffolders are generally small because rice plants at the vegetative growth stage have a large capacity to compensate for damage to foliage, leaffolders caused 24–32% yield loss in some rice paddies when no control was applied (22). Further, leaffolder damage is highly visible to farmers and is often the most important stimulus for insecticide applications (63), which adds additional control costs and environmental damage.

Aside from stem borers, the brown planthopper, *Nilaparvata lugens*, and the white-backed planthopper, *Sogatella furcifera* (both Homoptera: Delphacidae), which once were minor insect pests in China prior to 1968, are now primary pests of rice, with 10 disastrous outbreaks in China since 1975. The most recent outbreak occurred in 2005, causing an estimated loss of 2.8 billion kilograms (24). The planthopper outbreaks have been attributed to high adoption of hybrid rice, increased use of insecticides and chemical fertilizers, insecticide resistance, climate change resulting in elevated autumn temperatures, more intense and higher frequency of typhoons that transport hoppers over wider areas, and changes in cropping systems in southern China that may have affected migration patterns (24, 25).

In addition to the primary insect pests listed above, there are 27 common secondary insect pests of rice (**Table 1**). These insects are distributed in different rice regions in China, with some species sporadically causing high localized losses.

Present Pest Management Strategies

Various control strategies for rice pest management have been practiced during the thousands of years of rice cultivation in China, and these include mechanical (e.g., deep plowing for stem borer control and digging ditches for burying locusts), biological, chemical, and cultural control (52). Some of these strategies continue today, although there is more emphasis on chemical control. The development of insect pest management in "New China" (since 1949) has been divided into three stages as the concept of IPM has evolved (67). Most recently, China's Ministry of Agriculture has promulgated the concept for rice pest management as "public plant protection and green plant protection" (103).

Although different control theories and strategies have been developed for rice pests, in practice farmers in many rice-growing regions still rely heavily on synthetic insecticides because of a lack of education and an incomplete understanding of modern IPM concepts and the quick visible control effects of some effective, cheap and easily accessible synthetic insecticides. The use of all pesticides (including insecticides, fungicides, and herbicides) for agricultural crops including rice in China in 2005 (1.5 billion kg) has doubled since 1990 (65). Consequently, severe insect outbreaks, environmental pollution, insecticide-related food poisoning, and farmer illnesses are frequently reported. Thus, newer and safer pest management strategies are sorely needed for rice production in China.

IRGM RICE DEVELOPMENT IN CHINA

The Development of IRGM Rice

Because of the prominent pest status of stem borers and leaffolders, the limited sources of

_Bt_ rice plants: transgenic plants with insecticidal gene(s) from _B. thuringiensis_

CpTI: cowpea trypsin inhibitor

GMO: genetically modified organism

Pilot field testing: first stage of field testing for agricultural GMOs in China, also known as restricted field testing; conducted on a small scale (up to 0.26 ha)

Environmental release testing: second stage of field testing for agricultural GMOs in China, also known as enlarged field testing; conducted on a middle scale (0.26–2.0 ha)

Preproduction testing: last stage of field testing for agricultural GMOs in China, also known as productive testing; conducted on a larger scale (2.1–66.7 ha)

resistance to these pests in rice germplasm (27), and the success of *Bt* cotton in China and elsewhere (62, 64, 103), the Chinese government, research institutes, and academic researchers have devoted large efforts to finding newer and safer tactics to control rice pests. Agricultural biotechnology has been extensively explored in China as a source for creating such tactics. Since 1985, with the government's support, many National Key laboratories have been established across the country in the general areas of agricultural biotechnology and crop genetics, which formed an infrastructure for Chinese biotechnology researchers to test their ideas on rice IPM (114). In 1989, scientists from the CAAS generated *Bt* rice plants (108), which, to our knowledge, was the earliest successful *Bt* rice transformation in the world. Public research expenditures on GM rice in China increased from 8 million yuan (US$1.18 million) in 1986 to 195 million yuan (US$28.68 million) in 2003 (43). In late 2008, China started a 26 billion yuan (US$3.5 billion) research and development (R&D) initiative on GM plants, which paved the way for commercialization of IRGM rice (19). Since 1989, IRGM rice lines expressing insecticidal genes with lepidopteran activity [e.g., *cry1Aa*, *cry1Ab*, *cry1Ac*, *cry1Ab/Ac*, *cry1C*, *cry2A*, *CpTI* (cowpea trypsin inhibitor)] or hemipteran activity (e.g., *Galanthus nivalis* agglutinin, *gna*, and *Pinellia ternata* agglutinin, *pta*) under control of various promoters have been developed and tested at various stages based on the regulatory process for agricultural genetically modified organisms (GMOs) in China (**Supplemental Table 1**; follow the **Supplemental Material link** from the Annual Reviews home page at **http://www.annualreviews.org**). Although IRGM rice research in China has been given great opportunities both in terms of the fast development of research infrastructure and in terms huge input of public research funds, issues of intellectual property rights, government regulations on GM plants, educational outreach programs for farmers about IRGM crops, and effective and well-regulated seed distribution systems have all delayed IRGM rice from reaching full commercialization and reflected some of the weaknesses of the Chinese research and regulatory systems. For instance, there have been reports of illegal planting of unapproved IRGM rice in central China in 2005 (101, 125) and intellectual property rights issues involving foreign-owned patents used in several IRGM rice lines (123).

First Release and Biosafety Certificate Approval for *Bt* Rice

For the past 20 years in China, numerous IRGM rice lines have been developed (**Supplemental Table 1**) and the first field tests took place in 1998 (111). Based on the regulation policy for agricultural GMOs in China, GM crops go through three tiers of field testing (pilot field testing, environmental release testing, and preproduction testing) before being submitted to the Office of Agricultural Genetic Engineering Biosafety Administration (OAGEBA) to apply for the "Biosafety Certificate" for commercialization. Each year hundreds of applications are submitted to the OAGEBA for different tiers of field testing. By the first half of 2009, the OAGEBA had approved 357 applications for field testing. These included 228 applications for pilot field testing (or restricted field testing), 95 applications for environmental release testing (or enlarged field testing), and 34 applications for preproduction testing (or productive testing). On October 22, 2009, China's Ministry of Agriculture issued two biosafety certificates for commercial production of *Bt* rice lines Huahui No. 1 and *Bt* Shanyou 63 in Hubei Province. Huahui No. 1 is a CMS (cytoplasmic male sterile) restorer line and *Bt* Shanyou 63 is a hybrid of Huahui No. 1 and Zhenshan 97A (the CMS line). Both lines express a *cry1Ab/Ac* fusion gene. China is now poised to become the first nation in the world to commercialize IRGM rice, which will likely result in a positive influence on global acceptance and the speed at which biotech food and feed crops are adopted (45).

Other IRGM Rice in Development

Although two *Bt* rice lines have been issued biosafety certificates, many other IRGM rice lines have been extensively tested and are now waiting for approval. In addition to the two *Bt* rice lines, the National Biosafety Committee approved in August 2009 five other IRGM rice lines after preproduction testing, although no biosafety certificate has been issued. These IRGM rice lines are KMD (expressing *cry1Ab* gene), T1c-9 (expressing *cry1C*), T2A-1 (expressing *cry2A* gene), and Kefeng 6 and 8 (both expressing *cry1Ac+CpTI* genes) (**http://www.stee.agri.gov.cn/biosafety/spxx/t20091022_819217.htm**).

Although *cry1* and *cry2* are the primary genes used for IRGM rice, the *Bt* vip gene (*vip3H*) (31), plant-derived insect-resistant lectin genes (e.g., *gna*, *pta*), protease inhibitor genes [e.g., *CpTI*, *pinII* (potato inhibitor II) and *SbTI* (soybean trypsin inhibitor)], and animal-derived insect-resistant gene (e.g., spider toxin gene, *SpI*) are also being used for IRGM rice (**Supplemental Table 1**). New *cry* genes (e.g., *cry4Cc1*, *cry30Ga1*, and *cry56Aa1*) have been identified as having insecticidal activity on stem borers (53, 121). Therefore, it may be possible to substitute genes for *cry1* and *cry2* or they could be used for gene pyramiding to broaden activity and delay the evolution of resistance. In addition, DNA shuffling has been used to construct novel insecticidal genes from existing *cry* genes for rice transformation (92).

In addition to the insecticidal genes, promoters play a vital role in determining where and when the genes are expressed in the plant (6a). Thus, promoters can influence the environmental fate of insecticidal proteins and the evolution of resistance (6a, 35). Constitutive promoters generally allow the genes to be expressed continuously in most parts of the plant. An alternative is to have them expressed only in certain tissues attacked by insects. The tissue-specific promoter *rbsc* (ribulose-1,5-bisphosphate carboxylase/oxygenase) was used for *cry1C* rice to reduce potential ecological and food risks (112). IRGM rice with stacked traits of herbicide resistance, disease resistance, or both has also been explored (**Supplemental Table 1**). More recently, suppressing the expression of key genes for insect development or biochemical metabolism through RNA interference (RNAi) using gene fragments from a target pest has been achieved with IRGM corn (8) and cotton (61). Such second-generation IRGM plants hold promise for future work with rice.

BIOSAFETY REGULATION AND POLICIES ON GENETICALLY MODIFIED RICE

The first biosafety regulation for GMOs in China was issued in 1993 by the Chinese Ministry of Science and Technology. Since then, the regulations have been updated and revised (44), with the latest version issued by the State Council in 2001, followed by three additional regulations focusing on agricultural GMOs by the Ministry of Agriculture and one food hygiene regulation by the Ministry of Public Health in 2002 (**http://www.agri.gov.cn/xzsp/xgzl/**).

The National Biosafety Committee was formed under OAGEBA to evaluate all biosafety assessment applications relating to agricultural GMOs and to provide positive or negative recommendations to OAGEBA based on the results of biosafety assessments. However, OAGEBA is responsible for the final decision. Safety assessment for agricultural GMOs including GM rice in China is conducted on a case-by-case scientific examination using safety regulations appropriate to the testing stage. Safety assessment of GM plants is divided into five stages: (*a*) laboratory research, (*b*) pilot field testing, (*c*) environmental release field testing, (*d*) preproduction testing, and (*e*) application for biosafety certificates. In addition, after being issued biosafety certificates, GM rice lines along with other GM plants need to pass seed variety testing standards regulated by The Seed Law prior to entering production and marketing (44, 93).

Although the current regulations on agricultural GMOs in China are comprehensive and elaborate, criticisms and challenges exist. For instance, the biosafety decision making process for agricultural GMOs in China relies primarily on the National Biosafety Committee, which has 75 members, of which the majority are biotechnologists (44). Thus, different but well-informed voices may be needed to achieve a balanced decision on the biosafety of agricultural GMOs. In addition, appropriate biosafety assessment practices and approaches, such as tier-based risk assessment (71), proper test species selection, and statistical analysis (69), are needed to reduce some repeated and unnecessary studies; to ensure sound experimental design for risk assessments, good quality data, and interpretation; and to harmonize biosafety assessment processes in China with those in other countries.

LABORATORY AND FIELD TESTING OF IRGM RICE

Summary of Peer-Reviewed Publications

Since the first *Bt* rice plant was developed in China in 1989 (108), numerous laboratory and field studies have been conducted on its environmental and food safety. Although many peer-reviewed papers on IRGM rice in China were published in English language journals, most were published in Chinese journals that are unknown or inaccessible to many scientists in the western world. Two searching methods were used to summarize the publications: papers published in Chinese on IRGM rice were found using China Academic Journals Full-Text Database under the China National Knowledge Infrastructure (**http://www.cnki.net/**), which is the largest searchable full-text Chinese database in the world; papers published in English on China's IRGM rice were found using the Web of Science® (including Science Citation Index Expanded, Social Sciences Citation Index, and Arts & Humanities Citation Index databases). From January 1995 to December 2009, there were 378 and 108 peer-reviewed papers published in Chinese (including MS and PhD dissertations) and in English, respectively (**Supplemental Figure 1**). Those papers included laboratory and field studies on GM rice development; on the efficacy on target insects; and the effects on nontarget insects including soil biota, gene flow, transgene detection, food safety, and agronomic traits. Publications on field studies of IRGM rice accounted for 24% of all papers published in Chinese sources and 36% of papers published in English sources. From 1995 to 2009, an average of 34.7 peer-reviewed papers were published annually on IRGM rice, with a maximum of 70 papers in 2005. This exceeds the publication record of both IRGM cotton and corn, and this large source of references serves as the source for the various topics on *Bt* rice discussed in this manuscript.

Environmental Risk Assessment

Interactions of IRGM rice with biological control agents. Rice insecticides accounted for nearly 15% of the global crop insecticide market value (98). Farmers tend to overreact to slight infestations caused by leaf-feeding pests, such as rice leaffolders, and make routine preventive applications, which usually removes most natural enemies from the system and leaves the field open for pest buildup (63). The evolution of resistance to the major classes of insecticides in rice stem borers decreased their efficacy and often led farmers to compensate by increasing the amounts used (46, 68). These factors are detrimental to biological control and pest management in the rice ecosystem.

Although field surveys indicated that there were approximately 200 species of insect pests injurious to rice in China, in rice-growing region I alone more than 228 species of natural enemies of insect pests were found, including 111 parasitoid species and 117 predator species (23, 52). Because of the critical role biological control has in rice IPM and the previous deleterious experiences with indiscriminate use of another insect control technology (i.e., broad-spectrum insecticides), the need to carefully evaluate the ecological effects of IRGM rice before its release has been generally recognized in China. Many trials have assessed the potential impacts of IRGM rice on parasitoids and predators under laboratory and field conditions (summarized in Reference 18). These trials include studies of direct toxicity of purified insecticidal proteins or IRGM rice conducted in the laboratory and ecological studies conducted in the field. The field studies have examined populations of target and nontarget herbivores and their natural enemies using various insect sampling methods (e.g., vacuum sampling, sweep nets, and sticky traps). In general, negative effects of IRGM rice on natural enemies have not been observed, as measured by indicators of fitness, population density and dynamics, and biodiversity indices (18). For instance, the fitness of *Propylea japonica* (Coleoptera: Coccinellidae) and *Chrysoperla sinica* (Neuroptera: Chrysopidae) was not negatively affected by *Bt* rice through direct or indirect feeding (2, 3, 5).

Several multiple-year per site field studies indicated that the population dynamics of *Cyrtorhinus lividipennis* (Hemiptera: Miridae) and of five common spider species was similar in *Bt* and non-*Bt* rice fields (15, 16, 57). As expected, some negative effects on parasitoids have been observed when *Bt*-susceptible herbivores are used as hosts (47, 48, 87), but this is most likely attributed to poor host quality than to toxic effects of the parasitoid (72). However, the dispersal dynamics of the parasitoids (18) and the overall temporal dynamics of species richness, diversity, and evenness of the parasitoid communities (50, 56) were similar between *Bt* and non-*Bt* rice fields. More conclusively, 14 parasitic arthropod families and 26 predatory arthropod families belonging to Hemiptera, Neuroptera, Coleoptera, Diptera, Hymenoptera, and Aranaea were collected from both *Bt* and non-*Bt* rice fields, with no consistent differences in population structure in the two rice ecosystems (58). Although such studies were usually conducted on fields less than 1 ha in size, the results provide initial evidence that changes in natural enemy populations on a landscape level would be minimal, if at all. From our review of the Chinese and English literature, we believe that the results from studies with IRGM rice on natural enemies are consistent with those from other *Bt* crops, as reviewed in individual studies (72) or meta-analyses (64).

Nontarget Herbivores

Because the deployment of lepidopteran-resistant GM rice in China may potentially release competition pressure for planthoppers in the rice ecosystem and further worsen their already severe pest status (24, 25), planthoppers and leafhoppers have been identified as a key group of nontarget herbivores for IRGM rice research in China. In tests of direct toxicity, Bai et al. (5) reported that *N. lugens* ingested Cry proteins from *Bt* rice lines, but that they had no detectable negative effects on its fitness. Similarly, *Bt* rice had no significant effect on the feeding and oviposition behavior of planthoppers and leafhoppers (18, 85). Multiple-year per site studies indicated that populations of planthoppers and leafhoppers in *Bt* and non-*Bt* rice fields were similar (15, 17). A recent two-year field trial indicated that *cry1Ab/Ac Bt* Shanyou 63 rice harbored higher planthopper populations than did non-*Bt* rice at the late growth stage of rice, although not at early and middle growth stages (94). However, this result may have been caused by migration from nearby non-*Bt* rice fields where non-*Bt* rice leaf tissues were severely damaged by rice stem borers and leaffolders. A field study with six *Bt* rice lines indicated that *Bt* rice posed no risk of causing higher *Stenchaetothrips biformis* (Thysanoptera: Thripidae) populations in the field compared with non-*Bt* rice (1).

The nontarget effects on storage pests have also been studied. IRGM rice grains did not cause negative effects on four nonlepidopteran storage pests but did result in less damage by *Sitotroga cerealella* (Lepidoptera: Gelechiidae) compared with non-GM rice grains (10, 11, 51).

Effect of IRGM Rice Pollen on the Silkworm

The culture of silkworms, *Bombyx mori* (Lepidoptera: Bombycidae), or sericulture, has a long history in China. Silkworm larvae feed exclusively on fresh mulberry leaves. In southeast China, mulberry trees are generally planted near or around the edges of paddy fields, so-called mulberry-rice mixed cropping (30). Thus, mulberry leaves could be contaminated with IRGM rice pollen. Because silkworms and stem borers belong to the same order, this could be problematic. Fan et al. (30) reported that rice pollen escaped to nearby mulberry trees and contaminated mulberry leaves with an average concentration of 93 pollen grains per cm^2, which was slightly lower than a threshold concentration of 109 *Bt* rice pollen grains per cm^2, under which development of silkworm larvae could be negatively affected.

Different effects of IRGM rice lines expressing different Cry proteins on silkworm larvae under laboratory conditions have been reported; these effects might be due to a different insecticidal spectrum of the Cry protein or to different expression levels in rice pollen (95, 96, 113). Yao et al. (110) reported that *Bt* rice line TT9-3 expressing a *cry1Ab/Ac* gene had no significant adverse effects on young silkworm larvae, even after the neonates had been exposed to *Bt* pollen at the highest density of 3395.0 grains per cm^2 for 48 h. Such pollen density is more than twofold greater than the highest pollen density on mulberry leaves, 1635.9 grains per cm^2, naturally occurring in the field. However, in a worst-case-scenario laboratory feeding bioassay, pollen from *cry1Ab* rice lines (KMD1 and B1) was toxic to silkworm larvae and caused pathological midgut changes (109).

The data suggest that some IRGM rice pollen may be toxic and therefore a hazard to silkworm larvae. However, risk is a function of hazard $\times$ exposure (77), and the deposition of rice pollen on mulberry leaves is very limited under field conditions and appears to pose minimal risk to silkworms (110). Furthermore, because the silkworm has been completely domesticated, routine colony maintenance practices (including leaf cleaning prior to feeding) dramatically decrease the amount of IRGM rice pollen on mulberry leaves. This, in conjunction with some environmental factors in southeast China (such as more rainfall) and the physical requirement of temporal and spatial overlap between silkworm larval occurrence and plant anthesis, makes IRGM rice pollen negligible on silkworm (109).

Soil Biota

Soil-dwelling detritivores, such as collembolans, play an important role in rice ecosystems (36, 37). Bai et al. (4) found that Cry1Ab could be detected in *Entomobrya griseoolivata* (Collembola: Entomobryidae) feeding on *Bt* rice tissue in the laboratory. However, field studies indicated the populations of common collembolan families (e.g., Scatopsidae, Sminthuridae, and Tomoceridae) and detritivorous dipteran families (e.g., Ceratopogonidae, Mycetophilidae, Phoridae, and Psychodidae) were similar in *Bt* and non-*Bt* rice fields (6, 58).

Wang et al. (91) found that degradation of Cry1Ab from *Bt* rice occurred in soils under aerobic conditions with half-lives ranging from 19.6 to 41.3 days. However, under water-flooded conditions, the half-life of Cry1Ab was prolonged to 45.9–141 days, indicating that soil microbial organisms may be exposed to Cry proteins for longer periods in flooded *Bt* rice fields than in a dry *Bt* cotton or *Bt* corn field. Under laboratory conditions, *Bt* rice straw could significantly increase the number of hydrolytic-fermentative and anaerobic nitrogen-fixing bacteria in flooded paddy soil (106). However, the numbers of anaerobic fermentative bacteria, denitrifying bacteria, hydrogen-producing acetogenic bacteria, methanogenic bacteria, and colony-forming units of culturable bacteria and actinomycetes were similar between soil amended with *Bt* rice straw and non-*Bt* rice straw (70, 104). Yang et al. (107) identified 303 bacteria strains belonging to 20 genera from two *cry1Ab* rice fields and one non-*Bt* isoline rice field, with no statistical differences in Shannon-Wiener, Simpson, and Pielou indexes for the total bacterial community among the three rice fields. Based on published laboratory and field studies of 1–2 years, the results to date indicate that IRGM rice has not caused significant changes to soil biota in China.

Outcrossing of Insect Resistance Transgenes

In general, cultivated rice is primarily self-pollinating with very little cross-pollination between GM and non-GM rice cultivars (crop-to-crop) under field conditions (59). Field studies with *Bt+CpTI* rice lines indicated that risk of pollen-mediated crop-to-crop gene flow from IRGM rice to non-GM-cultivated rice in China is at a manageable level (73, 75). However, transgenic outcrossing from IRGM rice varieties to weedy rice and wild species (e.g., *Oryza rufipogon* and *O. nivara*) (crop-to-wild species) could occur at a higher level because these plants are present in and around cultivated rice fields where pollen from cultivated rice is at a high level (59). Pollen from cultivated rice can fertilize weedy rice (14) and *O. rufipogon* (14, 83) and produce fertile progeny. The rate of outcrossing declines rapidly with distance, but weedy and wild rice could occur within and around rice fields (59). Song et al. (81, 83) found that the maximum frequency of gene flow from cultivated rice to adjacent wild rice was less than 3%. Some fitness costs (producing fewer seeds) of the F1 hybrid of cultivated rice and wild rice could reduce the rate of transgene introgression into wild populations (82).

A recent field study (12) in China compared the field performances of three weedy rice strains and their six F1 hybrids with two IRGM rice lines (*CpTI* and *Bt+CpTI*), and the results indicated an enhanced relative performance of the crop-weed hybrids (e.g., taller plants and more tillers and panicles). Such results call for careful evaluation of the potential consequence of crop-to-wild gene outcrossing. Because seed markets in China are still not fully developed, seed trading is common in households, counties, and provinces. This suggests that seed-mediated gene flow should also be closely evaluated.

To better control gene flow from IRGM rice, Lin et al. (55) developed a built-in strategy for containing transgenes in GM rice. In addition, Rong et al. (74) recently constructed a model that takes into account the outcrossing rates of recipients and cross-compatibility between rice and its wild relatives to better predict pollen-mediated crop-to-wild gene flow. Based on such studies, strategies such as developing future IRGM rice lines with limited or no gene flow or releasing lines in areas where wild rice is absent can be incorporated into management programs that reduce the risk of outcrossing.

Food Safety Assessment for IRGM Rice

Food safety of IRGM rice in China has been studied primarily on the basis of the principle of substantial equivalence, the method used in the United States. Various biochemical methods have been used to compare the nutritional components of IRGM and non-GM parental rice grains. In addition, feeding experiments have been conducted on small animals. No significant differences in major nutritional components (e.g., crude protein, crude lipid, free amino acids, and mineral elements) and physicochemical properties (e.g., amylose content, alkali spreading value, and starch viscosity) were found between *cry1Ab* or *CpTI+cry1Ac* rice lines and their non-GM counterparts (54, 99). However, a compositional difference (three amino acids, two fatty acids, and two vitamins) between disease-resistant and insect-resistant GM rice grains and non-GM controls was recently reported (49). A 90-day laboratory feeding test on rats indicated that *cry1Ab* rice flour had no effects on the development of the rats. Necropsy indicated that neither pathological lesions nor histopathological abnormalities were present in liver, kidneys, and intestines of rats in either the IRGM rice group or the non-GM rice group (97). Similar results were reported for mice (21), rats (20, 124), and pigs (38) fed *CpTI* rice grains. Different proactive measures, such as cooking (99) or gamma irradiation (100), have also been tested to further reduce insecticidal proteins in IRGM rice grains.

A recent laboratory study indicated that *cry1Ab* rice could accumulate more heavy metals in grain and straw than non-*Bt* rice could (90). This study calls for more attention to IRGM rice food safety in areas with heavy-metal pollution, which is not uncommon in China (90).

SOCIOECONOMIC IMPACTS OF IRGM RICE

Yield and Gross Income

The rice-growing area in China has been decreasing in the last decade due to a lack of agricultural labor (migration into cities), water shortages, and poor profitability of rice production (105), which is in striking contrast to the increasing trend of growing cotton (**Supplemental Figure 2**). From 2002 to 2004, two IRGM rice lines (*cry1Ab/Ac Bt* Shanyou 63 and *CpTI* GM II-youming 86), as a part of preproduction trials, were tested at 17 villages located in eight different counties in Hubei and Fujian Provinces. A three-year survey was conducted with rice farmers to address whether IRGM rice could increase rice yields, reduce insecticide use, and increase farmers' incomes by adopting the new technology.

In a small-scale field trial conducted in Hubei Province in 1999, no insecticides were applied to both *Bt* and non-*Bt* rice fields, and the former rice field yielded 29% more rice (88). In the preproduction field trials in Hubei and Fujian Provinces, based on a survey of 330 households, *Bt* rice increased yield by up to 9% compared with non-*Bt* rice (42, 43). IRGM rice farmers spent only 31 yuan per hectare per season on insecticides (US$4.56), whereas non-GM rice farmers spent 243 yuan per hectare per season (US$35.74). Moreover, 3–11% of non-GM rice farmers reported insecticide poisonings, whereas there were no such reports from IRGM rice farmers (41, 43). Recently, a two-year field trial with *cry1Ab/Ac Bt* Shanyou 63 in Wuhan indicated that *Bt* rice could increase rice yield by 60–65% compared with non-*Bt* rice without insecticide applications (94). Clearly, IRGM rice will help rice farmers save on labor and insect control costs and increase their profit.

Insecticide Use

In the preproduction trials in Hubei and Fujian Provinces based on over 500 individual fields (GM and non-GM rice), IRGM rice farmers applied 0.6 insecticide applications per season, while non-GM rice farmers applied 3.7 applications per season (41–43). On a per hectare basis, 3.0 kg of insecticides were used on IRGM rice, which starkly contrasts with 23.5 kg of insecticides used on non-GM rice, and IRGM rice produced a higher yield (6688 kg ha^{-1}) than non-GM rice (6457 kg ha^{-1}). Recent field trials indicated that *cry1Ab/Ac* rice could reduce insecticide applications up to 60% compared with the non-*Bt* control rice (94). Reduced insecticide use from adopting IRGM rice has been clearly demonstrated in China (41–43) and is similar to trends reported for cotton and corn (9).

Impact of IRGM Rice on Farmers' Pest Management Practices

Preproduction trials on IRGM rice indicated that IRGM rice could substantially reduce insecticide use while increasing rice yield (42, 43, 94), which may lead to a change in farmers' attitudes regarding insecticide use, reduce unnecessary insecticide use, and result in increased profits. However, the impact of IRGM rice on farmers' pest management practices may be more complicated. Current IRGM rice lines developed in China are primarily first-generation biotech crops with one insecticidal gene. Even if two genes (e.g., *CpTI+Cry1Ac* rice) are used, both genes are often targeting the same group of lepidopteran pests (stem borers and leaffolders) (**Supplemental Table 1**). Although other insecticidal genes with different pest spectrums, such as *gna*, were also used for IRGM rice, unsatisfactory control of the target planthopper species made them less desirable for commercial release. Considering the complex of rice pest species in China, enhanced varieties are needed to address the other insects and pest organisms, including planthoppers. This is similar to the situation with *Bt* cotton in China, which effectively suppressed the target pest, cotton bollworm, but required increasing amounts of insecticide applications for sucking insect species (116). Taking into account the additional cost that an IRGM rice farmer needs to pay for higher-priced seed and control of sucking insect pests, the impact of current IRGM rice lines on farmers' pest management practices in China needs careful evaluation. Developing future IRGM rice lines with stacked traits targeting multiple groups of insect pests will likely have a more profound effect on farmers' pest management practices. In addition, proper training and education in agricultural biotechnology and IPM will be crucial to achieve a positive impact on Chinese rice farmers' pest management practices.

INSECTICIDE RESISTANCE MANAGEMENT

Challenges to IRM for IRGM Rice

Although transgenic plants offer many unique opportunities for the management of pest populations, one major concern regarding long-term use of IRGM plants is the potential for insect resistance (7, 35). Among the various options that have been considered for insecticide resistance management (IRM) for IRGM crops, especially *Bt* crops, the high dose/refuge and gene pyramiding strategies have strong theoretical support (33, 35, 118) and have been broadly implemented in the United States, Canada, and Australia. After more than a decade of widespread IRGM crops, there have only been two clear-cut cases of resistance involving *cry1F and cry1Ab* corn (62, 84). However, the few reported resistance incidences in IRGM crops does not mean the failure of the high dose/refuge strategy; instead, insecticidal proteins not expressed at a high dose level in the IRGM crops, plus an insufficient refuge area, are probably the key reasons (84).

IRM: insecticide resistance management

The high dose/refuge strategy calls for high expression of insecticidal protein in IRGM plants. The Bt protein expression level in *Bt* rice is much lower than that in *Bt* corn (88), and most of the current IRGM rice lines developed in China could not achieve 100% kill of late instars of target pests (18). Hu et al. (40) reported that the mortality of first to sixth instar *C. suppressalis* after feeding on *Bt+CpTI* rice for 7 days was 89.6, 87.1, 72.4, 50, 26, and 0%, respectively. Having <99% mortality is not ideal and can lead to a more rapid evolution of resistance, as was seen with *Spodoptera frugiperda* (Lepidoptera: Noctuidae) (84).

On-farm refuges are not required for *Bt* cotton in China because its principal target pest *Helicoverpa armigera* is highly polyphagous and natural refuges can function as unstructured refuges for this pest (102, 120). It is not clear whether there will be a refuge requirement for IRGM rice, but there are no significant alternative wild or cultivated host plants to serve as natural refuges for rice stem borers in most rice-growing regions (28). Thus, it will be difficult to have a highly effective IRM program for IRGM rice in China based on natural refuges alone. In addition, a mixture of single-gene and dual-gene IRGM rice lines is currently under various testing stages in China, and this may result in sequential or concurrent planting of single-gene and dual-gene IRGM rice lines in the fields, a practice that may further challenge IRM for IRGM rice in China (118).

Lastly, in rice fields most stem borer larvae move from plant to plant (29). Such movement can decrease the dose to which pests are exposed and decrease the effective size of refuges in a seed mix IRM strategy (33, 78, 86). Due to routine farming practices such as planting and trading self-kept rice seeds by rice farmers, deliberate or inadvertent mixing of IRGM and non-GM rice seeds could occur and challenge the sustainable use of IRGM rice in China.

Options to Increase Durability

Rice stem borers, the key target pests of the current IRGM rice lines in China, have multiple generations per year [e.g., up to seven generations for *Sesamia inferens* (Noctuidae) and *Scirpophaga incertulas* in rice region I]. Moreover, the number of generations of stem borers per year on rice in China has increased owing to a warming climate (60). Insecticidal proteins could become ineffective within a few years of deployment in IRGM rice unless a proper IRM program is deployed (27). For instance, under laboratory conditions, after 17 generations of selection on *CpTI* rice plants, the percentage striped stem borers that survived on *CpTI* rice plants increased from 10.7% to 42.7% (122). To help rice farmers fully benefit from IRGM rice, the following practical actions need to be considered.

First, for the longer term it is vital to develop and release IRGM rice lines with pyramided insecticidal genes in which each gene has a different mode of action. IRGM crops with pyramided genes require smaller refuges than do one-toxin lines (7, 76) and are more durable (119). Aside from *CpTI+cry1Ac* and *gna+SbTI* rice, more IRGM rice lines with pyramided genes in China are being actively investigated, including *cry1Ab+cry1C*, *cry1Ab+cry2A*, *cry1Ac+cry1C*, *cry1Ac+cry2A*, *cry2A+cry1C*, and *cry1Ab+vip3H+G6-epsps* (13, 31). These pyramided lines will certainly benefit IRM programs in the future.

Second, given the difficulty of implementing structured refuges in China, regulatory authorities may concurrently release GM rice lines with different traits (e.g., insect resistant, disease resistant, herbicide resistant, and drought tolerant). Thus, disease-resistant or herbicide-resistant GM rice can serve as refuges for IRGM rice while saving farmers additional costs for disease or weed control. In addition, ensuring adequate seed supplies of popular non-GM varieties may help maintain a certain amount of non-GM rice in the field (27).

Finally, developing education programs on agricultural biotechnology and basic understanding of IPM and IRM for rice farmers will certainly help achieve a sustainable use of IRGM rice. However, because a large number of Chinese farmers are illiterate, on-site or audio-visual interactions will be essential.

FUTURE DIRECTIONS AND RESEARCH NEEDS

In the past 20 years, a tremendous amount of research has been conducted on IRGM rice; however, to meet the demand for food from the increasing population in China and to fully benefit from the technology, knowledge gaps on IRGM rice need to be better understood and it should be recognized that additional challenges are yet to come.

With the majority of first-generation IRGM rice lines targeting stem borers, urgent attention should be given to identifying new insecticidal genes with different modes of action targeting different groups of insect pests. Furthermore, emphasis should be placed on developing rice lines with pyramided genes for IRM and GM lines with stacked traits to battle the various rice yield constraints in the field. The new 26 billion yuan (US$3.5 billion) R&D initiative on GM plants in China is helping Chinese scientists work on these aspects (19). Environmental and food safety assessments have been conducted primarily on *Bt* rice in comparison with other IRGM rice (e.g., *gna* rice and *gna+SbTI* rice), but research is needed on other non-*Bt* insecticidal genes that are or will be used in IRGM rice lines because some of them have broader insecticidal spectrums than *cry* genes. In addition, some studies have been conducted with unfocused research objectives and unclear hypotheses that have little use in risk assessments. An appropriate evaluation approach, such as tier-based risk assessment on nontarget organisms (71) and proper species selection (69), is needed to optimize risk assessments for IRGM rice. Strengthening regulations on IRGM rice seed distribution to prevent seed-mediated gene flow and studying its potential ecological and social impacts are urgently needed.

Rice is of great cultural importance throughout Asia and is the predominant staple food for over 3 billion people worldwide. Near and long-term effects of commercial release of IRGM rice in China on the rice trade among different countries will be profound. Although it is clear that Chinese farmers and the Chinese public will benefit from IRGM rice that decreases environmental and food safety risks, the release of IRGM rice may affect rice exports from China to some trading partners (27). China is in a difficult position of balancing its own production needs with the evolving regulations of international trade of GM crops. However, as two biosafety certificates for commercial production of *cry1Ab/Ac Bt* Shanyou 63 and Huahui No. 1 have been issued, this suggests that China sees IRGM rice as an important part of the future.

SUMMARY POINTS

1. China is the largest producer and consumer of rice in the world, with a >7,000-year history of rice cultivation and six rice-growing regions.
2. There are over 200 insect pests of rice in China. Stem borers and planthoppers are the two major groups of insects that cause losses in rice production totaling billions of yuan annually.
3. In 1989, scientists from the CAAS generated the first *Bt* rice line, and the first field tests of insect-resistant genetically modified rice took place in 1998 in China.
4. On October 22, 2009, China's Ministry of Agriculture issued its first two biosafety certificates for commercial production of *Bt* rice (*cry1Ab/Ac* Huahui No. 1 and *cry1Ab/Ac Bt* Shanyou 63) for Hubei Province.
5. A new generation of GM rice with pyramided genes and stacked traits is needed in China to battle the complex of insect pests of rice and other yield constraints in rice.
6. Effective IRM strategies for IRGM rice are needed to sustain their effectiveness and continued benefits.
7. Education programs on agricultural biotechnology and basic understanding of IPM and IRM will help achieve a positive impact on rice farmers' pest management practices and sustainable use of IRGM rice.

DISCLOSURE STATEMENT

The authors are not aware of any affiliations, memberships, funding, or financial holdings that might be perceived as affecting the objectivity of this review.

ACKNOWLEDGMENTS

We are grateful to Dr. J.Z. Zhao and Ms. H.L. Collins for their helpful comments on earlier drafts of the manuscript and Dr. J.C. Tian for his help with graphics. We also acknowledge the financial support from the National Program on Key Basic Research Projects (973 Program, 2007CB109202), the Ministry of Science and Technology of China, the Special Research Projects for Developing Transgenic Plants (2008ZX08011-01), and the National Natural Science Foundation of China (30671377).

LITERATURE CITED

1. Akhtar ZR, Tian JC, Chen Y, Fang Q, Hu C, et al. 2010. Impacts of six *Bt* rice lines on nontarget rice feeding thrips under laboratory and field conditions. *Environ. Entomol.* 39:715–26
2. Bai YY, Jiang MX, Cheng JA. 2005. Effects of transgenic *cry1Ab* rice pollen on the oviposition and adult longevity of *Chrysoperla sinica* Tjeder. *Acta Phytophylacica Sin.* 32:225–30
3. Bai YY, Jiang MX, Cheng JA. 2005. Effects of transgenic *cry1Ab* rice pollen on fitness of *Propylea japonica* (Thunberg). *J. Pest Sci.* 78:123–28
4. Bai YY, Jiang MX, Cheng JA. 2005. Impacts of transgenic *cry1Ab* rice on two collembolan species and predation of *Microvelia horvathi* (Hemiptera: Veliidae). *Acta Entomol. Sin.* 48:42–47
5. Bai YY, Jiang MX, Cheng JA, Wang D. 2006. Effects of Cry1Ab toxin on *Propylea japonica* (Thunberg) (Coleoptera: Coccinellidae) through its prey, *Nilaparvata lugens* Stål (Homoptera: Delphacidae), feeding on transgenic Bt rice. *Environ. Entomol.* 35:1130–36
6. Bai YY, Jiang MX, Cheng JA, Wang D. 2006. Effects of transgenic Bt *cry1Ab* rice on collembolan population in paddy field. *Chin. J. Appl. Ecol.* 17:903–6

6a. Bates SL, Cao J, Zhao JZ, Earle ED, Roush RT, Shelton AM. 2005. Evaluation of a chemically inducible promoter for developing a within-plant refuge for resistance management. *J. Econ. Entomol.* 98:2188–94

7. Bates SL, Zhao JZ, Roush RT, Shelton AM. 2005. Insect resistance management in GM crops: past, present, and future. *Nat. Biotechnol.* 23:57–62
8. Baum JA, Bogaert T, Clinton W, Heck GR, Feldmann P, et al. 2007. Control of coleopteran insect pests through RNA interference. *Nat. Biotechnol.* 10:1038–59
9. Brookes G, Barfoot P. 2009. GM crops: global socio-economic and environmental impacts 1996–2007. *May 2009 Rep.* Dorchester, UK: PG Economics Ltd.
10. Cai WL, Shi SB, Yang CJ, Peng YF, Quan MG. 2006. The effect of transgenic *Bt* rice grain on the several main storage pests. *Biotechnol. Bull.* 22(Suppl. 1):268–71
11. Cai WL, Zhang HY, Yang S, Yang CJ, Hua HX, Peng YF. 2008. Impact of *Bt* rice grain on the development of red flour beetle, *Tribolium castaneum* (Coleoptera: Tenebrionidae). *Acta Phytophylacica Sin.* 35:471–72
12. Cao QJ, Xia H, Yang X, Lu BR. 2009. Performance of hybrids between weedy rice and insect-resistant transgenic rice under field experiments: implication for environmental biosafety assessment. *J. Integr. Plant Biol.* 51:1138–48
13. Chen H, Lin YJ, Zhang QF. 2009. Review and prospect of transgenic rice research. *Chin. Sci. Bull.* 54:4049–68
14. Chen LJ, Lee DS, Song ZP, Suh HS, Lu BR. 2004. Gene flow from cultivated rice (*Oryza sativa*) to its weedy and wild relatives. *Ann. Bot.* 93:67–73
15. Chen M, Liu ZC, Ye GY, Shen ZC, Hu C, et al. 2007. Impacts of transgenic *cry1Ab* rice on non-target planthoppers and their main predator *Cyrtorhinus lividipennis* (Hemiptera: Miridae)—a case study of the compatibility of Bt rice with biological control. *Biol. Control* 42:242–50
16. Chen M, Ye GY, Liu ZC, Fang G, Hu C, et al. 2009. Analysis of Cry1Ab toxin bioaccumulation in a food chain of Bt rice, an herbivore and a predator. *Ecotoxicology* 18:230–38
17. Chen M, Ye GY, Liu ZC, Yao HW, Chen XX, et al. 2006. Field assessment of the effects of transgenic rice expressing a fused gene of *cry1Ab* and *cry1Ac* from *Bacillus thuringiensis* Berliner on non-target planthoppers and leafhoppers. *Environ. Entomol.* 35:127–34
18. **Chen M, Zhao JZ, Ye GY, Fu Q, Shelton AM. 2006. Impact of insect-resistant transgenic rice on target insect pests and non-target arthropods in China. *Insect Sci.* 13:409–20**

18. Summarizes the laboratory and field studies with IRGM rice in China on target insect pests and nontarget arthropods.

19. Chen X, Jia HP. 2008. China plans $3.5 billion GM crops initiative. *Science* 321:1279
20. Chen XP, Zhuo Q, Gu LZ, Piao JH, Yang XG. 2004. The nutritional evaluation of transgenic rice. *Acta Nutr. Sin.* 26:199–23
21. Chen XP, Zhuo Q, Piao JH, Yang XG. 2004. Immunotoxicologic assessment of transgenetic rice. *J. Hyg. Res.* 33:77–80
22. Chen Y, Deng YM. 2008. Field survey on rice yield losses caused by rice leafroller. *Anhui Agric. Bull.* 14:173–74
23. **Cheng JA, ed. 1996. *Insect Pests of Rice*. Beijing: Chin. Agric. Publ. Press. 213 pp.**

23. Lists insect pests of rice in China and their economic importance.

24. Cheng JA. 2009. *Ricehopper problems intensify in China*. **http://ricehoppers.net/2009/01/01/ricehopper-problems-intensify-in-china/**
25. Cheng JA, Zhu ZR. 2006. Analysis on the key factors causing the outbreak of brown planthopper in Yangtze Area, China in 2005. *Plant Prot.* 32:1–4
26. Cheng SH, Cao LY, Zhuang JY, Chen SG, Zhan XD, et al. 2007. Super hybrid rice breeding in China: achievements and prospects. *J. Integr. Plant Biol.* 49: 805–10
27. **Cohen M, Chen M, Bentur JS, Heong KL, Ye GY. 2008. *Bt* rice in Asia: potential benefits, impact, and sustainability. In *Integration of Insect-Resistant GM Crops within IPM Programs*, ed. J Romeis, AM Shelton, G Kennedy, pp. 223–48. Dordrecht, The Nether.: Springer**

27. Introduces research and development on *Bt* rice in Asia.

28. Cuong NL, Cohen MB. 2002. Field surveys and greenhouse evaluation of non-rice host plants of the striped stem borer, *Chilo suppressalis* (Lepidoptera: Pyralidae), as refuges for resistance management of rice transformed with *Bacillus thuringiensis* toxin genes. *Bull. Entomol. Res.* 92:265–68
29. Dirie AM, Cohen MB, Gould F. 2000. Larval dispersal and survival of two stem borer (Lepidoptera: Crambidae) species on *cry1Ab*-transformed and non-transgenic rice. *Environ. Entomol.* 29:972–78
30. Fan LJ, Wu YY, Pang HQ, Wu JG, Shu QY, et al. 2003. Bt rice pollen distribution on mulberry leaves near ricefields. *Acta Ecol. Sin.* 23:202–9
31. Fang J. 2008. *The vegetative insecticidal protein genes of Bacillus thuringiensis and expression in transgenic rice*. PhD thesis. Zhejiang Univ., Hangzhou, China. 86 pp.
32. FAO. 2009. *FAOSTAT*. **http://faostat.fao.org/site/567/default.aspx**
33. Ferré J, Van Rie J, MacIntosch SC. 2008. Insecticidal genetically modified crops and insect resistance management. In *Integration of Insect-Resistant GM Crops within IPM Programs*, ed. J Romeis, AM Shelton, G Kennedy, pp. 41–86. Dordrecht, The Nether.: Springer
34. Gong ZT, Chen HZ, Yuan DG, Zhao YG, Wu YJ, Zhang GL. 2007. The temporal and spatial distribution of ancient rice in China and its implications. *Chin. Sci. Bull.* 52:1071–79
35. Gould F. 1998. Sustainability of transgenic insecticidal cultivars-integrating pest genetics and ecology. *Ann. Rev. Entomol.* 43:701–26
36. Guo YJ, Wang NY, Jiang JW, Chen JW, Tang J. 1995. Ecological significance of neutral insects as nutrient bridge for predators in irrigated rice arthropod community. *Chin. J. Biol. Control* 11:5–9
37. Haimi J. 2000. Decomposer animals and bioremediation of soils. *Environ. Pollut.* 107:233–38
38. Han JH, Yang YX, Men JH, Bian LH, Yang XG, Zhu YZ. 2004. The comparison of ileal digestibility of protein and amino acids in GM rice and parental rice. *Acta Nutr. Sin.* 26:362–65
39. Herdt RW. 1991. Research priorities for rice biotechnology. In *Rice Biotechnology*, ed. GS Khush, GH Toenniessen, pp. 19–54. Los Banos, The Philippines: Wallingford/Int. Rice Res. Inst., CAB Int.
40. Hu YQ, Li YR, Zheng Y, Hu XB, Zhang MJ, Li BJ. 2005. Insecticidal activity of transgenic rice expressing *CpTI* or *CpTI+Bt* to *Chilo suppressalis* Walker. *J. Fujian Agric. Forest. Uni. (Nat. Sci.)* 34:185–88
41. Huang JK, Hu RF. 2007. Impacts of GM rice on rice farmers. *J. Agric. Sci. Technol.* 9:13–17
42. **Huang JK, Hu RF, Rozelle S, Pray C. 2005. Insect-resistant GM rice in farmers' fields: assessing productivity and health effects in China. *Science* 308:688–90**

42. First comprehensive survey of the performance of IRGM rice in the field in China based on preproduction testing.

43. Huang JK, Hu RF, Rozelle S, Pray C. 2008. Genetically modified rice, yields, and pesticides: assessing farm-level productivity effects in China. *Econ. Dev. Cult. Change* 56:241–63
44. Huang JK, Wang QF. 2002. Agricultural biotechnology development and policy in China. *AgBioForum* 5:122–35
45. James C. 2009. China approves biotech rice and maize in landmark decision. *Crop Biotech Update, 4 Dec. 2009*. **http://www.isaaa.org/kc/cropbiotechupdate/online/default.asp?Date=12/4/2009**
46. Jiang XH, Zhang QH, Hu SM, Xie SJ, Xu SG. 2001. The status of pesticide resistance of rice stalk borer in Zhejiang Province and their management tactics. *Plant Prot. Technol. Ext.* 21:27–29
47. Jiang YH, Fu Q, Cheng JA, Ye GY, Bai YY, Zhang ZT. 2004. Effects of transgenic *Bt* rice on the biological characteristics of *Apanteles chilonis* (Munakata) (Hymenoptera: Braconidae). *Acta Entomol. Sin.* 47:124–29
48. Jiang YH, Fu Q, Cheng JA, Zhu ZR, Jiang MX, et al. 2005. Effect of transgenic *sck+cry1Ac* rice on the survival and growth of *Chilo suppressalis* (Walker) and its parasitoid *Apanteles chilonis* (Munakata). *Acta Entomol. Sin.* 48:554–60

49. Jiao Z, Si XX, Li GK, Zhang ZM, Xu XP. 2010. Unintended compositional changes in transgenic rice seeds (*Oryza sativa* L.) studied by spectral and chromatographic analysis coupled with chemometrics methods. *J. Agric. Food Chem.* 58:1746–54
50. Li FF, Ye GY, Wu Q, Peng YF, Chen XX. 2007. Arthropod abundance and diversity in *Bt* and non-*Bt* rice fields. *Environ. Entomol.* 36:646–54
51. Li GT, Cao Y, Ye GY, Liu GL, He KL. 2008. Evaluation of transgenic Bt rice resistance to *Sitotroga cerealella*. *Acta Phytophylacica Sin.* 35:205–8
52. Li LY. 1982. Integrated rice pest control in the Guangdong province of China. *Entomophaga* 27:81–88
53. Li P, Zheng AP, Zhu J, Tan FR, Wang LX, et al. 2009. Bt *cry4Cc1* gene, encode protein and its application. *China Pat. No. CN101,497,658*
54. Li X, Huang K, He X, Zhu B, Liang Z, et al. 2007. Comparison of nutritional quality between Chinese *indica* rice with *sck* and *cry1Ac* genes and its nontransgenic counterpart. *J. Food Sci.* 72:S420–S24
55. **Lin CY, Fang J, Xu XL, Zhao T, Cheng JA, et al. 2008. A built-in strategy for containment of transgenic plants: creation of selectively terminable transgenic rice. *PLoS One* 3:e1818. doi:10.1371/journal.pone.0001818**

55. Introduces a built-in strategy for preventing gene flow from transgenic rice.

56. Liu YF, He L, Wang Q, Hu SQ, Liu WH, et al. 2006. Evaluation of the effects of insect-resistant *cry1Ac/sck* transgenic rice on the parasitoid communities in paddy fields. *Acta Entomol. Sin.* 49:955–62
57. Liu ZC, Ye GY, Hu C, Datta SK. 2002. Effects of *Bt* transgenic rice on population dynamics of main non-target insect pests and dominant spider species in rice paddies. *Acta Phytophylacica Sin.* 29:138–44
58. Liu ZC, Ye GY, Hu C, Datta SK. 2003. Impact of transgenic *indica* rice with a fused gene of *cry1Ab/1AC* on the rice paddy arthropod community. *Acta Entomol. Sin.* 46:454–65
59. Lu BR. 2008. Potential commercialization of genetically modified rice in China: key questions for environmental biosafety assessments. *J. Agric. Biotechnol.* 16:547–54
60. Lu ZQ, Du YZ, Zhou FC, Chen LF, Wu YF, et al. 2005. Dynamics of the rice borers, *Chilo suppressalis* (Walker) and *Tryporyza incertula* (Walker) and some problems involved in evidence-based control against the rice borers. *Plant Prot.* 31:47–51
61. Mao YB, Cai WJ, Wang JW, Hong GJ, Tao XY et al. 2007. Silencing a cotton bollworm P450 monooxygenase gene by plant-mediated RNAi impairs larval tolerance of gossypol. *Nat. Biotechnol.* 25:1307–13
62. Matten SR, Head GP, Quemada HD. 2008. How government regulations can help or hinder the integration of Bt crops within IPM programs. In *Integration of Insect-Resistant Genetically Modified Crops within IPM Programs*, ed. J Romeis, AM Shelton, GG Kennedy, pp. 27–40. Dordrecht, The Nether.: Springer
63. Matteson PC. 2000. Insect pest management in tropical Asian irrigated rice. *Annu. Rev. Entomol.* 45:549–74
64. Naranjo SE. 2009. Impacts of *Bt* crops on non-target organisms and insecticide use patterns. *CAB Rev. Perspect. Agric. Vet. Sci. Nutr. Nat. Resour.* 4:No.011 (doi: 10.1079/PAVSNNR20094011)
65. NBSC (Natl. Bur. Statistics China). 2007. *China Rural Statistical Yearbook 2007*. Beijing: China Stat. Press
66. NBSC (Natl. Bur. Statistics China). 2008. *China Statistical Yearbook 2008*. Beijing: China Stat. Press
67. Pan CX. 1988. The development of integrated pest control in China. *Agric. Hist.* 62:1–12
68. Peng Y, Chen CK, Han ZJ, Wang YC. 2001. Resistance measurement of *Chilo suppressalis* from Jiangsu Province and its resistance mechanism to methamidophos. *Acta Phytophylacica Sin.* 28:173–77
69. Prasifka JR, Hellmich RL, Dively GP, Higgins LS, Dixon PM, Duan JJ. 2007. Selection of nontarget arthropod taxa for field research on transgenic insecticidal crops: using empirical data and statistical power. *Environ. Entomol.* 37:1–10
70. Ren X, Wu WX, Ye QF, Chen YX, Thies JE. 2004. Effect of *Bt* transgenic rice straw on the bacterial community composition in a flooded paddy soil. *Acta Sci. Circumst.* 24:871–78
71. **Romeis J, Bartsch D, Bigler F, Candolfi MP, Gielkens MMC, Hartley SE, et al. 2008. Assessment of risk of insect-resistant transgenic crops to nontarget arthropods. *Nat. Biotechnol.* 26:203–8**

71. Introduces tier-based risk assessments for nontarget organisms for IRGM crops.

72. Romeis J, Meissle M, Bigler F. 2006. Transgenic crops expressing *Bacillus thuringiensis* toxins and biological control. *Nat. Biotechnol.* 24:63–71
73. Rong J, Lu BR, Song ZP, Su J, Snow AA, et al. 2006. Dramatic reduction of crop-to-crop gene flow within a short distance from transgenic rice fields. *New Phytol.* 173:346–53

74. Rong J, Song ZP, de Jong TJ, Zhang XS, Sun SG, et al. 2010. Modelling pollen-mediated gene flow in rice: risk assessment and management of transgene escape. *Plant Biotechnol. J.* 8:1–13
75. Rong J, Song ZP, Su J, Xia H, Lu BR, Wang F. 2005. Low frequency of transgene flow from *Bt/CpTI* rice to its nontransgenic counterparts planted at close spacing. *New Phytol.* 168:559–66
76. Roush RT. 1997. Two-toxin strategies for management of insect resistant transgenic crops: Can pyramiding succeed where pesticide mixtures have not? *Philos. Trans. R. Soc. Lond. Biol.* 353:1777–86
77. Sears MK, Hellmich RL, Stanley-Horn DE, Oberhauser KS, Pleasants JM, et al. 2001. Impact of *Bt* corn pollen on monarch butterfly populations: a risk assessment. *Proc. Natl. Acad. Sci. USA* 98:12328–30
78. Shelton AM, Tang JD, Roush RT, Metz TD, Earle ED. 2000. Field tests on managing resistance to *Bt*-engineered plants. *Nat. Biotechnol.* 18:339–42
79. Sheng CF, Wang HT, Gao LD, Xuan JW. 2003. The occurrence status, damage cost estimate and control strategies of stem borers in China. *Plant Prot.* 29:37–39
80. Sheng CF, Wang HT, Sheng SY, Gao LD, Xuan JW. 2003. Pest status and loss assessment of crop damage caused by the rice borers, *Chilo suppressalis* and *Tryporyza incertulas* in China. *Entomol. Knowl.* 40:289–94
81. Song ZP, Lu BR, Chen JK. 2004. Pollen flow of cultivated rice measured under experimental conditions. *Biodivers. Conserv.* 13:579–90
82. Song ZP, Lu BR, Wang B, Chen JK. 2004. Fitness estimation through performance comparison of F1 hybrids with their parental species *Oryza rufipogon* and *O. sativa*. *Ann. Bot.* 93:311–16
83. Song ZP, Lu BR, Zhu YG, Chen JK. 2003. Gene flow from cultivated rice to the wild species *Oryza rufipogon* under experimental field conditions. *New Phytol.* 157:657–65
84. Tabashnik BE, Van Rensburg JBJ, Carrière Y. 2009. Field-evolved insect resistance to *Bt* crops: definition, theory and data. *J. Econ. Entomol.* 102:2011–25
85. Tan H, Ye GY, Shen JH, Peng YF, Hu C. 2006. Effects of transgenic *indica* rice expressing a gene of *cry1Ab* with insect resistance on the development and reproduction of nontarget pest, *Sogatella furcifera* (Homoptera: Delphacidae). *Acta Phytophylacica Sin.* 33:251–6
86. Tang JD, Collins HL, Metz TD, Earle ED, Zhao J, et al. 2001. Greenhouse tests on resistance management of Bt transgenic plants using refuge strategies. *J. Econ. Entomol.* 94:240–47
87. Tian JC, Liu ZC, Yao HW, Ye GY, Peng YF. 2008. Impact of transgenic rice with a *cry*1Ab gene on parasitoid subcommunity structure and the dominant population of parasitoid wasps in rice paddy. *J. Environ. Entomol.* 30:1–7
88. Tu JM, Zhang GA, Datta K, Xu CG, He YQ, et al. 2000. Field performance of transgenic elite commercial hybrid rice expressing *Bacillus thuringiensis* δ-endotoxin. *Nat. Biotechnol.* 18:1101–4
89. Vaeck M, Reynaerts A, Höfte H, Jansens S, De Beuckeleer M, et al. 1987. Transgenic plants protected from insect attack. *Nature* 328:33–37
90. Wang HY, Huang JZ, Ye QF, Wu DX, Chen ZY. 2009. Modified accumulation of selected heavy metals in Bt transgenic rice. *J. Environ. Sci.* 21:1607–12
91. Wang HY, Ye QF, Gan J, Wu LC. 2007. Biodegradation of Cry1Ab protein from *Bt* transgenic rice in aerobic and flooded paddy soils. *J. Agric. Food Chem.* 55:1900–4
92. Wang J, Zhang XQ, Lin YJ. 2008. The study on rice transformation of recombined *Bt* genes derived from DNA shuffling library. *Mol. Plant Breed.* 6:1005–10
93. Wang Y, Johnson S. 2007. The status of GM rice R&D in China. *Nat. Biotechnol.* 25:717–18
94. Wang YM, Zhang GA, Du JP, Liu B, Wang MC. 2010. Influence of transgenic hybrid rice expressing a fused gene derived from *cry1Ab* and *cry1Ac* on primary insect pests and rice yield. *Crop Prot.* 29:128–33
95. Wang ZH, Ni XQ, Xu MK, Shu QY, Xia YW. 2001. The effect on the development of silkworm larvae of transgenic rice pollen with a synthetic *cry1Ab* gene from *Bacillus thuringiensis*. *Hereditas* 23:463–66
96. Wang ZH, Shu QY, Cui HR, Xu MK, Xie XB, Xia YW. 2002. The effect of *Bt* transgenic rice flour on the development of silkworm larvae and the sub-microstructure of its midgut. *Sci. Agric. Sin.* 35:714–18
97. Wang ZH, Yin W, Cui HR, Xia YW, Altosaar I, Shu QY. 2002. Toxicological evaluation of transgenic rice flour with a synthetic *cry1Ab* gene from *Bacillus thuringiensis*. *J. Sci. Food Agric.* 82:738–44
98. Woodburn AT. 1990. The current rice agrochemicals market. In *Proc. Conf. Pest Manag. Rice*, ed. BT Grayson, MB Green, pp. 15–30. London: Elsevier Appl. Sci.

99. Wu DX, Shu QY, Ye QF, Zhang L, Ma CX, Xia YW. 2003. Comparative studies on major nutritional components and physicochemical properties of the transgenic rice with a synthetic *cry1Ab* gene from *Bacillus thuringiensis*. *J. Food Biochem.* 27:295–308
100. Wu DX, Ye QF, Wang ZH, Xia YW. 2004. Effect of gamma irradiation on nutritional components and Cry1Ab protein in the transgenic rice with a synthetic *cry1Ab* gene from *Bacillus thuringiensis*. *Radiat. Phys. Chem.* 69:79–83
101. Wu G, Wu YH, Nie SJ, Zhang L, Ling X, et al. 2009. Real-time PCR method for detection of the transgenic rice event TT51-1. *Food Chem.* 119:417–22
102. Wu KM. 2007. Monitoring and management strategy for *Helicoverpa armigera* resistance to Bt cotton in China. *J. Invertebr. Pathol.* 95:220–23
103. Wu KM, Lu YH, Wang ZY. 2009. Advance in integrated pest management of crops in China. *Chin. Bull. Entomol.* 46:831–36
104. Wu WX, Ye QF, Min H, Chen HE. 2003. Effect of Cry1Ab toxin released from straw of *Bt*-transgenic rice on microflora and enzymatic activities in upland soil. *Acta Pedol. Sin.* 40:606–12
105. Xin LJ, Li XB. 2009. Changes of multiple cropping in double cropping rice area of southern China and its policy implications. *J. Nat. Resour.* 24:58–65
106. Xu XY, Ye QF, Wu WX, Min H. 2004. Incorporation of the straw of transgenic *Bt* rice into the soil on anaerobic microbial populations and enzyme activity in paddy soil. *Plant Nutr. Fertil. Sci.* 10:63–67
107. Yang BJ, Tang J, Jang YZ, Peng YF. 2009. The effects of *crylAb* transgenic rice on the culturable bacterial flora in the rhizospheres. *Acta Ecol. Sin.* 29:3036–43
108. **Yang H, Li JX, Guo SD, Chen XJ, Fan YL. 1989. Transgenic rice plants produced by direct uptake of δ-endotoxin protein gene from *Bacillus thuringenesis* into rice protoplasts. *Sci. Agric. Sin.* 22:1–5**

108. Reports the first IRGM rice transformation in China.

109. Yao HW, Jiang CY, Ye GY, Hu C, Peng YF. 2008. Toxicological assessment of pollen from different *Bt* rice lines on *Bombyx mori* (Lepidoptera: Bombyxidae). *Environ. Entomol.* 37:825–37
110. Yao HW, Ye GY, Jiang CY, Fan LJ, Datta K, et al. 2006. Effect of the pollen of transgenic rice line, TT9-3 with a fused *cry1Ab/1AC* gene from *Bacillus thuringiensis* Berliner on non-target domestic silkworm, *Bombyx mori* Linnaeus (Lepidoptera: Bombyxidae). *Appl. Entomol. Zool.* 41:339–48
111. **Ye GY, Shu QY, Yao HW, Cui HR, Cheng XY, et al. 2001. Field evaluation of resistance of transgenic rice containing a synthetic *cry1Ab* gene from *Bacillus thuringiensis* Berliner to two stem borers. *J. Econ. Entomol.* 94:271–76**

111. Documents the first field test of IRGM rice in China.

112. Ye RJ, Huang HQ, Yang Z, Chen TY, Liu L, et al. 2009. Development of insect-resistant transgenic rice with Cry1C*-free endosperm. *Pest Manag. Sci.* 65:1015–20
113. Yuan ZD, Yao HW, Ye GY, Hu C. 2006. Survival analysis of the larvae from different hybrids of silkworm, *Bombyx mori* exposure to *Bt* rice pollen. *Bull. Sericult.* 37:23–27
114. Zhang QF. 2000. China: agricultural biotechnology opportunities to meet the challenges of food production. In *Agricultural Biotechnology and the Poor*, ed. GJ Persley, MM Lantin, pp. 45–50. Washington, DC: Consul. Group Int. Agric. Res. U.S. Natl. Acad. Sci.
115. **Zhang QF. 2007. Strategies for developing green super rice. *Proc. Natl. Acad. Sci. USA* 104:16402–9**

115. Introduces the concept of green super rice.

116. Zhang WJ, Pang Y. 2009. Impact of IPM and transgenics in the Chinese agriculture. In *Integrated Pest Management: Dissemination and Impact*, ed. R Peshin, AK Dhawan, pp. 525–53. Dordrecht, The Nether.: Springer
117. Zhang XF, Wang DY, Fang FP, Zhen YK, Liao XY. 2005. Food safety and rice production in China. *Res. Agric. Mod.* 26:85–88
118. Zhao JZ, Cao J, Collins HL, Bates SL, Roush RT, et al. 2005. Concurrent use of transgenic plants expressing a single and two *Bacillus thuringiensis* genes speeds insect adaptation to pyramided plants. *Proc. Natl. Acad. Sci. USA* 102:8426–30
119. Zhao JZ, Cao J, Li YX, Collins HL, Roush RT, et al. 2003. Transgenic plants expressing two *Bacillus thuringiensis* toxins delay insect resistance evolution. *Nat. Biotechnol.* 21:1493–97
120. Zhao JZ, Rui CH, Lu MG, Fan XL, Meng XQ. 2000. Monitoring and management of *Helicoverpa armigera* resistance to transgenic *Bt* cotton in northern China. *Resist. Pest Manag.* 11:28–31

121. Zheng AP, Li P, Zhu J, Wang LX, Wang SQ, et al. 2009. Bt *cry30Ga1* gene, encode protein and its application. *China Pat. No. CN101,531,712*
122. Zheng Y, Hu QY, Li BJ, Zhang XJ, Li YR. 2007. Development and risk assessment of *CpTI* transgenic rice resistance in *Chilo suppressalis*. *J. Fujian Agric. For. Univ.* (*Nat. Sci.*) 2:113–16
123. Zhou YJ, Xue K, Liu Q, Hou LT, Xue DY. 2008. The investigation on the patents and properties of transgenic *Bt/CpTI* rice. *J. CUN* (*Nat. Sci.*) 17:33–39
124. Zhuo Q, Chen XP, Piao JH, Gu LZ, Yang XG. 2004. Study on food safety of genetically modified rice which expressed cowpea trypsin inhibitor by 90 day feeding test on rats. *J. Hyg. Res.* 33:176–80
125. Zi X. 2005. GM rice forges ahead in China amid concerns over illegal planting. *Nat. Biotechnol.* 23:637
126. Zong Y, Chen Z, Innes JB, Chen C, Wang Z, Wang H. 2007. Fire and flood management of coastal swamp enabled first rice paddy cultivation in east China. *Nature* 449:459–63

Energetics of Insect Diapause

Daniel A. Hahn[1] and David L. Denlinger[2]

[1]Department of Entomology and Nematology, University of Florida, Gainesville, Florida 32611; email: dahahn@ufl.edu

[2]Departments of Entomology and Evolution, Ecology, and Organismal Biology, Ohio State University, Columbus, Ohio 43210; email: denlinger.1@osu.edu

Annu. Rev. Entomol. 2011. 56:103–21

First published online as a Review in Advance on August 2, 2010

The *Annual Review of Entomology* is online at ento.annualreviews.org

This article's doi:
10.1146/annurev-ento-112408-085436

0066-4170/11/0107-0103$20.00

Key Words

dormancy, metabolic depression, energy reserves, nutrient homeostasis, costs of diapause, global warming

Abstract

Managing metabolic resources is critical for insects during diapause when food sources are limited or unavailable. Insects accumulate reserves prior to diapause, and metabolic depression during diapause promotes reserve conservation. Sufficient reserves must be sequestered to both survive the diapause period and enable postdiapause development that may involve metabolically expensive functions such as metamorphosis or long-distance flight. Nutrient utilization during diapause is a dynamic process, and insects appear capable of sensing their energy reserves and using this information to regulate whether to enter diapause and how long to remain in diapause. Overwintering insects on a tight energy budget are likely to be especially vulnerable to increased temperatures associated with climate change. Molecular mechanisms involved in diapause nutrient regulation remain poorly known, but insulin signaling is likely a major player. We also discuss other possible candidates for diapause-associated nutrient regulation including adipokinetic hormone, neuropeptide F, the cGMP-kinase *For*, and AMPK.

INTRODUCTION

Diapause: an environmentally preprogrammed period of arrested development, characterized by metabolic depression that can occur during any stage of insect development

Photoperiodic control of diapause: a response such as diapause that is evoked in response to predictable, seasonal changes in daylength

Surviving long periods without eating is a challenge, and this is precisely the challenge most diapausing insects confront. Diapause offers a tremendous adaptive advantage by allowing survival in seasonal environments that could not otherwise be tolerated and permits life cycle synchronization with periods suitable for growth, development, and reproduction. But to succeed in this venture implies an impressive capacity for managing energy reserves. Arrests of 9–10 months are common and in a few cases diapause may stretch to several years. What resources are sequestered? How are they parsed out during diapause, preserving enough nutrients to complete postdiapause challenges such as metamorphosis and flight, energy-intensive activities that are frequent prerequisites before feeding can replenish lost reserves?

Insects use two strategies to mitigate the energetic costs of diapause: accumulation of reserves and metabolic depression. How does an insect know when it has stored adequate reserves? Do diapausing insects have an energy-sensing mechanism that signals depletion of a critical mass of reserves? Is metabolic depression a simple turning down of the metabolic furnace? We argue that diapause is a metabolically dynamic state that may involve shifts from one energy source to another as diapause progresses, and in some cases diapause is characterized by dramatic pulses of metabolic activity that spike with a frequency of several days. We are just beginning to understand mechanisms regulating such decisions and processes in insects, and the goal of this review is to provide the basis for what we think are critical questions for understanding how resources are managed during the extreme energy deprivation that characterizes diapause.

Previous reviews provide a good context for understanding the ecological implications of insect diapause (26, 69, 107), photoperiodic control of diapause (78, 92), hormonal (30) and molecular regulation of diapause (36), dynamics of the diapause state (64), and nutrient storage and utilization (48). Although our previous review (48) is closely related to the issues presented here, the earlier paper placed more emphasis on the ecological context of nutrient issues associated with diapause, and this review focuses more extensively on physiological mechanisms governing diapause energetics. We also recommend several reviews on the insect fat body and energetic homeostasis as background (5, 45, 102, 103).

CONSEQUENCES OF ENERGY SHORTFALLS AND ABUNDANCE

The energy reserves an insect sequesters can affect the decision to enter diapause, the decision to terminate diapause, and fitness during the postdiapause period. We discussed these implications more fully in our earlier review (48) and only briefly summarize this literature by citing a few recent papers.

Insects that have not sequestered sufficient reserves to survive a lengthy diapause have four options: die during diapause or postdiapause development when all reserves have been depleted; opt to avert diapause, gambling that an attempt to produce one more generation is a better option than dying; terminate diapause prematurely when energy reserves become dangerously low; or compensate for this deficiency by feeding during diapause. All four of these responses do occur. Diapausing insects with low energy reserves do indeed have higher mortality during diapause, as reported in numerous arthropods (48). The blow fly *Calliphora vicina* is a species capable of averting diapause if it is too small: Undersized larvae, although exposed to diapause-inducing conditions of short daylength and low temperature, fail to enter diapause if they are below a certain size (90). The third option, breaking diapause early, is also evident in the larval diapause of *C. vicina*; undersized larvae that do enter diapause terminate diapause much earlier than heavier larvae (90). The final option is available only to insects that retain the ability to feed during larval or adult diapause. Overwintering larvae of the damselfly *Lestes eurinus* that enter winter with poor energy reserves compensate by feeding more than their well-fed cohorts (28).

Extra reserves that remain with an insect after diapause termination can be an asset for postdiapause development and reproduction, enhancing postdiapause performance. Lipid reserves remaining after termination of adult reproductive diapause in the mosquito *Culex pipiens* are readily used for subsequent egg production (120). Diapause-destined larvae of the corn stalk borer, *Sesamia nonagrioides*, feed more than nondiapausing larvae, resulting in larger adults that produce more eggs (38). Similarly, seasonal field trapping data of bivoltine moths in Estonia show that adults derived from pupae or larvae that overwintered in diapause were larger than adults from subsequent nondiapause generations, despite lower host plant quality (108). However, gathering additional reserves needed to ensure a successful diapause in itself can exact a toll. More time spent feeding can make an insect more vulnerable to predation and parasitism, and as demonstrated in the butterfly *Aglais urticae*, the added weight acquired by feeding in preparation for winter can impair the butterfly's escape ability by decreasing its flight muscle ratio, i.e., the ratio of thorax muscle mass to total body mass (3).

Although we usually think of the diapause program driving the accumulation of energy stores, the reverse is also possible, as seen in *Polistes* paper wasps (55). In this scenario, it is proposed that accumulation of hexameric storage proteins leads to suppression of juvenile hormone (JH) production, a prerequisite for diapause. Wasps that accumulate an abundance of hexamerins thus produce less JH and are channeled into diapause, suggesting an evolutionary integration of nutritional and endocrine mechanisms in diapause.

NUTRIENT STORAGE IN PREPARATION FOR DIAPAUSE

Both diapausing and nondiapausing insects store metabolic reserves of the same three macronutrient groups: lipids, carbohydrates, and amino acids, as well as essential micronutrients such as vitamins and minerals. Alterations in both the quantity and quality of nutrient stores are often apparent in diapause-destined individuals. During diapause, reserves are used in cellular maintenance for both catabolic energy production and anabolic processes, including protein turnover and cell membrane maintenance and remodeling. Diapausing insects are simply not running slower than nondiapausing insects of the same life stage; they have entered an alternative developmental pathway that has its own metabolic demands (64). For example, diapausing flesh fly pupae enhance cold hardiness by upregulating production of glycerol and several classes of heat shock proteins as a component of the diapause program. Thus, diapausing pupae require substantial quantities of amino acids and carbohydrate precursors to support synthesis of these protective molecules (33).

Because of their high caloric content, low hydration state, and perhaps relatively high yield of metabolic water, triacylglyceride fat stores are the most important energy reserve in most diapausing insects, often accounting for as much as 80–95% of total lipid content. The fat body is the primary site of fatty acid synthesis, triacylglyceride production, and triacylglyceride storage in insects, although all cells can store some triacylglycerides and substantial stores can occur in tissues such as the large, metabolically active flight muscles (118). Diapause-destined individuals of some species accumulate greater triacylglyceride stores than nondiapausing individuals, and increased triacylglyceride storage is thought to be an important factor mitigating the metabolic demands of diapause (26, 48, 107). To accumulate greater reserves, diapausing individuals must eat more, increase their digestive efficiency, divert nutrients away from somatic growth to storage, or use a combination of these, nominating the prediapause preparatory period as an excellent developmental window for examining the regulation of feeding and nutrient processing. Although the dogma is that diapause-destined insects accumulate greater reserves, this is not

Juvenile hormone (JH): a family of sesquiterpenoids produced by the corpora allata of insects that act together with ecdysteroids to regulate insect molting, metamorphosis, and reproduction

Triacylglyceride: the dominant storage form of lipids in insects synthesized and stored primarily in the fat body

always the case (48). What ecological and physiological factors determine whether a species accumulates greater reserves as part of its diapause preparatory program is a wide-open question.

The fatty acid composition of triacylglycerides accumulated prior to diapause can differ qualitatively and quantitatively between nondiapause and diapause-destined individuals within a species. Fatty acid composition of triacylglyceride stores is highly influenced by the fatty acid composition of insect diets. Despite feeding on the same diet, however, the triacylglyceride stores of diapausing individuals from several species contain more unsaturated fatty acids, whereas the triacylglyceride stores of nondiapausing individuals contain more saturated fatty acids (17, 40, 63). Triacylglyceride stores can only be enzymatically mobilized from intracellular lipid droplets when they are in liquid form and when the lipid droplet surface proteins can appropriately interact with lipolytic enzymes; therefore, greater unsaturation in diapause-destined individuals may be important for fat mobilization at low temperatures. This idea receives support from both the literature on mammalian hibernators, showing that ground squirrels and chipmunks fed diets high in saturated fats are more likely to avoid hibernation and have shorter hibernation periods with higher body temperatures (27, 89), and the literature on invertebrate cold tolerance, showing a substantial increase in fatty acid unsaturation associated with cold acclimation and adaptation to cold climates (52, 65, 79). In addition to triacylglycerides, alterations in lipid metabolism associated with modifications of cell membrane fluidity and deposition of cuticular waxes and hydrocarbons have been noted during the diapause preparatory period (76, 110, 117).

Like most animals, the primary carbohydrate reserve in insects is the polysaccharide glycogen. Insect muscles, particularly flight muscles, can store substantial quantities of glycogen that are catabolized during exercise, but the fat body is the primary site of glycogen synthesis and storage, and fat body glycogen stores can be rapidly converted to glucose or the disaccharide trehalose for transport to other tissues (5). Diapause is associated with qualitative and quantitative shifts in glycogen metabolism. Prior to diapause many insects accumulate large glycogen reserves, and as with fat stores, diapausing individuals often accumulate greater glycogen reserves than their nondiapausing counterparts, although this is not always the case (26, 107, 120).

Fat body glycogen reserves play two major roles in diapausing insects: They are converted to glucose or trehalose for transport out of the fat body to tissues for fueling catabolism, and they are metabolized to produce a variety of sugar-alcohol and sugar-based cryoprotectant molecules. Glycerol and sorbitol are the most common cryoprotectant molecules found in insects, but some species synthesize other polyols such as ethylene glycol, erythritol, mannitol, ribitol, and threitol, as well as sugars with a cryoprotective role including trehalose and glucose (33). The production of cryoprotectants from glycogen can occur either as part of the diapause preparatory program without prior exposure to low temperatures or in response to cold exposure in diapausing or nondiapausing individuals. For example, when held at constant temperatures, diapausing pupae of the flesh fly, *Sarcophaga crassipalpis*, contain greater quantities of glycerol and are more cold tolerant than nondiapausing individuals, illustrating that increased glycerol levels are part of the diapause program (33).

Increased amino acid concentrations are present in the blood of some diapausing insects, but it is unclear whether these additional amino acids primarily play a role in nutrient storage or cold and desiccation resistance (71, 77). Many diapausing insects store amino acids in the form of specialized storage proteins, initially termed diapause proteins, but most belong to the storage hexamerin family of insect proteins, which are also accumulated in lesser quantities in nondiapausing insects (16, 36). Storage proteins are typically accumulated prior to diapause, and during diapause their constituent amino acids may be used for both anabolic activities in maintaining turnover of active protein pools, such as molecular chaperones, and catabolic respiratory metabolism, as well as supporting anabolic and catabolic roles in postdiapause functions such as resumption of development (36). Several studies indicate that storage proteins may continue to be synthesized at low levels during diapause (42, 100). Whether storage proteins produced during diapause are playing roles in amino acid storage and intermediary metabolism or some other function, such as humoral defense or cold tolerance, is not known, but their rapid disappearance following diapause termination suggests a role in postdiapause tissue remodeling.

METABOLIC DEPRESSION AND NUTRIENT MOBILIZATION

In addition to accumulating reserves during the preparatory period, diapausing individuals save on metabolic costs by suppressing intermediary and respiratory metabolism. Although metabolic depression is a near universal aspect of diapause, the degree of depression can vary widely among species and diapause strategies from a mild 15% suppression in the flight-capable diapausing adults of the monarch butterfly (19) to the extreme 90% suppression observed in diapausing flesh fly pupae (35, 85). Metabolic depression in diapause results from a combination of ecological and physiological factors (48). Metabolism is proportional to temperature in diapausing insects, as in all ectotherms, and careful site selection can reduce metabolic stress (56, 57, 97, 113). The low temperatures of winter greatly favor metabolic depression, but summer and tropical diapauses are also common and metabolic depression is particularly important in these species (31).

Most diapausing insects are not actively growing or reproducing, and cell cycle arrest, decreased synthesis of proteins and other molecules, and genomic remodeling associated with reduced transcription all contribute to decreased requirements for energy or anabolic substrates (36). Whole-organism maintenance costs can be reduced by allowing unnecessary tissues, such as gut or flight muscle, to atrophy (29). However, diapausing animals must actively maintain some tissues, such as the brain and imaginal discs, that are critical to surviving diapause and performing postdiapause functions. Therefore, metabolic depression is not just a reduction of all cellular processes, but is a selective reduction of some processes, such as growth and reproduction, and an increase in others, such as stress resistance mechanisms, yielding a net decrease in overall metabolic demand (36). An important question is, Which modules of intermediary metabolism are decreased during diapause, and which are maintained or enhanced?

In addition to resource conservation, metabolism may be rearranged during diapause to promote stress resistance. Hypoxia/anoxia stress is a common risk for diapausing insects, especially those in soil that may become inundated with rain, snow, or ice. Some diapausing insects show substantial resistance to hypoxia/anoxia stress, such as diapausing flesh fly pupae that survive six days of anoxia, whereas nondiapausing pupae perish after one day (66). In addition to resource conservation, metabolic depression may enhance resistance to hypoxia/anoxia stress. Indeed, metabolic suppression has been associated with hypoxic overwintering states in numerous ectotherms including the roundworm, *Caenorhabditis elegans*, turtles, fish, and snails (39, 74, 103). In these examples, animals decrease aerobic metabolism and shift largely to anaerobic metabolism, favoring the activity of glycolysis and gluconeogenesis, the pentose phosphate shunt, and the phosphoenolpyruvate carboxykinase (PEPCK)-succinate pathway to generate ATP and reduce equivalents. Are diapausing insects following this same shift away from aerobic metabolism during diapause? A series of recent transcriptomic and metabolomics studies of adult reproductive diapause in *Drosophila melanogaster* (7), larval diapause in the pitcher plant mosquito, *Wyeomyia smithii* (37), and pupal diapause of the flesh fly *S. crassipalpis* (84) show diapause enrichment in key transcripts for glycolysis and gluconeogenesis. Diapausing flesh fly pupae have increased glucose, glycerol,

Respiratory quotient (RQ): the ratio of an animal's oxygen consumption to carbon dioxide release, reflecting specific forms of energy utilization

and pyruvate (75, 76), a difference in metabolites consistent with a shift away from aerobic metabolism and towards anaerobic metabolism. In the above mentioned studies, individuals were never exposed to hypoxia or anoxia, suggesting that a preponderance for anaerobic metabolism at the time of metabolic suppression is a preprogrammed component of diapause development.

Further work is needed to determine whether diapausing insects generally attain metabolic depression by specifically suppressing aerobic metabolism while maintaining anaerobic pathways, how it relates to their diapause life histories, and what biochemical mechanisms regulate the shift toward anaerobic metabolism in diapause. One promising mechanism for coordinating the shift from aerobic to anaerobic metabolism in diapausing insects is the transcriptional suppressor *hairy*, revealed by a recent study of hypoxia-selected lines in *D. melanogaster* (119). *hairy* transcription was upregulated in hypoxia-selected lines, and important aerobic respiration genes in the citric acid cycle that had lower expression in hypoxia-selected lines also contained *hairy* binding elements in their regulatory regions. *hairy* loss-of-function mutants were more susceptible to hypoxia stress and did not show coordinated downregulation of TCA cycle transcripts in response to hypoxia, further reinforcing the role of *hairy* as an important metabolic switch. Could *hairy* also act as a metabolic switch in diapausing insects? Diapausing pupae of the flesh fly, *S. crassipalpis*, suppress aerobic metabolism in favor of anaerobic metabolism during diapause-induced metabolic depression, and *hairy* transcript abundance is substantially greater in diapausing pupae than in nondiapausing pupae (84). However, *hairy* expression levels apparently do not differ between diapausing and reproductive *D. melanogaster* adults (7), necessitating more work to establish the generality and importance of *hairy* as a metabolic switch in diapausing insects.

Although metabolic depression conserves nutrient reserves, reserves are substantially depleted during diapause. Many studies show depletion of one reserve material or another, but few carefully track the utilization of multiple reserve classes throughout diapause. Studies that do track multiple nutrient classes indicate reserve mobilization is dynamic and not uniform throughout diapause. For example, in the adult diapause of the mosquito *Cx. pipiens*, females deplete glycogen stores during the first month of diapause and then switch to use lipids as shown by both radiotracer tracking (120) and expression studies of β-oxidation genes (96). Flesh flies show the opposite pattern where lipids are depleted early in diapause followed by a shift to other substrates (1). Further, a respiratory quotient (RQ) study in diapausing leafcutter bee larvae suggests lipid catabolism (RQ near 0.7) early in diapause, followed by a shift to catabolism of amino acids, glycogen, or a mixture later in diapause (RQ values from 0.8 to 1.0) (116). Specific nutrient subclasses may also be selectively utilized during diapause. For example, within the larger fatty acid pool stored as triacylglycerides, unsaturated fatty acids decrease more rapidly than saturated fatty acids during diapause (17, 40, 63). Although selective mobilization of unsaturated fats has been suggested as an adaptation to low temperatures, mammals and birds also show preferential mobilization of unsaturated fatty acids across a range of conditions, suggesting that preferential depletion of unsaturated fatty acids may be a general pattern of mobilization, not a diapause-specific pattern (82, 83).

In addition to longer-term shifts in nutrient utilization, rapid changes also occur. For example, diapausing larvae of the goldenrod gall fly, *Eurosta solidaginis*, convert glycogen stores to glycerol in response to low temperature in the fall (104). However, conversion of glycogen to glycerol is reversible and temperature dependent, so that as insects warm in the spring, much of the glycerol is recovered and stored again as glycogen. Glycerol production remains environmentally sensitive during diapause wherein short-term low-temperature spells are accompanied by increased glycerol production and rewarming drives reconversion to glycogen. The dynamic conversion and reconversion between glycogen and glycerol are regulated by a simple, elegant mechanism modifying the activity of glycogen phosphorylase, a key glycogen-mobilizing enzyme. Temperature-dependent activation of glycogen phosphorylase occurs under cold conditions because low-temperature exposure in the 0°C–5°C range inactivates glycogen phosphorylase phosphatase, while the reduction in glycogen phosphorylase kinase activity at 0°C–5°C is less pronounced and in line with Q_{10} predictions for the enzyme (20). Therefore, glycogen phosphorylase is more likely to be activated and provide glucose substrates for glycerol synthesis at low temperatures, providing direct environmental control of metabolic flux via protein phosphorylation reactions. Whereas short-term alterations in the intermediary metabolism of other substrates and products are not well known, it is clear that other mechanisms, such as some heat shock proteins and desaturases that modify membrane lipids (21, 33, 36), can respond rapidly to environmental stress during diapause and must presumably mobilize materials from reserves to synthesize new proteins and membrane lipids.

Cyclic Bouts of Metabolism during Diapause

The metabolic rate during diapause is not constant in all insects but instead may be characterized by periodic bouts of high metabolic activity, lasting 1–2 days at 25°C, interspersed between 4 and 5 days of low metabolic activity that is barely detectable. These cycles were first noted in diapausing pupae of sarcophagid flies (35) but have also been reported in diapausing pupae of several lepidopteran species (24, 25). Not all species exhibit such cycles, and it remains unclear why some species are cyclic and others are not. These cycles should not be confused with the better known cyclic discharge patterns of CO_2 of saturniid and other large lepidopteran pupae (51, 93). During the saturniid CO_2 bursts, O_2 consumption remains constant, but the cycles observed in *Sarcophaga* are indeed cycles of O_2 consumption that thus represent distinct pulses of metabolic activity that persist throughout diapause.

Metabolic cycles in flesh fly pupae are close together in early diapause, become further apart in mid-diapause, and then become closer together again near the end of diapause (35). The onset of the metabolic pulse is abrupt (98), suggesting a rapid switching mechanism that rapidly turns on the metabolic machinery. Several lines of experimental evidence suggest that these metabolic cycles are driven by JH (32, 34).

The interesting question is why diapausing flesh fly pupae display cyclic bouts of metabolic activity rather than maintain a constant low metabolic rate. Protein synthesis peaks during O_2 consumption pulses (59), and select genes, such as the transcript encoding apurinic/apyrimidinic (AP) endonuclease, a protein that repairs DNA damage in oxidation and hypoxia injury, show peak expression during the trough of the cycle (23). Presumably it is metabolically more efficient for flesh flies to dramatically suppress metabolism most of the time and to just briefly stoke up the metabolic furnace when needed. Possibly the fly uses these bouts of metabolic activity to generate ATP, replenish other key metabolites, boost its stress defense, and repair damage while burning off metabolic end products that may accumulate during the trough of the cycle. Such cycles underscore the dynamic nature of energy utilization that can operate in some species during diapause. These cycles of metabolism have a striking resemblance to the regular periods of arousal noted in mammalian hibernators (18, 111), suggesting a common advantage for cyclic metabolic activity during diverse types of dormancy.

Oxygen (O_2) consumption bouts: a cyclic infradian pattern of metabolism that persists throughout diapause in certain Diptera and Lepidoptera pupae

ENERGY SENSING

We argue that some sort of energy-sensing mechanism operates in diapausing insects, enabling them to assess levels of reserves in their body. Such information could be used during the diapause preparatory period to affect the decision to enter diapause (have enough reserves been stored?) or during diapause to affect the decision to terminate diapause (are reserves running low?). Indeed, small larvae of the blow fly *Calliphora vicina* avert diapause or terminate diapause earlier than well-fed larvae (90), suggesting that such an energy sensor is operating. Experimental manipulations of the rate of metabolic activity in diapausing flesh fly pupae also suggest this possibility (32). High temperature or application of a JH analog elevates the diapause metabolic rate, resulting in a shorter diapause. A comparison of metabolic rates and diapause durations under these diverse conditions suggests that diapause could be terminated when reserves reach some critical set point. When the metabolic rate is high the available energy is consumed quickly, and when the metabolic rate is low this set point is reached much later, yielding a longer diapause. Presumably, most insects enter diapause with more than adequate reserves, and a limit on such reserves is not what normally determines the duration of diapause, yet a modest adjustment in either the amount of stores or rate of utilization could shorten or lengthen the duration of diapause in some species.

Energy sensing: refers to the ability of an animal to accurately assess its energetic state based on reserves and current feeding state

Insulin signaling: a signal transduction pathway affecting growth, reproduction, and carbohydrate and lipid metabolism that is activated by the presence of insulin or an ILP

ILP: insulin-like peptide

Dauer larvae: a dormant state in nematodes equivalent to diapause in insects

Although it was once thought that fat body cells were just depots for accumulating stores, our understanding of adipocytes has blossomed in the past two decades so that in both mammals and invertebrates adipocytes are recognized to actively participate in regulating metabolism both through localized subcellular mechanisms and in endocrine interactions with the gut and brain. This recognition has led to a flurry of studies on *D. melanogaster* and a few on other insects such as mosquitoes and the hawk moth, *Manduca sexta*, reinforcing the idea that fat body cells play critical roles in nutrient sensing, nutritional homeostasis, feeding behaviors, and coordinating growth and reproduction with nutrition (5, 8, 10). Similarly, a recent high-throughput study of 350 transgenic *D. melanogaster* lines that either suppress or hyperactivate specific neuronal groups in the brain revealed two sets of neurons that when silenced produced very fat flies and when overactivated produced very lean flies (2). This finding crystallizes the view that the insect brain has distinct groups of neurons that regulate feeding and nutrient homeostasis similar to the vertebrate brain-hypothalamic axis. Unfortunately, few experiments have directly explored energy sensing, feeding and digestive physiology, and nutrient mobilization in diapausing insects, and the mechanisms we propose are simply possibilities that have not yet been critically examined experimentally. Below we discuss insulin signaling, which has been implicated as an important pathway in nutrient regulation, growth, and dormancy in insects and nematode worms, and we suggest some additional candidate mechanisms for nutrient sensing and metabolic reorganization in diapausing insects.

Insulin Signaling

The insulin signaling pathway, best known for its role in regulating carbohydrate and fat metabolism in mammals, is highly conserved and a promising candidate for regulating reserves in insect diapause. As in mammals, insulin signaling has roles in metabolism and growth in nematodes (39) and insects (8, 10, 15, 114). Although mammals have relatively few insulin-like peptides (ILPs), many members of this family are present in insects (114): Eight ILPs are known from the mosquito *Ae. aegypti*, but more than 30 are present in the silkmoth *B. mori*. In spite of this wealth of ILPs, all or most of the ILPs appear to exert their effect through a single receptor (114). What it means for insects to have such an array of ILPs is not clear but is matched by diverse physiological functions linked to ILPs, including growth and development, metabolism, reproduction, and caste determination in social insects (114).

The centrality of insulin signaling to energy storage and utilization in mammals suggests that this pathway could be central to energy issues during insect diapause. A link to diapause was first suggested by elegant work defining molecular events in dauer larva formation (a dormant state equivalent to diapause) in the nematode *C. elegans*. Like insects in diapause, *C. elegans* dauer larvae accumulate fat reserves and enter an arrested state that has most of the attributes of insect diapause (39). Several key genes linked to dauer formation in *C. elegans*, referred to as *dauer formation* (*daf*) genes, proved to be orthologs of genes in the insulin signaling pathway. In *D. melanogaster*, disrupting insulin signaling shuts down reproduction and increases energy stores, inducing a physiologic state reminiscent of the natural adult diapause in this species (106). JH terminates reproductive diapause in *D. melanogaster*, and the inhibition of reproduction and accumulation of reserves can be reversed by applying exogenous JH mimics (105).

Insulin signaling is apparently also involved in both shutting down reproduction and regulating fat accumulation associated with adult diapause in the mosquito *Cx. pipiens* (94). Knocking down the insulin receptor (InR) in nondiapausing females with dsInR blocks ovarian development, simulating the diapause state. This blockage can be reversed with JH, a hormone essential for the initiation of follicle development, and the response is similar to the JH mitigation of impaired insulin signaling on aging in *D. melanogaster* (105). Many of the effects of insulin signaling on diapause-like states in *C. elegans* and *D. melanogaster* are mediated by the fork-head transcription factor (FOXO), a critical downstream member of the pathway. FOXO is suppressed in the presence of insulin, but in the absence of insulin FOXO is activated and translocated from the cytoplasm into the nucleus where it initiates a number of responses including fat accumulation in *D. melanogaster* (8). During the prediapause period, adult females of *Cx. pipiens* eschew blood feeding and increase sugar feeding, changing their feeding behavior and digestive physiology to accumulate much greater fat reserves compared with nondiapausing females (88). If FOXO expression is knocked down using RNAi in diapause-destined females of *Cx. pipiens*, the females fail to accumulate the huge fat reserves characteristic of diapause (94), thus suggesting a critical role for the insulin signaling pathway in fat reserve accumulation. Not all ILPs appear to contribute to diapause regulation in *Cx. pipiens*; ILP 1, but not ILP 2 or ILP 5, regulates the response (95).

Across insects, diapause states are often maintained by modulating the titers of the major developmental hormones JH and ecdysteroids, and insulin may be a critical signaling system upstream of both JH and ecdysteroids. Insulin signaling has been implicated in maintaining and eventually terminating the adult reproductive diapause of *D. melanogaster* and *Cx. pipiens* mosquitoes by modulating JH titers, and adding exogenous JH to diapausing adults terminates diapause and rescues the nondiapause reproductive phenotype. Similarly, many pupal diapauses and some larval diapauses are characterized by decreased ecdysteroid production and terminated by application of exogenous ecdysteroids (36). Insulin signaling stimulates ecdysteroid production by ovaries in mosquito reproduction (15); could it play an important role in terminating larval and pupal diapause by stimulating ecdysteroid production? Pupal diapause of the butterfly *Pieris brassicae* is characterized by the absence of ecdysteroid production, and application of exogenous ecdysteroids can terminate diapause and initiate adult morphogenesis (4). Injecting bovine insulin into diapausing pupae of *P. brassicae* stimulates ecdysteroid production and terminates diapause (4). Enhanced insulin signaling is also involved in terminating the larval dauer state in *C. elegans*, wherein insulin signaling stimulates production of dafachronic acids, a blend of steroid hormones that act via the DAF-12 nuclear hormone receptor to resume larval development (39). Therefore, we expect that insulin signaling plays an important role in both metabolic suppression during diapause and the resumption of development at diapause termination.

FOXO: Forkhead transcription factor

TOR Signaling

Target of rapamycin (TOR) signaling is a sister pathway that interacts with insulin signaling to regulate body size and nutritional status in insects and vertebrates (44, 68, 101). The fat body acts as a nutrient-sensing organ in the

TOR: target of rapamycin

AKH: adipokinetic hormone

PKA: cyclic AMP-dependent protein kinase A

LSD-1: lipid storage droplet surface protein-1

context of TOR signaling in *D. melanogaster*, wherein amino acid uptake by the fat body during larval feeding enhances TOR activity and stimulates larval growth (22, 41). Similarly, the fat body acts as an important nutrient-sensing organ controlling the initiation of vitellogenesis in adult mosquitoes by regulating the flow of amino acids derived from either a recent blood meal or protein stores in the fat body into newly synthesized yolk proteins (6, 49). It is unknown how the fat body interprets and transduces amino acid availability into enhanced TOR signaling to coordinate growth and reproduction, but several fat body amino acid transporters have been implicated in activating TOR signaling (22, 41). The role of TOR in coordinating both growth and reproduction in a nutrition-dependent manner suggests it could play important roles in three contexts: differential nutrient accumulation during the diapause preparatory period, nutrient utilization during diapause, and the shift from developmental arrest during diapause to active growth or reproduction at diapause termination. Similarly, other mechanisms of nutrient homeostasis may also exert their effects, at least partially, by modulating insulin signaling.

Adipokinetic Hormones

The insect adipokinetic hormones, AKHs, are a family of small peptides that regulate mobilization of lipids, glycogen, and even amino acids (72, 112). Originally characterized for their role in mobilizing flight fuels, AKH function in *D. melanogaster* and the silkworm *B. mori* is important for nutrient homeostasis, including accumulation and mobilization of lipid and glycogen reserves (58, 70). Although evidence for the involvement of AKH in diapause is currently limited, one study of AKH in the firebug *Pyrrhocoris apterus* shows that females in adult reproductive diapause mobilize approximately twice as much lipid into the blood as nondiapausing reproductive females when injected with synthetic migratory locust *Locusta migratoria* AKH-I peptide or extracts of the *P. apterus* corpora cardiacum, the gland that secretes endogenous AKH (99). This finding suggests that diapausing adults have greater sensitivity to the lipid-mobilizing effects of AKH, but differences in AKH sensitivity as a general pattern in diapause are unknown.

Differences in AKH sensitivity for nutrient mobilization have been shown across life stages in several insects (72, 112). For example, the same AKH peptide prompts glycogen mobilization in larvae but mobilizes lipid in adults of the hawk moth, *M. sexta* (5). The migratory locust *L. migratoria* produces multiple distinct AKH peptides that differ in substrate mobilization, perhaps revealing specialized functions for each different peptide in nutritional homeostasis. Although all locust AKH peptides mobilize triacylglycerides, AKH-I shows greater ability to mobilize lipids than AKH-II or AKH-III, while AKH-II shows a greater affinity for mobilizing glycogen (112). In addition to differences in the dose-dependent propensity to mobilize lipids, AKH-I mobilizes diacylglycerides with a fatty acid profile distinct from AKH-II and AKH-III, further suggesting specialized roles for each AKH peptide in mobilizing different fuel classes (109).

What factors promote specific mobilization of discrete nutrient classes or subclasses by AKH between different life stages or among different AKH peptides? In *M. sexta*, AKH activates cAMP-dependent protein kinase A (PKA) in adult fat body cells, which then phosphorylates triacylglyceride lipase and lipid storage droplet surface protein-1 (LSD-1), a protein that localizes to the surface of lipid droplets in fat body cells, similar to the vertebrate perilipins (5, 67). In *M. sexta*, phosphorylation of LSD-1 at the surface of the lipid droplet appears necessary for triacylglyceride lipase to mobilize stored triacylglycerides from lipid droplets. LSD-1 is present only in adult fat body cells, which mobilize lipids in response to AKH, and is absent from larval fat body cells, which mobilize glycogen rather than lipids in response to AKH, prompting the suggestion that LSD-1 controls stage-specific differences in energy mobilization in *M. sexta* (5). Investigation of lipid droplet surfaces in *D. melanogaster* shows a complex mixture of dozens of proteins, some of which are implicated in metabolism and lipid mobilization, concordant with studies of mammalian lipid droplets (5, 9, 10, 67). Changes in composition of lipid droplet surface proteins could contribute to the differences in fuel mobilization observed between diapausing and nondiapausing insects, as well as shorter-term differences in fuel mobilization during diapause, either through their ability to modulate AKH action or through other undescribed routes.

Other Mechanisms of Energy Sensing and Nutrient Mobilization

AMP-activated protein kinase (AMPK) is another promising candidate. AMPK is often regarded as part of the cellular energy-sensing system because this enzyme becomes activated by phosphorylation in response to high ratios of AMP:ATP, and then AMPK itself further phosphorylates a series of downstream proteins that control flux through metabolic pathways affecting appetite, ATP production, and the synthesis/degradation of lipid and carbohydrate reserves (50). In mammals, suppression of AMPK activity leads to increased lipid biosynthesis and storage, whereas activation leads to lipid and carbohydrate catabolism. Furthermore, recent work on AMPK suggests that it may bind directly with glycogen, essentially acting as a cellular fuel gauge for glycogen (73), and that it interacts with insulin/TOR signaling to regulate synthetic activity underlying growth, thereby directly linking growth with energy supplies (46). Although AMPK function has been best studied in mammals, AMPK has similar functions in *D. melanogaster* (80). AMPK activity is implicated in developmental suppression and metabolic rate depression in hypoxic *Trachemys scripta* turtles (87) and estivating *Otala lactea* land snails (86), but it does not appear to be important in hibernating *Spermophilus tridecemlineatus* squirrels (53). A recent microarray study of diapausing *S. crassipalpis* flesh fly pupae showed that transcripts matching the AMPK α, β, and γ subunits were upregulated in diapausing pupae compared with nondiapausing pupae (84). AMPK is activated posttranslationally by phosphorylation, so further work on protein abundance and activity in diapausing flesh fly pupae is clearly needed, but this regulatory enzyme is a likely candidate for regulating reserves in diapause.

In addition, several other signaling pathways stand out as promising candidates for mediating feeding and nutritional homeostasis in diapause-destined and diapausing insects. The invertebrate neuropeptide F (NPF), homologous to the vertebrate hypothalamic neuropeptide Y, is produced in medial neurosecretory cells of the *D. melanogaster* brain and plays roles in both feeding behavior and insulin signaling (14, 115). Overexpression of NPF causes prolonged feeding in *D. melanogaster* larvae and NPF loss-of-function mutants feed less. Similarly, higher expression of the cGMP-dependent protein kinase, *For*, has been associated with enhanced feeding and foraging behavior in *D. melanogaster*, *C. elegans*, honey bees, and ants (62). In *D. melanogaster*, individuals carrying a naturally occurring *For* allele associated with higher PKG expression and enhanced feeding also accumulate greater lipid reserves and have apparently greater glucose mobilization during starvation (60, 61). Could diapause-destined individuals become hyperphagic and accumulate greater nutrient reserves by upregulating NPY, *For*, or other feeding-behavior genes that alter the brain–gut–fat body axis?

The greater nutrient reserves accumulated in diapause-destined insects may be facilitated by hypertrophy of fat body cells, with each cell storing more, greater proliferation of fat body cells so there are more cells available for storage, or both. Several candidates for regulating fat body hypertrophy and proliferation have been identified in *D. melanogaster*. For example, flies with partial-loss-of-function mutations in the transcriptional repressor gene *adipose* show substantial triacylglyceride accumulation by fat cell hypertrophy, but overexpression of *adipose* in wild-type flies leads to decreased triacylglyceride accumulation in fat body cells (47). To our knowledge, direct comparisons of the number

AMPK: AMP-activated protein kinase

Neuropeptide F (NPF): the invertebrate ortholog of vertebrate neuropeptide Y

Foraging (*For*): a locus coding for a conserved cGMP-dependent protein kinase

of fat body cells in diapausing and nondiapausing insects have not been made, but many insects that diapause as fully developed larvae or pupae initiate the prediapause developmental trajectory as embryos; thus, there is ample time for enhanced fat cell proliferation during the diapause-preparatory program (30). Illustrating the importance of fat cell number for storage, a recent RNAi screen for obesity phenotypes in *D. melanogaster* identified several dozen genes associated with enhanced triacylglyceride or glycogen storage (81). Among these fat body obesity genes were *hedgehog*-signaling pathway members that enhanced fat body cell proliferation, leading to phenotypes with greater triacylglyceride stores. Could the *hedgehog*-signaling pathway and perhaps even some of these same genes be involved in the greater accumulation of nutrient reserves in diapause-destined insects? Comparative studies of fat body cell hypertrophy and number are needed to untangle the relative roles of these two processes in diapause-induced reserve accumulation.

CHALLENGES OF CLIMATE CHANGE

Global warming has already had an impact on insect populations, as demonstrated in the Northern hemisphere by northward invasions of more southerly species and shifts in photoresponsiveness and voltinism to accommodate longer growing seasons (12, 13, 54). There is enough naturally segregating genetic variation in photoperiodic responses in some insects to provide the grist for rapid evolutionary shifts in critical photoperiod for diapause. In the pitcher plant mosquito, *Wyeomyia smithii*, a shift toward a shorter, more southern critical daylength corresponding to a longer growing season was detectable within a 30-year period (11), and the critical photoperiod in a Japanese population of the fall webworm, *Hyphantria cunea*, was shortened by 13–19 min over a 7-year period, resulting in a concomitant shift from bivoltinism to trivoltinism (43).

Implicit in this warming trend is an issue critical to energy utilization during diapause, a feature of global warming that is perhaps underappreciated. As discussed above, insects rely on low temperatures of winter to help suppress metabolism and conserve energy stores. A warmer winter could thus mean that an insect with a tight energy budget may indeed deplete its reserves too early and thus jeopardize survival and/or postdiapause fitness. For example, overwintering populations of the monarch butterfly, *Danaus plexippus*, appear to be on a tight energy budget, with fat content dropping over a two-month period from 71% lean dry weight in late November to 36% in late January (19). The monarch female must still retain enough fat at the end of diapause to migrate to a suitable larval habitat for oviposition. Cool nights are critical for suppressing the metabolic rate at these monarch overwintering sites, and just a slight temperature elevation would appear to jeopardize the overwintering success of the butterfly, especially at some of the current overwintering sites in southern California. Just how critical a slight temperature elevation may be is demonstrated in observations comparing survival and reproductive success in individuals of the goldenrod gall fly, *Eurosta solidaginis*, that overwinter above the snow in standing goldenrod stems or below the snow in broken stems (57). Although the mean temperature in the better-insulated galls held below the snow was only 0.6°C higher than in galls found above the snow, fewer adults emerged from the warmer galls and their reproductive output dropped 18%, a difference possibly attributable to the more rapid depletion of reserves in the warmer individuals. Thus, what appears to be a small temperature change can be expected to have a considerable impact on the depletion of energy reserves and consequently on postdiapause energetics and fecundity.

Alternatively, we can anticipate energy-related benefits of a warming planet for certain species. As discussed earlier, the blow fly, *C. vicina*, does not appear to adjust its fat storage or utilization rate as a function of diapause or latitude (91), and warmer winters are translated into a shorter larval diapause in this multivoltine, facultative-diapausing species. This, in turn, enables the fly to carry over more of its larval reserves into adulthood, possibly enhancing fecundity. Impacts on diapause energetics, both positive and negative, as well as phenotypic plasticity and generic variation in the insect's ability to respond to climate change, are likely to have far-reaching implications for insect populations in our changing world.

FINAL PERSPECTIVES

By understanding how insects manage their energy budgets during diapause we can identify one of the critical dimensions defining the limits of insect survival in seasonal environments. We anticipate that such information could help explain current species distributions and will be essential for predicting success of invasive species, the movement of native species into new habitats, and new species interactions such as insects shifting onto new host plants, predictions that are especially important as we look ahead to global warming trends that will likely compromise current insect energy budgets. Because many insects appear to be on a tight energy budget, the potential of capitalizing on this vulnerability by searching for new agents that could selectively prevent fat accumulation, elevate diapause metabolism, or deplete energy reserves prematurely would appear to be a tactic that could be exploited to manipulate pest populations.

Although management of an energy budget is critical for understanding the narrow question of how insects survive adverse conditions and synchronize development, diapause also offers a good model for probing more basic questions related to energy storage and utilization. For example, we show that many insects make a clear decision to store fat at a particular time during their prediapause development. How is such a decision made? With the challenges of obesity and diabetes plaguing the developed world, knowing how such physiologic decisions are made is a question of huge significance to human health. Are there lessons that could be learned from diapause? Similarly, during diapause, insects switch between the utilization of lipid and nonlipid sources. How are such processes regulated? The ability to turn down the metabolic furnace is a trait that is especially impressive during insect diapause. Understanding how this is done may offer useful tips for developing strategies of tissue storage that ameliorate hypoxia-reperfusion injury, a high priority goal for maintaining healthy tissues for human organ transplants. Although studies of nutrient homeostasis in *D. melanogaster* and other model organisms, facilitated by large community resources for novel screening technologies, have revolutionized our understanding of conserved mechanisms of energy homeostasis, we strongly advocate a place in this field for nontraditional models. Although nontraditional models, such as the flesh fly, *S. crassipalpis*, may not have the genetic resources of *D. melanogaster*, diapausing flesh fly pupae display much more drastic metabolic depression during diapause. By choosing to study nutrient homeostasis in organisms from the extremes of the diapause spectrum, we may find new mechanisms that are highly conserved but overlooked in model organisms that show less pronounced diapause responses.

SUMMARY POINTS

1. Many, but not all, insects store additional energy reserves during the preparatory phase of diapause. Triacylglycerides are the dominant form of energy storage, but glycogen reserves and hexameric storage proteins are also frequently accumulated.
2. Metabolic depression, facilitated by low winter temperatures, is an essential mechanism for conserving energy stores. Metabolic depression is especially important in summer and tropical diapauses that must also contend with high temperatures.

3. Utilization of reserves during diapause is a dynamic process. Switching from one type of reserve to another as diapause progresses appears to be common. Infradian cycles of metabolism, akin to the arousal periods of mammalian hibernators, are evident in some diapausing Diptera and Lepidoptera.
4. Insects are capable of evaluating the abundance of their energy stores and in some cases use this information to avert diapause or terminate diapause prematurely if stores are low, but the mechanism of energy-sensing remains unknown.
5. Insulin signaling appears to be an important pathway regulating storage of diapause energy reserves, but the mammalian hibernation literature and other studies suggest additional pathways that are likely involved in diapause energy management.
6. Modest temperature elevation can dramatically affect energy budgets that are already tight during diapause. Thus, rising global temperatures are likely to significantly affect seasonal and geographic distributions of insects.

FUTURE ISSUES

1. At the organismal level, we know surprisingly little about how the management of diapause energy reserves dictates patterns of seasonal distribution. This issue becomes especially important as the temperature of our planet increases and thus places new constraints on energy budgets. Although there is some plasticity and underlying generic variation in this response, it is unclear what the limits of tolerance may be.
2. At the suborganismal level, much work remains in defining the fine details of the metabolic switches and pathways that lead to metabolic depression and the dynamic changes in energy utilization that characterize diapause.
3. At the molecular level, the regulatory pathways used by insects for accumulating reserves and for regulating the utilization of those reserves as diapause progresses remain largely unknown. How does an insect make the decision to switch from one type of reserve to another midway through diapause? How does it know when its energy reserves are becoming critically low and how does it use this information to trigger the end of dormancy?

DISCLOSURE STATEMENT

The authors are not aware of any affiliations, memberships, funding, or financial holdings that might be perceived as affecting the objectivity of this review.

ACKNOWLEDGMENTS

We acknowledge that the work of many of our colleagues could not be included because of space restrictions. While preparing this review we were supported by funds from NSF IOS-641505, USDA-TSTARC-09051246, and the Florida State Agricultural Experiment Station to D.A.H. and by NSF IOS-0840772, USDA NRI 2006-35607-16582, and NIH RO1-AI1058279 to D.L.D.

LITERATURE CITED

1. Adedokun TA, Denlinger DL. 1985. Metabolic reserves associated with pupal diapause in the flesh fly, *Sarcophaga crassipalpis*. *J. Insect Physiol.* 31:229–33
2. Al-Anzi B, Sapin V, Waters C, Zinn K, Wyman RJ, Benzer S. 2009. Obesity-blocking neurons in *Drosophila*. *Neuron* 63:329–41
3. Almbro M, Kullberg C. 2008. Impaired escape flight ability in butterflies due to low flight muscle ratio prior to hibernation. *J. Exp. Biol.* 211:24–28
4. Arpagaus M. 1987. Vertebrate insulin induces diapause termination in *Pieris brassicae* pupae. *Roux Arch. Dev. Biol.* 196:527–30
5. Arrese EL, Soulages JL. 2010. Insect fat body: energy, metabolism, and regulation. *Annu. Rev. Entomol.* 55:207–25
6. Arsic D, Guerin PM. 2008. Nutrient content of diet affects the signaling activity of the insulin/target of rapamycin/p70 S6 kinase pathway in the African malaria mosquito *Anopheles gambiae*. *J. Insect Physiol.* 54:1226–35
7. Baker D, Russell S. 2009. Gene expression during *Drosophila melanogaster* egg development before and after reproductive diapause. *BMC Genomics* 10:242
8. Baker KD, Thummel CS. 2007. Diabetic larvae and obese flies—emerging studies of metabolism in *Drosophila*. *Cell Metab.* 6:257–66
9. Beller M, Riedel D, Jansch L, Dieterich G, Wehland Jr, et al. 2006. Characterization of the *Drosophila* lipid droplet subproteome. *Mol. Cell. Proteomics* 5:1082–94
10. Bharucha KN. 2009. The epicurean fly: using *Drosophila melanogaster* to study metabolism. *Pediatr. Res.* 65:132–37
11. Bradshaw WE, Holzapfel CM. 2001. Genetic shift in photoperiodic response correlated with global warming. *Proc. Natl. Acad. Sci. USA* 98:14509–11
12. Bradshaw WE, Holzapfel CM. 2007. Evolution of animal photoperiodism. *Annu. Rev. Ecol. Syst.* 38:1–25
13. Bradshaw WE, Holzapfel CM. 2010. Insects at not so low temperature: climate change in the temperate zone and its biotic consequences. In *Low Temperature Biology of Insects*, ed. DL Denlinger, RE Lee, pp. 242–75. Cambridge, UK: Cambridge Univ. Press
14. Brown MR, Crim JW, Arata RC, Cai HN, Chun C, Shen P. 1999. Identification of a *Drosophila* brain-gut peptide related to the neuropeptide Y family. *Peptides* 20:1035–42
15. Brown MR, Sieglaff DH, Rees HH. 2009. Gonadal ecdysteroidogenesis in arthropoda: occurrence and regulation. *Annu. Rev. Entomol.* 54:105–25
16. Burmester T. 1999. Evolution and function of the insect hexamerins. *Eur. J. Entomol.* 96:213–25
17. Cakmak O, Bashan M, Kocak E. 2008. The influence of life-cycle on phospholipid and triacylglycerol fatty acid profiles of *Aelia rostrata* Boheman (Heteroptera: Pentatomidae). *J. Kans. Entomol. Soc.* 81:261–75
18. Carey HV, Andrews MT, Martin SL. 2003. Mammalian hibernation: cellular and molecular responses to depressed metabolism and low temperature. *Physiol. Rev.* 83:1153–81
19. Chaplin SB, Wells PH. 1982. Energy reserves and metabolic expenditures of monarch butterflies overwintering in Southern California. *Ecol. Entomol.* 7:249–56
20. Churchill TA, Storey KB. 1989. Metabolic consequences of rapid cycles of temperature change for freeze-avoiding versus freeze-tolerant insects. *J. Insect Physiol.* 35:579–85
21. Clark MS, Worland MR. 2008. How insects survive the cold: molecular mechanisms—a review. *J. Comp. Physiol. B* 178:917–33
22. Colombani J, Raisin S, Pantalacci S, Radimerski T, Montagne J, Leopold P. 2003. A nutrient sensor mechanism controls *Drosophila* growth. *Cell* 114:739–49
23. Craig TL, Denlinger DL. 2000. Sequence and transcription patterns of 60S ribosomal protein P0, a diapause-regulated AP endonuclease in the flesh fly, *Sarcophaga crassipalpis*. *Gene* 255:381–88
24. Crozier AJG. 1979. Diel oxygen uptake rhythms in diapausing pupae of *Pieris brassicae* and *Papilio machaon*. *J. Insect Physiol.* 25:647–52
25. Crozier AJG. 1979. Supraradian and infraradian cycles in oxygen uptake of diapausing pupae of *Pieris brassicae*. *J. Insect Physiol.* 25:575–82

26. Danks HV. 1987. *Insect Dormancy: An Ecological Perspective*. Ottowa: Biol. Surv. Can.
27. Dark J. 2005. Annual lipid cycles in hibernators: integration of physiology and behavior. *Annu. Rev. Nutr.* 25:469–97
28. de Block M, McPeek MA, Stoks R. 2007. Winter compensatory growth under field conditions partly offsets low energy reserves before winter in a damselfly. *Oikos* 116:1975–82
29. de Kort CAD. 1990. Thirty-five years of diapause research with the Colorado potato beetle. *Entomol. Exp. Appl.* 56:1–14
30. Denlinger DL. 1985. Hormonal control of diapause. In *Comprehensive Insect Physiology, Biochemistry, and Pharmacology*, ed. GA Kekurt, LI Gilbert, pp. 353–412. Oxford, UK: Pergamon
31. Denlinger DL. 1986. Dormancy in tropical insects. *Annu. Rev. Entomol.* 31:239–64
32. Denlinger DL, Giebultowicz J, Adedokun T. 1988. Insect diapause: dynamics of hormone sensitivity and vulnerability to environmental stress. In *Endocrinological Frontiers in Physiological Insect Ecology*, ed. F Sehnal, A Zabaza, DL Denlinger, pp. 309–23. Wroclaw, Poland: Wroclaw Tech. Univ. Press
33. Denlinger DL, Lee RE. 2010. *Low Temperature Biology of Insects*. Cambridge, UK: Cambridge Univ. Press. 390 pp.
34. Denlinger DL, Tanaka S. 1989. Cycles of juvenile hormone esterase activity during the juvenile hormone driven cycles of oxygen consumption in pupal diapause of flesh flies. *Experientia* 45:474–76
35. Denlinger DL, Willis JH, Fraenkel G. 1972. Rates and cycles of oxygen consumption during pupal diapause in *Sarcophaga* flesh flies. *J. Insect Physiol.* 18:871–82
36. Denlinger DL, Yocum GD, Rinehart JL. 2005. Hormonal control of diapause. In *Comprehensive Molecular Insect Science*, ed. LI Gilbert, K Iatrou, SS Gill, pp. 615–50. Amsterdam: Elsevier
37. Emerson KJ, Bradshaw WE, Holzapfel CM. 2010. Microarrays reveal early transcriptional events during the termination of larval diapause in natural populations of the mosquito, *Wyeomyia smithii*. *PLoS ONE* 5:e9574
38. Fantinou AA, Perdikis DC, Zota KF. 2004. Reproductive responses to photoperiod and temperature by diapausing and nondiapausing populations of *Sesamia nonagrioides* Lef. *Physiol. Entomol.* 29:169–75
39. Fielenbach N, Antebi A. 2008. *C. elegans* dauer formation and the molecular basis of plasticity. *Genes Dev.* 22:2149–65
40. Foster DR, Crowder LA. 1980. Diapause of the pink bollworm, *Pectinophora gossypiella* (Saunders), related to dietary lipids. *Comp. Biochem. Physiol. B* 65:723–26
41. Geminard C, Rulifson EJ, Leopold P. 2009. Remote control of insulin secretion by fat cells in *Drosophila*. *Cell Metab.* 10:199–207
42. Godlewski J, Kludkiewicz B, Grzelak K, Cymborowski B. 2001. Expression of larval hemolymph proteins (Lhp) genes and protein synthesis in the fat body of greater wax moth (*Galleria mellonella*) larvae during diapause. *J. Insect Physiol.* 47:759–66
43. Gomi T, Nagasaka M, Fukuda T, Hagihara H. 2007. Shifting of the life cycle and life-history traits of the fall webworm in relation to climate change. *Entomol. Exp. Appl.* 125:179–84
44. Grewal SS. 2009. Insulin/TOR signaling in growth and homeostasis: a view from the fly world. *Int. J. Biochem. Cell Biol.* 41:1006–10
45. Guppy M, Withers P. 1999. Metabolic depression in animals: physiological perspectives and biochemical generalizations. *Biol. Rev. Camb. Philos. Soc.* 74:1–40
46. Gwinn DM, Shackelford DB, Egan DF, Mihaylova MM, Mery A, et al. 2008. AMPK phosphorylation of raptor mediates a metabolic checkpoint. *Mol. Cell* 30:214–26
47. Hader T, Muller S, Aguilera M, Eulenberg KG, Steuernagel A, et al. 2003. Control of triglyceride storage by a WD40/TPR-domain protein. *EMBO Rep.* 4:511–16
48. Hahn DA, Denlinger DL. 2007. Meeting the energetic demands of insect diapause: nutrient storage and utilization. *J. Insect Physiol.* 53:760–73
49. Hansen IA, Attardo GM, Park JH, Peng Q, Raikhel AS. 2004. Target of rapamycin-mediated amino acid signaling in mosquito anautogeny. *Proc. Natl. Acad. Sci. USA* 101:10626–31
50. Hardie DG. 2008. AMPK: a key regulator of energy balance in the single cell and the whole organism. *Int. J. Obes.* 32:S7–S12
51. Hetz SK, Bradley TJ. 2005. Insects breathe discontinuously to avoid oxygen toxicity. *Nature* 433:516–19

52. Holmstrup M, Sorensen LI, Bindesbol AM, Hedlund K. 2007. Cold acclimation and lipid composition in the earthworm *Dendrobaena octaedra*. *Comp. Biochem. Physiol. A* 147:911–19
53. Horman S, Hussain N, Dilworth SM, Storey KB, Rider MH. 2005. Evaluation of the role of AMP-activated protein kinase and its downstream targets in mammalian hibernation. *Comp. Biochem. Physiol. B* 142:374–82
54. Huey RB. 2010. Evolutionary physiology of insect thermal adaptation to cold environments. In *Low Temperature Biology of Insects*, ed. DL Denlinger, RE Lee, pp. 223–41. Cambridge, UK: Cambridge Univ. Press
55. Hunt JH, Kensinger BJ, Kossuth JA, Henshaw MT, Norberg K, et al. 2007. A diapause pathway underlies the gyne phenotype in *Polistes* wasps, revealing an evolutionary route to caste-containing insect societies. *Proc. Natl. Acad. Sci. USA* 104:14020–25
56. Irwin JT, Lee RE. 2000. Mild winter temperatures reduce survival and potential fecundity of the goldenrod gall fly, *Eurosta solidaginis* (Diptera: Tephritidae). *J. Insect Physiol.* 46:655–61
57. Irwin JT, Lee RE. 2003. Cold winter microenvironments conserve energy and improve overwintering survival and potential fecundity of the goldenrod gall fly, *Eurosta solidaginis*. *Oikos* 100:71–78
58. Isabel G, Martin JR, Chidami S, Veenstra JA, Rosay P. 2005. AKH-producing neuroendocrine cell ablation decreases trehalose and induces behavioral changes in *Drosophila*. *Am. J. Physiol. Reg. Integr. Comp. Physiol.* 288:R531–R38
59. Joplin KH, Yocum GD, Denlinger DL. 1990. Diapause specific proteins expressed by the brain during the pupal diapause of the flesh fly, *Sarcophaga crassipalpis*. *J. Insect Physiol.* 36:775–84
60. Kaun KR, Chakaborty-Chatterjee M, Sokolowski MB. 2008. Natural variation in plasticity of glucose homeostasis and food intake. *J. Exp. Biol.* 211:3160–66
61. Kaun KR, Riedl CAL, Chakaborty-Chatterjee M, Belay AT, Douglas SJ, et al. 2007. Natural variation in food acquisition mediated via a *Drosophila* cGMP-dependent protein kinase. *J. Exp. Biol.* 210:3547–58
62. Kaun KR, Sokolowski MB. 2009. cGMP-dependent protein kinase: linking foraging to energy homeostasis. *Genome* 52:1–7
63. Khani A, Moharramipour S, Barzegar M, Naderi-Manesh H. 2007. Comparison of fatty acid composition in total lipid of diapause and nondiapause larvae of *Cydia pomonella* (Lepidoptera: Tortricidae). *Insect Sci.* 14:125–31
64. Kostál V. 2006. Eco-physiological phases of insect diapause. *J. Insect Physiol.* 52:113–27
65. Kostál V, Simek P. 1998. Changes in fatty acid composition of phospholipids and triacylglycerols after cold-acclimation of an aestivating insect prepupa. *J. Comp. Physiol. B* 168:453–60
66. Kukal O, Denlinger DL, Lee RE. 1991. Developmental and metabolic changes induced by anoxia in diapausing and nondiapausing flesh fly pupae. *J. Comp. Physiol. B* 160:683–89
67. Lay SL, Dugail I. 2009. Connecting lipid droplet biology and the metabolic syndrome. *Prog. Lipid Res.* 48:191–95
68. Layalle S, Arquier N, Leopold P. 2008. The TOR pathway couples nutrition and developmental timing in *Drosophila*. *Dev. Cell* 15:568–77
69. Leather SR, Walters KFA, Bale JS. 1993. *The Ecology of Insect Overwintering*. Cambridge, UK: Cambridge Univ. Press
70. Lee GH, Park JH. 2004. Hemolymph sugar homeostasis and starvation-induced hyperactivity affected by genetic manipulations of the adipokinetic hormone-encoding gene in *Drosophila melanogaster*. *Genetics* 167:311–23
71. Lefevere KS, Koopmanschap AB, Dekort CAD. 1989. Changes in the concentrations of metabolites in hemolymph during and after diapause in female Colorado potato beetle, *Leptinotarsa decemlineata*. *J. Insect Physiol.* 35:121–28
72. Lorenz MW, Gade G. 2009. Hormonal regulation of energy metabolism in insects as a driving force for performance. *Integr. Comp. Biol.* 49:380–92
73. McBride A, Hardie DG. 2009. AMP-activated protein kinase—a sensor of glycogen as well as AMP and ATP? *Acta Physiol.* 196:99–113
74. McElwee JJ, Schuster E, Blanc E, Thornton J, Gems D. 2006. Diapause-associated metabolic traits reiterated in long-lived daf-2 mutants in the nematode *Caenorhabditis elegans*. *Mech. Ageing Dev.* 127:458–72

75. Michaud MR, Denlinger DL. 2007. Shifts in the carbohydrate, polyol, and amino acid pools during rapid cold-hardening and diapause-associated cold-hardening in flesh flies (*Sarcophaga crassipalpis*); a metabolomic comparison. *J. Comp. Physiol. B.* 177:753–63
76. Michaud MR, Denlinger DL. 2006. Oleic acid is elevated in cell membranes during rapid cold-hardening and pupal diapause in the flesh fly, *Sarcophaga crassipalpis*. *J. Insect Physiol.* 52:1073–82
77. Morgan TD, Chippendale GM. 1983. Free amino acids of the hemolymph of the Southwestern corn borer and the European corn borer in relation to their diapause. *J. Insect Physiol.* 29:735–40
78. Nelson RJ, Denlinger DL, Somers DE. 2010. *Photoperiodism: The Biological Calendar*. Oxford, UK: Oxford Univ. Press
79. Ohtsu T, Katagiri C, Kimura MT, Hori SH. 1993. Cold adaptations in *Drosophila*—qualitative changes of triacylglycerols with relation to overwintering. *J. Biol. Chem.* 268:1830–34
80. Pan DA, Hardie DG. 2002. A homologue of AMP-activated protein kinase in *Drosophila melanogaster* is sensitive to AMP and is activated by ATP depletion. *Biochem. J.* 367:179–86
81. Pospisilik JA, Schramek D, Schnidar H, Cronin SJF, Nehme NT, et al. 2010. *Drosophila* genome-wide obesity screen reveals *hedgehog* as a determinant of brown versus white adipose cell fate. *Cell* 140:148–60
82. Price ER, Krokfors A, Guglielmo CG. 2008. Selective mobilization of fatty acids from adipose tissue in migratory birds. *J. Exp. Biol.* 211:29–34
83. Raclot T. 2003. Selective mobilization of fatty acids from adipose tissue triacylglycerols. *Prog. Lipid Res.* 42:257–88
84. Ragland GJ, Denlinger DL, Hahn DA. 2010. Mechanisms of suspended animation are revealed by transcript profiling of diapause in the flesh fly. *Proc. Natl. Acad. Sci. USA.* 107:14909–14
85. Ragland GJ, Fuller J, Feder JL, Hahn DA. 2009. Biphasic metabolic rate trajectory of pupal diapause termination and postdiapause development in a tephritid fly. *J. Insect Physiol.* 55:344–50
86. Ramnanan CJ, McMullen DC, Groom AG, Storey KB. 2010. The regulation of AMPK signaling in a natural state of profound metabolic rate depression. *Mol. Cell. Biochem.* 335:91–105
87. Rider MH, Hussain N, Dilworth SM, Storey KB. 2009. Phosphorylation of translation factors in response to anoxia in turtles, *Trachemys scripta elegans*: role of the AMP-activated protein kinase and target of rapamycin signaling pathways. *Mol. Cell. Biochem.* 332:207–13
88. Robich RM, Denlinger DL. 2005. Diapause in the mosquito *Culex pipiens* evokes a metabolic switch from blood feeding to sugar gluttony. *Proc. Natl. Acad. Sci. USA* 102:15912–17
89. Ruf T, Arnold W. 2008. Effects of polyunsaturated fatty acids on hibernation and torpor: a review and hypothesis. *Am. J. Physiol. Reg. Integr. Comp. Physiol.* 294:R1044–R52
90. Saunders DS. 1997. Under-sized larvae from short-day adults of the blow fly, *Calliphora vicina*, side-step the diapause program. *Physiol. Entomol.* 22:249–55
91. Saunders DS. 2000. Larval diapause duration and fat metabolism in three geographical strains of the blow fly, *Calliphora vicina*. *J. Insect Physiol.* 46:509–17
92. Saunders DS, Vafopoulou X, Lewis RD. 2002. *Insect Clocks*. Cambridge, UK: Elsevier
93. Schneiderman HA, Williams CM. 1955. An experimental analysis of the discontinuous respiration of the *Cecropia* silkworm. *Biol. Bull.* 109:123–43
94. Sim C, Denlinger DL. 2008. Insulin signaling and FOXO regulate the overwintering diapause of the mosquito *Culex pipiens*. *Proc. Natl. Acad. Sci. USA* 105:6777–81
95. Sim C, Denlinger DL. 2009. A shut-down in expression of an insulin-like peptide, ILP-1, halts ovarian maturation during the overwintering diapause of the mosquito *Culex pipiens*. *Insect Mol. Biol.* 18:325–32
96. Sim C, Denlinger DL. 2009. Transcription profiling and regulation of fat metabolism genes in diapausing adults of the mosquito *Culex pipiens*. *Physiol. Genomics* 39:202–09
97. Sinclair BJ. 2001. Field ecology of freeze tolerance: interannual variation in cooling rates, freeze-thaw and thermal stress in the microhabitat of the alpine cockroach *Celatoblatta quinquemaculata*. *Oikos* 93:286–93
98. Slama K, Denlinger DL. 1992. Infradian cycles of oxygen consumption in diapausing pupae of the flesh fly, *Sarcophaga crassipalpis*, monitored by a scanning microrespirographic method. *Arch. Insect Biochem. Physiol.* 20:135–43
99. Socha R, Kodrik D. 1999. Differences in adipokinetic response of *Pyrrhocoris apterus* (Heteroptera) in relation to wing dimorphism and diapause. *Physiol. Entomol.* 24:278–84

100. Sonoda S, Fukumoto K, Izumi Y, Ashfaq M, Yoshida H, Tsumuki H. 2007. Expression profile of arylphorin gene during diapause and cold acclimation in the rice stem borer, *Chilo suppressalis* Walker (Lepidoptera: Crambidae). *Appl. Entomol. Zool.* 42:35–40
101. Soulard A, Cohen A, Hall MN. 2009. TOR signaling in invertebrates. *Curr. Opin. Cell Biol.* 21:825–36
102. Staples JF, Buck LT. 2009. Matching cellular metabolic supply and demand in energy-stressed animals. *Comp. Biochem. Physiol. A* 153:95–105
103. Storey KB. 2007. Anoxia tolerance in turtles: metabolic regulation and gene expression. *Comp. Biochem. Physiol. A* 147:263–76
104. Storey KB, Storey JM. 1990. Metabolic rate depression and biochemical adaptation in anaerobiosis, hibernation and estivation. *Q. Rev. Biol.* 65:145–74
105. Tatar M, Kopelman A, Epstein D, Tu MP, Yin CM, Garofalo RS. 2001. A mutant *Drosophila* insulin receptor homolog that extends life-span and impairs neuroendocrine function. *Science* 292:107–10
106. Tatar M, Yin CM. 2001. Slow aging during insect reproductive diapause: why butterflies, grasshoppers and flies are like worms. *Exp. Gerontol.* 36:723–38
107. Tauber MJ, Tauber CA, Masaki S. 1986. *Seasonal Adaptations of Insects*. New York: Oxford Univ. Press
108. Teder T, Esperk T, Remmel T, Sang A, Tammaru T. 2010. Counterintuitive size patterns in bivoltine moths: late-season larvae grow larger despite lower food quality. *Oecologia* 162:117–25
109. Tomcala A, Bártu I, Simek P, Kodrík D. 2010. Locust adipokinetic hormones mobilize diacylglycerols selectively. *Comp. Biochem. Physiol. B* 156:26–32
110. Tomcala A, Tollarova M, Overgaard J, Simek P, Kostal V. 2006. Seasonal acquisition of chill tolerance and restructuring of membrane glycerophospholipids in an overwintering insect: triggering by low temperature, desiccation and diapause progression. *J. Exp. Biol.* 209:4102–14
111. van Breukelen F, Martin SL. 2002. Reversible depression of transcription during hibernation. *J. Comp. Physiol. B* 172:355–61
112. van der Horst DJ. 2003. Insect adipokinetic hormones: release and integration of flight energy metabolism. *Comp. Biochem. Physiol. B* 136:217–26
113. Williams JB, Shorthouse JD, Lee RE. 2003. Deleterious effects of mild simulated overwintering temperatures on survival and potential fecundity of rose-galling *Diplolepis* wasps. *J. Exp. Zool. A* 298:23–31
114. Wu Q, Brown MR. 2006. Signaling and function of insulin-like peptides in insects. *Annu. Rev. Entomol.* 51:1–24
115. Wu Q, Wen TQ, Lee G, Park JH, Cai HN, Shen P. 2003. Developmental control of foraging and social behavior by the *Drosophila* neuropeptide Y-like system. *Neuron* 39:147–61
116. Yocum GD, Kemp WP, Bosch J, Knoblett JN. 2005. Temporal variation in overwintering gene expression and respiration in the solitary bee *Megachile rotundata*. *J. Insect Physiol.* 51:621–29
117. Yoder JA, Blomquist GJ, Denlinger DL. 1995. Hydrocarbon profiles from puparia of diapausing and nondiapausing flesh flies (*Sarcophaga crassipalpis*) reflect quantitative rather than qualitative differences. *Arch. Insect Biochem. Physiol.* 28:377–85
118. Zera AJ, Denno RF. 1997. Physiology and ecology of dispersal polymorphism in insects. *Annu. Rev. Entomol.* 42:207–30
119. Zhou D, Xue J, Lai JCK, Schork NJ, White KP, Haddad GG. 2008. Mechanisms underlying hypoxia tolerance in *Drosophila melanogaster*: *hairy* as a metabolic switch. *PLoS Genet.* 4:e1000221
120. Zhou G, Miesfeld RL. 2009. Energy metabolism during diapause in *Culex pipiens* mosquitoes. *J. Insect Physiol.* 55:40–46

NOTE ADDED IN PROOF

We would also like to point out a very useful recent review of diapause molecular biology published while this article was in press.

MacRae TH. 2010. Gene expression, metabolic regulation and stress tolerance during diapause. *Cell Mol. Life Sci.* 67:2405–24

Arthropods of Medicoveterinary Importance in Zoos

Peter H. Adler,[1] Holly C. Tuten,[1] and Mark P. Nelder[2]

[1]Department of Entomology, Soils & Plant Sciences, Clemson University, Clemson, South Carolina 29634; email: padler@clemson.edu; htuten@gmail.com

[2]Enteric, Zoonotic & Vector-Borne Diseases Unit, Ministry of Health and Long-Term Care, Toronto, Ontario M2M 4K5, Canada; email: mark.nelder@ontario.ca

Annu. Rev. Entomol. 2011. 56:123–42

First published online as a Review in Advance on August 16, 2010

The *Annual Review of Entomology* is online at ento.annualreviews.org

This article's doi:
10.1146/annurev-ento-120709-144741

0066-4170/11/0107-0123$20.00

Key Words

diseases, ectoparasites, exotic species, insects, mosquitoes, vectors

Abstract

Zoos present a unique assemblage of arthropods, captive vertebrates, free-roaming wildlife, humans, and plants, each with its own biota of symbiotic organisms. Arthropods of medicoveterinary importance are well represented in zoos, and an ample literature documents their influence in these animal-rich environments. Mosquitoes are of greatest significance because of the animal and human pathogens they transmit, followed by ectoparasites, many of which are exotic and present health risks to captive and native animals. Biting flies, cockroaches, filth flies, and triatomid bugs represent additional concerns. Integrated management programs for arthropods in zoos are commonplace. Zoos can play a role in biosurveillance, serving as an advanced guard for detecting exotic arthropods and vector-borne diseases. We provide the first review of arthropods of medicoveterinary importance in zoos. A case is made for the value of collaborations between entomologists and zoo personnel as a means of enhancing research and public education while safeguarding the health of captive animals and the public.

Medicoveterinary: portmanteau for medical and veterinary

Vector: an organism capable of transmitting a disease agent from one organism to another

INTRODUCTION

The hackneyed view of a world becoming smaller is often recited in reference to human travel; more than 90% of the world's land is now accessible within 48 hours from the nearest city (143). Often overlooked, however, is the international movement of animals, an enterprise that runs deep in recorded human history to the early civilizations of China, Egypt, Latin America, and Mesopotamia. Zoos, as public institutions, are a recent incarnation of the acquisition and display of exotic animals, the word "zoo" dating from 1847 as a nickname for zoological garden (70). The concentration of hundreds to thousands of animals from all zoogeographic regions of the world in novel surroundings now provides not only entertainment and public education, but also opportunities for animal research and conservation.

The first modern zoos date from the late eighteenth century when the Ménagerie du Jardin des Plantes was established in Paris in 1793, followed by the London Zoological Garden in 1828, and eventually the first zoo in the United States, the Philadelphia Zoological Garden, which opened in 1874 (70). Today, the global number of facilities exhibiting exotic animals is unknown, although in the United States alone, more than 2,300 animal operations represent themselves as zoos (108), while an untold number function without license. The Association of Zoos and Aquariums (4) estimated that its approximately 220 accredited facilities house nearly 150,000 captive amphibians, birds, mammals, and reptiles.

Interactions of humans, captive animals, and free-roaming wildlife in zoos often are mediated by arthropods such as cockroaches, filth flies, fleas, lice, mosquitoes, and ticks (**Table 1**). Among the early records of arthropods of medicoveterinary importance in zoos is a case from the early 1800s of scabies contracted by zoo personnel from a captive wombat (77). A modern-day survey of zoo veterinarians indicated that 3.2% of respondents had acquired scabies, 3.6% had been bitten by insects, and 14.2% had allergies to insects (55). About 25 arthropod-borne diseases have been reported in zoo animals during the past 30 years. Zoos, therefore, present arthropod problems in need of solutions.

The evolution of the zoo over the past two centuries from a "living natural history cabinet" to the modern-day "environmental resource center" (60) has brought new threats from arthropods and vector-borne diseases. Although research in zoos has provided insights into wildlife epidemics through the application of veterinary medicine in characterizing diseases (66), the awareness of arthropods of medicoveterinary importance has lagged, leaving their role in zoos understudied. Deaths of cheetahs (*Acinonyx jubatus*) from anemia caused by fleas in a zoo, for example, are recognized as preventable mortality (108), but only if zoo personnel are aware of the potential for arthropod problems.

We review the scattered literature on arthropods of medicoveterinary importance associated with zoos. We show that zoos are ideal settings for (*a*) detecting arthropods and arthropod-borne diseases of potential threat to animals and humans, (*b*) conducting research on arthropods of medicoveterinary importance, (*c*) developing management programs for arthropods, (*d*) building collaborations between entomologists and zoo personnel, and (*e*) educating the public and zoo personnel about medically important arthropods. Although arthropods have become prominent as the focus of exhibits (e.g., insect zoos), we are concerned here only with those of unintentional presence and potential medical or veterinary importance.

THE ZOO ENVIRONMENT

All zoo-animal spaces on the planet could fit within New York's borough of Brooklyn (27). Yet, zoos represent one of the few environments in which exotic and native vertebrates, arthropods, and humans routinely comingle—where radically different genomes intersect and selection pressures transcend traditional boundaries. Although uniquely free of predators,

Table 1 Examples of novel records of arthropods of medicoveterinary importance in zoos

Arthropod[a]	Pathogen	Associated vertebrates	Zoo	Situation (Reference)
Blattaria				
(Cockroaches)	*Gongylonema macrogubernaculum*[b]	African squirrels (*Funisciurus substriatus*, *Xerus erythropus*), 3 primate species	Baltimore Zoo, MD	Deaths of 13 animals, possibly by ingesting infected cockroaches (28)
Blattella germanica	*Pterygodermatites nycticebi*[b?]	Golden lion tamarin (*Leontopithecus rosalia*)	National Zoological Park, Washington, DC	1 death; cockroaches implicated as intermediate hosts (89)
Blattella germanica, *Supella longipalpa*, *Acheta domesticus* (cricket)	*Geopetitia aspiculata*[b]	24 bird species	Lincoln Park Zoo, IL	Deaths among 24 species; experimental infection of intermediate hosts (48)
Nauphoeta cinerea[c]	*Oxyspirura conjunctivalis*[b]	9 primate species	Moscow Zoo, Russia	Ocular disease (61)
Hemiptera				
Triatoma longipennis	*Trypanosoma cruzi*[b]	Polar bear (*Ursus maritimus*)	Guadalajara Zoo, Mexico	Death of bear (62)
Triatoma sp.	*Trypanosoma* sp.[b]	Sugar glider (*Petaurus breviceps*), hedgehog (*Atelerix albiventris*)	Unidentified zoo	Deaths of 6 animals; transmission possibly by ingesting bugs (76)
Siphonaptera				
Echidnophaga sp.[d]		Agoutis (*Dasyprocta mexicana*), maras (*Dolichotis* sp.)	Africam Safari, Mexico	Deaths of 2 agoutis from anemia; milder cases in maras (31)
Orchopeas sp.	*Francisella tularensis*[b]	Black and red tamarins (*Saguinus nigricollis*), talapoins (*Cercopithecus talapoin*)	Assiniboine Park Zoo, Canada	Deaths of 4 animals; rodents with fleas and tularemia (93, 105)
Pulex simulans[d]		Giant anteaters (*Myrmecophaga tridactyla*), aardvarks (*Orycteropus afer*)	Rolling Hills Wildlife Adventure, KS	Fleas controlled with imidacloprid and bedding replacement (92)
Diptera: Ceratopogonidae				
Culicoides	Blue tongue virus[b]	East African greater kudu (*Tragelaphus strepsiceros*), Reeve's muntjac (*Muntiacus reevesi*)	San Diego Zoo and Wild Animal Park, CA	Deaths of kudu and muntjac (57)
Diptera: Culicidae				
Mosquitoes	Eastern equine encephalitis virus[b]	African penguins (*Spheniscus demersus*)	Mystic Aquarium, CT	64% of 22 penguins infected (132)
	Plasmodium spp.[b]	4 bird species	National Zoological Park, Washington, DC	Deaths of most affected birds (136)
	Usutu virus[b]	Great gray owls (*Strix nebulosa*)	Tiergarten Schönbrunn Vienna Zoo, Austria	Deaths of 5 birds (141)
	West Nile virus[b]	14 bird species	Bronx Zoo, NY	Deaths of 27 birds (122)
	West Nile virus[b]	Polar bear (*Ursus maritimus*)	Toronto Zoo, Canada	Paraparesis followed by euthanization (40)

(Continued)

Table 1 (*Continued*)

Arthropod[a]	Pathogen	Associated vertebrates	Zoo	Situation (Reference)
	Dirofilaria immitis[b]	Humboldt penguin (*Spheniscus humboldti*)	Akita Municipal Omoriyama Zoo, Japan	First avian record of heartworm; death possibly caused by *D. immitis* (113)
Diptera: Psychodidae				
Sand flies	*Leishmania* sp.[b]	Red kangaroo (*Macropus rufus*)	Unidentified zoo, Australia	Locally acquired leishmaniasis; vectors putatively absent from Australia (110)
Phlebotomus sp.	*Leishmania* prob. *donovani*[b]	Fennec fox (*Vulpes zerda*)	Lincoln Park Zoo, IL	Visceral leishmaniasis; fox originally from Sahara Desert (26)
Diptera: Calliphoridae				
Chrysomya bezziana[d]		21 mammal species	Zoo Negara, Malaysia	91 myiasis cases, 21 deaths (120)
Acari (Mites)				
Audycoptid mites[d]		Polar bears (*Ursus maritimus*), black bear (*U. americanus*)	Sacramento Zoo, CA	Mange (47)
Hirstiella dioli[c]		4 iguana species	Taronga Zoo, Australia	46 mites on hosts (42)
Pneumonyssus procavians[c]		African rock hyrax (*Procavia capensis*)	Kansas City Zoo, MO	Mite from deceased host (133)
Sarcoptes scabiei[d]		5 ruminant species, humans	5 zoological gardens, Israel	Mange; 107 deaths (148)
Acari (Ticks)				
Amblyomma variegatum	*Ehrlichia ruminantium*[b]	Sitatunga (*Tragelaphus spekei*)	Jos Zoo, Nigeria	Heartwater; 1 death (101)
Aponomma komodoense[c?]		Komodo dragon (*Varanus komodoensis*)	Miami Metro Zoo, FL	3 animals infested (18)
Rhipicephalus evertsi[c]		Eland (*Taurotragus* sp.)	Africa USA, Florida	Eradication with DDT (16)
Rhipicephalus praetextatus[c]		Aardvark (*Orycteropus afer*)	Busch Gardens, FL	1 male tick (133)

[a]Taxa in parentheses refer to probable vector, but connection not investigated or demonstrated.
[b]New vertebrate-host record for pathogen.
[c]New country record for arthropod.
[d]New host record for arthropod.

these concentrations of conspecifics and mixed species are prone to parasitic infections and arthropod-borne diseases (65). Many zoos are located in areas that historically were hotbeds for arthropod-borne diseases of humans (1), and they continue to sit in vulnerable areas, often near population centers and international nodes such as airports, military bases, ports, and universities. The annual flow of more than 600 million visitors through zoos worldwide (60) brings the equivalent of nearly 10% of the global human population in contact with exotic animals.

Zoos are human-maintained, urbanized islands that harbor a mix of free-ranging opportunists and captive animals in indoor and outdoor exhibits, often with year-round climate control. The animal displays feature disparate habitats (e.g., arctic and tropical) in remarkable proximity, complemented by exotic and native flora, public amenities, storage areas, and refuse—all relatively stable over space and time and all favorable to arthropods of medicoveterinary importance. Although zoos are ecologically insular, their exotic animals remaining within the perimeter, they are connected with the surrounding environment through daily bidirectional movement of humans, wildlife, and arthropods. Ectoparasites, for instance, were associated with 91% of 96 free-roaming animals of 14 species, including humans, in two South Carolina zoos (98). Giant anteaters (*Myrmecophaga tridactyla*) infested with fleas probably acquired the parasites from free-roaming wildlife (92). *Heterodoxus longitarsus*, a louse native to Australia, spread to dogs after kangaroos were introduced to zoos and circuses in the United States (43). Molecular analyses of 52 blood meals in mosquitoes from South Carolina's Riverbanks Zoo indicated that about 40% of the meals came from free-roaming wildlife (95).

The two-way flow of free-living arthropods and opportunistic wildlife with their ectoparasites also facilitates the movement of vector-borne pathogens. The most notorious example of vector-borne disease in the history of zoos involved the mosquito-borne West Nile virus in the United States, which infected and killed exotic animals in about 100 zoos from its discovery in 1999 through the end of 2002 (140). During the 1999 West Nile outbreak, 34% of 368 birds and 8% of 117 mammals in the Bronx Zoo/Wildlife Conservation Park tested positive for the virus (80). An outbreak of *Culicoides*-borne African horsesickness in Spain was traced to the importation of zebras from Namibia to the Safari Park in Aldea del Fresno (109). Six bacterial agents, including species of *Bartonella* and *Rickettsia*, were detected in ectoparasites from free-roaming animals in South Carolina zoos (98).

The stark cages and metal bars that characterized zoos of recent memory have given way to more natural environments (30). Today's zoos emphasize the authenticity of the environment in which animals are exhibited. In 1972, the San Diego Wild Animal Park opened, ushering in a new era of exhibiting animals, with 730 hectares of diverse habitats for uncaged, mixed-species assemblages of exotic animals. Although habitat authenticity might permit animals to display more natural behaviors and defenses against arthropods, it also increases habitat heterogeneity, carrying with it the potential for greater arthropod problems, such as those from ticks requiring multiple hosts. Bamboo, a common planting in zoos, imparts an exotic ambience to animal facilities, but the stumps provide habitat for container-breeding mosquitoes such as the invasive Asian tiger mosquito, *Aedes albopictus* (95). Bromeliads in Florida's Parrot Jungle Island provided habitat for the yellow fever mosquito, *Aedes aegypti* (116), and about 15% of exotic bromeliads in New Zealand's Auckland Zoological Park supported larvae of another invasive mosquito, *Aedes notoscriptus* (33). Covering the traditional bare floors of monkey exhibits with wood chips or straw improves the health of monkeys, but increases the level of insect nuisance (108). The introduction of plants and mulch increases the probability that ticks will be introduced into a zoo (7, 124). Introduction of landscaping materials into South Carolina's Greenville Zoo is believed to have brought the exotic ant *Pachycondyla chinensis* into the zoo, where it is considered one of the most pestiferous insects to humans (96).

The arrangement of exhibits and the feeding and housing of animals can subject the animals to stresses and associations to which they are not adapted and can influence the associated arthropod community. Proximity of taxonomically related species of animals fosters transfer of ectoparasites or movement of free-living arthropods bearing parasites and pathogens. The presence of the large chicken louse, *Goniodes gigas*, on peafowl (*Pavo cristatus*) in a Michigan zoo was attributed to sheltering the birds with chickens (58). African ticks on a captive Asian water monitor (*Varanus salvator*)

Reservoir: an organism that serves as a host for a parasite or pathogen and is a source of infection, typically without manifesting acute disease

Exotic arthropod: an organism imported into an area where it does not occur naturally or typically does not have feral populations

probably were acquired from African reptiles in captivity (144). German cockroaches (*Blattella germanica*) bearing acanthocephalans moved freely among cages of various species of great apes in Utah's Hogle Zoo, causing an epizootic of acanthocephaliasis (90). The introduction of mammalian prey into a royal python (*Python regius*) exhibit in Nigeria's Ibadan Zoo resulted in transfer of the tick *Amblyomma hebraeum* from the mammal to the reptile (2). Some evidence indicates that mosquitoes rest in habitats adjacent to the hosts from which they acquire blood (95), suggesting that taxonomically related hosts, logically grouped for care and exhibition, would be at risk for transfer of pathogens. Increased vigilance, therefore, is needed to diagnose subclinical infections of arthropod-borne diseases in zoo animals, particularly if refractory hosts serve as pathogen reservoirs (53).

TAXONOMIC TRENDS AMONG ARTHROPODS IN ZOOS

Little is known of how the arthropod faunas of zoos compare with the surrounding biodiversity, largely because surveys of arthropods in zoos are limited in number and taxonomic scope and because the arthropod fauna of the surrounding area is often poorly known. About 20 surveys of arthropods of medicoveterinary importance have been conducted in zoos (**Table 2**), and about one-quarter of those were incidental to prospection for disease agents. More than half of the arthropod surveys focused on ectoparasites and about one-third on Diptera.

Free-living arthropods in zoos should represent a subset of the regional fauna. This pattern holds for Diptera in German zoos (72), ceratopogonids in South Carolina zoos (99), mosquitoes in the Baltimore Zoo (10), and fleas in the Los Angeles Zoo (114). Three of the 16 species of mosquitoes in New Zealand have been found in the Wellington Zoo (33), where the mosquito *Ae. notoscriptus* was more common than in nearby native forests (36). Adult densities of mosquitoes were greater in New Zealand's Auckland Zoo than in native habitats (33).

Ectoparasites in zoos would be expected to include both exotic and native representatives. The importation of exotic animals also involves the transport of symbiotic organisms, including those that threaten animal welfare and human health. The problem is not insignificant; the most frequent mode of entry for pests is through importation of host animals and their products (14). The United States brought in nearly 1.5 billion animals from 2000 to 2006, often with minimal disease surveillance; the vast majority of animals were designated for the pet trade, with a proportion funneled to zoos (119). The major importers of reptiles in Florida, for example, had animals infested with exotic ticks, some of which were transferred to zoos (17). Based on USDA interceptions at ports of entry into the United States from 1961 through 1993, 69% of exotic arthropods were found on equids, especially zebras, and another 14% on various species of antelopes, with a few additional records from giraffes, rhinoceroses, and tortoises (14). Of 58 exotic ticks detected on premises in the United States from 1958 through 1990, 26% were on rhinoceroses that typically entered legally from Africa (14). The red tick, *Rhipicephalus evertsi*, a vector of the agent of heartwater, was reported for the first time in the United States at Africa USA in Florida and was believed to have arrived from Africa on elands or zebras (16). The discovery of exotic ectoparasites on imported hosts can lead to government prohibition of further imports of the hosts, as with the 1989 ban on importation of ratites (e.g., ostriches) into the United States (88).

Most surveys of ectoparasites in zoos reveal exotic species. Ixodid ticks are among the best-represented ectoparasites on hosts imported from Africa into zoos (14). Exotic ticks also are collected from reptiles in zoos (45, 98). Exotic chewing lice have been found on ostriches in Brazilian zoos (134), the exotic elephant louse, *Haematomyzus elephantis*, on its namesake host in North American zoos (121), and the imported hippoboscid fly *Hippobosca longipennis*

Table 2 Surveys of arthropods of medicoveterinary importance in zoos

Zoo	Taxa surveyed	Number of arthropod taxa	Reference
Auckland Zoological Park, New Zealand	Mosquitoes	3 species	33
Amsterdam and Rotterdam Zoos, Netherlands	Ectoparasites of birds, mammals, reptiles	14 species	127
Aviarios Sloth Sanctuary, Costa Rica	Ectoparasites of sloths	2 species	117
Baltimore Zoo, MD	Mosquitoes attracted to African penguins (*Spheniscus demersus*)	7 species	10
Chicago Zoological Park, IL, and St. Louis Zoological Park, MO	Ectoparasites of birds, mammals, and reptiles during necropsy	10 species	137
Greenville and Riverbanks Zoos, SC	Ectoparasites of captive and free-roaming animals[a,b]	46 species	98
Greenville and Riverbanks Zoos, SC	Ceratopogonidae	16 species	99
Konya Zoo, Turkey	Chewing lice of 15 bird species[b]	7 species	37
Los Angeles Zoo, CA	Fleas on roof rats (*Rattus rattus*)	3 species	114
Maiduguri Zoological Garden, Nigeria	Ectoparasites, biting Diptera	5 species (fleas, ticks), 5 taxa (Diptera)	84
Moscow Zoo, Russia	Cockroaches	4 species	61
National Zoological Gardens, Sri Lanka	Ticks of reptiles[a,b]	3 species	45
National and Quilpué Zoos, Chile	Lice of raptors[a,b]	4 species	51
Quinzinho de Barros Zoo, Brazil	Chewing lice[a,b]	8 species	118
São Paulo Zoo, Brazil	Chewing lice[a,b]	28 species	134
Unidentified zoo, Kanagawa, Japan	Mosquitoes	11 species	41
Whipsnade Wild Animal Park, England	Chewing lice on peafowl (*Pavo cristatus*)	2 species	123
Zoological Gardens, Germany	Diptera	159 species	72
Zoos and wildlife theme parks, FL	Ticks of reptiles	2 species	17

[a]New host record for arthropod.
[b]New country record for arthropod.

on cheetahs in zoos in the United States (67). In most cases, flea infestations of zoo animals probably are acquired locally (6, 31, 92).

Mites and other small ectoparasites are common on zoo animals but typically go unnoticed unless they produce conspicuous problems, such as mange or death of the host (47, 50, 148). Small ectoparasites can be difficult to detect without destructive techniques that involve the dissolution of the entire host skin, a technique rarely used in zoos (123). Internal arthropod parasites also prove elusive unless symptoms alert zoo personnel to their presence or they are discovered during autopsies, as with lung mites in primates (63, 126).

Among the more common arthropods in zoos are cockroaches and filth flies, the latter breeding in bedding, dung, and food of the animals. Few reports document the importation of either of these groups. Imported cosmopolitan arthropods, such as house flies (*Musca domestica*), are likely to be detected infrequently, blending imperceptibly with local, resident populations. A significant danger, however, is that cosmopolitan arthropods could introduce exotic pathogens into zoos (98). More than 100 species of filth flies have been documented in a German zoo (72) and more than 50 species in two South Carolina zoos (95), but whether any of these flies was of extralimital origin is unknown.

Insects that rely on stealth to exploit hosts probably tend to be overlooked in zoos. Black flies, for example, rarely have been documented

Zoonoses: diseases of animals, of which the causal agents can be transmitted to humans

in zoos (72) but are likely to be present, feeding for example in the ears of mammals, particularly if large rivers are nearby. Records of phlebotomine sand flies in zoos are wanting, although phlebotomine-borne leishmaniasis is not uncommon in zoos (110), particularly in South America (83).

Although mosquitoes can be furtive in their quests for blood meals, they are a widely recognized problem in zoos. Their presence is often signaled by attacks on humans (33), frequent manifestation of mosquito-borne diseases among captive animals (10, 128), or discovery of pathogens, such as the heartworm, *Dirofilaria immitis*, at necropsy (113). Mosquitoes lead the list of arthropods responsible for vector-borne diseases of zoo animals, both in types and case numbers.

Some arthropods pose minimal risk of importation or establishment in zoos, although they might bear diseases that could enter zoos via infected animals. In most regions, zoos are at low risk for tsetse. In areas where tsetse are native, however, particularly Africa south of the Sahara, they have invaded zoos (131), and tsetse-borne trypanosomiasis has entered zoos beyond the native range through the importation of infected animals (85). Some widespread and conspicuous arthropods, such as screwworms, stable flies, and tabanids, are rarely reported in zoos but can pose nuisance or disease problems (84, 111, 120).

ZOOS AS NATURAL EXPERIMENTS AND LABORATORIES

Zoos are natural experiments. Their unique ecology can be used to explore novel and well-established relationships among vertebrates, arthropods, humans, and pathogens, providing insight into ectoparasite-host associations and zoonoses. Yet, less than half of zoos responding to a 1986 survey reported a formal research relationship with a university (19).

Most research on medically important arthropods in zoos or the pathogens they transmit is reactive, based on finding a treatment for illness in animals. Even when arthropods are incriminated, they often are not sought, entomologists are rarely involved, and the literature typically is not disseminated beyond a zoo audience. Some investigations of medically important arthropods have been undertaken in response to epizootics in zoos, including a survey of mosquitoes at the Auckland Zoo after an outbreak of avian malaria and avian pox (33).

Considerable work has been conducted in zoos on metazoan parasites mechanically transmitted by arthropods (e.g., cockroaches). Zoo animals typically are assumed to acquire the parasites by eating an arthropod host (46). Studies in zoos have identified the vector or potential vector (48, 61), have shown that the parasite can be transmitted among animals (49) and zoos (89), have revealed that some animals act as reservoirs of the parasites (28), and have documented the diversity of hosts and associated pathology (28, 48, 61). The potential for mechanical transmission of bacteria or viruses is understudied, although agents, such as those causing Q fever and tularemia, have been found on arthropods in zoos (93, 97, 105).

Arthropods of medicoveterinary importance can provide ingenious solutions to questions of host specificity and the diagnosis of animal diseases. To demonstrate the adaptation of lice to their hosts, Ewing (43) removed lice from hosts in the National Zoological Park, allowed them to feed on him, and recorded the time until their death. The snake mite *Ophionyssus serpentium* was collected from snakes at Ohio's Cleveland and Columbus Zoos and used to show that it is a mechanical vector of a bacterium that causes hemorrhagic septicemia in snakes (20). The mosquito *Aedes aegypti* was used to xenodiagnose hemogregarine infections in captive reptiles (146), and triatomid bugs were used as living syringes in a survey of bluetongue virus in European zoos, a technique pioneered at the Wuppertal Zoo in Germany (82).

The observational nature of animal care enhances the utility of zoos as sentinel sites for emerging diseases (87). Observations during the West Nile epidemic at the Bronx Zoo established a record of behavioral and pathological changes in exotic and native birds and the diversity of birds that could be infected, before West Nile virus infected birds beyond New York and surrounding states (122). The investigation also provided new molecular protocols for diagnosing West Nile infections in wildlife. A subsequent serosurvey documented susceptible birds and mammals in the collection, indicating potential wild reservoirs, and demonstrated that they were not the source of the virus's introduction into the United States (80). Novel diseases in animals or serosurveys of zoo animals can provide insight into potential wild- and captive-animal reservoirs of arthropod-borne pathogens such as bluetongue virus (57, 112), Lyme spirochetes (124), *Leishmania* spp. (26, 81, 83, 110), and *Trypanosoma cruzi* (56). Zookeepers themselves can be the subjects of serosurveys; in two zoos, for example, employees have been tested for Lyme spirochetes (29, 107).

Translocated or reintroduced animals can have a wild or captive origin and destination. Because of ecological differences between environments, animals should be screened for pathogens before movement, destinations should be surveyed for pathogens to which the animals might not have been exposed, and medical records, including vouchers of ectoparasites, should be kept before and after release (8). Zoos frequently nurse wild animals to health (e.g., raptor clinics) or run captive-breeding and reintroduction or translocation programs. An animal in these circumstances can become a source of ectoparasites or arthropod-borne pathogens; animals, therefore, should be quarantined not only on arrival at a zoo, but also on exit (15, 71). Ectoparasite burdens were one of the metrics used at the Adelaide Zoo to assess the health of yellow-footed rock wallabies (*Petrogale xanthopus*) under consideration for a reintroduction program (25). Of 485 translocation cases, 24% of the animals were not examined for diseases, injuries, or parasites before release, and only 22% of the releases had protocols adequate for estimating the number of animals lost to disease after release (52).

Diseases in zoo animals and the documentation of behavioral and pathological changes can help conservation biologists anticipate future occurrences of wildlife diseases and prepare treatments for diseases and responses to outbreaks. Currently, a polar bear (*Ursus maritimus*) can contract Chagas disease or West Nile illness only in a zoo or similar setting (40, 62). With global climate change and increased international movements of animals, interactions among previously disparate hosts, parasites, and vectors could increase. Zoos, thus, can be used as an early-warning or prediction system for what could happen beyond the zoo environment (75). For example, an eastern equine encephalitis outbreak in captive African penguins (*Spheniscus demersus*) provided documentation of physiological changes and mortality in the birds, although the disease has not occurred in penguins outside captivity (132).

Serosurvey: serological survey; a screening of the serum of animals for evidence of infection by pathogens

Many animals react differently to the same pathogens, and emerging wildlife diseases can be anticipated by observing diseases in captive animals. Tick toxicity has been a problem in Australia's Taronga Zoo, and a series of case reports presented a comparison of differing responses to tick bites in the zoo's animals (7). A wealth of data was gathered in zoos on the acquisition, transmission, and treatment of avian malaria in penguins (9, 10, 41, 86, 128) before the disease was discovered in wild populations (78). An outbreak of avian malaria in endangered yellowheads (*Mohoua ochrocephala*) at the Oruna Wildlife Park in New Zealand gave researchers a glimpse into the severity of illness should the birds encounter the pathogen in the wild (3). Another pathogen that can be explored in zoos is *D. immitis*, the causal agent of dog heartworm, which has been found in a variety of zoo animals, many for the first time (94, 113).

Most novel observations and disease studies that have emerged from zoos occurred almost accidentally, but they emphasize that zoos are excellent arenas for conducting research to understand the epidemiology of arthropod-borne pathogens transmitted to wildlife. Some studies suggest that high host heterogeneity facilitates pathogen transmission and emergence of

Biosurveillance: biological surveillance; monitoring of early-warning sites, such as zoos, to identify possible threats to animal and public health from invasive organisms and disease outbreaks

epizootics (68), whereas others suggest that high host heterogeneity has a dilution effect that lowers parasite transmission (13). Zoos provide environments for the study of host choice and pathogen presence in blood-feeding arthropods, parasite diversity in areas with high host heterogeneity, and development of diagnostic tests to detect pathogens in wild animals. Zoos and entomologists will benefit from collaborative research on host preferences of blood-feeding arthropods; testing primers for blood meal identification in arthropod vectors, using banked zoo sera; and serosurveys for understanding arbovirus transmission, mutation, and historical distribution.

The unique faunal attributes of zoos benefit entomology by providing material for taxonomic descriptions, systematic revisions, population genetics, and other areas of investigative biology. The value of zoos in illuminating the morphology, life histories, and taxonomy of parasites, including arthropods, was recognized as early as 1923 (139). The National Zoological Park in Washington, DC, was instrumental in providing material for the first comprehensive monograph on the sucking lice of American monkeys (44). In a survey of filth flies in South Carolina zoos, 10% of 51 species were new to science, including muscid and sarcophagid flies reared from elephant and lion dung, respectively (95). The parasitic mite *Hirstiella diolii* was described originally from material collected from the rhinoceros iguana (*Cyclura cornuta*) in the London Zoo (5). Ectoparasites, such as lice and mites, are difficult to collect from feral animals but can be surveyed more readily in zoos (123). Nineteen of 28 species of chewing lice collected from birds in the São Paulo Zoo had not been found previously in Brazil, even though the distributional ranges of most of the birds include Brazil (134). Ticks (*Amblyomma dissimile*) collected from snakes in four Venezuelan zoos provided more than half of the source material for an investigation of the tick's population genetics (74). The tick *Aponomma varanensis* from a king cobra (*Ophiophagus hannah*) in the Bronx Zoo was the subject of a study on chromosomes and reproductive biology (102). The deposition of arthropods (voucher specimens) collected in zoos, whether during animal wellness exams and necropsies or through routine surveillance and research, ensures a wealth of future information.

BIOSURVEILLANCE IN ZOOS

The accidental importation of exotic arthropods on legally imported animals puts animal caretakers and zoo veterinarians on the front line of vigilance and protection (14). A proactive approach to biosurveillance is economically beneficial to zoos, considering the costs to mitigate arthropod problems. Zoo animals provide a readily available and affordable vanguard alert to potential zoonoses. Animals are the source of more than 70% of emerging infections worldwide (73). Zoo biosurveillance is a relatively new approach to pathogen detection, whereby captive animals are monitored for health-related events that pose threats to animals and the public. An international system for monitoring infectious diseases in free-ranging and captive wildlife was suggested as early as 1993 (91).

The 1999 outbreak of West Nile disease at the Bronx Zoo and its subsequent investigation marked a seminal event in recognizing the role that zoos can play as sentinels for emerging diseases (87). Tracey McNamara, credited with associating deaths of crows (*Corvus brachyrhynchos*) at the Bronx Zoo with concurrent encephalitis cases at the city hospital, conceived a national consortium of zoos and public health officials—the National Zoological West Nile Virus Surveillance Group (138). Dominic Travis of the Chicago Lincoln Park Zoo, who set up the project's database, considered zoos to be perfect sentinels for tracking the disease (100). Zoo animals lend themselves to biosurveillance because they are broadly susceptible, located countrywide near humans, and associated with veterinary expertise and archival samples of serum and tissue (64). Less clearly articulated in proposals to integrate zoos into public health surveillance networks, however, is the beneficial role that entomologists could play in diagnosing sources of infection in zoos.

Other arthropod-borne diseases with potential for rapid spread reinforce the role of zoos in biosurveillance. The threat of another West Nile scenario has been claimed for the mosquito-borne African Usutu virus, discovered in 2001 among birds in Europe, including some in an Austrian zoo (141). Of 146 European zoos responding to a survey, 49 were within 20 km of cases of *Culicoides*-borne bluetongue virus in a 2006–2007 epizootic, and 13 zoos had 62 cases of clinical disease (112). In Canada's Assiniboine Park Zoo, four monkeys succumbed to tularemia and a zoo veterinarian acquired the disease after being bitten by an infected monkey; the mode of transmission to the monkeys was not determined but could have been through fleas on free-roaming ground squirrels that tested positive for the causal agent (105).

The groundwork is in place to develop more inclusive disease surveillance systems, such as a system for all animals that depend on humans but are not used as food (125). The Zoo Animal Health Network, managed by Chicago's Lincoln Park Zoo, with support from the USDA Animal and Plant Health Inspection Service (APHIS), integrates with national animal and public health monitoring to provide leadership in education and research on infectious diseases of zoo animals and visitors (21). BioPortal, an infectious disease–informatics project, integrates datasets for West Nile and other diseases to provide predictive models and data analyses (149). The Zoological Information Management System is a step toward a comprehensive disease database equipped with animal diagnoses and treatment data in real time (22). The One Health Initiative seeks to integrate animal- and human-disease monitoring programs (69). A new role for zoos lies in biosecurity. Many Category A and B bioterrorism agents recognized by the Centers for Disease Control are zoonotic agents, including some that are arthropod borne.

ARTHROPOD MANAGEMENT IN ZOOS

The first step in managing arthropods in zoos involves a survey of relevant taxa, extended to import centers that temporarily house animals destined for zoos, with subsequent monitoring of species deemed potential risks. The idea of a zoo free of arthropods of medicoveterinary importance, however, is neither realistic nor sensible. For each case, entomologists and zoo personnel must determine how much reduction is needed and possible, recognizing that animals in the wild have evolved mechanisms for coexisting with arthropods and the pathogens they transmit. Infections with some pathogens are more prevalent and virulent in species of zoo animals that do not have contact with the vector in their native environment (124, 136). Zoo personnel have discussed the possibility of increasing the tolerance of zoo animals to parasites by judiciously infesting them (30). The complexities of interactions between zoo animals and arthropods are illustrated by the case of an aardwolf (*Proteles cristatus*) that would not eat the food provided, yet survived by eating free-living cockroaches (108). Cockroach consumption is common in zoo animals such as primates (38), but any benefit must be weighed against the risk of ingesting individuals with vertebrate parasites and pathogens (28, 48, 61).

Although cockroaches and ectoparasites can be problematic in zoos year round, other arthropods, such as hematophagous flies, are more seasonal. Mosquitoes cause problems during warm weather, and outbreaks of avian malaria tend to be confined to the summer, prompting personnel at the San Diego Zoo to experiment with a curtain of air passing across penguin exhibits to discourage entry of mosquitoes (54). Restricting the importation and quarantine of animals to periods of reduced arthropod activity could reduce problems of acquiring vector-borne pathogens before establishment of the animals in a zoo. The Netherlands's Rotterdam Zoo initiated a study to investigate avian malaria in African penguins

(*Spheniscus demersus*) and puffins (*Fratercula* sp.) and determined that a mosquito management plan was needed, with biological control providing the backbone (128). Additional measures to prevent avian malaria in zoos include housing birds indoors, covering outdoor exhibits with mosquito netting, and developing integrated mosquito management programs.

Early efforts to control arthropods in exotic-animal exhibits typically involved chemicals including DDT (16). Because of collateral risks, chemical applications now are considered a last resort in zoos (24), or if used, they are of a more benign nature. Insecticidal ear tags and permethrin pour-on insecticides, for example, have been used to control myiasis in musk oxen (*Ovibos moschatus*) at the Minnesota Zoological Garden (106, 147). One of the first approaches to treating zoo animals for ectoparasites in a manner sensitive to the needs of zoos (e.g., low toxicity) involved the use of a silica aerogel at the San Diego Zoo (129). Oral doses of ivermectin often are administered to control parasites such as mites that cause mange (148). Certain arthropod control programs require specific approaches to accommodate the needs of endangered or vulnerable species, such as komodo dragons (*Varanus komodoensis*) infested with exotic ticks (18). Some zoo animals are trained to receive insecticide applications, whereas others must be hand searched for ectoparasites (7).

Alternatives to chemical control, such as classic biological control, are limited only by the creativity of those designing the programs. The potential of the emerald cockroach wasp, *Ampulex compressa*, to control cockroaches was investigated at the Artis Zoo in The Netherlands (135), and chickens and quail were introduced into high-risk exhibits to eat scorpions (*Centruroides sculpturatus*) at the Phoenix Zoo in Arizona (103). Control of stable flies (*Stomoxys calcitrans*) by trapping at Australia's Taronga Zoo, although effective, required too much maintenance to be sustainable (111). Food containers for animals can be designed to exclude insects, and trash disposal can be managed on a daily schedule (24). Larval mosquitoes often develop in discarded containers (34), which can be minimized by informed zoo personnel (35) through routine disposal and tipping. Features that are integral to the animals and exhibits can be ameliorated by introducing mosquito-feeding fish (136), regularly flushing exhibit ponds and drinking troughs, applying *Bacillus thuringiensis israelensis* (35, 116), filling tree holes with utility foam (115), and drilling holes in the bottoms of rubbish containers to prevent water accumulation. Zoos also are ideal long-term surveillance sites for mosquitoes and can be included in routine monitoring programs that are in place in many areas.

An integral step toward preventing zoonoses in zoos is quarantine. New arrivals are placed in quarantine, typically for 2–4 weeks, where they undergo medical examination that includes screening for parasites (108). Quarantine, however, also places stresses on animals and can increase susceptibility to arthropod-borne diseases (145). Movement of infected animals among zoos remains a concern even after quarantine. *Plasmodium* infections, probably acquired locally, occurred in 36% of captive pisittacine birds in three Brazilian zoos, necessitating tighter controls on the movement of birds (11).

Vector control is recognized as a viable solution for minimizing parasitic infections, beginning in holding facilities (65, 91). Yet, vector control sometimes is recommended before specific vectors have been identified or incriminated. Of 29 papers reporting diseases actually or probably arthropod-borne in zoo animals, published in the *Proceedings of the American Association of Zoo Veterinarians* (1968–2008), only 21% sought or investigated the vectors, while 31% implemented or recommended control strategies against arthropods before vectors were confirmed or identified. Proactive vector control needs to begin in the design of exhibits, and requires partnerships with local control programs, pest-control operators, public health agencies, and zoo management and veterinary services.

The Animal Welfare Act requires that facilities exhibiting animals be licensed with the

USDA-APHIS and that they maintain an effective management program for pests, including arthropods. Before the implementation of a new pest control program at the Baltimore Zoo, employees brought in their own ant and wasp controls (32). Many zoos now have well-run, legal pest control programs (23, 24, 32). Few entomologists, however, are employed by zoos to address arthropod problems, although the National Zoo created a position for a pesticide program manager in 2003 and hired an entomologist with IPM experience (24). Other zoos contract with private pest control companies or work with local vector control programs (23, 32). Disney's Animal Programs initiated a computer-based training program for zoo employees that included a component on avoiding insect/vector exposure (130). Given the modest funding typically available for zoos to conduct a dizzying variety of operations, collaborations with entomologists provide viable options for arthropod management and vector control.

ARTHROPOD CONSERVATION AND EDUCATION IN ZOOS

Zoos can serve as public platforms for disseminating information about medicoveterinary entomology. Jewel wasps were exhibited in The Netherlands's Artis Zoo to educate the public about biological control of cockroaches by parasitoids (135). Zoos with active management programs can not only educate their own personnel (24), but also improve their public image by educating visitors on proactive approaches to safeguarding animal and public health. Efforts by zoos to inform the public about arthropods of medicoveterinary importance are nascent, but local entomologists could contribute by presenting public talks and demonstrations at zoos while training and involving zoo personnel.

Entomologists also can serve as conservation consultants. With increasing emphasis on the importance of zoos in animal conservation (150), each vertebrate species should be recognized as a collection of miniature ecosystems supporting unique communities of symbiotic arthropods, many of which are host specific. If the host faces extinction, so too does the community of species-specific arthropods (coextinction). The loss of these symbiotes, some of which are saddled with the stigma of being parasites, represents a diminution of biodiversity, in most cases before we know the role, pro and con, they might play in the ecology of their hosts. A poignant example is the loss of the chewing louse *Colpocephalum californici* when the last surviving members of its only host, the California condor (*Gymnogyps californianus*), were dusted with pesticide when they entered California zoos (39).

Because zoos tend to focus on rare, threatened, and endangered vertebrates, they also can collect and conserve affiliated ectoparasites. Zoos are ideal arenas for the conservation and breeding of threatened invertebrates, and groups exist that could oversee captive invertebrate care and educational outreach (59). Parasite conservation can be included in recovery plans for endangered species (12). Ectoparasites of animals could be conserved to serve as seed-stock during reintroductions, and animals gradually could be introduced to parasites they will encounter after reintroduction (71). If the last individuals of a vertebrate species are taken into captivity, rid of their ectoparasites, and subsequently reintroduced, they might be susceptible to generalist parasites or freed of constraints previously imposed by their parasites (142). If coevolved ectoparasites of captive animals are maintained in captivity, they could provide clues to paternity in promiscuous group collections (123), serve as components of educational exhibits on ecology, and provide genetic material to elucidate the evolutionary history of their hosts (142).

CONCLUDING REMARKS

The modern zoo of the past 200 years has changed in mission, philosophy, and structure from an operation shackled by ignorance of the connection between disease and vectors to a progressive leader in conservation biology and endangered species management. The importance of arthropod-borne diseases of captive animals is now widely recognized in zoos, and veterinary staffs are versed in recognizing the symptoms; yet, the vectors remain underreported. Although the importance of pathogen screening for zoo animals is recognized (79), the role of entomologists and the value of pathogen screening of arthropods associated with zoos often are not.

As zoos assume greater roles in wildlife conservation (27), proactive awareness of the vectors, not simply the diseases, becomes imperative. Eruptions of arthropod-borne diseases, such as West Nile encephalitis, focus attention on the problem, leading to monitoring networks (87, 138), but in the absence of outbreaks or monitoring programs, potential problems remain hidden. The unprecedented deaths of 12 tigers (*Panthera tigris*) at India's Nandankanan Zoo, caused by trypanosomiasis, were recognized as a tragedy requiring attention through "interaction and partnership between veterinarians, entomologists, and medical microbiologists" (104). Today's zoo thus represents a focus where interactions benefit entomologists, zoo veterinarians and staff, and the public. The entomologist is afforded specimens and a venue for conducting collaborative research, the zoo receives solutions to problems and prophylactic advice, and the public gains an added level of protection from vector-borne diseases and an appreciation for arthropods.

SUMMARY POINTS

1. Zoos are unique environments where captive and free-roaming animals, arthropods, humans, and pathogens interact in novel ways.
2. Zoos can serve as biosurveillance sites for invasive arthropods and arthropod-borne pathogens such as West Nile virus.
3. Zoos are excellent arenas for research collaborations among conservation biologists, entomologists, and veterinarians.
4. More research on arthropods of medicoveterinary importance in zoos is needed to ensure animal and human welfare.
5. A substantial literature on arthropods of medicoveterinary importance is disseminated largely to specialized groups; a need exists for more widespread exchange of this information.

DISCLOSURE STATEMENT

The authors are not aware of any affiliations, memberships, funding, or financial holdings that might be perceived as affecting the objectivity of this review.

ACKNOWLEDGMENTS

We are grateful to the staffs of South Carolina's Greenville Zoo and Riverbanks Zoo, especially K. Benson, J. Bullock, B. Foster, and K. Gilchrist for welcoming research in their facilities and fostering collaborations with entomologists. We thank K. Benson, B. Foster, J. Shimonski, A.G. Wheeler, and W. Wills for reviewing the manuscript. This is Technical Contribution No. 5777 of the Clemson University Experiment Station and is based, in part, on work supported by NIFA/USDA under project number SC-1700276.

LITERATURE CITED

1. Adler PH, Wills W. 2003. Legacy of death: the history of arthropod-borne human diseases in South Carolina. *Am. Entomol.* 49:216–28
2. Ajuwape ATP, Sonibare AO, Adedokun RA, Adedokun OA, Adejinmi JO, Akinboye DG. 2003. Infestation of royal python (*Python regius*) with ticks *Amblyomma hebraeum* in Ibadan Zoo, Nigeria. *Trop. Vet.* 21:38–41
3. Alley MR, Fairley RA, Martin DG, Howe L, Atkinson T. 2008. An outbreak of avian malaria in captive yellowheads/mohua (*Mohoua ochrocephala*). *N. Z. Vet. J.* 56:247–51
4. Association of Zoos and Aquariums. 2009. *Zoo and aquarium statistics.* **http://www.aza.org/zoo-aquarium-statistics/**
5. Baker AS. 1998. A new species of *Hirstiella* Berlese (Acari: Pterygosomatidae) from captive rhinoceros iguanas *Cyclura cornuta* Bonnaterre (Reptilia: Iguanidae). *Syst. Appl. Acarol.* 3:183–92
6. Baker RT, Beveridge I. 2001. Imidacloprid treatment of marsupials for fleas (*Pygiopsylla hoplia*). *J. Zoo Wildl. Med.* 32:391–92
7. Barnes J, Vogelnest L, Hulst F, Rose K, Bryant B. 2001. Tick toxicity in zoo animals in eastern Australia. *Proc. Am. Assoc. Zoo Vet.* (2001):140–45
8. Beck B, Cooper M, Griffith B. 1993. Infectious disease considerations in reintroduction programs for captive wildlife. *J. Zoo Wildl. Med.* 24:394–97
9. Beier JC, Stoskopf MK. 1980. The epidemiology of avian malaria in black-footed penguins (*Spheniscus demersus*). *J. Zoo Anim. Med.* 11:99–105
10. **Beier JC, Trpis M. 1981. Incrimination of natural culicine vectors which transmit *Plasmodium elongatum* to penguins at the Baltimore Zoo. *Can. J. Zool.* 59:470–75**

    10. Investigates the epidemiology of avian malaria in African penguins at the Baltimore Zoo.
11. Belo NO, Passos LF, Júnior LM, Goulart CE, Sherlock TM, Braga EM. 2009. Avian malaria in captive psittacine birds: detection by microscopy and 18S rRNA gene amplification. *Prev. Vet. Med.* 88:220–24
12. Blasier MW, Mashima TY. 2003. Endangered species recovery plans: considering parasite conservation. *Proc. Am. Assoc. Zoo Vet.* (2003):29–30
13. Bradley CA, Altizer S. 2007. Urbanization and the ecology of wildlife diseases. *Trends Ecol. Evol.* 22:95–102
14. Bram RA, George JE. 2000. Introduction of nonindigenous arthropod pests of animals. *J. Med. Entomol.* 37:1–8
15. Brossy JJ, Plös AL, Blackbeard JM, Kline A. 1999. Diseases acquired by captive penguins: What happens when they are released into the wild? *Mar. Ornithol.* 27:185–86
16. Bruce WG. 1962. Eradication of the red tick from a wild animal compound in Florida. *J. Wash. Acad. Sci.* 52:81–85
17. Burridge MJ, Simmons L-A, Allan SA. 2000. Introduction of potential heartwater vectors and other exotic ticks into Florida on imported reptiles. *J. Parasitol.* 86:700–4
18. Burridge MJ, Simmons L-A, Condie T. 2004. Control of an exotic tick (*Aponomma komodoense*) infestation in a komodo dragon (*Varanus komodoensis*) exhibit at a zoo in Florida. *J. Zoo Wildl. Med.* 35:248–49
19. Burton MS. 1993. Scientific study in a small zoo. *Proc. Am. Assoc. Zoo Vet.* (1993):293–300
20. Camin JH. 1948. Mite transmission of a hemorrhagic septicemia in snakes. *J. Parasitol.* 34:345–54
21. Chosy J. 2009. *NAHLN aids in zoo surveillance.* **http://archive.constantcontact.com/fs038/1102142667536/archive/1102655766931.html**
22. Cohn JP. 2006. New at the zoo: ZIMS. *BioScience* 56:564–66
23. Collins D, Powell D. 1996. Applied pest control at Woodland Park Zoological Gardens. *Proc. Am. Assoc. Zoo Vet.* (1996):290–95
24. **Committee on a Review of the Smithsonian Institution's National Zoological Park. 2004. Pest management. In *Animal Care and Management at the National Zoo: Interim Report*, pp. 55–58. Washington, DC: Natl. Acad. Press**

    24. Outlines elements and steps for successful integrated pest management in zoos, featuring the National Zoo in Washington, DC.
25. Conaghty S, Schultz DJ. 1998. Veterinary investigation of the yellow-footed rock-wallaby (*Petrogale xanthopus xanthopus*) for reintroduction. *Proc. Am. Assoc. Zoo Vet.* (1998):294–99
26. Conroy JD, Levine ND, Small E. 1970. Visceral leishmaniosis in a fennec fox (*Fennecus zerda*). *Pathol. Vet.* 7:163–70

27. Conway W. 2007. Entering the 21st century. See Ref. 150, pp. 12–21
28. Craig LE, Kinsella JM, Lodwick LJ, Cranfield MR, Strandberg JD. 1998. *Gongylonema macrogubernaculum* in captive African squirrels (*Funisciurus substriatus* and *Xerus erythropus*) and lion-tailed macaques (*Macaca silenus*). *J. Zoo Wildl. Med.* 29:331–37
29. Cranfield MR, Schwartz BS, Hofmeister E, Glass GE, Arthur RR, Childs JE. 1990. Potential zoonotic risk of Lyme disease at the Baltimore zoo. *Proc. Am. Assoc. Zoo Vet.* (1990):328–29
30. Croke V. 1997. *The Modern Ark; the Story of Zoos: Past, Present and Future*. NY: Scribner. 272 pp.
31. Cucchi-Stefanoni K, Juan-Sallés C, Parás A, Garner MM. 2008. Fatal anemia and dermatitis in captive agoutis (*Dasyprocta mexicana*) infested with *Echidnophaga* fleas. *Vet. Parasitol.* 155:336–39
32. Denver M. 2008. Evolution of a pest control program at the Maryland Zoo in Baltimore. *Proc. Am. Assoc. Zoo Vet.* (2008):77–78
33. Derraik JB. 2004. A survey of the mosquito (Diptera: Culicidae) fauna of Auckland Zoological Park. *N. Z. Entomol.* 27:51–55
34. Derraik JB. 2004. Mosquitoes (Diptera: Culicidae) breeding in artificial habitats at the Wellington Zoo. *Weta* 28:28–31
35. Derraik JB. 2005. Recommendations for mosquito control in zoological parks to reduce disease transmission risk. *Weta* 29:16–20
36. Derraik JB, Slaney D, Weinstein P. 2008. Effect of height on the use of artificial larval container habitats by *Aedes notoscriptus* in native forest and zoo in Wellington, New Zealand. *N. Z. Entomol.* 31:31–34
37. Dik B, Uslu U. 2009. Chewing-lice (Phthiraptera: Amblycera, Ischnocera) occurring on birds in the Konya Zoo. *Turk. Parazitol. Derg.* 33:43–49
38. Duncan M, Tell L, Montaili RJ. 1993. Lingual spirurids and pasteurellosis in Goeldi's monkeys (*Callimico goeldi*). *Proc. Am. Assoc. Zoo Vet.* (1993):194–95
39. Dunn RR. 2009. Coextinction: anecdotes, models, and speculation. In *Holocene Extinctions*, ed. ST Turvey, pp. 167–80. New York: Oxford Univ. Press
40. Dutton CJ, Quinnell M, Lindsay R, DeLay J, Barker IK. 2009. Paraparesis in a polar bear (*Ursus maritimus*) associated with West Nile virus infection. *J. Zoo Wildl. Med.* 40:568–71
41. Ejiri H, Sato Y, Sawai R, Sasaki E, Matsumoto R, et al. 2009. Prevalence of avian malaria parasite in mosquitoes collected at a zoological garden in Japan. *Parasitol. Res.* 105:629–33
42. Evans DE, Shaw M. 2002. First record of the mite *Hirstiella diolii* Baker (Prostigmata: Pterygosomatidae) from Australia, with a review of mites found on Australian lizards. *Aust. J. Entomol.* 41:30–34
43. Ewing HE. 1933. Some peculiar relationships between ectoparasites and their hosts. *Am. Nat.* 67:365–73
44. Ewing HE. 1938. The sucking lice of American monkeys. *J. Parasitol.* 24:13–33
45. Fernando SP, Udagama-Randeniya PV. 2009. Parasites of selected reptiles of the National Zoological Garden, Sri Lanka. *J. Zoo Wildl. Med.* 40:272–75
46. Ferrell ST, Pope KA, Gardiner C, Bradway DS, Ambrose DL, et al. 2009. Proventricular nematodiasis in wrinkled hornbills (*Aceros corrugatus*). *J. Zoo Wildl. Med.* 40:543–50
47. Fowler ME, Lavoipierre M, Schulz T. 1979. Audycoptid mange in bears. *Proc. Am. Assoc. Zoo Vet.* (1979):104–5
48. French RA, Todd KS, Meehan TP, Zachary JF. 1994. Parasitology and pathogenesis of *Geopetitia aspiculata* (Nematoda: Spirurida) in zebra finches (*Taeniopygia guttata*): experimental infection and new host records. *J. Zoo Wildl. Med.* 25:403–22
49. Glover GJ, Swendrowski M, Cawthorn RJ. 1990. An epizootic of besnoitiosis in captive caribou (*Rangifer tarandus caribou*), reindeer (*Rangifer tarandus tarandus*) and mule deer (*Odocoileus hemionus hemionus*). *J. Wildl. Dis.* 26:186–95
50. Goldman L, Feldman MD. 1949. Human infestation with scabies of monkeys. *Arch. Dermatol. Syphilol.* 59:175–78
51. González-Acuña D, Ardiles K, Figueroa RA, Barrientos C, González P, et al. 2008. Lice of Chilean diurnal raptors. *J. Raptor Res.* 42:281–86
52. Griffith B, Scott JM, Carpenter JW, Reed C. 1993. Animal translocations and potential disease transmission. *J. Zoo Wildl. Med.* 24:231–36
53. Grim KC, McCutchan T, Sullivan M, Cranfield MR. 2008. Unidentified *Plasmodium* species in Australian black swans (*Cygnus atratus*) hatched and raised in North America. *J. Zoo Wildl. Med.* 39:216–20

54. Griner LA. 1974. Some diseases of zoo animals. *Adv. Vet. Sci. Comp. Med.* 18:251–71
55. Hill DJ, Langley RL, Morrow WM. 1998. Occupational injuries and illnesses reported by zoo veterinarians in the United States. *J. Zoo Wild. Med.* 29:371–85
56. Hoare CA. 1963. Does Chagas' disease exist in Asia? *J. Trop. Med. Hyg.* 66:297–99
57. Hoff GL, Griner LA, Trainer DO. 1973. Bluetongue virus in exotic ruminants. *J. Am. Vet. Med. Assoc.* 163:565–67
58. Hollamby S, Sikarskie JG, Stuht J. 2003. Survey of peafowl (*Pavo cristatus*) for potential pathogens at three Michigan zoos. *J. Zoo Wildl. Med.* 34:375–79
59. Hughes DG, Bennett PM. 1991. Captive breeding and the conservation of invertebrates. *Int. Zoo Yearb.* 30:45–51
60. IUDZG/CBSG (IUCN/SSC). 1993. *Executive Summary, The World Zoo Conservation Strategy; The Role of the Zoos and Aquaria of the World in Global Conservation*. Brookfield: Chicago Zool. Soc. 12 pp.
61. **Ivanova E, Spiridonov S, Bain O. 2007. Ocular oxyspirurosis of primates in zoos: intermediate host, worm morphology, and probable origin of the infection in the Moscow Zoo. *Parasite* 14:287–98**

61. Model study of disease in zoo animals, using arthropod surveys and molecular and morphological techniques to establish the vector and original source of infection.

62. Jaime-Andrade GJ, Avila-Figueroa D, Lozano-Kasten FJ, Hernández-Gutiérrez RJ, Magallón-Gastélum E, et al. 1997. Acute Chagas' cardiopathy in a polar bear (*Ursus maritimus*) in Guadalajara, México. *Rev. Soc. Brasil. Med. Trop.* 30:337–40
63. Janssen DL. 1994. Morbidity and mortality of douc langurs (*Pygathrix nemaeus*) at the San Diego Zoo. *Proc. Assoc. Reptil. Amphib. Vet.* (1994):195–200
64. Kahn LH. 2007. *Animals: the world's best (and cheapest) biosensors.* **http://www.thebulletin.org/node/146**
65. Kalter SS. 1989. Infectious diseases of nonhuman primates in a zoo setting. *Zoo Biol. Suppl.* 1:61–76
66. Karesh WB, Cook RA. 1995. Application of veterinary medicine to *in-situ* conservation efforts. *Oryx* 29:653–58
67. **Keh B, Hawthorne RM. 1977. The introduction and eradication of an exotic ectoparasitic fly, *Hippobosca longipennis* (Diptera: Hippoboscidae), in California. *J. Zoo Anim. Med.* 8:19–24**

67. Provides a follow-through investigation, from detection to eradication of exotic arthropod, in collaboration with zoo personnel.

68. Kilpatrick AM, Daszak P, Jones MJ, Marra PP, Kramer LD. 2006. Host heterogeneity dominates West Nile virus transmission. *Proc. Biol. Sci. R. Soc. B* 273:2327–33
69. King LJ, Anderson LR, Blackmore CG, Blackwell MJ, Lautner EA, et al. 2008. Executive summary of the AVMA One Health Initiative Task Force report. *J. Am. Vet. Med. Assoc.* 233:259–61
70. Kisling VN Jr. 2001. *Zoo and Aquarium History: Ancient Collections to Zoological Gardens*. Boca Raton, FL: CRC Press. 415 pp.
71. Kock RA, Soorae PS, Mohammed OB. 2007. Role of veterinarians in reintroductions. *Int. Zoo Yearb.* 41:24–37
72. **Kühlhorn F. 1981. Über Dipteren im Säugetier-Bereich zoologishcer Gärten unter Berücksichtigung infektionsmedizinischer Gesichtspunkte. *Angew. Parasitol.* 22:92–103**

72. Provides a comprehensive survey of Diptera in a German zoo.

73. Kuiken T, Leighton FA, Fouchier RAM, LeDuc JW, Peiris JSM, et al. 2005. Pathogen surveillance in animals. *Science* 309:1680–81
74. Lampo M, Rangel Y, Mata A. 1998. Population genetic structure of a three-host tick, *Amblyomma dissimile*, in eastern Venezuela. *J. Parasitol.* 84:1137–42
75. Lasry JE, Sheridan BW. 1965. Chagas' myocarditis and heart failure in the red uakari. *Int. Zoo Yearb.* 5:182–84
76. Latas P, Reavill DR, Nicholson D. 2004. *Trypanosoma* infection in sugar gliders (*Petaurus breviceps*) and a hedgehog (*Atelerix albiventris*) from Texas. *Proc. Am Assoc. Zoo Vet.* (2004):179–80
77. Latreille M. 1818. *Nouveau Dictionnaire d'Histoire Naturelle.* 21:222–23
78. Levin II, Outlaw DC, Vargas FH, Parker PG. 2009. *Plasmodium* blood parasite found in endangered Galapagos penguins. *Biol. Conserv.* 142:3191–95
79. Lewis JCM. 2007. Conservation medicine. See Ref. 150, pp. 192–204
80. Ludwig GV, Calle PP, Mangiafico JA, Raphael BL, Danner DK, et al. 2002. An outbreak of West Nile virus in a New York City captive wildlife population. *Am. J. Trop. Med. Hyg.* 67:67–75
81. Luppi MM, Malta MCC, Silva TMA, Silva FL, Motta ROC, et al. 2008. Visceral leishmaniasis in captive wild canids in Brazil. *Vet. Parasitol.* 155:146–51

82. MacKenzie D. 2008. Mexican kissing bugs tell all about bluetongue. *New Sci.* 198(2659):14
83. Malta MCC, Tinoco HP, Xavier MN, Vieira ALS, Costa EA, Santos RL. 2010. Naturally acquired visceral leishmaniasis in nonhuman primates in Brazil. *Vet. Parasitol.* 169:193–97
84. Mbaya AW, Aliyu MM, Nwosu CO, Ibrahim UI. 2008. Captive wild animals as potential reservoirs of haemo and ectoparasitic infections of man and domestic animals in the arid-region of northeastern Nigeria. *Vet. Archiv.* 78:429–40
85. Mbaya AW, Aliyu MM, Ibrahim UI. 2009. The clinic-pathology and mechanisms of trypanosomosis in captive and free-living wild animals: a review. *Vet. Res. Commun.* 33:793–809
86. McConkey GA, Li J, Rogers MJ, Seeley DC, Graczyk TK, et al. 1996. Parasite diversity in an endemic region for avian malaria and identification of a parasite causing penguin mortality. *J. Eukaryot. Microbiol.* 43:393–99
87. **McNamara T. 2007. The role of zoos in biosurveillance. *Int. Zoo Yearb.* 41:12–15**

87. Seminal paper, resulting from 1999 West Nile virus outbreak, calling for the use of zoos as epidemiological monitoring sites.

88. Mertins JW, Schlater JL. 1991. Exotic ectoparasites of ostriches recently imported into the United States. *J. Wildl. Dis.* 27:180–82
89. Montali RJ, Bush M. 1992. Some diseases of golden lion tamarins acquired in captivity and their impact on reintroduction. *Proc. Am. Assoc. Zoo Vet.* (1992):14–16
90. Moore JG. 1970. Epizootic of acanthocephaliasis among primates. *J. Am. Vet. Med. Assoc.* 157:699–705
91. Murphy FA, Brooks DL, Boyce WM, Lasley W, Lowenstine L, et al. 1993. An international system for the prevention and control of infectious diseases in free-ranging and captive wild animals. *J. Zoo Wildl. Med.* 24:365–73
92. Mutlow AG, Dryden MW, Payne PA. 2006. Flea (*Pulex simulans*) infestation in captive giant anteaters (*Myrmecophaga tridactyla*). *J. Zoo Wildl. Med.* 37:427–29
93. Nayar GP, Crawshaw GJ, Neufeld JL. 1979. Tularemia in a group of nonhuman primates. *J. Am. Vet. Med. Assoc.* 175:962–63
94. Neiffer DL, Klein EC, Calle PP, Linn M, Terrell SP, et al. 2002. Mortality associated with melarsomine dihydrochloride administration in two North American river otters (*Lontra canadensis*) and a red panda (*Ailurus fulgens fulgens*). *J. Zoo Wildl. Med.* 33:242–48
95. Nelder MP. 2007. *Arthropods at the interface of exotic and native wildlife: a multifaceted approach to medical and veterinary entomology in zoos of South Carolina*. PhD thesis. Clemson Univ., Clemson, SC. 251 pp.
96. Nelder MP, Paysen ES, Zungoli PA, Benson EP. 2006. Emergence of the introduced ant *Pachycondyla chinensis* (Formicidae: Ponerinae) as a public health threat in the southeastern United States. *J. Med. Entomol.* 43:1094–98
97. Nelder MP, Lloyd JE, Loftis AD, Reeves WK. 2008. *Coxiella burnetii* in wild-caught filth flies. *Emerg. Infect. Dis.* 14:1002–4
98. Nelder MP, Reeves WK, Adler PH, Wozniak A, Wills B. 2009. Ectoparasites and associated pathogens of free-roaming and captive animals in zoos of South Carolina. *Vector-Borne Zoonot. Dis.* 9:469–77
99. Nelder MP, Swanson DA, Adler PH, Grogan WL. 2010. Biting midges of the genus *Culicoides* in South Carolina zoos. *J. Insect Sci.* 10:1–9
100. Nolen RS. 2001. Nation's zoos and aquariums help track West Nile virus. *J. Am. Vet. Med. Assoc.* 219:1327–30
101. Okoh AEJ, Oyetunde IL, Ibu JO. 1986. Naturally contracted fatal *Cowdria ruminantium* infection in a captive sitatunga in Nigeria. *J. Zoo Anim. Med.* 17:72–74
102. Oliver JH, Owsley MR, Claiborne CM. 1988. Chromosomes, reproductive biology, and developmental stages of *Aponomma varaensis* (Acari: Ixodidae). *J. Med. Entomol.* 25:73–77
103. Orr K. 2002. Envenomation of small primates by bark scorpions (*Centruroides sculpturatus*): symptoms and treatment. *Proc. Am. Assoc. Zoo Vet.* (2002):374–76
104. Parija SC, Bhattacharya S. 2001. The tragedy of tigers: lessons to learn from Nandankanan episode. *Ind. J. Med. Microbiol.* 19:116–18
105. Preiksaitis JK, Crawshaw GJ, Nayar GSP, Stiver HG. 1979. Human tularemia at an urban zoo. *Can. Med. Assoc. J.* 121:1097–99
106. Rasmussen JM, Wünschmann A, Guo J, Concha-Bermejillo A. 2004. A multi-species outbreak of orf within a zoological collection. *Proc. Am. Assoc. Zoo Vet.* (2004):388–92

107. Rees DHE, Axford JS. 1994. Evidence for Lyme disease in urban park workers: a potential new health hazard for city inhabitants. *Br. J. Rheumatol.* 33:123–28
108. Robinson PT. 2004. *Life at the Zoo: Behind the Scenes with the Animal Doctors*. New York: Columbia Univ. Press. 293 pp.
109. Rodriguez M, Hooghuis H, Castaño M. 1992. African horse sickness in Spain. *Vet. Microbiol.* 33:129–42
110. Rose K, Curtis J, Baldwin T, Mathis A, Kumar B, et al. 2004. Cutaneous leishmaniasis in red kangaroos: isolation and characterization of the causative organisms. *Int. J. Parasitol.* 34:655–64
111. Rugg D. 1982. Effectiveness of Williams traps in reducing the numbers of stable flies (Diptera: Muscidae). *J. Econ. Entomol.* 75:857–59
112. Sanderson S, Garn A-K, Kaandorp J. 2008. Species susceptibility to bluetongue in European zoos during the Bluetongue Virus subtype 8 (BTV8) epizootic August 2006–December 2007. *Proc. Am. Assoc. Zoo Vet.* (2008):232–33
113. Sano Y, Aoki M, Takahashi H, Miura M, Komatsu M, et al. 2005. The first record of *Dirofilaria immitis* infection in a Humboldt penguin (*Spheniscus humboldti*). *J. Parasitol.* 91:1235–37
114. Schwan TG, Thompson D, Nelson BC. 1985. Fleas on roof rats in six areas of Los Angeles County, California: their potential role in the transmission of plague and murine typhus to humans. *Am. J. Trop. Med. Hyg.* 34:372–79
115. Shimonski J. 2009. An experiment in arboriculture and mosquito control. *City Trees* 45(2):6–10
116. Shimonski J. 2010. Integrated pest management in a zoological theme park. *Wing Beats* (Spring):37–41
117. Sibaja-Morales KD, Oliveira JB, Rocha AEJ, Gamboa JH, Gamboa JP, et al. 2009. Gastrointestinal parasites and ectoparasites of *Bradypus variegatus* and *Choloepus hoffmanni* sloths in captivity from Costa Rica. *J. Zoo Wildl. Med.* 40:86–90
118. Silva OS, Oliveira HH, Friccielo RH, Serra-Freire NM. 2004. Malófagos parasitas de aves campestres cativas do Zoológico Municipal Quinzinho de Barros, Sorocaba, Estado de São Paulo, Brasil. *Entomol. Vect.* 11:333–39
119. Smith KF, Behrens M, Schloegel LM, Marano N, Burgiel S, Daszak P. 2009. Reducing the risks of the wildlife trade. *Science* 324:594–95
120. Spradbery JP, Vanniasingham JA. 1980. Incidence of the screw-worm fly, *Chrysomya bezziana*, at the Zoo Negara, Malaysia. *Malays. Vet. J.* 7:28–32
121. Stadler CK, Burns EA. 2007. Treatment of a louse (*Haematomyzus elephantis*) infestation in a captive herd of African elephants (*Loxodonta africana*). *Proc. Am. Assoc. Zoo Vet.* (2007):152
122. Steele KE, Linn MJ, Schoepp RJ, Komar N, Geisbert TW, et al. 2000. Pathology of fatal West Nile virus infections in native and exotic birds during the 1999 outbreak in New York City, New York. *Vet. Pathol.* 37:208–24
123. Stewart IRK, Clark F, Petrie M. 1996. Distribution of chewing lice upon the polygynous peacock *Pavo cristatus*. *J. Parasitol.* 82:370–72
124. Stoebel K, Schoenberg A, Streich WJ. 2003. The seroepidemiology of Lyme borreliosis in zoo animals in Germany. *Epidemiol. Infect.* 131:975–83
125. Stone AB, Hautala JA. 2008. Meeting report: panel on the potential utility and strategies for design and implementation of a national companion animal infectious disease surveillance system. *Zoon. Publ. Health* 55:378–84
126. Stone WB, Hughes JA. 1969. Massive pulmonary acariasis in the pig-tailed macaque. *Bull. Wildl. Dis. Assoc.* 5:20–22
127. Swierstra D, Jansen J, Van Den Broek E. 1959. Parasites of zoo-animals in the Netherlands. *Tijdschr. Diergeneesk* 84:1301–5
128. Takken W, Huijben S, Wijsman A, Paaijmans K, Schmidt H, Schaftenaar W. 2003. Avian malaria in Rotterdam Zoo. *Dutch Wildl. Health Cent. Newsl.* 2:3–4
129. Tarshis IB, Penner LR. 1960. Treatment of ectoparasites in captive wild animals. *Int. Zoo Yearb.* 2:107–9
130. Terrell SP, Pace P. 2007. Worms, germs, and you: use of an on-line interactive program for zoonotic disease training in the zoo environment. *Proc. Am. Assoc. Zoo Vet.* (2007):78–79
131. Tongué LK, Diabakana PM, Bitsindou P, Louis FJ. 2009. Have tsetse flies disappeared from Brazzaville town? *PanAfrican Med. J.* 3:3

132. Tuttle AD, Andreadis TG, Frasca S, Dunn JL. 2005. Eastern equine encephalitis in a flock of African penguins maintained at an aquarium. *J. Am. Vet. Med. Assoc.* 226:2059–62
133. USDA APHIS. 2008. United States Animal Health Report. *Agric. Inform. Bull.* 803:1–178
134. Valim MP, Teixeira RHF, Amorim M, Serra-Freire NM. 2005. Malófagos (Phthiraptera) recolhidos de aves silvestres no Zoológico de São Paulo Zoo, SP, Brasil. *Rev. Brasil. Entomol.* 49:584–87
135. Veltman J, Wilhelm W. 1991. Husbandry and display of the jewel wasp *Ampulex compressa* and its potential value in destroying cockroaches. *Int. Zoo Yearb.* 30:118–26
136. Viner TC, Nichols D, Montali RJ. 2001. Malaria in birds at the Smithsonian National Zoological Park. *Proc. Am. Assoc. Zoo Vet.* (2001):68–70
137. Wallach JD, Williamson WM. 1968. Partial list of parasites found at necropsy at the Chicago and St. Louis Zoological Parks. *Proc. Am. Assoc. Zoo Vet.* (1968):7–8
138. Watanabe M. 2002. News profile: Tracey McNamara, veterinary pathologist. *Scientist* 16(5):60
139. Weidman FD. 1923. The animal parasites, their incidence and significance. In *Disease in Captive Wild Mammals and Birds*, ed. H Fox, pp. 614–59. Philadelphia: Lippincott
140. Weiss R. 2002. West Nile's widening toll: impact on North American wildlife far worse than on humans. *Washington Post*, Dec. 28, p. 2
141. Weissenböck H, Kolodziejek J, Url A, Lussy H, Rebel-Bauder B, Nowotny N. 2002. Emergence of *Usutu virus*, an African mosquito-borne *Flavivirus* of the Japanese encephalitis virus group, Central Europe. *Emerg. Infect. Dis.* 8:652–56
142. Whiteman NK, Parker PG. 2005. Using parasites to infer host population history: a new rationale for parasite conservation. *Anim. Conserv.* 8:175–81
143. Williams C. 2009. It's a small world. *New Sci.* 202(2704):40–43
144. Wilson N, Barnard SM. 1985. Three species of *Aponomma* (Acari: Ixodidae) collected from imported reptiles in the United States. *Fla. Entomol.* 68:478–80
145. Woodford MH, Rossiter PB. 1993. Disease risks associated with wildlife translocation projects. *Rev. Sci. Tech. Off. Int. Epizoot.* 12:115–35
146. Woziak EJ, McLaughlin GL. 1993. A molecular epidemiologic study of hemogregarine infections in captive reptiles. *Proc. Am. Assoc. Zoo Vet.* (1993):223
147. Wright FH. 1983. Contagious ecthyma in a herd of musk oxen (*Ovibos moschatus*). *Proc. Am. Assoc. Zoo Vet.* (1983):116–19
148. Yeruham I, Rosen S, Hadani A, Nyska A. 1996. Sarcoptic mange in wild ruminants in zoological gardens in Israel. *J. Wildl. Dis.* 32:57–61
149. Zeng D, Chen H, Lynch C, Eidson M, Gotham I. 2005. Infectious disease informatics and outbreak detection. In *Medical Informatics: Knowledge Management and Data Mining in Biomedicine*, ed. H Chen, SS Fuller, C Friedman, W Hersh, pp. 360–95. New York: Springer-Verlag
150. Zimmermann A, Gatchwell M, Dickie LA, West C, eds. 2007. *Zoos in the 21st Century: Catalysts for Conservation?* Cambridge, UK: Cambridge Univ. Press. 373 pp.

Climate Change and Evolutionary Adaptations at Species' Range Margins

Jane K. Hill, Hannah M. Griffiths, and Chris D. Thomas

Department of Biology, University of York, YO10 5DD, United Kingdom; email: jkh6@york.ac.uk

Annu. Rev. Entomol. 2011. 56:143–59

First published online as a Review in Advance on August 30, 2010

The *Annual Review of Entomology* is online at ento.annualreviews.org

This article's doi: 10.1146/annurev-ento-120709-144746

0066-4170/11/0107-0143$20.00

Key Words

dispersal, flight morphology, geographic distribution, habitat selection, invasions, postglacial expansion

Abstract

During recent climate warming, many insect species have shifted their ranges to higher latitudes and altitudes. These expansions mirror those that occurred after the Last Glacial Maximum when species expanded from their ice age refugia. Postglacial range expansions have resulted in clines in genetic diversity across present-day distributions, with a reduction in genetic diversity observed in a wide range of insect taxa as one moves from the historical distribution core to the current range margin. Evolutionary increases in dispersal at expanding range boundaries are commonly observed in virtually all insects that have been studied, suggesting a positive feedback between range expansion and the evolution of traits that accelerate range expansion. The ubiquity of this phenomenon suggests that it is likely to be an important determinant of range changes. A better understanding of the extent and speed of adaptation will be crucial to the responses of biodiversity and ecosystems to climate change.

INTRODUCTION

Ice age refugia: locations to which species retracted to during glacial periods and from which species subsequently spread during warmer, interglacial periods

Genetic erosion: loss of genetic diversity and reduced heterozygosity within populations

Recent global climate warming has caused species to shift their ranges poleward (90, 91). Insects are poikilothermic (cold-blooded) organisms and so are particularly sensitive to temperature changes, especially those species that have narrow thermal tolerances (1, 31). There is now a substantial literature documenting that insects are expanding at their high-latitude and high-elevation cool range margins (2, 58, 90, 91, 115) and retracting at their low-latitude and low-elevation warm margins (2, 37, 88, 118). Regardless of any climate change mitigation that may occur in the future to reduce the release of greenhouse gases, there is a commitment to future warming due to continuing emissions from existing industrial and agricultural activities, and there are lags in the climate system resulting from thermal inertia in the oceans. Even the most optimistic scenarios mean that there is the inevitability of future warming for most if not all of this century (72, 73, 81, 116). Thus, understanding the factors affecting the pattern and rate of change in species' ranges is crucial for making reliable predictions of the potential future distribution of biodiversity as climates continue to warm.

Responding to global climate change is not a new phenomenon for insects. For example in Britain, the distributions of many butterfly species have fluctuated over the past two centuries (5), probably related to climate fluctuations (65). Many species occupy temperate regions because they colonized these regions after the last ice age, and evidence from postglacial range expansions has been proposed as indicative of a species' capacity to respond to current climate change (6, 24, 34, 35, 70). Species can respond to climate change by adapting to climate changes in situ, by migrating to new climatically suitable areas and tracking climate changes, by dying out, or by some combination of these. For beetles, evidence from the fossil record suggests that adaptation of species was not observed following postglacial range expansion, but that species shifted their ranges to track the changing climate (23). However, fossil evidence may not preserve evidence of adaptation, and current data indicate that evolutionary changes in response to current warming are commonly observed in insects.

During recent climate warming, species have been faced with shifting their distributions across heavily fragmented, human-modified landscapes and many species are failing to colonize new areas and thus failing to track climate changes (82, 115). For these species, their potential for adaptation to new climatic conditions may be crucial to their longer-term persistence. Those insects that are tracking climate changes may also undergo evolutionary changes. Theoretical studies support the notion of increased dispersal evolving during range expansion (33, 68), and increased dispersal ability has been observed in species at their expanding range margins. In this paper, we review empirical evidence for such evolutionary changes occurring in insects.

Shifts at the leading edge of range expansion may often be achieved by small numbers of long-distance migrant individuals, potentially resulting in founder bottlenecks. Bottlenecks associated with postglacial range expansions have resulted in latitudinal gradients of genetic diversity across species' ranges, from ice age refugia to high-latitude range margins (57). These clines in genetic diversity may affect the ability of populations to adapt to future climate warming. In addition, anthropogenic habitat fragmentation may alter dispersal rates (60) and gene flow among populations, further affecting patterns of genetic diversity across species' ranges. Changes to the distributions of species in the future are also likely to result in the loss of genetic diversity at low-latitude and low-elevation warm range boundaries where climate will become unsuitable (i.e., too hot and/or dry). In these locations, which are often the genetically richest parts of species' ranges, populations are likely to become fragmented, leading to genetic bottlenecks and genetic erosion as a result of reduced gene flow among isolated populations (57). However, survival of populations at the trailing edges of species shifting distributions can be particularly important to the persistence and evolution of species (47). Thus, examining patterns of genetic diversity across insect ranges, and considering how these patterns might affect species' potential to evolve and adapt to climate changes, is important for biodiversity conservation.

SCOPE OF REVIEW

In this review, we consider the impacts of climate warming on insect distributions, and we focus on examining evolutionary changes at the margins of species ranges. The review covers three main aspects: (*a*) We describe current geographical patterns of genetic diversity and evolutionary adaptations at range margins arising from postglacial range shifts. (*b*) We review evolutionary adaptations at recently expanding range margins. (*c*) We discuss the implications of these adaptations for species' abilities to respond to future climate warming, and highlight current knowledge gaps.

Geographical Patterns in Genetic Diversity

Evidence from plant and animal fossil remains suggests that at the height of the Last Glacial Maximum (LGM; around 25,000 years ago; 22) many members of Northern Hemisphere terrestrial biota had distribution ranges that were significantly smaller, as well as farther south, than those observed today (56). The climate began warming after the LGM, then cooled again during the Younger Dryas period, followed by warming around 12,000 years ago (104) as climatic conditions shifted rapidly toward the current, full interglacial (Holocene), warm conditions. Many species rapidly expanded their range boundaries northward out of these southern refugia, tracking the retreating ice sheets and favorable climatic conditions (56).

In species with long-distance leptokurtic (i.e., with a long "tail") dispersal patterns, colonization of new habitats may come about from a small number of founder individuals dispersing over long distances, which may lead to a loss of alleles and reduced heterozygosity in these newly colonized sites (55, 56, 71, 97, 106). By contrast, species with a lower incidence of long-distance dispersal tend not to show such marked clines in genetic diversity; in these species, many individuals repeatedly colonize sites over relatively short distances, such that there is greater gene flow and mixing of alleles among subpopulations (55, 71). Nonetheless, a commonly observed consequence of rapid postglacial expansions is the genetic impoverishment of populations inhabiting newly colonized areas compared to those residing in areas of persistently suitable habitat (55). This has been demonstrated in a wide variety of insect taxa, including Lepidoptera, phasmids, weevils, pitcher-plant mosquitoes, grasshoppers, caddis flies, bees, Megaloptera, and cicadas (3, 4, 11, 17, 30, 45, 54, 55, 79, 83, 94, 117). The patterns are indicative of rapid postglacial range expansion in many regions of the globe including Europe (11), North America (36), New Zealand (16, 79), and Asia (3). However, in some studies these latitudinal cline effects are relatively minor (98) or not evident at all (75, 99, 112). Variation among species is likely to reflect different patterns of dispersal and habitat availability, coupled with differences in life-history traits (e.g., affecting population growth) and different numbers and locations of refugial populations. In the tropics and subtropics, variation in moisture availability is likely to have been as important as temperature in determining fluctuations in insect distributions in response to glacial and interglacial climates. In Australia, the caddisfly *Tasimia palpata* appears to have expanded during interglacial periods, associated with the expansion of subtropical rainforests (85); dung beetle phylogeographies and speciation also reflect the historical dynamics of rainforests (8).

The loss of within-population variation associated with range expansion is sometimes, but not always, accompanied by losses of large-scale among-population variation. For example, populations with different refugial populations from the Iberian, Italian, and Balkan peninsulas of Europe (as well as farther east, within Asia) are often regarded as separate races or subspecies (77, 111). In many species, not

Range shifts: expansion poleward at species' leading-edge, cold, high-latitude range boundaries and retractions poleward at trailing-edge, warm, low-latitude range boundaries

Leptokurtic dispersal pattern: the distribution of dispersal distances traveled by a sample of individuals can be plotted on a graph as a dispersal curve or kernel

Quaternary: the geological period encompassing the past 2.6 million years

all of these refugia contributed to the northward spread that took place under warmer Holocene conditions, such that northern populations commonly represent only a subset of the southern refugial lineages (57, 77, 114).

Some studies have shown that spatial patterns of genetic diversity reflect the existence of more ancient refugia that predate the last glacial period (78). This is particularly the case for species that are adapted to cool environments and would, in many cases, have been more widespread during glacial climates; their current distributions may represent Holocene (interglacial) refugia. For example, the cold-adapted European leaf beetle *Gonioctena pallida* has a hotspot of genetic diversity in the Carpathian Mountains in central Europe, which implies that current patterns of genetic variation in this species reflect a Pleistocene climatic event that resulted in population isolation >90,000 years ago, well before the end of the last ice age (78). Other studies demonstrate evidence of repeated glacial/interglacial cycles on current patterns of genetic diversity, with likely range expansion during glacial periods (25).

Closer focus on butterflies. One well-studied taxon in terms of phylogeography and regional patterns in genetic diversity is butterflies. Genetic differentiation among three strongly divergent lineages of the satyrid butterfly *Maniola jurtina* across Europe demonstrates the genetic consequences of past isolation of populations associated with different glacial refugia (44). Similar genetic patterns, showing distinct lineages within currently continuously distributed populations, have also been observed in several other butterfly species (11, 46, 100, 102). Studies of *Coenonympha arcania* (11) that compared the genetic structure of populations across the species' European range have demonstrated a latitudinal cline in allele frequencies from south to north (56, 71). Isolated peripheral populations at northern range boundaries displayed significantly lower levels of genetic variation and high differentiation, while central and southern populations (in areas of previous glacial refugia) showed much higher genetic diversity and no significant differentiation. These findings correspond to a reduction in polymorphic loci from 64% in populations of *C. arcania* in southeastern Europe to 27% in populations at the northern periphery of its range in Sweden (11). By contrast, no south-north decline in genetic diversity was observed in Europe for two sibling species of butterfly, the marbled white *Melanargia galathea* and the Iberian marbled white *M. lachesis* (46), and high genetic diversity in *M. galathea* has been proposed as one reason for its recent range expansion at its northern range limits (46). However, genetic differentiation is commonly high between different population refugias, compared to areas of postglacial expansion. This is because only a subset of refugial populations may give rise to the new range, and this will be combined with the loss of variation associated with range expansion itself (57).

Comparison of three species of satyrid butterfly in Britain (61) showed that these species differed in their genetic diversity. The species with more extensive current ranges in Britain had greater genetic diversity (*Maniola jurtina* > *Pyronia tithonus* > *Pararge aegeria*). Greater genetic diversity may reflect longer occupancy of these regions by species since the LGM, greater habitat availability, and therefore more gene flow, or it may reflect more recent fluctuations in ranges over the past two centuries (63). A genetic study of the host-specific butterfly *Parnassius smintheus* (Papilionidae) and its larval host plant *Sedum lanceolatum* (Crassulaceae) showed that the two species responded to postglacial climate warming in similar ways (through repeated uphill contractions and downhill retractions during climate fluctuations) but independently (patterns of within-population genetic variation and among-population differentiation showed no congruence) (26).

Implications for population persistence in the future. Evidence from a wide range of insect groups leads us to conclude that differences in dispersal potential, along with the climatic events of the past two million years (the Quaternary period), have had major influences on the current patterns of genetic diversity evident within the distributions of many insects (56). Reduction in genetic diversity as a consequence of postglacial expansion means that, for many species, the lowest genetic diversity is found in populations that inhabit regions at the cool, leading margins of species' ranges. Hence, recent climate-driven range expansions will involve individuals from populations at range boundaries that already exhibit reduced allelic diversity arising from postglacial range expansions.

However, there is relatively limited knowledge of when these genetic effects might accelerate or limit distribution changes, relative to that expected in the absence of such genetic effects. Genetic bottlenecks represent a double-edged sword during range expansion. On the one hand, they may result in the reduced capacity for subsequent adaptation and potentially in the fixation of deleterious alleles; on the other hand, they may result in the rapid fixation of traits that facilitate fast range expansion and hence survival during a period of rapid climate change. However, loss of genetic diversity has been associated with reduced fitness leading to population declines in some insects (101). In *Polyommatus coridon* (Lycaenidae), populations with lower genetic diversity have reduced adult lifetime expectancy, reduced dispersal, and smaller population size (113). Given that previous studies have demonstrated a link between lower genetic diversity and increased local extinction of populations (96), the impact of current range expansion and population bottlenecks as species spread through human-dominated landscapes could be expected to result in reduced fitness, and this could reduce the rate of expansion. In contrast, although the recent range expansion of *Battus philenor* (Papilionidae) into California has resulted in populations with reduced genetic diversity, this has been associated with shifts onto a novel host plant, laying significantly larger egg clutches and loss of mimics in these California populations. This implies that losses in genetic variation may be compensated by increases in adaptation to conditions in the areas that are being invaded (36). Loss of genetic variation associated with recent range expansions is expected to continue in modern landscapes and may be even greater than in the past, given the fragmented nature of many modern landscapes. As with *B. philenor* (36), genetic losses are frequently associated with species spreading across landscapes that contain patchily distributed habitats, for example, in gall wasps associated with patches of planted host plants (107).

Anthropogenic climate change: refers to recent climate warming attributable to human activities and influences on greenhouse gas emissions

Further range changes associated with climate warming are expected to result in erosion of genetic variation within species (30, 108). Expansions at high-latitude boundaries will be initiated from populations that already contain relatively low levels of variation, and further founder effects will reduce this even more. Meanwhile extinctions of populations at their hot, lower-latitude trailing boundaries are expected (64, 103) and observed (90, 118) as a result of climate warming. These local extinctions occur within parts of species' distributions that fall within their historical refugial distributions, where within-population genetic variability is greatest. For example, the butterfly *Coenonympha arcania* shows high genetic diversity in parts of its distribution (11) that are projected to become extinct in response to future climate warming (103). Hence, within-species genetic variation may be eroded even when the species as a whole is not threatened by climate change. It is unknown whether such genetic attrition will affect the expected persistence times and speciation capacities of species that have been so-affected.

Evolutionary Adaptations at Expanding Range Boundaries

There is now a considerable body of evidence documenting the degree to which animal species are responding to recent anthropogenic climate change and shifting their distributions to high latitudes (89, 91). This body of evidence covers many insect species, but most analysis of climate-driven range expansion has focused on a few well-studied groups such as butterflies (90, 93, 115) and Odonata (59), for which there are long-term, fine-scale spatial and temporal datasets recording species' distributions and abundances. Insects colonizing newly available habitats are not a random subset of the original population and usually share a suite of life-history traits associated with increased colonization ability. In wing-dimorphic species, these effects are evident in terms of increased proportions of strong-flying macropterous individuals within newly colonized populations (86, 105, 109). In wing-monomorphic species, such as butterflies, these effects are evident as more subtle changes in adult flight morphology (62) and flight metabolic rates (43) in newly colonized sites. Many additional traits, such as habitat or host selection or reproductive strategy, may come under selection during range shifts and thereby also influence the rate of range expansion. In this section, we review the types of adaptations evident at expanding range boundaries and discuss how these changes might affect species' abilities to respond to climate change. We include discussion of evolutionary adaptations arising from climate-driven range expansions as well as changes occurring through range expansion attributable to other nonclimatic factors.

Our review focuses primarily on expanding range boundaries that involve expansion to higher latitudes. Although uphill range shifts are recorded in insects (21, 119), we do not consider uphill expansions in this review, and in any case these may provide little opportunity for evolutionary adaptation given that colonization distances are much shorter, resulting in greater gene flow between core and margin sites. We also do not consider evolutionary changes at contracting boundaries for which empirical studies are lacking in insects, although there is some such evidence for other taxa (e.g., beech trees; 74). Evolutionary shifts in photoperiod sensitivity, and hence phenologies, have taken place throughout the distribution of the pitcher plant mosquito, *Wyeomyia smithii*, in the eastern United States since 1972, but responses have been strongest at higher latitudes, with limited responses at the southern boundary (13). Museum specimens of the stream-dwelling beetle *Gyretes sinuatus* showed an 8% increase in body size and a 6% increase in the length/width ratio of the body over a 60-year period in the southern United States. These changes were attributed to cooling local temperatures over the same period (7), and this is a possible explanation for the lack of phenology evolution in southern populations of *W. smithii*. There is little or no evidence of the evolution of increased tolerances of high temperatures under climate warming (14).

Predictions for evolutionary changes that might occur at contracting range boundaries may mirror changes observed in populations that are declining (regardless of where populations are located within their geographical range) and becoming increasingly fragmented. For example, studies of museum specimens of the garden tiger moth, *Arctia caja*, which has declined in abundance by 85% in Britain over the past 30 years, have provided evidence of a significant decline in genetic diversity and of changes in wing morphology during the twentieth century (2). The changes in wing morphology imply evolution of increased dispersal ability over time, which may be due to evolutionary responses to increased habitat fragmentation (2, 10, 68), although nongenetic effects due to environmental factors may also be responsible for these changes. By contrast, with more extreme levels of habitat fragmentation, evolution of reduced dispersal might be expected in isolated island-like populations, as typified by true oceanic island endemic species (19, 68). Analysis of the morphology of museum specimens showed likely reduction in flight capacity of the swallowtail butterfly, *Papilio machaon britannicus*, as populations contracted over time (27), but as with all studies of wild-caught material it is difficult to disentangle genetic changes from phenotypic responses to environmental factors. Common-environment rearing of the Glanville fritillary butterfly, *Melitaea cinxia*, showed reduced dispersal in long-established populations (48). Studies at contracting range boundaries are required to determine the relative effects of population size, genetic diversity, and habitat availability on evolutionary changes.

Morphological changes during climate-driven range expansion. Evidence of evolutionary changes in insect populations comes from two types of study: long-term temporal studies at specific locations and snapshot comparisons of populations from different locations. Longitudinal temporal studies examining morphological changes over time come from studies of museum specimens (7, 27). However, such data are rarely available for recently colonized populations that have existed only for a few years or decades.

Snapshot studies of evolutionary changes during range expansion have compared populations at different times-since-colonization by comparing populations at increasing distances from the expanding range margin. These studies have examined evolutionary changes in adult flight morphology, as well as changes in larval diet. Climate-driven range expansion has been associated with the exploitation of a wider range of host plants in some species. The comma butterfly, *Polygonia c-album*, has expanded its northern range margin northward in Britain by ~200 km in the past 60 years. This range expansion has been associated with an apparent change in its preferred larval host plant from *Humulus lupulus* (hop) to include other plant species, especially *Urtica dioica* (nettle) and *Ulmus glabra* (wych elm), host plants utilized in other parts of its European range and by other closely related *Polygonia* species (15). These new larval host plants are more widely available in Britain compared with *H. lupulus*, and so changes in larval diet may have facilitated the recent range expansion this species has displayed, especially because butterfly growth and survival were higher and adults were larger on the new plant hosts (15). A similar pattern has been observed in expanding populations of the brown argus butterfly, *Aricia* (*Plebeius*) *agestis*, where climate warming has been associated with a shift in host plant selection by ovipositing females from exploiting common rockrose, *Helianthemum nummularium*, on chalk and limestone grassland habitats to exploiting *Erodium cicutarium* and *Geranium molle* in cooler microclimates in a wider range of grasslands (5, 109). As with *P. c-album*, the potential to utilize more widely available host plants that grow on a variety of geological substrates has increased the potential breeding habitats available to *A. agestis* and thus increased its ability to expand its range in Britain as the climate warms. These within-species shifts mirror between-species differences that have been observed, in which habitat generalists and species that utilize widespread host plants and habitats are expanding their ranges faster than those with more specialized requirements (115).

Theoretical studies predict that dispersal rates will often evolve at expanding range boundaries, and there is considerable empirical evidence to support this notion. In many cases, this empirical evidence is based on indirect measures of insect dispersal based on adult flight morphology. Measures of thorax size and shape (which measure flight muscle mass), measures of wing size and shape (aspect ratio; $wingspan^2/area$), and measures of wingloading (insect mass/wing area) are associated with insect flight speed and maneuverability (9, 20). Differences in morphological traits associated with flight and reproduction have been shown in the speckled wood butterfly, *Pararge aegeria* (62). More recently colonized sites generally have populations with larger adults, greater thorax mass, and broader thorax shape (62, 68, 69). Differences in wing aspect ratio are also evident, although patterns are not consistent, and both higher (=narrower wings; 68) and lower (=broader wings; 62) aspect ratios have been recorded at range margin sites. There is also evidence that selection pressures may vary between sexes, with evolutionary changes more evident in females than in males during responses to climate change (68) and habitat fragmentation (48). Given that butterflies use flight for activities other than dispersal, for example, mate location, nectaring, and oviposition, it is perhaps not surprising that evolutionary adaptations to range expansion do not always predominate over other selection pressures. Hence, evolutionary changes in flight morphology are not always evident in recently colonized butterfly populations (48, 84). The studies cited above for *P. aegeria* are from studies of F1 offspring reared under common environmental conditions (62, 68, 69), consistent with a genetic basis for the change. Studies where differences in dispersal morphology among populations are evident only in wild-caught material and not in F1 offspring imply that environmental factors at range margin sites, rather than any genetic changes, are affecting dispersal morphology, perhaps through effects of larval host plant quality and/or temperature on insect development (15, 92).

In studies where changes in flight morphology are not evident in newly colonized sites, individuals nonetheless have higher dispersal ability compared with older populations. In expanding populations of the map butterfly, *Araschnia levana*, butterflies in newly colonized sites have higher levels of PGi alleles (84), which have been associated with superior flight metabolic rates and population growth rates (43, 51), even though colonizing individuals show no changes in wing or thorax measures (84). Similarly, Glanville fritillary butterflies, *Melitaea cinxia*, have greater dispersal ability in newly colonized sites but show no change in their flight morphology (48). Empirical measures of dispersal from mark-release-recapture field studies, as well as physiological measures of the ADP/ATP ratio in butterfly flight muscles (49), provide further support that increased dispersal capability is present in newly colonized populations of *M. cinxia* in the absence of any changes in adult flight morphology.

Increased dispersal ability has also been demonstrated in range margin populations of other insect taxa (105, 109). A comparison of three British damselfly species currently expanding their distributions northward showed that *Calopteryx splendens*, the most rapidly expanding of the three species, had higher wing aspect ratios (=narrower wings) in individuals from the range margin compared with those closer to the core of the range (52). In Orthoptera, the frequency of dispersive (long-winged macropterous) morphs was approximately eight and four times greater for *Conocephalus discolor* and *Metrioptera roeselii*, respectively, at expanding range margins than in the core of their ranges (105). Once established, populations of these bush crickets returned to low levels of macroptery within 5 to 10 years (generations) after colonization. This finding highlights that evolutionary adaptations can be rapid; for example, photoperiodic adaptation in pitcher plant mosquitoes, *W. smithii*, was evident after 5 years (13).

In many studies, trade-offs between flight and reproduction have been evident (120), and these trade-offs are particularly strong in taxa that disperse by flight but are capable of completing other adult activities by different means. Captive-reared macropterous females of *C. discolor* showed a threefold reduction in fecundity, relative to brachypterous individuals, and this trade-off is likely to have driven the decline in macroptery in populations once they had established. The trade-off is more complex in insects that use flight for a variety of purposes. Nonetheless, in the speckled wood butterfly, *P. aegeria*, the increased dispersal ability of individuals at range margin sites was associated with reduced reproductive investment; range margin females have relatively larger and broader thoraxes but produced significantly fewer eggs compared with those from core sites (69). This suggests that any benefits to species expanding their ranges in response to climate warming from evolutionary increase in dispersal ability are balanced by reduced fecundity and lower population growth rates in margin sites. Populations developing under cooler conditions in range margin sites also show different development rates (53) and multivoltine species develop through fewer generations (40), further contributing to reduced population growth rates at range margin sites. Thus, the benefits to expanding species of increased dispersal at range margins may be relatively short-lived but essential in enabling species to shift their distributions.

Evolutionary changes during latitudinal range expansions have hardly begun to be explored for other traits that could be important for insect population persistence. Following Quaternary range expansion, many species developed genetically based latitudinal clines in photoperiod and temperature responses (12, 80, 87), and in body size and melanism (28, 29, 76, 121), within the last ~10,000 years. As expanding populations again spread poleward in response to anthropogenic climate warming, the selection pressures that generated these clines will presumably be renewed, but such data are currently lacking.

Morphological changes during invasions and metapopulation dynamics. The studies discussed above are for species expanding their ranges primarily in response to climate warming. However, range expansions of insects colonizing and invading new areas commonly occur in the absence of climate warming. There is a large literature on invasive species showing evolutionary adaptations during range expansion, and evolutionary changes are also evident as a consequence of colonization-extinction metapopulation dynamics in fragmented landscapes. For example, studies of *M. cinxia* (50) showed altered life-history traits in newly colonized sites within metapopulations. A meta-analysis of the Mediterranean fruit fly, *Ceratitis capitata* (a globally successful invader), showed that life-history traits such as longevity and fecundity differed in newly colonized sites (32).

By contrast to climate-driven range expansions, many invasive species that are expanding their ranges are not necessarily limited by suitable climate conditions and so may not experience the same environmental challenges and selection pressures as species expanding in response to climate warming. Nonetheless, the invasive hemlock woolly adelgid, *Adelges tsugae*, has expanded its range northward in areas with cooler climates, and this has been associated with the evolution of greater cold hardiness (18). Local climate conditions may also have resulted in changes in voltinism during the invasion of Japan by the North American moth *Hyphantria cunea* (39). A temperature-related latitudinal gradient in body size evolved in *Drosophila subobscura* within two decades of its introduction to North America (67).

These studies suggest that many of the evolutionary adaptations of species expanding their ranges owing to climate warming may mirror evolutionary responses in expanding species regardless of the drivers of range changes. However, for other traits and evolutionary changes, the colonization of newly available climatically suitable habitats may result in selection pressures and environmental challenges different from other types of range changes. This may mean that the traits under selection, the strength of selection pressures, and the types of evolutionary adaptation that are evident may not be comparable between climate-driven and nonclimate-driven range changes.

Implications of Evolutionary Adaptations

Climate change and the resulting community and distribution changes of species mean that virtually all populations of all species will have experienced changes in selection pressures as a result of recent climate change, and this will continue for the foreseeable future. The evolutionary shifts driven by these changes affect the ecology of species and communities, potentially resulting in the emergence of new pest species and the extinction of other taxa, as flexible, mobile generalist species and phenotypes come to dominate biological communities (109, 115). Whereas the ecological shifts generated by climate change are increasingly well documented (73, 89, 95), the evolutionary responses are far less well understood, as are the feedbacks between the evolutionary and ecological changes that are presumably affecting most populations of species in the world.

Knowledge gaps and future research. Recent anthropogenic climate change provides an opportunity to evaluate the role of evolution in ecology, and vice versa, and the role of evolution in permitting species to survive or go extinct. In particular, the feedbacks between ecology and evolution might be expected to be strongest at range boundaries, where the strength of selection may be greatest and population bottlenecks common. The extent and speed of adaptation could be crucial to the responses of biodiversity and ecosystems to climate change, and several areas of research require further attention.

There is no information about whether the rate of environmental change might affect the likelihood and speed of evolutionary responses at contracting and expanding range margins. At range boundaries where conditions are deteriorating (i.e., trailing edge margins), novel adaptations will presumably be rare (increasingly limited by low mutation rates) because most of these marginal populations are likely to have been under historical directional selection to survive in hot/dry conditions in the recent past; the failure to adapt sufficiently rapidly under such conditions is likely to result in population decline and extinction if conditions become more extreme (41, 42). The degree to which populations may or may not adapt to new environmental conditions has not been examined at the trailing edges of insect species' distributions. Thus, new empirical and theoretical work is needed. Laboratory selection experiments often show the capacity of populations to adapt to new conditions, but it is unclear how such findings might translate to field conditions where trailing-edge populations have experienced past directional selection and where they must compete and interact with newly arriving thermophilous species that may already be better adapted to the new physical conditions. Given that evolutionary feedbacks are expected to reduce rates of population decline when environmental conditions are deteriorating (41, 42), reduced rates of climate change presumably increase the likelihood of sufficient evolution taking place to enable populations to persist; but this has not been empirically tested.

At expanding range margins, rapid climate change may increase the likelihood of the evolution of increased dispersal capacity because species lagging behind climate changes will experience strong directional selection, generation after generation, whereas those that are keeping track of climate changes will experience periodic setbacks in cooler years. Our review has highlighted that the evolution of dispersal is commonly observed in expanding insect species. Many studies have highlighted that the dispersal ability of insects is crucial for enabling them to respond to environmental changes (60, 115). Some measures of dispersal, such as distance traveled and incidence of flight, are relatively well understood (through mark-release-recapture studies), as is the evolution of dispersal (through theoretical simulation models and some empirical examples). However, evolutionary changes to the behavioral and physiological processes involved in flight in the context of climate change and range shifts are less well known. One option for conservation management to help species respond to climate changes is to develop landscapes that have better habitat connectivity and thus aid species' dispersal and tracking of climate, but the potential success of such a strategy is largely unknown (66). A better understanding of the behavioral inclination of individuals to disperse, their behavior at habitat boundaries, and their likelihood of traveling across unsuitable habitats is required. Investigation into behavioral adaptations of colonizing individuals and newly established populations compared with ancestral populations and individuals in the core of the range may help in developing more resilient landscapes that better conserve biodiversity.

In general, far less information is available on the evolutionary responses of populations in regions where conditions are deteriorating for those species, at least in insects, compared with evolutionary responses at expanding range margins (38). Such information is important because evolution in refugial populations may be necessary if species are to survive in the long term and then recolonize areas at lower latitudes, assuming cooler climates occur again at some point in the future. If populations lack the capacity to adapt, then projections of future range retreats and extinctions are more likely to be realized (110).

Relationships between species' genetic variability and their patterns of range expansion are not well understood. For example, it is not clear whether species showing the most rapid range expansions during recent climate change are those that have high genetic variation because of their high evolutionary capacity. Alternatively, species with attributes that facilitate rapid expansion may have lost genetic variation in the past (during rapid postglacial expansions) but are capable of responding rapidly to recent climate change. More information on the habitat associations and life-history traits of species that exhibit reductions in genetic variability as species shift their ranges would help in understanding which species may or may not show adaptations to climate change in the future.

In this review, we have focused on evolutionary adaptations along latitudinal gradients, and studies examining evolutionary responses along elevation gradients are lacking. The distances insects need to move to track climate change along elevation gradients are far shorter than those required to track climate with latitude (89). Thus, greater gene flow might prevent local adaptation at low-elevation range boundaries because of gene swamping from higher altitude populations but accelerate the rate of evolution at high-elevation boundaries through the immigration of warm-adapted genes from lower altitude populations. In a similar manner, it is unknown whether the patterns that have been observed in species with large continental-scale distributions are repeated in more narrowly distributed species, but at a smaller scale. Again, it is possible that increased capacity for gene flow between the leading and trailing range boundaries of narrowly distributed species results in patterns different from those discussed in this review. In the field of evolutionary adaptations to recent climate change, there are still far more unknowns than established generalities. Notwithstanding the serious negative impacts of recent climate change on biodiversity and ecosystems, we do currently have opportunities to examine and understand the feedbacks between ecological and evolutionary processes during a period of rapid environmental change.

SUMMARY POINTS

1. Many species are responding to recent climate change and shifting their distributions to higher latitudes and higher altitudes. Evolutionary adaptations have been observed at species' expanding high-latitude range margins.
2. Recent range changes mirror those observed during climate warming after the LGM about 25,000 years ago. These postglacial expansions have resulted in latitudinal declines in genetic diversity from low to higher latitudes in many species from a wide range of insect taxa.
3. Declines in genetic diversity arise when colonization of new habitats is by a small number of founder individuals dispersing over long distances, which leads to loss of alleles and reduced heterozygosity.
4. It is unclear whether these postglacial genetic effects might accelerate or limit current range changes, but loss of genetic variation is likely to continue as species spread across modern landscapes with habitats that are heavily fragmented.
5. Evolutionary changes occur in populations at expanding range boundaries. The most commonly observed adaptation is increased dispersal, e.g., changes in flight morphology (wing and thorax), and increased metabolic rates. Changes in habitat associations are also observed.
6. Trade-offs between flight and reproduction are commonly observed such that range margin sites have individuals with reduced fecundity.

7. A better understanding of the extent and speed of adaptation is crucial to the responses of biodiversity and ecosystems to climate change.

FUTURE ISSUES

1. How does the rate of environmental change affect the likelihood and speed of evolutionary responses in expanding species?
2. What is the likelihood of expanding species adapting to new environmental conditions?
3. What are the relative roles of changes in (*a*) physiological processes underlying dispersal ability and (*b*) the behavioral inclination of individuals to disperse in determining colonization rates?
4. Are evolutionary adaptations evident at contracting range boundaries, and under what circumstances are they fast enough to prevent population extinction?
5. Are evolutionary adaptations observed on latitudinal gradients also evident along elevational gradients, and in species with small geographical ranges?
6. What is the relationship between the genetic diversity of species and their rate of range expansion?

DISCLOSURE STATEMENT

The authors are not aware of any affiliations, memberships, funding, or financial holdings that might be perceived as affecting the objectivity of this review.

LITERATURE CITED

1. Addo-Bediako A, Chown SL, Gaston KJ. 2000. Thermal tolerance, climatic variability and latitude. *Proc. R. Soc. Biol. Sci. Ser. B* 267:739–45
2. Anderson S, Conrad K, Gillman M, Woiwod I, Freeland J. 2008. Phenotypic changes and reduced genetic diversity have accompanied the rapid decline of the garden tiger moth (*Arctia caja*) in the U.K. *Ecol. Entomol.* 33:638–45
3. Aoki K, Kato M, Murakami N. 2008. Glacial bottleneck and postglacial recolonization of a seed parasitic weevil, *Curculio hilgendorfi*, inferred from mitochondrial DNA variation. *Mol. Ecol.* 17:3276–79
4. Armbruster P, Bradshaw WE, Holzapfel CM. 1998. Effects of postglacial range expansion on allozyme and quantitative genetic variation of the pitcher-plant mosquito, *Wyeomyia smithii*. *Evolution* 52:1697–704
5. Asher J, Warren M, Fox R, Harding P, Jeffcoate G, Jeffcoate S. 2001. *The Millennium Atlas of Butterflies in Britain and Ireland*. Oxford: Oxford Univ. Press. 433 pp.
6. Atkinson TC, Briffa KR, Coope GR. 1987. Seasonal temperatures in Britain during the past 22,000 years, reconstructed using beetle remains. *Nature* 325:587–92
7. Babin-Fenske J, Anand M, Alarie Y. 2008. Rapid morphological change in stream beetle museum specimens correlates with climate change. *Ecol. Entomol.* 33:646–51
8. Bell KL, Moritz C, Moussalli A, Yeates DK. 2007. Comparative phylogeography and speciation of dung beetles from the Australian Wet Tropics rainforest. *Mol. Ecol.* 16:4984–98
9. Berwaerts K, Van Dyck H, Aerts P. 2002. Does flight morphology relate to flight performance? An experimental test with the butterfly *Pararge aegeria*. *Funct. Ecol.* 16:484–91

10. Berwaerts K, Van Dyck H, Van Dongen S, Matthysen E. 1998. Morphological and genetic variation in the speckled wood butterfly (*Pararge aegeria* L.) among differently fragmented landscapes. *Neth. J. Zool.* 48:241–53
11. Besold J, Schmitt T, Tammaru T, Cassel-Lundhagen A. 2008. Strong genetic impoverishment from the center of distribution in southern Europe to peripheral Baltic and isolated Scandinavian populations of the pearly heath butterfly. *J. Biogeogr.* 35:2090–101
12. Blanckenhorn WU, Fairbairn DJ. 1995. Life-history adaptation along a latitudinal cline in the water strider *Aquarius remigis* (Heteroptera, Gerridae). *J. Evol. Biol.* 8:21–41
13. Bradshaw WE, Holzapfel CM. 2001. Genetic shift in photoperiodic response correlated with global warming. *Proc. Natl. Acad. Sci. USA* 98:14509–11
14. Bradshaw WE, Holzapfel CM. 2008. Genetic response to rapid climate change: It's seasonal timing that matters. *Mol. Ecol.* 17:157–66
15. Braschler B, Hill JK. 2007. Role of larval host plants in the climate-driven range expansion of the butterfly *Polygonia c-album*. *J. Anim. Ecol.* 76:415–23
16. Buckley TR, Marske K, Attanayake D. 2010. Phylogeography and ecological niche modelling of the New Zealand stick insect *Clitarchus hookeri* (White) support survival in multiple coastal refugia. *J. Biogeogr.* 37:682–95
17. Buckley TR, Marske KA, Attanayake D. 2009. Identifying glacial refugia in a geographic parthenogen using palaeoclimate modelling and phylogeography: the New Zealand stick insect *Argosarchus horridus* (White). *Mol. Ecol.* 18:4650–63
18. Butin E, Porter AH, Elkinton J. 2005. Adaptation during biological invasions and the case of *Adelges tsugae*. *Evol. Ecol. Res.* 7:887–900
19. Cassel-Lundhagen A, Tammaru T, Windig JJ, Ryrholm N, Nylin S. 2009. Are peripheral populations special? Congruent patterns in two butterfly species. *Ecography* 32:591–600
20. Chai P, Srygley RB. 1990. Predation and the flight, morphology, and temperature of neotropical rain-forest butterflies. *Am. Nat.* 135:748–65
21. Chen IC, Shiu HJ, Benedick S, Holloway JD, Cheye VK, et al. 2009. Elevation increases in moth assemblages over 42 years on a tropical mountain. *Proc. Natl. Acad. Sci. USA* 106:1479–83
22. Clark PU, Dyke AS, Shakun JD, Carlson AE, Clark J, et al. 2009. The Last Glacial Maximum. *Science* 325:710–14
23. Coope G. 1977. Quaternary Coleoptera as aids in the interpretation of environmental history. In *British Quaternary Studies: Recent Advances*, ed. FW Shotton, pp. 55–68. Oxford: Clarendon. 298 pp.
24. Coope G. 1995. The effects of Quaternary climatic changes on insect populations: lessons from the past. In *Insects in a Changing Environment*, ed. R Harrington, N Stork, pp. 29–48. London: Academic. 535 pp.
25. DeChaine EG, Martin AP. 2004. Historic cycles of fragmentation and expansion in *Parnassius smintheus* (Papilionidae) inferred using mitochondrial DNA. *Evolution* 58:113–27
26. DeChaine EG, Martin AP. 2006. Using coalescent simulations to test the impact of Quaternary climate cycles on divergence in an alpine plant-insect association. *Evolution* 60:1004–13
27. Dempster JP, King ML, Lakhani KH. 1976. Status of the swallowtail butterfly in Britain. *Ecol. Entomol.* 1:71–84
28. Dennis RLH. 1993. *Butterflies and Climate Change*. Manchester, UK: Manchester Univ. Press. 302 pp.
29. Dennis RLH, Shreeve TG. 1989. Butterfly wing morphology variation in the British Isles—the influence of climate, behavioural posture and the hostplant-habitat. *Biol. J. Linn. Soc.* 38:323–48
30. Descimon H, Zimmermann M, Cosson E, Barascud B, Neve G. 2001. Genetic variation, geographic variation and gene flow in some French butterfly species. *Genet. Sel. Evol.* 33:S223–S49
31. Deutsch CA, Tewksbury JJ, Huey RB, Sheldon KS, Ghalambor CK, et al. 2008. Impacts of climate warming on terrestrial ectotherms across latitude. *Proc. Natl. Acad. Sci. USA* 105:6668–72
32. Diamantidis AD, Carey JR, Papadopoulos NT. 2008. Life-history evolution of an invasive tephritid. *J. Appl. Entomol.* 132:695–705
33. Dytham C. 2009. Evolved dispersal strategies at range margins. *Proc. R. Soc. Biol. Sci. Ser. B* 276:1407–13
34. Elias SA. 1991. Insects and climate change. *Bioscience* 41:552–59
35. Elias SA. 2006. Quaternary beetle research: the state of the art. *Quat. Sci. Rev.* 25:1731–37

36. Fordyce JA, Nice CC. 2003. Contemporary patterns in a historical context: phylogeographic history of the pipevine swallowtail, *Battus philenor* (Papilionidae). *Evolution* 57:1089–99
37. Franco AMA, Hill JK, Kitschke C, Collingham YC, Roy DB, et al. 2006. Impacts of climate warming and habitat loss on extinctions at species' low-latitude range boundaries. *Glob. Change Biol.* 12:1545–53
38. Gilchrist GW, Lee CE. 2006. All stressed out and nowhere to go: Does evolvability limit adaptation in invasive species? *Genetica* 129:127–32
39. Gomi T. 2007. Seasonal adaptations of the fall webworm *Hyphantria cunea* (Drury) (Lepidoptera: Arctiidae) following its invasion of Japan. *Ecol. Res.* 22:855–61
40. Gomi T, Adachi K, Shimizu A, Tanimoto K, Kawabata E, Takeda M. 2009. Northerly shift in voltinism watershed in *Hyphantria cunea* (Drury) (Lepidoptera: Arctiidae) along the Japan Sea coast: evidence of global warming? *Appl. Entomol. Zool.* 44:357–62
41. Gomulkiewicz R, Holt RD. 1995. When does evolution by natural-selection prevent extinction? *Evolution* 49:201–7
42. Gomulkiewicz R, Holt RD, Barfield M, Nuismer SL. 2010. Genetics, adaptation, and invasion in harsh environments. *Evol. Appl.* 3:97–108
43. Haag CR, Saastamoinen M, Marden JH, Hanski I. 2005. A candidate locus for variation in dispersal rate in a butterfly metapopulation. *Proc. R. Soc. Biol. Sci. Ser. B* 272:2449–56
44. Habel JC, Dieker P, Schmitt T. 2009. Biogeographical connections between the Maghreb and the Mediterranean peninsulas of southern Europe. *Biol. J. Linn. Soc.* 98:693–703
45. Habel JC, Meyer M, El Mousadik A, Schmitt T. 2008. Africa goes Europe: the complete phylogeography of the marbled white butterfly species complex *Melanargia galathea/lachesis* (Lepidoptera: Satyridae). *Organ. Divers. Evol.* 8:121–29
46. Habel JC, Schmitt T, Muller P. 2005. The fourth paradigm pattern of postglacial range expansion of European terrestrial species: the phylogeography of the marbled white butterfly (Satyrinae, Lepidoptera). *J. Biogeogr.* 32:1489–97
47. Hampe A, Petit RJ. 2005. Conserving biodiversity under climate change: the rear edge matters. *Ecol. Lett.* 8:461–67
48. Hanski I, Breuker CJ, Schops K, Setchfield R, Nieminen M. 2002. Population history and life history influence the migration rate of female Glanville fritillary butterflies. *Oikos* 98:87–97
49. Hanski I, Eralahti C, Kankare M, Ovaskainen O, Siren H. 2004. Variation in migration propensity among individuals maintained by landscape structure. *Ecol. Lett.* 7:958–66
50. Hanski I, Saastamoinen M, Ovaskainen O. 2006. Dispersal-related life-history trade-offs in a butterfly metapopulation. *J. Anim. Ecol.* 75:91–100
51. Hanski I, Saccheri I. 2006. Molecular-level variation affects population growth in a butterfly metapopulation. *PLoS Biol.* 4:719–26
52. Hassall C, Thompson DJ, Harvey IF. 2009. Variation in morphology between core and marginal populations of three British damselflies. *Aquat. Insects* 31:187–97
53. Hassall M, Telfer MG. 1999. Ecotypic differentiation in the grasshopper *Chorthippus brunneus*: Life history varies in relation to climate. *Oecologia* 121:245–54
54. Heilveil JS, Berlocher SH. 2006. Phylogeography of postglacial range expansion in *Nigronia serricornis* Say (Megaloptera: Corydalidae). *Mol. Ecol.* 15:1627–42
55. Hewitt GM. 1996. Some genetic consequences of ice ages, and their role in divergence and speciation. *Biol. J. Linn. Soc.* 58:247–76
56. Hewitt GM. 1999. Post-glacial recolonization of European biota. *Biol. J. Linn. Soc.* 68:87–112
57. Hewitt GM. 2000. The genetic legacy of the Quaternary ice ages. *Nature* 405:907–13
58. Hickling R, Roy DB, Hill JK, Fox R, Thomas CD. 2006. The distributions of a wide range of taxonomic groups are expanding polewards. *Glob. Change Biol.* 12:450–55
59. Hickling R, Roy DB, Hill JK, Thomas CD. 2005. A northward shift of range margins in British Odonata. *Glob. Change Biol.* 11:502–6
60. Hill JK, Collingham YC, Thomas CD, Blakeley DS, Fox R, et al. 2001. Impacts of landscape structure on butterfly range expansion. *Ecol. Lett.* 4:313–21
61. Hill JK, Hughes CL, Dytham C, Searle JB. 2006. Genetic diversity in butterflies: interactive effects of habitat fragmentation and climate-driven range expansion. *Biol. Lett.* 2:152–54

62. Hill JK, Thomas CD, Blakeley DS. 1999. Evolution of flight morphology in a butterfly that has recently expanded its geographic range. *Oecologia* 121:165–70
63. Hill JK, Thomas CD, Fox R, Moss D, Huntley B. 2001. Analyzing and modelling range changes in UK butterflies. In *Insect Movement: Mechanisms and Consequences*, ed. IP Woiwod, DR Reynolds, CD Thomas, pp. 415–41. Wallingford, UK: CABI. 458 pp.
64. Hill JK, Thomas CD, Fox R, Telfer MG, Willis SG, et al. 2002. Responses of butterflies to twentieth century climate warming: implications for future ranges. *Proc. R. Soc. Biol. Sci. Ser. B* 269:2163–71
65. Hill JK, Thomas CD, Huntley B. 1999. Climate and habitat availability determine twentieth century changes in a butterfly's range margin. *Proc. R. Soc. Biol. Sci. Ser. B* 266:1197–206
66. Hodgson JA, Thomas CD, Wintle BA, Moilanen A. 2009. Climate change, connectivity and conservation decision making: back to basics. *J. Appl. Ecol.* 46:964–69
67. Huey RB, Gilchrist GW, Carlson ML, Berrigan D, Serra L. 2000. Rapid evolution of a geographic cline in size in an introduced fly. *Science* 287:308–9
68. Hughes CL, Dytham C, Hill JK. 2007. Modelling and analyzing evolution of dispersal in populations at expanding range boundaries. *Ecol. Entomol.* 32:437–45
69. Hughes CL, Hill JK, Dytham C. 2003. Evolutionary trade-offs between reproduction and dispersal in populations at expanding range boundaries. *Proc. R. Soc. Biol. Sci. Ser. B* 270:147–50
70. Huntley B. 1991. How plants respond to climate change–migration rates, individualism and the consequences for plant-communities. *Ann. Bot.* 67:15–22
71. Ibrahim KM, Nichols RA, Hewitt GM. 1996. Spatial patterns of genetic variation generated by different forms of dispersal during range expansion. *Heredity* 77:282–91
72. Intergovernmental Panel on Climate Change. 2001. *Climate Change 2001: Impacts, Adaptation, and Vulnerability*. Cambridge, UK: Cambridge Univ. Press. 1032 pp.
73. Intergovernmental Panel on Climate Change. 2007. *Climate Change 2007: Impacts, Adaptation and Vulnerability*. Cambridge, UK: Cambridge Univ. Press. 976 pp.
74. Jump AS, Hunt JM, Martinez-Izquierdo JA, Penuelas J. 2006. Natural selection and climate change: temperature-linked spatial and temporal trends in gene frequency in *Fagus sylvatica*. *Mol. Ecol.* 15:3469–80
75. Kohnen A, Wissemann V, Brandl R. 2009. No genetic differentiation in the rose-infesting fruit flies *Rhagoletis alternata* and *Carpomya schineri* (Diptera: Tephritidae) across central Europe. *Eur. J. Entomol.* 106:315–21
76. Loeschcke V, Bundgaard J, Barker JSF. 2000. Variation in body size and life history traits in *Drosophila aldrichi* and *D. buzzatii* from a latitudinal cline in eastern Australia. *Heredity* 85:423–33
77. Lunt DH, Ibrahim KM, Hewitt GM. 1998. mtDNA phylogeography and postglacial patterns of subdivision in the meadow grasshopper *Chorthippus parallelus*. *Heredity* 80:633–41
78. Mardulyn P, Mikhailov YE, Pasteels JM. 2009. Testing phylogeographic hypotheses in a Euro-Siberian cold-adapted leaf beetle with coalescent simulations. *Evolution* 63:2717–29
79. Marshall DC, Hill KBR, Fontaine KM, Buckley TR, Simon C. 2009. Glacial refugia in a maritime temperate climate: cicada (*Kikihia subalpina*) mtDNA phylogeography in New Zealand. *Mol. Ecol.* 18:1995–2009
80. McColl G, McKechnie SW. 1999. The *Drosophila* heat shock hsr-omega gene: an allele frequency cline detected by quantitative PCR. *Mol. Biol. Evol.* 16:1568–74
81. Meehl GA, Washington WM, Collins WD, Arblaster JM, Hu A, et al. 2005. How much more global warming and sea level rise? *Science* 307:1769–72
82. Menendez R, Megias AG, Hill JK, Braschler B, Willis SG, et al. 2006. Species richness changes lag behind climate change. *Proc. R. Soc. Biol. Sci. Ser. B* 273:1465–70
83. Miguel I, Iriondo M, Garnery L, Sheppard WS, Estonba A. 2007. Gene flow within the M evolutionary lineage of *Apis mellifera*: role of the Pyrenees, isolation by distance and postglacial recolonization routes in the western Europe. *Apidology* 38:141–55
84. Mitikka V, Hanski I. 2010. Pgi genotype influences flight metabolism at the expanding range margin of the European map butterfly. *Ann. Zool. Fenn.* 47:1–14

85. Murria C, Hughes JM. 2008. Cyclic habitat displacements during Pleistocene glaciations have induced independent evolution of *Tasimia palpata* populations (Trichoptera: Tasimiidae) in isolated subtropical rain forest patches. *J. Biogeogr.* 35:1727–37
86. Niemela J, Spence JR. 1991. Distribution and abundance of an exotic ground-beetle (Carabidae)—a test of community impact. *Oikos* 62:351–59
87. Noda H. 1992. Geographic-variation of nymphal diapause in the small brown planthopper in Japan. *Jpn. Agric. Res. Q.* 26:124–29
88. Parmesan C. 1996. Climate and species' range. *Nature* 382:765–66
89. Parmesan C. 2006. Ecological and evolutionary responses to recent climate change. *Annu. Rev. Ecol. Evol. Syst.* 37:637–69
90. Parmesan C, Ryrholm N, Stefanescu C, Hill JK, Thomas CD, et al. 1999. Poleward shifts in geographical ranges of butterfly species associated with regional warming. *Nature* 399:579–83
91. Parmesan C, Yohe G. 2003. A globally coherent fingerprint of climate change impacts across natural systems. *Nature* 421:37–42
92. Pellegroms B, Van Dongen S, Van Dyck H, Lens L. 2009. Larval food stress differentially affects flight morphology in male and female speckled woods (*Pararge aegeria*). *Ecol. Entomol.* 34:387–93
93. Poyry J, Luoto M, Heikkinen RK, Kuussaari M, Saarinen K. 2009. Species traits explain recent range shifts of Finnish butterflies. *Glob. Change Biol.* 15:732–43
94. Rich KA, Thompson JN, Fernandez CC. 2008. Diverse historical processes shape deep phylogeographical divergence in the pollinating seed parasite *Greya politella*. *Mol. Ecol.* 17:2430–48
95. Rosenzweig C, Karoly D, Vicarelli M, Neofotis P, Wu QG, et al. 2008. Attributing physical and biological impacts to anthropogenic climate change. *Nature* 453:353–57
96. Saccheri I, Kuussaari M, Kankare M, Vikman P, Fortelius W, Hanski I. 1998. Inbreeding and extinction in a butterfly metapopulation. *Nature* 392:491–94
97. Salle A, Arthofer W, Lieutier F, Stauffer C, Kerdelhue C. 2007. Phylogeography of a host-specific insect: Genetic structure of *Ips typographus* in Europe does not reflect past fragmentation of its host. *Biol. J. Linn. Soc.* 90:239–46
98. Schmitt T, Giessl A, Seitz A. 2003. Did *Polyommatus icarus* (Lepidoptera: Lycaenidae) have distinct glacial refugia in southern Europe? Evidence from population genetics. *Biol. J. Linn. Soc.* 80:529–38
99. Schmitt T, Habel JC, Zimmermann M. 2006. Genetic differentiation of the marbled white butterfly, *Melanargia galathea*, accounts for glacial distribution patterns and postglacial range expansion in southeastern Europe. *Mol. Ecol.* 15:1889–901
100. Schmitt T, Haubrich K. 2008. The genetic structure of the mountain forest butterfly *Erebia euryale* unravels the late Pleistocene and postglacial history of the mountain coniferous forest biome in Europe. *Mol. Ecol.* 17:2194–207
101. Schmitt T, Hewitt GM. 2004. The genetic pattern of population threat and loss: a case study of butterflies. *Mol. Ecol.* 13:21–31
102. Schmitt T, Hewitt GM, Muller P. 2006. Disjunct distributions during glacial and interglacial periods in mountain butterflies: *Erebia epiphron* as an example. *J. Evol. Biol.* 19:108–13
103. Settele J, Kudrna O, Harpke A, Kuhn I, Van Swaay C, et al. 2008. *Climatic Risk Atlas of European Butterflies*. Moscow: Pensoft. 710 pp.
104. Severinghaus JP, Sowers T, Brook EJ, Alley RB, Bender ML. 1998. Timing of abrupt climate change at the end of the Younger Dryas interval from thermally fractionated gases in polar ice. *Nature* 391:141–46
105. Simmons AD, Thomas CD. 2004. Changes in dispersal during species' range expansions. *Am. Nat.* 164:378–95
106. Smith CI, Farrell BD. 2005. Phylogeography of the longhorn cactus beetle *Moneilema appressum* LeConte (Coleoptera: Cerambycidae): Was the differentiation of the Madrean sky islands driven by Pleistocene climate changes? *Mol. Ecol.* 14:3049–65
107. Sunnucks P, Stone GN. 1996. Genetic structure of invading insects and the case of the Knopper gallwasp. In *Frontiers of Population Ecology Conference to Celebrate the Centenary of the Birth of the Population Ecologist AJ Nicholson (1895–1969)*, ed. RB Floyd, AW Sheppard, PJ DeBarro, pp. 485–95. East Melbourne: CSIRO

108. Thomas CD. 2005. Recent evolutionary effects of climate change. In *Climate Change and Biodiversity*, ed. TE Lovejoy, L Hannah, pp. 75–88. New Haven, CT: Yale Univ. Press
109. Thomas CD, Bodsworth EJ, Wilson RJ, Simmons AD, Davies ZG, et al. 2001. Ecological and evolutionary processes at expanding range margins. *Nature* 411:577–81
110. Thomas CD, Cameron A, Green RE, Bakkenes M, Beaumont LJ, et al. 2004. Extinction risk from climate change. *Nature* 427:145–48
111. Tolman T. 1997. *Butterflies of Britain and Europe*. London: HarperCollins. 320 pp.
112. Vandewoestijne S, Neve G, Baguette M. 1999. Spatial and temporal population genetic structure of the butterfly *Aglais urticae* L. (Lepidoptera, Nymphalidae). *Mol. Ecol.* 8:1539–43
113. Vandewoestijne S, Schtickzelle N, Baguette M. 2008. Positive correlation between genetic diversity and fitness in a large, well-connected metapopulation. *BMC Biol.* 6:46
114. Wahlberg N, Saccheri I. 2007. The effects of Pleistocene glaciations on the phylogeography of *Melitaea cinxia* (Lepidoptera: Nymphalidae). *Eur. J. Entomol.* 104:675–84
115. Warren MS, Hill JK, Thomas JA, Asher J, Fox R, et al. 2001. Rapid responses of British butterflies to opposing forces of climate and habitat change. *Nature* 414:65–69
116. Wigley TML. 2005. The climate change commitment. *Science* 307:1766–69
117. Wilcock HR, Hildrew AG, Nichols RA. 2001. Genetic differentiation of a European caddisfly: past and present gene flow among fragmented larval habitats. *Mol. Ecol.* 10:1821–32
118. Wilson RJ, Gutierrez D, Gutierrez J, Martinez D, Agudo R, Monserrat VJ. 2005. Changes to the elevational limits and extent of species ranges associated with climate change. *Ecol. Lett.* 8:1138–46
119. Wilson RJ, Gutierrez D, Gutierrez J, Monserrat V. 2007. An elevational shift in butterfly species richness and composition accompanying recent climate change. *Glob. Change Biol.* 13:1873–87
120. Zera AJ, Denno RF. 1997. Physiology and ecology of dispersal polymorphism in insects. *Annu. Rev. Entomol.* 42:207–30
121. Zwaan BJ, Azevedo RBR, James AC, Van 't Land J, Partridge L. 2000. Cellular basis of wing size variation in *Drosophila melanogaster*: a comparison of latitudinal clines on two continents. *Heredity* 84:338–47

Ecological Role of Volatiles Produced by Plants in Response to Damage by Herbivorous Insects

J. Daniel Hare

Department of Entomology, University of California, Riverside, California 92521; email: daniel.hare@ucr.edu

Annu. Rev. Entomol. 2011. 56:161–80

The *Annual Review of Entomology* is online at ento.annualreviews.org

This article's doi:
10.1146/annurev-ento-120709-144753

0066-4170/11/0107-0161$20.00

Key Words

HIPVs, indirect defenses, tritrophic interactions, natural enemies, habitat location, evolution, variation

Abstract

Plants often release a blend of volatile organic compounds in response to damage by herbivorous insects that may serve as cues to locate those herbivores by natural enemies. The blend of compounds emitted by plants may be more variable than is generally assumed. The quantity and the composition of the blends may vary with the species of the herbivore, the plant species and genotype within species, the environmental conditions under which plants are grown, and the number of herbivore species attacking the plant. Although it is often assumed that induced emission of these compounds is an adaptive tactic on the part of plants, the evidence that such responses minimize fitness losses of plants remains sparse because the necessary data on plant fitness rarely have been collected. The application of techniques of evolutionary quantitative genetics may facilitate the testing of widely held hypotheses about the evolution of induced production of volatile compounds under natural conditions.

Herbivore-induced plant volatile (HIPV): volatile compounds induced in plants following damage by herbivores

VOC: volatile organic compound

Fitness: the average lifetime contribution of individuals of a particular genotype to future generations

Type I error: rejection of the null hypothesis when it is true

INTRODUCTION

Plant odors are used as cues by the parasitoids and predators of plant-feeding insects to aid in the location of their prey (77, 123, 124). The role of the host plant upon the interaction between the plant's herbivores and the natural enemies of herbivores, however, was largely ignored until Price et al. (84) reviewed how interactions between herbivorous insects and natural enemies varied across different plant species. They suggested the potential role that natural enemies could play as a component of plant defense against herbivorous insects. Implicit in this idea is the possibility that natural selection may be imposed on plants by natural enemies for traits that enhance the effectiveness of those natural enemies. In the early 1990s, it was discovered that many of the volatile compounds produced by plants that can be used as cues by natural enemies of herbivores were inducible and only released by plants after the plants had been damaged by herbivores (18, 21, 107, 112, 120). These compounds are termed herbivore-induced plant volatiles (HIPVs) and are a subset of the more general class of volatile organic compounds (VOCs) emitted by plants.

Hundreds of studies examining the effect of HIPVs on the behavior and effectiveness of natural enemies have been published in the past two decades, and it is common to find natural enemies to be attracted to odors of damaged plants, especially under the controlled laboratory conditions under which such investigations are usually performed (2, 50). This field has been reviewed numerous times. Whereas the benefits to natural enemies in utilizing plant-produced cues to more efficiently locate their hosts seems clear, the benefits to plants are less certain. Most reviewers consistently call for additional research under field conditions and on undomesticated plant species in their natural environments (19, 39, 54, 108, 113, 117), but these calls have largely gone unheeded. Thus, we still know little about the significance of variation in the production of HIPVs on variation in plant fitness, and we can draw few inferences on the evolution of HIPV production by plants as a mechanism to reduce damage by herbivores through the enhanced activities of natural enemies.

The first goal of this review is to focus on current questions about the adaptive value of HIPVs as components of plant defense against herbivores in light of relatively recent research showing substantial variation in HIPV production within plant species. The second goal is to emphasize the need to adopt a Darwinian perspective in this field that focuses on the potential variation in HIPV production among plant genotypes and the potential for natural selection on that variation in order to understand the evolutionary ecology of HIPV production by plants. Interested readers may wish to consult other reviews with different perspectives (15, 16, 43, 55).

STATISTICAL TESTING OF BLENDS

The array of HIPVs may be composed of hundreds of compounds (20) and should be properly considered as an intercorrelated, multivariate suite of traits. Because many of these compounds share common precursors, patterns of correlated variation within groups of compounds may be the result of variation in the quantity of their common precursor. Moreover, in some cases, particular ratios of several compounds can be the product of a single enzyme (92). Thus, it is incorrect to assume that all HIPVs vary independently. Although the individual HIPV components are conventionally analyzed statistically as though each compound is free to vary independently of others, such analyses run the risk of inflating the Type I error and concluding that treatments differ with respect to individual compounds even though the differences are due simply to correlations with other compounds (89). To avoid such problems, multivariate statistical techniques are required that take into account the patterns of correlations of variables in the determination of statistically significant variation. This raises its own set of problems, especially if state-of-the art analytical equipment is used to quantify tens, if not hundreds, of individual compounds from small numbers of plants. The problem is that each individual variable consumes one degree of freedom, so even a minimal statistical test requires that the number of subjects (individual plants) minimally exceeds the number of variables (compounds measured), and even greater replication would be preferred. At present, our chemical analytical abilities may exceed our abilities to grow and sample a sufficient number of plants under uniform conditions for an appropriate statistical analysis, and compromises may be needed between the goals of exhaustively quantifying all measurable compounds on the one hand and properly replicating the observations for a meaningful statistical analysis on the other hand.

In addition to the lack of independence among individual variables, another problem with multivariate data is that general patterns of variation may be difficult to discern from the variation of those variables. In such cases, it often is helpful to utilize other statistical techniques to capture the general patterns of variation. This is actually an application of multivariate techniques from the rapidly growing fields of metabolomics and/or chemometrics, and the application of such techniques has so far been notably absent from the analysis of HIPVs (115).

The overall goals of such techniques are to reduce the dimensionality of the data in search of general patterns and their underlying causes. One general approach is to calculate a new, smaller, and unbiased set of components, or factors, that are linear combinations of the original variables. Perhaps the most widely used technique is principal components analysis (PCA), in which each new component accounts for a progressively smaller proportion of the variance of the original data. Most importantly, the new components are independent and uncorrelated. The component loadings (relationships between the original variables and the new component) show how much each of the original variables contributes to each new component, and scores can be calculated for each original subject using the new components. Statistical analyses then can be performed on these new components (82), but some limitations may exist in statistical testing (52). Newer statistical techniques are being developed to overcome these limitations (53), but they may have their own limitations as well (24). Perhaps the best recommendation at this point is to evaluate several different techniques and determine to what extent techniques lead to different conclusions (67).

Principal components analysis (PCA): an early but still widely used multivariate statistical method that operates upon a number of possibly correlated variables and calculates a smaller number of uncorrelated variables called principal components

These techniques also may require large sample sizes to develop stable patterns of association that are not unduly influenced by outliers. Several rules of thumb regarding sufficient sample size appear in the literature (34), although most of these may have little justification. The actual samples sizes required for stability depend on the strength of the patterns, with fewer numbers of observations required for stronger patterns than for weaker patterns (11, 33).

In principle, these techniques should allow one to determine how blends from different induction treatments might differ statistically far better than the conventional approach of conducting a series of independent univariate tests on the array of individual compounds. It should not be forgotten that statistical discrimination is not the same as identifying causality, and the behavioral significance of statistically significant variation in multivariate blends needs to be determined independently. The development of techniques for metabolomics or chemometrics analyses is a rapidly growing field and one from which the statistical and behavioral analysis of HIPVs would benefit (115).

GENETIC VARIATION IN HIPV PRODUCTION

Variation in the production of HIPVs has been observed among cultivars of numerous crops, including apple, *Malus* sp. (103); maize, *Zea mays* (46); cotton, *Gossypium hirsutum* (66); rice, *Oryza sativa* (65); carrot, *Daucus carota* (76); pear, *Pyrus* spp. (94); gerbera, *Gerbera jamesonii* (61); and cruciferous crops such as cabbage, *Brassica oleracea* (5). The range of variation in HIPV production among cultivars can be

Inducible direct defenses: the change in plant traits in response to damage by herbivores that reduces herbivore preference or performance and minimizes the fitness losses of plants to those herbivores

considerable, and the differences in individual compounds are too numerous to summarize here. In one of the more extensive analyses, there was an approximate 70-fold variation in the quantities of volatiles emitted and in the odor profile of 31 lines of maize. In some cases, the quantities of odors emitted by undamaged plants of some lines were higher than the quantities emitted by damaged plants of other lines (12).

Although of potential interest in biological control (46, 65), such variation is most assuredly merely a fortuitous consequence of plant breeding for desirable agronomic traits that are incidental to HIPV production, for none of the crops listed above was specifically bred for particular HIPV blends to improve the biological control of their pests. This, when coupled with the fact that most of the well-studied crop-based systems are far removed from their original ecological and evolutionary context, makes it difficult, if not impossible, to draw strong evolutionary inferences from crop-based systems. To draw such inferences, it is necessary to study natural systems in their native habitats (39).

Variation in HIPV production among genotypes of wild species has been rarely studied compared with that in cultivated species. Up to eightfold variation in total quantities of HIPVs as well as significant variation in the proportions of several sesquiterpenes occurred among five accessions of teosinte, the progenitor of cultivated maize (31). More recently, qualitative and quantitative variation in HIPV production has been reported in the uncultivated solanaceous species, sacred datura, *Datura wrightii* (40); horsenettle, *Solanum carolinense* (14); and coyote tobacco, *Nicotiana attenuata* (93). In the first case, either treating plants with methyl jasmonate or allowing the plant's primary herbivore, *Lema daturaphila* (40), to remove 10%–15% of foliage by feeding induced the production of 17 compounds, the most abundant of which was (*E*)-β-caryophyllene. Eight genetic lines of *D. wrightii* were examined (two lines from each of the four southern California populations selected) and the lines varied in inducibility, but the patterns of variation showed as much or more variation within populations as among them. The total production of volatiles increased from nearly 4- to 16-fold, depending on the genetic line, in response to insect damage, and from 3- to 32-fold in response to methyl jasmonate. The proportion of (*E*)-β-caryophyllene in induced plants varied from 17% to 88% of all volatiles, depending on inducer and genetic line (40). Significant quantitative and qualitative variation was also observed among 12 genotypes of *S. carolinense*, with up to 10-fold variation in total HIPV production among lines. The composition of the blends was relatively complex in this study, but one index of variation in the composition of the blends is the proportion of (*E*)-β-caryophyllene, which varied from 4% to 38% among lines (14).

Early research on the HIPV blends of *N. attenuata* showed substantial variation in HIPV production among widely separated populations in the American Southwest. The compound, linalool, was inducible in one population, not inducible in a second, and completely absent in a third. The populations also differed in the absolute and relative concentrations of induced *E*-β-ocimene and α-bergamotene, with two populations highly inducible for *E*-β-ocimene and the third more inducible for α-bergamotene; the production of *E*-β-ocimene varied up to 80-fold and production for α-bergamotene up to 10-fold within these three populations (35). In addition, a population from Arizona showed little if any change in HIPV production compared to a population from Utah, which suggests that the adaptive value of inducible indirect defenses may vary among plant populations (27).

In contrast to these previous studies on *N. attenuata*, which focused on differences among widely distant plant populations, considerable variation has been shown within a single population in Utah as well. Progenies from four plants from a single population were studied in a greenhouse, and there were significant differences in the relative concentration of 20 compounds, although not all compounds were produced by all accessions (93). One accession produced no α-bergamotene, for example, a compound that attracts *Geocoris pallens*, a predator of *N. attenuata*'s primary herbivore, *Manduca sexta* (36, 56).

There is substantial variation in HIPV production among genotypes of both cultivated and wild plant species, and perhaps more than might have been expected under the assumption that the blends of extant genotypes are the product of strong natural selection for the blends most attractive to natural enemies. Nevertheless, the fact that such variation exists is a prerequisite to determining if such variation might be subject to natural selection.

ENVIRONMENTAL VARIATION IN HIPV BLENDS

The relative composition of HIPV blends of a single plant genotype and induction treatment can vary with environmental conditions. In maize, for example, the quantities of several components varied up to sixfold depending on the experimental level of light intensity, fertilization, and soil moisture used in different experiments (32). With regard to fertilization, the lowest HIPV levels were observed from unfertilized plants, with higher levels observed from plants receiving half-strength or full-strength applications of a nutrient solution. The relative proportions of compounds such as (*E*)-4,8,dimethyl-1,3,7-nonatriene (DMNT), indole, geranyl acetate, β-caryophyllene, and (*E*,*E*)-4,8,12 trimethyl-1,3,7,11-tridecatetraene (TMTT) increased with fertilization level, whereas the relative concentration of other compounds, such as β-myrcene, α-bergamotene, and (*E*,*E*)-α-farnesene, showed no effect of variation in fertilization level, and still others, such as (*Z*)-3-hexenyl acetate and linalool, showed highest relative concentrations at intermediate fertilization levels. In contrast to these findings, others report that maximal amounts of sesquiterpenes were emitted in maize plants growing under the lowest nitrogen fertilization regimes (91).

High nitrogen levels also reduced the HIPV production of cotton. Surprisingly, the emission of some sesquiterpenes, notably β-caryophyllene, (*E*)-β-farnesene, α-bergamotene, α-humulene, and γ-bisabolene, was significantly lower after insect damage than prior to damage (7). In another study, however, the highest HIPV concentrations were observed in cotton plants receiving an intermediate level of nitrogen fertilization compared with plants receiving higher or lower levels (80). In contrast, HIPV production was independent of nitrogen fertilization level in *N. attenuata* (64).

The availability of water affects both the quantity and composition of HIPV blends under controlled conditions. In maize, the total HIPVs declined significantly with increasing soil humidity, and the relative composition of the blend varied in a complex way with soil humidity (32). Some compounds of the HIPV blend of maize, particularly (*Z*)-3-hexenyl acetate, β-caryophyllene, nerolidol, and TMTT, were relatively more abundant when collections were carried out at high temperatures (37°C) than at lower temperatures, although the relative concentrations of most other major compounds were unaffected by temperature (32).

The limited studies available provide variable results and may be difficult to reconcile. In some cases, the differences in the results may be due to differences in experimental protocol and/or to differences in the range over which fertilization and other factors are manipulated relative to the mineral nutritional needs of the plants. For domesticated crop species, the differences in results may be due to correlated differences among the cultivars studied. The response of natural enemies to the HIPV production of stressed plants has rarely been examined, and no general patterns have yet been obtained. In some cases, the differences in HIPV production resulted in differences in attraction, but in others cases not (7, 80).

In one of the few studies to date to directly compare the production of HIPVs in the laboratory and the field, substantial differences both in composition and in response to herbivory were found. Fewer individual components were detected in field-grown than in laboratory-grown plants, and the emission of some compounds decreased in response to increasing herbivory in the field but increased in response to herbivory in the laboratory. Although there may be numerous factors that contribute to the different patterns in the laboratory and field, the results reinforce the importance of the environment on both the total quantity and the composition of HIPV blends emitted by individual plants (58).

The ability of plants to respond to elicitors may also vary with plant growth stage. In maize, HIPV production in seedlings is inducible, but HIPV production in mature leaves apparently is not (60). In soybean (*Glycine max*), feeding by the fall armyworm, *Spodoptera frugiperda*, induced approximately 10 times the total HIPVs from leaves of five-week-old nonflowering vegetative plants compared with 10-week-old flowering plants (88). Such results suggest that the ability of plants to attract natural enemies in response to herbivory may be constrained to the early stages of plant development.

Although the number of studies is limited, they document substantial variation in both the quantity and the composition of HIPV blends of plants grown under varying environmental conditions. The significance of such variation as a factor influencing the ability of plants to consistently attract natural enemies of the plant's herbivores is largely unexplored, especially under field conditions.

SINGLE VERSUS MULTIPLE HERBIVORES AND HIPV BLENDS

Most studies of HIPVs have been conducted primarily in the simple ecological food webs of a single genotype or cultivar of a plant species, a single species of herbivore, and a single species of natural enemy in a tightly controlled laboratory environment. Previous sections have documented substantial variation in HIPVs induced by a single elicitor from different plant genotypes or from a single plant genotype depending on environmental conditions. Yet another potential source of variation in HIPVs is attack by other herbivore species. To date, relatively little research has been completed in this area. Most, if not all such research simply adds a second species of herbivore to the plant to determine how this may affect the behavioral choices of natural enemies when presented choices between HIPV blends induced by one versus two herbivores.

One of the earliest studies utilized cultivated Brussels sprouts that were damaged either by larvae of the imported cabbageworm, *Pieris rapae*, or of the diamondback moth, *Plutella xylostella*. *P. rapae* is a host for the parasitoid, *Cotesia glomerata*, but *P. xylostella* is not. Nevertheless, the parasitoid was attracted nonspecifically to the HIPV blends induced by each caterpillar species and was unable to discriminate between the HIPV blends elicited by each. When given choices between plants damaged by both caterpillar species compared to damage by the nonhost, the parasitoid failed to discriminate, thus leading to essentially random choices (125). Although the authors argued that this may increase the persistence of parasitoid communities because parasitoids are attracted to plants damaged by herbivores that they cannot utilize, such inefficient searching also may lead to rates of parasitism too low for the parasitoid population to persist.

In a similar experiment using the same two herbivore species on cultivated cabbage plants but two parasitoid species, *C. glomerata* and *Cotesia plutellae*, the latter of which attacks *P. xylostella* but not *P. rapae*, *C. glomerata* once again failed to discriminate between HIPV blends induced by each herbivore species, whereas *C. plutellae* preferred HIPV blends induced by its host (*P. xylostella*) over the nonhost (*P. rapae*) (95). In contrast to the previous study, *C. glomerata* actually preferred the HIPV blends induced by simultaneous feeding by both caterpillars, whereas *C. plutellae* preferred the HIPV blends induced by *P. xylostella* over the blends induced by both species. Analyses of the HIPV blends from plants damaged by each herbivore alone or both simultaneously showed slight though statistically significant variation in the proportional concentration of several components. It would appear that the presence of damage by a nonhost may be relatively unimportant for *C. plutellae*, but for *C. glomerata*, the presence of a nonhost may increase the detectability of a nonspecific cue. Similarly, the parasitoid, *Diaeretiella rapae*, which successfully attacks any of a number of generalist and specialist aphids that attack crucifers, also discriminates between plants infested with aphids and uninfested plants or plants damaged by *P. xylostella*. The parasitoid did not discriminate between plants infested only with aphids and plants infested with both aphids and caterpillars, suggesting that damage by a nonhost was relatively unimportant for foraging by *D. rapae* (1).

Several studies have been performed on the variation of HIPV blends induced by one versus two herbivores of lima bean, *Phaseolus lunatus*, and on the responses of predatory mites. Lima beans infested with the phytophagous twospotted spider mite, *Tetranychus urticae*, or the beet armyworm, *Spodoptera exigua*, produced generally the same HIPV components but in different amounts. Of 27 compounds, 8 were produced in greater abundance when caterpillars and mites were feeding simultaneously on plants compared with the sum of each species damaging plants alone. No compounds were produced in significantly lower abundance, but there was substantial variability in the data such that the total abundance of HIPVs of doubly damaged plants was not significantly different from that predicted as the sum of each herbivore species acting alone (9).

Similar experiments were reported in the same study for cucumber, *Cucumis sativus*, either singly or doubly damaged by the same two herbivores, but in this case two of nine components were produced at significantly lower abundance than would be predicted under an additive model. Once again, the data were highly variable, and the total abundance of HIPVs of doubly damaged plants was not significantly different from that predicted as the sum of each herbivore species acting alone (9). In both cases, individuals of the predatory mite *Phytoseiulus persimilis* preferred the blends from both species of plants infested with both herbivores simultaneously to plants infested with either herbivore species singly.

The generalist mirid predator *Macrolophus caliginosus* also preferred pepper plants infested by either of two of its prey species, the spider mite, *T. urticae*, or the green peach aphid, *Myzus persicae*, to uninfested plants, and preferred plants infested with both species to plants infested with either species alone. The total HIPV production of doubly infested plants was significantly greater than the production of plants singly infested with either herbivore, and the ratios of some components varied as well. Although the possible significance of changes in relative composition cannot be overlooked, a simple explanation for these results is that the predator prefers the more concentrated blend produced by doubly infested plants (74).

Such may not be the case for pairs of herbivore species that differ more markedly in mode of feeding. When lima bean was infested with spider mites and silver-leaf whitefly, *Bemisia argentifolii*, predatory mites preferred plants damaged only by spider mites to plants damaged by both (130). The main difference in the blends was in the proportion of (*E*)-β-ocimene, and the attractiveness of doubly infested plants was restored by supplementing the blend of doubly infested plants with this single compound. The absence of (*E*)-β-ocimene was the result of suppressing the synthesis of jasmonic acid and signaling by whitefly infestations, ultimately causing a reduction in the emission of (*E*)-β-ocimene. The suppression of HIPV production by suppressing jasmonic acid signaling may be adaptive in the context of suppressing both direct and indirect defenses against whiteflies (127), but these results suggest that such suppression may also benefit the interspecific competitors of whiteflies on the same plants.

Multiple herbivore species attacking plants can be physically displaced on different plant tissues and still affect the behavioral choices of natural enemies. The activities of belowground and aboveground herbivores on the same plant may interact to alter the HIPV blends produced by doubly infested plants, compared to the blends produced by plants attacked by either herbivore alone, although the few studies to date do not lend themselves to simple generalizations. Larvae of the cabbage maggot, *Delia radicum*, induced the production of a HIPV blend by *Brassica nigra* different from that of uninfested plants or of plants infested only by larvae of the large cabbage white butterfly, *Pieris brassicae*, and plants infested by both herbivores produced a still different blend as well. The blends of plants infested by the root herbivore were richer in repellent sulfides and poorer in terpenoids than plants infested by caterpillars alone, and the parasitoid, *C. glomerata*, avoided plants suffering root herbivory by *D. radicum* (102).

The responses of adults of the generalist parasitoid, *Cotesia marginiventris*, to the volatiles produced by maize when attacked by larvae of *Spodoptera littoralis* aboveground and/or larvae of the western corn rootworm, *Diabrotica virgifera*, belowground varied with the level of experience, or training, of the parasitoid prior to bioassay. Naïve adults, or those allowed to oviposit while exposed to the volatile blends of maize induced only by *S. littoralis*, preferred the HIPV blends induced by *S. littoralis* in subsequent tests, whereas wasps allowed to oviposit while exposed to the HIPV blends of maize induced by both herbivore species preferred the HIPV blends of maize induced by both species subsequently (85). These results point out the necessity of carefully considering and controlling for the type of preassay experience of natural enemies in evaluating their responses to HIPV blends of a particular plant species when attacked by different combinations of herbivore species. In particular, it may be premature to assume that host location by natural enemies is impeded on plants damaged by multiple herbivore species without specifically testing a cohort of natural enemies that had prior experience with their hosts on plants that were damaged by multiple herbivore species.

In summary, there is a substantial range of effects of damage by an additional herbivore species on host location of a focal herbivore by its natural enemy. The change in the potential benefits of HIPV production by any plant species or genotype in response to variation in the structure of communities of herbivores attacking those plants may be difficult to predict, and I know of no studies to date that have examined communities of herbivores that comprise more than two species. Nevertheless, these few studies document yet another source of environmental variation in HIPV production based on ecological context.

SPECIFICITY

One area of controversy in the production of HIPVs that attract the natural enemies of herbivores is whether different herbivore species induce consistently different, or specific, blends, and how specific a blend might need to be to attract the natural enemy species most appropriate to suppress a particular herbivore species. Several factors contribute to this controversy. One factor is that all researchers do not use uniform criteria to determine if blends from different inducers do indeed differ statistically (see Statistical Testing of Blends, above). If blends differ qualitatively, i.e., if some compounds are produced in response to one elicitor but not to another, then it is reasonable to conclude that the HIPV blends differ between inducers. On the other hand, if two inducers result in plants producing HIPV blends having the same compounds but in slightly different proportions, then without multivariate statistical tests of the whole blend, it is difficult to conclude if the blends differ.

Another factor is the type and intensity of the prior experience that natural enemies might have had with their hosts and host plants prior to their testing in bioassays, because the ability of natural enemies to discriminate often increases with the amount of experience (72, 111, 114, 121). The volume of literature in this area is vast, and the following is necessarily a selective review to illustrate the controversial nature of this area.

Nonspecific Responses

When the regurgitant from four different caterpillar species, the beet armyworm, *S. exigua*; the cabbage looper, *Trichoplusia ni*; the velvetbean caterpillar, *Anticarsia gemmatalis*; and the corn earworm, *Helicoverpa zea*, was applied to maize seedlings that were damaged mechanically, similar HIPV blends were elicited (110). Regurgitant from the American grasshopper, *Schistocerca americana*, also elicited HIPV production. Both the specialist *Microplitis croceipes* and the generalist *C. marginiventris* were more attracted to plants treated with regurgitant than to untreated plants, but neither parasitoid discriminated between plants treated with regurgitant from caterpillars or grasshoppers. Moreover, *M. croceipes*, which does not attack *S. exigua* but does attack *H. zea*, was attracted to plants treated with regurgitant from *S. exigua*. Results such as this suggested that, although the detectability of HIPVs by natural enemies was high, the reliability of these cues to allow natural enemies to identify particular herbivore species was limited (109, 120, 122).

Another series of studies in which HIPV production was relatively nonspecific involves the wild tobacco species, *N. attenuata*, and its community of herbivores. Species in three different feeding guilds (87), the strip-feeding lepidopteran *Manduca quinquemaculata*, the pit-feeding flea beetle *Epitrix hirtipennis*, and the piercing-sucking bug *Tupiocoris notatus* (=*Dicyphus minimus*), induced plants to produce the same suite of HIPVs but in slightly different proportions. The similarity of the HIPV blends elicited by the different herbivore species suggests that these HIPVs may serve as "universal signs of herbivore damage" to generalist predators (56). Indeed, the HIPV blend elicited by *T. notatus* was sufficiently similar to that elicited by *Manduca* spp. that the predator, *Geocoris pallens*, was attracted to plants damaged by *T. notatus* in sufficiently high numbers to inflict 4.6-fold-higher mortality of *Manduca* eggs in a field experiment (57).

Similarly, two noctuid lepidopteran species, *Helicoverpa armigera* and *Pseudaletia separata*, induced different blends and quantities of HIPVs from maize, but naïve adults of the generalist parasitoid species *Campoletis chlorideae* did not discriminate between the HIPV blends elicited by the two herbivore species (128). Naïve adults of *Cotesia flavipes*, which attacks numerous species of stem borers, were similarly attracted to sorghum plants damaged by both host and nonhost species compared to undamaged plants despite the qualitative and quantitative differences in the HIPV blends that were elicited by the different herbivore species (78). This parasitoid species may be unable to discriminate suitable from unsuitable hosts even after probing, and the attractive but unsuitable hosts may be a reproductive sink for *C. flavipes* (78).

In crucifers, herbivore-damaged plants and undamaged plants produce similar arrays of compounds, but quantities are greater in herbivore-damaged plants (25). Naïve adults of both *Cotesia rubecula* and *C. glomerata* preferred the HIPV blends of Brussels sprouts damaged by caterpillars of host and nonhost species to HIPV blends of undamaged plants, but neither discriminated among the HIPV blends elicited by either host or nonhost species (26). After multiple oviposition experiences, however, *C. glomerata* learned to prefer the HIPV blends from Brussels sprouts infested with *Pieris brassicae* to HIPV blends from Brussels sprouts infested with *P. rapae*. *C. rubecula* was unable to learn any preferences, however (4).

Adults of *C. rubecula* also were unable to discriminate HIPV blends from *A. thaliana* elicited by the wasp's host, *P. rapae*, from blends elicited by the nonhost caterpillar species, *P. xylostella* and *S. exigua* (119). The absence of discrimination by *C. rubecula* also extends to herbivore species that damage plants in substantially different ways. When assayed against undamaged plants, *C. rubecula* was attracted not only to plants damaged by the nonhost, *S. exigua*, but also to plants damaged by the spider mite, *T. urticae*, and no discrimination was observed between plants damaged by *P. rapae* or spider mites (119). In general, whereas HIPVs permit natural enemies to discriminate crucifers damaged by herbivorous insects from undamaged plants, it appears that only a few of the natural enemies of those herbivores utilize those HIPV blends to discriminate between host and nonhost herbivores (28).

Another controversy in this area focuses on whether the HIPV blends produced by plants in response to damage by herbivores are anything more than a simple response to mechanical injury. In most studies, damage by herbivores elicits HIPV blends containing a greater range of compounds and in greater concentrations compared with HIPV blends elicited by mechanical damage, although mechanical damage that is most similar to damage by herbivores in terms of the mechanism of damage as well as its duration elicits HIPV blends most similar to those elicited by natural herbivores (8, 73). In many species, similar HIPV blends can be induced by treating plants with the plant hormones jasmonic acid or methyl jasmonate and by damage by particular herbivores (13, 17, 40, 49). The jasmonate signaling pathway can be activated by numerous stresses, and it is certainly possible that wounding or abiotic stress (e.g., ozone exposure) may induce emissions of many of the same HIPV components that are induced by herbivore damage (126). Plant species also may differ on how closely the volatile profile of plants induced by mechanical damage resembles that induced by herbivore damage, however (68).

Herbivore-Specific Responses

When given a learning experience as short as 20 s, adults of the generalist parasitoid *Cotesia marginiventris* learned to recognize the odors from four different host and host plant combinations (111). Plants of cultivated tobacco, cotton, and maize not only produce quantitatively different HIPV blends when attacked by either *Heliothis virescens* or *Helicoverpa zea*, but the HIPV blends induced by each herbivore also differed among the three plant species. In field tests on each of the plant species, adults of the specialist parasitoid *Cardiochiles nigriceps* preferred to visit plants damaged by its host, *H. virescens*, to plants damaged by the nonhost, *H. zea* (10). These results show a high level of behavioral plasticity, likely involving learning of host-associated HIPV blends from each of the three different host plant species, in order for *C. nigriceps* to locate its polyphagous host. The authors noted that this strategy of host location and identification by specialists based on learning contrasted with earlier views of rigid, innate, or hard-wired responses to key stimuli for specialist natural enemies (10).

In rare cases, HIPV blends may differ not only in response to feeding by different herbivore species, but also in response to feeding by different life stages of a single herbivore species. Naïve adults of the parasitic wasp *Cotesia kariyai* differentiated between the qualitatively and quantitatively different HIPV blends of maize induced by regurgitant or feeding by different instars of the caterpillar *Pseudaletia separata* (104). This suggests a particularly high level of differential HIPV induction that indicates not only the identity of its herbivore, but also the most susceptible life stages that the parasitoid can attack. The differences in plant responses to different life stages of herbivores are intriguing, but the phenomenon may not be general. For example, another wasp, *Microplitis rufiventris*, which can attack only the early instars of its host, *S. littoralis*, did not differentiate among HIPVs produced by plants damaged by different instars of *S. littoralis* either before or after oviposition experience (30). The proportional composition of the HIPV blends did not differ significantly following damage by different instars of *S. littoralis* but did differ significantly with the total amount of damage independent of instar. The absence of discrimination by *M. rufiventris* for damage inflicted by suitable and unsuitable instars of *S. littoralis* likely follows from the absence of any instar-specific variation in the HIPV blends. Similarly, neither naïve nor experienced adults of *C. glomerata* were able to discriminate among odors produced by suitable and unsuitable instars of its host, *P. brassicae* (69). Discrimination among instars can occur after landing on the plant by utilizing other cues, however (70). Although the evidence is limited, the idea that HIPVs induced by different life stages of a particular herbivore species can provide specific information on the age or developmental life stage of the herbivore is far from universal at this point.

In summary, there are cases in which plants under attack by different herbivore species produce the same or sufficiently similar blends that natural enemies fail to discriminate among them. In other cases, plants produce statistically different blends that also fail to result in discrimination by natural enemies. Still, other cases exist in which different herbivore species induce sufficiently different HIPV blends that do indeed lead to discrimination by natural enemies.

When Should Plants Produce Host-Specific HIPV Blends?

If specificity in HIPV production is the result of natural selection based on the activities of natural enemies, then it is because plant genotypes that produced specific blends had greater fitness (i.e., survived at greater frequency and/or left more offspring) than genotypes that produced less specific blends. Some have argued that plants would benefit most from natural enemies that quickly kill herbivores or cause them to immediately cease feeding (22, 117). By this line of reasoning, predators and idiobiont parasitoids that immediately terminate feeding by herbivores may have a greater impact on plant fitness than koinobiont parasitoids, in which the parasitized host continues to feed before eventually dying. Because predators generally have broader diet ranges than parasitoids do, this argument also suggests that natural selection may favor the evolution of nonspecific cues indicative of damage by any number of potential hosts (e.g., the "universal sign of herbivore damage"; 56). Such cues should potentially attract generalist natural enemies, especially predators, that can quickly suppress damage by several herbivore species. On the other hand, many specialist herbivores accumulate and sequester plant toxins and become toxic to generalist natural enemies (6, 75, 79). If so, only the natural enemies sufficiently specialized to cope with such toxins may effectively reduce damage by specialized herbivores (3).

Although the problem of sequestration of plant toxins by specialist herbivores would seem to suggest that plants might benefit more by attracting specialist rather than generalist natural enemies (116, 117), an analysis of the literature on life tables of herbivorous insects suggests that such a benefit may be infrequently realized in natural systems (42). Whereas a single specialist parasitoid species often is a key mortality factor on herbivore populations in simplified habitats composed of exotic plant and herbivore species (e.g., crops and their invasive herbivores regulated by introduced parasitoids), no cases were found in which native parasitoids acted as key factors regulating native herbivore populations on native plants in their natural habitats. Instead, suites of generalist predators imposed top-down suppression of herbivores in natural systems more frequently than specialist parasitoids did (42). These results suggest that successful biological control programs that utilize single species of specialized parasitoids to regulate populations of exotic pests on cultivated crops may be an inappropriate model to understand how herbivore populations are regulated in natural systems (41, 51). By extension, simple systems of single species of plant, herbivore, and natural enemy may also be inappropriate models to infer how HIPV-mediated indirect defenses of those plants might have evolved.

Congenitally fixed responses to herbivore-specific cues have been predicted for cases involving extreme specialization by natural enemies on particular species of herbivore and similar specialization of herbivores on a particular species of plant; progressively greater levels of learned responses are predicted at less extreme levels of specialization (22, 120). In view of the substantial potential for environmental and intrapopulational genetic variation in HIPV blends from a single inducer, it is unclear if HIPV blends from different inducers differ consistently over a range of plant-growing conditions and genotypes. Extremely rapid learning may provide a solution to cope with such variation, and such learning may inhibit and obviate the need for the evolution of congenital or innate responses (81). Sequentially learning the HIPV blends of different individual plants over time may be an optimal behavioral strategy even for natural enemies with narrow host ranges. Such temporary specificity may facilitate reproduction and increase the fitness of the natural enemies. The impact of such learning on plant fitness and on the evolution of HIPV production by plants is less clear, because the plasticity in the responses of natural enemies to different HIPV blends is not expected to result in consistent selection on any plant genotype to produce a particular HIPV blend from generation to generation.

EVOLUTION OF HIPVs IN CONTEMPORARY TIME

Others have pointed out the difficulties in answering whether the activity of natural enemies can consistently impose natural selection on plant traits that enhance plant fitness through differential effectiveness of those natural enemies on different plant genotypes (2, 16, 22, 38, 117). The answer to this question still is not clear because the critical questions have not yet been addressed in any system. The little information available only partially documents the potential impact of HIPV production on plant fitness via the activity of natural enemies. Numerous laboratory studies show that natural enemies may be attracted to plants damaged by herbivores (113), but relatively few studies demonstrate that such attraction actually reduces populations of herbivores in the field (36, 56, 57, 105), and in none of these studies was plant fitness measured. Parasitized caterpillars, when placed on plants, may sometimes consume less foliage and therefore cause a smaller fitness loss than unparasitized herbivores on plants (47, 101, 118), but the experimental designs in these studies precluded adult parasitoids from finding and parasitizing their hosts on their own. Similarly, manipulative experiments differentially exposing plants only to herbivores or to herbivores and natural enemies can show the value of natural enemies in suppressing herbivore populations and improving plant fitness (29, 106), but actual HIPV-mediated attraction of natural enemies to plants in the field in these studies has not been demonstrated. Researchers therefore disagree on the strength of this incomplete evidence that HIPV production benefits plants and is a product of natural selection. Progress in answering this question has been slow because there is little information available on the differences in plant fitness that differentially attract effective natural enemies.

Natural selection requires differences in relative fitness among individuals, but evolution also requires that such variation be heritable. A major advance in the study of the evolution of direct mechanisms of plant resistance to herbivory in contemporary time was the application of quantitative genetics to study the variation in plant resistance to herbivores in natural populations (97, 99, 100).

HERITABILITY

Heritability is a measure of the resemblance among relatives. The total phenotypic variance of a population is defined as V_P and is composed of genetic (V_G) and nongenetic, or environmental (V_E), components such that $V_P = V_G + V_E$. The genetic variance can be decomposed into additive and nonadditive components such as dominance and epistasis ($V_G = V_A + V_D + V_I$). Heritability in the broad sense (H^2) is calculated as the ratio of V_G divided by V_P (V_G/V_P), which shows the proportion of the phenotypic variation that has a genetic basis. More useful in quantitative genetics is heritability in the narrow sense (h^2), which is calculated as the ratio of the additive genetic variation to the total phenotypic variation (V_A/V_P). For a univariate trait, the narrow sense heritability then can be used to calculate the response to selection as $R = h^2S$, where S is the selection differential and is equal to the covariance of the trait and relative fitness and R is the response to selection. This equation also can be written as $\Delta\bar{z} = V_A \cdot (S/V_p)$, where $\Delta\bar{z}$ is the change in the population mean of the character under selection. The latter equation emphasizes the importance of determining V_A in order to predict responses of traits to natural selection.

This methodology utilized precise mating designs to partition the phenotypic variation in resistance of plants to herbivores into its genetic and nongenetic components in order to understand the process and constraints on the evolution of resistance to those herbivores.

Such studies have yet to be applied to the evolution of indirect defense mechanisms mediated by HIPV production. Only recently has genetically based variation in HIPV production been documented (see Genetic Variation in HIPV Production, above), but in these experiments, the designs were sufficient only to demonstrate that HIPV blends were heritable in the broad sense. More detailed experimental designs are required to isolate the additive component of genetic variation (V_A) to calculate any change in HIPV production due to natural selection.

Because of the multivariate nature of HIPV blends, the most relevant approach toward understanding the evolution of HIPV production is multivariate selection analysis. This technique also has proven useful in understanding the evolution of direct mechanism of plant resistance to herbivores and groups of herbivore species in natural systems (71, 96, 98, 129). Although such an approach has the potential to detect selection on the production of individual HIPVs, or suites of HIPVs that may vary together, numerous statistical difficulties need to be considered. The power of the statistical tests to determine if selection coefficients differ significantly from zero depends greatly on the sample size. Moreover, increasing the number of traits in the statistical model not only diminishes available degrees of freedom for any statistical test, but increases more rapidly the number of correlational terms in the model. As a consequence, it may be difficult to detect selection without utilizing a far greater level of replication than is conventionally used in the analysis of HIPVs (45, 59). In such situations it may be advisable to reduce the dimensionality of the dataset prior to performing multivariate selection analysis (see Statistical Testing of Blends, above). Numerous treatments of multivariate selection analysis to address contemporary problems in evolutionary genetics are available (23, 37, 44, 86, 90, 100).

MULTIVARIATE SELECTION ANALYSIS

Multivariate selection analysis is a technique of quantitative genetics used to determine the intensity and direction of natural selection on a group of traits. The methodology is based on the equation $\Delta\bar{z} = GP^{-1}S$ or $\Delta\bar{z} = G\beta$, where $\Delta\bar{z}$ is a column vector representing the change in trait means across generations; G is the additive genetic variance covariance matrix, which is the multivariate analog of V_A; and P^{-1} is the inverse of the phenotypic variance covariance matrix and is the multivariate analog of $1/V_P$. S is a vector of selection differentials, and β, which is equal to $P^{-1}S$, is the multivariate vector of linear selection gradients (62). The selection gradients can be estimated from multivariate regression using the relative fitness of an individual as the dependent variable and the values of the traits in question as the independent variables. The method can be expanded to estimate not only linear (directional) selection but also nonlinear (stabilizing, disruptive) selection and correlational selection by including the squares and cross products of the independent variables in the multivariate regression (63).

In addition, at least some volatile components from plants may influence plant fitness by mechanisms not involving the attraction of natural enemies. These may include attracting or deterring herbivores, attracting pollinators, defending against pathogens, quenching of ozone, and coping with heat or drought stress (16, 20, 48, 83). Therefore, more experiments may be required to isolate the potential selection due only to natural enemies on the evolution of HIPVs.

Finally, such experiments can only provide a single snapshot of selection on HIPV production at a particular place and time, and such experiments likely need to be replicated in both time and space to better understand the net, long-term advantage that may accrue to plant genotypes differing in HIPV production. We should not be surprised if different populations of a single plant species follow different evolutionary trajectories (27). Although such analyses are far from trivial, they may provide the best hope of resolving whether and to what extent the HIPV blends either may be signals to natural enemies resulting from natural selection on both the emitter and receiver or are merely damage-related cues that natural enemies learn to associate with damage caused by particular herbivore species for a relatively short period of time (2).

PERSPECTIVE AND FUTURE ISSUES

The majority of research on the ecological significance of HIPVs has been performed largely in the laboratory and largely from the perspective of the natural enemy. Although there is abundant evidence that natural enemies can learn to exploit HIPVs to locate the habitats of their hosts, that should not imply that there has been natural selection by natural enemies on plants to produce specific HIPV blends. To better understand the ecological significance of HIPVs, additional field experiments are needed, but such experiments should be more than laboratory experiments performed outside. Field experiments should be designed to determine if effects seen in the laboratory are sufficiently strong to persist in the face of an increased level of environmental variation. The substantial genetic variation in HIPV production recently found within populations of undomesticated plant species is exciting because it provides the raw material to test widely held hypotheses about how natural enemies may influence the evolution of HIPV production by plants.

SUMMARY POINTS

1. Multivariate statistical tests should be used routinely to determine whether different HIPV blends do indeed differ statistically.
2. There is substantial variation in the HIPV blends produced by different genotypes of both cultivated and wild species. HIPV blends also vary quantitatively and in composition owing to variation in the environment, variation in plant development, and variation in ecological context.
3. The data supporting the hypothesis that different herbivore species induce different and specific blends remain equivocal, and further consideration may need to be given to the ecological conditions favoring the evolution of herbivore-specific blends over nonspecific blends.
4. Whether the activities of natural enemies can consistently impose natural selection on plant traits that enhance plant fitness through differential effectiveness of those natural enemies on plant genotypes remains to be addressed.
5. Future research in this field would benefit from exploiting intraspecific genetic variation in HIPV production and adopting a Darwinian perspective that emphasizes genetic variation and natural selection to determine whether such variation might be adaptive to plants.
6. Techniques of quantitative genetics have the potential to isolate and measure the intensity and direction of natural selection on HIPV production and to provide a better understanding of how contemporary patterns of HIPV production might have evolved.

DISCLOSURE STATEMENT

The author is not aware of any affiliations, memberships, funding, or financial holdings that might be perceived as affecting the objectivity of this review.

ACKNOWLEDGMENTS

I thank the Editorial Committee for the invitation to prepare this review and J. Allison, J. Sun, and M. Turcotte for comments on previous drafts. Funding support for the preparation of this review was provided by the National Science Foundation under grant number 0414181.

LITERATURE CITED

1. Agbogba BC, Powell W. 2007. Effect of the presence of a nonhost herbivore on the response of the aphid parasitoid *Diaeretiella rapae* to host-infested cabbage plants. *J. Chem. Ecol.* 33:2229–35
2. Allison JD, Hare JD. 2009. Learned and naïve natural enemy responses and the interpretation of volatile organic compounds as cues or signals. *New Phytol.* 184:768–82
3. Barbosa P. 1988. Natural enemies and herbivore-plant interactions: influence of plant allelochemicals and host specificity. In *Novel Aspects of Insect-Plant Interactions*, ed. P Barbosa, DK Letourneau, pp. 201–29. New York: Wiley
4. Brodeur J, Geervliet JBF, Vet LEM. 1998. Effects of *Pieris* host species on life history parameters in a solitary specialist and gregarious generalist parasitoid (*Cotesia* species). *Entomol. Exp. Appl.* 86:145–52
5. Bukovinszky T, Gols R, Posthumus MA, Vet LEM, van Lenteren JC. 2005. Variation in plant volatiles and attraction of the parasitoid *Diadegma semiclausum* (Hellen). *J. Chem. Ecol.* 31:461–80
6. Camara MD. 1997. Predator responses to sequestered plant toxins in buckeye caterpillars: Are tritrophic interactions locally variable? *J. Chem. Ecol.* 23:2093–106
7. Chen Y, Schmelz EA, Wäckers F, Ruberson JR. 2008. Cotton plant, *Gossypium hirsutum* L., defense in response to nitrogen fertilization. *J. Chem. Ecol.* 34:1553–64
8. Connor EC, Rott AS, Samietz J, Dorn S. 2007. The role of the plant in attracting parasitoids: response to progressive mechanical wounding. *Entomol. Exp. Appl.* 125:145–55
9. De Boer JG, Hordijk CA, Posthumus MA, Dicke M. 2008. Prey and nonprey arthropods sharing a host plant: effects on induced volatile emission and predator attraction. *J. Chem. Ecol.* 34:281–90
10. De Moraes CM, Lewis WJ, Pare PW, Alborn HT, Tumlinson JH. 1998. Herbivore-infested plants selectively attract parasitoids. *Nature* 393:570–73
11. de Winter JCF, Dodou D, Wieringa PA. 2009. Exploratory factor analysis with small sample sizes. *Multivar. Behav. Res.* 44:147–81
12. Degen T, Dillmann C, Marion-Poll F, Turlings TCJ. 2004. High genetic variability of herbivore-induced volatile emission within a broad range of maize inbred lines. *Plant Physiol.* 135:1928–38
13. Degenhardt DC, Lincoln DE. 2006. Volatile emissions from an odorous plant in response to herbivory and methyl jasmonate exposure. *J. Chem. Ecol.* 32:725–43
14. **Delphia CM, Rohr JR, Stephenson AG, De Moraes CM, Mescher MC. 2009. Effects of genetic variation and inbreeding on volatile production in a field population of horsenettle. *Int. J. Plant Sci.* 170:12–20**

14. Documents genetic variation in HIPV production within an uncultivated species in the field.

15. Dicke M. 2009. Behavioural and community ecology of plants that cry for help. *Plant Cell Environ.* 32:654–65
16. Dicke M, Baldwin IT. 2010. The evolutionary context for herbivore-induced plant volatiles: beyond the 'cry for help'. *Trends Plant Sci.* 15:167–75
17. Dicke M, Gols R, Ludeking D, Posthumus MA. 1999. Jasmonic acid and herbivory differentially induce carnivore-attracting plant volatiles in lima bean plants. *J. Chem. Ecol.* 25:1907–22
18. Dicke M, Sabelis MW. 1988. How plants obtain predatory mites as bodyguards. *Neth. J. Zool.* 38:148–65
19. Dicke M, van Loon JJA. 2000. Multitrophic effects of herbivore-induced plant volatiles in an evolutionary context. *Entomol. Exp. Appl.* 97:237–49
20. **Dudareva N, Negre F, Nagegowda DA, Orlova I. 2006. Plant volatiles: recent advances and future perspectives. *Crit. Rev. Plant Sci.* 25:417–40**

20. Reviews the broad subject of the production of volatile compounds by plants.

21. Elzen GW, Williams HJ, Vinson SB. 1983. Response by the parasitoid *Campoletis sonorensis* (Hymenoptera: Ichneumonidae) to chemicals (synomones) in plants: implications for host habitat location. *Environ. Entomol.* 12:1872–76

22. Faeth SH. 1994. Induced plant responses: effects on parasitoids and other natural enemies of phytophagous insects. In *Parasitoid Community Ecology*, ed. BA Hawkins, W Sheehan, pp. 245–60. Oxford, UK: Oxford Univ. Press
23. Fairbairn DJ, Reeve JP. 1999. Natural selection. In *Evolutionary Ecology: Concepts and Case Studies*, ed. CW Fox, DA Roff, DJ Fairbairn, pp. 29–43. Oxford, UK: Oxford Univ. Press
24. Francois N, Govaerts B, Guyot-Declerck C. 2007. Inferential non-centred principal curve analysis of time-intensity curves in sensory analysis: the methodology and its application to beer astringency evaluation. *J. Chemom.* 21:187–97
25. Geervliet JBF, Posthumus MA, Vet LEM, Dicke M. 1997. Comparative analysis of headspace volatiles from different caterpillar-infested or uninfested food plants of *Pieris* species. *J. Chem. Ecol.* 23:2935–54
26. Geervliet JBF, Vet LEM, Dicke M. 1996. Innate responses of the parasitoids *Cotesia glomerata* and *C. rubecula* (Hymenoptera, Braconidae) to volatiles from different plant-herbivore complexes. *J. Insect Behav.* 9:525–38
27. Glawe GA, Zavala JA, Kessler A, van Dam NM, Baldwin IT. 2003. Ecological costs and benefits correlated with trypsin protease inhibitor production in *Nicotiana attenuata*. *Ecology* 84:79–90
28. Gols R, van Dam NM, Raaijmakers CE, Dicke M, Harvey JA. 2009. Are population differences in plant quality reflected in the preference and performance of two endoparasitoid wasps? *Oikos* 118:733–43
29. Gomez JM, Zamora R. 1994. Top-down effects in a tritrophic system: Parasitoids enhance plant fitness. *Ecology* 75:1023–30
30. Gouinguene S, Alborn H, Turlings TCJ. 2003. Induction of volatile emissions in maize by different larval instars of *Spodoptera littoralis*. *J. Chem. Ecol.* 29:145–62
31. **Gouinguene S, Degen T, Turlings TCJ. 2001. Variability in herbivore-induced odour emissions among maize cultivars and their wild ancestors (teosinte). *Chemoecology* 11:9–16**
32. **Gouinguene SP, Turlings TCJ. 2002. The effects of abiotic factors on induced volatile emissions in corn plants. *Plant Physiol.* 129:1296–307**
33. Guadagnoli E, Velicer WF. 1988. Relation of sample-size to the stability of component patterns. *Psychol. Bull.* 103:265–75
34. Hair JF, Anderson RE, Tatham RL. 1987. *Multivariate Data Analysis with Readings*. New York: Macmillan. 449 pp.
35. Halitschke R, Kessler A, Kahl J, Lorenz A, Baldwin IT. 2000. Ecophysiological comparison of direct and indirect defenses in *Nicotiana attenuata*. *Oecologia* 124:408–17
36. Halitschke R, Stenberg JA, Kessler D, Kessler A, Baldwin IT. 2008. Shared signals:- 'Alarm calls' from plants increase apparency to herbivores and their enemies in nature. *Ecol. Lett.* 11:24–34
37. Hansen TF, Houle D. 2008. Measuring and comparing evolvability and constraint in multivariate characters. *J. Evol. Biol.* 21:1201–19
38. Hare JD. 1992. Effects of plant variation on herbivore-natural enemy interactions. In *Plant Resistance to Herbivores and Pathogens: Ecology, Evolution and Genetics*, ed. RS Fritz, EL Simms, pp. 278–98. Chicago: Univ. Chicago Press
39. Hare JD. 2002. Plant genetic variation in tritrophic interactions. In *Multitrophic Level Interactions*, ed. T Tscharntke, BA Hawkins, pp. 8–43. Cambridge, UK: Cambridge Univ. Press
40. Hare JD. 2007. Variation in herbivore and methyl jasmonate-induced volatiles among genetic lines of *Datura wrightii*. *J. Chem. Ecol.* 33:2028–43
41. Hawkins BA. 1992. Parasitoid-host food webs and donor control. *Oikos* 65:159–62
42. **Hawkins BA, Mills NJ, Jervis MA, Price PW. 1999. Is the biological control of insects a natural phenomenon? *Oikos* 86:493–506**
43. Heil M. 2008. Indirect defense via tritrophic interactions. *New Phytol.* 178:41–61
44. Hereford J, Hansen TF, Houle D. 2004. Comparing strengths of directional selection: How strong is strong? *Evolution* 58:2133–43
45. Hersch EI, Phillips PC. 2004. Power and potential bias in field studies of natural selection. *Evolution* 58:479–85
46. Hoballah MEF, Tamo C, Turlings TCJ. 2002. Differential attractiveness of induced odors emitted by eight maize varieties for the parasitoid *Cotesia marginiventris:* Is quality or quantity important? *J. Chem. Ecol.* 28:951–68

31. Documents genetic variation in HIPV production within species.

32. Documents environmental variation in HIPV production within species.

42. This analysis of published life tables contrasts the role of specialist parasitoids and suites of generalist predators as factors regulating populations of herbivores in exotic and native plant communities.

47. Hoballah MEF, Turlings TCJ. 2001. Experimental evidence that plants under caterpillar attack may benefit from attracting parasitoids. *Evol. Ecol. Res.* 3:553–65
48. Holopainen JK. 2004. Multiple functions of inducible plant volatiles. *Trends Plant Sci.* 9:529–33
49. Hopke J, Donath J, Blechert S, Boland W. 1994. Herbivore-induced volatiles: The emission of acyclic homoterpenes from leaves of *Phaseolus lunatus* and *Zea mays* can be triggered by a beta-glucosidase and jasmonic acid. *FEBS Lett.* 352:146–50
50. Hunter MD. 2002. A breath of fresh air: beyond laboratory studies of plant volatile-natural enemy interactions. *Agric. For. Entomol.* 4:81–86
51. Hunter MD, Price PW. 1992. Playing chutes and ladders—heterogeneity and the relative roles of bottom-up and top-down forces in natural communities. *Ecology* 73:724–32
52. Jackson JE. 1991. *A User's Guide to Principal Components*. New York: Wiley. 569 pp.
53. Jansen JJ, Hoefsloot HCJ, van der Greef J, Timmerman ME, Westerhuis JA, Smilde AK. 2005. ASCA: analysis of multivariate data obtained from an experimental design. *J. Chemom.* 19:469–81
54. Janssen A, Sabelis MW, Bruin J. 2002. Evolution of herbivore-induced plant volatiles. *Oikos* 97:134–38
55. Kant MR, Bleeker PM, Van Wijk M, Schuurink RC, Haring MA. 2009. Plant volatiles in defense. *Adv. Bot. Res.* 51:613–66
56. Kessler A, Baldwin IT. 2001. Defensive function of herbivore-induced plant volatile emissions in nature. *Science* 291:2141–44
57. **Kessler A, Baldwin IT. 2004. Herbivore-induced plant vaccination. Part I. The orchestration of plant defenses in nature and their fitness consequences in the wild tobacco *Nicotiana attenuata*. *Plant J.* 38:639–49**

57. Shows that plants may benefit by producing a nonspecific HIPV blend that is attractive to the shared natural enemy of two dissimilar species of herbivore.

58. Kigathi RN, Unsicker SB, Reichelt M, Kesselmeier J, Gershenzon J, Weisser WW. 2009. Emission of volatile organic compounds after herbivory from *Trifolium pratense* (L.) under laboratory and field conditions. *J. Chem. Ecol.* 35:1335–48
59. Kingsolver JG, Hoekstra HE, Hoekstra JM, Berrigan D, Vignieri SN, et al. 2001. The strength of phenotypic selection in natural populations. *Am. Nat.* 157:245–61
60. Köllner TG, Schnee C, Gershenzon J, Degenhardt J. 2004. The sesquiterpene hydrocarbons of maize (*Zea mays*) form five groups with distinct developmental and organ-specific distributions. *Phytochemistry* 65:1895–902
61. Krips OE, Willems PEL, Gols R, Posthumus MA, Gort G, Dicke M. 2001. Comparison of cultivars of ornamental crop *Gerbera jamesonii* on production of spider mite-induced volatiles, and their attractiveness to the predator *Phytoseiulus persimilis*. *J. Chem. Ecol.* 27:1355–72
62. Lande R. 1979. Quantitative genetic-analysis of multivariate evolution, applied to brain: body size allometry. *Evolution* 33:402–16
63. Lande R, Arnold SJ. 1983. The measurement of selection on correlated characters. *Evolution* 37:1210–26
64. Lou Y, Baldwin IT. 2004. Nitrogen supply influences herbivore-induced direct and indirect defenses and transcriptional responses to *Nicotiana attenuata*. *Plant Physiol.* 135:496–506
65. Lou Y, Hua X, Turlings TCJ, Cheng JA, Chen X, Ye G. 2006. Differences in induced volatile emissions among rice varieties result in differential attraction and parasitism of *Nilaparvata lugens* eggs by the parasitoid *Anagrus nilaparvatae* in the field. *J. Chem. Ecol.* 32:2375–87
66. Loughrin JH, Manukian A, Heath RR, Tumlinson JH. 1995. Volatiles emitted by different cotton varieties damaged by feeding beet armyworm larvae. *J. Chem. Ecol.* 21:1217–27
67. Luciano G, Naes T. 2009. Interpreting sensory data by combining principal component analysis and analysis of variance. *Food Qual. Pref.* 20:167–75
68. Maffei ME, Mithofer A, Boland W. 2007. Insects feeding on plants: rapid signals and responses preceding the induction of phytochemical release. *Phytochemistry* 68:2946–59
69. Mattiacci L, Dicke M. 1995. Host-age discrimination during host location by *Cotesia glomerata*, a larval parasitoid of *Pieris brassicae*. *Entomol. Exp. Appl.* 76:37–48
70. Mattiacci L, Dicke M. 1995. The parasitoid *Cotesia glomerata* (Hymenoptera: Braconidae) discriminates between first and fifth larval instars of its host *Pieris brassicae*, on the basis of contact cues from frass, silk, and herbivore-damaged leaf tissue. *J. Insect Behav.* 8:485–98
71. Mauricio R, Rausher MD. 1997. Experimental manipulation of putative selective agents provides evidence for the role of natural enemies in the evolution of plant defense. *Evolution* 51:1435–44

72. McCall PJ, Turlings TCJ, Lewis WJ, Tumlinson JH. 1993. Role of plant volatiles in host location by the specialist parasitoid *Microplitis croceipes* Cresson (Braconidae: Hymenoptera). *J. Insect Behav.* 6:625–39

73. Mithöfer A, Wanner G, Boland W. 2005. Effects of feeding *Spodoptera littoralis* on lima bean leaves. II. Continuous mechanical wounding resembling insect feeding is sufficient to elicit herbivory-related volatile emission. *Plant Physiol.* 137:1160–68

73. Shows that mechanical damage that closely mimics the time course of feeding by herbivores is sufficient to induce a blend of HIPVs qualitatively similar to that induced by those herbivores.

74. Moayeri HRS, Ashouri A, Poll L, Enkegaard A. 2007. Olfactory response of a predatory mirid to herbivore induced plant volatiles: multiple herbivory versus single herbivory. *J. Appl. Entomol.* 131:326–32

75. Nishida R. 2002. Sequestration of defensive substances from plants by Lepidoptera. *Annu. Rev. Entomol.* 47:57–92

76. Nissinen A, Ibrahim M, Kainulainen P, Tiilikkala K, Holopainen JK. 2005. Influence of carrot psyllid (*Trioza apicalis*) feeding or exogenous limonene or methyl jasmonate treatment on composition of carrot (*Daucus carota*) leaf essential oil and headspace volatiles. *J. Agric. Food Chem.* 53:8631–38

77. Nordlund DA, Lewis WJ, Altieri MA. 1988. Influences of plant-produced allelochemicals on the host/prey selection behavior of entomophagous insects. In *Novel Aspects of Insect-Plant Interactions*, ed. P Barbosa, DK Letourneau, pp. 65–90. New York: Wiley

78. Obonyo M, Schulthess F, Gerald J, Wanyama O, Le Ru B, Calatayud PA. 2008. Location, acceptance and suitability of lepidopteran stemborers feeding on a cultivated and wild host-plant to the endoparasitoid *Cotesia flavipes* Cameron (Hymenoptera: Braconidae). *Biol. Control* 45:36–47

79. Ode PJ. 2006. Plant chemistry and natural enemy fitness: effects on herbivore and natural enemy interactions. *Annu. Rev. Entomol.* 51:163–85

80. Olson DM, Cortesero AM, Rains GC, Potter T, Lewis WJ. 2009. Nitrogen and water affect direct and indirect plant systemic induced defense in cotton. *Biol. Control* 49:239–44

81. Papaj DR. 1993. Automatic behavior and the evolution of instinct: lessons from learning in parasitoids. In *Insect Learning: Ecological and Evolutionary Perspectives*, ed. DR Papaj, AC Lewis, pp. 243–72. New York: Chapman & Hall

82. Pareja M, Mohib A, Birkett MA, Dufour S, Glinwood RT. 2009. Multivariate statistics coupled to generalized linear models reveal complex use of chemical cues by a parasitoid. *Anim. Behav.* 77:901–9

83. Peñuelas J, Llusia J. 2004. Plant VOC emissions: making use of the unavoidable. *Trends Ecol. Evol.* 19:402–4

84. Price PW, Bouton CE, Gross P, McPheron BA, Thompson JN, Weis AE. 1980. Interactions among three trophic levels: influence of plants on interactions between insect herbivores and natural enemies. *Annu. Rev. Ecol. Syst.* 11:41–65

85. Rasmann S, Turlings TCJ. 2007. Simultaneous feeding by aboveground and belowground herbivores attenuates plant-mediated attraction of their respective natural enemies. *Ecol. Lett.* 10:926–36

86. Roff DA. 1997. *Evolutionary Quantitative Genetics*. New York: Chapman & Hall. 493 pp.

87. Root RB. 1973. Organization of a plant-arthropod association in simple and diverse habitats: the fauna of collards (*Brassica oleracea*). *Ecol. Monogr.* 43:95–124

88. Rostas M, Eggert K. 2008. Ontogenetic and spatio-temporal patterns of induced volatiles in *Glycine max* in the light of the optimal defense hypothesis. *Chemoecology* 18:29–38

89. Scheiner SM. 2001. MANOVA: multiple response variables and multispecies interactions. In *Design and Analysis of Ecological Experiments*, ed. SM Scheiner, J Gurevitch, pp. 99–155. New York: Oxford Univ. Press

90. Scheiner SM, Donohue K, Dorn LA, Mazer SJ, Wolfe LM. 2002. Reducing environmental bias when measuring natural selection. *Evolution* 56:2156–67

91. Schmelz EA, Alborn HT, Engelberth J, Tumlinson JH. 2003. Nitrogen deficiency increases volicitin-induced volatile emission, jasmonic acid accumulation, and ethylene sensitivity in maize. *Plant Physiol.* 133:295–306

92. Schnee C, Köllner TG, Held M, Turlings TCJ, Gershenzon J, Degenhardt J. 2006. The products of a single maize sesquiterpene synthase form a volatile defense signal that attracts natural enemies of maize herbivores. *Proc. Natl. Acad. Sci. USA* 103:1129–34

93. Schuman MC, Heinzel N, Gaquerel E, Svatos A, Baldwin IT. 2009. Polymorphism in jasmonate signaling partially accounts for the variety of volatiles produced by *Nicotiana attenuata* plants in a native population. *New Phytol.* 183:1134–48

93. Documents genetic variation in HIPV production within single population over a small spatial scale in an undomesticated plant species.

94. Scutareanu P, Bruin J, Posthumus MA, Drukker B. 2003. Constitutive and herbivore-induced volatiles in pear, alder and hawthorn trees. *Chemoecology* 13:63–74
95. Shiojiri K, Takabayashi J, Yano S, Takafuji A. 2001. Infochemically mediated tritrophic interaction webs on cabbage plants. *Popul. Ecol.* 43:23–29
96. Shonle I, Bergelson J. 2000. Evolutionary ecology of the tropane alkaloids of *Datura stramonium* L. (Solanaceae). *Evolution* 54:778–88
97. Simms E, Rausher MD. 1987. Costs and benefits of plant resistance to herbivory. *Am. Nat.* 130:570–81
98. Simms EL. 1990. Examining selection on the multivariate phenotype: plant resistance to herbivores. *Evolution* 44:1177–88
99. Simms EL, Rausher MD. 1989. The evolution of resistance to herbivory in *Ipomoea purpurea* II. Natural selection by insects and costs of resistance. *Evolution* 43:573–85
100. **Simms EL, Rausher MD. 1992. Uses of quantitative genetics for studying the evolution of plant resistance. In *Plant Resistance to Herbivores and Pathogens: Ecology, Evolution and Genetics*, ed. RS Fritz, EL Simms, pp. 42–68. Chicago: Univ. Chicago Press**

100. A well-written introduction to the application of evolutionary quantitative genetics to the study of plant resistance against herbivores.

101. Smallegange RC, van Loon JJA, Blatt SE, Harvey JA, Dicke M. 2008. Parasitoid load affects plant fitness in a tritrophic system. *Entomol. Exp. Appl.* 128:172–83
102. Soler R, Harvey JA, Kamp AFD, Vet LEM, van der Putten WH, et al. 2007. Root herbivores influence the behavior of an aboveground parasitoid through changes in plant-volatile signals. *Oikos* 116:367–76
103. Takabayashi J, Dicke M, Posthumus MA. 1994. Volatile herbivore-induced terpenoids in plant mite interactions: variation caused by biotic and abiotic factors. *J. Chem. Ecol.* 20:1329–54
104. Takabayashi J, Takahashi S, Dicke M, Posthumus MA. 1995. Developmental stage of herbivore *Pseudaletia separata* affects production of herbivore-induced synomone by corn plants. *J. Chem. Ecol.* 21:273–87
105. Thaler JS, Fidantsef AL, Duffey SS, Bostock RM. 1999. Trade-offs in plant defense against pathogens and herbivores: a field demonstration of chemical elicitors of induced resistance. *J. Chem. Ecol.* 25:1597–609
106. Tooker JF, Hanks LM. 2006. Tritrophic interactions and reproductive fitness of the prairie perennial *Silphium laciniatum* Gillette (Asteraceae). *Environ. Entomol.* 35:537–45
107. Tumlinson JH, Turlings TCJ, Lewis WJ. 1992. The semiochemical complexes that mediate insect parasitoid foraging. *Agric. Zool. Rev.* 5:221–52
108. Turlings TCJ, Benrey B. 1998. Effects of plant metabolites on the behavior and development of parasitic wasps. *Ecoscience* 5:321–33
109. Turlings TCJ, Loughrin JH, McCall PJ, Rose USR, Lewis WJ, Tumlinson JH. 1995. How caterpillar-damaged plants protect themselves by attracting parasitic wasps. *Proc. Natl. Acad. Sci. USA* 92:4169–74
110. Turlings TCJ, McCall PJ, Alborn HJ, Tumlinson JH. 1993. An elicitor in caterpillar oral secretions that induces corn seedlings to emit chemical signals attractive to parasitic wasps. *J. Chem. Ecol.* 19:411–25
111. Turlings TCJ, Scheepmaker JWA, Vet LEM, Tumlinson JH, Lewis WJ. 1990. How contact foraging experiences affect preferences for host-related odors in the larval parasitoid *Cotesia marginiventris* (Cresson) (Hymenoptera: Braconidae). *J. Chem. Ecol.* 16:1577–89
112. Turlings TCJ, Tumlinson JH, Eller FJ, Lewis WJ. 1991. Larval-damaged plants: Source of volatile synomones that guide the parasitoid *Cotesia marginiventris* to the microhabitat of its host. *Entomol. Exp. Appl.* 58:75–82
113. Turlings TCJ, Wäckers F. 2004. Recruitment of predators and parasitoids by herbivore-injured plants. In *Advances in Insect Chemical Ecology*, ed. RT Cardé, JG Millar, pp. 21–75. Cambridge, UK: Cambridge Univ. Press
114. Turlings TCJ, Wäckers FL, Vet LEM, Lewis WJ, Tumlinson JH. 1993. Learning of host-finding cues by hymenopterous parasitoids. In *Insect Learning: Ecological and Evolutionary Perspectives*, ed. DR Papaj, AC Lewis, pp. 51–78. New York: Chapman & Hall
115. **van Dam NM, Poppy GM. 2008. Why plant volatile analysis needs bioinformatics: detecting signal from noise in increasingly complex profiles. *Plant Biol.* 10:29–37**

115. Makes a compelling argument to utilize multivariate statistical techniques to analyze HIPV blends as a whole instead of as a series of independent components.

116. van der Meijden E. 1996. Plant defense, an evolutionary dilemma: contrasting effects of (specialist and generalist) herbivores and natural enemies. *Entomol. Exp. Appl.* 80:307–10
117. van der Meijden E, Klinkhamer PGL. 2000. Conflicting interests of plants and the natural enemies of herbivores. *Oikos* 89:202–8

118. van Loon JJA, de Boer JG, Dicke M. 2000. Parasitoid-plant mutualism: Parasitoid attack of herbivore increases plant reproduction. *Entomol. Exp. Appl.* 97:219–27
119. van Poecke RMP, Roosjen M, Pumarino L, Dicke M. 2003. Attraction of the specialist parasitoid *Cotesia rubecula* to *Arabidopsis thaliana* infested by host or nonhost herbivore species. *Entomol. Exp. Appl.* 107:229–36
120. Vet LEM, Dicke M. 1992. Ecology of infochemical use by natural enemies in a tritrophic context. *Annu. Rev. Entomol.* 37:141–72
121. Vet LEM, Groenewold AW. 1990. Semiochemicals and learning in parasitoids. *J. Chem. Ecol.* 16:3119–35
122. Vet LEM, Wäckers FL, Dicke M. 1991. How to hunt for hiding hosts: the reliability-detectability problem in foraging parasitoids. *Neth. J. Zool.* 41:202–13
123. Vinson SB. 1976. Host selection by insect parasitoids. *Annu. Rev. Entomol.* 21:109–33
124. Vinson SB. 1981. Habitat location. In *Semiochemicals: Their Role in Pest Control*, ed. DA Nordlund, pp. 51–77. New York: Wiley
125. Vos M, Berrocal SM, Karamaouna F, Hemerik L, Vet LEM. 2001. Plant-mediated indirect effects and the persistence of parasitoid-herbivore communities. *Ecol. Lett.* 4:38–45
126. Vuorinen T, Nerg AM, Holopainen JK. 2004. Ozone exposure triggers the emission of herbivore-induced plant volatiles, but does not disturb tritrophic signaling. *Environ. Pollut.* 131:305–11
127. Walling LL. 2008. Avoiding effective defenses: strategies employed by phloem-feeding insects. *Plant Physiol.* 146:859–66
128. Yan ZG, Wang CZ. 2006. Similar attractiveness of maize volatiles induced by *Helicoverpa armigera* and *Pseudaletia separata* to the generalist parasitoid *Campoletis chlorideae*. *Entomol. Exp. Appl.* 118:87–96
129. Zangerl AR, Stanley MC, Berenbaum MR. 2008. Selection for chemical trait remixing in an invasive weed after reassociation with a coevolved specialist. *Proc. Natl. Acad. Sci. USA* 105:4547–52
130. Zhang PJ, Zheng SJ, van Loon JJA, Boland W, David A, et al. 2009. Whiteflies interfere with indirect plant defense against spider mites in lima bean. *Proc. Natl. Acad. Sci. USA* 106:21202–7

Native and Exotic Pests of *Eucalyptus*: A Worldwide Perspective

Timothy D. Paine,[1] Martin J. Steinbauer,[2] and Simon A. Lawson[3]

[1]Department of Entomology, University of California, Riverside, California 92521; email: Timothy.Paine@ucr.edu

[2]Department of Zoology, La Trobe University, Melbourne, Victoria, 3086 Australia; email: M.Steinbauer@latrobe.edu.au

[3]Horticulture and Forestry Science, Department of Employment, Economic Development and Innovation, Indooroopilly, Queensland, 4068 Australia; email: simon.lawson@deedi.qld.gov.au

Annu. Rev. Entomol. 2011. 56:181–201

First published online as a Review in Advance on August 30, 2010

The *Annual Review of Entomology* is online at ento.annualreviews.org

This article's doi: 10.1146/annurev-ento-120709-144817

0066-4170/11/0107-0181$20.00

Key Words

invasive species, host plant shift, host range expansion, plantation forestry

Abstract

Eucalyptus species, native to Australia, Indonesia, the Philippines, and New Guinea, are the most widely planted hardwood timber species in the world. The trees, moved around the globe as seeds, escaped the diverse community of herbivores found in their native range. However, a number of herbivore species from the native range of eucalypts have invaded many *Eucalyptus*-growing regions in North America, Europe, Africa, Asia, and South America in the last 30 years. In addition, there have been shifts of native species, particularly in Africa, Asia, and South America, onto *Eucalyptus*. There are risks that these species as well as generalist herbivores from other parts of the world will invade Australia and threaten the trees in their native range. The risk to *Eucalyptus* plantations in Australia is further compounded by planting commercially important species outside their endemic range and shifting of local herbivore populations onto new host trees. Understanding the mechanisms underlying host specificity of Australian insects can provide insight into patterns of host range expansion of both native and exotic insects.

INTRODUCTION

Endemic: native to a specific region or environment and not occurring naturally anywhere else

Invasive: the capacity of an organism to both colonize and expand its geographic distribution once it establishes in a new environment

Host plant specificity: the suite of plant species and leaf types utilized by an insect in the wild at endemic population densities

Exotic: not native, introduced from or originating in a foreign area

More than 700 species in the genera *Eucalyptus*, *Angophora*, and *Corymbia* (all formerly classified in the single genus *Eucalyptus*) are native to Australia, Indonesia, the Philippines, and New Guinea (28). *Eucalyptus* species (in the broadest sense) are now planted around the Mediterranean, in southern Africa, South America, and Asia as one of the most important sources of commercial cellulose fiber. *Eucalyptus* have been widely planted as ornamental trees in North America (42) and small plantations have also been established for the production of ornamental foliage (36). Even in Australia, eucalypt plantations have become increasingly important as harvesting of native forests is restricted.

As the trees have been established around the world, Australian insect herbivores of *Eucalyptus* have also colonized the new environments. Most of these movements have been relatively recent and the patterns of colonization have not been uniform. Although some species that have colonized virtually all of the *Eucalyptus*-growing regions cause significant problems, some guilds have successfully established on one continent but not on others. However, the flow of insect pests has been, essentially, unidirectional. Despite a large number of insects in the Southern Hemisphere and Asia that have shifted onto *Eucalyptus*, few of these insects have been detected in other parts of the world, including Australia.

The Australian timber industry and agencies focused on natural resource protection must deal not only with insects that damage plantations, but also with the community-level impacts in native forests. The most commercially important *Eucalyptus* species are endemic to restricted ranges in Australia but may have been widely planted in commercial plantations throughout many parts of the country, often with widely different environmental conditions. Movement of both native and invasive insect herbivores, with or without their natural enemies, into the different environments can result in significantly different pest complexes and damage than that observed in the endemic range.

Our objective is to examine the global movement of insect herbivores of *Eucalyptus*. These movements include colonization of insects native to Australia in different parts of the world where *Eucalyptus* have been planted. In addition, there are *Eucalyptus*-feeding insects from many parts of the world that have been introduced or are at great risk of being introduced into Australia. Plantations of commercial *Eucalyptus* timber species have been established in Australia outside their endemic ranges, and we explore the establishment of new geographic distributions of native herbivores. We also discuss the basis for host plant specificity and limitations in host range expansion among the native herbivores with regard to the implications for host range expansion among exotic *Eucalyptus* herbivores.

NORTH AMERICA AND EUROPE

North America

Approximately 90 species of *Eucalyptus* have been introduced into North America, particularly into California and Florida, over the past 150 years (42). Eucalypts were first brought into North America as seeds for at least a century; they were not associated with the insects and diseases that utilize the trees in their native range. The single exception was a galling wasp, *Quadrastichodella nova*, infesting seed capsules reported in 1957 (135). However, the eucalyptus psyllid, *Blastopsylla occidentalis*, the tristania psyllid, *Ctenarytaina longicauda*, and the eucalyptus longhorned borer, *Phoracantha semipunctata*, were introduced into southern California in 1983–1984 (56). Since that time, at least 15 different Australian eucalyptus-feeding insect species from at least four different feeding guilds (2 borer species, 3 leaf-eating beetle species, 4 gall wasp species, and at least 8 psyllid species) have been introduced into California, Florida, or Hawaii.

Although considered to be of minor economic importance in eastern Australia (44), the eucalyptus longhorned borer, *P. semipunctata*, has been accidentally introduced into virtually all of the *Eucalyptus*-growing regions of the world and has caused significant tree mortality in many of those areas (43, 111). Interactions between the beetle and its host trees (63, 64, 66, 67), as well as intraspecific competitive interactions, are important in population regulation in native and novel environments. The beetles are also subject to mortality from natural enemies (111). The egg parasitoid *Avetianella longoi*, introduced into California from Australia, has had a significant impact on the population dynamics of the beetle (61, 65). However, a second species, *Phoracantha recurva*, was detected in southern California in 1995 (56). Although the congeneric borers appear to have similar ecological requirements, the behavior of *A. longoi* in relation to the eggs of the two species is markedly different; the wasp parasitizes a greater proportion of *P. semipunctata* eggs than *P. recurva* eggs and *P. recurva* eggs encapsulate the parasitoid eggs (88, 89, 113). Escape from extensive parasitism of eggs and earlier seasonal activity, which may permit *P. recurva* to colonize available resources before emergence of *P. semipunctata*, can help explain the replacement of *P. semipunctata* by *P. recurva* in southern California now that the latter has established (23, 24).

The eucalyptus snout beetle, *Gonipterus scutellatus*, was discovered defoliating eucalyptus in California in 1994 (32). This insect has been introduced into several eucalyptus-growing regions around the world from Australia and has caused extensive damage (79, 112). Fortunately, it has proven relatively easy to control by the introduction of a specific egg parasitoid, *Anaphes nitens*. By 1997, densities of beetle larvae had dropped to barely detectable levels in California (62).

The eucalyptus tortoise beetle, *Trachymela sloanei*, was first collected in southern California in March of 1998 (56). Adults and larvae feed on leaves and young stems, but of particular importance, adults will clip off young, tender leaf shoots as the tree refoliates. A second chrysomelid beetle, *Chrysophtharta m-fuscum*, was introduced into southern California in 2005 (14). The combined feeding effects of the leaf beetle and the psyllids can cause significant damage to commercial eucalyptus foliage production (14).

At least eight Australian psyllid species have been introduced into North America (*Ctenarytaina eucalypti*, *Glycaspis brimblecombei*, *Eucalyptolyma maideni*, *Blastopsylla occidentalis*, *Ctenarytaina longicauda*, *Ctenarytaina spatulata*, *Acizzia uncatoides*, and *Cryptoneossa triangular*) (56). The blue gum psyllid, *C. eucalypti*, seriously damages the foliage of members of the blue gum species group including several species used for ornamental foliage production. Release of the parasitoid *Psyllaephagus pilosus* resulted in complete biological control (36). The free-living psyllids, *B. occidentalis*, *C. longicauda*, *C. spatulata*, and *A. uncatoides*, are present on a variety of *Eucalyptus* hosts but typically do not cause significant damage. However, the lerp (coverings formed of sugar and wax produced by the insect)-forming psyllids, *G. brimblecombei* and *E. maideni*, threaten red and lemon-scented gum species.

The red gum lerp psyllid, *G. brimblecombei*, has killed thousands of host trees in California (18), typically on sites lacking supplementary irrigation. An encyrtid parasitoid, *Psyllaephagus bliteus*, has established good biological control in many parts of California (34, 35). Variability in the biological control could be a result of infection of the psyllid with an endosymbiotic bacterium that appears to confer some resistance to parasitism (68).

The spotted gum lerp psyllid, *E. maideni*, colonizes leaves of both lemon-scented gum and spotted gum. Like other lerp-forming psyllids, this species produces a shelter on the leaf surface constructed primarily of sugars. In California, occupied or abandoned lerps may be colonized by the free-living lemon gum psyllid, *C. triangular*. Although infestations can be severe and honeydew production can be heavy, no tree mortality has been reported.

Five species of gall-forming wasps have been introduced into North America. Following the early introduction of *Q. nova*, a leaf petiole-galling *Aprostocetus* species was introduced into Hawaii and subsequently into

Polyphagous: having many host plants, particularly across genera and families

California (56). *Epichrysocharis burwelli* produces small, dark, pustule-like galls on the leaves of lemon-scented gums in California (116). *Selitrichodes globulus* was described galling twigs and small branches from California in 2008, but the authors ascribe it to an Australian origin (81). *Leptocybe invasa* produces galls that swell stems, petioles, and leaf midribs (142). This species has been recently reported from Florida but had been widely distributed throughout other regions of the world (99).

Europe and the Mediterranean Basin

Introductions of several eucalyptus-feeding insects into Mediterranean Europe and North Africa occurred slightly earlier than those into North America. *P. semipunctata* was reported infesting plantations of eucalypts in Spain before 1982 (95) and the congener *P. recurva* was discovered in Tunisia about a decade later (12). The eucalyptus snout weevil was reported in France in 1979 (112), with one chrysomelid recently having been reported in Ireland in 2008. Only four psyllid species have been reported from Europe: *C. spatulata* was reported from France and Italy in 2003 (31), *C. eucalypti* appeared in central Europe by 1998 (22), *G. brimblecombei* was recorded from Spain and Portugal in 2007 (138), and *C. peregrina* was described from the United Kingdom and Ireland in 2007 (70). Of these four, only *C. peregrina* is not found in North America. Five species comprise the European fauna of galling wasps: *Ophelimus maskelli* (7), *Leptocybe invasa* (99), *Q. nova* (39), *Leprosa milga* (96), *Megastigmus eucalypti* (139), and an *Aprostocetus* sp. (139). The only species common to both the Mediterranean basin and North America is *L. invasa*.

SOUTH AMERICA AND SOUTH AFRICA

Introduced Australian Insects

Extensive plantations of *Eucalyptus* have been established in Chile, Argentina, and Brazil for the production of cellulose. There have been accidental introductions of at least eight Australian insects that feed on the trees (78), including four psyllids (*B. occidentalis*, *C. eucalypti*, *C. spatulata*, and *G. brimblecombei*) (115), one leaf-eating weevil (*G. scutellatus* = *Gonipterus gibberus*) (54, 79), two borers (*P. semipunctata* and *P. recurva*) (79), two seed-galling wasps (*Q. nova*, 80; and *Moona spermophaga*, 76), and the hemipteran *Thaumastocoris peregrinus* (25). Seven of these species are also found in North America.

The list of Australian insects that feed on eucalyptus in South Africa is very similar. In addition to the two *Phoracantha* species, *G. scutellatus* and *T. peregrinus* are also found in South Africa (143). Two chrysomelid beetles, *Trachymela tincticollis* and *T. sloanei*, are established in southern Africa (137), as are the seed-galling wasps *L. milgra* (96), *Q. nova* (80), *M. spermophaga* (75), and *Megastigmus zebrinus* (58).

Host Shifts of Native Insects onto *Eucalyptus*

Africa, Australia, and South America have a diversity of native myrtaceous plants that is much greater than that in the Northern Hemisphere. Consequently, insect herbivores that feed on plants related to *Eucalyptus* may be preadapted to shift onto them (78, 143). Brazil has the most diverse fauna, with more than 224 indigenous species recorded as having eucalypts as hosts (13), 10% of which were considered to be pests. These include highly polyphagous species such as ants in the genera *Atta*, *Acromyrmex*, *Sericomyrmex*, *Mycocepurus*, *Trachymyrmex*, and subterranean termites in the families Kalotermitidae, Rhinotermitidae, and Termitidae (78, 150). Termites have also been commonly recorded as problems in eucalypt plantations in southern Africa (6, 57, 100, 101).

Many eucalypt plantations in Brazil abut native vegetation or may include strips of native vegetation within them. The vegetation may include many myrtaceous species, enhancing the opportunity for host-switching from these native hosts to the exotic *Eucalyptus*. This has been especially prevalent among the lepidopteran fauna of Brazil (15). For example, the geometrid moth *Thyrinteina arnobia* has guava (*Psidium guajava*) and several other Myrtaceae as its native hosts (59, 121). Its pest status appears to be driven by significantly higher larval survival on eucalypt hosts compared to its original host (71). The mechanisms for this are unclear, but *T. arnobia* may have been released from predation, parasitism, and pathogen pressure by obtaining natural enemy–free space on its new host (10, 11, 59).

Other indigenous Lepidoptera that defoliate eucalypt plantations in Brazil include *Sarsina violascens* and a complex of *Glena* spp. (15). Hosts of *S. violascens* include Myrtaceae (*Psidium* spp.), Asteraceae, and Oleaceae (78). *Glena* spp. also have a wide host range (15) that may include *Pinus* spp. Adults of the chrysomelid beetle *Costalimaita ferruginea* feed on the foliage of a wide range of species, including *Eucalyptus*. Aside from defoliators, a range of wood- and bark-boring beetles, particularly scolytids and platypodids, have been recorded on eucalypts in Brazil (49).

ASIA AND NEW ZEALAND

Large-scale plantings of eucalypts for a variety of purposes have occurred throughout Asia, from India to Indonesia, Thailand, Malaysia, the Philippines, Vietnam, and China, with much smaller-scale plantings in New Zealand. Herbivores occurring on eucalypts in Asia and New Zealand present widely contrasting trends. Asia has had very few introductions of Australian insects, but large numbers of endemic insects utilize eucalypts as hosts. More than 60 species of insects were recorded in India in 1983 as associated with eucalypts (118), while in China 207 species across 10 orders and 50 genera were recorded in 2000 (27), increasing from 96 species in 1987 and 167 in 1992 (148). This would appear to be a common theme throughout Southeast Asia where eucalypts have been grown.

The only insect of Australian origin to have been introduced into Asia and to have caused significant damage is *L. invasa*, with introductions occurring between 2002 and 2007 (119, 133). Prior to this, the leafroller *Strepsicrates* sp. nr. *semicanella* was recorded as potentially the only Australian insect to have been introduced into China (147), while in India *Icerya purchasii* was recorded as the only Australian introduction (106). Species such as *P. semipunctata*, which has colonized all other continents where eucalypts are grown, is notable by its absence in Asia.

New Zealand, due to its proximity to Australia and the large trade volume between the two countries, has seen many Australian insects become established (144). Between 1869 and 1999, 57 species of eucalypt-feeding insects had established in New Zealand, including 26 eucalypt specialists. Of these, a number, including *P. semipunctata* and *G. scutellatus*, are common introductions into the Americas, Africa, and Europe. However, since 1999, only two new species of Australian origin have established in New Zealand, the psyllids *Creiis liturata* and *Anoeconeossa communis* (145). Small sap-suckers dominate the fauna of insect introductions into New Zealand, suggesting that aerial dispersal from Australia may be the dominant pathway there (145) in contrast to other regions where movement of goods and people are the most likely pathways.

The movement of Australian insects onto *Eucalyptus* plantations around the world appears to have come in two phases. The first reports come from parts on the British Commonwealth (e.g., South Africa, New Zealand, and India), some as early as 1869 (144). These invasions were probably associated with movements of people and commerce within the former British Empire. The second phase of large-scale movement was on a much broader geographic scale beginning in the latter half of the twentieth century. This period of invasion coincided with significant increases in plantation-grown *Eucalyptus* destined for production of high-quality paper. The insects may have originated in Australia, but there has undoubtedly been redistribution among eucalypt-growing regions. Molecular ecological techniques could be used to elucidate the pattern of these movements.

EXOTIC INSECTS ON EUCALYPTUS IN AUSTRALIA

The continent of Australia has a large community of insect herbivores adapted to feeding on *Eucalyptus* species. However, a significant threat to the native stands of timber, as well as plantations of trees, can come from the introduction of herbivores from outside the continent. In the absence of natural control agents, these have the potential to cause significant losses. In addition, species of *Eucalyptus* that have important commercial characteristics are being planted in many different parts of Australia well outside their endemic geographic ranges. If native herbivores accompany the movement of their host trees to these new habitats, there is also the possibility for pest outbreaks and tree injury. Both of these possibilities need to be explored to assess potential risks to plantation and native eucalypts.

Established Exotic Insects

Few insects exotic to Australia damage eucalypts in native forests or plantations. Those that do are highly polyphagous. The most significant is the African black beetle (*Heteronychus arator*), a grasslands melolonthine scarab accidentally introduced into Western Australia from southern Africa in 1938 (98), which is now a significant pest of *Eucalyptus globulus* plantations established for pulp production (87). Belowground stem girdling by adult beetles occurs three to six months after planting, reducing seedling growth and survival.

Other exotic insects recorded on eucalypts in Australia include the coccids *Ceroplastes sinensis* and *Coccus hesperidum* (91, 93), the diaspidid *Diaspidiotus perniciosus* (92), the scolytines *Xyleborinus saxeseni* and *Hypothenemus birmanus* (140), and the platypodid *Platypus parallelus* (141). Eucalypts are minor hosts for all these species. The scolytines and platypodid are usually associated only with dead or dying trees, and the hemipterans are highly polyphagous.

Eucalypts in nurseries host common nursery pests including whiteflies (*Trialeurodes vaporariorum* and *Bemisia tabaci*), the two-spotted spider mite, *Tetranychus urticae*, and various aphids including *Aphis gossypii* (21). All exotic insects recorded on eucalypts in Australia are generalists with a diverse host range and in most cases are associated with juvenile eucalypts; none has adapted to feed in any significant way on mature eucalypt foliage or other aboveground tissues. Juvenile eucalypt foliage and other tissues have various chemical and physical characteristics that may enable utilization by insects otherwise not well adapted to feeding on adult plant parts.

Given that the ubiquity of Australia's eucalypts (most of the 164 million hectares of Australian forest is dominated by eucalypts) provides a highly abundant resource, and represents potential natural enemy–free space for successful colonizers, why more exotic insects have not become significant pests of eucalypts in Australia is unclear. Possible explanations may relate to (*a*) competitive exclusion of less-well-adapted exotics by the diversity of endemic insects occupying existing niches; (*b*) lack of exposure of introduced insects to eucalypts and/or other Myrtaceae in their endemic range, with consequent low-risk of host shifting; and (*c*) Australia's history of strong quarantine procedures limiting the introduction/establishment of exotic herbivores. The greatest exotic insect threats to Australia's eucalypts are thus (*a*) highly polyphagous species with hosts from diverse phylogenetic groups; (*b*) insects that have adapted to eucalypts grown overseas; and (*c*) insects on endemic eucalypts in Australia's near neighbors, Papua New Guinea, Timor, and the Philippines.

Polyphagous Threats

Twenty-six exotic insects are listed as threats to the Australian forest plantation industry in the National Biosecurity Plan for the Plantation Timber Industry (1), five of which are associated with eucalypts: *Coryphodema tristis*, *Chilecomadia valdiviana*, *Lymantria dispar*, *Orgyia thyellina*, and *Anoplophora glabripennis*. *L. dispar*, *O. thyellina* and *A. glabripennis* are highly polyphagous in their native and introduced ranges, while *C. tristis* and *C. valdiviana* represent recent host shifts in their regions of origin (55, 79).

The lymantriids *L. dispar* and *O. thyellina* pose significant environmental and commercial threats to eucalypts in Australia. The gypsy moth (*L. dispar*) has a host range of over 650 species in 53 families (86). Matsuki et al. (97) assessed the risk posed by *L. dispar* to endemic Australian and New Zealand tree species and recorded larval performance on five eucalypt species comparable to that on *Quercus pubescens* and *Q. robur*, the preferred hosts in Europe. Australia has a diverse lymantriid fauna of at least 74 species, including four *Lymantria* species (46). Native natural enemies may therefore adapt to invasive lymantriids in Australia, although there is likely to be a lag period during which considerable damage may be done and in which spread is not hindered by lowered rates of reproduction. Pathways into Australia, especially via northeast Asia on imported containers, vehicles, and machinery, are well known, with several recorded egg-mass interceptions at Australian ports. No moth detections using the sex pheromone disparlure at major Australian ports have been made to date; however, the capture of a single male *L. dispar* ssp. *praeterea* moth in Hamilton, New Zealand, in a similar trapping program in March 2003 triggered a successful eradication campaign using aerial spraying and mating disruption (2) costing around NZ$5.4 million (114). Australia's response plan for an *L. dispar* incursion (149) suggests that eradication in an urban area (the most likely scenario) would require political and public acceptance of an aerial spraying campaign similar to that conducted in New Zealand.

The Asian longhorned beetle borer, *A. glabripennis*, native to East Asia, is a threat to hardwood species worldwide. It is highly polyphagous and has expanded its exotic distribution significantly, being detected in New York in 1997 (60), Austria in 2001, France in 2003, and Germany in 2004 (136). Although eucalypts are yet to be recorded as hosts, the species' wide and expanding host range for other hardwoods suggests that it has the potential to utilize this host should it become established in Australia.

Indigenous Insects on Eucalypts Outside Australia

The cossid moths *Chilecomadia valdiviana* in Chile (79) and *Coryphodema tristis* in South Africa (55) are polyphagous species, facilitating potential for host-switching. The host range of *C. valdiviana* in Chile includes *Salix*, *Nothofagus*, and a range of fruit, ornamental, and forest trees (26, 79). It was reported to attack *Eucalyptus nitens* in 1992 and thereafter expanded its host range to occasionally include *E. gunnii*, *E. camaldulensis*, and *E. delegatensis* (79), although *E. globulus* stands adjacent to infested *E. nitens* were not attacked. Usually no more than 5% of trees in a stand are infested (78) and attack is not generally associated with tree stress.

In South Africa, *Coryphodema tristis* feeds on a range of native and exotic trees in the Ulmaceae, Vitaceae, Rosaceae, Scrophulariaceae, Myoporaceae, Malvaceae, and Combretaceae (55). It was found attacking *E. nitens* in 2004 in Mpumalanga Province. Surveys in affected stands showed severe damage, with up to 90% of trees attacked (17, 55). This host switch may have been in response to extreme tree stress (55) and expansion into enemy-free space: Parasitoids have rarely been reared from *C. tristis* in *E. nitens* logs.

At present there are no established pathways for these cossids into Australia, although untreated logs in which immature stages can survive are a possible route. Risk assessment of eucalypt log and chip imports into the United States from South America rates *C. valdiviana* high on *E. nitens* and moderate on *E. camaldulensis* and *E. gunnii* logs (78). For untreated eucalypt logs sourced from Australia (77), several species of hepialid ghost moths and cossids

Extralimital: used in the context of species not normally found in a specific geographic area

Kairomone: a chemical from one species that elicits an adaptively favorable response in an unrelated species

were likewise rated as high overall pest risk potential, with pest risk potential on chips rating low. The biology of *C. tristis* is similar to that of *C. valdiviana* and some of the Australian cossids evaluated, so the pathways and risk ratings are likely to be similar. However, importation of eucalypt timber into Australia from South Africa and South America is minimal, so the risk from importation of cossid-infested material from either source is presently low.

The Pathway from Neighboring Countries

The highest risk for exotic insects associated with eucalypts to enter Australia is from south and Southeast Asia and Papua New Guinea into northern Australia, particularly via trade between the islands (50). Pest risk analysis prioritized 10 insect species posing the greatest threat, including two generalist termite species (50). Others included sap-suckers (the coreid *Amblypelta cocophaga* and the mirids *Helopeltis* spp.), stem borers (the buprestids *Agrilus opulentus* and *Agrilus sexsignatus*, the cerambycids *Celosterna scabrator* and *Oxymagis horni*, and the cossid *Zeuzera coffeae*), and one defoliator, the tortricid *Strepsicrates rothia*. Most have wide host ranges and are likely to have switched hosts to eucalypts, while *A. opulentus* and *A. sexsignatus* seem to have coevolved with *E. deglupta* in Papua New Guinea and the Philippines, respectively (50).

NATIVE INSECTS AND EUCALYPTS PLANTED EXTRALIMITALLY WITHIN AUSTRALIA

As awareness of the biological traits and environmental tolerances of different species of eucalypt has increased, so has their use within Australia in locations extralimital to their regions' of endemism. As nonendemic species of eucalypt have been introduced into new areas, utilization by locally occurring native insects has risen accordingly. Nevertheless, the biological significance of these events has rarely been appreciated and it was not until the seminal paper by Strauss (131) that host expansion by native insects was placed within a broader ecological context.

Eucalyptus nitens was introduced into Tasmania in the 1970s and trees were exposed to aseasonal and severe defoliation of juvenile leaves by the larvae of the geometrid *Mnesampela privata* (9, 90). The first serious outbreak of *M. privata* on *E. nitens* occurred in northwest Tasmania in the summer of 1993 to 1994 (90). Since this time, populations of *M. privata* that have warranted insecticide control have been recorded on juvenile *E. globulus* ssp. *globulus* planted extralimitally on the Australian mainland and on juvenile *E. grandis* planted extralimitally in northwest Victoria (3, 4, 33, 87, 126, 128). In all these instances, the introduced eucalypts have been planted in locations within the geographic distribution of *M. privata* and resident moths have begun utilizing them in addition to the local host species.

Following its introduction into Tasmania, *E. nitens* was also quickly utilized by five species of endemic chrysomelids: *Paropsisterna* (=*Chrysophtharta*) *agricola*, *Pt.* (=*Chrysophtharta*) *bimaculata*, *Paropsis charybdis*, *P. delittlei* and *P. porosa* (132). Both *Pt. agricola* and *Pt. bimaculata* are now considered to be economically significant pests of *E. nitens*, with all life cycle stages of the former species utilizing both juvenile and adult foliage and adults of the latter species using only adult *E. nitens* foliage in the wild (103). The defoliation of *E. nitens* by *Pt. bimaculata* was considered unusual because the insect was thought to be host specific for species of eucalypt belonging to the subgenus *Monocalyptus*, whereas *E. nitens* is a *Symphyomyrtus* species (38). When given a choice, adult *Pt. bimaculata* do not discriminate between a *Monocalyptus* (*E. regnans*) and a *Symphyomyrtus* (*E. nitens*) species when the leaves available to them are of the same age (125). This insect apparently ignores plant kairomones that indicate host identity and selects oviposition sites on the basis of a female's ability to grasp opposing leaf margins when depositing eggs on leaf tips (72).

E. globulus ssp. *globulus* has been widely planted outside its endemic range of Tasmania and coastal regions of mainland southeastern Australia. The first commercial plantation of this species was established in southwestern Western Australia in 1980 and there followed a period of about 10 to 15 years before insect defoliation became of concern (87). In addition to *M. privata*, a suite of defoliating beetles, most notably *G. scutellatus* and *Pt. variicollis*, are now considered important pests of young *E. globulus* ssp. *globulus* plantations (33, 87). Recent molecular and morphological studies suggest that *G. scutellatus* was most likely introduced into Western Australia from Tasmania (or perhaps from northeastern New South Wales), probably not long after *E. g. globulus* was introduced into Western Australia (94). Whatever its origins, populations of *G. scutellatus* have benefited greatly from the abundance of same-age foliage that occurs in plantations (33). A survey of the oviposition host preferences of *G. scutellatus* in a native forest in Tasmania found that *E. g. globulus*, a gum, is less preferred to three species of eucalypt belonging to the peppermint group (29).

Extralimital plantings of *E. globulus* ssp. *globulus*, in locations within southeastern mainland Australia, are also attacked and have occasionally been significantly defoliated by various *Anoplognathus* and *Heteronyx* species of scarab beetles (51). In southwestern Western Australia, adult *Cadmus excrementarius* have caused significant damage to juvenile and adult leaves of *E. g. globulus*, which represents a novel host for this insect (40, 41). *C. excrementarius* was first recorded from *E. g. globulus* in Western Australia in 1993 (3).

In addition to associations of native insects on eucalypts planted extralimitally in temperate regions of southern Australia, there has been a significant expansion of hardwood eucalypt plantation establishment in tropical and subtropical regions of Australia. Here again, chrysomelid beetles have been among the first insects to utilize these plantings (102). Of these, the record of the collection of the paropsine *Paropsis atomaria* from *Corymbia citriodora* ssp. *variegata* is novel (102). Typically, the hosts of *P. atomaria* include *E. camaldulensis*, *E. cloeziana*, *E. dunnii*, *E. grandis*, *E. pilularis* (120), and only recently some *Corymbia* hybrids.

UNDERSTANDING HOST EXPANSIONS OF NATIVE INSECTS AND IMPLICATIONS FOR EXOTIC INSECTS

Eucalyptus trees were introduced into North America (California) approximately 150 years ago (42). Despite the length of exposure to native herbivores, reports of colonization are rare. In contrast, *Eucalyptus* species have been grown in South America for a similar length of time and a large number of insects have shifted onto the trees (3, 13, 78, 79). It is not yet clear whether host shifts will occur in Australia as plantations of nonendemic *Eucalyptus* species become widely established. Therefore, it is important to understand the

BIOSECURITY THREAT TO EUCALYPTS REQUIRES PREEMPTIVE RESEARCH

As the area of plantation eucalypts has increased worldwide, so has the number of insects utilizing them as hosts. Many of these insects now pose biosecurity threats to eucalypts in Australia and in regions where eucalypts are grown as exotics. Such insects generally have other Myrtaceae as hosts and/or are highly polyphagous. In addition, the number of native Australian insects established as pests overseas has increased markedly and is likely to continue to do so; many of these were not significant pests, or were previously unrecorded or poorly known, in Australian forests and plantations prior to their establishment overseas.

Research on interactions between eucalypt herbivores and their hosts will enable better prediction and assessment of threats posed by exotic and endemic insects and elucidate mechanisms by which they adapt to new hosts. Advances in host physiology (particularly host chemistry and resistance mechanisms), chemical ecology (including host selection), population and community dynamics (population modeling and biological control), and pathway and risk analysis will be essential to achieve this. Outcomes from eucalypt genome-sequencing projects may also enable new approaches to identifying and deploying resistance. Australia, as the origin of eucalypt diversity, has an essential role to play in this research.

factors associated with host specificity to assess the potential for host shifts.

Physical and Chemical Aspects of Host Specificity

The specificity of some insects for certain species of host may not apply to the entire life of an individual plant. This is probably quite widespread within *Eucalyptus* due to the prominence of heterophylly within the genus. Heterophylly is the production of different leaf types, which differ markedly in toughness and potential kairomones, during the life of an individual tree. Insects such as *Ctenarytaina eucalypti* and *Mnesampela privata* are host specific for juvenile leaves of *E. globulus* ssp. *globulus* (20, 123), whereas *C. spatulata* and *Pt. agricola* are host specific for adult leaves of the same species (20, 82). Populations of pest insects that are specific for juvenile leaves will have less time to exploit resources the faster their hosts undergo the transition to adult foliage (suggested in Reference 107). Host specificity based on leaf type is distinct from an insect's preference for leaves according to their age, e.g., young or expanding leaves versus old or fully expanded leaves. Many eucalypt-feeding insects have a preference for young leaves of either juvenile or adult leaf types.

Perhaps the simplest mechanisms determining an insect's host specificity are those that are mediated by physical leaf characteristics. For many species of chrysomelid leaf beetles and some psyllids, nonstructural waxes prevent adherence to leaves of certain eucalypt species (see 19, 103, 108 and references therein). Nonstructural waxes can be physically abraded and give leaves a whitish bloom; β-diketones in the waxes cause them to form compound, acutely branching tubes (**Figure 1**). By contrast, structural waxes cannot be physically abraded and give leaves a dull matte appearance; the absence of β-diketones in the waxes leads to the formation of platelets (**Figure 1**). When nonstructural waxes are intact, they prevent all but specialist insects from adhering to leaves. In the case of *M. privata*, nonstructural epicuticular waxes do not provide a physical defense against adherence and are considered to provide the final cue necessary for host acceptance (129). At low population densities, *M. privata* will not oviposit on species that produce structural waxes.

Eucalypts with tough or dense leaves require that insects exert mechanical forces greater than those of the leaves themselves before they can bite into them. This has limited the suite of mandibulate insect species that can use such

Eucalyptus nitens a 10 µm

Corymbia eximia

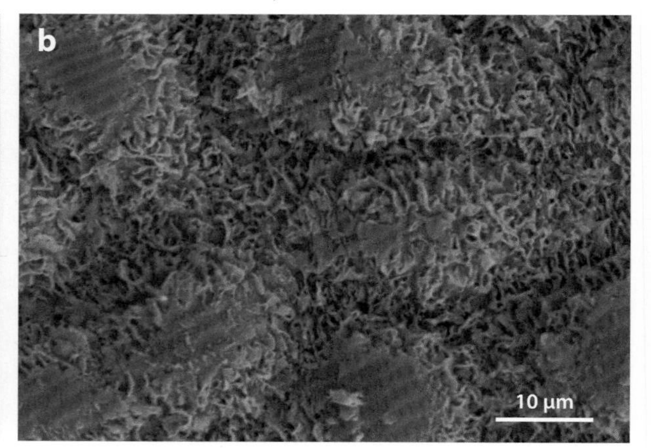

Figure 1

Scanning electron micrographs of different types of eucalypt epicuticular waxes (110). (*a*) Nonstructural waxes on the abaxial surface of juvenile *Eucalyptus nitens* leaves. (*b*) Structural waxes on intermediate *Corymbia eximia* leaves (leaves of this type are isobilateral).

hosts. The larvae of these insects, whose survival and performance are reduced on tough or dense leaves, often feed in groups; species with comparatively large neonates feed on leaf margins (73, 105), whereas those with small neonates feed on leaf lamellae (127). Specific leaf weight and lamellar thickness are measurements that can be used to compare the relative susceptibility of different species' leaves to feeding by chewing insects as determined by their toughness (105, 122). Four species of lerp-forming psyllids utilize dense leaves by inserting their mouthparts through the stomata to the tissues on which they feed (146). Consequently, no measure of leaf toughness or density provides any insight into the susceptibility of eucalypts to such insects, but knowing the distribution and density of stomata in conjunction with the insect behavior may.

More difficult to determine are kairomonal mechanisms in insect host specificity. de Little (37) suggested that similarities in foliar terpenes may partly explain host expansion of *Pt. bimaculata* onto *E. nitens* in Tasmania. A similar suggestion was used to explain the utilization of an exotic tree species common in Australia by the normally eucalypt-feeding *Anoplognathus montanus* and *A. pallidicollis* (130). Experimental evidence has shown that *A. chloropyrus* and *M. privata* perceive terpene odors and that *M. privata* may use a mosaic (or "fingerprint" sensu 8) of electrophysiologically active volatiles for host recognition (52, 129). The new expansion hosts of *M. privata* share many terpenes in common with natural hosts (**Figure 2**). Although similarities in terpene profiles may facilitate the utilization of novel species by *M. privata*, subtle differences will nevertheless affect oviposition. Natural and novel hosts with higher concentrations of α-pinene received fewer eggs than did hosts with lower concentrations of this terpene, and eucalypts with higher concentrations of α-terpineol received more eggs than those with lower concentrations (110).

Although the phylogenetic relatedness of plant families and genera may influence the degree of similarity in the types of plant metabolites produced, unrelated plants may produce similar compounds. Similarities in plant metabolites, perhaps more so than phylogenetic relatedness, could influence host shifts of exotic insects onto eucalypts planted outside Australia. Whenever apparent host shifts occur, effort should be made to determine the possible contribution the relatedness of plant metabolites has had on the exchange so that generalities can begin to be inferred and documented. Similarly, evaluation of the contributions of plant metabolites may reveal possible instances of host race formation.

Foliar terpenes: in insect-eucalypt interactions, the suite of volatile mono- (C10) and sesquiterpene (C15) plant secondary metabolites (PSM) with kairomonal activity

Host plant suitability: the relative capacities of different species of plant and leaf type to sustain the development of immature insects

Intraspecific Variation in Host Suitability and Insect Population Growth

Antibiosis as a means to select defoliation-resistant genotypes of eucalypt has received the majority share of research. The research is based on evidence that the expression of various plant metabolites that determine host plant suitability is genetically based (5, 53, 69, 74). Although this is potentially useful, there are only two examples in which variation in the suitability of hosts to a particular insect herbivore has been attributed to a specific metabolite in eucalypts. The phenol quercetin decreases larval performance of *Uraba lugens* by reducing nitrogen assimilation (48). Taylor (134) implicated increases in foliar phenolics with the collapse of populations of *Cardiaspina albitextura* on *E. camaldulensis* but did not identify quercetin as the toxic compound. In the second example, 1,8-cineole has repellent activity against five species of *Anoplognathus*, but the response is mediated because sideroxylonal (which co-occurs with 1,8-cineole) has antifeedant activity (5, 47, 52). Although research on a number of other insect species has narrowed the potential field of biologically active metabolites, none has advanced beyond correlations between putative causal agents and various insect responses. If significant advances in this area are to be achieved, correlative studies must be followed by insect bioassays with plant extracts and synthetic compounds.

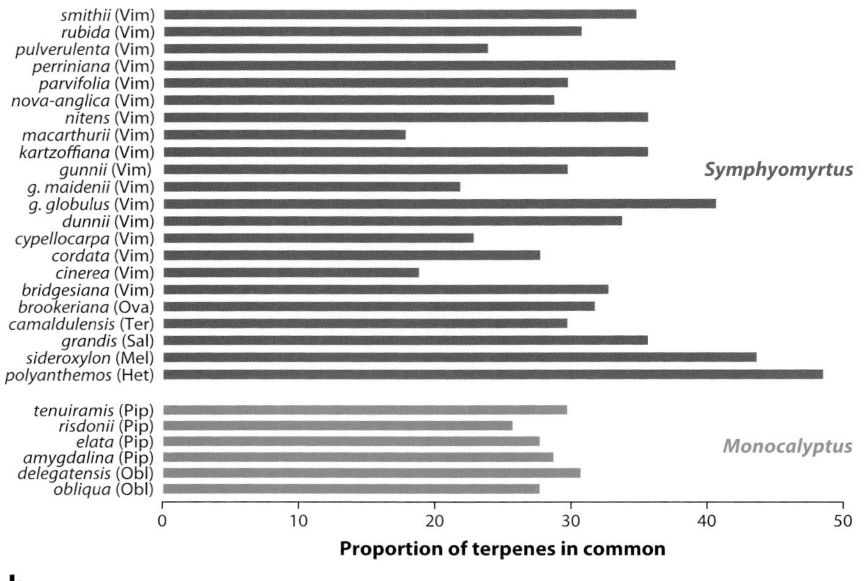

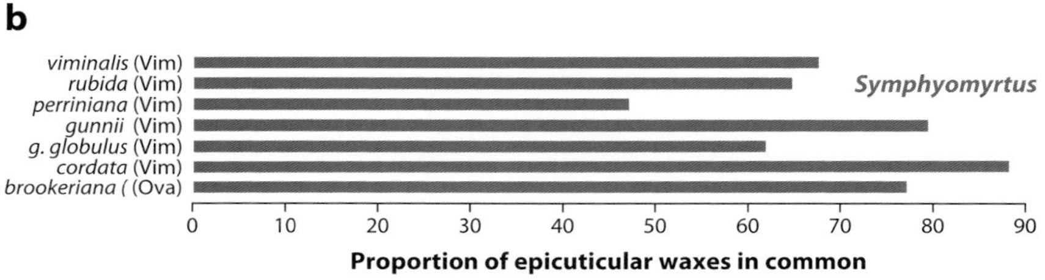

Figure 2

Similarities in terpenes and waxes of natural and novel hosts of *Mnesampela privata* (see 125 and references therein; collection records for *Eucalyptus kartzoffiana* and *E. pulverulenta* from M.J. Steinbauer). (*a*) Foliar terpenes of adult leaves grouped by subgenus and series (adapted from References 16, 83, and 84). (*b*) Epicuticular waxes of juvenile leaves of *Symphyomyrtus* species grouped by series (adapted from Reference 85). Abbreviations of series classifications (from Reference 41) given in parentheses as follows: Het, Heterophloiae; Mel, Melliodorae; Obl, Obliquae; Ova, Ovatae; Pip, Piperitae; Sal, Salignae; Ter, Tereticornes; Vim, Viminales. Data given in Reference 16 are assumed to be for adult leaves; data from References 83 and 84 are also for adult leaves and provided for consistency with Reference 16. Note, however, that female *M. privata* prefer to oviposit on the juvenile leaves of their most preferred host, i.e., *Eucalyptus globus* ssp. *globulus*.

Insect Life-History Traits and Host Expansion

The life-history traits of the native Australian insects that have utilized species of eucalypt planted off-site within Australia share only a few characteristics with one another, i.e., all have good dispersal ability and most lay eggs in clutches. Populations of paropsine chrysomelid beetles are rather transient; some species disperse soon after ovipositing (30, 104, 131). It has been suggested that oviposition followed by the location of new hosts by *Pt. bimaculata* limits the potential for intraspecific larval competition (30). This oviposition strategy is driven by a preference for high-quality foliage, the occurrence of which is spatially and temporally limited. Nevertheless, because some eggs are laid on less suitable foliage and eucalypt leaves can mature rapidly, even neonate *Pt. bimaculata* larvae are capable of dispersing to find expanding foliage (73). Large areas of plantation hosts at the same stage of development have been linked to an increase in intergenerational reproductive fitness in *M. privata* and oviposition on nearby novel hosts, possibly as oviposition mistakes (124, 127).

Leaf beetles and *M. privata*, as well as many other eucalypt-feeding insects, lay their eggs in clutches and upon hatching, and for varying lengths of time thereafter, feed gregariously. Factors that reduce the size of larval aggregations help limit the severity of defoliation suffered by trees. Egg parasitoids are especially effective in this regard. For example, species of *Telenomus* egg parasitoid have been reported to kill between 5% and 69% of *M. privata* eggs per clutch (109, 117) and a *Neopolycystus* species kills an average of 30% of *Paropis atomoria* eggs per clutch (45). Egg parasitoids have the advantage of locating herbivores before they disperse and before the larvae are able to defend themselves or hide in leaf shelters. For these reasons, and because they can be mass-reared and often use highly specific kairomones to guide their oviposition, they appear to be particularly attractive for use in classical biological control programs.

CONCLUSIONS

Eucalyptus species have become an important source of short-fiber pulp demanded for the production of high-quality paper used in modern office copiers and printers. The trees have a rapid growth rate, have short rotation times, can be grown in coppiced production, and are extraordinarily well suited for large-scale plantation production in diverse parts of the world, with distributions limited at the present by low temperatures (42). However, contemporary breeding practices and expanded variety testing could reduce the susceptibility of the trees to frost damage and yield trees adapted to new growing regions (e.g., China and southeastern North America). In addition to the commercial value, there are additional opportunities to exploit members of the genus for carbon sequestration. However, as *Eucalyptus* plantations are expanded worldwide, international movement of insects and diseases threaten growth and productivity. It is critical to foster research to understand the international routes of movement of the herbivores and to develop ways of detecting and preventing the accidental invasions. Research leading to a more advanced understanding of patterns of host suitability, host susceptibility, and host selection will improve evaluation and mitigation of those invasive risks.

SUMMARY POINTS

1. Eucalypts are important plantation timber species that are widely planted throughout regions of the world with suitable growing conditions. Many of these areas were free of insect herbivores on the trees until recently. Australian herbivores have been introduced into many parts of the world in the last three decades and can significantly affect plant health.
2. A number of species have been introduced into all eucalypt-growing regions of the world, but the patterns are not identical. For example, the leaf beetles and psyllids were introduced into North America before they appeared in Europe. A greater diversity of hymenopteran galling wasps was introduced into Europe before establishing in other parts of the world.

3. Few introduced exotic insects have become pests on eucalypts in Australia perhaps because of strong competitive exclusion by endemic insect herbivores, lack of preexisting adaptations to feeding on eucalypts, and Australia's history of strong quarantine procedures limiting numbers of introductions.
4. An increasing number of insect herbivores overseas have shifted onto eucalypts and achieved pest status. These now pose a significant biosecurity threat to Australia's native and planted eucalypt forests, as do insects on endemic eucalypts in Australia's near neighbors, Papua New Guinea, Timor, and the Philippines.
5. Native insects that have become pests on eucalypts outside of Australia are generally either highly polyphagous or have native Myrtaceae as their natural hosts. Insects in the latter group may be preadapted to shift to eucalypts as hosts.
6. Similarities in plant metabolites between species of *Eucalyptus* and noneucalypts are likely to be influential to the utilization of novel hosts by herbivorous insects.
7. The long-standing premise that foliar terpenes have antibiotic activity does not account for their possible activity as host location and selection kairomones.
8. Nonstructural epicuticular waxes not only limit adherence of nonspecialist insects but also likely act as oviposition-stimulating kairomones.

DISCLOSURE STATEMENT

The authors are not aware of any affiliations, memberships, funding, or financial holdings that might be perceived as affecting the objectivity of this review.

ACKNOWLEDGMENTS

We thank Dr. Helen Nahrung and Dr. Ross Wylie for content and editorial assistance and Dr. Mamoru Matsuki for provision of unpublished data.

LITERATURE CITED

1. 2007. National Biosecurity Plan for the Plantation Timber Industry. Plant Health Australia. **http://www.planthealthaustralia.com.au**
2. 2008. Biosecurity New Zealand. *Gypsy moth.* **http://www.biosecurity.govt.nz/pests/gypsy-moth**
3. Abbott I. 1993. Insect pest problems of eucalypts plantations in Australia. 6. Western Australia. *Aust. For.* 56:381–84
4. Abbott I, Wills A, Burbidge T. 1999. The impact of canopy development on arthropod faunas in recently established *Eucalyptus globulus* plantations in Western Australia. *For. Ecol. Manag.* 121:147–58
5. Andrew RL, Wallis IR, Harwood CE, Henson M, Foley WJ. 2007. Heritable variation in the foliar secondary metabolite sideroxylonal in *Eucalyptus* confers cross-resistance to herbivores. *Oecologia* 153:891–901
6. Atkinson PR, Nixon KM, Shaw MJP. 1992. On the susceptibility of *Eucalyptus* species and clones to attack by *Macrotermes natalensis* Haviland (Isoptera, Termitidae). *For. Ecol. Manag.* 48:15–30
7. Badmin J. 2008. Spread of *Ophelimus maskelli* Ashmead (Hymenoptera: Eulophidae) in south-east England. *Br. J. Entomol. Nat. Hist.* 21:147 (Abstr.)
8. Barata EN, Mustaparta H, Pickett JA, Wadhams LJ, Araujo J. 2002. Encoding of host and non-host plant odors by receptor neurons in the *Eucalyptus* woodborer, *Phoracantha semipunctata* (Coleoptera: Cerambycidae). *J. Comp. Physiol. A* 188:121–33

9. Bashford R. 1993. Insect pest problems of eucalypt plantations in Australia. 4. Tasmania. *Aust. For.* 56:375–77
10. Batista-Pereira LG, Marques EN, Groke PH Jr, da Silva MJ, Pereira Neto SD. 1994. Percentage mortality of larvae of *Thyrinteina arnobia* (Stoll, 1782) (Lepidoptera: Geometridae) collected from the edge and interior of plantations of *Eucalyptus grandis* W. Hill ex Maiden. *Rev. Setor Cienc. Agrar.* 13:233–38
11. Batista-Pereira LG, Marques EN, da Silva MJ, Groke Junior PH, Pereira Neto SD. 1995. Mortality rate of *Thyrinteina arnobia* (Stoll, 1782) (Lepidoptera: Geometridae) by parasitoids and entomopathogens. *Rev. Arvore* 19:396–404
12. Ben Jamaa ML, Villemant C, M'Nar S. 2002. *Phoracantha recurva* Newman, 1840: nouveau ravageur des eucalyptus en Tunisie [Coleoptera, Cerambycidae]. *Rev. Fr. Entomol.* 24:19–21
13. Berti Filho E. 1985. Insects associated to eucalypt plantations in Brazil. In *Noxious Insects to Pine and Eucalypt Plantations in the Tropics*, pp. 162–78. Curitiba, Braz.: IUFRO Work. Party S2.07.07. Protection of Forests in the Tropics. (24-30 Nov. 1985, Curitiba). Univ. Fed. Parana
14. Bethke JA. 2007. *Chrysophtharta* control. *CAPCA Advisor.* June 2007. 10:20
15. Bittencourt MAL, Boaretto L, Serafim I, Berti Filho E. 2003. Fauna of Lepidoptera associated to a natural ecosystem of the state of São Paulo, Brazil. *Arq. Inst. Biol.* 70:85–87
16. Boland DJ, Brophy JJ, House APN, eds. 1991. Eucalyptus *Leaf Oils: Use, Chemistry, Distillation and Marketing*. Melbourne, Aust.: Inkata Press. 252 pp.
17. Boreham GR. 2006. A survey of cossid moth attack in *Eucalyptus nitens* on the Mpumalanga highveld of South Africa. *S. Afr. For. J.* 206:23–26
18. Brennan EB, Gill RJ, Hrusa GF, Weinbaum SA. 1999. First record of *Glycaspis brimblecombei* (Moore) (Homoptera: Psyllidae) in North America: initial observations and predator associations of a potentially serious new pest of eucalyptus in California. *Pan-Pac. Entomol.* 75:55–57
19. Brennan EB, Weinbaum SA. 2001. Effect of epicuticular wax on adhesion of psyllids to glaucous juvenile and glossy adult leaves of *Eucalyptus globulus* Labillardière. *Aust. J. Entomol.* 40:270–77
20. **Brennan EB, Weinbaum SA, Rosenheim JA, Karban R. 2001. Heteroblasty in *Eucalyptus globulus* (Myricales: Myricaceae) affects ovipositional and settling preferences of *Ctenarytaina eucalypti* and *C. spatulata* (Homoptera: Psyllidae). *Environ. Entomol.* 30:1144–49**

20. Shows that host plant specificity can be intra- as well as interspecific and is determined by an insect's ability to adhere to different types of epicuticular wax.

21. Brown BN, Wylie FR. 1991. Diseases and pests of Australian forest nurseries: past and present. In *Proc. First Meet. IUFRO Work. Party S.2.07–09 (Diseases and Insects in Forest Nurseries)*, ed. JR Sutherland, SG Glover, pp. 3–15. Victoria, B.C. Aug. 22–30, 1990. For. Can., Pac. For. Cent., Victoria, B.C.
22. Burckhardt D. 1998. *Ctenarytaina eucalypti* (Maskell) (Hemiptera, Psylloidea) neu fur Mitteleuropa mit Bemerkungen zur Blattflohfauna von *Eucalyptus*. *Mitt. Entomol. Gesell. Basel* 48:59–67
23. Bybee LF, Millar JG, Paine TD, Campbell K, Hanlon CC. 2004. Effects of temperature on fecundity and longevity of *Phoracantha recurva* and *P. semipunctata* (Coleoptera: Cerambycidae). *Environ. Entomol.* 33:138–46
24. Bybee LF, Millar JG, Paine TD, Campbell K, Hanlon CC. 2004. Seasonal development of *Phoracantha recurva* and *P. semipunctata* (Coleoptera: Cerambycidae) in Southern California. *Environ. Entomol.* 33:1232–41
25. Carpintero DL, Dellape PM. 2006. A new species of *Thaumastocoris* Kirkaldy from Argentina (Heteroptera: Thaumastocoridae: Thaumastocorinae). *Zootaxa* 1228:61–68
26. Cerda ML. 1996. *Chilecomadia valdiviana* (Philippi) (Lepidoptera: Cossidae). Insecto Taladrador de la Madera Asociado al Cultivo del E*ucalyptus* spp. en Chile. *Nota Técnica Año 16. de la CONAF (Corp. Nac. For.)*
27. Chen P, Gu M. 2000. Research on the fauna of *Eucalyptus* pests in China. *For. Res.* 13:51–56
28. Chippendale GM. 1988. *Flora of Australia. Volume 19. Myrtaceae-Eucalyptus*, Angophora. Canberra, Australia: Aust. Gov. Publ. Serv. 542 pp.
29. Clarke AR, Paterson S, Pennington P. 1998. *Gonipterus scutellatus* Gyllenhal (Coleoptera: Curculionidae) oviposition on seven naturally co-occurring *Eucalyptus* species. *For. Ecol. Manag.* 110:89–99
30. Clarke AR, Zalucki MP, Madden JL, Patel VS, Paterson SC. 1997. Local dispersion of the *Eucalyptus* leaf-beetle *Chrysophtharta bimaculata* (Coleoptera: Chrysomelidae), and implications for forest protection. *J. Appl. Ecol.* 34:807–16

31. Costanzi M, Malausa JC, Cocquempot C. 2003. Un nouveau psylle sur les *Eucalyptus* de la Riviera Ligure et de la Cote d'Azur: premieres observations de *Ctenarytaina spatulata* Taylor dans le Bassin mediterraneen occidental. *Phytoma* 566:48–51
32. Cowles RS, Downer JA. 1995. Eucalyptus snout beetle detected in California. *Calif. Agric.* 49:38–40
33. Cunningham SA, Floyd RB, Weir TA. 2005. Do *Eucalyptus* plantations host an insect community similar to remnant *Eucalyptus* forest? *Aust. Ecol.* 30:103–17
34. Daane KM, Sime KR, Dahlsten DL, Andrews JW Jr, Zuparko RL. 2005. The biology of *Psyllaephagus bliteus* Riek (Hymenoptera: Encyrtidae), a parasitoid of the red gum lerp psyllid (Hemiptera: Psylloidea). *Biol. Control* 32:228–35
35. Dahlsten DL, Daane KM, Paine TD, Sime KR, Lawson AB, et al. 2005. Imported parasitic wasp helps control red gum lerp psyllid. *Calif. Agric.* 59:229–34
36. Dahlsten DL, Hansen EP, Zuparko RL, Norgaard RB. 1998. Biological control of the blue gum psyllid proves economically beneficial. *Calif. Agric.* 52:35–40
37. de Little DW. 1989. Paropsine chrysomelid attack on plantations of *Eucalyptus nitens* in Tasmania. *N.Z. J. For. Sci.* 19:223–27
38. de Little DW, Madden JL. 1975. Host preference in the Tasmanian eucalypt defoliating Paropsini (Coleoptera: Chrysomelidae) with particular reference to *Chrysophtharta bimaculata* (Olivier) and *C. agricola* (Chapuis). *J. Aust. Entomol. Soc.* 14:387–94
39. Doganlar O, Doganlar M. 2008. First record of the eucalyptus seed gall wasp, *Quadrastichodella nova* Girault, 1922 (Eulophidae: Tetrastichinae) from Turkey. *Turk. J. Zool.* 32:457–59
40. dos Anjos N, Majer JD, Loch AD. 2002. Occurrence of the eucalypt leaf beetle, *Cadmus excrementarius* Suffrian (Coleoptera: Chrysomelidae: Cryptocephalinae), in Western Australia. *J. R. Soc. West. Aust.* 85:161–64
41. dos Anjos N, Majer JD, Loch AD. 2002. Spatial distribution of a chrysomelid leaf beetle (*Cadmus excrementarius* Suffrian) and potential damage in a *Eucalyptus globulus* subsp. *globulus* plantation. *Aust. For.* 65:227–31
42. Doughty RW. 2000. *The Eucalyptus: A Natural and Commercial History of the Gum Tree*. Baltimore, MD: The Johns Hopkins Univ. Press. 237 pp.
43. Drinkwater TW. 1975. The present pest status of eucalyptus borers *Phoracantha* spp. in South Africa. In *Proc. 1st Congr. Entomol. Soc. South. Afr.*, pp. 119–29. Pretoria: Entomol. Soc. S. Afr.
44. Duffy EAJ. 1963. A monograph of the immature stages of Australasian timber beetles (*Cerambycidae*). London: Br. Mus. Nat. Hist. 135 pp.
45. Duffy MP, Nahrung HF, Lawson SA, Clarke AR. 2008. Direct and indirect effects of egg parasitism by *Neopolycystus* Girault sp. (Hymenoptera: Pteromalidae) on *Paropsis atomaria* Olivier (Coleoptera: Chrysomelidae). *Aust. J. Entomol.* 47:195–202
46. Edwards ED. 1996. Lymantriidae. In *Checklist of the Lepidoptera of Australia. Monogr. Australian Lepidoptera*, ed. ES Nielsen, ED Edwards, TV Rangs, pp. 275–77. Canberra, Aust.: CSIRO Publ. 529 pp.
47. Edwards PB, Wanjura WJ, Brown WV. 1993. Selective herbivory by Christmas beetles in response to intraspecific variation in *Eucalyptus* terpenoids. *Oecologia* 95:551–57
48. Farr JD. 1985. *The performance of* Uraba lugens *Walker (Lepidoptera: Nolidae) in relation to nitrogen and phenolics in its food*. PhD thesis. Univ. Adel., Aust. 189 pp.
49. Flechtmann CAH, Ottati ALT, Berisford CW. 2001. Ambrosia and bark beetles (Scolytidae: Coleoptera) in pine and eucalypt stands in southern Brazil. *For. Ecol. Manag.* 142:183–91
50. Floyd R, Wylie R, Old K, Dudzinski M, Kile G. 1998. *Pest risk analysis of Eucalyptus spp. at risk from incursions of plant pests and pathogens through Australia's northern border*. Contract. Rep. No. 44. Canberra, Aust.: CSIRO Publ.
51. Floyd RB, Farrow RA, Matsuki M. 2002. Variation in insect damage and growth in *Eucalyptus globulus*. *Agric. For. Entomol.* 4:109–15
52. Floyd RB, Foley WJ. 2001. Identifying pest-resistant eucalypts using near-infrared spectroscopy. *Rural Ind. Res. Dev. Corp. Publ. No. 01/112*. Canberra, Aust.: RIRDC
53. Freeman JS, O'Reilly-Wapstra JM, Vaillancourt RE, Wiggins N, Potts BM. 2008. Quantitative trait loci for key defensive compounds affecting herbivory of eucalypts in Australia. *New Phytol.* 178:846–51

54. de Freitas S. 1991. Biologia de *Gonipterus gibberus* (Boisduval 1835) (Coleoptera: Curculionidae) uma praga de eucaliptos. *An. Soc. Entomol. Brasil.* 20:339–44
55. Gebeyehu S, Hurley BP, Wingfield MJ. 2005. A new lepidopteran insect pest discovered on commercially grown *Eucalyptus nitens* in South Africa. *S. Afr. J. Sci.* 101:26–28
56. Gill RJ. 1998. Recently introduced pests of *Eucalyptus*. *Calif. Plant Pest Dis. Rep.* 17:21–24
57. Govender P, Nair KSS, Sharma JK, Varma RV. 1996. *Soil pest complex and its control in the establishment of commercial plantations in South Africa*. Presented at Impact of diseases and insect pests in tropical forests. Proc. IUFRO Symp., Peechi, India, 23–26 Nov. 1993, pp. 406–15
58. Grissell EE. 2006. A new species of *Megastigmus* Dalman (Hymenoptera: Torymidae), galling seed capsules of *Eucalyptus camaldulensis* Denhardt (Myrtacae) in South Africa and Australia. *Afr. Entomol.* 14:87–94
59. Grosman AH, van Breemen M, Holtz A, Pallini A, Rugama AM, et al. 2005. Searching behaviour of an omnivorous predator for novel and native host plants of its herbivores: a study on arthropod colonization of eucalyptus in Brazil. *Entomol. Exp. Appl.* 116:135–42
60. Haack RA, Law KR, Mastro VC, Ossenbruggen HS, Raimo BJ. 1997. New York's battle with the Asian long-horned beetle. *J. For.* 95:11–15
61. Hanks LM, Gould JR, Paine TD, Millar JG. 1995. Biology and host relations of *Avetianella longoi*, an egg parasitoid of the eucalyptus longhorned borer. *Ann. Entomol. Soc. Am.* 88:666–71
62. Hanks LM, Millar JG, Paine TD, Campbell CD. 2000. Classical biological control of the Australian weevil *Gonipterus scutellatus* (Coleoptera: Curculionidae) in California. *Environ. Entomol.* 29:369–75
63. Hanks LM, Paine TD, Millar JG. 1991. Mechanisms of resistance in *Eucalyptus* against the larvae of the eucalyptus longhorned borer (Coleoptera: Cerambycidae) in California. *Environ. Entomol.* 20:1583–88
64. Hanks LM, Paine TD, Millar JG. 1993. Host species preference and larval performance in the wood-boring beetle *Phoracantha semipunctata* F. *Oecologia* 95:22–29
65. Hanks LM, Paine TD, Millar JG. 1996. Tiny wasp helps protect eucalypts from eucalyptus longhorned borer. *Calif. Agric.* 50:14–16
66. Hanks LM, Paine TD, Millar JG, Campbell CD, Schuch UK. 1999. Water relations of host trees and resistance to the phloem-boring beetle *Phoracantha semipunctata* F. (Coleoptera: Cerambycidae). *Oecologia* 119:400–7
67. Hanks LM, Paine TD, Millar JG, Hom JL. 1995. Variation among *Eucalyptus* species in resistance to eucalyptus longhorned borer in southern California. *Entomol. Exp. Appl.* 74:185–94
68. **Hansen AK, Jeong G, Paine TD, Stouthamer R. 2007. Frequency of secondary symbiont infection in an invasive psyllid relates to parasitism pressure on a geographic scale in California. *Appl. Environ. Microbiol.* 73:7531–35**

68. Reports the first example from a field system of secondary endosymbiont infection of a hemipteran conferring protection from parasitism by a hymenopteran parasitoid.

69. Henery ML, Moran GL, Wallis IR, Foley WJ. 2007. Identification of quantitative trait loci influencing foliar concentrations of terpenes and formylated phloroglucinol compounds in *Eucalyptus nitens*. *New Phytol.* 176:82–95
70. Hodkinson ID. 2007. A new introduced species of *Ctenarytaina* (Hemiptera, Psylloidea) damaging cultivated *Eucalyptus parvula* (=*parvifolia*) in Europe. *Dtsch. Entomol. Z.* 54:27–33
71. Holtz AM, Oliveira HGd, Pallini A, Marinho JS, Zanuncio JC, Oliveira CL. 2003. Adaptation of *Thyrinteina arnobia* to a new host and herbivore induced defense in eucalyptus. *Pesqui. Agropecu. Bras.* 38:453–58
72. Howlett BG, Clarke AR. 2003. Role of foliar chemistry versus leaf-tip morphology in egg-batch placement by *Chrysophtharta bimaculata* (Olivier) (Coleoptera: Chrysomelidae). *Aust. J. Entomol.* 42:144–48
73. Howlett BG, Clarke AR, Madden JL. 2001. The influence of leaf age on the oviposition preference of *Chrysophtharta bimaculata* (Olivier) and the establishment of neonates. *Agric. For. Entomol.* 3:121–27
74. Jones TH, Potts BM, Vaillancourt RE, Davies NW. 2002. Genetic resistance of *Eucalyptus globulus* to autumn gum moth defoliation and the role of cuticular waxes. *Can. J. For. Res.* 32:1961–69
75. Kim I-K, LaSalle J. 2008. A new genus and species of tetrastichinae (Hymenoptera: Eulophidae) inducing galls in seed capsules of *Eucalyptus*. *Zootaxa* 1745:63–68
76. Kim I-K, McDonald MW, La Salle J. 2005. *Moona*, a new genus of tetrastichine gall inducers (Hymenoptera: Eulophidae) on seeds of *Corymbia* (Myrtaceae) in Australia. *Zootaxa* 989:1–10

77. Kliejunas JT, Burdsall HH Jr, DeNitto GA, Eglitis A, Haugen DA, et al. 2003. *Pest risk assessment of the importation into the United States of unprocessed logs and chips of eighteen eucalypt species from Australia*. Gen. Tech. Rep. FPL-GTR-137. USDA For. Serv., For. Prod. Lab., Madison, WI. 206 pp.
78. Kliejunas JT, Tkacz BM, Burdsall HH Jr, DeNitto GA, Eglitis A, et al. 2001. *Pest risk assessment of the importation into the United States of unprocessed Eucalyptus logs and chips from South America*. Gen. Tech. Rep. FPL-GTR-124. USDA For. Serv. For. Prod. Lab., Madison, WI. 134 pp.
79. Lanfranco D, Dungey HS. 2001. Insect damage in *Eucalyptus*: a review of plantations in Chile. *Aust. Ecol.* 26:477–81
80. La Salle J. 2005. Biology of gall inducers and evolution of gall induction in Chalcidoidea (Hymenoptera: Eulophidae, Eurytomidae, Pteromalidae, Tanaostigmatidae, Torymidae). In *Biology, Ecology, and Evolution of Gall-Inducing Arthropods*, ed. A Raman, CW Schaefer, TM Withers, pp. 503–33. Enfield, UK: Science
81. La Salle J, Arakelian G, Garrison RW, Gates MW. 2009. A new species of invasive gall wasp (Hymenoptera: Eulophidae: Tetrastichinae) on blue gum (*Eucalyptus globulus*) in California. *Zootaxa* 2121:35–43
82. Lawrence R, Potts BM, Whitham TG. 2003. Relative importance of plant ontogeny, host genetic variation, and leaf age for a common herbivore. *Ecology* 84:1171–78
83. Li H, Madden JL, Potts BM. 1995. Variation in volatile leaf oils of the Tasmanian *Eucalyptus* species I. Subgenus *Monocalyptus*. *Biochem. Syst. Ecol.* 23:299–318
84. Li H, Madden JL, Potts BM. 1996. Variation in volatile leaf oils of the Tasmanian *Eucalyptus* species II. Subgenus *Symphyomyrtus*. *Biochem. Syst. Ecol.* 24:547–69
85. Li H, Madden JL, Potts BM. 1997. Variation in leaf waxes of the Tasmanian *Eucalyptus* species I. Subgenus *Symphyomyrtus*. *Biochem. Syst. Ecol.* 25:631–57
86. Liebhold AM, Gottschalk KW, Muzika RM, Montgomery ME, Young R, et al. 1995. Suitability of North American tree species to the gypsy moth: a summary of field and laboratory tests. *USDA For. Serv., Gen. Tech. Rep. NE-211*. Northeast. For. Exp. Stn., Radnor, Pa. 34 pp.
87. Loch AD, Floyd RB. 2001. Insect pests of Tasmanian blue gum, *Eucalyptus globulus globulus*, in southwestern Australia: history, current perspectives and future prospects. *Aust. Ecol.* 26:458–66
88. Luhring KA, Millar JG, Paine TD, Reed D, Christiansen H. 2004. Ovipositional preferences and progeny development of the egg parasitoid *Avetianella longoi*: factors mediating replacement of one species by a congener in a shared habitat. *Biol. Control* 30:382–91
89. Luhring KA, Paine TD, Millar JG, Hanks LM. 2000. Suitability of the eggs of two species of eucalyptus longhorned borers (*Phoracantha recurva* and *P. semipunctata*) as hosts for the encyrtid parasitoid *Avetianella longoi*. *Biol. Control* 19:95–104
90. Lukacs Z. 1999. *Phenology of autumn gum moth*, Mnesampela privata (*Guenée*) (*Lepidoptera: Geometridae*). PhD thesis. Univ. Tasmania, Hobart. 252 pp.
91. Malipatil M, Wainer J. 2007. Chinese wax scale (*Ceroplastes sinensis*). **http://www.padil.gov.au**
92. Malipatil M, Wainer J. 2007. San José scale (*Diaspidiotus perniciosus*). **http://www.padil.gov.au**
93. Malipatil M, Wainer J. 2007. Soft brown scale (*Coccus hesperidum*). **http://www.padil.gov.au**
94. Mapondera TS. 2008. *Molecular phylogenetics and phylogeography of the cryptic species complex* Gonipterus scutellatus (*Coleoptera, Curculionidae*). Honours thesis. Murdoch Univ., Perth. 80 pp.
95. Martinez Egea JM. 1982. *Phoracantha semipunctata* Fab. en el Suroeste espanol. Resumen de la campana de colocacion de arboles cebo. *Bol. Estac. Cent. Ecol.* 11:57–69
96. Marzo L de. 2008. Ulteriori dati sulla presenza di *Leprosa milga* Kim & La Salle su eucalipto in Sud Italia (Hymenoptera Eulophidae). *Boll. Zool. Agrar. Bachicol.* 40:227–32
97. **Matsuki M, Kay M, Serin J, Floyd R, Scott JK. 2001. Potential risk of accidental introduction of Asian gypsy moth (*Lymantria dispar*) to Australasia: effects of climatic conditions and suitability of native plants. *Agric. For. Entomol.* 3:305–20**
98. Matthiessen JN, Ridsdill-Smith TJ. 1991. Populations of African black beetle, *Heteronychus arator* (Coleoptera: Scarabaeidae), in a Mediterranean climate region of Australia. *Bull. Entomol. Res.* 81:85–91
99. Mendel Z, Protasov A, Fisher N, LaSalle J. 2004. Taxonomy and biology of *Leptocybe invasa* gen. & sp. n. (Hymenoptera: Eulophidae), an invasive gall inducer on *Eucalyptus*. *Aust. J. Entomol.* 43:101–13

97. An analysis of the potential threat of a highly polyphagous exotic insect to *Eucalyptus* in Australia.

100. Mitchell JD. 2002. Termites as pests of crops, forestry, rangeland and structures in southern Africa and their control. *Sociobiology* 40:47–69
101. Mitchell MR. 1989. Susceptibility to termite attack of various tree species planted in Zimbabwe. *ACIAR Monogr. Ser.* 10:215–27
102. Nahrung HF. 2006. Paropsine beetles (Coleoptera: Chrysomelidae) in south-eastern Queensland hardwood plantations: identifying potential pest species. *Aust. For.* 69:270–74
103. Nahrung HF, Allen GR. 2003. Intra-plant host selection, oviposition preference and larval survival of *Chrysophtharta agricola* (Chapuis) (Coleoptera: Chrysomelidae: Paropsini) between foliage types of a heterophyllous host. *Agric. For. Entomol.* 5:155–62
104. Nahrung HF, Allen GR. 2004. Sexual selection under scramble competition: mate location and mate choice in the eucalypt leaf beetle *Chrysophtharta agricola* (Chapuis) in the field. *J. Insect Behav.* 17:353–66
105. Nahrung HF, Dunstan PK, Allen GR. 2001. Larval gregariousness and neonate establishment of the eucalypt-feeding beetle *Chrysophtharta agricola* (Coleoptera: Chrysomelidae: Paropsini). *Oikos* 94:358–64
106. Nair KSS, Mathew G, Varma RV, Sudheendrakumar VV, Sharma JK, et al. 1986. Insect pests of eucalypts in India. In *Eucalypts in India. Past, Present and Future. Proc. Natl. Semin. held at Kerala For. Res. Inst., Peechi, Kerala, India,. Jan. 30–31, 1984*, pp. 325–35
107. Neumann FG, Collett NG. 1997. Insecticide trials for control of the autumn gum moth (*Mnesampela privata*), a primary defoliator in commercial eucalypt plantations prior to canopy closure. *Aust. For.* 60:130–37
108. Ohmart CP, Edwards PB. 1991. Insect herbivory on *Eucalyptus*. *Annu. Rev. Entomol.* 36:637–57
109. Östrand F, Elek JA, Steinbauer MJ. 2008. Monitoring autumn gum moth (*Mnesampela privata*): relationships between pheromone and light trap catches and oviposition in eucalypt plantations. *Aust. For.* 70:185–91
110. Östrand F, Wallis IR, Davies NW, Matsuki M, Steinbauer MJ. 2008. Causes and consequences of host expansion by *Mnesampela privata* (Lepidoptera: Geometridae). *J. Chem. Ecol.* 34:153–67
111. Paine TD, Millar JG, Bellows TS, Hanks LM. 1997. Enlisting an under-appreciated clientele: public participation in distribution and evaluation of natural enemies in urban landscapes. *Am. Entomol.* 43:163–72
112. Rabasse JM, Perrin H. 1979. Introduction in France of the eucalyptus snout beetle *Gonipterus scutellatus* Gyll. *Acta Oecol. Oecol. Appl.* 1:21–28
113. Reed DA, Luhring KA, Stafford CA, Hansen AK, Millar JG, et al. 2007. Host defensive response against an egg parasitoid involves cellular encapsulation and melanization. *Biol. Control* 41:214–22
114. Ross M. 2005. $80 million pest eradicated. *Biosecurity* 61:18–19
115. de Santana DLQ, Burckhardt D. 2007. Introduced *Eucalyptus* psyllids in Brazil. *J. For. Res.* 12:337–44
116. Schauff ME, Garrison R. 2000. An introduced species of *Epichrysocharis* (Hymenoptera: Eulophidae) producing galls on *Eucalyptus* in California with notes on the described species and placement of the genus. *J. Hymenopt. Res.* 9:176–81
117. Schumacher RK. 1997. *A study of the parasitoids of eggs and larvae of the autumn gum moth,* Mnesampela privata *(Guen.) (Lepidoptera: Geometridae), on* Eucalyptus globulus *Labill. subsp.* bicostata *(Maiden, Blakely & J. Simm.) Kirkpatr. in Canberra, Australian Capital Territory*. Graduate Diploma in Science thesis. Aust. Natl. Univ., Canberra. 84 pp.
118. Sen-Sarma PK, Thakur ML. 1983. Insect pests of *Eucalyptus* and their control. *Indian Forest.* 109:864–81
119. Shivesh K, Sharma SK, Tarun K, Emmanuel CJSK. 2007. Emergence of gall inducing insect *Leptocybe invasa* (Hymenoptera: Eulophidae) in *Eucalyptus* plantations in Gujarat, India. *Indian For.* 133:1566–68
120. Simmul TL, de Little DW. 1999. Biology of the Paropsini (Chrysomelidae: Chrysomelinae). In *Advances in Chrysomelidae Biology*, ed. ML Cox, pp. 463–77. Leiden: Backhuys
121. Speight MR, Wylie FR. 2001. *Insect Pests of Tropical Forestry*. Wallingford, UK: CABI. 307 pp.
122. Steinbauer MJ. 2001. Specific leaf weight as an indicator of juvenile leaf toughness in Tasmanian bluegum (*Eucalyptus globulus* ssp. *globulus*): implications for insect defoliation. *Aust. For.* 64:32–37
123. Steinbauer MJ. 2002. Oviposition preference and neonate performance of *Mnesampela privata* in relation to heterophylly in *Eucalyptus dunnii* and *E. globulus*. *Agric. For. Entomol.* 4:245–53

110. Shows that oviposition on novel hosts is facilitated by presence of waxes and terpenes characteristic of preferred species.

113. An encapsulation response in an insect egg against a hymenopteran egg parasitoid is documented and explains host selection behavior by the wasp.

121. Excellent introduction to pests of tropical forestry with extensive references to eucalypts.

124. Steinbauer MJ. 2005. How does host abundance affect oviposition and fecundity of *Mnesampela privata* (Lepidoptera: Geometridae)? *Environ. Entomol.* 34:281–91

125. Steinbauer MJ, Clarke AR, Madden JL. 1998. Oviposition preference of a *Eucalyptus* herbivore and the importance of leaf age on interspecific host choice. *Ecol. Entomol.* 23:201–6

126. Steinbauer MJ, Kriticos DJ, Lukacs Z, Clarke AR. 2004. Modelling a forest lepidopteran: phonological plasticity determines voltinism which influences population dynamics. *For. Ecol. Manag.* 198:117–31

127. Steinbauer MJ, Matsuki M. 2004. Suitability of *Eucalyptus* and *Corymbia* for *Mnesampela privata* (Guenée) (Lepidoptera: Geometridae) larvae. *Agric. For. Entomol.* 6:323–32

128. Steinbauer MJ, McQuillan PB, Young CJ. 2001. Life history and behavioral traits of *Mnesampela privata* that exacerbate population responses to eucalypt plantations: comparisons with Australian and outbreak species of forest geometrid from the Northern hemisphere. *Austr. Ecol.* 26:525–34

129. Steinbauer MJ, Schiestl FP, Davies NW. 2004. Monoterpenes and epicuticular waxes help female autumn gum moth differentiate between waxy and glossy *Eucalyptus* and leaves of different ages. *J. Chem. Ecol.* 30:1117–42

129. Shows that epicuticular waxes are perceived by antennal sensilla and that they elicit oviposition.

130. Steinbauer MJ, Wanjura WJ. 2002. Christmas beetles (*Anoplognathus* spp., Coleoptera: Scarabaeidae) mistake peppercorn trees for eucalypts. *J. Nat. Hist.* 36:119–25

131. Strauss SY. 2001. Benefits and risks of biotic exchange between *Eucalyptus* plantations and native Australian forests. *Aust. Ecol.* 26:447–57

131. Considers host range expansion by native insects onto eucalypts planted outside their endemic ranges.

132. Strauss SY, Morrow PA. 1988. Movement patterns of an Australian chrysomelid beetle in a stand of two *Eucalyptus* host species. *Oecologia* 77:231–37

133. Tang C, Wan X, Wan F, Ren S, Peng Z. 2008. The bluegum chalcid, *Leptocybe invasa*, invaded Hainan province. *Chin. Bull. Entomol.* 45:967–71

134. Taylor GS. 1997. Effect of plant compounds on the population dynamics of the lerp insect, *Cardiaspina albitextura* Taylor (Psylloidea: Spondyliaspididae) on eucalypts. In *Ecology and Evolution of Plant-Feeding Insects in Natural and Man-Made Environments*, ed. A Raman, pp. 37–57. Leiden, The Neth.: Backhuys

135. Timberlake PH. 1957. A new entedontine chalcid-fly from seed capsules of eucalyptus in California (Hymenoptera: Eulophidae). *Pan-Pac. Entomol.* 33:109–10

136. Tomiczek C, Hoyer-Tomiczek U. 2007. Asian longhorned beetle (*Anoplophora glabripennis*) and citrus longhorned beetle (*Anoplophora chinensis*) in Europe—actual situation. *Forst. Akt.* 38:2–5

137. Tribe GD, Cillie JJ. 1997. Biology of the Australian tortoise beetle *Trachymela tincticollis* (Blackburn) (Chrysomelidae: Chrysomelini: Paropsina), a defoliator of *Eucalyptus* (Myrtaceae), in South Africa. *Afr. Entomol.* 5:109–23

138. Valente C, Hodkinson I. 2009. First record of the red gum lerp psyllid, *Glycaspsis brimblecombei*, in Europe. *J. Appl. Entomol.* 133:315–17

139. Viggiani G, Laudonia S, Bernardo U. 2002. Aumentano gli insetti dannosi agli eucalipti. *Inform. Agric.* 58:86–87

140. Walker J. 2008. *Kiawe scolytid* (Hypothenemus birmanus). **http://www.padil.gov.au**

141. Walker K. 2006. *Common ambrosia beetle* (Platypus parallelus). **http://www.padil.gov.au**

142. Wiley J, Skelley P. 2008. A *Eucalyptus* pest, *Leptocybe invasa* Fisher and LaSalle (Hymenoptera: Eulophidae), genus and species new to Florida and North America. Fla. Dep. Agric. Consum. Serv. 2 pp. **http://www.doacs.state.fl.us/pi/enpp/ento/leptocybe_invasa.html**

143. Wingfield MJ, Slippers B, Hurley BP, Coutinho TA, Wingfield BD, Roux J. 2008. Eucalypt pests and diseases: growing threats to plantation productivity. *South. For.* 70:139–44

144. Withers TM. 2001. Colonization of eucalypts in New Zealand by Australian insects. *Aust. Ecol.* 26:467–76

144. A useful history and breakdown of the trends of introductions of eucalypt insects in Australia's near neighbor.

145. Withers TM, Bain J. 2009. Reducing rate of Australian *Eucalyptus* insects invading New Zealand. *NZ Plant Prot.* 62:411

146. Woodburn TL, Lewis EE. 1973. A comparative histological study of the effects of feeding by nymphs of four psyllid species on the leaves of eucalypts. *J. Aust. Entomol. Soc.* 12:134–38

147. Wylie FR. 1992. *A comparison of insect problems in Eucalypt plantations in Australia and in Southern China*. Presented at XIX Int. Congr. Entomol., Beijing, China, 28 June–5 July

148. Wylie FR, Floyd RB. 2002. The insect threat to eucalypt plantations in tropical areas of Australia and Asia. *FORSPA Publ.* No. 30/2002, pp. 11–17
149. Wylie R, Simpson J, Bashford D. 2007. A response plan and strategy for gypsy moth, *Lymantria dispar*. Aust. Gov., Dep. Agric., Fish. For. 54 pp.
150. Zanetti R, Zanuncio JC, Vilela EF, Leite HG, Jaffe K, Oliveira AC. 2003. Level of economic damage for leaf-cutting ants (Hymenoptera: Formicidae) in *Eucalyptus* plantations in Brazil. *Sociobiology* 42:433–42

Urticating Hairs in Arthropods: Their Nature and Medical Significance

Andrea Battisti,[1] Göran Holm,[2] Bengt Fagrell,[2] and Stig Larsson[3]

[1]Department of Environmental Agronomy, University of Padova, Legnaro I-35020 Italy; email: andrea.battisti@unipd.it

[2]Karolinska Institutet, Karolinska University Hospital Solna, L08:00, Stockholm, S-17176 Sweden; email: goran.holm@ki.se; fag.ben@telia.com

[3]Department of Ecology, Swedish University of Agricultural Sciences, Uppsala, S-75007 Sweden; email: stig.larsson@ekol.slu.se

Annu. Rev. Entomol. 2011. 56:203–20

First published online as a Review in Advance on August 30, 2010

The *Annual Review of Entomology* is online at ento.annualreviews.org

This article's doi:
10.1146/annurev-ento-120709-144844

0066-4170/11/0107-0203$20.00

Key Words

seta, Lepidoptera, spider, ecology, epidemiology, immunology

Abstract

The ecological phenomenon of arthropods with defensive hairs is widespread. These urticating hairs can be divided into three categories: true setae, which are detachable hairs in Lepidoptera and in New World tarantula spiders; modified setae, which are stiff hairs in lepidopteran larvae; and spines, which are complex and secretion-filled structures in lepidopteran larvae. This review focuses on the true setae because their high density on a large number of common arthropod species has great implications for human and animal health. Morphology and function, interactions with human tissues, epidemiology, and medical impact, including inflammation and allergy in relation to true setae, are addressed. Because data from epidemiological and other clinical studies are ambiguous with regard to frequencies of setae-caused allergic reactions, other mechanisms for setae-mediated disease are suggested. Finally, we briefly discuss current evidence for the adaptive and ecological significance of true setae.

Integument: multilayer structure forming the external part of the body of an arthropod, replaced at each molt with the shedding of the exuvia

INTRODUCTION

The arthropod integument is often richly endowed with hairs. There are many different types of hairs, which have several functions. In this review we focus on hairs with a putative protective function against invertebrate and vertebrate enemies (113). Defensive hairs from several taxa, primarily lepidopterans (32) and spiders (18), but also coleopterans (83) and millipedes (27), have been described.

The ecological significance of hairs as a defensive agent is unclear and inferred mainly from studies on humans and domesticated vertebrates, for which the hairs constitute great medical problems (4, 78, 112). An important task for entomological, medical, and veterinary research is to explain the mechanisms behind the symptoms occurring after exposure to hairs. Previous reviews have addressed urticating hairs, primarily from lepidopterans as well as spiders (18, 104), from the perspective of both entomology (21, 32, 52, 75, 88, 106, 112) and medicine (4, 26, 40, 41, 42, 78).

The aim of this review is to present basic and applied aspects of arthropod urticating hairs and to combine entomological and medical knowledge on urticating hairs into an integrated framework for future research. Although urticating hairs are discussed in general, the focus is on true urticating setae because of their complex mechanisms and great impact on humans.

CLASSIFICATION AND OCCURRENCE OF URTICATING HAIRS

Urticating hairs are derived from the typical arthropod hair, which is formed by at least two cells [trichogen (or hair-forming cell) and tormogen (or auxiliary cell)] embedded in the epidermal cells and is connected to one or more neurons for the transmission of sensorial information (16) (**Figure 1*a***). We divide urticating hairs into three main categories: (*a*) true setae, (*b*) modified setae, and (*c*) spines (**Table 1**).

True Setae

True setae are modified by the loss of the neural connection and the detachment of the proximal end of the hair from the integument (**Figure 1*b***). The base of each seta is attached to an integument stalk (18, 31), or inserted into a socket (52, 81, 82), and can easily be removed with any kind of mechanical stimulation. The setae are short (generally 100–500 μm long, 3–7 μm in diameter), have barbs along the shaft, and occur in special parts of the insect body called mirrors or setae fields. The density of setae can be very high [60,000 setae/mm^2 in the larvae of the pine processionary moth (58); 10,000–12,000 setae/mm^2 in tarantula spiders (18)]. The seta enters the human skin at the proximal end (37).

True setae occur in Lepidoptera larvae and adults (32, 52, 75, 88, 106) and in some tarantula spider taxa (18) (**Table 1**) (**Figure 1*b*** and **Supplemental Figure 1**; follow the **Supplemental Material link** from the Annual Reviews home page at **http://www.annualreviews.org**). The setae are a distinct feature of the larval stage of processionary (Thaumetopoeinae, Notodontidae) and tussock moths (Lymantriidae) and occur in the adult stage of a few species [e.g., the African *Anaphe* spp. (Notodontidae) (93) and *Euproctis* spp. (Lymantriidae) (54, 108)]. A few species of Saturniidae in South America (88, 98) and Zygaenidae in Australia carry setae only as adults (101).

Early lepidopteran instars are without urticating hairs, although the cellular apparatus that produces them is present (56). In processionary moths, setae start to develop in the third instar, and the numbers increase at successive molts (**Supplemental Figure 2**). The larval exuvia left after the molt may carry the setae that were not dispersed during the previous larval instar. Similarly, exuviae left inside the cocoon at pupation are covered with numerous setae. Some *Euproctis* species may pick up such setae at emergence from cocoons and use them to protect the eggs alone or in combination with other true setae that they produce for this purpose (54). Female moths

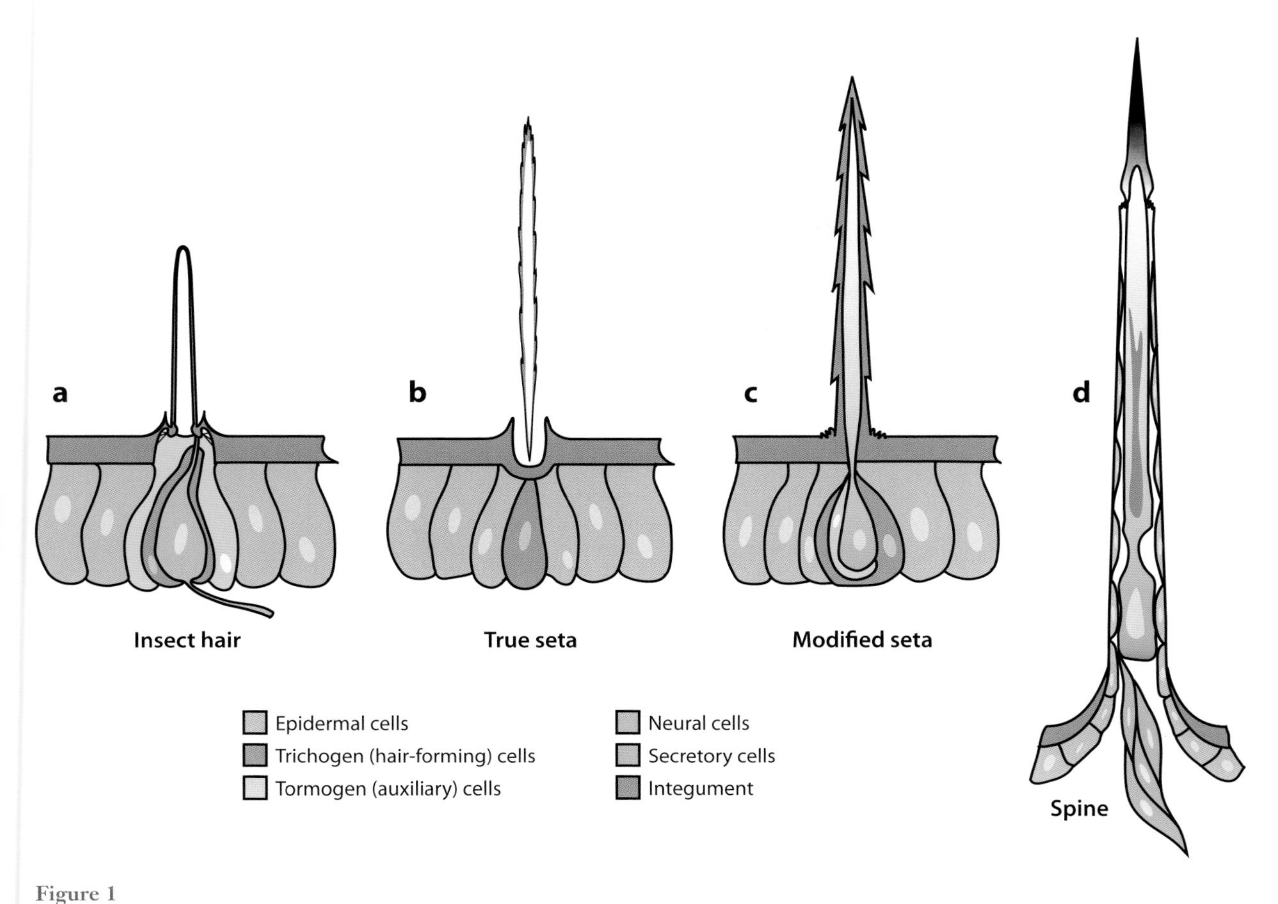

Figure 1

Schematic representation of (*a*) an insect hair, (*b*) a true seta, (*c*) a modified seta, and (*d*) a spine. Note the different scale between (*a*), (*b*), (*c*), and (*d*) by comparing the epidermal cell size.

of *Hylesia* spp. and *Anaphe* spp. carry setae on the ventral part of the abdomen and use them to protect their eggs, presumably from mammalian and avian predators (59, 60, 62, 92, 96). Neonate larvae may carry on their body some of the setae originally protecting the eggs and become urticating even though they cannot produce the setae by themselves (88).

In tarantula spiders, true setae occur in the large family Theraphosidae, but only in New World species in the subfamilies Aviculariinae (tropical, tree-dwelling tarantulas, with less than 100 species), Grammostolinae, and Theraphosinae (terrestrial tarantulas, with a few hundred ground-dwelling species) (18). Spider setae have been classified into a number of subtypes (18) (**Supplemental Figure 1**) and are used in taxonomic classifications and in phylogenetic analyses (86, 87).

Spider setae are situated on the abdomen (in one species also on the pedipalps) (31, 73) and start to develop in the third instar, roughly corresponding to the time when the spider leaves the parental burrows. Subsequently, setae are renewed at every molt, including when the spider molts to the adult stage (18). The number of setae increases at every molt; some types of seta appear earlier than others during development and in a gender-specific pattern (87).

Hairs similar to true urticating setae occur in other arthropods. The larvae of dermestid beetles and some millipedes carry

Table 1 **Urticating hair types in relation to taxonomy, distribution, life-history traits, and previous classification types**

Hair type	Group	Family	Distribution	Classification type		
				Gilmer (1925)	Maschwitz & Kloft (1971)	Kawamoto & Kumada (1984)
Seta	Spiders	Theraphosidae	America			
	Lepidoptera larvae	Lymantriidae	Cosmopolitan	Modified hair	Pointed base	1, 2, 6
		Notodontidae				
	Lepidoptera moths	Lymantriidae	Cosmopolitan		Anal tuft	7
		Notodontidae	Africa			
		Saturniidae	America			
		Zygaenidae	Australia			
Modified seta	Lepidoptera larvae	Zygaenidae	Cosmopolitan	Primitive	Blunt base	3, 4, 5, 8, 9, 10
		Limacodidae	Cosmopolitan			
		Nolidae	Australia			
		Arctiidae	Cosmopolitan			
		Anthelidae	Indo-Australia			
		Eupterotidae	Africa, Australia			
		Lasiocampidae	Cosmopolitan			
		Lymantriidae	Cosmopolitan			
Spine	Lepidoptera larvae	Saturniidae	America	Spine	Poison spine	11
		Noctuidae	Cosmopolitan			
		Megalopygidae	America			
		Limacodidae	Cosmopolitan			
		Nymphalidae				

hastate hairs, or hastisetae, that are detachable and may entangle small predators like ants (27, 83). The adults of the caddisfly *Eubasilissa avalokhita* (Trichoptera) carry urticating setae on the thorax and forewings (111).

Modified Setae

Modified setae have a blunt base and are connected to the integument; the neural connection is lost (**Figure 1*c***). However, another cell type is connected to the seta and is hypothesized to have a secretory function (52). Modified setae vary in size but are generally longer than true setae (up to 1 mm), are rather stiff, have a very sharp distal end, and have barbs of various sizes along the shaft or at the base. Modified setae are clustered at a much lower density than true setae on different parts of the body and can easily break off at the connection with the integument. Modified seta enters the skin at the distal end.

Modified setae are exclusive to Lepidoptera and occur in a large number of taxa. Less is known about the structure and function of modified setae compared with true setae. Many of the hair types described by Kawamoto & Kumada (52) belong to this category because they have the same anatomical origin, i.e., they are formed by one or a few cells, and differ substantially from the complex structure of the spine. Modified setae occur on larvae of Lymantriidae (e.g., the gypsy moth, *Lymantria dispar*) (22), Lasiocampidae (e.g., pine caterpillars *Dendrolimus* spp.) (52), Arctiidae (52), Anthelidae (6), and in several other small families (**Table 1**). The mature larvae can detach the setae from the body at pupation and incorporate them into the cocoon as an external protection (6). Many species have transformed normal hairs into defensive setae simply by hardening them [stiff end, primitive type described by Gilmer (32)] and in some cases by adding secretions that can cause specific reactions. First-instar gypsy moth larvae carry small balloon setae that contain, among other compounds, nicotine (1, 23).

Spines

Compared with true or modified setae, spines have a more complex structure (**Figure 1*d***). They can be seen as an outgrowth of the integument and involve a high number of specialized cells and a sensory function. Spines are stiff, are filled with a secretion, and respond to mechanical stimulation; the spine's contents can be injected into the skin from the broken tip (98). The size of the spines may vary considerably but their diameter is generally larger than that of true and modified setae. Spines are often grouped around an integument outgrowth (scolus). There is a large body of literature on spines (14), but in this review we do not discuss details of their function.

Spines are a common type of urticating hair in a number of Lepidoptera, particularly Megalopygidae, Limacodidae, and Saturniidae, and are well described (32, 52, 98). Spines occur to a lesser extent in larvae of other lepidopteran families (Noctuidae and Nymphalidae) as well, but specific information is scarce. Large spiders of the Sparassidae carry spine-like structures on the legs (43), although their morphology has yet to be described.

The number and size of the scoli that host the spines increase with larval instar (e.g., *Lonomia obliqua*; 98). Spines may occur alone or in combination with modified setae (e.g., Limacodidae), but not with true setae. Certain Saturniidae species that bear spines in larval stages however, may produce true setae as adults (98). The occurrence of spines in lepidopteran larvae is commonly associated with aposematic traits, such as bright color, unusual shape, or alarm posture (98).

GENERAL MEDICAL ASPECTS

Human exposure to arthropods carrying setae, modified setae, or spines is a worldwide concern. Arthropod density is the key to hair dynamics and risk of exposure to humans, with complications related to repeated exposure (see Tissue Reactions to True Setae, below). Outbreaks of defoliating lepidopterans result in dispersal of true setae in the environment (33, 52, 57, 69, 85). Modified setae and spines are generally present at a much lower density on the insect body compared with true setae and are normally not released into the air. Modified setae can become a major problem at high population densities as observed during occupational exposure, such as forest workers exposed to *Dendrolimus* sp. (Lasiocampidae) in China and Siberia (40) and to *Premolis semirufa* (Arctiidae) in Brazil (19). Spider populations generally do not reach outbreak densities comparable to those of lepidopterans, and their urticating hairs are therefore less of a problem to humans and animals. Pest control operations have been used worldwide to reduce the abundance of defoliating lepidopterans carrying urticating hairs (8, 74, 95).

Erucism, lepidopterism: syndromes associated with larval and adult Lepidoptera, respectively, often confused and of limited clinical relevance

Urticating hairs are harmful to humans in a number of ways and different syndromes have been proposed in the literature (e.g., erucism and lepidopterism). The use of these terms, however, is problematic because erucism and lepidopterism refer to different life stages of the insect (larvae and adults, respectively) rather than to reactions in humans (78). Furthermore, the distinction of these syndromes based on local and general reactions is complicated because true setae can be released by the insects, either larva or adult, and distributed in the environment by air (56, 88). Thus, true setae are more likely to cause reactions as they may affect several parts of the body simultaneously. As a result, it must be stressed that these definitions are of limited value when the effect of true setae on humans is described in clinical practice. Because larval modified setae and spines are normally not released into the air, they are more likely associated with reactions from direct contact, thus leading to specific syndromes (10, 13, 14, 19, 39, 100, 105).

Urticating hairs can also affect domestic animals (78). Ingestion of caterpillars, or urticating hairs, seems to be more common in animals and may have dramatic consequences, such as tongue necrosis in dogs (3, 36, 44, 71, 89). In addition, setae coming into contact with tissues of the mouth, pharynx, or intestine may provoke severe symptoms and sometimes have life-threatening consequences (12).

Urticaria: itchy skin eruption with wheal-like swelling and redness (erythema) in the skin

Dermatitis: skin inflammation due either to direct contact with an irritating substance or to an allergic reaction

TRUE SETAE IN LEPIDOPTERA

Nature and Dynamics

Most of what is known about true setae in Lepidoptera comes from studies on *Thaumetopoea* spp. (24, 28, 58, 90, 97, 106) and *Euproctis* spp. (7, 47, 48, 49, 50, 52, 53). The irritating nature of *Thaumetopoea* larvae has been known since the ancient Greeks (Dioscorides, reported in Reference 76). Setae from both species groups share similar traits; the setae appear during larval development on the dorsal part of the abdomen in specific areas called mirrors, owing to their brightness and light reflectivity. The mirrors increase in number and size as the larva molts. In the mature larva a full set of mirrors consists of several hundred thousand setae, e.g., 630,000 in *T. processionea* (97) and up to one million in *T. pityocampa* (58). New setae are produced at every molt (**Supplemental Figure 2**); old setae remain with the exuvia.

The setae of *Thaumetopoea* larvae are packed close together on the mirror, whereas *Euproctis* larval setae are situated on small outgrowths of the mirror called papillae (75). In both cases, the sharp basal end of the seta is loosely inserted into a socket. In *Euproctis* larvae, a hole situated close to the basal end of the seta acts as a bridge between the epidermal cell and the seta, although the nature of the connection is not clear (52). Such an opening has not been observed in *Thaumetopoea* larvae (81). Distal barbs of *Euproctis* are more conspicuous than those of *Thaumetopoea*, making identification possible (52).

The mechanism by which larvae release the setae has been studied in *T. pityocampa* (24). In this species, the mirrors are kept folded under normal conditions and only the distal end of the setae is visible. When disturbed, the larva opens the mirror (**Supplemental Figure 3**), releasing the setae, a process that is further facilitated by the action of a few normal hairs that are mixed with the setae in the mirror (82). Accidental release of setae by *Euproctis* adults has been observed (54, 65, 108).

Once in the air, the setae can be carried by the wind far from the source. In pollen trap studies setae have been found several kilometers away from the source for both *Euproctis* (108, 109) and *Thaumetopoea* (107). Models have predicted high densities of setae of *Thaumetopoea processionea* (20–30% of the initial density) as far as several hundred meters from the source, depending on weather conditions (30).

In general, setae can persist in the environment for a long time. In silk shelters used by larvae, for example, a seta's reactive property can last for at least one year (69). Setae can remain active for long periods in the soil where larvae have pupated (25), in collection material (37), and in contaminated clothes (37), although no precise estimates are available. Thus, humans can be exposed to setae long after the active insect has disappeared.

Medical Aspects

Few large epidemiological studies have been published describing the frequency and distribution of clinical symptoms following exposure to true setae. Vega et al. (103) found that 10% of children exposed to *Thaumetopoea pityocampa* had symptoms, predominantly urticaria and dermatitis. In a study of 1,025 persons living within 500 m of three oak trees infested with *T. processionea*, about 6% had symptoms (69).

In order to depict the frequency of setae-associated symptoms, a questionnaire-based study was performed covering 70% of the population living in an area heavily infested with *T. pinivora* in the southern part of the Swedish island Gotland in the Baltic Sea. Approximately 18% of 4,300 persons reported symptoms likely to be caused by setae (38). The frequency of symptoms varied between 4% and 87%, mainly depending on the local density of larvae.

The importance of how much setae one is exposed to is further illustrated by large

Euproctis outbreaks reported in Japan and China, with some hundred thousand persons affected (52). Most studies, however, imply that a considerable proportion of a human population does not show symptoms in spite of exposure to setae. The parts of the body exposed to setae, as well as genetic or other factors such as sweating, may be important for determining the severity and manifestations of clinical symptoms.

In general, symptoms from exposure to setae are often highly nonspecific, and consequently, accurate diagnosis must be based on a careful reconstruction of events. Hossler (40) summarized certain clinical findings, especially in children, that suggest contact with caterpillars or moths. Microscopic evaluation in vivo of the affected areas may be an excellent tool for verifying that setae are present in the affected areas (5, 29, 114). With the use of microscopy the pathophysiologic changes of affected skin areas can be followed for weeks (29).

Local skin symptoms vary extensively from one person to another. The first symptoms and signs of urticaria include itching, local swelling (edema and wheal formation), and red discoloration (erythema) (29, 40). Extended edema of skin or mucous membranes, and sometimes fever and malaise, may be present. In rare cases, life-threatening acute anaphylactic reactions have been described (34, 35, 102).

The pathophysiology of skin changes following setae attachment has also been studied experimentally. Setae from *Euproctis chrysorrhoea* were applied epicutaneously and intracutaneously to volunteers (45, 46, 47, 48). Within a few hours almost 70% of the individuals exhibited positive skin reactions including erythema with varying degrees of edema. These results are similar to the reactions in a recent study in which a drop of setae suspension from *Thaumetopoea pinivora* was applied onto the skin of the forearm of volunteers (29). The local blood perfusion, measured by laser Doppler fluxmetry, increased in all individuals within the first two days. In two individuals with a history of severe symptoms, microscopic vesicles developed around the setae, followed by desquamation and severe local symptoms (**Figure 2*b***). The remaining individuals, who experienced only mild symptoms during previous exposure, exhibited only mild reactions that disappeared within three weeks (**Figure 2*a***).

Eyes may also be affected by true setae. Hase (37) described technicians acquiring severe conjunctivitis during laboratory work with setae from *T. pityocampa*. Watson & Sevel (105) report patients with severe and long-lasting eye symptoms, including keratitis, conjunctivitis, and damage to the inner eye and lens, that were produced by setae from *T. pityocampa* and *E. chrysorrhoea*. Oral exposure to setae may occur in humans and animals by unintentional ingestion of larvae or contaminated feed. Lee et al. (68) reported 26 patients who had oropharyngeal exposure to lepidopterans, with 8 patients actually ingesting the larvae. The lips, tongue, and buccal mucosa were most frequently affected, but sometimes also the esophagus was involved. The small size of setae allows for inhalation into the airways. The ensuing symptoms may include coughing, shortness of breath, and asthma-like symptoms (2, 40, 99).

Anaphylactic: describes an acute, severe multisystem, and sometimes life-threatening hypersensitivity reaction usually mediated by IgE antibodies

TRUE SETAE IN SPIDERS

Nature and Dynamics

The details about the dynamics and mode of action of spider setae are less known compared with those of lepidopteran setae. Unlike lepidopteran setae, spider setae are carried on stalks (31). Spider setae release occurs at a junction between the stalk and the shaft, with the proximal end of the released seta resembling a hypodermic needle. The shaft is hollow, with a central lumen of 1 μm and a wall thickness of 2 μm. The density of setae in the parts of the integument carrying them (field) is much lower than that observed in the mirrors of lepidopterans.

There are two functional patterns of seta systems in spiders. In Aviculariinae, the spiders cannot flick the setae into the air when disturbed; instead the setae are pressed into an enemy upon contact, except in the genus *Ephebopus*, which carry setae on the palps and actively release them (31). The second pattern is found

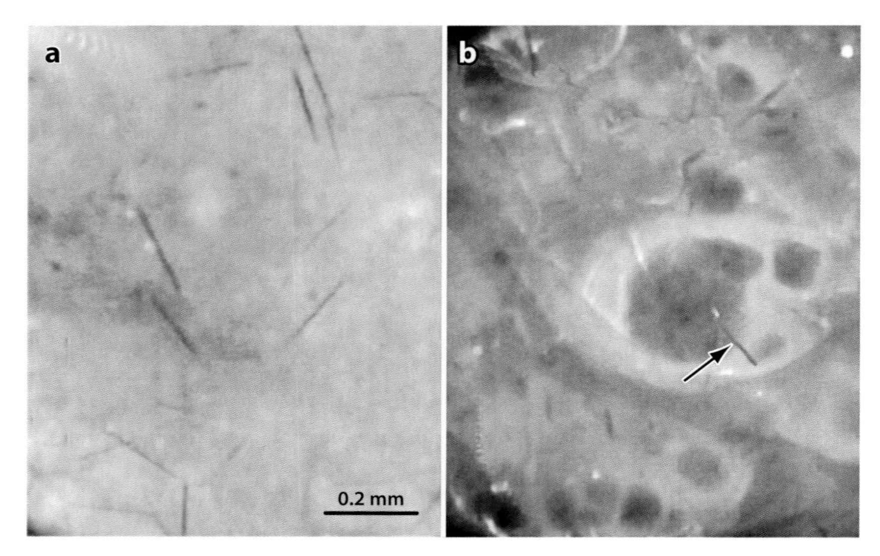

Figure 2

Microscopic images (×100) of skin areas experimentally exposed to setae of *Thaumetopoea pinivora*. Both pictures are taken 7 days after exposure. (*a*) A subject with mild symptoms. (*b*) A subject with severe skin symptoms. In panel *a*, setae are attached to the outer skin layer and there are no obvious signs of inflammation. In panel *b*, several small and large vesicles are present, indicating a pronounced inflammatory reaction. In one vesicle a seta is clearly visible (*arrow*).

in Theraphosinae, which actively release the setae as well. The setae are dislodged from their supporting stalks by rapid vibrations of one or both hindlegs, which are lifted onto the dorsum of the abdomen. It appears that the ventral metatarsal spines play an important role in combing off the setae (18, 73). Setae fields also include normal hairs (see figure 24 in Reference 18), in addition to true setae, which may play a role in the release of setae, similar to that in certain Lepidoptera. The released setae may adhere to the silk tunnels, egg sacs, and shedding mats of Theraphosidae (72). When actually released by the spiders, setae cause significant symptoms in mice and attach to mucous membranes of the eye, nose, airways, and outer skin layer, as documented through microscopy (18).

Medical Aspects

Spider setae cause reactions in humans when in direct contact with skin or mucous membranes (17) or when released close to sensitive parts such as the eyes (80). Thus, symptoms are recognized mainly in persons handling spiders in places such as museums and zoos. Eye injuries seem to be an emerging problem as keeping tarantulas as pets has increased in popularity (80, 104), although the size of the clinical problem is still poorly known (9). The symptoms vary from red eyes to severe intraocular inflammation that may persist for several weeks (80). The reactions caused by setae to skin may be local (17, 18) or general (15). General symptoms were observed in sensitized persons mainly because of occupational exposure in a center for tarantula study in Brazil. The typical symptoms, varying only in intensity, suggest a hypersensitivity reaction mediated by IgE (15).

Local symptoms differ somewhat depending on the type of seta. Only type I, II, and III setae (**Supplemental Figure 1**) have been tested so far (17, 18). Type I setae seem to have limited effects, as experimental exposure to lips and eyelids cause mild reactions. Type II setae from *Avicularia* spiders (nonflicking) effect strong reactions in humans after contact with the spider or with its arboreal silk nests. Type II setae seem to penetrate deeper into the tissues than Type I setae and provoke itching and other symptoms of intermediate severity. Type III setae penetrate more deeply into the skin (up to 2 mm for a seta with an average length of 0.6 mm) and commonly cause skin reactions associated with inflammation and itching. Microscopy of skin biopsies taken within 30 min after exposure revealed only local edema and the results were interpreted as mechanical irritation (17, 18).

TISSUE REACTIONS TO TRUE SETAE

Setae contain proteins and other molecules that the mammalian immune system recognizes as foreign, causing allergy and other types of immune reactions. However, as discussed above, data from epidemiological and case studies are ambiguous with regard to the frequency of allergic reactions to setae, and a thorough understanding of tissue reactions to setae remains to be developed. We next review the evidence for IgE-mediated allergic reactions to setae and suggest a model that includes additional mechanisms to explain host responses at the cellular level. Most of current understanding of setae-tissue interactions refers to exposure to *Thaumetopoea* spp.; thus, we restrict the discussion to this system.

Allergic and Toxic Reactions

Two different antigens reactive with patient IgE antibodies have been described in larval extracts and in setae from *T. pityocampa*. Lamy et al. (61, 63) found a soluble 28-kDa dimeric protein called thaumetopoein in extracts of setae that caused skin reactions in people as well as guinea pigs and that degranulated mast cells (51, 64). Thaumetopoein-binding IgE antibodies have been described in about 20% of highly exposed wood workers (110). Moneo et al. (77) purified a 15-kDa protein from extracts of *T. pityocampa* larvae (Tha p 1). The protein was recognized by IgE antibodies from allergic patients. Although the nature of the antigen is not clear, amino acid sequences homologous to chemosensory proteins have been noted (66).

There is no conclusive evidence for the participation of other types of allergic immune reactions mediated by IgG antibodies or thymus-derived (T) lymphocytes (cell-mediated immune reactions). Note that a majority of patients report that their skin reactions to *T. pinivora* setae, manifested by itching, flare, and rash, do not appear until 12–24 h after exposure (38). Similarly, volunteers with or without a history of previous exposure, when provoked by setae applied to the skin of the forearm, developed skin reactions that were delayed (29). These findings suggest that mechanisms other than those mediated by IgE are at play.

Setae are sometimes referred to as toxic (115). Histamine-like effects exerted by extracted compounds from setae have been described and molecules with enzymatic actions have been identified in extracts (11, 64). The slow incitement of skin reactions after intracutaneous exposure to setae, however, contradicts the notion that rapid reactants such as histamine are at play (29). Pretreating setae with chemicals or heating to prevent symptoms did not remove their inflammatory action (37). Thus, the current literature does not support the notion that setae can exert toxicity in mammals. The important issue therefore is to define other possible immune, or nonimmune, reactions that contribute to the reactions brought about by setae.

Immune reactions: defense reactions, allergies, and other immune diseases and tissue reactions induced by T-lymphocytes and antibodies

Mast cells: cells in blood and tissues containing vesicles loaded with histamine and other vasoactive reactants

Lymphocytes: white mononuclear cells of the immune system present in blood and lymphatic organs

Toxic: containing a poisonous substance produced by living organisms that causes disease after contact with body tissues

Inflammatory Reactions Induced by Setae

Setae constitute a special case for the interaction between insects and mammals. Setae, like the general insect integument, are built up by a chitin skeleton with a matrix of proteins and are covered by layers of tannin-bound lipoproteins, wax, and mucopolysaccharides (16). Not only are these constituents foreign to mammals but chitin may also act as a strong inducer of inflammation and potentiator of immune responses. When introduced via the skin or mucous membranes into mammalian tissues, setae become

Macrophages: white blood and tissue cells that can engulf and usually also degrade small particles

Cytokines: hormone-like proteins that participate in the cooperation between cells

exposed to cells and other reactants of the local and circulating defense systems. As a result, the introduction of setae into the mammalian body is expected to lead to inflammatory or immunological defense reactions of varying severities causing local as well as systemic symptoms. Although the details concerning the mechanism of induction of inflammation by chitin are not fully understood, the following outline, based mainly on studies of bacterial chitin in mammals, is offered as a hypothesis for chitin involvement in setae-human interactions.

Chitin is a polysaccharide biopolymer composed of *N*-acetyl-β-D-glucosamine that constitutes the rigid structures of insects, helminths, fungi, and crustaceans. Chitin is not present in mammalian tissues, but it can be degraded by chitinases expressed in mammalian tissues (55). Among a number of mammalian chitinase-like proteins, only two have enzymatic activity, e.g., chitotriosidase and acid mammalian chitinase.

Chitin particles can bind to receptors on macrophages and activate them to produce chitinases together with proinflammatory cytokines and other inflammatory and immunoregulatory mediators. In this way the breakdown process of chitin may contribute to an inflammatory process (20).

In addition to its proinflammatory properties, chitin, or its breakdown fragments, executes complex interactions with cells of the immune system. Chitin compounds may modulate cellular as well as antibody-mediated immune responses by direct and indirect interactions with T-lymphocytes (67). Thus, chitin may also act as an adjuvant to promote immune responses mediated by T-helper cells. Chitin upregulates immunity against *Mycobacterium bovis* bacillus as well as HIV and influenza virus (67). Moreover, low-molecular-weight chitin particles may play a role in asthma (84). Chitin also seems to have an important role in the immune response against some helminth infections (91). It is conceivable that setae chitin and its breakdown products exert similar effects on inflammatory and immune processes. If so, the levels of chitinases and hence the levels of chitin fragments may be an important factor for the ensuing biological effects. Recent observations have revealed functional acid mammalian chitinase gene polymorphisms to be associated with asthma and other immune-mediated diseases (84). Thus, genetically determined variations of chitinase activity could be a contributory explanation for the variable human sensitivity to setae exposure observed in epidemiological studies.

In conclusion, chitin and its breakdown products are potent promoters and regulators of inflammatory and immune reactions. It is therefore conceivable to attribute a similar role to such compounds in setae to explain the symptoms caused by setae in the host (**Figure 3**). This hypothesis implies that chitin components of setae may promote harmful allergic and other immune responses against setae antigens. However, direct studies of the cellular and molecular mechanisms behind the clinical outcome of exposure to setae chitin need to be performed.

ADAPTIVE SIGNIFICANCE

Protection of arthropods from predation has been invoked as the only function of urticating hairs (18, 32, 52), although experimental evidence in support of this hypothesis is scarce (but see Reference 79). It is conceivable that spines, and possibly modified setae, have a defensive function because these hairs contain, at least in some species (23), chemicals that can deter an enemy on contact. It is less clear, however, how true setae can function in protection against predators. In the following, we summarize setae-related life-history traits in larval Lepidoptera, based mainly on our own experience with *Thaumetopoea* spp., and offer tentative suggestions how setae may function in an adaptive way.

Based on what is known from reactions to setae in humans, it is likely that setae act primarily against vertebrate predators; there is no evidence to suggest that arthropod enemies are affected. Two important issues need to be emphasized. First, setae are released into the

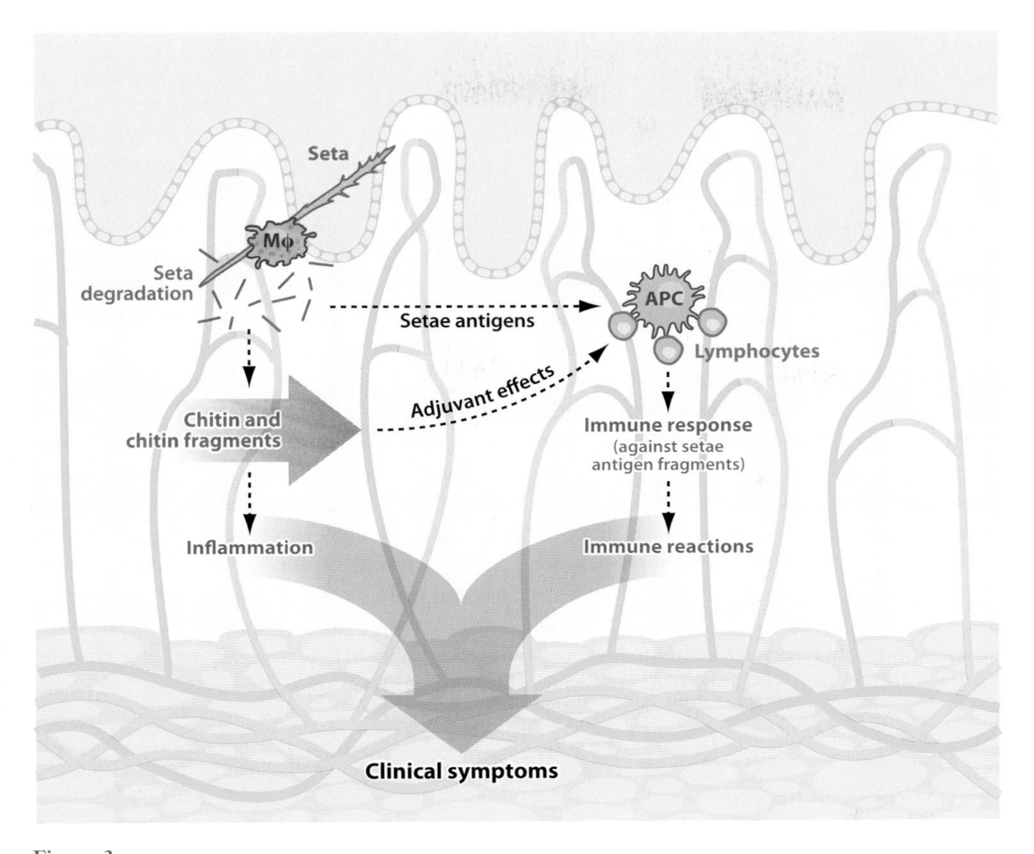

Figure 3

Tentative role of setae as inducers of inflammation and immune reactions in the skin. When introduced into the skin tissues, setae induce inflammatory and immunological processes; macrophages (Mφ) and other inflammatory cells accumulate and become activated. Chitinases synthesized by such cells break down the chitin skeleton of setae. Chitin fragments, proteins, and other antigenic components are released. Setae proteins are taken up and processed by antigen presenting cells (APCs) and presented to lymphocytes, thereby initiating lymphocyte proliferation and specific immune responses. Chitin fragments promote inflammation and further support proliferation of lymphocytes (adjuvant effects) leading to allergic and other kinds of immune reactions. In addition, other setae components may contribute to inflammation and immune system stimulation. The final outcome is tissue damage, clinical symptoms, and disease.

environment upon the attack or disturbance and are then dispersed away from the larva. Presumably, the release is directed toward the attacking enemy but potential attackers farther away may also be affected, immediately and for longer periods following the attack. Second, the predator's reaction to the setae is unlikely to be instantaneous; at least in humans, and probably in other vertebrates, the reaction starts to appear after a considerable delay, sometimes up to 12 h after exposure. For these two reasons, assuming that vertebrate predators respond similarly to setae as humans do, it would be difficult for the predator to immediately associate symptoms with the attack on the prey armed with setae. Thus, it is difficult to see at first how releasing setae during the attack contributes to the fitness of that individual.

The defensive benefits of setae may be contingent on the particular aspects of the life history of the insect. *Thaumetopoea* larvae do not have distinct warning colors, but the

Aposematism: association of conspicuous traits such as bright color, shape, and posture with defensive traits

larvae are highly social and spend their entire life in tight groups, most often probably composed of closely related individuals. Like many other social larvae, they have a synchronized defensive behavior. Furthermore, the individuals are covered with normal hairs, in addition to setae, that grow longer as larvae mature. In combination, these traits give the larval group a very conspicuous appearance, likely resulting in aposematism based on urticating hairs instead of toxic chemicals. Thus, the signaling effect of the colony may be an important factor in maintaining the defense once it has evolved, similar to that in other bad-tasting group-living larvae (94). Conspicuousness may thus reinforce learning of avoidance behavior in vertebrate predators.

The key issue of how predators can relate illness to attacked prey, however, is still puzzling. In other aposematic prey the attacking predator most often receives an immediate negative feedback. It is possible that vertebrate predators are more sensitive to setae than humans and thus respond with an immediate reaction to setae, similar to predators attacking aposematic prey with a chemical defense. Nothing is known about the mechanisms underlying the responses of putative vertebrate predators to setae; it is unlikely, however, that they are different from responses in humans or pets because the immune system and basic inflammatory processes are similar. On the other hand, the delayed effects are expected to be more costly to the predator than the immediate effects because of the cumulative action of the setae (70), possibly amplified by sensitization processes, which would select for avoidance behavior of setae-carrying larvae by the predators.

Although there is a lack of experimental data to convincingly argue for an adaptive advantage of setae, their very high numbers in larvae strongly point to their importance. No functions other than defense against natural enemies have so far been suggested for setae.

SUMMARY POINTS

1. The large body of literature on various types of urticating hairs in arthropods is synthesized, and hairs are divided into three categories (true seta, modified seta, spine) on the basis of morphological and functional features.
2. The medical relevance of major syndromes (erucism and lepidopterism) associated with the three hair categories is discussed. The use of the syndrome terms, however, is problematic because erucism and lepidopterism refer to different life stages of the insect (larvae and adults, respectively) rather than to reactions in humans; it is concluded that these syndromes have limited value in clinical practice.
3. The nature of true setae (high density on the arthropod, easy release into the air, severe but delayed reactions in humans) makes them particularly interesting; a coherent view of true setae is attempted that includes setae morphology and function, epidemiology of reactions in humans, and cellular and tissue-level responses to setae penetration.
4. Epidemiological data are limited in terms of the incidence of human reactions to true setae and are hampered by the sparse information on the degree of individual exposure. The clinical picture is dominated by local tissue symptoms but may include complex inflammatory and/or allergic reactions.
5. Inflammatory reactions may occur in most individuals after exposure to setae independent of previous exposure. In sensitized persons, IgE-mediated processes contribute to the observed patterns of reactions.

FUTURE ISSUES

1. Recent data suggest that setae inserted into human tissue may induce reactions in addition to IgE-mediated allergy. Chitin is a key component in the insect integument, including setae. An intriguing possibility is that chitins, or their breakdown products, may interfere with cells of the inflammatory and immune systems and thus induce a cascade of events eventually leading to tissue damage and clinical symptoms. A challenge to future research will be to combine entomological and immunological efforts to obtain model systems in which the chitin hypothesis can be tested in mechanistic detail.
2. The massive investment by larvae in setae strongly points to an advantage of some sort, presumably protection against natural enemies. Unequivocal evidence from controlled experiments, however, is still missing. The fitness benefits of setae could be primarily at the cocoon stage; the soil around the cocoons is extremely rich in setae, deposited by larvae during pupation. In all these cases it is likely that the benefit of setae is protection from a putative vertebrate enemy, based on the type of reactions documented in humans, but this is also in need of experimental support.

DISCLOSURE STATEMENT

The authors are not aware of any affiliations, memberships, funding, or financial holdings that might be perceived as affecting the objectivity of this review.

ACKNOWLEDGMENTS

We would like to acknowledge Geoffrey Isbister, Carlos Lopez-Vaamonde, and Michael Stastny for valuable comments on earlier drafts of the manuscript, and two anonymous reviewers for useful suggestions. Thanks also to Paolo Paolucci, Edoardo Petrucco Toffolo and Daniel Zovi for drawing the figures. The work was partly supported by The Swedish Research Council for Environment, Agricultural Sciences and Spatial Planning, the project URTICLIM from ANR France, the San Michele Foundation, and Torsten and Ragnar Söderbergs foundations.

LITERATURE CITED

1. Aldrich JR, Schaefer PW, Oliver JE, Puapoomchareon P, Lee C-J, Vander Meer RK. 1997. Biochemistry of the exocrine secretion from gypsy moth caterpillars (Lepidoptera: Lymantriidae). *Ann. Entomol. Soc. Am.* 90:75–82
2. Aparicio VF, Fernandez MB, Sotes MR, Paredes AR, Molero MIM, et al. 2004. Non-occupational allergy caused by the pine processionary caterpillar (*Thaumetopoea pityocampa*). *Allergol. Immunopathol.* 32:69–75
3. Arditti J, David JM, Jean P, Jouglard J. 1988. Injuries induced by the pine processionary caterpillar in Provence. *J. Toxicol. Clin. Exp.* 8:247–51
4. Arlian LG. 2002. Arthropod allergens and human health. *Annu. Rev. Entomol.* 47:395–433
5. Bakos RM, Rezende RL, Bakos L, Cartell A. 2006. Spider spines detected by dermoscopy. *Arch. Dermatol.* 142:1517–18
6. Balit CR, Geary MJ, Russell RC, Isbister GK. 2004. Clinical effects of exposure to the Whitestemmed gum moth (*Chelepteryx collesi*). *Emerg. Med. Australas.* 16:74–81
7. Balit CR, Ptolemy HC, Geary MJ, Russell RC, Isbister GK. 2001. Outbreak of caterpillar dermatitis caused by airborne hairs of the mistletoe browntail moth (*Euproctis edwardsi*). *Med. J. Aust.* 175:641–43

8. Battisti A, Longo S, Tiberi R, Triggiani O. 1998. Results and perspectives in the use of *Bacillus thuringiensis* Berl. var. *kurstaki* and other pathogens against *Thaumetopoea pityocampa* (Den. et Schiff.) in Italy (Lep., Thaumetopoeidae). *J. Pest Sci.* 71:72–76
9. Belyea DA, Tuman DC, Ward TP, Babonis TR. 1998. The red eye revisited: ophthalmia nodosa due to tarantula hairs. *South. Med. J.* 91:565–67
10. Bessler E, Bildner E, Yassur Y. 1987. *Thaumetopoea wilkinsoni* (toxic pine caterpillars) blepharoconjunctivitis. *Am. J. Ophthalmol.* 103:117–18
11. Bleumink E, Jong MCJM, Kawamoto F, Meyer GT, Kloosterhuis AJ, Slijperpal IJ. 1982. Protease activities in the spicule venom of *Euproctis* caterpillars. *Toxicon* 20:607–13
12. Bruchim Y, Ranen E, Saragusty J, Aroch I. 2005. Severe tongue necrosis associated with pine processionary moth (*Thaumetopoea wilkinsoni*) ingestion in three dogs. *Toxicon* 45:443–47
13. Cadera W, Pachtman MA, Fountain JA, Ellis FD, Wilson FM. 1984. Ocular lesions caused by caterpillar hairs (ophthalmia nodosa). *Can. J. Ophthalmol.* 19:40–44
14. Carrijo-Carvalho LC, Chudzinski-Tavassi AM. 2007. The venom of the *Lonomia* caterpillar: an overview. *Toxicon* 49:741–57
15. Castro FFM, Antila MA, Croce J. 1995. Occupational allergy caused by urticating hair of Brazilian spider. *J. Allergy Clin. Immunol.* 95:1282–85
16. Chapman RF. 1998. *The Insects. Structure and Function*. Cambridge, UK: Cambridge Univ. Press. 770 pp.
17. Cooke JA, Miller FH, Grover RW, Duffy JL. 1973. Urticaria caused by tarantula hairs. *Am. J. Trop. Med. Hyg.* 22:130–33
18. Cooke JA, Roth VD, Miller FH. 1972. The urticating hairs of theraphosid spiders. *Am. Mus. Novit.* 2498:1–43

18. The only overview of the true seta system in tarantula spiders.

19. Costa RM, Atra E, Ferraz MB, da Silva NP, Batista Júnior J, et al. 1993. "Pararamose": an occupational arthritis caused by Lepidoptera (*Premolis semirufa*). An epidemiological study. *São Paulo Med. J.* 111:462–65
20. Da Silva CA, Chalouni C, Williams A, Hartl D, Lee CG, Elias JA. 2009. Chitin is a size-dependent regulator of macrophage TNF and IL-10 production. *J. Immunol.* 182:3573–82
21. Delgado QA. 1978. Venoms of Lepidoptera. In *Arthropod Venoms*, ed. S Bettini, pp. 555–611. Berlin: Springer
22. Deml R, Dettner K. 1995. 'Balloon hairs' of gypsy moth larvae (Lep., Lymantriidae): morphology and comparative chemistry. *Comp. Biochem. Physiol.* 112:673–81
23. Deml R, Dettner K. 2003. Comparative morphology and secretion chemistry of the scoli in caterpillars of *Hyalophora cecropia*. *Naturwissenschaften* 90:460–63
24. Démolin G. 1963. Les 'miroirs' urticants de la processionnaire du pin (*Thaumetopoea pityocampa* Schiff.). *Rev. Zool. Agric.* 10–12:107–14

24. A careful description of the structure of the urticating apparatus in *Thaumetopoea*.

25. Démolin G. 1971. Incidences de quelques facteurs agissant sur le comportement social des chenilles de *Thaumetopoea pityocampa* Schiff. (Lepidoptera) pendant la période des processions de nymphose. Répercussion sur l'efficacité des parasites. *Ann. Zool. Ecol. Anim.* Num. hors série: 33–56
26. Diaz JH. 2005. The evolving global epidemiology, syndromic classification, management, and prevention of caterpillar envenoming. *Am. J. Trop. Med. Hyg.* 72:347–57
27. Eisner T, Eisner M, Deyrup M. 1996. Millipede defense: use of detachable bristles to entangle ants. *Proc. Natl. Acad. Sci. USA* 93:10848–51
28. Fabre J-H. 1900. *Souvenirs Entomologiques, Études sur l'Instinct et les Moeurs des Insectes*. Paris: Delagrave
29. Fagrell B, Jörneskog G, Salomonsson AC, Larsson S, Holm G. 2008. Skin reactions induced by experimental exposure to setae from larvae of the northern pine processionary moth (*Thaumetopoea pinivora*). *Contact Dermat.* 59:290–95

29. An application of skin microscopy and skin blood perfusion to the study of reactions to setae.

30. Fenk L, Vogel B, Horvath H. 2007. Dispersion of the bio-aerosol produced by the oak processionary moth. *Aerobiologia* 23:79–87

30. An experimental and modeling approach to the dispersal of the setae in a natural habitat.

31. Foelix R, Rast B, Erb B. 2009. Palpal urticating hairs in the tarantula *Ephebopus*: fine structure and mechanism of release. *J. Arachnol.* 37:292–98

31. A thorough scanning microscope analysis of the structure of setae in a tarantula spider.

32. Gilmer PM. 1925. A comparative study of the poison apparatus of certain lepidopterous larvae. *Ann. Entomol. Soc. Am.* 18:203–39

32. The first comprehensive review of the defensive hairs of lepidopteran larvae.

33. Glasser CM, Cardoso JL, Carreribruno GC, Domingos MD, Moraes RHP, Ciaravolo RMD. 1993. Epidemic outbreaks of dermatitis caused by *Hylesia* (Lepidoptera, Hemileucidae), in Sao-Paulo State, Brazil. *Rev. Saude Publ.* 27:217–20
34. Gottschling S, Meyer S. 2006. An epidemic airborne disease caused by the oak processionary caterpillar. *Pediatr. Dermatol.* 23:64–66
35. Gottschling S, Meyer S, Dill-Mueller D, Wurm D, Gortner L. 2007. Outbreak report of airborne caterpillar dermatitis in a kindergarten. *Dermatology* 215:5–9
36. Grundmann S, Arnold P, Montavon P, Schraner EM, Wermelinger B, Hauser B. 2000. Toxic tongue necrosis from processional pine caterpillar (*Thaumetopoea pityocampa* Schiff.). *Kleintierpraxis* 45:45–50
37. Hase A. 1939. Über den Pinienprozessionsspinner und über die Gefährlichkeit seiner Raupenhaare (*Thaumetopoea pityocampa* Schiff.). *J. Pest Sci.* 15:133–42
38. **Holm G, Sjoberg J, Ekstrand C, Bjorkholm M, Granath F, et al. 2009. Tallprocessionsspinnare—stort hälsoproblem på södra Gotland. *Lakartidningen* 106:1891–94**
39. Horng C-T, Chou P-I, Liang J-B. 2000. Caterpillar setae in the deep cornea and anterior chamber. *Am. J. Ophthalmol.* 129:384–85
40. Hossler EW. 2009. Caterpillars and moths. *Dermatol. Ther.* 22:353–66
41. Hossler EW. 2010. Caterpillars and moths. Part I. Dermatologic manifestations of encounters with Lepidoptera. *J. Am. Acad. Dermatol.* 62:1–10
42. Hossler EW. 2010. Caterpillars and moths. Part II. Dermatologic manifestations of encounters with Lepidoptera. *J. Am. Acad. Dermatol.* 62:13–28
43. Isbister GK, Hirst D. 2002. Injuries from spider spines, not spider bites. *Vet. Hum. Toxicol.* 44:339–42
44. Jans HWA, Franssen AEM. 2008. The urticating hairs of the oak processionary caterpillar (*Thaumetopoea processionea* L.), a possible problem for animals? *Tijdschr. Diergeneesk.* 133:424–29
45. Jong MCJM, Bleumink E. 1977. Investigative studies of dermatitis caused by larva of brown-tail moth, *Euproctis chrysorrhoea* (Lepidoptera, Lymantriidae). 3. Chemical analysis of skin reactive substances. *Arch. Dermatol. Res.* 259:247–62
46. Jong MCJM, Bleumink E. 1977. Investigative studies of dermatitis caused by larva of brown-tail moth, *Euproctis chrysorrhoea* L. (Lepidoptera, Lymantriidae). 4. Further characterization of skin reactive substances. *Arch. Dermatol. Res.* 259:263–81
47. Jong MCJM, Bleumink E, Nater JP. 1975. Investigative studies of dermatitis caused by larva of brown-tail moth (*Euproctis chrysorrhoea* Linn.). 1. Clinical and experimental findings. *Arch. Dermatol. Res.* 253:287–300
48. Jong MCJM, Hoedemaeker PJ, Jongeblod WL, Nater JP. 1976. Investigative studies of dermatitis caused by larva of brown-tail moth (*Euproctis chrysorrhoea* Linn.). 2. Histopathology of skin lesions and scanning electron-microscopy of their causative setae. *Arch. Dermatol. Res.* 255:177–91
49. Jong MCJM, Kawamoto F, Bleumink E, Kloosterhuis AJ, Meijer GT. 1982. A comparative study of the spicule venom of *Euproctis* caterpillars. *Toxicon* 20:477–85
50. Jongeblod WL, Jong MCJM. 1984. Three-dimensional structure of the skin-irritating spicules of the caterpillar of *Euproctis chrysorrhoea* (L.) or brown tail moth. *Ultramicroscopy* 15:382–83
51. Kalender Y, Kalender S, Uzunhisarcikli M, Ogutcu A, Acikgoz F. 2004. Effects of *Thaumetopoea pityocampa* (Lepidoptera: Thaumetopoeidae) larvae on the degranulation of dermal mast cells in mice; an electron microscopic study. *Folia Biol. Krakow* 52:13–17
52. **Kawamoto F, Kumada N. 1984. Biology and venoms of Lepidoptera. In *Handbook of Natural Toxins. Vol. 2. Insect Poisons, Allergens, and other Invertebrate Venoms*, ed. AT Tu, pp. 291–330. New York: Dekker**
53. Kawamoto F, Suto C, Kumada N. 1978. Studies on venomous spicules and spines of moth caterpillars. 1. Fine-structure and development of venomous spicules of *Euproctis* caterpillars. *Jpn. J. Med. Sci. Biol.* 31:291–99
54. Kemper H. 1955. Experimentelle Untersuchungen uber die durch Afterwolle *Euproctis chrysorrhoea* (Lepidoptera) erzeugte Dermatitis, verglichen mit der Wirkung von Arthropodenstichen. *Z. Angew. Zool.* 42:37–59
55. Kzhyshkowska J, Gratchev A, Goerdt S. 2007. Human chitinases and chitinase-like proteins as indicators for inflammation and cancer. *Biomarker Insights* 2:128–46

38. The largest epidemiological study on humans reacting to setae.

52. An attempt to classify the defensive structures of Lepidoptera.

56. Lamy M. 1990. Contact dermatitis (erucism) produced by processionary caterpillars (genus *Thaumetopoea*). *J. Appl. Entomol.* 110:425–37
57. Lamy M. 1990. Urticating caterpillars and moths: an unrecognised form of "pollution". *Recherche Paris* 21:896–900
58. Lamy M, Ducombs G, Pastureaud MH, Vincendeau P. 1982. Productions tégumentaires de la processionnaire du pin (*Thaumetopoea pityocampa* Schiff.) (Lépidoptères). Appareil urticant et appareil de ponte. *B. Soc. Zool. Fr.* 107:515–29
59. Lamy M, Lemaire C. 1983. Contribution à la systématique de *Hylesia*: étude au microscope électronique à balayage des "fléchettes urticantes". *Bull. Soc. Entomol. Fr.* 88:176–92
60. Lamy M, Michel M, Pradinaud R, Ducombs G, Maleville J. 1982. The urticant apparatus of the moths *Hylesia urticans* Floch & Abonnenc and *H. umbrata* Schaus (Lepidoptera: Saturniidae) caused by lepidopterans in French Guiana. *Int. J. Insect Morphol.* 11:129–35
61. Lamy M, Pastureaud MH, Ducombs G. 1985. Thaumetopoein, an urticating protein of the pine processionary caterpillar (*Thaumetopoea pityocampa* Schiff.) (Lepidoptera, Thaumetopoeidae). *C. R. Acad. Sci. III-Vie* 30:173–76
62. Lamy M, Pastureaud MH, Novak F, Ducombs G. 1984. Papillons urticants d'Afrique et d'Amérique du sud (*g. Anaphae* et *g. Hylesia*): contribution du microscope électronique à balayage à l'étude de leur appareil urticant et à leur mode l'action. *Bull. Soc. Zool. Fr.* 109:163–77
63. Lamy M, Pastureaud MH, Novak F, Ducombs G, Vincendeau P, et al. 1986. Thaumetopoein: an urticating protein from the hairs and integument of the pine processionary caterpillar (*Thaumetopoea pityocampa* Schiff. Lepidoptera Thaumetopoeidae). *Toxicon* 24:347–56
64. Lamy M, Vincendeau P, Ducombs G, Pastureaud MH. 1983. Irritating substance extracted from the *Thaumetopoea pityocampa* caterpillar; mechanism of action. *Experientia* 39:299
65. Lamy M, Werno J. 1989. Le papillon du *Bombyx* "cul-brun" *Euproctis chrysorrhoea* L. (Lepidoptera) responsable de papillonite en France: interpretation biologique. *C. R. Acad. Sci. Paris* 309:605–10
66. Larsson S, Backlund A. 2009. Regarding the putative identity of a moth (*Thaumetopoea* spp.) allergen. *Allergy* 64:493
67. Lee CG, Da Silva CA, Lee J-Y, Hartl D, Elias JA. 2008. Chitin regulation of immune responses: an old molecule with new roles. *Curr. Opin. Immunol.* 20:684–89
68. Lee DL, Pitetti RD, Casselbrant MD. 1999. Oropharyngeal manifestations of lepidopterism. *Arch. Otolaryngol. Head Neck Surg.* 125:50–52
69. **Maier H, Spiegel W, Kinaciyan T, Krehan H, Cabaj A, et al. 2003. The oak processionary caterpillar as the cause of an epidemic airborne disease: survey and analysis. *Br. J. Dermatol.* 149:990–97**
70. Mappes J, Marples N, Endler JA. 2005. The complex business of survival by aposematism. *Trends Ecol. Evol.* 20:598–603
71. Maronna A, Stache H, Sticherling M. 2008. Lepidopterism—oak processionary caterpillar dermatitis: appearance after indirect out-of-season contact. *J. Dtsch. Dermat. Gesell.* 6:1–4
72. Marshall SD, Uetz GW. 1990. Incorporation of urticating hairs into silk: a novel defense mechanism in two neotropical tarantulas (Araneae: Theraphosidae). *J. Arachnol.* 18:143–50
73. Marshall SD, Uetz GW. 1990. The pedipalpal brush of *Ephebopus* sp. (Araneae: Theraphosidae): evidence of a new site for urticating hairs. *Bull. Br. Arachnol. Soc.* 8:122–24
74. Martin JC, Mazet R. 2001. Winter control of pine processionary caterpillar. *Phytoma* 540:32–35
75. Maschwitz WJ, Kloft W. 1971. Morphology and function of the venom apparatus of insects—bees, wasps, ants, and caterpillars. In *Venomous Animals and Their Venoms*, ed. W Bucherl, EE Buckley, pp. 1–60. New York: Academic
76. Matthioli PA. 1568. *I Discorsi di M. Pietro Andrea Matthioli, Medico Cesareo nelli sei Libri di Pedacio Dioscoride Anazarbeo*. Venezia: Valgrisi
77. **Moneo I, Vega JM, Caballero ML, Vega J, Alday E. 2003. Isolation and characterization of Tha p 1, a major allergen from the pine processionary caterpillar *Thaumetopoea pityocampa*. *Allergy* 58:34–37**
78. **Mullen GR. 2009. Moths and butterflies (Lepidoptera). In *Medical and Veterinary Entomology*, ed. GR Mullen, LA Durden, pp. 353–70. Amsterdam: Elsevier**

69. A thorough epidemiological and clinical study of an outbreak in an urban area.

77. A study pointing at identifying the antigen role of proteins associated with urticating larvae.

78. A synthesis of the knowledge on Lepidoptera of medical importance.

79. Murphy SM, Leahy SM, Williams LS, Lill JT. 2010. Stinging spines protect slug caterpillars (Limacodidae) from multiple generalist predators. *Behav. Ecol.* 21:153–60
80. Norris JH, Carrim ZI, Morrel AJ. 2010. Spiderman's eye. *Lancet* 375:92
81. Novak F, Lamy M. 1987. Etude ultrastructurale de la glande urticante de la chenille processionnaire du pin, *Thaumetopoea pityocampa* Schiff (Lepidoptera: Thaumetopoeidae). *Int. J. Insect Morphol.* 16:263–70
82. Novak F, Pelissou V, Lamy M. 1987. Comparative morphological, anatomical and biochemical studies of the urticating apparatus and urticating hairs of some Lepidoptera: *Thaumetopoea pityocampa* Schiff., *Th. processionea* L. (Lepidoptera, Thaumetopoeidae) and *Hylesia metabus* Cramer (Lepidoptera, Saturniidae). *Comp. Biochem. Physiol.* 88:141–46
83. Nutting WL, Spangler HG. 1969. The hastate setae of certain dermestid larvae: an entangling defense mechanism. *Ann. Entomol. Soc. Am.* 62:763–69
84. Ober C, Chupp G. 2009. The chitinase and chitinase-like proteins: a review of genetic and functional studies in asthma and immune-mediated diseases. *Curr. Opin. Allergy Cl.* 9:401–8
85. Ooi PL, Goh KT, Lee HS, Goh CL. 1991. Tussockosis—an outbreak of dermatitis caused by tussock moths in Singapore. *Contact Dermat.* 24:197–200
86. Pérez-Miles F. 1998. Notes on the systematics of the little known theraphosid spider *Hemirrhagus cervinus*, with a description of a new type of urticating hair. *J. Arachnol.* 26:120–23
87. Pérez-Miles F. 2002. The occurrence of abdominal urticating hairs during development in Theraphosinae (Araneae, Theraphosidae): phylogenetic implications. *J. Arachnol.* 30:316–20
88. Pesce H, Delgado QA. 1971. Poisoning from adult moths and caterpillars. In *Venomous Animals and Their Venoms*, Vol. III: *Venomous Invertebrates*, ed. W Bürchel, EE Buckley, pp. 119–56. New York: Academic
89. Poisson L, Boutet JP, Paillassou P, Fuhrer L. 1994. Four cases of pine caterpillar poisoning in dog. *Point Vet.* 25:992–1002
90. Reamur RAF. 1736. *Mémoire pour Servir à l'Histoire des Insectes*. Paris: Imprimerie Royale
91. Reese TA, Liang H-E, Tager ANM, Luster AD, van Rooijen N, et al. 2007. Chitin induces tissue accumulation of innate immune cells associated with allergy. *Nature* 447:92–96
92. Rodriguez J, Hernandez JV, Fornes L, Lundberg U, Pinango CLA, Osborn F. 2004. External morphology of abdominal setae from male and female *Hylesia metabus* adults (Lepidoptera: Saturniidae) and their function. *Fla. Entomol.* 87:30–36
93. Rothschild M, Reichstein T, Lane NJ, Parsons J, Prince W, Swales SW. 1970. Toxic Lepidoptera. *Toxicon* 8:293–99
94. Ruxton G, Sherratt T, Speed MP. 2004. *Avoiding Attack: The Evolutionary Ecology of Crypsis, Warning Signals and Mimicry*. Oxford, UK: Oxford Univ. Press
95. Salomon OD, Simon D, Rimoldi JC, Villaruel M, Perez O, et al. 2005. Lepidopterism due to the butterfly *Hylesia nigricans*. Preventive research-intervention in Buenos Aires. *Med. B. Aires* 65:241–46
96. Schabel HG. 2006. *Forest Entomology in East Africa. Forest Insects of Tanzania*. Dordrecht: Springer
97. Scheidter F. 1934. Auftreten der "Gifthaare" bei den Prozession-spinnerraupen in den einzelnen Stadien. *Z. Pflanzenk. Pflanzen.* 44:223–26
98. Specht A, Corseuil E, Barreto Abella H. 2008. *Lepidopteros de Importancia Medica. Principais Especies no Rio Grande do Sul*. Pelotas, Bras.: USEB
99. Spiegel W, Maier H, Maier M. 2004. A non-infectious airborne disease. *Lancet* 363:1438
100. Steele C, Lucas DR, Ridgway AE. 1984. Endophthalmitis due to caterpillar setae: surgical removal and electron microscopic appearances of the setae. *Br. J. Ophthalmol.* 68:284–88
101. Tarmann GM. 2004. *Zygaenid Moths of Australia: A Revision of the Australian Zygaenidae*, Vol. 9. Australia: CSIRO
102. Vega JM, Moneo I, Armentia A, Lopez-Rico R, Curiel G, et al. 1997. Anaphylaxis to a pine caterpillar. *Allergy* 52:1244–45
103. Vega ML, Vega J, Vega JM, Moneo I, Sanchez E, Miranda A. 2003. Cutaneous reactions to pine processionary caterpillar (*Thaumetopoea pityocampa*) in pediatric population. *Pediatr. Allergy Immunol.* 14:482–86
104. Vetter RS, Isbister GK. 2008. Medical aspects of spider bites. *Annu. Rev. Entomol.* 53:409–29
105. Watson PG, Sevel D. 1965. Ophthalmia nodosa. *Br. J. Ophthalmol.* 50:209–17

106. Weidner H. 1937. Beiträge zur einer Monographie der Raupen mit Gifthaaren. *Z. Angew Entomol.* 23:433–84
107. Werno J, Lamy M. 1990. Animal atmospheric pollution: the urticating hairs of the pine processionary caterpillar (*Thaumetopoea pityocampa* Schiff.) (Insecta, Lepidoptera). *C. R. Acad. Sci.* 310:325–31
108. Werno J, Lamy M. 1991. Urticating hairs of "brown-tail" moth (*Euproctis chrysorrhoea* L.) (Lepidoptera): preliminary studies in urban and laboratory. *C. R. Acad. Sci. III Vie* 312:455–59
109. Werno J, Lamy M. 1994. Daily cycles for emission of urticating hairs from the pine processionary caterpillar (*Thaumetopoea pityocampa* S.) and the brown tail moth (*Euproctis chrysorrhoea* L.) (Lepidoptera) in laboratory conditions. *Aerobiologia* 10:147–51
110. Werno J, Lamy M, Vincendeau P. 1993. Caterpillars hairs as allergens. *Lancet* 342:936–37
111. Wiggins GB. 1998. *The Caddisfly Family Phryganeidae (Trichoptera).* Toronto: Univ. Toronto Press. 306 pp.
112. Wirtz RA. 1984. Allergic and toxic reactions to non-stinging arthropods. *Annu. Rev. Entomol.* 29:47–69
113. Witz BW. 1990. Antipredator mechanisms in arthropods: a twenty year literature survey. *Fla. Entomol.* 73:71–99
114. Zalaudek I, Glacomel J, Gabo H, Di Stefani A, Ferrara G, et al. 2008. Entodermoscopy: a new tool for diagnosing skin infections and infestations. *Dermatology* 216:14–23
115. Ziprkowski L, Rolant F. 1972. Study of the toxin from poison hairs of *Thaumetopoea wilkinsoni* caterpillars. *J. Invest. Dermatol.* 58:274–77

The Alfalfa Leafcutting Bee, *Megachile rotundata*: The World's Most Intensively Managed Solitary Bee*

Theresa L. Pitts-Singer and James H. Cane

USDA ARS Bee Biology & Systematics Laboratory, Utah State University, Logan, Utah 84322; email: Theresa.Pitts-Singer@ars.usda.gov, Jim.Cane@ars.usda.gov

Annu. Rev. Entomol. 2011. 56:221–37

First published online as a Review in Advance on August 30, 2010

The *Annual Review of Entomology* is online at ento.annualreviews.org

This article's doi: 10.1146/annurev-ento-120709-144836

0066-4170/11/0107-0221$20.00

Key Words

Apoidea, bivoltinism, chalkbrood, Fabaceae, pollination, Megachilidae

Abstract

The alfalfa leafcutting bee (ALCB), *Megachile rotundata* F. (Megachildae), was accidentally introduced into the United States by the 1940s. Nest management of this Eurasian nonsocial pollinator transformed the alfalfa seed industry in North America, tripling seed production. The most common ALCB management practice is the loose cell system, in which cocooned bees are removed from nesting cavities for cleaning and storage. Traits of ALCBs that favored their commercialization include gregarious nesting; use of leaves for lining nests; ready acceptance of affordable, mass-produced nesting materials; alfalfa pollination efficacy; and emergence synchrony with alfalfa bloom. The ALCB became a commercial success because much of its natural history was understood, targeted research was pursued, and producer ingenuity was encouraged. The ALCB presents a model system for commercializing other solitary bees and for advancing new testable hypotheses in diverse biological disciplines.

Alfalfa a.k.a. lucerne: *Medicago sativa* L. (Fabaceae). Originating in southwest Asia, the tetraploid is now widely naturalized in Europe and North America

Alfalfa leafcutting bee (ALCB): *Megachile* (*Eutricharaea*) *rotundata* Fab. (Apoidea: Megachilidae)

INTRODUCTION

Fifty-three years ago in the second volume of the *Annual Review of Entomology*, George "Ned" Bohart summarized the wild bees that pollinate the world's leading forage legume, alfalfa (*Medicago sativa*). He concluded that only honey bees, although mediocre pollinators, could satisfy this crop's need for an abundant bee (7). At that time, U.S. alfalfa seed production was shifting west to California, where fields packed with hived honey bees increased seed yields fivefold to 450 kg ha^{-1}. Fifteen years later, Bohart again wrote for these pages about an unheralded new agricultural pollinator from Eurasia, the alfalfa leafcutting bee (ALCB) (8). Although detected in the United States without fanfare by the 1940s (122), the ALCB revolutionized the alfalfa seed industry, boosting yields to a remarkable 1300 kg ha^{-1} (95). No other solitary bee is produced and managed so intensively, although several species are propagated to fill regional niche markets. What are this bee's attributes that have made its management uniquely successful, but only where it and alfalfa are not native? Is ALCB management a model for other solitary bee pollinators or, like honey beekeeping, is it peculiar and unlikely to be replicated? In reviewing the ALCB's life history, management, and ecological impacts, we highlight the factors that enable successful solitary bee management for crop pollination.

TAXONOMY AND BIOGEOGRAPHY

The bee genus *Megachile* is massive (ca. 1,478 described species, or one-third of all megachilids) and cosmopolitan (76); most cut leaf pieces to line their nests, a behavior to which they owe their common name (76, 144). This genus is subdivided into 52 subgenera (76), in part reflecting their collective diverse behaviors and morphologies. The largest subgenus, *Eutricharaea*, has more than 230 described species, all from the Eastern Hemisphere (37, 76). It includes the ALCB, which has been introduced widely (e.g., the Americas, Australia) to pollinate alfalfa.

Fossil *Megachile* are unknown. However, trace fossils have been reported periodically for more than a century. These are leaf imprints that bear the characteristic semicircular marginal notches that remain after females clip leaf discs for nest building (see Life Cycle, below). Most examples come from the Eocene (34–55 Ma) (66); the oldest one dates to the Paleocene (55–65 Ma) (144).

POLLINATION

Owing to its use in alfalfa, the value of the ALCB is surpassed only by the honey bee for pollination of field crops. For example, ALCB use yielded 46,000 metric tons of alfalfa seed in North America in 2004, two-thirds of world production (80, 95). Planted worldwide for hay, alfalfa is fed to livestock, especially dairy cows. Alfalfa seed and resultant hay constitute one-third of the $14 billion value ascribed to honey bees pollinating U.S. crops (139); managed ALCBs account for an additional 50% of alfalfa seed production in the northwestern United States (139) and central Canada. Paradoxically, the ALCB remains uncommon and irrelevant for pollinating alfalfa in its native range. It constituted just 0.03% of the 8,168 wild bees taken in 27 Hungarian alfalfa fields (78). In wild regions of Spain where alfalfa is also grown, the ALCB was absent from the 59 species sampled (82), and even in southern France, large populations are difficult to sustain for alfalfa (126).

Bees pollinate alfalfa flowers when they trip the staminal column, for which ALCBs are effective (29) (**Figure 1**). Rates of pod and seed set reflect primarily the frequencies of tripping (15), with lesser benefits from cross-pollination (124). Females of the ALCB and the alkali bee (*Nomia melanderi*; Halictidae) excel at pollinating alfalfa, tripping 80% of visited flowers (15), comparable to alfalfa's effective but unmanaged European pollinators (23). In North America, diverse bee species—including native *Megachile* species—also pollinate alfalfa well (8, 45, 68), but their abundances are never adequate to satisfy modern seed yield expectations.

Rates and distances of alfalfa gene flow cannot be reliably extrapolated from the flight range of ALCBs. ALCBs readily fly more than 100 m from the nest but then forage locally in a field, moving pollen only short distances of about 4 m (117). Among isolated alfalfa plots or plants, however, minor pollen-mediated gene flow occurs over much greater distances, 8% between plots 200 m apart (10) and 0.5% between plots 330 m apart (139).

The name alfalfa leafcutting bee belies its moderate foraging and pollination versatility. ALCBs have varying success in pollinating several other North American field crops. They are reportedly in extensive use for producing hybrid seed of canola (*Brassica napus*) in western Canada. In cage studies, they pollinated several annual clovers well (*Trifolium* spp.) (72, 104), but not vetches (*Vicia* spp.) (107). They pollinate some native legumes farmed for wildland restoration seed (16). ALCBs pollinate the small flowers of lowbush blueberries (*Vaccinium angustifolium*) grown commercially in the Canadian Maritimes (54) but are exposed to lethally cold nights there during bloom (113). ALCBs foraged at and pollinated cranberry in field cages (19), but in open bogs they foraged little and dispersed (69).

The ALCB can also be amenable to pollinating in confinement. Along with mason bees (*Osmia lignaria*) and honey bees, ALCBs are useful in cages to increase or regenerate germplasm accessions stored at crop seed repositories (21). Caged ALCBs effectively pollinated carrot (130) and canola (116) for hybrid seed, but they eschewed bloom of field-grown carrot (129). Greenhouse pollination of vegetables or fruits by ALCBs receives scant attention; they forage readily in glasshouses (15, 70, 136) but orient poorly under some plastic films (133).

LIFE CYCLE

Adult ALCBs naturally emerge and nest during the hot days of summer. Females mate once, soon after emergence, and then consume nectar and pollen as their first eggs mature (106); within a week females begin constructing and provisioning cells sequentially. Like most other *Megachile* species, they nest in existing holes above ground, fashioning nest walls, partitions, and plugs from strips and disks of leaves that are transported singly. Each nest cell requires 14–15 leaf pieces, both for wall strips and partition disks (58, 70) (**Figure 1**). They cut these pieces using the opposing beveled edges of their mandibles like scissors (**Figure 1**). At the nest, the female thoroughly chews the edges of each new leaf piece. The resultant sticky pulp binds the new leaf piece to the others (136). An ALCB can line and later cap one cell with leaf pieces on average in 81 min (70) to 2.5 h (58). Leaf piece dimensions and cell architecture represent precisely measurable physical manifestations of the bee's complex behaviors (44).

Scopa: a brush of setae (hairs) beneath the female bee's abdomen for transporting dry pollen

A female spends from 5 to 6 h per day foraging (58, 70), returning from flowers with both dry pollen in her scopa and nectar in her crop. The female enters the cavity headfirst to regurgitate her crop full of nectar. She then backs out, turns, and backs into the nest, using her metatarsal hair combs to sweep her abdominal scopa clean of its pollen load (32). Early in the provisioning sequence, the female carries mostly pollen (ca. 80%), but with each subsequent trip, she returns with proportionally more nectar (59, 136). On her final foraging trip, a female invariably returns with just nectar (58, 59, 70). Regurgitated atop the provision mass, this nectar constitutes the young larva's largely liquid diet (136). The final provision mass has a wet pasty consistency, weighs about 90–94 mg, and consists of 33%–36% pollen and 64%–67% nectar by weight (18, 59). It contains 1.3 million pollen grains and is 47% sugar by weight (18). Male progeny receive 17% less provision than do females (59). Larval provision masses of ALCBs contain diverse aerobic bacteria, filamentous fungi, and yeasts, but neither their removal by irradiation (50) nor individual restoration to sterile provisions (49) affected larval performance.

Under ideal greenhouse conditions with abundant sweetclover (*Melilotus officinalis*),

bloom for forage, ALCB females sometimes laid two eggs per day, and each female completed on average 57 cells with eggs over their 7–8 week life spans (71). Far fewer cells result when floral resources become limited (92, 96). Interference between females, or excessive male harassment of females, slows the pace of many nesting activities and curtails reproductive output, at least in cages (70, 112).

The immature ALCB develops rapidly (**Figure 1**). Embryogenesis takes 2–3 days, followed by five larval instars (136). The first instar is spent inside the egg chorion from which the second instar larva hatches to begin feeding

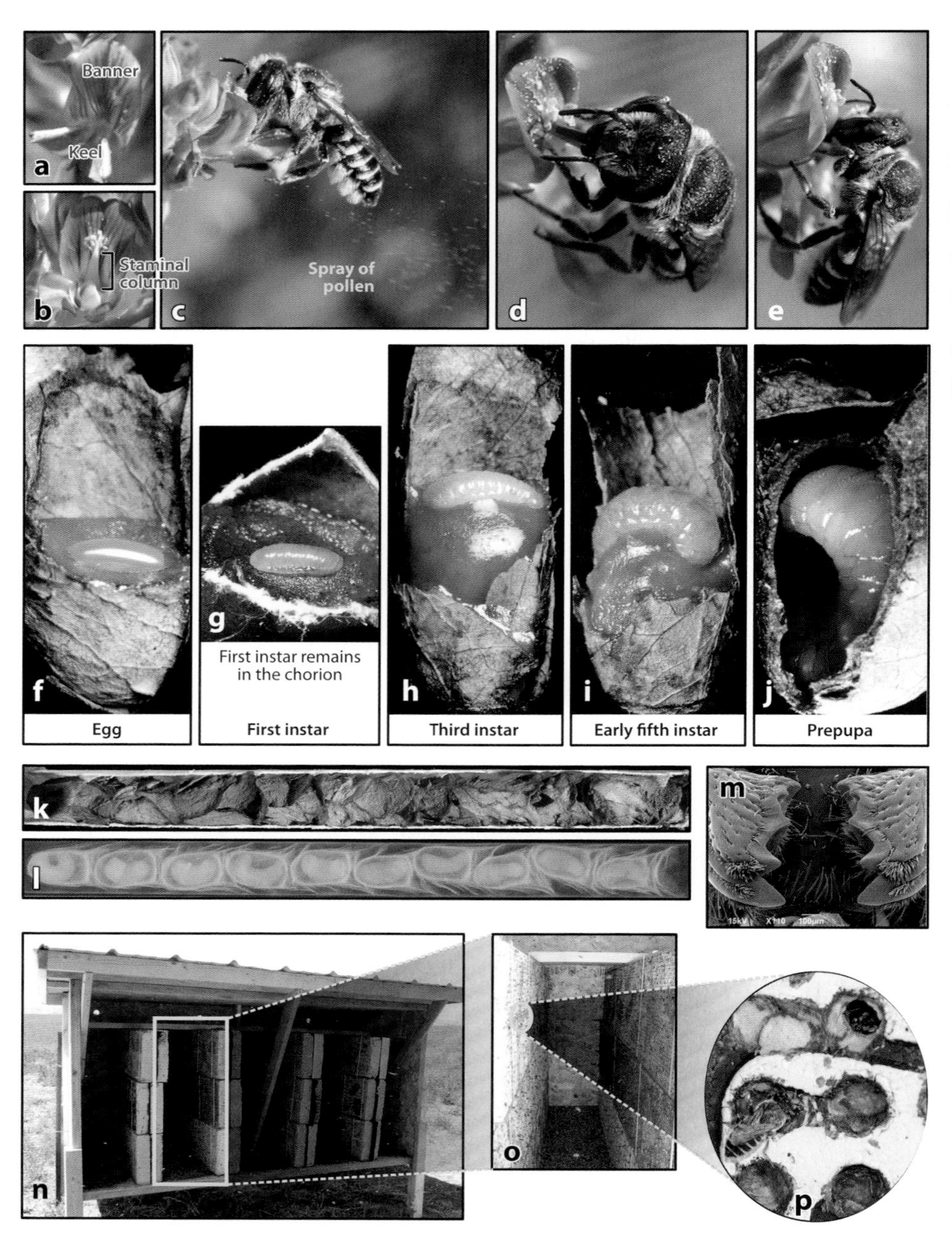

(136). As with *A. mellifera*, young ALCB larvae consume a liquid diet, but they imbibe it from a shallow trough that the larva carves atop its provision mass (136). After 3–4 more days of development, the larva molts to its final fifth instar. In just the next three days, it eats the entire pollen provision. Few insects rival ALCB nitrogen and energy assimilation efficiencies (146); moreover, only 2% of the provision is left uneaten. Only when the larva is done feeding does it defecate (76). As with other megachilids and *Apis* species (among other long-tongued bees) (76), the larva then weaves a tough multilayered cocoon of secreted silk of unknown composition.

Typical of other *Megachile* and many summer-flying bees (76), the ALCB brood overwinters as a postfeeding, diapausing larva, the prepupa. Diapause terminates with warming conditions of late spring or early summer, or when they are artificially incubated (see Commercial Management, below). Alternatively, instead of diapause, up to half of the early ALCB summer brood completes development to yield a second generation before late summer (48, 62, 126). This facultative bivoltinism is irreversible (131). The cues and mechanism(s) that induce diapause or bivoltinism are unknown, but suspected factors that elicit summer emergence include maternal and/or larval responses to long photoperiods, excessive heat, and poor larval nutrition, as well as maternal inheritance (57, 84, 86, 101, 127, 132).

Facultative bivoltinism: refers to some progeny emerging as a second generation in the same reproductive season

MATING

The courtship behaviors and mating biology of the ALCB resemble those of many solitary bees (24). Male ALCBs are usually smaller than females (83), having received smaller provisions as larvae (32). Females vary more in size than males (83), perhaps because daughters are more likely to receive less food when forage is sparse (92). Like most bees (111), the ALCB is protandrous. The first adult males emerge 1–3 days before females regardless of thermal regime (56, 109). Protandry facilitates an orderly exodus from a linear cavity nest whose male cells are nearer the entrance (**Figure 1**), as well as early access to receptive females. Sib-mating within intact ALCB nests is not reported. Confined to desktop cages, some female ALCBs mated and their spermathecae received sperm within hours of emergence (61). However, when newly emerged marked females were released in the field, only one-third of recaptured females were inseminated within 48 h;

Figure 1

Tripping alfalfa flower, alfalfa leafcutting bee (ALCB) life cycle, and ALCB field domicile. (*a*) Banner petal and keel of untripped alfalfa flower. (*b*) Staminal column snaps upward to banner petal when tripped by a bee. (*c*) The ALCB female lands on a flower and applies pressure to keel; the flower is tripped; pollen is exposed and released. (*d*) Extended proboscis of ALCB female while taking nectar. (*e*) ALCB female always probes flower for nectar. (*f*) ALCB egg laid atop provision mass of pollen and nectar that is enveloped in leaf pieces. (*g*) First instar usually remains inside the chorion; second instar then hatches to imbibe nectar pooled atop the provision. (*h*) Third and fourth instars continue feeding but do not defecate. (*i*) Fifth instar consumes final portions of the provision and defecates before spinning a cocoon. (*j*) In the cocoon stage, the postfeeding late fifth instar is called a prepupa. (*k*) ALCB cells are arranged linearly within a nest, with female cells made before male cells (left to right = back to front). (*l*) An X-radiograph of the same nest. (*k*) reveals healthy diapausing prepupae, with larger females in the first (*left*) cells. (*m*) Scanning electron micrograph of female ALCB mandibles. Beveled, chisel-like opposing mandibular edges are used to cut leaf ovals and discs for lining nest cavities and capping individual nest cells, respectively. (Micrograph courtesy of N. Dakota State Univ., Fargo, North Dakota.) (*n*) ALCB domicile aside a blooming alfalfa field. (*o*) Crowded polystyrene nesting boards with thousands of ALCBs (*black dots*) seeking nest cavities. (*p*) Close-up view of ALCB nesting board surface showing a female completing a nest plug made of leaf discs; the face of another female peers out from a cavity from which second-generation bees chewed through the leaf plug and have already emerged. (Except where noted, all images are property of USDA ARS Bee Biology & Systematic Laboratory.)

Loose cell bee management system: nest cells are removed from cavities of bees' nesting boards and then are separated, cleaned, and sorted using specialized equipment, rather than leaving cells within nests in boards

all were inseminated within a week (106). Mating is commonly seen at commercial nesting shelters (32), but its frequency at flowers is not known, particularly for bees emerging from intact natal nests at natural nesting densities. Territoriality is not evident, but males do pounce on perched females and other males.

ALCB males have simple genitalia (85), and copulation for ALCBs is apparently brief (8–10 s) (112). Copulation is preceded by male wing flipping and accompanied by pulsed buzzing and lateral sweeps of his antennae (32). Males firmly grip females in a stereotypical embrace that employs distinctive forecoxal spines and basal mandibular protrusions (147) but without the paddle-like foretarsi of some other *Megachile* species (147). Putative female sex pheromones of ALCBs are reported among their cuticular alkenes (88). Glandular regions of the male's foretarsi are pressed against the female's antennae during mating (147), but the possible exudates have not been isolated or studied for their functionality.

Males undoubtedly inseminate multiple females when they can (polygyny), as seems ubiquitous with non-*Apis* bees (24). As for many solitary bees (89), female ALCBs seem to be monogamous (31) and, in confinement at least, physically reject later suitors (61). Allozyme markers (74) and early genetic studies using RAPD markers indicated singular paternity among nestmates (6). Oogenesis can begin without mating in ALCBs, but as is the case for many other insects (145), eggs will not mature unless the female eats pollen (106). Overall, the mating biology of ALCBs seems unexceptional for *Megachile* (147) and solitary bees in general.

COMMERCIAL MANAGEMENT

Several works over the decades have described ALCB management for alfalfa seed production (8, 28, 46, 47, 105, 119, 120). Unless otherwise cited, most of the following information is from comprehensive guides by Richards (105) and Frank (28) written for alfalfa pollination in Canada. U.S. bee managers tailor Canadian guidelines for their climate to maximize alfalfa seed yield. Depending on the age, quality, and production potential of the alfalfa, growers seek to optimize the ALCB stocking rate for pollination. Because of their higher seed yield potential, U.S. producers use more bees (100,000–150,000 bees per hectare) than do Canadian producers (50,000–75,000 bees per hectare) (96). A loose cell bee management system is most commonly practiced in North America, consisting of four sequential phases: spring/summer incubation, summer brood production, fall/winter cleanup, and winter storage.

ALCB prepupae spend the winter in cold storage (at 4°C–5°C for 7–10 months), usually as cells removed from their nests (i.e., loose cells). In the spring, nest cells bearing prepupae are poured into large, shallow trays to incubate at constant 30°C and 50%–60% relative humidity for safe, synchronous, and timely adult emergence. Emergence is timed for when alfalfa bloom is expected to be at 25%–50% (3, 28). Prior to bee emergence, wasp parasitoids emerge and must be controlled (see Parasitoids and Predators, below). Male ALCBs emerge by day 17–20, and females emerge a few days later (97). Once ~75% of females have emerged, bees are taken to domiciles in blooming alfalfa fields for release from incubation trays. However, if weather is unfavorable, alfalfa bloom is delayed, or an insecticide application is needed for alfalfa pests, deliberate cooling (15°C–20°C) during incubation will slow adult emergence.

Prior to bee release, field domiciles containing nesting boards are readied. Occurring in many shapes, sizes, and materials, domiciles can be uniformly dispersed throughout the fields or placed around field borders. The domicile opening is oriented southeastward (121) so that bees and brood are warmed by the morning sun but shaded during the heat of the day. Nest materials are wood or polystyrene boards containing evenly spaced holes (hole length is 95–150 mm; diameter is 5–7 mm) (32) (**Figure 1**). Bees will only fly once the sun has risen and ambient temperature has warmed to 21°C; flight ceases at twilight, regardless of temperature (65, 125).

Most bees immediately fly out of the opened trays into the blooming field. However, bees that have not eclosed that continue to develop under field conditions emerge slowly and may even die. Nesting begins a few days after release and continues for about 11 weeks, although later activity may represent only the second-generation bees. After bee activity subsides, brood-filled nesting boards are moved from the field to a shop or incubator for storage.

Conscientious bee management using controlled conditions during the fall is practiced more in Canada than in the United States. Nests are allowed to dry in boards, with some of the larvae maturing to the prepupal stage if kept warm enough (>15°C); parasitoids may need to be controlled. Bees are gradually cooled for winter storage (about 5°C), arresting development. During fall and winter, bee nests are removed from boards by stripping them from grooved wooden boards or "punching" them from polystyrene molded boards. This process is followed by cell tumbling, breaking, and separation, which are mechanized techniques for minimizing extraneous matter and helping to rid the bee stock of chalkbrood (see Chalkbrood Disease, below). During winter storage, stocks are sampled for the percent of live progeny and their sex ratio. Healthy, diseased, parasitized, and other dead bees can be diagnosed from X-radiographs of cells or manual cell dissections (123).

ALCB management is strikingly different from honey bee management. Migratory honey beekeepers move hives to follow bloom, but moving ALCB domiciles or boards disorients most nesting females and increases dispersal. Because ALCB adults are active for only about two months, most ALCB care focuses on prepupae in nest cells. Farmers can manage ALCBs amid regular duties or can contract with specialists to handle all ALCB management, in the same way farmers can rent honey bee hives.

To control *A. mellifera* hive pests, pesticides are placed in direct contact with adult bees. For ALCBs, most pests are controlled while larvae are protected in their cells and no adults are present (see Parasitoids and Predators, below). Nevertheless, ALCB adults and brood may be exposed to pesticides and fungicides applied to the crop plant during the nesting season. Bees may be killed or sublethally affected if toxins are sprayed during daylight or if active residues persist (1, 110). Poisoned adult bees are evidenced by their corpses or irregular behavior, but it is difficult to ascertain if brood succumb to toxins that have reached them directly in their nests, as provision contaminants, or through transovarial transport to eggs from mother bees after their topical or oral exposure to active chemicals (79). Generally, pesticides toxic to honey bees are toxic to ALCBs, bumble bees, and wild bees, although LD_{50} levels can differ (64, 110, 134). Toxicological screening of new pesticide formulations and classes are vital for safekeeping all bees whose different life histories and foraging behaviors may render some more susceptible than others (e.g., ALCBs handling contaminated leaves) (1, 79, 90, 118, 134).

CHALKBROOD DISEASE

ALCBs are attacked by disease pathogens just like other bees (5, 52). The most common pathogen that infects ALCB brood is the fungus *Ascosphaera aggregata* (Ascomycete), which causes chalkbrood (34, 35). The related *A. apis* causes chalkbrood in honey bees (20), but the two *Ascosphaera* species do not cross-infect. In the 1970s, chalkbrood disease was devastating to ALCB production; controls used today have lessened disease impact (≤20% of U.S. brood) (36, 98). In Canada, chalkbrood is stringently controlled (≤2%) (28), and Canadian bees are sold to the United States, but not vice versa.

Larval cadavers filled with *A. aggregata* spores are the source of future disease. When emerging bees are trapped in nests behind chalkbrood cadavers, they must chew through the cadavers and thus are dusted with fungal spores. Spore-laden adults may then contaminate provisions of their own brood and also may contaminate other bees through physical contact or by depositing spores in tunnels when investigating future nest cavities

(102, 141). Although loose cell management keeps adults from chewing through cadavers (105), spores erupting from cadavers during cell removal and processing may contaminate loose cells (53). In addition, where ALCB bivoltinism is prevalent in the United States, the spread of chalkbrood disease is exacerbated by summer-emerging bees that must chew through any spore-filled siblings (140).

Canadian managers use paraformaldehyde fumigation to kill fungal spores on loose cells, trays, nesting boards, and any bee equipment (35). Paraformaldehyde is not registered for this use in the United States, so U.S. managers have tried to cleanse nesting boards with methyl bromide fumigation, heat (wood boards only), and chlorine dips (51). New chalkbrood controls are desired because paraformaldehyde is carcinogenic, methyl bromide is now banned, and heat and chlorine treatments are difficult and labor-intensive.

PARASITOIDS AND PREDATORS

ALCBs host many natural enemies native to both North America and Eurasia (8, 25). Fortunately, no pest that accompanied ALCBs from Europe has affected North American native bees. Most knowledge of ALCB parasitism and predation comes from managed populations. Where the ALCB has been used around the world, the same groups of parasitoids and predators have followed or adapted (63, 108, 126, 148).

Up to 20% of ALCB cells produced in U.S. western states can be parasitized by wasps (98), which are known natural enemies of other *Megachile* and *Osmia* spp. (25, 60) in their native ranges. The prevalent parasitoids are the European *Pteromalus venustus* (Pteromalidae) (25) and the native minute *Tetrastichus megachilidis* (Eulophidae). Also, the European *Monodontomerus aeneus* (Torymidae) (40, 60) once, but no longer, devastated ALCB populations. Adult parasitoids are typically killed with dichlorvos pest strips (73) that are placed in storage or incubator chambers, but that do not harm ALCBs enclosed in nest cells. Ultraviolet lamps above liquid traps also can be effective. During nesting, parasitoids are physically deterred from nests by the thick walls of artificial cavities and by felt cloth tightly affixed to the back of nesting boards.

Sapyga pumila (Sapygidae) is a North American wasp that attacks native bees in several genera and quickly adopted the ALCB as a host (25, 60). A trap designed for this wasp (2, 25, 135) is seldom used because *S. pumila* parasitism is no longer a major concern. A trap also exists, but is rarely used, for the checkered flower beetle, *Trichodes ornatus* (Cleridae) (22, 67). Larvae of these beetles are pests during the nesting season and while bees are stored (25). They attack other megachilids, bees in other families, and some wasps (67).

Less persistent or problematic parasitoids, predators, and pests include the wasp parasitoids *Melittobia chalybii* and *M. acasta* (Eulophidae) (73), six native *Coelioxys* species (60), one *Stelis* species (cuckoo bees) (Megachilidae), the parasitic beetle *Nemognatha lutea* (Meloidae), and the beetles *Trogoderma* spp. (Dermestidae) and *Tribolium* spp. (Tenebrionidae), which eat immature bees and their provisions (25). Predators also include yellowjacket wasps (Vespidae: *Vespula* spp.), earwigs (Forficulidae: *Forficula auricularia*), ants (Formicidae), birds, and rodents (25).

As with honey bees, nest and brood destroyers can be problematic year-round. Unlike honey bees, ALCB adults are present only in summer with their brood sealed inside leafy capsules. Thus, pests can be treated with insecticides while ALCBs are protected within their cells.

UNEXPLAINED MORTALITY

ALCB populations are difficult to sustain in the United States, so U.S. producers import ALCBs from Canada to supplement or completely supply their bees (95). Although all bee stocks suffer pestilence (chalkbrood, parasites, and predators), unexplained mortality occurs more often in U.S. than Canadian populations.

Poor management may account for some U.S. bee mortality. Compared to Canadian samples, more U.S. ALCBs died over the winter and during incubation, with most bees dying as prepupae (97). Although optimal durations and temperatures for winter storage and incubation are reliable (56, 100, 109, 137, 138), less is known about potential bee losses due to conditions during the handling of completed nests prior to winter storage (57, 99). The duration of prewinter storage at the recommended temperature of 16°C can subtly affect bee survival to adulthood (99), but effects of excessive heat for various storage durations need study. Storage differences may underlie the superior survival of Canada-raised bees.

Unexplained brood mortality in the United States includes pollen ball (at times 40%–60% of brood), which are cells containing the pollen and nectar provision at a time when the provision should have been consumed by a larva. Many pollen ball cells lack eggs or larvae; others contain collapsed eggs, dead larvae, or fungi (93). Causes of pollen balls probably include unsuitable microclimates and overly dense bee populations (96, 98). Overstocked bees used in U.S. alfalfa fields quickly drain floral resources, suffer crowded nesting sites, and may increase the spread of disease (41, 52). Depleted resources constrain ALCB reproduction (91, 92, 96), and adults likely perish or depart crowded sites. Often under such conditions during the hot U.S. summers, less than half of ALCBs released into fields are replaced through reproduction. Other suspected causes of larval mortality include molds, viruses, other as yet identified pathogens, and pesticides (34, 35).

Second-generation emergence in the United States can reach 90% for the first nests made in the season, tapering to none for the season's last nests (55); over an entire season, summer emergence can total about 40% progeny loss. In contrast, summer adult emergence in Alberta, Canada, is quite low (e.g., <5%) (62). Reproduction by second-generation bees is constrained by sparse floral resources on farms and their progeny's race to become prepupae before cool weather arrests their development.

VISUAL AND CHEMICAL CUES FOR ALFALFA LEAFCUTTING BEES

Just as honey bees orient to a hive, ALCBs must orient to their nests after foraging bouts. Large field domiciles can serve as long-distance visual cues for ALCBs flying over seed fields (32, 105, 121). Once at the domicile, a female bee finds her nest among thousands of holes using three-dimensionality, color contrast, and color patterns (27, 41, 42). If visual cues are manipulated, females become temporarily disoriented. Poor local orientation can result in wasted time, an increase in bee collisions, dropped nest-building materials, or further spread of disease that diminishes efficient, productive nesting.

Pollen ball: intact pollen and nectar provision mass remaining at end of the field season

Olfactory cues detected in close proximity or upon antennation seem to be important when initiating or recognizing nests. Females prefer once-used nesting boards over new ones, implicating olfactory attractants at nest initiation (14, 26, 87, 142). Isolating effective cues from the old nests has proven difficult. In laboratory Y-tube assays, ALCB females were attracted to a complete nest cell (after the adult emerged), larval feces, and solvent-extracted components from leaf pieces that once lined a cell (94). However, field bioassays were less conclusive, finding that only cavities cued with feces and a cocoon were somewhat more attractive than other components (142). Fractionation of chemicals from old nest components and testing them in field assays are needed for understanding chemical mediation of nest initiation and aggregation.

A nesting ALCB female distinguishes her own nest from others. ALCB females mark their nests by dragging the abdomen throughout the cavity (32, 43). Replacing an outer bee-marked section of a nest cavity with a clean section causes a returning female to be confused and hesitant to enter, indicating an individual nest recognition pheromone. Cuticular waxes and/or the Dufour's gland may be the origin of the nest recognition cue (13). For honey bees, in which many individuals need to recognize the colony odor to discern kin from nonkin, recognition requires them to learn odors both from cuticles of nestmates and from the colony's wax comb (11). Solitary bees need to recognize only their own nests.

ALFALFA LEAFCUTTING BEES AS ECOLOGICAL DISRUPTORS

Exotic species can become ecologically disruptive invaders. Feral honey bees have penetrated far into wildlands (77), as have introduced European *Bombus terrestris* that are now escaping around the world (39). Among the few surveys of U.S. wildland bee communities where the ALCB might be expected, it has been rare or absent. In Wyoming short-grass prairie, none of the 13 species of *Megachile* (52 species of Megachilidae) were ALCBs (128). ALCB was similarly absent from the 125 bee species netted in tallgrass prairie remnants and restorations (103). In the shrub-steppe of eastern Oregon, 12 species of Megachilidae were trap-nested, but no ALCB was found (30). At a reserve amid intensive agriculture in central California, the ALCB was more common, but occupied only 3%–4% of available nesting cavities; it was outnumbered both by native cavity-nesters and by another escaped Eurasian species of *Eutricharaea*, *M. apicalis*, known to aggressively compete for nesting sites (4, 122). The ALCB is also rare or absent in southern Europe, even in alfalfa seed fields. The ALCB seems not to venture far from agricultural or otherwise disturbed sites in North America and so appears to be an inconsequential competitor with native bees for nesting sites or floral resources.

Some nonnative bees prefer and pollinate nonnative forbs, contributing to the invasions of weedy species that can outcompete native wildflowers for pollination services (38). Feral honey bees and feral bumble bees have been widely implicated in pollinating invasive Eurasian weeds in the Americas, New Zealand, Tasmania, and Australia (39). ALCBs also avidly visit several Eurasian weeds in North America, notably sweetclovers (*Melilotus alba* and *M. officinalis*) (58, 81) and purple loosestrife (*Lythrum salicaria*), preferring both to alfalfa in choice tests (114). However, these and other exotic and invasive weeds are eagerly sought and pollinated by honey bees and bumble bees too (12).

A MODEL SOLITARY BEE

Management of ALCBs is an exemplary system for solitary bee commercial pollination. Fortuitously, ALCBs are efficient crop pollinators that also thrive as aggregating cavity-nesters in anthropogenic habitats. In contrast, the solitary ground-nesting alkali bee is restricted in its use as a managed alfalfa pollinator because it requires specific soil moisture, temperature, and alkalinity. Few growing areas naturally or artificially can satisfy such requirements (95).

Where farming is intense and an abundance of pollinating bees is needed, managed cavity-nesting bees are advantageous. Cavity-nesters are transportable as packaged nests or loose cells; some species tolerate nesting in large aggregations. With progeny in nests or as loose cells, dead or parasitized cells can be culled, cells and management equipment can be treated for diseases, and the bee number and sex ratio can be determined. Other megachilids (e.g., *Osmia* spp. for orchard and berry crops) can be used as pollinators (9), but today they are not available at the scale of ALBCs (9, 17).

Just as honey bees and bumble bees represent the social bees for studies in evolution, genetics, behavior, learning, neurophysiology, and ecology (e.g., 33, 75), the readily available ALCB can serve as a model for solitary bees. ALCB laboratory maintenance is relatively simple, and adults can be kept to emerge through most of the year. Beyond management research, ALCB studies already include the effects of male harassment on reproductive success and the lack of response to proboscis extension reflex elicitation used in learning and conditioning experiments (112, 143). Recent reports have identified ALCB genes related to diapause regulation and bee immunity (149, 150). Future ALCB studies might reveal solitary bee commonalities in recognition, aggregation behavior, mating strategies and courtship behaviors, and pheromone use. On the horizon are genome sequences of ALCB and other bees from which ancestral and shared traits in solitary, subsocial, and primitively eusocial bees may be found and thus contribute to the understanding of the mechanisms and evolutionary origins of bee sociality (115).

SUMMARY POINTS

1. Accidentally introduced into the United States, the ALCB has become the world's most effectively used and intensely managed solitary bee. Its use transformed the alfalfa seed industry in North America. The most common management practice is the loose cell system.
2. Traits of ALCBs that contribute to their commercialization include their gregarious nature; philopatry; use of leaves for lining nests; ready acceptance of cheap, mass-produced nesting materials; and pollination efficacy at and emergence synchrony with alfalfa bloom.
3. The ALCB became a commercial success because its natural history was studied, targeted research was performed, and producer ingenuity was encouraged. ALCB management is a model system for commercializing other solitary bees and for advancing new testable hypotheses in diverse biological disciplines.

DISCLOSURE STATEMENT

The authors are not aware of any affiliations, memberships, funding, or financial holdings that might be perceived as affecting the objectivity of this review.

ACKNOWLEDGMENTS

We thank Gordon Frank for insights into ALCB pollination of canola. Vincent Tepedino and James Pitts kindly provided helpful reviews.

LITERATURE CITED

1. Abbott VA, Nadeau JL, Higo HA, Winston ML. 2008. Lethal and sublethal effects of imidacloprid on *Osmia lignaria* and clothianidin on *Megachile rotundata* (Hymenoptera: Megachilidae). *J. Econ. Entomol.* 101:784–96
2. Arnett WH. 1980. Alfalfa leafcutting bee parasite *Sapyga pumula* trap: effects of color and domicile placement on catch. *Environ. Entomol.* 9:783–84
3. Baird CR, Bitner RM. 1991. *Loose cell management of alfalfa leafcutting bees in Idaho*. Moscow: Univ. Idaho Coll. Agric., Curr. Inf. Ser. 588
4. Barthell JF, Frankie GW, Thorp RW. 1998. Invader effects in a community of cavity nesting megachilid bees (Hymenoptera: Megachilidae). *Environ. Entomol.* 27:240–47
5. Batra LR, Batra SWT, Bohart GE. 1973. The mycoflora of domesticated and wild bees (Apoidea). *Mycopathol. Mycol. Appl.* 49:13–44
6. Blanchetot A. 1992. DNA fingerprinting analysis in the solitary bee *Megachile rotundata*: variability and nest mate genetic relationships. *Genome* 35:681–88
7. Bohart GE. 1957. Pollination of alfalfa and red clover. *Annu. Rev. Entomol.* 2:355–80
8. Bohart GE. 1972. Management of wild bees for the pollination of crops. *Annu. Rev. Entomol.* 17:287–312
9. Bosch J, Sgolastra F, Kemp WP. 2008. Life cycle ecophysiology of *Osmia* mason bees used as crop pollinators. In *Bee Pollination in Agricultural Ecosystems*, ed. RR James, TL Pitts-Singer, 6:83–104. New York: Oxford Univ. Press. 232 pp.

10. Bradner NR, Frakes RV, Stephen WP. 1965. Effects of bee species and isolation distance on possible varietal contamination in alfalfa. *Agron. J.* 57:247–48
11. Breed MD. 1998. Chemical cues in kin recognition: criteria for identification, experimental approaches, and the honey bee as an example. In *Pheromone Communication in Social Insects*, ed. RK Vander Meer, MD Breed, ML Winston, KE Espelie, 3:57–78. Boulder, CO: Westview. 368 pp.
12. Brown BJ, Mitchell RJ, Graham SA. 2002. Competition for pollination between an invasive species (purple loosestrife) and a native congener. *Ecology* 83:2328–36
13. Buckner JS, Pitts-Singer TL, Guédot C, Hagen MM, Fatland CL, Kemp WP. 2009. Cuticular lipids of female solitary bees, *Osmia lignaria* Say and *Megachile rotundata* (F.) (Hymenoptera: Megachilidae). *Comp. Biochem. Physiol. B* 153:200–5
14. Buttery RG, Parker FD, Teranishi R, Mon TR, Ling LC. 1981. Volatile components of alfalfa leaf-cutter bee cells. *J. Agric. Food Chem.* 29:955–58
15. Cane JH. 2002. Pollinating bees (Hymenoptera: Apiformes) of U.S. alfalfa compared for rates of pod and seed set. *J. Econ. Entomol.* 95:22–27
16. Cane JH. 2006. An evaluation of pollination mechanisms for purple prairie-clover, *Dalea purpurea* (Fabaceae: Amorpheae). *Am. Midl. Nat.* 156:193–97
17. Cane JH. 2008. Pollinating bees crucial to farming wildflower seed for U.S. habitat restoration. In *Bee Pollination in Agricultural Ecosystems*, ed. RR James, TL Pitts-Singer, 4:48–64. New York: Oxford Univ. Press. 232 pp.
18. Cane JH, Gardner D, Harrison P. 2011. Nectar and pollen sugars constituting larval provisions of the alfalfa leaf-cutting bee (*Megachile rotundata*) (Hymenoptera: Apiformes: Megachilidae). *Apidologie*. In press
19. Cane JH, Schiffhauer D. 2003. Dose-response relationships between pollination and fruiting refine pollinator comparisons for cranberry (*Vaccinium macrocarpon* [Ericaceae]). *Am. J. Bot.* 90:1425–32
20. Christensen M, Gilliam M. 1983. Notes on the *Ascosphaera* species inciting chalkbrood in honey bees. *Apidologie* 14:291–97
21. Cox RL, Abel C, Gustafson E. 1996. A novel use of bees: controlled pollination of germplasm collections. *Am. Bee J.* 136:709–12
22. Davis HG, George DA, McDonnough LM, Tamaki G, Burditt AK. 1983. Checkered flower beetle (Coleoptera: Cleridae) attractant: development of an effective bait. *J. Econ. Entomol.* 76:674–75
23. Dylewska M, Jablonski B, Sowa S, Bilinski M, Wrona S. 1970. An attempt of determination of the number of bees (*Hym., Apoidea*) needed for adequate pollination of alfalfa. *Pol. Pismo Entomol.* 40:371–98
24. Eickwort GC, Ginsberg HS. 1980. Foraging and mating behavior in Apoidea. *Annu. Rev. Entomol.* 25:421–46
25. Eves JD, Mayer DF, Johansen CA. 1980. Parasite, predators, and nest destroyers of the alfalfa leafcutting bee, *Megachile rotundata*. *West. Reg. Ext. Publ.* 32:1–15
26. Fairey DT, Lieverse JAC. 1986. Cell production by the alfalfa leafcutting bee (*Megachile rotundata*) in new and used wood and polystyrene nesting materials. *J. Appl. Entomol. Z. Angew. Entomol.* 102:148–53
27. Fauria K, Campan R, Grimal A. 2004. Visual marks learned by the solitary bee *Megachile rotundata* for localizing its nest. *Anim. Behav.* 67:523–30
28. Frank G. 2003. *Alfalfa Seed and Leafcutter Bee Production & Marketing Manual*. Brooks, Alberta, Can.: Irrig. Alfalfa Seed Prod. Assoc. 160 pp.
29. Free JB. 1993. *Insect Pollination of Crops*. New York: Academic. 684 pp.
30. Frohlich DR, Clark WH, Parker FD, Griswold TL. 1988. The xylophilous bees and wasps of a high, cold desert: Leslie Gulch, Oregon (Hymenoptera: Apoidea, Vespoidea). *Pan-Pac. Entomol.* 64:266–69
31. **Gerber HS, Klostermeyer EC. 1970. Sex control by bees: a voluntary act of egg fertilization during oviposition. *Science* 167:82–84**

31. First documentation of female control of progeny sex in a solitary bee.

32. Gerber HS, Klostermeyer EC. 1972. Factors affecting the sex ratio and nesting behavior of the alfalfa leafcutter bee. *Wash. Agric. Exp. Stn. Tech. Bull.* 73:1–11
33. Giurfa M. 2007. Behavioral and neural analysis of associative learning in the honeybee: a taste from the magic well. *J. Comp. Physiol. A* 193:801–24
34. Goerzen DW. 1989. *Foliar Moulds in Alfalfa Leafcutting Bees*. Can.: Saskatchewan Agric. Dev. Fund Ext. Bull. (ISBN 0-88656-534-0) 4 pp.

35. Goerzen DW. 1991. Microflora associated with the alfalfa leafcutting bee, *Megachile rotundata* (Fab.) (Hymenoptera: Megachilidae) in Saskatchewan, Canada. *Apidologie* 22:553–61
36. Goerzen DW, Watts TC. 1991. Efficacy of the fumigant paraformaldehyde for control of microflora associated with the alfalfa leafcutting bee, *Megachile rotundata* (Fabricius) (Hymenoptera: Megachilidae). *Bee Sci.* 1:212–18
37. Gonzalez VH, Engel MS, Hinojosa-Diaz IA. 2010. A new species of *Megachile* from Pakistan, with taxonomic notes on the subgenus *Eutricharaea* (Hymenoptera: Megachilidae). *J. Kans. Entomol. Soc.* 83:58–67
38. Goodell K. 2008. Invasive exotic plant-bee interactions. In *Bee Pollination in Agricultural Ecosystems*, ed. RR James, TL Pitts-Singer, 10:166–83. New York: Oxford Univ. Press. 232 pp.
39. Goulson D. 2003. Effects of introduced bees on native ecosystems. *Annu. Rev. Ecol. Evol. Syst.* 34:1–26
40. Grissell EE. 2007. Torymidae (Hymenoptera: Chalcidoidea) associated with bees (Apoidea), with a list of chalcidoid bee parasitoids. *J. Hymenopt. Res.* 16:234–65
41. Guédot C, Bosch J, James RR, Kemp WP. 2006. Effects of three-dimensional and color patterns on nest location and progeny mortality in alfalfa leafcutting bee (Hymenoptera: Megachilidae). *J. Econ. Entomol.* 99:626–33
42. Guédot C, Bosch J, Kemp WP. 2005. The relative importance of vertical and horizontal visual cues in nest location by *Megachile rotundata*. *J. Apic. Res.* 44:109–15
43. Guédot C, Pitts-Singer TL, Buckner JS, Bosch J, Kemp WP. 2005. Use of olfactory cues for individual nest recognition in the solitary bee *Megachile rotundata* (F.) (Hymenoptera: Megachilidae). Proc. Winter Seed Sch. Conf. Northwest Alfalfa Seed Growers Assoc. Meet., 30 Jan. – 1 Feb. pp. 45
44. Hasenkamp KR. 1974. Ökophysiologische und ökethologische Untersuchungen an Blattschneiderbienen. *Forma Funct.* 7:139–78
45. Hobbs GA. 1956. Ecology of the leaf-cutter bee *Megachile perihirta* Ckll. (Hymenoptera: Megachilidae) in relation to production of alfalfa seed. *Can. Entomol.* 87:625–31
46. Hobbs GA. 1965. Importing and managing the alfalfa leaf-cutter bee. *Publ. No. 1209*. Ottawa, Ontario: Can. Dep. Agric.
47. Hobbs GA. 1967. Domestication of alfalfa leaf-cutter bees. *Publ. No. 1313*. Ottawa, Ontario: Can. Dep. Agric.
48. Hobbs GA, Richards KW. 1976. Selection for a univoltine strain of *Megachile* (Euthricharea) *pacifica* (Hymenoptera: Megachilidae). *Can. Entomol.* 108:165–67
49. Inglis GD, Goettel MS, Sigler L. 1993. Influence of microorganisms on alfalfa leafcutter bee (*Megachile rotundata*) larval development and susceptibility to *Ascosphaera aggregata*. *J. Invertebr. Pathol.* 61:236–43
50. Inglis GD, Goettel MS, Sigler L, Borsa J. 1992. Effects of decontamination of eggs and γ-irradiation of provisions on alfalfa leafcutter bee (*Megachile rotundata*) larvae. *J. Apic. Res.* 31:15–21
51. James RR. 2005. Impact of disinfecting nesting boards on chalkbrood control in the alfalfa leafcutting bee. *J. Econ. Entomol.* 98:1094–100
52. James RR. 2008. The problem of disease when domesticating bees. In *Bee Pollination in Agricultural Ecosystems*, ed. RR James, TL Pitts-Singer, 8:124–41. New York: Oxford Univ. Press. 232 pp.
53. James RR, Pitts-Singer TL. 2005. *Ascosphaera aggregata* contamination on alfalfa leafcutting bees in a loose cell incubation system. *J. Invertebr. Pathol.* 89:176–78
54. Javorek SK, MacKenzie KE, Vander Kloet SP. 2002. Comparative pollination effectiveness among bees (Hymenoptera: Apoidea) on lowbush blueberry (Ericaceae: *Vaccinium angustifolium*). *Ann. Entomol. Soc. Am.* 95:345–51
55. Johansen CA, Eves J. 1973. Effects of chilling, humidity and seasonal conditions on emergence of the alfalfa leafcutting bee. *Environ. Entomol.* 2:23–26
56. Kemp WP, Bosch J. 2000. Development and emergence of the alfalfa pollinator *Megachile rotundata* (Hymenoptera: Megachilidae). *Ann. Entomol. Soc. Am.* 93:904–11
57. Kemp WP, Bosch J. 2001. Postcocooning temperatures and diapause in the alfalfa pollinator *Megachile rotundata* (Hymenoptera: Megachilidae). *Ann. Entomol. Soc. Am.* 94:244–50
58. Klostermeyer EC, Gerber HS. 1969. Nesting behavior of *Megachile rotundata* (Hymenoptera: Megachilidae) monitored with an event recorder. *Ann. Entomol. Soc. Am.* 62:1321–26

59. Klostermeyer EC, Mech SJ Jr, Rasmussen WB. 1973. Sex and weight of *Megachile rotundata* (Hymenoptera: Megachilidae) progeny associated with provision weights. *J. Kans. Entomol. Soc.* 46:536–48
60. Krombein KV, Hurd PD, Smith DR, Burks BD. 1979. *Catalog of Hymenoptera in America North of Mexico*. Washington, DC: Smithson. Inst. Press. 2731 pp.
61. Krunic MD. 1971. Mating of alfalfa leafcutter bee, *Megachile rotundata* (F.), in the laboratory. *Can. J. Zool.* 49:778–79
62. Krunic MD. 1972. Voltinism in *Megachile rotundata* (Megachilidae: Hymenoptera) in Southern Alberta. *Can. Entomol.* 104:185–88
63. Krunic MD, Tasei JN, Pinzauti M. 1995. Biology and management of *Megachile rotundata* Fabricius under European conditions. *Apicoltura* 10:71–97
64. Ladurner E, Bosch J, Kemp WP, Maini S. 2005. Assessing delayed and acute toxicity of five formulated fungicides to *Osmia lignaria* Say and *Apis mellifera*. *Apidologie* 36:449–60
65. Lerer H, Bailey WG, Mills PF, Pankiw P. 1982. Pollination activity of *Megachile rotundata* (Hymenoptera: Apoidea). *Environ. Entomol.* 11:997–1000
66. Lewis SE. 1994. Evidence of leaf-cutting bee damage from the Republic sites (Middle Eocene) of Washington. *J. Paleontol.* 68:172–73
67. Linsley EG, MacSwain JW. 1943. Observations on the life history of *Trichodes ornatus* (Coleoptera, Cleridae), a larval predator in the nests of bees and wasps. *Ann. Entomol. Soc. Am.* 36:589–601
68. Linsley EG, MacSwain JW. 1947. Factors influencing the effectiveness of insect pollinators of alfalfa in California. *J. Econ. Entomol.* 40:349–57
69. MacKenzie K, Javorek S. 1997. The potential of alfalfa leafcutter bees (*Megachile rotundata* L.) as pollinators of cranberry (*Vaccinium macrocarpon* Aiton). *Acta Hortic.* 437:345–51
70. Maeta Y, Adachi K. 2005. Nesting behaviors of the alfalfa leaf-cutting bee, *Megachile* (*Eutricharaea*) *rotundata* (Fabricius) (Hymenoptera, Megachilidae). *Chugoku Kontyu* 18:5–21
71. Maeta Y, Kitamura T. 2005. On the number of eggs laid by one individual of females in the alfalfa leaf-cutting bee, *Megachile* (*Eutricharaea*) *rotundata* (Fabricius) (Hymenoptera, Megachilidae). *Chugoku Kontyu* 19:39–43
72. Maeta Y, Kitamura T. 2007. Efficiency of seed production in breeding of ladino clover by three species of wild bees, *Ceratina flavipes* Smith, *Megachile rotundata* (Fabricius) and *Megachile spissula* Cockerell (Hymenoptera, Apidae and Megachilidae). *Chugoku Kontyu* 21:35–53
73. Matthews RW, González JM, Matthews JR, Deyrup LD. 2009. Biology of the parasitoid *Melittobia* (Hymenoptera: Eulophidae). *Annu. Rev. Entomol.* 54:251–66
74. McCorquodale DB, Owen RE. 1997. Allozyme variation, relatedness among progeny in a nest, and sex ratio in the leafcutter bee, *Megachile rotundata* (Fabricius) (Hymenoptera: Megachilidae). *Can. Entomol.* 129:211–19
75. Menzel R. 2001 Behavioral and neural mechanisms of learning and memory as determinants of flower constancy. In *Cognitive Ecology of Pollination: Animal Behavior and Floral Evolution*, ed. L Chittka, JD Thomson, 2:21–40. Cambridge, UK: Cambridge Univ. Press. 343 pp.
76. Michener CD. 2007. *The Bees of the World*. Baltimore, MD: Johns Hopkins Univ. Press. 953 pp.
77. Minckley RL, Cane JH, Kervin L, Yanega D. 2003. Biological impediments to measures of competition among introduced honey bees and desert bees (Hymenoptera: Apiformes). *J. Kans. Entomol. Soc.* 76:306–19
78. Móczár L. 1961. The distribution of wild bees in the lucerne fields of Hungary (Hymenoptera: Apoidea). *Ann. Hist. Nat. Mus. Natl. Hungar.* (*Pars Zool.*) 53:451–61
79. Mommaerts V, Sterk G, Smagghe G. 2006. Hazards and uptake of chitin synthesis inhibitors in bumblebees *Bombus terrestris*. *Pest Manag. Sci.* 62:752–58
80. Morse RA, Calderone NW. 2000. The value of honey bees as pollinators of U.S. crops in 2000. *Bee Cult.* 128:1–15
81. O'Neill KM, O'Neill RP, Blodgett S, Fultz J. 2004. Composition of pollen loads of *Megachile rotundata* in relation to flower diversity (Hymenoptera: Megachilidae). *J. Kans. Entomol. Soc.* 77:619–25
82. Ortiz-Sanchez FJ, Aguirre-Segura A. 1991. Composition and seasonal patterns of a community of Apoidea in Almeria. *EOS* 67:3–22

83. Owen RE, McCorquodale DB. 1994. Quantitative variation and heritability of postdiapause development time and body size in the alfalfa leafcutting bee (Hymenoptera: Megachilidae). *Ann. Entomol. Soc. Am.* 87:922–27

84. Pankiw P, Lieverse JAC, Siemens B. 1980. The relationship between latitude and the emergence of alfalfa leafcutter bees *Megachile rotundata* (Hymenoptera: Megachilidae). *Can. Entomol.* 112:555–58

85. Parker FD. 1978. An illustrated key to alfalfa leafcutter bees *Eutricharaea* (Hymenoptera: Megachilidae). *Pan-Pac. Entomol.* 54:61–64

86. Parker FD, Tepedino VJ. 1982. Maternal influence on diapause in the alfalfa leafcutting bee (Hymenoptera: Megachilidae). *Ann. Entomol. Soc. Am.* 75:407–10

87. Parker FD, Teranishi R, Olson AC. 1983. Influence of attractants on nest establishment by the alfalfa leafcutting bee (Hymenoptera: Megachilidae) in styrofoam and rolled paper. *J. Kans. Entomol. Soc.* 56:477–82

88. Paulmier I, Bagnères AG, Afonso CMM, Dusticier G, Rivière G, Clément JL. 1999. Alkenes as a sexual pheromone in the alfalfa leaf-cutter bee *Megachile rotundata*. *J. Chem. Ecol.* 25:471–90

89. Paxton RJ. 2005. Male mating behaviour and mating systems of bees: an overview. *Apidologie* 36:145–56

90. Peach ML, Alston DG, Tepedino VJ. 1995. Sublethal effects of carbaryl bran bait on nesting performance, parental investment, and offspring size and sex ratio of the alfalfa leafcutting bee (Hymenoptera, Megachilidae). *Environ. Entomol.* 24:34–39

91. Peterson JH, Roitberg BD. 2006. Impacts of flight distance on sex ratio and resource allocation to offspring in the leafcutter bee, *Megachile rotundata*. *Behav. Ecol. Sociobiol.* 59:589–96

92. Peterson JH, Roitberg BD. 2006. Impact of resource levels on sex ratio and resource allocation in the solitary bee, *Megachile rotundata*. *Environ. Entomol.* 35:1404–10

93. Pitts-Singer TL. 2004. Examination of 'pollen balls' in nests of the alfalfa leafcutting bee, *Megachile rotundata*. *J. Apic. Res.* 43:40–46

94. Pitts-Singer TL. 2007. Olfactory response of megachilid bees, *Osmia lignaria*, *Megachile rotundata*, and *M. pugnata*, to individual cues from old nest cavities. *Environ. Entomol.* 36:402–8

95. Pitts-Singer TL. 2008. Past and present management of alfalfa bees. In *Bee Pollination in Agricultural Ecosystems*, ed. RR James, TL Pitts-Singer, 7:105–23. New York: Oxford Univ. Press. 232 pp.

95. Review of alfalfa pollination by two managed solitary bees. Other chapters in book cover different aspects of solitary bees in agriculture.

96. Pitts-Singer TL, Bosch J. 2010. Nest establishment, pollination efficiency, and reproductive success of *Megachile rotundata* (Hymenoptera: Megachilidae) in relation to resource availability in field enclosures. *Environ. Entomol.* 39:149–58

97. Pitts-Singer TL, James RR. 2005. Emergence success and sex ratio of commercial alfalfa leafcutting bees from the United States and Canada. *J. Econ. Entomol.* 98:1785–90

98. Pitts-Singer TL, James RR. 2008. Do weather conditions correlate with findings in failed, provision-filled nest cells of *Megachile rotundata* (Hymenoptera: Megachilidae) in western North America? *J. Econ. Entomol.* 101:674–85

99. Pitts-Singer TL, James RR. 2009. Prewinter management affects *Megachile rotundata* (Hymenoptera: Megachilidae) prepupal physiology and adult emergence and survival. *J. Econ. Entomol.* 102:1407–16

100. Rank GH, Goerzen DW. 1982. Effect of incubation temperatures on emergence of *Megachile rotundata* (Hymenoptera: Megachilidae). *J. Econ. Entomol.* 75:467–71

101. Rank GH, Rank FP. 1989. Diapause intensity in a French univoltine and a Saskatchewan commercial strain of *Megachile rotundata* (Fab.). *Can. Entomol.* 121:141–48

102. Rank GH, Rank FP, Watts R. 1990. Chalkbrood (*Ascosphaera aggregata* Skou.) resistance of a univoltine strain of the alfalfa leafcutting bee, *Megachile rotundata* F. *J. Appl. Entomol.* 109:524–27

103. Reed CC. 1995. Species richness of insects on prairie flowers in southeastern Minnesota. In *Proc. 14th Annu. N. Am. Prairie Conf.*, ed. DC Hartnett, pp. 103–15. Manhattan: Kans. State Univ.

104. Richards K. 1995. The alfalfa leafcutter bee, *Megachile rotundata*: a potential pollinator for some annual forage clovers. *J. Apic. Res.* 34:115–21

105. Richards KW. 1984. Alfalfa leafcutter bee management in Western Canada. *Agric. Can.* 1495/E:1–53

105. A comprehensive guide to using ALCB for alfalfa seed production that includes ALCB life history and is largely based on author's research. A needed update to this guide is Reference 28.

106. Richards KW. 1994. Ovarian development in the alfalfa leafcutter bee, *Megachile rotundata*. *J. Apic. Res.* 33:199–203

107. Richards KW. 1997. Potential of the alfalfa leafcutter bee, *Megachile rotundata* (F.) (Hym., Megachilidae) to pollinate hairy and winter vetches (*Vicia* spp.). *J. Appl. Entomol. Z. Angew. Entomol.* 121:225–29
108. Richards KW, Krunic MD. 1990. Introduction of alfalfa leafcutter bees to pollinate alfalfa in Yugoslavia. *Entomologist* 109:130–35
109. Richards KW, Whitfield GH. 1988. Emergence and survival of leafcutter bees, *Megachile rotundata*, held at constant incubation temperatures (Hymenoptera: Megachilidae). *J. Apic. Res.* 27:197–204
110. Riedl H, Johansen E, Brewer L, Barbour J. 2006. *How to reduce bee poisoning from pesticides. Rep. PWN 591*. Pac. Northwest Ext. Publ., Oregon State Univ., Corvallis
111. Robertson C. 1930. Proterandry and flight of bees (Hymen.: Apoidea). III. *Entomol. News* 41:331–36
112. Rossi BH, Nonacs P, Pitts-Singer TL. 2010. Sexual harassment by males reduces female fecundity in the alfalfa leafcutting bee, *Megachile rotundata*. *Anim. Behav.* 79:165–71
113. Sheffield CS. 2008. Summer bees for spring crops? Potential problems with *Megachile rotundata* (Fab.) (Hymenoptera: Megachilidae) as a pollinator of lowbush blueberry (Ericaceae). *J. Kans. Entomol. Soc.* 81:276–87
114. Small E, Brookes B, Lefkovitch LP, Fairey DT. 1997. A preliminary analysis of the floral preferences of the alfalfa leafcutting bee, *Megachile rotundata*. *Can. Field Nat.* 111:445–53
115. Smith CR, Toth AL, Suarez AV, Robinson GE. 2008. Genetic and genomic analyses of the division of labour in insect societies. *Nat. Rev. Genet.* 9:735–48
116. Soroka JJ, Goerzen DW, Falk KC, Bett KE. 2001. Alfalfa leafcutting bee (Hymenoptera: Megachilidae) pollination of oilseed rape (*Brassica napus* L.) under isolation tents for hybrid seed production. *Can. J. Plant Sci.* 81:199–204
117. St. Armand PC, Skinner DZ, Peaden RN. 2000. Risk of alfalfa transgene dissemination and scale-dependent effects. *Theor. Appl. Genet.* 101:107–14
118. Stark JD, Banks JE. 2003. Population-level effects of pesticides and other toxicants on arthropods. *Annu. Rev. Entomol.* 48:505–19
119. Stephen WP. 1961. Artificial nesting sites for the propagation of the leaf-cutter bee, *Megachile* (*Eutricharaea*) *rotundata*, for alfalfa pollination. *J. Econ. Entomol.* 54:989–93
120. Stephen WP. 1962. Propagation of the leaf-cutter bee for alfalfa seed production. *Agric. Exp. Stn. Bull.* 586:1–16. Oregon State Univ., Corvallis
121. Stephen WP. 1981. The design and function of field domiciles and incubators for leafcutting bee management (*Megachile rotundata* (Fabricius)). *Oregon State Coll. Agric. Exp. Stn. Bull.* 654:1–13
122. **Stephen WP. 2003. Solitary bees in North American agriculture: a perspective. In *For Non-native Crops, Whence Pollinators of the Future?*, ed. K Strickler, JH Cane, 3:41–66. Lanham, MD: Entomol. Soc. Am. 204 pp.**

122. Comprehensive review and history of establishment of solitary bees for pollination. Other chapters in book cover different aspects of alternative pollinators in agriculture.

123. Stephen WP, Undurraga JM. 1976. X-radiography, an analytical tool in population studies of the leaf-cutter bee *Megachile pacifica*. *J. Apic. Res.* 15:81–87
124. Strickler K. 1999. Impact of flower standing crop and pollinator movement on alfalfa seed yield. *Environ. Entomol.* 28:1067–76
125. Szabo TI, Smith MV. 1972. The influence of light intensity and temperature on the activity of the alfalfa leaf-cutter bee *Megachile rotundata* under field conditions. *J. Apic. Res.* 11:157–65
126. Tasei JN. 1975. Le problème de l'adaptation de *Megachile* (Eutricharea) *pacifica* Panz. (Megachilidae) Américain en France. *Apidologie* 6:1–57
127. Tasei JN, Masure MM. 1978. Sur quelques facteurs influencant le développement de *Megachile pacifica* Panz (Hymenoptera, Megachilidae). *Apidologie* 9:273–90
128. Tepedino VJ. 1982. Flower visitation and pollen collection records for bees of high altitude shortgrass prairie in southeastern Wyoming, USA. *Southwest. Entomol.* 7:16–25
129. Tepedino VJ. 1983. An open-field test of *Megachile rotundata* as a potential pollinator in hybrid carrot fields. *J. Apic. Res.* 22:64–68
130. Tepedino VJ. 1997. A comparison of the alfalfa leafcutting bee (*Megachile rotundata*) and the honey bee (*Apis mellifera*) as pollinators for hybrid carrot seed in field cages. *Acta Hortic.* 437:457–61
131. Tepedino VJ, Parker FD. 1986. Effect of rearing temperature on mortality, second-generation emergence, and size of adult in *Megachile rotundata* (Hymenoptera: Megachilidae). *J. Econ. Entomol.* 79:974–77

132. Tepedino VJ, Parker FD. 1988. Alternation of sex ratio in a partially bivoltine bee, *Megachile rotundata* (Hymenoptera: Megachilidae). *Ann. Entomol. Soc. Am.* 81:467–76
133. Tezuka T, Maeta Y. 1993. Effect of UVA film on extranidal activities of three species of bees. *Jpn. J. Appl. Entomol. Zool.* 37:175–80
134. Torchio PF. 1973. Relative toxicity of insecticides to the honey bee, alkali bee, and alfalfa leafcutting bee (Hymenoptera: Apidae, Halictidae, Megachilidae). *J. Kans. Entomol. Soc.* 46:446–53
135. Torchio PF. 1979. An eight-year field study involving control of *Sapyga pumila* Cresson (Hymenoptera: Sapygidae), a wasp parasite of the alfalfa leafcutter bee, *Megachile pacifica* Panzer. *J. Kans. Entomol. Soc.* 52:412–19
136. Trostle G, Torchio PF. 1994. Comparative nesting behavior and immature development of *Megachile rotundata* (Fabricius) and *Megachile apicalis* Spinola (Hymenoptera, Megachilidae). *J. Kans. Entomol. Soc.* 67:53–72

136. Describes nesting behavior and immature development of *Megachile* species.

137. Undurraga JM, Stephen WP. 1980. Effects of temperature on development and survival in post-diapausing alfalfa leafcutting bee prepupae and pupae (*Megachile rotundata* (F.): Hymenoptera: Megachilidae). I. High temperatures. *J. Kans. Entomol. Soc.* 53:669–76
138. Undurraga JM, Stephen WP. 1980. Effects of temperature on development and survival in post-diapausing alfalfa leafcutting bee prepupae and pupae (*Megachile rotundata* (F.): Hymenoptera: Megachilidae). II. Low temperatures. *J. Kans. Entomol. Soc.* 53:677–82
139. Van Deynze AE, Fitzpatrick SM, Hammon B, McCaslin MH, Putnam DH, et al. 2008. Gene flow in alfalfa: biology, mitigation, and potential impact on production. *CAST Spec. Publ. 28*. Counc. Agric. Sci. Technol., Ames, IA. 30 pp.
140. Vandenberg JD, Fichter BL, Stephen WP. 1980. Spore load of *Ascosphaera* spp. on emerging adults of the alfalfa leafcutting bee, *Megachile rotundata*. *Appl. Environ. Microbiol.* 39:650–55
141. Vandenberg JD, Stephen WP. 1982. Etiology and symptomatology of chalkbrood in the alfalfa leafcutting bee, *Megachile rotundata*. *J. Invertebr. Pathol.* 39:133–37
142. Vorel CA. 2010. *Learning ability and factors influencing nest establishment of the solitary bees* Osmia lignaria *and* Megachile rotundata *(Hymenoptera: Megachilidae)*. PhD thesis. Utah State Univ. 126 pp.
143. Vorel CA, Pitts-Singer TL. 2010. The proboscis extension reflex not elicited in megachilid bees. *J. Kans. Entomol. Soc.* 83:80–83
144. Wedmann S, Wappler T, Engel MS. 2009. Direct and indirect fossil records of megachilid bees from the Paleogene of Central Europe (Hymenoptera: Megachilidae). *Naturwissenschaften* 96:703–12
145. Wheeler D. 1996. The role of nourishment in oogenesis. *Annu. Rev. Entomol.* 41:407–31
146. Wightman JA, Rogers VM. 1978. Growth, energy and nitrogen budgets and efficiencies of the growing larvae of *Megachile pacifica* (Hymenoptera: Megachilidae). *Oecologia* 36:245–57
147. Wittmann D, Blochtein B. 1995. Why males of leafcutter bees hold the females' antennae with their front legs during mating. *Apidologie* 26:181–95
148. Woodward DR. 1994. Predators and parasitoids of *Megachile rotundata* (F.) (Hymenoptera: Megachilidae) in South Australia. *J. Aust. Entomol. Soc.* 33:13–15
149. Xu J, James R. 2009. Genes related to immunity, as expressed in the alfalfa leafcutting bee, *Megachile rotundata*, during pathogen change. *Insect Mol. Biol.* 18:785–94
150. Yocum GD, Kemp WP, Bosch J, Knoblett JN. 2005. Temporal variation in overwintering gene expression and respiration in the solitary bee *Megachile rotundata*. *J. Insect Physiol.* 51:621–29

Vision and Visual Navigation in Nocturnal Insects

Eric Warrant and Marie Dacke

Department of Biology, University of Lund, S-22362 Lund, Sweden,
email: Eric.Warrant@cob.lu.se, Marie.Dacke@cob.lu.se

Annu. Rev. Entomol. 2011. 56:239–54

First published online as a Review in Advance on September 3, 2010

The *Annual Review of Entomology* is online at ento.annualreviews.org

This article's doi:
10.1146/annurev-ento-120709-144852

0066-4170/11/0107-0239$20.00

Key Words

apposition compound eye, superposition compound eye, nocturnal vision, neural summation, homing

Abstract

With their highly sensitive visual systems, nocturnal insects have evolved a remarkable capacity to discriminate colors, orient themselves using faint celestial cues, fly unimpeded through a complicated habitat, and navigate to and from a nest using learned visual landmarks. Even though the compound eyes of nocturnal insects are significantly more sensitive to light than those of their closely related diurnal relatives, their photoreceptors absorb photons at very low rates in dim light, even during demanding nocturnal visual tasks. To explain this apparent paradox, it is hypothesized that the necessary bridge between retinal signaling and visual behavior is a neural strategy of spatial and temporal summation at a higher level in the visual system. Exactly where in the visual system this summation takes place, and the nature of the neural circuitry that is involved, is currently unknown but provides a promising avenue for future research.

Optical sensitivity: the ratio of the number of photons absorbed by a photoreceptor to the number emitted per steradian of solid angle from a unit area of an extended source (i.e., a measure of the light-gathering capacity of an eye)

INTRODUCTION

Despite their tiny size and comparatively few neurons, the visual systems of insects are remarkably sophisticated. With the help of their compound eyes, insects recognize and react to conspecifics; distinguish and avoid predators; locate food sources and intercept prey; navigate to and from a nest using learned visual landmarks; and walk, swim, or fly through a complicated three-dimensional habitat. Even in the bright sunlit world in which many insects are active, these visual tasks would be demanding enough. But for the vast numbers of insects that are active exclusively at night, when light levels can be up to 11 orders of magnitude lower, such tasks may at first glance seem impossible. However, over the past two decades we have begun to realize that nocturnal insects perform the same visual tasks as their diurnal relatives—and with the same precision and accuracy—despite the difficult light conditions they face. This remarkable fact has led to a new appreciation of the visual abilities of nocturnal insects and to an impetus to understanding how these abilities are realized in the nervous system.

How is such impressive visual performance achieved in nocturnal insects? Part of the answer clearly lies in behavioral modifications (such as slower locomotion) that enhance the reliability of visual information that reaches the insect at night. Part of the answer also lies in the visual system itself. The optical designs of most nocturnal compound eyes are well suited to high optical sensitivity (e.g., the superposition compound eyes of nocturnal moths and beetles) (**Figure 1*a***), but in some rare cases they are not (e.g., the apposition compound eyes of some exceptional species of nocturnal bees and wasps) (**Figure 1*b***). Compared to their diurnal relatives, in nocturnal insects optical sensitivity can typically be improved by one (apposition eyes) to three (superposition eyes) orders of magnitude, but rarely more. In addition, the photoreceptors of nocturnal insects tend to respond more slowly and have higher visual gain (roughly five

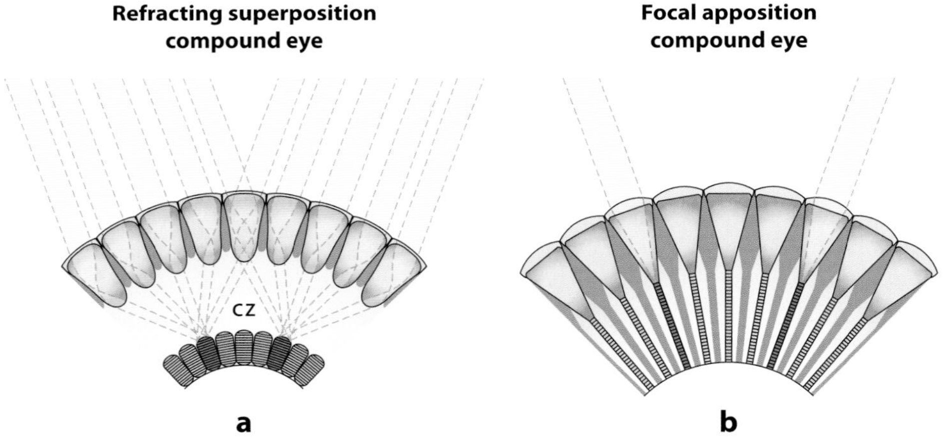

Figure 1

Two compound eye designs in insects. (*a*) A refracting superposition compound eye. A large number of corneal facet lenses and bullet-shaped crystalline cone lenses (each possessing a powerful internal gradient of refractive index) collect and focus light across the clear zone of the eye (*cz*) toward single photoreceptors in the retina. Several hundred, or even thousands, of facets service a single photoreceptor. Not surprisingly, many nocturnal insects have refracting superposition eyes and benefit from the significant improvement in sensitivity. (*b*) A focal apposition compound eye. Each ommatidium is isolated from its neighbors by a sleeve of light-absorbing screening pigment, thus preventing light from reaching the photoreceptive rhabdom from all but its own small corneal lens, significantly limiting light capture. This eye design is therefore typical of diurnal insects. Diagrams courtesy of Dan-Eric Nilsson.

times higher), physiological adaptations that also improve visual reliability. However, these optical and physiological gains are relatively modest and suggest that peripheral visual mechanisms on their own are unable to explain behavioral visual performance. Higher visual mechanisms—including spatial and temporal summation, and other mechanisms—are thus likely to be responsible. Where this takes place in the visual system, and how the necessary circuitry works, remains largely unknown.

This gap between our knowledge of what nocturnal insects can actually see and our understanding of how it is achieved neurally is thus large. In this review we describe our current understanding of vision and visual processing in nocturnal insects and how the gains in visual performance afforded by these mechanisms appear to fall short of the actual visual performance of freely behaving nocturnal insects. The results of this synthesis suggest several future avenues of fruitful research.

WHAT CAN NOCTURNAL INSECTS SEE?

From the outset it is important to point out that the nocturnal visual world is essentially identical to the diurnal visual world. The contrasts of objects are identical and so (or nearly so) are their colors. The only distinguishing difference is the mean level of light intensity, which can be up to 11 orders of magnitude dimmer at night (58, 88), depending on the presence or absence of moonlight and clouds and whether the habitat is open or closed (i.e., beneath the canopy of a forest). It is this difference that severely limits the ability of a visual system to distinguish the colors and contrasts of the nocturnal world. Indeed, many animals, especially diurnal animals, distinguish little at all. In the end, the greatest challenge for an eye that views a dimly illuminated object is to absorb sufficient photons of light to reliably discriminate it from other objects (51, 85, 86, 88).

Nonetheless, many nocturnal insects have evolved sufficiently sensitive visual systems to orient themselves and navigate at night. Some navigate under the open sky, and take full advantage of the celestial cues available there. Others navigate in more difficult conditions, such as through the understory of a dense tropical rainforest, analyzing the optic flow of the passing world generated by their own movements in order to hold a stable course, discriminate landmarks, or control flight. Many nocturnal insects even experience the world in color. In all these nocturnal insects, the apparent disadvantages for visual orientation and navigation imposed by low light levels, in particular a decreased visual reliability arising from greater visual noise and lower visual contrast, have been overcome. These facts, inextricably linked to their highly sensitive visual systems, suggest that the visual world experienced by nocturnal arthropods is essentially no different than that experienced by their diurnal relatives. This surprising fact is underscored by an examination of a few case examples, notably the ability of nocturnal insects to see color and their ability to navigate using celestial cues and terrestrial landmarks. Fuller accounts of the impressive visual abilities of nocturnal insects can be found elsewhere (85, 86, 88–90).

Nocturnal Color Vision

The world is as equally colorful at night as during the day, and the usefulness of color for object discrimination (e.g., flower identification; 40) does not decrease with falling light intensities. The nocturnal nectar-feeding hawk moth *Deilephila elpenor* has superposition eyes (**Figure 1*a***) with three different spectral classes of photoreceptors, centered in the UV, the violet, and the green parts of the spectrum (33, 73). This moth can not only be trained to associate a sugar reward with a blue disc at starlight levels of illumination (39), but it can also discriminate this blue disk from other discs in various shades of gray with a choice frequency of at least 80%. This hawk moth, despite its tiny eyes and brain, thus has color vision at light levels 100 times dimmer than the dimmest level at which the human visual system can distinguish color. Moreover, color vision in *D. elpenor* is color-constant, meaning that color discrimination is not affected by moderate shifts in the spectrum of illumination (2). This is a feature of all advanced color vision systems (40). Remarkably, color vision has also been demonstrated in the nocturnal Indian carpenter bee *Xylocopa tranquebarica* (74, 75), which has apposition eyes, the design apparently unsuited to life in dim light (**Figure 1*b***). The discovery of nocturnal trichromatic color vision in two widely separated taxa of nocturnal pollinators suggests that color is equally visible and important for both nocturnal and diurnal insects.

Nocturnal Navigation and Orientation

To navigate back to the safety of a nest after a long and tortuous foraging trip, or to optimally orient themselves in order to efficiently escape from rivals or predators, insects require reliable detection of both terrestrial and celestial visual cues (reviewed for insects in References 95 and 98), and this is equally true at night as it is during the day. Most insects rely on hierarchies of both kinds of cues. Thus, in the examples that follow, even though we separately review selected case studies concerning how nocturnal insects use terrestrial and celestial cues for navigation and orientation, in any one species both types of cues are likely to be used in concert.

Navigation and orientation using celestial cues. At night, the brightest and most easily discernable cue in the sky is undoubtedly the moon. Because of its variable rise time and prominence, the moon is a much more complicated orientation cue than the sun (94), but nevertheless its bright disk is used for orientation and navigation in a number of different nocturnal insects, including ants (37, 44), earwigs (82), moths (77), and beetles (8).

A much dimmer and more subtle cue associated with the moon is its pattern of polarized light. This circular pattern, centered around the moon, arises because of the atmospheric scattering of moonlight as it travels toward Earth (17, 36). Light is most polarized around a circular celestial locus 90° from the moon, and the circular pattern of polarized light moves with the moon. In many day-active insects, such as ants and bees, the equivalent polarization pattern formed around the sun is used as a compass cue when returning to the nest (93, 95, 96, 97) and allows them to hold a constant bearing during their straight-line returns. A similar use of the moon's polarization pattern, which during a full moon is a million times dimmer, should only be limited by the sensitivity of nocturnal eyes.

After the sun sets over the savannah, the nocturnal dung beetle, *Scarabaeus zambesianus*, emerges from its temporary nest in the soil and takes flight in search of fresh dung. Once found, it rapidly makes a dung ball and rolls it away on foot (9), aiming toward a location well away from the ferocious competition of the dung pile, where the beetle can manipulate its ball safe from interference. The safest and most efficient way to escape is to move away from the dung pile along a straight line in any direction, as indeed the beetles do (5, 9, 10). To hold this course in a straight line, beetles rely on compass information from the dim moonlight polarization pattern. If a large linearly polarizing filter is placed above the ball-rolling beetle as it moves along such a straight path, the insect realigns itself as it passes underneath, in accordance with the dominant polarization axis of the filter. When it then rolls back out under the natural sky, it again realigns itself to the natural skylight polarization pattern (9). Thus, nocturnal dung beetles have sufficient visual sensitivity to detect dim polarized moonlight and to use it as a compass cue for orientation. This ability is likely shared by other nocturnal insects, including crickets (28, 71), tenebrionid beetles (4, 66), and bees (20, 42).

Navigation and orientation using terrestrial cues. Just as for diurnal insects, many nocturnal insects rely on terrestrial features and landmarks for orientation, for flight control, or for homing in on the exact location of a small nest entrance in the dark. The ability of an animal to use terrestrial cues for orientation at night is much more dependent on whether the cues have a sufficient contrast against the background than on their own particular luminance (38).

As shown in numerous excellent studies on diurnal flying insects, such as bees and flies, the ability to accurately analyze optic flow cues visually during flight is crucial for holding a stable course, hovering, landing, and homing (reviewed in References 100 and 101). Visual detection of optic flow is also clearly necessary for controlling nocturnal flight. For instance, in dim light gypsy moths (6) and mosquitoes (18) analyze optic flow cues to control their flight trajectories, and on brighter moonlit nights, nocturnally migrating locusts may be capable of doing the same thing (69).

For insects living in forests, where a clear view of the night sky is impossible, terrestrial features and landmarks seen silhouetted against the brighter night sky, such as the canopy or individual trees, are especially well suited for nocturnal orientation. As the animal moves under the tree canopy, the brighter sky in the gaps of the canopy, together with the darker area under the canopy, has the potential to form a spatial representation of the world above. Thus, an animal leaving its nest can memorize the structure of the canopy above and use it as a landmark for orientation (34), or use it to pinpoint the nest upon return from a longer trip. Indeed, both nocturnal (35, 79) and diurnal ants (1, 13, 34, 64) navigate to and from their nests using the rainforest canopy pattern as a navigational cue. Recently, the same ability has been demonstrated in a nocturnal shield bug, *Parastrachia japonensis* (29, 30, 31, 32).

Females of the Japanese subsocial shield bug *P. japonensis* forage on the floors of heavily wooded forests and look for food (stone fruits dropped from nearby trees) to take back to their nymphs in the home burrow. The outbound trip is typically tortuous, but the bug returns with fruit to the burrow along a straight course. To test if the shield bugs orient themselves with respect to the spatial pattern of gaps in the canopy, a nesting female was placed on the floor of a large closed box having a single round opening at one end of the box's dorsal lid (30). This opening, representing an artificial canopy gap, was always exposed to an overcast night sky to eliminate other celestial cues. For several days prior to testing, the bug was allowed to forage for fruit in this box and to home after finding it (**Figure 2*a,b***). During the test, the bug was allowed to forage for fruit as usual, but at the moment the fruit was found, the lid of the box was rapidly rotated by 180° so that the canopy gap was now positioned at the other end of the box. The shield bug, not realizing the change, homed in a direction exactly opposite the direction of the burrow (**Figure 2*a,c***). This shows that in nocturnal as well as in diurnal insects the canopy pattern provides a powerful

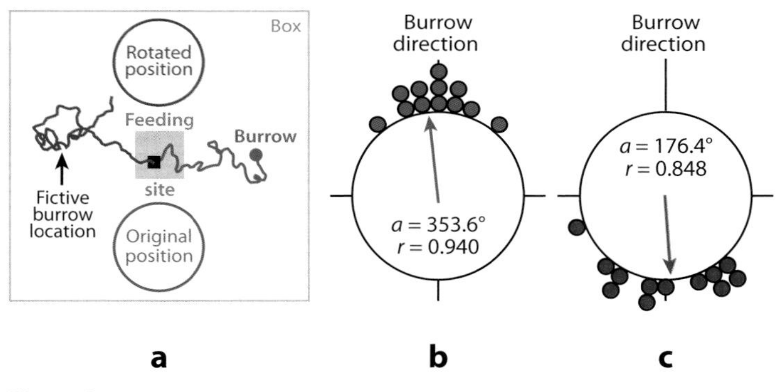

Figure 2

Nocturnal homing using canopy cues in the shield bug *Parastrachia japonensis*. (*a*) The typical homing route (*red line*) of a bug exposed to a sudden 180° rotation of an artificial canopy gap (circular hole, 90 cm in diameter; *blue circle* indicates original position, *red circle* indicates new position) in the roof of its box (*large square*: box measured 3 × 3 m, height 2 m). The bug was allowed to forage from its burrow to the feeding site where it located fruit (outbound foraging route shown by the *blue line*). At the moment the fruit was located, the canopy gap was rotated to its new position (*red circle*). The rotation caused the bug to home in the opposite direction to normal (*red line*) and to search for its burrow at its fictive location (*arrow*). (*b*) The distribution of homing directions in 12 bugs (*blue dots*) relative to the canopy gap (*blue circle* in panel *a* shows one of three original positions: four bugs were tested at each). In tests, all bugs oriented toward the real burrow. (*c*) The distribution of homing directions in 12 bugs (*red dots*) following a sudden 180° rotation of the canopy gap position (*red circle* in panel *a*). All bugs oriented 180° away from the real burrow. In panels *b* and *c*: *a* is the mean angle of orientation, *r* is the length of the mean resultant vector ($r = 1$ equates to perfect orientation at angle *a*). Adapted with kind permission from Reference 30.

Lamina: first optic ganglion of the optic lobe of insects

Lobula: third optic ganglion of the optic lobe of insects

compass cue for orientation. In shield bugs this ability seems particularly interesting: For an overcast nocturnal sky the contrast between the gap and the understory is at its minimum, implying that the bug possesses significant visual contrast sensitivity in dim light.

Visual landmarks in a forest, such as nearby bushes and tree trunks, typically have much lower visual contrast than bright gaps in the forest canopy. Nonetheless, two species of nocturnal bees, the Central American sweat bee (*Megalopta genalis*) and the Indian carpenter bee (*X. tranquebarica*), use such landmarks for homing after a foraging trip. Both species are active in extremely dim light and forage at long distances from their tiny nests at night, returning home without getting lost in the dark forest. While the involvement of canopy cues for this remarkable navigational ability remains to be tested, recent behavioral investigations reveal that both species perform orientation flights to visually learn landmarks around the nest entrance at night (74, 91). In *M. genalis* the nest is a hollowed-out stick. When a nest stick was placed in the middle of a row with four empty nest sticks (which could act as visual landmarks), the bee learned the spatial arrangement of these and would effortlessly return to its own stick in the center of the nest array after a foraging flight. If, when the bee was away, the positions of the bee's nest and an empty nest were swapped, the bee would still fly directly into the central unoccupied nest—the spatially correct nest—but after a couple of seconds would fly out again. This continued until the nest stick was returned to its original position in the middle of the array, after which the bee no longer emerged. This experiment shows that *M. genalis* is capable of using visually learned landmarks at night to find its way home, an ability that nocturnal carpenter bees also share (74). From studies of these homing behaviors at different light levels, it has been determined that *M. genalis* can find and land accurately on its nest when as few as five photons every second are absorbed by each photoreceptor of its apposition eye (80, 91). We return to this impressive performance below.

EYES AND VISION IN NOCTURNAL INSECTS

Most of our current knowledge of visual processing in nocturnal insects is confined to the optics and retina of the compound eye, although in recent years a small amount of data have also come from the first and third visual neuropiles of the optic lobe (the lamina ganglionaris of nocturnal bees and the lobula plate of nocturnal hawk moths). The following general conclusion can be drawn from these studies: Vision in nocturnal insects is made possible by having eyes with an enhanced optical sensitivity to light and visual neurons that sacrifice spatial and temporal resolution to improve visual reliability for the slower and coarser features of the world.

The Optical Designs of Nocturnal Compound Eyes

We have known for more than a century (14) that nocturnal insects typically possess superposition compound eyes (**Figure 1*a***) and thereby gain a considerable optical sensitivity to light. A superposition eye can have optical sensitivity 100–1000 times higher than that of an apposition compound eye (the design typically possessed by diurnal insects) of the same size (**Figure 1*b***).

In apposition compound eyes, each ommatidium is isolated from its neighbors by a sleeve of light-absorbing screening pigment, thus preventing light from reaching the photoreceptive rhabdom from all but its own small corneal lens. This tiny lens, typically some tens of micrometers across, represents the pupil of the apposition eye, and not surprisingly this eye design is typical of insects living in bright habitats. Remarkable exceptions do exist, including nocturnal mosquitoes (48, 49) and crane flies (52, 55), various species of nocturnal and crepuscular ants (e.g., within the bull ant genus *Myrmecia*; 21, 62), tropical nocturnal wasps (within the genera *Provespa* and *Apoica*; 19, 89), and nocturnal and crepuscular bees (41, 87, 89), including the central American sweat bee,

M. genalis (Halictidae), and the Indian carpenter bee, *X. tranquebarica* (Apidae). We discuss the latter two species below. Superposition eyes (**Figure 1*a***) are better known for their high sensitivity. In this eye design, typical of nocturnal insects, the pigment sleeve is withdrawn in the dark-adapted state, and a wide optically transparent area, the clear zone (**Figure 1*a***), is interposed between the lenses and the retina. The clear zone, and specially modified crystalline cones, allows light from a narrow region of space to be collected by a large number of ommatidia (comprising the superposition aperture) and to be focused onto a single rhabdom. Unlike the crystalline cones of most apposition eyes, those of superposition eyes of nocturnal insects have evolved powerful refractive index gradients (7, 14, 59) that allow as many as 2,000 lenses to collect light for a single photoreceptor (as in some large nocturnal moths). The width of this superposition aperture, which effectively acts as the pupil of the eye, is much larger than the width of a single corneal facet lens and represents a massive improvement in sensitivity.

The area of the pupil (diameter A) is not the only determinant of an eye's optical sensitivity (S) to a spatially extended source of broad-spectrum light. S, expressed in units of μm^2 sr, is given by (43, 47, 92):

$$S = \left(\frac{\pi}{4}\right)^2 A^2 \left(\frac{d}{f}\right)^2 \left(\frac{kl}{2.3 + kl}\right), \qquad 1.$$

where l is the length of the rhabdom, k is the peak absorption coefficient of the visual pigment, f is the focal length of the ommatidium, and d is the diameter of the rhabdom. This equation predicts that good sensitivity to a spatially extended scene results from a pupil of large area ($\pi A^2/4$) and photoreceptors that each view a large solid angle of visual space ($\pi d^2/4f^2$ steradians) and absorb a substantial fraction of the incident light ($kl/2.3 + kl$). The compound eyes of nocturnal insects tend to show all three trends.

To see how nocturnal life has affected the optical structure, and sensitivity, of insect compound eyes, consider the apposition eyes of the nocturnal sweat bee *M. genalis*. This bee has larger eyes and larger facets (diameters up to 36 μm) than the strictly day-active and similarly sized European honey bee, *Apis mellifera* (diameters up to 20 μm). Moreover, in *A. mellifera* the rhabdoms have a width of only 2 μm, whereas in *M. genalis* they reach an extraordinary 8 μm, resulting in a receptive field more than seven times greater in solid angular extent (23, 91). These differences in receptive field and facet size allow *M. genalis* an optical sensitivity that is roughly 27 times greater than that of *A. mellifera*: 2.7 μm^2 sr versus 0.1 μm^2 sr, a fact confirmed by measurements of responses to single photons in both species (15). Similar differences in sensitivity can be seen in the apposition eyes of nocturnal and diurnal carpenter bees (76), wasps (19), and ants (21, 60, 61). Even though nocturnal apposition eyes have a significantly higher optical sensitivity than diurnal apposition eyes, it is still modest compared to that found in a typical superposition eye, such as those of the nocturnal hawk moth *D. elpenor* ($S = 69\ \mu m^2$ sr). This finding exposes the inherent optical limitations of the apposition design for vision in dim light.

S: optical sensitivity

Rhabdom: in an ommatidium of a compound eye the light-sensitive structure formed by the microvilli of the retinula cells (contains the visual pigment rhodopsin)

Steradian (sr): one steradian is the solid angle subtended at the center of a sphere of radius r by a portion of the surface of the sphere whose area equals r^2

Photoreception and the Reliability of Vision in Dim Light

A general property of both vertebrate and invertebrate photoreceptors is their ability to respond to single photons of light with small but distinct electrical responses known as bumps (as they are called in the invertebrate literature) (**Figure 3*a,b***). At higher intensities, the bump responses fuse to create a graded response whose duration and amplitude are proportional to the duration and amplitude of the light stimulus. At very low light levels, a light stimulus of constant intensity is coded as a train of bumps generated in the retina at a particular rate, and at somewhat higher light levels the constant intensity is coded by a graded potential of particular amplitude. At the level of the photoreceptors, the reliability of vision is determined by the repeatability of this response: For repeated presentations of the stimulus, the reliability of vision is maximal if the rate of bump generation, or the amplitude of the graded response, remains exactly the same for each presentation. In practice this is never the case, especially in dim light.

Visual noise. Why is this so? The basic answer is that the visual response (and as a result its repeatability) is degraded by visual noise. Part of this noise arises from the stochastic nature of photon arrival and absorption (governed by Poisson statistics): Each sample of N absorbed photons (the signal) has a certain degree of uncertainty (or noise) associated with it ($\pm\sqrt{N}$ photons). The relative magnitude of this uncertainty is greater at lower rates of photon absorption, and these quantum fluctuations set an upper limit to the visual signal-to-noise ratio (i.e., $N/\sqrt{N} = \sqrt{N}$; 70, 83). As light levels fall, the fewer photons absorbed, the greater this shot noise relative to the signal and the less that can be seen. This is the famous Rose–de Vries or square root law of visual detection at low light levels: The visual signal-to-noise ratio, and thus the finest contrast that can be discriminated, improves as the square root of photon catch. Signal reliability in dim light can thus be

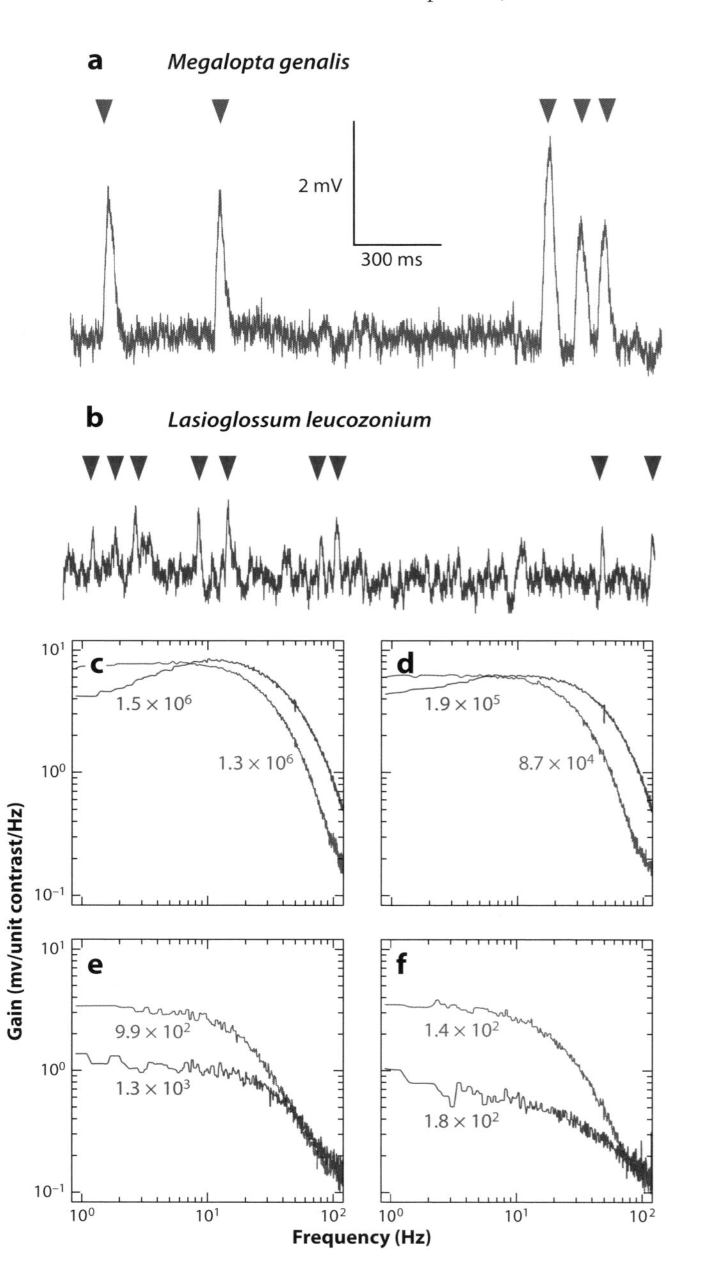

Figure 3

Adaptations for nocturnal vision in the photoreceptors of the nocturnal Central American sweat bee, *Megalopta genalis*, compared with photoreceptors in the closely related diurnal sweat bee *Lasioglossum leucozonium*. (*a, b*) Responses to single photons (or photon bumps, *red and blue arrowheads*) recorded from photoreceptors in (*a*) *M. genalis* and (*b*) *L. leucozonium*. Note that the bump amplitude is larger, and the bump time course much slower, in *M. genalis* than in *L. leucozonium*. (*c–f*). Average contrast gain as a function of temporal frequency in *M. genalis* (*blue curves*, $n = 8$ cells) and *L. leucozonium* (*red curves*, $n = 8$ cells) at different adapting intensities, indicated as effective photons per second in each panel for each species. [For each species, each stimulus intensity was calibrated in terms of effective photons, i.e., the number of photon bumps per second the light source elicited, thereby eliminating the effects of differences in the light-gathering capacity of the optics between the two species, which is about 27 times greater in *M. genalis* (57).] In light-adapted conditions (*c, d*), both species reach the same maximum contrast gain per unit bandwidth, although *L. leucozonium* has broader bandwidth and a higher corner frequency (the frequency at which the gain has fallen off to 50% of its maximum). In dark-adapted conditions (*e, f*), *M. genalis* has a much higher contrast gain per unit bandwidth. All panels adapted with kind permission from Reference 16.

improved with an eye design of higher sensitivity to light.

There are two other sources of noise that limit the reliability of nocturnal photoreception. The first source, transducer noise, arises because photoreceptors are incapable of producing identical bumps of fixed amplitude, latency, and duration to each (identical) photon of absorbed light. This source of noise, originating in the biochemical processes leading to signal amplification, degrades the reliability of vision (54, 56, 57). The second source of intrinsic noise, dark noise, arises because the biochemical pathways responsible for transduction are occasionally activated, even in perfect darkness (3). These activations are due either to spontaneous conversion of rhodopsin to metarhodopsin or to spontaneous activation of G-protein-coupled steps in the transduction chain. Irrespective of their origin, these activations produce dark events, electrical responses that are indistinguishable from those produced by real photons, and are more frequent at higher retinal temperatures. Even though dark noise is much lower in invertebrates than in vertebrates (11, 12, 25, 57), at very low light levels this dark noise could still significantly contaminate visual signals in insects.

Retinal adaptations for nocturnal vision. To investigate the effects of visual noise on the reliability of vision in dim light, a useful method is to measure the visual signal-to-noise ratio of the photoreceptors by electrophysiologically recording their responses to Gaussian-distributed white-noise light stimuli (45). As a result of such experiments, the photoreceptors of insects reveal several properties uniquely suited to a life in dim light.

First, photoreceptor responses to single photons (i.e., bumps) are much larger in nocturnal insects than in their closely related diurnal relatives. Large bumps have been demonstrated in nocturnal crane flies (55), cockroaches (27), and bees (16), as well as in other arthropods, notably spiders (53, 65). This trend can be seen in closely related nocturnal and diurnal sweat bees: The quantum bumps are much larger in the nocturnal *M. genalis* than in the diurnal *L. leucozonium* (**Figure 3a,b**) (16). The larger bumps of *M. genalis* and other nocturnal species indicate that the photoreceptor's gain of transduction is greater compared with diurnal species. This higher transduction gain manifests itself as a higher contrast gain, that is, in a greater photoreceptor voltage response per unit change in light intensity (or contrast). Contrast gain is plotted for the bees *M. genalis* and *L. leucozonium* as a function of temporal frequency in **Figure 3c–f**: At all levels of light and dark adaptation, at a frequency range both species can discriminate, the visual gain of *M. genalis* is always higher than that of *L. leucozonium*, and at the lowest intensities it is up to five times higher (**Figure 3e,f**). This higher gain results in greater signal amplification. Unfortunately, it also amplifies the noise, and thus on its own the higher gain does not alter the visual signal-to-noise ratio.

Contrast gain: amplitude of the electrical response of a photoreceptor per unit contrast

Second, as in slowly moving nocturnal ants (78), crane flies (55), and cockroaches (27), the dark-adapted photoreceptors of fast-flying nocturnal *M. genalis* are slow. In the frequency domain, this is equivalent to stating that the temporal corner frequency is low—this is the frequency at which the gain has fallen to 50% of its maximum value, and lower values indicate slower vision. In dark-adapted conditions, the corner frequency is around 7 Hz for *M. genalis*. In diurnal *L. leucozonium* it is about three times faster, at around 20 Hz (**Figure 3f**). Because both bees fly at similar speeds, their difference in temporal properties—most likely due to different photoreceptor sizes and to different numbers and types of ion channels in the photoreceptor membrane (52)—can be related only to the difference in light intensity experienced by the two species.

The slower vision of *M. genalis* in dim light, despite compromising temporal resolution, is beneficial because it increases the visual signal-to-noise ratio and improves contrast discrimination at lower temporal frequencies by suppressing photon noise at frequencies too high to be reliably resolved (26). Despite compromising temporal resolution, this low-pass filtering (which is evident in **Figure 3*c–f***) improves visual reliability in dim light. However, the narrower bandwidth possessed by nocturnal *M. genalis* (**Figure 3*c–f***) has a devastating effect on the visual information rate (calculated in units of bits per second): At all intensities, the intrinsic rate of visual information is significantly greater in the photoreceptors of *L. leucozonium* than in the photoreceptors of *M. genalis* (**Figure 4*a***). It is only when the approximately 27 times greater optical sensitivity of *M. genalis* apposition eyes is accounted for (**Figure 4*b***) that the rate of information is greater in *M. genalis* than in *L. leucozonium*, but then only at the lowest intensities. Information, it seems, has been sacrificed in *M. genalis* for a greater absolute sensitivity.

Information rate: the number of bits of information that are conveyed or processed per unit time

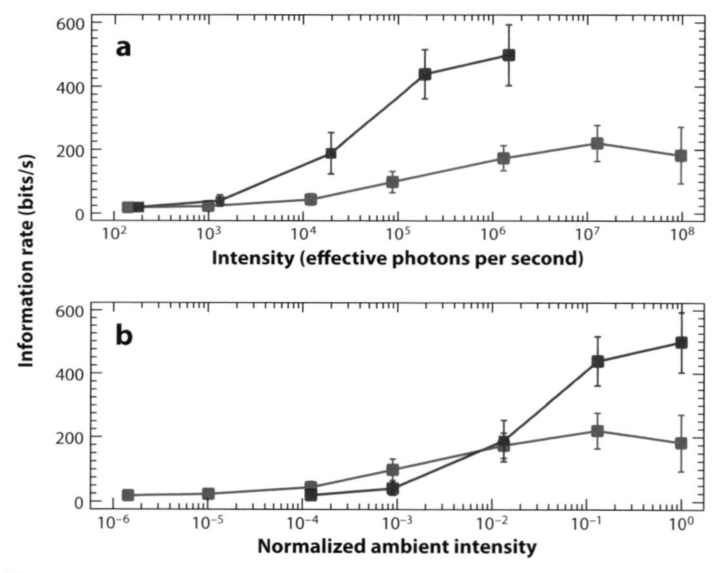

Figure 4

The average rates of information transmission (in bits per second) in the photoreceptors of the sweat bees *Megalopta genalis* (nocturnal) (*blue curves*, $n = 8$ cells) and *Lasioglossum leucozonium* (diurnal) (*red curves*, $n = 8$ cells). (*a*) When photoreceptors alone are considered (via a light source calibration in effective photons absorbed by the photoreceptor per second), it is evident that at all intensities *L. leucozonium* has a higher information rate than *M. genalis*. (*b*) When light sources are instead calibrated to external ambient intensities (a normalized intensity of 100 corresponds to the light intensity on an overcast day, or around 180 candela m^{-2}), *M. genalis* has a higher information rate in dim light. This results because the optics of *M. genalis* is 27 times more sensitive than that of *L. leucozonium*, not because an intrinsic adaptation is present within the photoreceptors. Error bars show ± standard deviation. Both panels adapted with kind permission from Reference 16.

FROM RETINA TO BEHAVIOR: A MISSING LINK

There is now little doubt that nocturnal insects have superb night vision, due to their ability (among other things) to see color, fly through complicated forests, and navigate using celestial cues and visual landmarks, all at light levels at least eight orders of magnitude lower than those experienced by their diurnal relatives when performing similar tasks. Neither is there any doubt that the compound eyes of nocturnal insects are well adapted to a life in dim light, with high optical sensitivity and photoreceptors that enhance visual reliability by having higher gains and slower responses. But is this the end of the story? Are the optical and neural adaptations of nocturnal insect eyes alone sufficient to explain nocturnal visual behavior?

To answer these questions, it is necessary to revisit our earlier case studies. Consider the nocturnal hawk moth *D. elpenor*, which can distinguish a blue disk from various gray disks in starlight. Despite having sensitive superposition eyes, its photoreceptors can absorb no more than 16 photons during each visual integration time when the moth performs this task. Theoretically (47), this photon catch is insufficient to reliably distinguish color (39). At similar intensities, each photoreceptor in the apposition eyes of the nocturnal bee *M. genalis* absorbs on average only a single photon every six visual integration times (91)! Yet at these light levels *M. genalis* lands accurately on its nest (80; **Supplemental Film**, follow the **Supplemental Material link** from the Annual Reviews home page at **http://www.annualreviews.org**) and distinguishes visual landmarks. For *M. genalis*, this level of photon absorption is theoretically (47) about 100 times too low for the bee to distinguish the dark entrance hole of its nest (91). Even diurnal house flies (*Musca domestica*), tethered within a rotating optomotor drum lined with vertical stripes, react to the movements of the stripes when as few as two or three photons reach each photoreceptor every second (11, 67, 68, 72).

Another anomaly concerns nocturnal photoreceptors themselves. Even though they respond more slowly than diurnal photoreceptors, thereby improving visual reliability in dim light (26), the narrower temporal bandwidth of nocturnal photoreceptors significantly reduces the intrinsic information rate (**Figure 4**) (16). Moreover, their higher contrast gain, while providing beneficial signal amplification, also amplifies the noise and thus leads to no improvement in the visual signal-to-noise ratio.

Spatial and Temporal Summation

How can this wide gap between the performance of the retina and the visual performance of freely behaving nocturnal insects be explained? Currently we have only one hypothesis, supported by growing indirect evidence but still lacking definitive proof: Visual signals leaving the retina might neurally be summed in space and time (50, 84). We have already discussed summation in time above: When light gets dim, the visual systems of nocturnal insects can improve visual reliability by integrating signals over longer periods of time (**Figure 3**) (26, 51). This can be achieved by having slower photoreceptors. Even slower vision could be obtained by neurally integrating (summing) signals at a higher level in the visual system. This temporal summation comes at a price: It can drastically degrade the perception of fast-moving objects, which is potentially disastrous for a fast-flying nocturnal animal (such as a nocturnal wasp or bee) that needs to negotiate obstacles. Not surprisingly, temporal summation is more likely to be employed by slowly moving animals (84).

Summation of photons in space can also improve visual reliability. Instead of each ommatidium collecting photons in isolation (as in bright light), the transition to dim light could activate specialized laterally spreading neurons that couple the outputs of the ommatidia together into groups. Each summed group—now comprising a super ommatidium—could collect considerably more photons over a much wider visual angle, albeit with a simultaneous and unavoidable loss of spatial resolution. Despite being much brighter, the image would become necessarily coarser. Evidence for such laterally spreading neurons has been found in the first optic ganglion (lamina ganglionaris) of nocturnal cockroaches, fireflies, and hawk moths (88), and these neurons have been interpreted as an adaptation for spatial summation (50). *M. genalis* also appears to have such neurons. The wide lateral branches of its laminar monopolar cells L2, L3, and L4, which spread to 12, 11, and 17 lamina cartridges, respectively, are considerably wider than the homologous cells of the diurnal honey bee *A. mellifera*, which spread to 2, 0, and 4 cartridges, respectively (22, 24).

By allowing the world to become coarser and slower, but more importantly much brighter, spatial and temporal summation could improve visual reliability of these coarser and slower details of the scene. Theoretically, this could explain the ability of nocturnal insects to see color (39), distinguish landmarks (81), and navigate using celestial patterns of polarized moonlight (46). It could also help explain the improved contrast gain of nocturnal photoreceptors (16, 89). The high contrast gain present in the photoreceptors of nocturnal bees amplifies both the signal and the noise. Because the noise is uncorrelated across ommatidia, spatial summation could effectively average out the noise and dramatically increase the visual signal-to-noise ratio in dim light, albeit for a lower range of spatial frequencies. Thus, a high visual gain, followed by spatial summation, could represent a significant strategy for vision in dim light.

Where in the Visual System Does Summation Occur?

The neural circuits responsible for spatial summation could very well be found in the lamina, the first optic ganglion of the optic lobe. Additional circuits might also be found in higher centers of the optic lobe, such as the medulla. Temporal summation, beyond the level already performed in the photoreceptors, could be performed by temporal integrating mechanisms (e.g., long membrane time constants and neural delays) at many locations in the optic lobe, although exactly where is hard to say and will most likely depend on the type of visual information being processed.

Thus, a promising future avenue of research is to determine the nature of the neural circuitry responsible for summation. A clear question concerns the role of the lamina monopolar cells in spatial summation, and this will require an electrophysiological investigation of their physiological properties in one or more model insect species. Another obvious question involves the nature of the neural mechanisms responsible for temporal summation. This question might best be addressed by studying higher-order cells in which the time domain can be easily investigated. The motion-sensitive cells of the lobula plate are good candidates: By understanding how the temporal properties of these cells change with light intensity, one may be able to decipher the type of temporal integrating mechanism employed. Other types of higher-order cells may also be useful in this respect, including those that analyze visual depth (99) or moving targets (63).

Although many questions remain unanswered, an undeniable fact remains: The visual systems of insects, despite their small size and comparatively few neurons, are nothing short of remarkable. With their exquisite visual sensitivity, nocturnal insects have evolved the impressive capacity to discriminate colors and navigate at night. It is a sobering thought indeed to realize that these abilities rival, and in some cases even exceed, those of many vertebrates despite the advantages afforded by their orders-of-magnitude larger eyes and brains.

SUMMARY POINTS

1. Nocturnal insects have excellent night vision, with the capacity to discriminate colors, orient themselves using faint celestial cues (such as polarized moonlight), fly unimpeded through a complicated habitat, and navigate to and from a nest using learned visual landmarks. These visual capacities are in all respects similar to those of diurnal insects.
2. Nocturnal insects with advanced visual behaviors possess either superposition compound eyes or apposition compound eyes. Apposition eyes (which are typical of diurnal insects) have an optical sensitivity up to around 1,000 times lower than superposition eyes.
3. At the dim light intensities at which nocturnal insects are visually active, each photoreceptor absorbs photons at extremely low rates.
4. The photoreceptors of nocturnal insects respond more slowly and have a higher contrast gain than those of diurnal insects, adaptations that potentially improve the reliability of vision, especially if visual signals are subsequently neurally summed.
5. A neural strategy of spatial and temporal summation at a higher level in the visual system is hypothesized as the necessary bridge between retinal signaling and visual behavior. Where in the visual system this summation takes place is currently unknown, although the lamina is a strong candidate. The elucidation of these summation strategies provides a fruitful avenue for future research.

DISCLOSURE STATEMENT

The authors are not aware of any affiliations, memberships, funding, or financial holdings that might be perceived as affecting the objectivity of this review.

ACKNOWLEDGMENTS

The authors are particularly grateful to the Swedish Research Council, the United States Air Force Office of Scientific Research, and the Royal Physiographic Society of Lund for their valuable and ongoing support. We also thank Dan-Eric Nilsson, Takahiko Hariyama, and Mantaro Hironaka for kindly allowing us to reproduce their figures.

LITERATURE CITED

1. Baader AP. 1996. The significance of visual landmarks for navigation of the giant tropical ant, *Paraponera clavata* (Formicidae: Ponerinae). *Insectes Soc.* 43:435–50
2. Balkenius A, Kelber A. 2004. Color constancy in diurnal and nocturnal hawkmoths. *J. Exp. Biol.* 207:3307–16
3. Barlow HB. 1956. Retinal noise and absolute threshold. *J. Opt. Soc. Am.* 46:634–39
4. Bisch SM. 1999. *Orientierungsleistungen des nachtaktiven Wüstenkäfers* Parastizopus armaticeps *Peringuey (Coleoptera: Tenebrionidae)*. PhD diss. Rheinischen Friedrich-Wilhems-Universität Bonn, Ger.
5. Byrne MJ, Dacke M, Nordström P, Scholtz CH, Warrant EJ. 2003. Visual cues used by ball-rolling dung beetles for orientation. *J. Comp. Physiol. A* 189:411–18
6. Cardé RT, Knols BGJ. 2000. Effects of light levels and plume structure on the orientation manoeuvres of male gypsy moths flying along pheromone plumes. *Physiol. Entomol.* 25:141–50
7. Caveney S, McIntyre P. 1981. Design of graded-index lenses in the superposition eyes of scarab beetles. *Philos. Trans. R. Soc. Lond. B* 294:589–32
8. Dacke M, Byrne M, Scholtz CH, Warrant EJ. 2004. Lunar orientation in a beetle. *Philos. Trans. R. Soc. Lond. B* 271:361–65
9. **Dacke M, Nilsson D-E, Scholtz CH, Byrne M, Warrant EJ. 2003. Animal behaviour: insect orientation to polarized moonlight. *Nature* 424:33**

9. First demonstration of an animal that orients itself using polarized moonlight.

10. Dacke M, Nordström P, Scholtz CH. 2003. Twilight orientation to polarised light in the crepuscular dung beetle *Scarabaeus zambesianus*. *J. Exp. Biol.* 206:1535–43
11. Dubs A, Laughlin SB, Srinivasan MV. 1981. Single photon signals in fly photoreceptors and first order interneurons at behavioural threshold. *J. Physiol.* 317:317–34
12. Doujak FE. 1985. Can a shore crab see a star? *J. Exp. Biol.* 166:385–93
13. Ehmer B. 1999. Orientation in the ant *Paraponera clavata*. *J. Insect Behav.* 12:711–22
14. Exner S. 1891. *Die Physiologie der facettirten Augen von Krebsen und Insecten*. Leipzig u. Wien: Franz Deuticke
15. Frederiksen R, Warrant EJ. 2008. The optical sensitivity of compound eyes—theory and experiment compared. *Biol. Lett.* 4:745–47
16. **Frederiksen R, Wcislo WT, Warrant EJ. 2008. Visual reliability and information rate in the retina of a nocturnal bee. *Curr. Biol.* 18:349–53**

16. Demonstrates that nocturnal insects can improve visual reliability at night by sacrificing photoreceptor signal-to-noise ratio and information rate for an increased contrast gain.

17. Gál J, Horváth G, Barta A, Wehner R. 2001. Polarization of the moonlit clear night sky measured by full-sky imaging polarimetry at full moon: comparison of the polarization of moonlit and sunlit skies. *J. Geophys. Res.* 106:22647–53
18. Gibson G. 1995. A behavioural test of the sensitivity of a nocturnal mosquito, *Anopheles gambiae*, to dim white, red and infra-red light. *Physiol. Entomol.* 20:224–28
19. Greiner B. 2006. Visual adaptations in the night-active wasp *Apoica pallens*. *J. Comp. Neurol.* 495:255–62
20. Greiner B, Cronin TW, Ribi WA, Wcislo WT, Warrant EJ. 2007. Anatomical and physiological evidence for polarisation vision in the nocturnal bee *Megalopta genalis*. *J. Comp. Physiol. A* 193:591–600
21. Greiner B, Narendra A, Reid SF, Dacke M, Ribi WA, Zeil J. 2007. Eye structure correlates with distinct foraging-bout timing in primitive ants. *Curr. Biol.* 17:R879–80
22. Greiner B, Ribi WA, Warrant EJ. 2004. Neuronal organisation in the first optic ganglion of the nocturnal bee *Megalopta genalis*. *Cell Tissue Res.* 318:429–37
23. Greiner B, Ribi WA, Warrant EJ. 2004. Retinal and optical adaptations for nocturnal vision in the halictid bee *Megalopta genalis*. *Cell Tissue Res.* 316:377–90

24. Greiner B, Ribi WA, Warrant EJ. 2005. A neural network to improve dim-light vision? Dendritic fields of first-order interneurons in the nocturnal bee *Megalopta genalis*. *Cell Tissue Res.* 323:313–20

24. Provides the possible neural substrate for spatial summation in the lamina of a nocturnal bee.

25. Hardie RC, Martin F, Cochrane GW, Juusola M, Georgiev P, Raghu P. 2002. Molecular basis of amplification in *Drosophila* phototransduction: roles for G protein, phospholipase C, and diacylglycerol kinase. *Neuron* 36:689–701
26. Hateren van JH. 1993. Spatiotemporal contrast sensitivity of early vision. *Vision Res.* 33:257–67
27. Heimonen K, Salmela I, Kontiokari P, Weckström M. 2006. Large functional variability in cockroach photoreceptors: optimization to low light levels. *J. Neurosci.* 26:13454–62
28. Herzmann D, Labhart T. 1989. Spectral sensitivity and absolute threshold of polarization vision in crickets: a behavioral study. *J. Comp. Physiol. A* 165:315–19
29. Hironaka M, Filippi L, Nomakuchi S, Horiguchi H, Hariyama T. 2007. Hierarchical use of chemical marking and path integration in the homing trip of a subsocial shield bug. *Anim. Behav.* 73:739–45
30. Hironaka M, Inadomi K, Nomakuchi S, Filippi L, Hariyama T. 2008. Canopy compass in nocturnal homing of the subsocial shield bug, *Parastrachia japonensis* (Heteroptera: Parastrachiidae). *Naturwissenschaften* 95:343–46
31. Hironaka M, Nomakuchi S, Filippi L, Tojo S, Horiguchi H, Hariyama T. 2003. The directional homing behavior of the subsocial shield bug, *Parastrachia japonensis* (Heteroptera: Cydnidae), under different photic conditions. *Zool. Sci.* 20:423–28
32. Hironaka M, Tojo S, Nomakuchi S, Filippi L, Hariyama T. 2007. Round-the-clock homing behavior of a subsocial shield bug, *Parastrachia japonensis* (Heteroptera: Parastrachiidae) using, path integration. *Zool. Sci.* 24:535–41
33. Höglund G, Hamdorf K, Rosner G. 1973. Trichromatic visual system in an insect and its sensitivity control by blue light. *J. Comp. Physiol.* 86:265–79
34. Hölldobler B. 1980. Canopy orientation: a new kind of orientation in ants. *Science* 210:86–88
35. Hölldobler B, Taylor RW. 1983. A behavioral study of the primitive ant *Nothomyrmecia macrops* Clark. *Insectes Soc.* 30:381–401
36. Horváth G, Varjú D. 2009. *Polarized Light in Animal Vision: Polarization Patterns in Nature*. Berlin/Heidelberg/New York: Springer. 447 pp.
37. Jander R. 1957. Die optische Richtungsorienterung der Roten Waldameise (*Formica rufa*). *Z. Vergl. Physiol.* 40:162–238
38. Kaul RM, Kopteva GA. 1982. Night orientation of ants *Formica rufa* (Hymenoptera: Formicidae) upon movement on routes. *Zool. Zh.* 61:1351–58

39. Kelber A, Balkenius A, Warrant EJ. 2002. Scotopic color vision in nocturnal hawkmoths. *Nature* 419:922–25

39. First demonstration of nocturnal color vision.

40. Kelber A, Balkenius A, Warrant EJ. 2003. Color vision in diurnal and nocturnal hawkmoths. *Integr. Comp. Biol.* 43:571–79
41. Kelber A, Warrant EJ, Pfaff M, Wallén R, Theobald JC, et al. 2006. Light intensity limits the foraging activity in nocturnal and crepuscular bees. *Behav. Ecol.* 17:63–72
42. Kerfoot WB. 1967. The lunar periodicity of *Sphecodogastra texana*, a nocturnal bee (Hymenoptera; Halictidae). *Anim. Behav.* 15:479–86
43. Kirschfeld K. 1974. The absolute sensitivity of lens and compound eyes. *Z. Naturforsch.* 29C:592–96
44. Klotz JH, Reid BL. 1993. Nocturnal orientation in the black carpenter ant *Camponotus pennsylvanicus* (De Geer) (Hymenoptera: Formicidae). *Insectes Soc.* 40:95–106
45. Kouvalainen E, Weckström M, Juusola M. 1994. A method for determining photoreceptor signal-to-noise ratio in the time and frequency domains with a pseudorandom stimulus. *Vis. Neurosci.* 11:1221–25
46. Labhart T, Petzold J, Helbling H. 2001. Spatial integration in polarization-sensitive interneurones of crickets, a survey of evidence, mechanisms and benefits. *J. Exp. Biol.* 204:2423–30
47. Land MF. 1981. Optics and vision in invertebrates. In *Handbook of Sensory Physiology*, ed. H Autrum, VII/6B:471–592. Berlin: Springer
48. Land MF, Gibson G, Horwood J. 1997. Mosquito eye design, conical rhabdoms are matched to wide aperture lenses. *Proc. R. Soc. Lond. B* 264:1183–87

49. Land MF, Gibson G, Horwood J, Zeil J. 1999. Fundamental differences in the optical structure of the eyes of nocturnal and diurnal mosquitoes. *J. Comp. Physiol. A* 185:91–103
50. Laughlin SB 1981. Neural principles in the peripheral visual systems of invertebrates. In *Handbook of Sensory Physiology*, ed. H Autrum, VII/6B:133–280. Berlin: Springer
51. Laughlin SB. 1990. Invertebrate vision at low luminances. In *Night Vision*, ed. RF Hess, LT Sharpe, K Nordby, pp. 223–250. Cambridge, UK: Cambridge Univ. Press
52. Laughlin SB. 1996. Matched filtering by a photoreceptor membrane. *Vision Res.* 36:1529–41
53. Laughlin SB, Blest AD, Stowe S. 1980. The sensitivity of receptors in the posterior median eye of the nocturnal spider, *Dinopis*. *J. Comp. Physiol.* 141:53–65
54. Laughlin SB, Lillywhite PG. 1982. Intrinsic noise in locust photoreceptors. *J. Physiol.* 332:25–45
55. **Laughlin SB, Weckström M. 1993. Fast and slow photoreceptors—a comparative study of the functional diversity of coding and conductances in the Diptera. *J. Comp. Physiol. A* 172:593–609**

55. Compares the responses and membrane properties of photoreceptors in 20 species of flies and demonstrates that temporal coding is matched to the visual ecology of the species.

56. Lillywhite PG. 1981. Multiplicative intrinsic noise and the limits to visual performance. *Vision Res.* 21:291–96
57. Lillywhite PG, Laughlin SB. 1979. Transducer noise in a photoreceptor. *Nature* 277:569–72
58. Martin GR. 1990. *Birds By Night.* London: Poyser. 227 pp.
59. McIntyre P, Caveney S. 1985. Graded-index optics are matched to optical geometry in the superposition eyes of scarab beetles. *Philos. Trans. R. Soc. Lond. B* 311:237–69
60. Menzi U. 1987. Visual adaptation in nocturnal and diurnal ants. *J. Comp. Physiol. A* 160:11–21
61. Moser JC, Reeve JD, Bento JMS, Della Lucia TMC, Cameron RS, Heck NM. 2004. Eye size and behavior of day- and night-flying leafcutting ant alates. *J. Zool. Lond.* 264:69–75
62. Narendra A, Reid SF, Hemmi JM. 2010. The twilight zone: ambient light levels trigger activity in primitive ants. *Proc. R. Soc. Lond. B* 277:1531–38
63. Nordström K, O'Carroll DC. 2009. Feature detection and the hypercomplex property in insects. *Trends Neurosci.* 32:383–91
64. Oliveira PS, Hölldobler B. 1989. Orientation and communication in the neotropical ant *Odontomachus bauri* Emery (Hymenoptera, Formicidae, Ponerinae). *Ethology* 83:154–66
65. Pirhofer-Walzl K, Warrant EJ, Barth FG. 2007. Adaptations for vision in dim light: impulse responses and bumps in nocturnal spider photoreceptor cells (*Cupiennius salei* Keys). *J. Comp. Physiol. A* 193:1081–87
66. Rasa OAE. 1990. Evidence for subsociality and division of labor in a desert tenebrionid beetles *Parastizopus armaticeps* (Puey). *Naturwissenschaften* 77:591–92
67. Reichardt WE. 1965. Quantum sensitivity of light receptors in the compound eye of the fly, *Musca*. *Cold Spring Harb. Symp. Quant. Biol.* 30:505–15
68. Reichardt WE, Braitenburg V, Weidel G. 1968. Auslösung von Elementarprozessen durch einzelne Lichtquanten im Fliegenauge. *Kybernetik* 5:148–69
69. Riley JR, Kreuger U, Addison CM, Gewecke M. 1988. Visual detection of wind-draft by high-flying insects at night: a laboratory study. *J. Comp. Physiol. A* 169:793–98
70. Rose A. 1942. The relative sensitivities of television pickup tubes, photographic film and the human eye. *Proc. Inst. Radio Eng. New York* 30:293–300
71. Rost R, Honegger HW. 1987. The timing of premating and mating behavior in a field population of the cricket *Gryllus campestris* L. *Behav. Ecol. Sociobiol.* 21:279–89
72. Scholes JH, Reichardt W. 1969. The quantal content of optomotor stimuli and the electrical responses of receptors in the compound eye of the fly *Musca*. *Kybernetik* 6:74–80
73. Schwemer J, Paulsen R. 1973. Three visual pigments in *Deilephila elpenor* (Lepidoptera, Sphingidae). *J. Comp. Physiol.* 86:215–29
74. **Somanathan H, Borges RM, Warrant EJ, Kelber A. 2008. Nocturnal bees learn landmark colours in starlight. *Curr. Biol.* 18:R996–97**

74. First demonstration of nocturnal color vision in an insect with apposition compound eyes, an eye design more typical of diurnal insects.

75. Somanathan H, Borges RM, Warrant EJ, Kelber A. 2008. Visual ecology of Indian carpenter bees I: light intensities and flight activity. *J. Comp. Physiol. A* 194:97–107
76. Somanathan H, Kelber A, Borges RM, Wallén R, Warrant EJ. 2009. Visual ecology of Indian carpenter bees II: adaptations of eyes and ocelli to nocturnal and diurnal lifestyles. *J. Comp. Physiol. A* 195:571–83
77. Sotthibandhu S, Baker BB. 1979. Celestial orientation by the large yellow underwing moth, *Noctua pronuba* L. *Anim. Behav.* 27:786–800

78. Souza de JM, Ventura DF. 1989. Comparative study of temporal summation and response form in hymenopteran photoreceptors. *J. Comp. Physiol. A* 165:237–45
79. Taylor RW. 2007. Bloody funny wasps! Speculations on the evolution of eusociality in ants. In *Advances in Ant Systematics (Hymenoptera: Formicidae): Homage to E.O. Wilson – 50 Years of Contributions*, Memoirs of the American Entomological Institute, ed. RR Snelling, BL Fisher, PS Ward, 80:580–609. Gainesville, FL: Am. Entomol. Inst.
80. Theobald JC, Coates MM, Wcislo WT, Warrant EJ. 2007. Flight performance in night-flying sweat bees suffers at low light levels. *J. Exp. Biol.* 210:4034–42
81. Theobald JC, Greiner B, Wcislo WT, Warrant EJ. 2006. Visual summation in night-flying sweat bees: a theoretical study. *Vision Res.* 46:2298–309
82. Ugolini A, Chiussi R. 1996. Astronomical orientation and learning in the earwig *Labidura riparia*. *Behav. Process.* 36:151–61
83. **Vries de H. 1943. The quantum character of light and its bearing upon threshold of vision, the differential sensitivity and visual acuity of the eye. *Physica* 10:553–64**

83. First demonstration of the relationship between the quantal nature of light and human visual threshold. Introduces the "square root law" of visual detection at low light levels.

84. Warrant EJ. 1999. Seeing better at night: life style, eye design and the optimum strategy of spatial and temporal summation. *Vision Res.* 39:1611–30
85. Warrant EJ. 2004. Vision in the dimmest habitats on earth. *J. Comp. Physiol. A* 190:765–89
86. Warrant EJ. 2006. Invertebrate vision in dim light. In *Invertebrate Vision*, ed. EJ Warrant, D-E Nilsson, pp. 83–126. Cambridge, UK: Cambridge Univ. Press
87. Warrant EJ. 2007. Nocturnal bees. *Curr. Biol.* 17:R991–92
88. Warrant EJ. 2008. Nocturnal vision. In *The Senses: A Comprehensive Reference, Vision II*, ed. T Albright, RH Masland, AI Basbaum, A Kaneko, GM Shepherd, G Westheimer, 2:53–86. Oxford: Academic
89. Warrant EJ. 2008. Seeing in the dark: vision and visual behavior in nocturnal bees and wasps. *J. Exp. Biol.* 211:1737–46
90. Warrant EJ, Dacke M. 2010. Visual orientation and navigation in nocturnal arthropods. *Brain Behav. Evol.* 75:156–73
91. **Warrant EJ, Kelber A, Gislén A, Greiner B, Ribi W, Wcislo WT. 2004. Nocturnal vision and landmark orientation in a tropical halictid bee. *Curr. Biol.* 14:1309–18**

91. First demonstration of an insect that navigates at night using visual landmarks.

92. Warrant EJ, Nilsson D-E. 1998. Absorption of white light in photoreceptors. *Vision Res.* 38:195–207
93. Waterman TH. 1981. Polarization sensitivity. In *Handbook of Sensory Physiology*, ed. H Autrum, VII/6B:281–469. Berlin: Springer
94. Wehner R. 1984. Astronavigation in insects. *Annu. Rev. Entomol.* 29:277–98
95. Wehner R. 1992. Arthropods. In *Animal Homing*, ed. F Papi, pp. 45–144. London: Chapman and Hall
96. Wehner R. 2001. Polarization vision—a uniform sensory capacity? *J. Exp. Biol.* 204:2589–96
97. Wehner R, Labhart T. 2006. Polarisation vision. In *Invertebrate Vision*, ed. EJ Warrant, D-E Nilsson, pp. 291–47. Cambridge, UK: Cambridge Univ.
98. Wehner R, Srinivasan M. 2003. Path integration in insects. In *The Neurobiology of Spatial Behavior*, ed. KJ Jeffery, pp. 9–30. Oxford, UK: Oxford Univ. Press
99. Wicklein M, Strausfeld NJ. 2000. Organization and significance of neurons that detect change of visual depth in the hawk moth *Manduca sexta*. *J. Comp. Neurol.* 424:356–76
100. Zeil J, Boeddeker N, Stürzl W. 2009. Visual homing in insects and robots. In *Flying Insects and Robots*, ed. D Floreano, pp. 87–100. Berlin: Springer-Verlag
101. Zeil J, Boeddeker N, Hemmi JM. 2009. Visually guided behaviour. In *Encyclopaedia of Neuroscience*, ed. LR Squire, 10:369–80. Oxford: Academic

The Role of Phytopathogenicity in Bark Beetle–Fungus Symbioses: A Challenge to the Classic Paradigm

Diana L. Six[1] and Michael J. Wingfield[2]

[1]Department of Ecosystem and Conservation Sciences, College of Forestry and Conservation, The University of Montana, Missoula, Montana 59812; email: diana.six@cfc.umt.edu

[2]Forestry and Agricultural Biotechnology Institute, University of Pretoria, Pretoria, Republic of South Africa 0002

Annu. Rev. Entomol. 2011. 56:255–72

First published online as a Review in Advance on September 3, 2010

The *Annual Review of Entomology* is online at ento.annualreviews.org

This article's doi:
10.1146/annurev-ento-120709-144839

0066-4170/11/0107-0255$20.00

Key Words

Scolytinae, ophiostomatoid fungi, mutualism, tree defenses, *Dendroctonus*, *Ips*

Abstract

The idea that phytopathogenic fungi associated with tree-killing bark beetles are critical for overwhelming tree defenses and incurring host tree mortality, herein called the classic paradigm (CP), has driven research on bark beetle–fungus symbiosis for decades. It has also strongly influenced our views of bark beetle ecology. We discuss fundamental flaws in the CP, including the lack of consistency of virulent fungal associates with tree-killing bark beetles, the lack of correspondence between fungal growth in the host tree and the development of symptoms associated with a successful attack, and the ubiquity of similar associations of fungi with bark beetles that do not kill trees. We suggest that, rather than playing a supporting role for the host beetle (tree killing), phytopathogenicity performs an important role for the fungi. In particular, phytopathogenicity may mediate competitive interactions among fungi and support survival and efficient resource capture in living, defensive trees.

INTRODUCTION

It has been more than 100 years since Von-Schrenk (107) first noted that trees killed by bark beetles often became stained by fungi within a few weeks of attack. Likewise, 80 years have passed since Craighead (24), observing this same relationship, speculated that the fungi may play an important role in the death of bark beetle–attacked trees or in the nutrition of the beetles. Although the latter possibility has received some attention over the years, the concept that tree-killing bark beetles require fungal pathogens to overcome tree defenses and to incur tree mortality has received the most attention. This hypothesis, which we hereafter refer to as the classic paradigm (CP), has formed the basis for the majority of research conducted on these interactions. However, despite numerous studies, no conclusive evidence exists supporting the CP. The common and self-perpetuating practice of citing the CP as fact in the literature has also meant that the CP is seldom questioned. As a result, few alternative hypotheses are considered when research is conducted on these systems.

Tree defenses: complex physical and chemical defenses against herbivorous insects and pathogens

Classic paradigm (CP): the main model that postulates that fungal associates of tree-killing bark beetles are responsible for overwhelming tree defenses and incurring host tree mortality

Phytopathogenicity: the ability to cause disease in plants

Scolytinae: a subfamily of weevils (Curculionidae) that includes the bark and ambrosia beetles

In this review, we question the validity of the CP. That some tree-killing bark beetles possess virulent fungal associates is not in question. It is well known that some do and that some of these fungi are capable of killing trees (20, 46, 102, 114). It is also not in question whether fungi elicit defensive reactions in conifers; an extensive literature exists documenting the form and process of these responses (29). Rather, we question the view that the fungi play a proximate role in aiding bark beetles to overwhelm trees. We hope this review provokes thought and initiates new avenues of investigation into these fascinating and complex interactions.

At the outset we provide a brief review of bark beetle–fungus associations. The focus here is primarily on conifer-infesting bark beetles, specifically because the CP arose from studies on these systems. For more in-depth treatments of these symbioses, we refer readers to several recent reviews (32, 47, 88, 91). Next, we describe the process of attack and colonization of trees by bark beetles. We then present evidence and arguments for and against the CP. Finally, we propose alternative explanations for the occurrence of phytopathogenicity in bark beetle–associated fungi and suggest some directions for future research.

BARK BEETLE–FUNGUS SYMBIOSES

Bark beetles (Curculionidae: Scolytinae) construct galleries under the bark in the phloem layer of woody plants, where they lay eggs and their brood feed and develop (**Figure 1**). Most are limited to colonizing weak or recently killed trees; however, a few species are capable of killing healthy trees or developing in living trees without causing mortality (113).

One of the most striking characteristics of bark beetles is their widespread association with fungi (8, 47, 88). Most species carry fungi, either in specialized structures of the integument

Figure 1

Generalized life cycle of a tree-killing bark beetle and its associated fungi. (1) Dispersal of adult beetles carrying fungi in mycangia and/or on exoskeleton. (2) Attack phase. (*a*) Tree choice by pioneer (first arriving) beetle. (*b*) Entry into tree and subsequent release of aggregation pheromones. Conspecifics of both sexes are attracted to the pheromone, enter the tree, and release additional pheromone. The pheromone-mediated mass attack typically occurs over a relatively short period (often 2–5 days). In some conifers, pitch tubes form as part of preformed defenses. (*c*) When tree defenses are overwhelmed (the point of no return), beetles switch from producing aggregation to antiaggregation pheromones to avoid overexploitation of the tree. (3) Colonization phase. (*d*) Initial egg gallery construction by parental adults, egg-laying, and inoculation of fungi into phloem. During the early stages of development of beetle larvae, there is low vertical spread of vegetative (hyphal) growth of fungi in phloem, and the beginnings of hyphal penetration into sapwood. Tree defensive chemistry and moisture levels are high and oxygen availability is low, limiting the growth of fungi at this stage. (*e*) Extensive larval tunneling. Phloem and sapwood begin to dry and defensive chemistry has declined, allowing extensive hyphal colonization by fungi. (*f*) Excavation of pupal chambers and pupation. Fungi begin to form spore layers in pupal chambers. (*g*) Spore feeding by teneral (newly emerged) adults, acquisition of fungi in mycangia or on exoskeleton.

called mycangia or phoretically on the exoskeleton (87). Most fungal partners are Ascomycetes in four sexual genera, *Ophiostoma*, *Ceratocystiopsis*, *Grosmannia*, and *Ceratocystis* (32, 33, 43, 47, 88, 111, 117). *Ophiostoma*, *Grosmannia*, and *Ceratocystiopsis* form a monophyletic group in the Ophiostomatales separate from *Ceratocystis*, which resides in the Microascales (103). A small

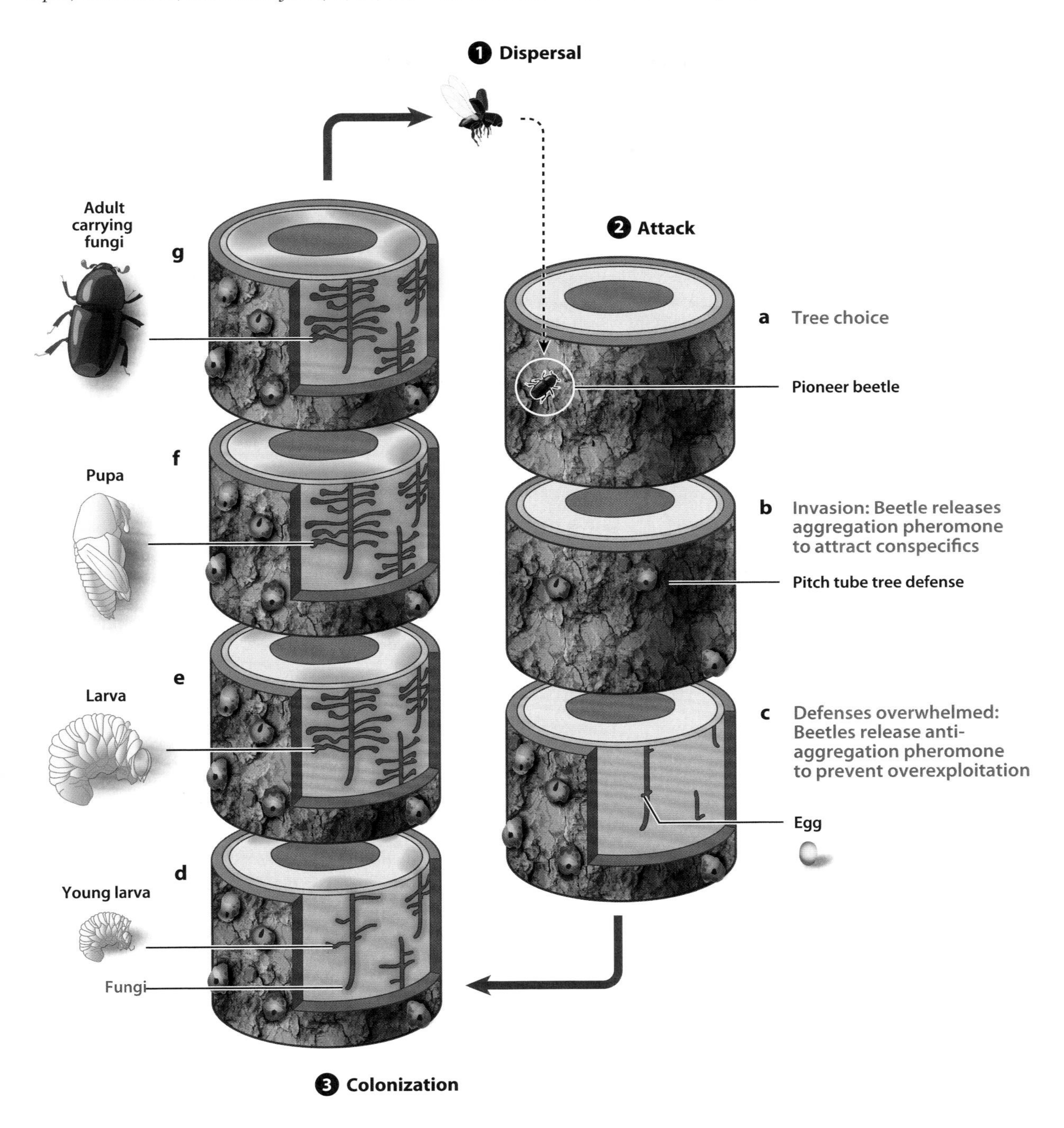

Mycangia: structures of the adult insect exoskeleton specialized for the transport of fungi

Ascomycetes: an extremely abundant and successful group of fungi (phylum Ascomycota)

Competition: an interaction between two organisms that results in a net negative effect on all interactants, often through the use or defense of similar resources

Mutualism: an interaction between two nonconspecific organisms that results in a net mutual benefit to both partners

Commensalism: an interaction between two nonconspecific organisms that results in a net positive benefit to one partner but no significant effect on the other

Antagonism: an interaction between two nonconspecific organisms where one or both partners exert a detrimental effect on the other

Ophiostomatoid fungi: a term used to denote a polyphyletic group of Ascomycete fungi that includes the genera *Ophiostoma*, *Grosmannia*, *Ceratocystiopsis* (Ophisotomatales), and *Ceratocystis* (Microascales)

number of bark beetles are also associated with Basidiomycetes in the genera *Entomocorticium* and *Phlebiopsis* (42, 108).

There has been a tendency to view all bark beetle–fungus symbioses as similar in function. However, there is actually a diversity of interaction types. Fungal associates benefit from the association through transport to ephemeral and otherwise inaccessible plant resources. Introduction into living or recently killed trees may also allow these fungi to avoid competition with later successional saprophytic fungi. Effects on the bark beetle host vary from beneficial to neutral to negative (88). Many of these associations are thought to be mutualisms based on the phytopathogenicity of the fungi (the CP). This idea is developed in subsequent sections. In contrast, a number appear to be obligate mutualisms in which the bark beetles rely on nutritional supplementation from fungi (5, 11, 15, 23, 93). At least some mutualistic partners exhibit parallel cladogenesis with their hosts, indicating long coevolutionary histories (87, 93). In these associations, larvae that feed on phloem colonized by mutualistic fungi are larger, more likely to complete development, have higher fecundity, and develop more rapidly than bark beetles that do not (5, 11, 23, 93). For at least one bark beetle species, feeding on fungal spores by new adults appears to be required for reproduction (93). Alternatively, some bark beetle–fungus mutualisms may be facultative, with hosts benefiting from feeding on fungi but not requiring it for survival (28, 54).

Some associations, especially those involving incidental fungi, are likely to be commensalisms with no measurable effect on the bark beetle host. In contrast, some fungi have strong negative effects on bark beetle development and survival (6, 37). The underlying cause of antagonism is not known but may be related to the inability of a fungus to provide critical nutrients (53). Whether a fungus is beneficial, commensal, or antagonistic is not strictly linked to taxonomy; fungi involved in all three types of relationships with bark beetles can be closely related congenerics. For example, two sister species interact with their hosts in different ways: *Ophiostoma montium* is a highly specific nutritional mutualist with *Dendroctonus ponderosae* (mountain pine beetle) (93), whereas *Ophiostoma ips* is considered a commensal with many bark beetles worldwide (83, 116).

Bark beetles carry complexes of fungi (44, 47, 57). Most of this diversity consists of incidental fungi likely to be of no importance to the insect host because of their variable and often low incidence. The symbiotic fungi (consistent associates) are less diverse and often include only two to three partners per bark beetle species. These partners can have either differing or redundant effects. For example, *Dendroctonus frontalis* (southern pine beetle) possesses three main associates, two nutritional mutualists, and one antagonist (6, 37).

While the general membership of the symbiont community associated with a host may remain constant, or nearly so, the relative prevalence of each symbiont may vary considerably over time and within and among locations (39, 44, 90, 104). This variability is due primarily to differences in the tolerances of the fungi to various environmental conditions. In nature, these differences translate to changes in the relative prevalence and competitiveness of each fungus, and thus its influence on the host, as conditions shift over time. For example, the two fungi associated with *D. ponderosae* (one cold tolerant, the other heat tolerant) shift in their relative prevalence on dispersing hosts as temperatures fluctuate over a season (90). *Grosmannia clavigera* dominates during cooler periods, but *O. montium* is dominant during warmer periods. Similar effects of temperature on fungal symbiont prevalence have also been observed in the *D. frontalis* system (38). Differences in fungal preferences for phloem or sapwood are likely dictated largely by different nutritional requirements and enzymatic capabilities (2).

Differences in virulence, in tolerances to host tree defensive chemistry, and in moisture and oxygen requirements also influence which fungus dominates and when (3, 12, 13, 44, 48, 97). The tree, as a resource for both fungi and bark beetles, changes considerably from the time of attack to the time when brood beetles emerge, often as long as one year later. At the time of attack, conditions are conducive to the proliferation of pathogens. However, over time as defensive compounds dissipate, tissues dry, oxygen increases, and nutritional content declines, conditions become more suitable for saprophytes (3, 12, 13, 44). As conditions within a tree change, so will the ability of a given fungus to grow and capture resources (48, 53). Depending on the particular requirements of a fungus, it may be optimally suited to the early, middle, or late stage of tree colonization, but not to conditions occurring over the entire period. Variability in local weather conditions, particularly precipitation and temperature, adds an additional layer of stochasticity to the structure of the symbiont community.

In summary, symbiont communities associated with bark beetles are diverse and dynamic, with variable structures strongly influenced by their environment. Although highly dynamic, many have likely been shaped and fine-tuned by long periods of coevolution.

BARK BEETLE COLONIZATION OF A TREE

Trees are not sitting ducks, but rather possess elaborate defense systems that function to protect them from a plethora of insects and pathogens. These include preformed defenses, in place regardless of whether an attack occurs, and induced defenses, which form only in response to attack (9, 29). Many bark beetles colonize recently killed or severely compromised trees. For these insects, tree defenses are of trivial concern as defenses in such trees are low to lacking. However, for bark beetles that attack, and kill healthy trees, overcoming host defenses is paramount for survival and reproduction. This is a matter of kill or be killed. The general attack, colonization, and developmental sequence for a tree-killing bark beetle and its associated fungi is presented in **Figure 1**.

The killing of a tree is initiated through a pheromone-mediated mass attack (112). The number of bark beetles are required to kill a tree in an attack varies depending on the vigor of the tree (21, 67, 76). In general, the more vigorous a tree, the more bark beetles are required to overcome its defenses (9, 19, 76). The sequence of a mass attack begins with a single bark beetle arriving at a tree and releasing an aggregation pheromone that attracts conspecifics of both sexes from the surrounding area. Arriving bark beetles, in turn, release additional pheromones that increase the attractive signal and the likelihood of a successful attack (112).

Virulence: the relative degree to which a pathogen affects a host

Conifer defense system: preformed defenses of conifers consist of resin that acts as a physical barrier to entry by the insect

The first obstacle bark beetles encounter is the preformed conifer defense system, which in conifers consists primarily of resin released as bark beetles bore into the tree (9, 29). Resin acts as a physical barrier to entry by repelling, and often drowning, bark beetles and can effectively halt an attack. However, if enough bark beetles are recruited within a sufficiently short time frame, resin can be depleted, allowing bark beetles access to the phloem layer beneath the outer bark. At this point, bark beetles may still have to contend with induced defenses (9, 29). In conifers, these consist of lesions that form in the immediate area surrounding the bark beetle. Lesions contain high concentrations of secondary chemicals, which can be toxic to the insects and halt the growth of symbiotic fungi (66, 80). A strong induced defense can end an attack, in which case the tree survives. When trees are overwhelmed, either no induced defense forms or it is terminated before bark beetle attack ceases. Note that we do not state that the tree dies at this point, merely that it reaches a point of no return where the eventual death of the tree is assured. This point occurs rapidly. In fact, the entire sequence of events, from the initiation of attack to the point of no return for a tree, typically occurs over just a few days (4, 58, 76, 98). Beetles appear able to detect when the point of no return occurs. Here, they respond accordingly by switching to releasing repellant antiaggregation pheromones that act to halt the attack (76, 112).

For more information on defense responses of conifers, we refer readers to reviews by Paine et al. (70), Lieutier (59), and Franceschi et al. (29). For additional information on the colonization sequence of trees by tree-killing bark beetles, readers are directed to the review by Raffa et al. (77).

THE CLASSIC PARADIGM

Broadly stated, the CP postulates that fungi associated with bark beetles play a critical role in overwhelming tree defenses and causing tree death. Two divergent hypotheses have been proposed: the tree killing hypothesis and the defense exhaustion hypothesis. The first hypothesis suggests that colonization of the tree by virulent fungi leads directly to tree death, primarily by blocking water conduction in the xylem (56, 70). The fungi do not appear to produce toxins (25, 34). The second hypothesis proposes that fungi lead indirectly to tree death by stimulating induced defenses in the phloem, which ultimately results in their exhaustion, allowing bark beetles to invade (60).

Both hypotheses have been tested primarily using artificial inoculations of living trees with symbiotic fungi; however, how the results of such studies are interpreted varies depending on the hypothesis. Studies employing low numbers of inoculations to living trees have been used to assess the length of lesions that result from fungal colonization of the phloem. Under the tree killing hypothesis, lesion length is used as a proxy for virulence and longer lesions are assumed to be produced by more virulent associates (64, 65, 80, 82). In this context, lesion length is used to assess the relative value of each fungus to the bark beetle. The fungal partner of a bark beetle that produces the longest lesion is considered to be the most virulent and thus the most beneficial because it is most likely to contribute to the death of the tree (102, 115). However, some symbionts are better suited to growth in phloem, whereas others are more aggressive in colonizing sapwood (52, 96). Because blockage of sapwood conduction is thought to be the primary mechanism by which the fungi incur tree death, assessments of sapwood penetration as well as measurements of lesion development are critical (14, 20, 41, 46, 74, 85, 102).

Under the defense exhaustion hypothesis, which postulates that the primary role of the fungi is to exhaust tree defenses, fungal activity in phloem is vital because induced defenses are initiated and form in tree phloem. Therefore, in this case, lesion length becomes the critical measure (60). Under this hypothesis, virulence of the fungus (ability to penetrate sapwood and block water conduction) is of minimal to no importance (60).

High-density inoculations have been used to investigate the tree killing hypothesis. In these studies, trees are inoculated with fungi at high densities (similar to those occurring under natural attack, or often, much higher densities) (46, 85, 102). The trees are then assessed weeks to months later for reductions in water conduction or mortality (14, 20, 41, 74).

Difficulties Testing the CP

The CP remains controversial because it is difficult to test directly through experimentation. The choice of a tree, and its subsequent attack and colonization by bark beetles and fungi, is a complex process and not one easily replicated in experiments. The cues used by bark beetles to choose an appropriate host are not known; thus, trees chosen by humans may or may not be those most suitable for attack. The logistics of both inoculating and punching holes into trees in a way that mimics bark beetle attack is daunting, especially when one factors in timing, the number of required treatment levels and replication, and the need for positive and negative controls. Such studies still lack other important factors that may influence a tree reaching the point of no return in a biologically meaningful way. These include proper dose of mimicked attacks to match the vigor of the tree and the effects of adult and larval tunneling (and consequently, effects of oxygen and moisture on fungi growth and the extent of their distribution within the tree). The fact that we have no way of precisely determining when the point of no return occurs, and that tree death is not a discrete event but rather one that occurs by compartment over an extended period, regardless of causal agent (27), is also problematic when it comes to assessing outcomes of our experiments.

Because of these and other difficulties in directly testing the CP, we have taken a more indirect approach to assess its validity. In the next section, we develop several arguments that show, alone and in combination, that the CP is fundamentally flawed.

ARGUMENTS AGAINST THE CLASSIC PARADIGM

Tree-Killing Bark Beetles Can Kill Trees Without Virulent Pathogens

Perhaps the most compelling argument against the role of fungi in killing trees is that tree-killing bark beetles can and often do kill trees in the absence of virulent pathogens. Most bark beetles are associated with more than one symbiotic fungus (44, 47, 88, 104). For tree-killing bark beetles that possess virulent associates (not all do), typically only one associate exhibits this characteristic and the prevalence of this associate in a population can vary considerably. Such variability is inconsistent with a critical direct role of fungi in tree killing. Under the CP, populations lacking the virulent associate would move rapidly toward extinction because they would not be able to kill trees. The success of a population would depend on the relative prevalence of the virulent fungus. If present with all or most bark beetles, the population should be robust and perhaps even expand. In contrast, when prevalence is low, populations should decline, and if environmental conditions do not shift rapidly in a manner to increase the prevalence of the virulent associate, those populations should also move rapidly toward extinction.

However, this is not what we observe. The first observations that trees can be killed by bark beetles in the absence of a virulent pathogen were made by Hetrick (35) and Bridges et al. (17). These authors observed pines that had been killed by *D. frontalis* but that were lacking *Ophiostoma minus*, the only pathogenic fungus commonly associated with this beetle. This phenomenon has subsequently been observed for *Dendroctonus brevicomis* (109). Another example is *Ips typographus*, the most serious killer of spruce in Europe. This beetle is associated with *Ceratocystis polonica*, a highly virulent pathogen capable of killing trees, and with several other weakly virulent or nonpathogenic fungi (44, 47). Because of its virulence, *C. polonica* is thought to be critical in causing tree mortality. However, the prevalence of this fungus with the beetle is highly variable, as it is found commonly in some portions of the host's range but only rarely or not at all in others (44, 47, 55, 86, 95, 104, 105). Yet, even in populations where the fungus is rare or apparently lacking, the bark beetle kills trees and is capable of developing outbreaks. Likewise, with *D. ponderosae* (the most serious pest of pines in the western United States and Canada) the presence of its more virulent associate is not required for tree death or the development of epidemics (88, 90).

Yet another example is *Dendroctonus rufipennis* (spruce beetle), which vectors *Leptographium abietinum* and *Ceratocystis rufipenni*. *C. rufipenni* is highly virulent and thus under the CP has been postulated to be an important mutualist (102). *L. abietinum*, on the other hand, is only weakly pathogenic, and its importance to the bark beetle has thus been downplayed. However, *L. abietinum* is by far the most prevalent fungus with this beetle (30, 81, 89). It is present with greater than 90% of bark beetles in all populations thus far surveyed (inclusive of most of its geographic range), whereas *C. rufipenni* is apparently lacking in most populations and is usually rare in those where it does occur (30, 81, 89). Although most *D. rufipennis* populations lack a virulent fungus, they remain capable of killing trees and causing extensive mortality whenever conditions support increases in bark beetles.

From an evolutionary perspective, the inconsistency of association of virulent fungi with their host bark beetles poses a severe dilemma for bark beetles under the CP. If a bark beetle required a virulent pathogen to overcome tree defenses and/or incur tree mortality, it would be inherently risky for it to enter a tree without carrying such a fungus. Any bark beetle without such a fungus would have an increased risk of being killed by the tree and of not reproducing with subsequent strong negative effects on fitness. This should result in strong selection pressure to maintain a highly consistent association with the virulent fungus. As previously noted, the virulent associates of aggressive bark beetles are actually some of those least consistently present. This suggests that if they benefit bark beetles at all through phytopathogenicity, it is a facultative effect at best. In fact, trees develop more pronounced induced defense responses when challenged by virulent associates. This finding suggests that that their association with a bark beetle may actually increase risk during attack, potentially decreasing the fitness of these bark beetles relative to those that enter carrying less virulent associates. Indeed, Raffa & Smalley (78) found that the presence of virulent bark beetle–associated fungi causes the accumulation of allelochemicals in trees in concentrations that adversely affect bark beetles.

Inconsistency of association also poses problems for the defense exhaustion hypothesis. Under this hypothesis, the presence of fungi that cause the most rapid and extensive stimulation of the induced defense is critical. If defenses are not exhausted rapidly, bark beetles are killed or repelled. Therefore, there should be a close correspondence between a bark beetle's aggressiveness and the ability of their main associated fungal species to rapidly stimulate the defenses of the host tree (60). However, in most systems this close correspondence does not occur. For example, the most consistent associates of *D. rufipennis*, *D. brevicomis*, and *D. frontalis* do not stimulate much in the way of an induced defense (69, 71, 72, 85), yet trees are still killed.

It has been proposed that a lack of consistency of association between a bark beetle and a single fungus highly efficient at stimulating defenses may be compensated for by the complex of fungi often carried by a bark beetle (60). It is true that most bark beetles are associated with a complex of fungal species, some of which can stimulate defenses. However, given that most fungi in such complexes are highly incidental and often present on only a low proportion of dispersing bark beetles in a population, the reliance of a bark beetle on such an undependable and variable group of fungi to fulfill such a critical function seems unlikely.

Although much has been made of the ability of some fungi associated with bark beetles to kill trees, few are capable of doing so. In most inoculation studies, trees survive inoculation with bark beetle symbiotic fungi (51, 102) unless inoculations have been done at high densities, with unnaturally high inoculum loads, or both (26, 47, 85).

The Point of No Return of the Tree is not Coincident with Fungal Colonization of Tree Tissues

For fungi to be the proximate cause of bark beetle attack, their successful growth in tree tissues and effects on tree function must occur within a short time frame. However, all studies indicate that fungi grow slowly within trees, especially during the critical initial stage of bark beetle colonization (36, 84, 98) (**Figure 1**). Likewise, effects on water conduction are not manifest until bark beetle brood development is substantially advanced (14, 115).

A successful attack on a tree is often complete within just a few days, and oviposition and brood development occur immediately thereafter (4, 76). Fungal colonization, however, occurs at a much slower rate. In some cases, particularly with less virulent associates, fungal growth is initially limited by high moisture and low oxygen (100). For more virulent associates, growth occurs more rapidly, but still relatively slowly compared with colonization of the tree by bark beetle hosts. For example, the fungal associates of *I. typographus* penetrate the sapwood to a depth of only about 20 mm after 4–5 weeks, a time when bark beetle larvae have nearly completed their development (98).

This slow rate of penetration indicates that any effect fungi have on conduction in the tree occurs long after trees have been overwhelmed, not before. In fact, effects on conduction may not be due wholly to fungal proliferation. Hobson et al. (36) found that fungal penetration of sapwood follows sapwood occlusion rather than preceding it. In any case, a substantial amount of sapwood must be affected relatively rapidly to induce the symptoms associated with bark beetle attack within a meaningful time frame. Vite (106), extrapolating from experimental work, estimated that more than two-thirds of the sapwood would need to be rapidly disrupted to cause symptoms in trees in a time frame similar to those which occur with bark beetle attack. In the case of bark beetle–associated fungi, growth in trees, even by virulent associates, occurs much too slowly to match this time frame. In fact, at natural attack densities, symbiotic fungi take several months to achieve similar effects, if they occur at all (73, 99). This contradiction in fungal colonization rate relative to the development of bark beetle brood and tree symptoms has been previously noted (60, 70) and has proven to be one of the most difficult dilemmas to resolve under the tree killing hypothesis.

From these studies, we can see that fungal colonization of tree tissues begins in earnest only after the point of no return has been reached and when bark beetle establishment is well underway. Thus, fungal colonization follows bark beetle colonization, not vice versa. Although some bark beetle–associated fungi are virulent, and a few have been shown to be capable of killing trees in inoculation studies, this virulence may not be biologically meaningful in the context of overwhelming tree defenses and causing tree death. Due to the slow growth rate of bark beetle–associated fungi within trees during the critical attack phase, fungus-caused mortality or a reduction in tree defenses would occur much too late to be of benefit to the insects. The phytopathogenicity exhibited by some of these fungi may play a different role as discussed below.

The Distribution of Virulent Fungal Associates is not Correlated with Bark Beetle Aggressiveness

Under the CP, one would predict that virulence in fungal associates would correlate with bark beetle life history; tree-killing beetles would possess virulent pathogenic fungal associates. Likewise, bark beetles that do not kill trees would have either no fungal associates or only incidental ones lacking virulence. However, this pattern is not observed in nature. For many aggressive bark beetles, the most consistent associates are nonpathogenic or weak pathogens. These include most of the obligate mycangial fungi involved in nutritional symbioses (40, 47, 69, 88, 91). In contrast, virulent fungi can often be found with bark beetles that do not typically kill trees. For example, *Dendroctonus murrayanae* (lodgepole pine beetle), *D. terebrans* (black turpentine beetle), and *D. valens* (red turpentine beetle) complete development in living trees (113). Although these beetles do not kill their tree hosts, they are often associated with *Leptographium terebrantis*, a fungus that is among the most virulent of all Ophiostomatales associated with bark beetles (7, 28, 45, 50, 64, 79, 110). Likewise, *L. wingfieldii* is highly virulent to pines (45, 64) but is carried by *Tomicus piniperda* (pine shoot beetle), which typically infests shoots without killing the host tree.

Non-Tree-Killing Bark Beetles also Have Fungal Associates

The CP arose from observations on tree-killing bark beetle systems, and it is on these systems that virtually all research has focused. However, tree killing is a rare strategy and not representative of the life histories of the vast majority of bark beetles. Worldwide, fewer than fifteen of the thousands of bark beetle species can be considered aggressive tree killers. Many, if not most, non-tree-killing species also possess ophiostomatoid fungal associates. This begs the question of why non-tree-killing bark beetles possess fungi similar to those associated with tree-killing bark beetles if they do not need them to kill trees. The presence of these fungi, as well as mycangia, with some non-tree-killing species indicates that for at least some species the fungi are important and likely play a role or roles other than those postulated under the CP. Unfortunately, little is known about these symbioses because they do not involve economic pests. For this reason, they have not engendered much interest from the forest entomology community. However, to avoid error due to bias in sampling from forming the basis for the hypotheses we use to investigate these systems, we need to study the composition and function of these symbioses over the entire range of bark beetle life strategies.

LESSONS LEARNED FROM THE *DENDROCTONUS FRONTALIS*–FUNGUS SYSTEM

The best studied of all bark beetle–fungus symbioses is the *D. frontalis* system. *D. frontalis* is the most important pine-killing species in the southern United States. In this region, the bark beetle is associated with three fungi. Two of these, *Ceratocystiopsis ranaculosus* and *Entomocorticium* sp. A, are specific to the bark beetle and carried consistently in mycangia (49). Both fungi appear coevolved with their host (94). Neither of these fungi is phytopathogenic, and defensive responses by the tree to them are minimal (40). Therefore, neither is capable of killing the tree, nor are they likely to contribute in any substantial way to the exhaustion of tree defenses. The third fungus, *O. minus*, is never carried in mycangia, but instead loosely and less consistently on the exoskeleton (49). In addition, this fungus is not specific to the bark beetle, as it is found with many bark beetles including species that colonize weakened, dying, or dead trees (47, 64, 101). This fungus initiates a moderate induced defense response in pines (22, 40, 51, 75).

Because of its pathogenicity, *O. minus* was long considered a critical mutualist of *D. frontalis* (14, 24, 68), overshadowing the beneficial nutritional roles of the mycangial fungi (5, 6, 15, 23). Gradually, however, perceptions of the relative importance and roles of the three fungi with their bark beetle host began to shift. Notably, observations that *O. minus* was not always present, or present only in small amounts, in trees killed by *D. frontalis* brought into question the role of this fungus in tree killing. Furthermore, inoculation studies indicated that *O. minus* is not capable of killing mature pines (51). Perhaps most revealing were observations that tunneling larvae of *D. frontalis* turn away from phloem colonized by *O. minus* and do not survive when they cannot avoid feeding in *O. minus*–colonized areas (6, 63). Together, these studies and observations suggested that not only is *O. minus* not responsible for overwhelming the tree, it is not a mutualist, but rather an antagonist.

Ironically, the notion that *O. minus* is a mutualist has remained firmly entrenched, and this has led to attempts to reconcile the antagonistic effects of the fungus with the CP. For example, it has been suggested that *O. minus* acts as a mutualist early in the colonization of a tree by aiding the bark beetle in overwhelming tree defenses, but that once the tree is overcome, the fungus acts as an antagonist during larval development (49, 54). This shift in roles by the fungus over a single insect generation has been presented as an example of context dependency (52). However, context dependency is more correctly defined as variation in net outcomes of an interaction due to stochastic shifts in biotic and abiotic conditions (18). This term is not inclusive of the normal predictable change in substrate conditions encountered by a host and its symbionts, such as that occurring in a tree over a generation of a bark beetle and its fungi.

Net outcomes are the sum total of effects of the interaction on partner fitness and determine whether an association can be considered a mutualism, a commensalism, or an antagonism. Even if a beneficial effect occurs at some point within a host generation, and if the net effect of the interaction overall is a reduction in either partner's fitness, the interaction cannot be considered a mutualism. In the case of

D. frontalis, *O. minus* is not required by the bark beetle and indeed is unlikely to contribute to tree mortality when it is present. In addition, its effects on *D. frontalis* reproduction and survival are strongly negative and increase with increasing prevalence of the fungus within a tree. Indeed, effects of this fungus are so severe that once a particular threshold of phloem area colonized by *O. minus* in the tree is reached, bark beetle populations collapse (37, 61).

O. minus is also unlikely to benefit the bark beetle through defense exhaustion. This fungus initiates only a small-to-moderate lesion response. Thus, for this fungus to rapidly stimulate defenses to the point that they are exhausted, high levels of fungal inoculation by attacking bark beetles are likely required. However, trees can be overwhelmed without the fungus and high levels of *O. minus* in a tree drastically reduce the host bark beetle's fitness, making this fungus an unlikely candidate to fulfill this role.

Additional indirect evidence that *O. minus* is not a mutualist of the bark beetle comes from studies on mites associated with *D. frontalis*. Tarsonemid mites phoretic on *D. frontalis* carry *O. minus* in structures called sporothecae (10, 16). The mites are involved in a nutritional mutualism with *O. minus*; feeding on the fungus results in high levels of mite productivity (62). The prevalence of *O. minus* in a tree colonized by *D. frontalis* is determined by the abundance of mites, which in turn is primarily driven by temperature (38, 61). When thermal conditions are favorable for mites, mite abundance increases and the prevalence of *O. minus* within the tree likewise increases (61). Once a particular threshold of area colonized by *O. minus* in the tree is reached, bark beetle populations decline. This fungus-driven decline in bark beetles can even result in the termination of outbreaks (37, 61).

A recent and fascinating discovery is that *D. frontalis* carries actinomycete bacteria that are antagonistic to *O. minus* (1, 87). Although the actinobacteria–bark beetle association is only peripherally understood, it would be difficult to reconcile the notion that *D. frontalis* would selectively carry a microbe antagonistic to a fungus on which it relies for tree killing.

The *D. frontalis*–fungus system includes one of the most important tree-killing bark beetles in the world. The insect carries a pathogenic fungus, but fungi do not appear to play a role in tree killing or in exhausting tree defenses. This exemplifies the fact that strong fidelity to the CP for many decades has frustrated our understanding of the manner in which these interactions truly function. The real story has only begun to unfold now that researchers have been willing to consider alternative roles for fungal partners.

WHY ARE SOME FUNGI ASSOCIATED WITH BARK BEETLES PHYTOPATHOGENIC?

If phytopathogenicity is not required to aid bark beetle hosts in overcoming tree defenses, why do some fungal associates possess this quality? Perhaps a useful place to begin to investigate this question is to inspect fungal lifestyles and strategies and to consider how phytopathogenicity may be important to the fungi rather than to the bark beetle.

The fungi associated with tree-killing bark beetles must initially face hostile conditions as the bark beetles deliver them into a tree that is still living and able to defend itself. Pathogenicity may allow these fungi to survive in a living tree until defenses decline and the environment becomes more conducive to growth. In addition, pathogenicity may play a role in competition among the fungal associates (31). Pathogens would be more competitive early in the colonization process when tree tissues are still living, whereas more saprophytic species may become dominant later on.

Relative differences in pathogenicity likely play a substantial role in determining fungal community dynamics within a tree over time. At first glance, the multiple symbionts associated with a bark beetle host appear to occupy the same niche (they occur in the same place at the same time, use similar resources, and compete for the same hosts for dispersal). This should result in strong direct competition and selection for whichever species is most competitive. However, even slight differences in environmental tolerances and intrinsic vital properties (such as virulence and resource use) can alter niche hyperspace to the degree that several fungi can occupy different realized niches within a limited resource base. This effectively reduces competition and allows for coexistence between a number of different symbionts. For example, we know that the two fungi associated with *D. ponderosae* possess different temperature tolerances (92, 100). These differences determine which fungus is vectored by dispersing host bark beetles as temperatures fluctuate over a season and which fungus dominates within a tree during the developmental period of a bark beetle (3, 12, 90). By growing at different temperatures, and thus at different times, the fungi may minimize competition with one another except within a narrow range of temperatures at which the growth of both fungi is equally supported. This separation in niches is most likely further attenuated through differential use of carbon and nitrogen sources within the tree (11).

Differences in relative virulence are also likely to play a role in the niche separation of these two fungi. *G. clavigera*, a common associate of *D. ponderosae*, is moderately virulent (100). It can grow in still-living host tissues containing defensive compounds and under the low-oxygen and high-moisture conditions that predominate in trees during the initial stages of bark beetle development (100). On the other hand, *O. montium*, also associated with this beetle, is only weakly virulent. It grows slowly during the initial stages of bark beetle development but proliferates rapidly once tree defenses have declined and oxygen content increases and moisture content decreases (11, 100).

Virulent fungi are often found with bark beetles that complete their entire development in living trees. In this case, high levels of virulence may be related to the need for these fungi to grow and survive in a defensive host that is continuously trying to kill them or to restrict their growth. These fungi must survive for up to a year under such conditions, until their vectors complete development and transport them to a new host. If these fungi did not exhibit relatively high levels of virulence, they would likely be killed or contained soon after entry into the tree.

SUMMARY POINTS

1. Symbioses between tree-killing bark beetles and ophiostomatoid fungi have been postulated to be mutualisms in which the fungi benefit through transport to new host trees and in return benefit the bark beetles by aiding in overwhelming tree defenses and/or killing the tree. The CP has driven most research on these symbioses, yet after decades of study, no conclusive evidence exists supporting this role for the fungi.
2. Several lines of indirect evidence strongly suggest that the CP is fundamentally flawed. These include the lack of consistency of virulent fungal associates with tree-killing bark beetles, the lack of correspondence between fungal growth in the host tree and the development of symptoms associated with a successful attack, and the ubiquity of similar associations of fungi with bark beetles that do not kill trees.
3. Nearly all focus on fungal phytopathogenicity has been on the importance of this characteristic to the bark beetle. However, we suggest that, rather than playing a supporting role for the host bark beetle (tree killing), phytopathogenicity performs an important role for the fungi that exhibit this characteristic, particularly in mediating competitive interactions with other fungi and supporting survival and efficient resource capture in living, defensive trees.

FUTURE ISSUES

1. An important first step will be to broaden our approach to include symbioses between fungi and non-tree-killing bark beetle species. It will be informative to investigate whether interactions among non-tree-killing bark beetles and fungi are similar or inherently different from those occurring between tree-killing species and fungi. In addition, it will be important to understand why species of related fungi can have profoundly different effects on the host insect. For instance, there is a need to understand why some *Ophiostoma* species are mutualists and others are commensals or antagonists.
2. An understanding of how biotic and abiotic factors affect the relative prevalence of the fungi with a host over time will be critically important in understanding the dynamics of these associations. Fungal dynamics must surely also affect bark beetle population dynamics, and understanding how this occurs will be important.
3. Powerful molecular tools are now available to aid in our understanding of the evolution and function of these symbioses. Also, mutualism theory has seen amazing advancements in just the past few years. Some of these powerful new tools should be used to investigate bark beetle–fungus systems.
4. In a review of bark beetle–fungus interactions, Stewart Whitney, an early pioneer of this topic, suggested that "Occam's razor might not be sharp enough to slice through the jungle of information and the simplest hypothesis may merely be simplest in a complex series of hypotheses" (107). Bark beetle–fungus symbioses are complex, and the role that the fungi play in the lives of the bark beetles remains substantially clouded. We have argued here that the CP, although a comfortable hypothesis, is strongly flawed. By seeking alternatives rather than a convenient explanation, we expect that many fascinating and previously unimagined roles for the fungi in their relationships with bark beetles will emerge.

DISCLOSURE STATEMENT

The authors are not aware of any affiliations, memberships, funding, or financial holdings that might be perceived as affecting the objectivity of this review.

ACKNOWLEDGMENTS

Many colleagues in various parts of the world, and too numerous to name individually, have contributed to interesting discussions and debates regarding the role of fungi in their symbiosis with bark beetles. We are grateful for the views that they have shared with us, and although some will not agree with all our arguments, we hope that these will stimulate a new wave of understanding of a fascinating and long-debated topic. We also appreciate the financial support of the DST/NRF Center of Excellence in Tree Health Biotechnology, South Africa.

LITERATURE CITED

1. Aanen DK, Slippers B, Wingfield MJ. 2009. Biological pest control in beetle agriculture. *Trends Microbiol.* 17:179–82
2. Abraham L, Hoffman B, Gao Y, Breuil C. 1998. Action of *Ophiostoma piceae* proteinase and lipase on wood nutrients. *Can. J. Microbiol.* 44:698–701

3. Adams AS, Six DL. 2007. Temporal variation in mycophagy and prevalence of fungi associated with developmental stages of the mountain pine beetle, *Dendroctonus ponderosae* (Coleoptera: Scolytinae, Curculionidae). *Environ. Entomol.* 36:64–72
4. Anderbrandt O. 1988. Survival of parent and brood bark beetles, *Ips typographus*, in relation to size, lipid content and reemergence or emergence day. *Physiol. Entomol.* 13:121–29
5. Ayres MP, Wilkens RT, Ruel JJ. 2000. Nitrogen budgets of phloem-feeding bark beetles with and without symbiotic fungi. *Ecology* 81:2198–210
6. Barras SJ. 1970. Antagonism between *Dendroctonus frontalis* and the fungus *Ceratocystis minor*. *Ann. Entomol. Soc. Am.* 63:1187–90
7. Barras SJ, Perry T. 1971. *Leptographium terebrantis* sp. nov. associated with *Dendroctonus terebrans* in loblolly pine. *Mycopathologia* 43:1–10
8. Beaver RA. 1989. Insect-fungus relationships in the bark and ambrosia beetles. In *Insect-Fungus Interactions*, ed. N Wilding, NM Collins, PM Hammond, JF Webber, pp. 121–43. London: Academic
9. **Berryman AA. 1972. Resistance of conifers to invasion by bark beetle–fungus associations. *Bioscience* 22:598–602**

9. Presents a seminal synthesis on conifer defenses against bark beetles and their fungal symbionts.

10. Blackwell M, Bridges JR, Moser JC, Perry TJ. 1986. Hyperphoretic dispersal of a *Pyxidiophora* anamorph. *Science* 232:993–95
11. Bleiker K, Six DL. 2007. Dietary benefits of fungal associates to an eruptive herbivore: potential implications of multiple associates on host population dynamics. *Environ. Entomol.* 36:1384–96
12. Bleiker KP, Six DL. 2008. Competition and coexistence in a multi-partner mutualism: interactions between two fungal symbionts of the mountain pine beetle in beetle-attacked trees. *Microbial Ecol.* 57:191–202
13. Bleiker KP, Six DL. 2009. Effects of water potential and solute on the growth and interactions of two fungal symbionts of the mountain pine beetle. *Mycol. Res.* 113:3–15
14. Bramble WC, Holst EC. 1940. Fungi associated with *Dendroctonus frontalis* in killing shortleaf pines and their effects on conduction. *Phytopathology* 30:881–99
15. Bridges R. 1983. Mycangial fungi of *Dendroctonus frontalis* (Coleoptera: Scolytidae) and their relationship to beetle population trends. *Environ. Entomol.* 12:858–61
16. Bridges RJ, Moser JC. 1983. Role of two phoretic mites in the transmission of bluestain fungus, *Ceratocystis minor*. *Ecol. Entomol.* 8:9–12
17. Bridges RJ, Nettleton WA, Connor MD. 1985. Southern pine beetle (Coleoptera: Scolytidae) infestations without the bluestain fungus, *Ceratocystis minor*. *J. Econ. Entomol.* 78:325–27
18. Bronstein JL. 1994. Conditional outcomes in mutualistic interactions. *Trends Ecol. Evol.* 9:214–17
19. Christiansen E. 1985. *Ips/Ceratocystis* infection of Norway spruce: What is a deadly dosage? *J. Appl. Entomol.* 99:6–11
20. Christiansen E, Solheim H. 1990. The bark beetle-associated blue stain fungus *Ophiostoma polonicum* can kill various spruces and Douglas fir. *Eur. J. For. Pathol.* 20:436–46
21. Christiansen E, Waring RH, Berryman AA. 1987. Resistance of conifers to bark beetle attack: searching for general relationships. *For. Ecol. Manag.* 22:89–106
22. Cook SP, Hain FP. 1985. Qualitative examination of hypersensitive response of loblolly pine, *Pinus taeda* L., inoculated with two fungal associates of the southern pine beetle, *Dendroctonus frontalis* Zimmermann (Coleoptera: Scolytidae). *Environ. Entomol.* 14:396–400
23. Coppedge BR, Stephen FM, Felton GW. 1995. Variation in female southern pine beetle size and lipid content in relation to fungal associates. *Can. Entomol.* 127:145–54
24. Craighead FC. 1928. The interrelation of tree-killing bark beetles (*Dendroctonus*) and blue stains. *J. For.* 26:886–87
25. DeAngelis JD, Hodges JD, Nebeker TE. 1986. Phenolic metabolites of *Ceratocystis minor* from laboratory cultures and their effects on transpiration in loblolly pine seedling. *Can. J. Bot.* 64:151–55
26. Fernandez MMF, Garcia AE, Lieutier F. 2004. Effects of various densities of *Ophiostoma ips* inoculations on *Pinus sylvestris* in north-western Spain. *For. Pathol.* 34:213–23
27. Filip GM, Schmitt CL, Scott DW, Fitzgerald SA. 2006. Understanding and defining mortality in western conifer forests. *West. J. Appl. For.* 22:105–15

28. Fox JW, Wood DL, Akers PP, Partmeter JR Jr. 1992. Survival and development of *Ips paraconfusus* Lanier (Coleoptera: Scolytidae) reared axenically and with tree pathogenic fungi vectored by cohabiting *Dendroctonus* species. *Can. Entomol.* 124:1157–67
29. Francheschi VR, Krokene P, Christiansen E, Krekling T. 2005. Anatomical and chemical defenses of conifer bark against bark beetle and other pests. *New Phytol.* 167:353–75
30. Haberkern KE, Illman BL, Raffa KF. 2002. Bark beetles and associates colonizing white spruce in the Great Lakes region. *Can. J. For. Res.* 32:1137–50
31. Harrington TC. 1993. Diseases of conifers caused by species of *Ophiostoma* and *Leptographium*. See Ref. 111, pp. 161–72
32. Harrington TC. 2005. Ecology and evolution of mycophagous bark beetles and their fungal partners. In *Insect-Fungal Associations: Ecology and Evolution*, ed. FE Vega, M Blackwell, pp. 257–91. Oxford: Oxford Univ. Press
33. Harrington TC, Wingfield MJ. 1998. The *Ceratocystis* species on conifers. *Can. J. Bot.* 17:1446–57
34. Hemingway RW, McGraw GW, Barras SJ. 1977. Polyphenols in *Ceratocystis minor*–infected *Pinus taeda*: fungal metabolites, phloem and xylem phenols. *Agric. Food Chem.* 25:717–22
35. Hetrick LA. 1949. Some overlooked relationships of the southern pine beetle. *J. Econ. Entomol.* 42:466–69
36. Hobson KR, Parmeter JR Jr, Wood DL. 1994. The role of fungi vectored by *Dendroctonus brevicomis* LeConte (Coleoptera: Scolytidae) in occlusion of ponderosa pine xylem. *Can. Entomol.* 126:277–82
37. Hofstetter RW, Cronin J, Klepzig KD, Moser JC, Ayres MP. 2006. Antagonisms, mutualisms and commensalisms affect outbreak dynamics of the southern pine beetle. *Oecologia* 147:679–91
38. Hofstteter RW, Dempsey TD, Klepzig KD, Ayres MP. 2007. Temperature-dependent effects on mutualistic, antagonistic and commensalistic interactions among insects, fungi and mites. *Community Ecol.* 8:47–56
39. Hofstetter RW, Klepzig KD, Moser JC, Ayres MP. 2006. Seasonal dynamics of mites and fungi and their effects on the southern pine beetle. *Environ. Entomol.* 35:22–30
40. Hofstetter RW, Mahfouz JB, Klepzig KD, Ayres MP. 2005. Effects of tree phytochemistry on the interactions among endophloedic fungi associated with the southern pine beetle. *J. Chem. Ecol.* 31:539–57
41. Hornvedt R, Christiansen E, Solheim H, Wand S. 1983. Artificial inoculation with *Ips typographus*–associated blue-stain fungi can kill healthy Norway spruce trees. *Medd. Nor. Inst. Skogforsk* 38:1–20
42. Hsiau P-TW, Harrington TC. 2003. Phylogenetics and adaptations of basidiomycetous fungi fed upon by bark beetles (Coleoptera: Scolytidae). *Symbiosis* 34:111–31
43. Jacobs K, Wingfield MJ. 2001. Leptographium *Species: Tree Pathogens, Insect Associates, and Agents of Blue-Stain*. St. Paul: Am. Phytopathol. Soc. Press
44. Jankowiak R. 2005. Fungi associated with *Ips typographus* on *Picea abies* in southern Poland and their succession into the phloem and sapwood of beetle-infested trees and logs. *For. Pathol.* 35:37–55
45. Jankowiak R. 2006. Fungi associated with *Tomicus piniperda* in Poland and assessment of their virulence using Scots pine seedlings. *Ann. For. Sci.* 63:801–8
46. Kim J-J, Plattner A, Lim YW, Breuil C. 2008. Comparison of two methods to assess the virulence of the mountain pine beetle associate, *Grosmannia clavigera*, to *Pinus contorta*. *Scand. J. For. Res.* 23:98–104
47. **Kirisits T. 2004. Fungal associates of European bark beetles with special emphasis on the ophiostomatoid fungi. In *Bark and Wood Boring Insects in Living Trees in Europe, A Synthesis*, ed. F Lieutier, KR Day, A Battisti, J-C Gregoire, HF Evans, pp. 181–236. Dordrecht, Nether.: Kluwer Acad.**

47. Reviews bark beetle-fungus symbioses.

48. Klepzig KD, Flores-Otero J, Hofstetter RW, Ayres MP. 2004. Effects of available water on growth and competition of southern pine beetle associated fungi. *Mycol. Res.* 108:183–88
49. Klepzig KD, Moser JC, Lombardero FJ, Hofstetter RW, Ayres MP. 2001. Symbiosis and competition: complex interactions among beetles, fungi and mites. *Symbiosis* 30:83–96
50. Klepzig KD, Raffa KF, Smalley EB. 1991. Association of an insect-fungal complex with red pine decline in Wisconsin. *For. Sci.* 37:1119–39
51. Klepzig KD, Robison DJ, Fowler G, Minchin PR, Hain FP, Allen HL. 2005. Effects of mass inoculation on induced oleoresin response in intensively managed loblolly pine. *Tree Physiol.* 25:681–88

52. Klepzig KD, Six DL. 2004. Context dependency in bark beetle-fungal symbioses: complex interactions in complex associations. *Symbiosis* 37:189–205
53. Klepzig KD, Wilkens RT. 1997. Competitive interactions among symbiotic fungi of the southern pine beetle. *Appl. Environ. Microbiol.* 63:621–27
54. Kopper BJ, Klepzig KD, Raffa KF. 2004. Components of antagonism and mutualism in *Ips pini*–fungal interactions: relationship to a life history of colonizing highly stressed and dead trees. *Environ. Entomol.* 33:28–34
55. Krokene P, Solheim H. 1996. Fungal associates of five bark beetles species colonizing Norway spruce. *Can. J. For. Res.* 26:2115–22
56. Langstrom B, Solheim H, Hellqvist C, Gref R. 1993. Effects of pruning young Scots pines on host vigor and susceptibility to *Leptographium wingfieldii* and *Ophiostoma minus*, two blue stain fungi associated with *Tomicus piniperda*. *Eur. J. For. Pathol.* 23:400–15
57. Lee S, Kim J-J, Breuil C. 2006. Fungal diversity associated with the mountain pine beetle, *Dendroctonus ponderosae* and infested lodgepole pines in British Columbia. *Fungal Divers.* 22:91–105
58. Lieutier F. 2002. Mechanisms of resistance in conifers and bark beetle attack strategies. In *Mechanism and Deployment of Resistance in Trees to Insects*, ed. MR Wagner, KM Clancy, F Lieutier, TD Paine, pp. 31–77. Dordrecht, Nether.: Kluwer Acad.
59. Lieutier F. 2004. Host resistance to bark beetles and its variations. In *Bark and Wood Boring Insects in Living Trees in Europe, A Synthesis*, ed. F Lieutier, KR Day, A Battisti, J-C Gregoire, HF Evans, pp. 135–80. Dordrecht, Nether.: Kluwer Acad.
60. Lieutier F, Yart A, Salle A. 2009. Stimulation of tree defenses by ophiostomatoid fungi can explain attack success of bark beetles in conifers. *Ann. For Sci.* 66:801
61. **Lombardero MJ, Ayres MP, Hofstetter RW, Moser JC, Klepzig KD. 2003. Strong indirect interactions of *Tarsonemus* mites (Acarina: Tarsonemidae) and *Dendroctonus frontalis* (Coleoptera: Scolytidae). *Oikos* 102:243–52**

61. Studies the roles and interactions of pathogenic and nonpathogenic fungi and phoretic mites with an important tree-killing beetle.

62. Lombardero MJ, Klepzig KD, Moser JC, Ayres MP. 2000. Biology, demography and community interactions of *Tarsonemus* (Acarina: Tarsonemidae) mites phoretic on *Dendroctonus frontalis* (Coleoptera: Scolytidae). *Agric. For. Entomol.* 2:193–202
63. Mathiesen-Kaarik A. 1960. Studies on the ecology, taxonomy, and physiology of Swedish insect-associated blue stain fungi, especially the genus *Ceratocystis*. *Oikos* 11:1–25
64. Matsuya H, Kaneko S, Yamaoka Y. 2003. Comparative virulence of blue-stain fungi isolated from Japanese red pine. *J. For. Res.* 8:83–88
65. Molnar AC. 1965. Pathogenic fungi associated with a bark beetle on alpine fir. *Can. J. Bot.* 43:563–70
66. Mullick DB. 1977. The nonspecific nature of defense in bark and wood during wounding, insect and pathogen attack. *Rec. Adv. Phytochem.* 11:359–441
67. Mulock P, Christiansen E. 1986. The threshold of successful attack by *Ips typographus* on *Picea abies*: a field experiment. *For. Ecol. Manag.* 14:125–32
68. Nelson RM. 1934. Effect of bluestain fungi on southern pines attacked by bark beetles. *Phytopathol. Z.* 7:327–53
69. Paine TD. 1984. Seasonal response of ponderosa pine to inoculation of the mycangial fungi of the western pine beetle. *Can. J. Bot.* 62:551–55
70. Paine TD, Raffa KF, Harrington TC. 1997. Interactions among scolytid bark beetles, their associated fungi, and live host conifers. *Annu. Rev. Entomol.* 42:179–206
71. Paine TD, Stephen FM. 1987. Response of loblolly pine to different inoculum doses of *Ceratocystis minor*, a blue-satin fungus associated with *Dendroctonus frontalis*. *Can. J. Bot.* 65:2093–95
72. Paine TD, Stephen FM, Cates RG. 1988. Phenology of an induced response in loblolly pine following inoculation of fungi associated with the southern pine beetle. *Can. J. For. Res.* 18:1556–62
73. Parmeter JR Jr, Slaughter GW, Chen M, Wood DL. 1992. Rate and depth of sapwood occlusion following inoculation of pines with bluestain fungi. *For. Sci.* 38:34–41
74. Parmeter JR Jr, Slaughter GW, Chen M-M, Wood DL, Stubbs HA. 1989. Single and mixed inoculations of ponderosa pine with fungal associates of *Dendroctonus* spp. *Phytopathology* 79:768–72
75. Popp MP, Johnson JD, Lesney MS. 1995. Characterization of the induced response of slash pine to inoculation with bark beetle vectored fungi. *Tree Physiol.* 15:619–23

76. Raffa KF, Berryman AA. 1983. Physiological aspects of lodgepole pine wound responses to a fungal symbiont of the mountain pine beetle. *Can. Entomol.* 115:723–34
77. Raffa KF, Phillips TW, Salom SM. 1993. Strategies and mechanisms of host colonization by bark beetles. In *Beetle-Pathogen Interactions in Conifer Forests*, ed. T Schowalter, G Filip, pp. 102–28. San Diego, CA: Academic
78. Raffa KF, Smalley EB. 1995. Interaction of preattack and induced monoterpene concentrations in conifer defense against bark beetle–fungal complexes. *Oecologia* 102:285–95
79. Rane KK, Tattar TA. 1987. Pathogenicity of blue stain fungi associated with *Dendroctonus terebrans*. *Plant Dis.* 71:879–83
80. Reid RW, Whitney HS, Watson JA. 1967. Reactions of lodgepole pine to attack by *Dendroctonus ponderosae* Hopkins and blue stain fungi. *Can. J. Bot.* 45:1115–26
81. Reynolds KM. 1992. Relations between activity of *Dendroctonus rufipennis* Kirby on Lutz spruce and blue stain associate *Leptograhium abietinum* (Peck) Wingfield. *For. Ecol. Manag.* 47:71–86
82. Rice A, Thormann MN, Langor DW. 2007. Virulence of and interactions among mountain pine beetle associated blue stain fungi on two pine species and their hybrids in Alberta. *Can. J. Bot.* 85:316–23
83. Romon P, Zhou X, Iturrondobeitia JC, Wingfield MJ, Goldarazena A. 2007. *Ophiostoma* species (Ascomycetes: Ophiostomatales) associated with bark beetles (Coleoptera: Scolytinae) colonizing *Pinus radiata* in northern Spain. *Can. J. Microbiol.* 53:756–67
84. Ross DW, Fenn P, Stephen FM. 1992. Growth of southern pine beetle associated fungi in relation to the induced wound response in loblolly pine. *Can. J. For. Res.* 22:1851–59
85. Ross DW, Solheim H. 1997. Pathogenicity of Douglas fir of *Ophiostoma pseudotsugae* and *Leptographium abietinum*, fungi associated with the Douglas fir beetle. *Can. J. For. Res.* 27:39–43
86. Salle A, Monclus R, Yart A, Garcia J, Romary P, Lieutier F. 2005. Fungal flora associated with *Ips typographus*: frequency, virulence, and ability to stimulate the host defense reaction in relation to insect population levels. *Can. J. For. Res.* 35:365–73
87. Scott JJ, Oh D-C, Yuceer MC, Klepzig KD, Clardy J, Currie CR. 2008. Bacterial protection of beetle-fungus mutualism. *Science* 322:63
88. Six DL. 2003. Bark beetle-fungus symbioses. In *Insect Symbiosis*, ed. K Bourtzis, T Miller, pp. 97–114. Boca Raton, FL: CRC Press
89. Six DL, Bentz BJ. 2003. Fungi associated with the North American spruce beetle, *Dendroctonus rufipennis*. *Can. J. For. Res.* 33:1815–20
90. Six DL, Bentz BJ. 2007. Temperature determines the relative abundance of symbionts in a multipartite bark beetle-fungus symbiosis. *Microb. Ecol.* 54:112–18
91. Six DL, Klepzig KD. 2004. *Dendroctonus* bark beetles as model systems for the study of symbiosis. *Symbiosis* 37:207–32
92. Six DL, Paine TD. 1997. *Ophiostoma clavigerum* is the mycangial fungus of the Jeffrey pine beetle, *Dendroctonus jeffreyi*. *Mycologia* 89:858–66
93. Six DL, Paine TD. 1998. Effects of mycangial fungi and host tree species on progeny survival and emergence of *Dendroctonus ponderosae* (Coleoptera: Scolytidae). *Environ. Entomol.* 27:1393–401
94. Six DL, Paine TD. 1999. A phylogenetic comparison of ascomycete mycangial fungi and *Dendroctonus* bark beetles (Coleoptera: Scolytidae). *Ann. Entomol. Soc. Am.* 92:159–66
95. Solheim H. 1986. Species of Ophiostomataceae isolated from *Pices abies* infested by the bark beetle *Ips typographus*. *Nord. J. Bot.* 6:199–207
96. Solheim H. 1988. Pathogenicity of some *Ips typographus*–associated blue-stain fungi to Norway spruce. *Medd. Nor. Inst. Skoforsk* 40:1–11
97. Solheim H. 1991. Oxygen deficiency and spruce resin inhibition of growth of blue stain fungi associated with *Ips typographus*. *Mycol. Res.* 95:1387–92
98. Solheim H. 1992. The early stages of fungal invasion in Norway spruce infested by the bark beetle *Ips typographus*. *Can. J. Bot.* 70:1–5
99. Solheim H. 1995. Early stages of blue-stain fungus invasion of lodgepole pine sapwood following mountain pine beetle attack. *Can. J. Bot.* 73:70–74
100. Solheim H, Krokene P. 1998. Growth and virulence of mountain pine beetle associated blue-stain fungi, *Ophiostoma clavigerum* and *Ophiostoma montium*. *Can. J. Bot.* 76:561–66

101. Solheim H, Langstrom B. 1991. Blue-stain fungi associated with *Tomicus piniperda* in Sweden and preliminary observations on their pathogenicity. *Ann. Sci. For.* 48:149–56
102. Solheim H, Safranyik L. 1997. Pathogenicity to Sitka spruce of *Ceratocystis rufipenni* and *Leptographium abietinum*, blue stain fungi associated with the spruce bark beetle. *Can. J. For. Res.* 27:1336–41
103. Spatafora JW, Blackwell M. 1994. The polyphyletic origins of ophiostomatoid fungi. *Mycol. Res.* 98:1–9
104. Viiri H. 1997. Fungal associates of the spruce bark beetle *Ips typographus* L. (Coleoptera: Scolytidae) in relation to different trapping methods. *J. Appl. Entomol.* 121:529–33
105. Viiri H, Lieutier F. 2004. Ophiostomatoid fungi associated with the spruce bark beetle, *Ips typographus*, in three areas in France. *Ann. For. Sci.* 61:215–19
106. Vite JP. 1961. Influence of water supply on oleoresin exudation pressure and resistance to bark beetle attack in *Pinus ponderosa*. *Contrib. Boyce Thompson Inst.* 21:37–66
107. VonSchrenk H. 1903. The "bluing" and the "red rot" of the western yellow pine with special reference to the Black Hills Forest Reserve. *US Bur. Plant. Ind. Bull.* 36:1–40
108. Whitney HS, Bandoni RJ, Oberwinkler F. 1987. *Entomocorticium dendroctoni* gen. et. sp. nov. (Basidiomycotina), a possible nutritional symbiote of the mountain pine beetle in lodgepole pine in British Columbia. *Can. J. Bot.* 65:95–102
109. Whitney HS, Cobb FW. 1972. Non-staining fungi associated with the bark beetle *Dendroctonus brevicomis* (Coleoptera: Scolytidae) on *Pinus ponderosa*. *Can. J. Bot.* 50:1943–45
110. Wingfield MJ. 1986. Pathogenicity of *Leptographium procerum* and *L. terebrantis* on *Pinus strobus* seedlings and established trees. *Eur. J. For. Pathol.* 16:299–308
111. Wingfield MJ, Seifert KA, Webber JF, eds. 1993. Ceratocystis *and* Ophiostoma: *Taxonomy Ecology and Pathogenicity*. St. Paul: Am. Phytopathol. Soc. Press
112. Wood DL. 1982. The role of pheromones, kairomones, and allomones in the host selection and colonization behavior of bark beetles. *Annu. Rev. Entomol.* 27:411–46
113. Wood S. 1982. *The Bark and Ambrosia Beetles of North and Central America (Coleoptera: Scolytidae): A Taxonomic Monograph*. Great Basin Nat. Mem. 6. Provo, UT: Brigham Young Univ. Press. 1359 pp.
114. Yamaoka Y, Hiratsuka Y, Maruyama PJ. 1995. The ability of *Ophiostoma clavigerum* to kill mature lodgepole pine trees. *Eur. J. For. Pathol.* 25:401–4
115. Yamaoka Y, Swanson RH, Hiratsuka Y. 1990. Inoculation of lodegepole pine with four blue stain fungi associated with mountain pine beetle, monitored by a heat pulse velocity (HPV) instrument. *Can. J. For. Res.* 20:31–36
116. Zhou XD, de Beer ZW, Wingfield BD, Wingfield MJ. 2001. Ophiostomatoid fungi associated with three pine-infesting bark beetles in South Africa. *Sydowia* 53:290–300
117. Zipfel RD, de Beer ZW, Jacobs K, Wingfield BD, Wingfield MJ. 2006. Multigene phylogenies define *Ceratocystiopsis* and *Grosmannia* distinct from *Ophiostoma*. *Stud. Mycol.* 55:75–97

Robert F. Denno (1945–2008): Insect Ecologist Extraordinaire

Micky D. Eubanks,[1] Michael J. Raupp,[2] and Deborah L. Finke[3]

[1]Department of Entomology, Texas A&M University, College Station, Texas 77843; email: m-eubanks@tamu.edu

[2]Department of Entomology, University of Maryland, College Park, Maryland 20742; email: mraupp@umd.edu

[3]Division of Plant Sciences, University of Missouri, Columbia, Missouri 65211; email: finked@missouri.edu

Annu. Rev. Entomol. 2011. 56:273–92

First published online as a Review in Advance on September 3, 2010

The *Annual Review of Entomology* is online at ento.annualreviews.org

This article's doi:
10.1146/annurev-ento-120709-144825

0066-4170/11/0107-0273$20.00

Key Words

competition, wing polymorphism, intraguild predation, omnivory, trophic cascade, food webs

Abstract

Robert F. Denno was widely recognized as one of the leading insect ecologists in the world. He made major contributions to the study of plant-insect interactions, dispersal, interspecific competition, predator-prey interactions, and food web dynamics. He was especially well known for his detailed and comprehensive study of the arthropods that inhabit salt marshes. Denno promoted a research approach that included detailed knowledge of the natural history of the study system, meticulous experiments that often pushed logistical possibilities, and a focus on important ecological questions of the day. He was an enthusiastic collaborator and excellent mentor who invested incredible amounts of time and energy in the training and placement of graduate students and postdoctoral associates. As a result, Denno's legacy will continue to shape the field of insect ecology for generations to come.

EARLY FOUNDATIONS OF AN INSECT ECOLOGIST

Robert Frederick Denno (1945–2008) was born in New York City but as a child moved to southern California. His love of natural history began in the foothills of California's coastal range near his home in Santa Barbara where much time was spent at La Cumbre Peak collecting insects. By the time he graduated from high school, Denno amassed 30 display drawers housing a large, meticulously curated collection of several hundred specimens of endemic butterflies, particularly checkerspots. This collection eventually swelled to over 36,000 specimens from around the world, housed in beautiful wooden cabinets that he crafted himself (**Figure 1**). His passion for science, ecology, and insects guided his curriculum choices through community college and as an undergraduate at the University of California at Davis. Denno graduated Phi Beta Kappa cum laude with a major in entomology and a minor in botany in 1967.

Figure 1

Robert Denno collecting butterflies in Arizona. Note the custom-built wooden specimen box and ever present diet Coke. Denno's lifelong passion for butterflies resulted in a meticulously curated personal collection of over 36,000 specimens.

Denno entered the doctoral program in the laboratory of an assistant professor, Dr. Warren Cothran, a dynamic member of the fledgling Insect Ecology group at Davis. Cothran's academic rigor, meticulous attention to detail, and rapport with students laid foundations that would become hallmarks of Denno's research and laboratory. Upon graduation in 1973, Denno began a postdoctoral appointment at Rutgers University, where he was first introduced to Atlantic coast salt marshes as part of a team studying the ecological effects of insecticides used in mosquito abatement programs. After only one year in this position, he was hired by Rutgers as an assistant professor of insect systematics, a testament to the breadth and depth of his knowledge of natural history. But his time at Rutgers was limited, as he was quickly lured away by the Department of Entomology at the University of Maryland, his home for the next three decades. Denno's original job description specified an emphasis on the management of insect pests of turfgrass, but he was encouraged to do the type of research that excited him. Over the course of his career, Denno became an internationally recognized figure in population and community ecology. He established a reputation for synthesis, with a talent for identifying broad patterns in nature and revealing their underlying mechanisms.

THE SEEDS OF COMPETITION

Denno's dissertation, completed in 1973, was entitled "Niche Relationships and Competitive Interactions of Carrion-Breeding Calliphoridae and Sarcophagidae" and resulted in two publications that provided the first examples of the comprehensive and in-depth writing style that was to become one of Denno's trademarks. Focusing on a foundational concept of competition theory, Denno addressed the Gaussian principle that no two species could occupy the same niche indefinitely (52). By manipulating periods of exposure and size of carcasses and measuring temporal patterns of exploitation, Denno & Cothran (14) provided evidence supporting Gaussian predictions that closely related species utilizing a similar resource diverged in patterns of utilization over several niche dimensions. Species of necrophagous flies differed in the size of carcasses used, seasonality, and stage of decomposition. Notably, in a theme that would recur in his monumental studies of insects inhabiting salt marshes, Denno & Cothran (14) found that the degree of specialization of carrion flies was linked to the stability of their food resources, with specialists utilizing stable resources and generalists utilizing less predictable resources. In a subsequent test, Denno & Cothran (15) found strong evidence of competitive release. Documenting and understanding patterns of resource utilization was a theme repeated by Denno in many systems including biting flies (109), aquatic insects (3, 4), spiders (39), tropical insects (16), and, most famously, salt marsh–inhabiting planthoppers (10, 12).

LIFE-HISTORY TRAITS AND WING POLYMORPHISMS

For more than three decades, Denno's primary focus was the community of arthropods inhabiting coastal salt marshes along the Atlantic Ocean and Gulf of Mexico. This ecosystem is dominated by relatively pure stands of the grasses *Spartina patens*, *S. alterniflora*, and *Distichlis spicata*. *D. spicata* and *S. patens* occupy higher elevations in the marsh, whereas *S. alterniflora* occurs at lower elevations and along creek banks. One striking feature of the herbivorous arthropods associated with these grasses is the widespread prevalence of wing dimorphism in guilds of sap-sucking Auchenorrhyncha, primarily leafhoppers and planthoppers. Many grasses are dominated by flightless, short-winged forms (brachypters), while some grasses house temporally and geographically shifting proportions of flight-capable, long-winged forms (macropters) (9, 10, 22).

Harlequin environment: insular patches of habitat that vary in quality or persistence occupied by species whose population dynamics and competitive interactions are governed by differential rates of colonization and extinction

Denno (9) predicted that brachyptery would prevail in denizens utilizing stable resources but that macroptery would characterize species occupying patchy or variable resources. Three species of planthoppers associated with the high marsh grass *S. patens* were strongly biased toward brachyptery. *S. patens* is a structurally uniform and spatially homogeneous resource with respect to patch quality. By contrast, at lower elevations the quality of *S. alterniflora* patches varied dramatically both spatially and temporally owing to variability in tidal inundation. Dimorphism in populations of *Prokelisia marginata* (**Figure 2**) was maintained by differing selective regimes in harlequin environments. Brachypters with limited dispersal capabilities were favored in stable patches, where they persisted and exploited *S. patens*, and macropters were favored in variable patches, escaping when resources declined in quality and colonizing others of higher quality. This model described shifting seasonal patterns in proportions of wing forms in populations of *P. marginata* during their annual interhabitat migrations between high- and low-quality patches of *S. alterniflora*, recolonization of defaunated patches of grass, and intraplant distributions as they exploited plant tissues of varying quality (9, 32).

Denno and colleagues (9, 11, 22) also proposed that a density-sensitive developmental switch triggered the production of macropters occupying variable host patches. Experiments designed to elucidate the mechanisms that trigger this switch found that developmental control of wing dimorphism in *P. marginata* was influenced by crowding, host nutrition, and their interaction, but crowding seemed to be the most important factor influencing wing development (17, 18, 26). Ideas spawned in *Spartina* systems culminated in a paradigm for

Figure 2

The Denno laboratory studied *Prokelisia* planthoppers and their cordgrass host plants in the genus *Spartina* for over 30 years. Here, two *P. marginata* planthoppers mate on *S. alterniflora*. Photographed by Dwight Kuhn.

wing dimorphism in planthoppers (35). In this study, levels of macroptery were strongly linked to habitat persistence in over 40 species of planthoppers. As habitat persistence declined, the proportion of migrants in the population increased.

Ecological forces underlying the evolution of migration in planthoppers were further revealed with the discovery of a second species of *Prokelisia*, *P. dolus*, inhabiting *S. alterniflora* (37). Unlike *P. marginata*, whose populations consisted of about 80% macropters, populations of *P. dolus* were composed of roughly 20% macropters. This species did not engage in annual interhabitat migrations on mid-Atlantic marshes. Different acoustic signaling and hybridization failure were two mechanisms isolating these sympatric species (57). This discovery and a lingering question regarding the adaptive significance of brachyptery were addressed in an important series of papers examining the cost of flight capability in planthoppers (29, 33). Using *P. dolus*, Denno et al. (29) demonstrated that female brachypters enjoyed enhanced fecundity, earlier age to first reproduction, and longer adult life than macropters. The pattern for greater fecundity and earlier reproduction in brachypters and enhanced dispersal abilities of macropters was documented in several species of planthoppers, confirming the trade-off between flight and reproduction (29, 33). Later studies confirmed that dispersal ability constrains reproduction in male *P. dolus* as well, with flight-capable macropterous males acquiring fewer matings and siring fewer offspring than their flight-incapable brachypterous counterparts (71). However, this dispersal polymorphism appears to be maintained in males, despite the fitness trade-off, because vegetation structure (sparseness) and low female density differentially favor macropterous over flightless males (72).

Interspecific competition: competition for resources among individuals of different species

These ideas were confirmed in an exhaustive review of the wing polymorphism literature (116). Zera & Denno (116) reviewed studies of insects from nine orders to convincingly demonstrate that habitat persistence selects for reduced dispersal capability in insects. They also demonstrated that increased fecundity of flightless females is a remarkably widespread fitness trade-off found in almost all major insect orders.

Denno's discovery of the coexistence of *P. marginata* and *P. dolus* provided an opportunity to return to questions regarding interspecific competition among closely related sympatric species. By rearing both species at different densities in isolation and together, Denno & Roderick found that interspecific crowding was as equally strong a stimulus for the production of migrants as intraspecific crowding (34). However, responses of the congeners to intraspecific crowding were asymmetrical. Intraspecific crowding resulted in delayed development, reduced body size, and lower survival in *P. marginata*, but not in *P. dolus*. Because of its profound effects on the reproduction and fecundity of *P. marginata*, interspecific competition emerged as an important selective force in this system by increasing emigration and decreasing population growth of planthoppers inhabiting *S. alterniflora* (34).

INTERSPECIFIC COMPETITION

Denno's growing awareness of the importance of interspecific competition in the population dynamics of *Prokelisia* planthoppers (34) led him to contemplate the broader importance of interspecific competition in ecology. During the 1960s and 1970s, interspecific competition was assumed to be an important force affecting the distribution, abundance, and community structure of herbivorous insects, although this view developed largely as a result of observational studies (5, 101). During the late 1970s and early 1980s, however, the presumed importance of competition on the structure of phytophagous insect communities was severely challenged and within a few brief years the prevailing view was that interspecific competition was weak and infrequent (76, 77, 105). These critiques of interspecific competition, in turn, stimulated many well-executed, experimental evaluations of the importance of competition among herbivorous insects. Many of these studies were published in the late 1980s and early 1990s. Denno et al. (28) immersed themselves in this literature and extracted data on 193 pairwise species interactions occurring in 104 different study systems. The majority of these pairwise interactions (148 in total) were experimental assessments of competition. Denno et al. found that interspecific competition was incredibly widespread: Seventy-six percent of the pairwise interactions showed strong evidence of competition, whereas only 18% of the interactions indicated an absence of competition and 6% supported facilitation. This remarkably comprehensive evaluation of the literature was extremely influential. The review has been cited 296 times to date and has definitively changed the way insect ecologists view competition.

In the ten years following Denno's original review of competition, the growing body of literature reflected a greater emphasis on indirect interactions including plants (i.e., induced defenses) and natural enemies (i.e., apparent competition). Given this shifting focus, Kaplan & Denno (62) revisited the issue of competition, this time using a quantitative meta-analytical approach to assess whether herbivorous insects conform to the traditional paradigm of interspecific competition. Notably, they found that indeed interspecific competition was common, but that many widely held assumptions did not hold. Specifically, there was no correlation between the amount of plant damage and the intensity of competition, competition occurred between distantly related species in different feeding guilds, it was not dampened by temporal or spatial resource partitioning, and it was highly asymmetric. These results point toward a significant contribution of indirect interactions, whereby plants and/or natural enemies mediate competitive interactions among herbivores. Kaplan & Denno concluded that, to understand how interspecific competition contributes to the organization of phytophagous insect communities, a new paradigm that accounts for indirect interactions and facilitation is required (23, 62).

Several studies emerged from Denno's laboratory that provided support for the prevalence of indirect interactions among herbivorous insects that partition their resources either temporally or spatially. For example, Denno et al. (31) found that early-season feeding by *P. dolus* on *S. alterniflora* grass in high marsh meadows induced changes in plant quality that negatively affected populations of *P. marginata* colonists that arrived later in the season. Lynch et al. (82) also found that previous feeding by potato leafhoppers, *Empoasca fabae*, resulted in adverse plant-mediated effects on later-feeding Colorado potato beetles, *Leptinotarsa decemlineata*, such as reduced survivorship and delayed development. This induced resistance to early-feeding leafhoppers in turn enhanced the

Indirect interactions: one species influences a second species by virtue of its impacts on an intermediary species

Induced defenses: dynamic plant defenses that are produced in response to attack by insects

Apparent competition: one prey species has an indirect negative effect on another prey species via its direct positive effect on the abundance of a shared predator

beetle's risk of predation because its slowed growth increased exposure to predators during susceptible stages (65). Spatially segregated herbivores that feed above- and belowground on the same host plant also interact, due to the physiological integration of root and shoot defense (63, 64, 66). Nematode herbivory of tobacco roots interferes with the defensive induction of the plant, positively influencing aboveground-feeding phytophagous insects. Likewise, aboveground-feeding herbivores facilitate plant parasitic nematodes by stimulating plants to allocate nutritional reserves to the roots.

VARIABLE PLANTS AND HERBIVORES

With recognition that planthoppers closely tracked high-quality resources at several spatial scales, plant nutrition as a determinant of herbivore behavior and population dynamics became an organizing theme in Denno's laboratory (32). Several graduate students embarked on studies of how variation in plant resources affects plant and herbivore interactions.

In one set of studies, phenological and sex-based variation in plant quality influenced herbivore feeding behavior. Along the upper fringes of the salt marsh grows the dioecious woody shrub *Baccharis halimifolia*. *B. halimifolia* has a specialized chrysomelid beetle, *Trirhabda bacharidis*. Kraft & Denno (68) found that young foliage was preferred by these beetles and that the beetles performed better when fed young foliage. Sex ratios of *B. halimifolia* were slightly female biased (~60%) in natural settings. Krischik & Denno (69, 70) found that *T. bacharidis* and a polyphagous flea beetle, *Paria thoracicai*, preferred to feed on male plants owing to a greater abundance of tender young leaves on rapidly growing male plants.

The defensive strategy of the host plant was also determined to affect the feeding behavior and host utilization of phytophagous insects. One widespread defensive syndrome in terrestrial plants is to arm secretory canals with noxious resins, gums, muscilages, or latexes. Some phytophagous insects are behaviorally adapted to preempt these defenses by cutting or trenching veins or canals that supply and distribute these defenses. By examining feeding behaviors of 33 species of caterpillars, beetles, and katydids on canal-bearing plants, Dussourd & Denno (40) found that distinct patterns emerged across widely disparate plant lineages. Regardless of the taxonomic affinity of the herbivores, there was a close link between feeding behavior and the architecture of secretory canals. Phytophagous insects that consumed plants with secretory canals prevented the flow of secretion to distal branches of the canals by cutting the major leaf veins, thereby rendering much of the leaf defenseless. Alternatively, herbivores that fed on plants with net-like anastomosing canals transected all strands of the network by cutting a trench in the leaf. Thus, diverse insects that successfully attack these plants employ the same behavior on similar canal architectures, even when their host plants otherwise differ in taxonomy and secondary chemistry (40).

In one of the earliest tests of the importance of inducible plant defenses, another chrysomelid, *Plagiodera versicolora*, was used to examine changes in host suitability in *Salix* following natural and artificial defoliation. Leaves damaged artificially by tearing or naturally by the feeding of mandibulate herbivores resulted in a ~20% reduction in the fecundity of *P. versicolora*. Furthermore, damage to one leaf reduced the suitability of adjacent undamaged ones as evidenced by reduced weight and slower development of leaf beetle larvae (96). Willow beetles were later shown by other investigators to deal with altered leaf quality by avoiding previously damaged leaves (97).

Denno's laboratory was also interested in how host plants mediated the effects of natural enemies on herbivores. Willow beetles provided an excellent model for examining patterns of host utilization and dietary breadth in the context of enemy-free space (24). Larvae of *Phratora vitellinae* produced defensive secretions from precursors found in leaves of their host, whereas larvae of another willow-feeding leaf beetle, *Galerucella lineola*, lacked this ability. Female *G. lineola* selected and oviposited on willow species where larval performance was greatest. By contrast, *P. vitellinae* oviposited on hosts of varying quality for larvae but preferred hosts rich in precursors for larval defensive secretions. In the presence of predators, the ability of these beetles to produce potent defensive secretions outweighed costs associated with slower development and lower survival on suboptimal hosts. The role of enemy-free space in shaping dietary breadth and specialization in these leaf beetles provides some of the clearest support for this hypothesis (24).

Benrey & Denno (1) further explored the effects of host plants on herbivore–natural enemy interactions by testing the slow-growth/high-mortality hypothesis. This hypothesis predicts that prolonged development on poor-quality hosts results in increased exposure to natural enemies and a subsequent increase in mortality. They conducted a series of experiments to determine how host plant species influenced the larval development of imported cabbageworm, *Pieris rapae*, and their vulnerability to attack by the parasitoid wasp *Cotesia glomerata*. They found that rapidly developing caterpillars reached instars that were invulnerable to attack faster and thus had a shorter "window of vulnerability" than slowly growing caterpillars. Consequently, caterpillars reared on relatively poor-quality host plants suffered significantly higher levels of parasitism. This was true, however, only within a given host plant species. Parasitism rates among plants were largely specific to the plant species and not explained by development time of caterpillars. These studies were extremely influential because they demonstrated that plants can increase the susceptibility of insect herbivores to natural enemies via multiple mechanisms and that these indirect effects can have important consequences for the ecological and evolutionary interactions of plants and insects.

Ultimately, Denno's laboratory documented that variability in host plants can have community-wide impacts on the trophic structure, composition, and diversity of an entire arthropod food web (115). Increasing *S. alterniflora* biomass production on the salt marsh by enhancing nutrient inputs resulted in an increase in the species richness of herbivores, detritivores, predators, and parasitoids, primarily due to an increase in the diversity of rare species. In addition, there were significant changes in arthropod species composition with increasing levels of production (115).

Although the vignettes described above provide a sample of his work, Denno's interests in the grand sweep of plant variation and the evolutionary and ecological responses of herbivorous insects crystallized in an early, remarkable treatise, *Variable Plants and Herbivores in Natural and Managed Systems* (27). Co-edited with his lifelong friend Mark McClure, the "green book," as it was known to many insect ecologists because of its bright green cover, is a classic synthesis of the evolutionary ecology of plant-herbivore interactions and remains a foundational work in plant-herbivore theory.

EVOLUTIONARY AND ECOLOGICAL TRADE-OFFS

Evolutionary trade-offs between reproduction and migration in planthoppers led to investigations of evolutionary trade-offs in other systems. In a series of fascinating studies with his student, Douglas Tallamy, Denno investigated another type of life-history trade-off: the trade-off between maternal care and reproduction in the eggplant lace bug, *Gargaphia solani*, a specialist on solanaceous weeds. Following the discovery that female *G. solani* defend their nymphs from predators but do not assist in feeding or locating resources, trade-offs between egg production, maternal care, and longevity were quantified (107, 108). Maternal care significantly reduced female fecundity, but the disadvantages of mothering were strongly offset by greatly enhanced survival of offspring when mothers protected progeny from predators. Thus, maternal care increased fitness and longevity of females (107, 108).

Interests in energetic trade-offs of differing life-history traits led to an examination of costs

Genetic isolation by distance: because neighboring populations are more likely to exchange genes, genetic exchange among populations of species with limited dispersal ability may be strongly correlated with geographic distance among populations

Plant stress hypothesis: drought promotes outbreaks of phytophagous insects because nitrogen availability is increased in water-stressed plants

and benefits of fecal shields in chrysomelid tortoise beetles. As larvae, many species of tortoise beetles collect feces and exuviae on modified caudal appendages. Three species of tortoise beetles, *Charidotella bicolor*, *Deloyala guttata*, and *Chelymorpha cassidea*, were raised under laboratory conditions with their fecal shield intact or removed. Removal of shields in the absence of predators did not affect performance of larvae and resulted in no compensatory feeding to build replacements. However, in laboratory settings, fecal shields deterred small predators with piercing mouthparts, such as *Geocoris*, and mandibulate predators, such as coccinellids. Tortoise beetles suffered high mortality in natural settings when fecal shields were removed, leading to the conclusion that fecal shields provided a low cost but potent defense to their chrysomelid bearers (86, 87).

EVOLUTIONARY CONSEQUENCES OF VARIATION IN DISPERSAL ABILITY

Denno's work with wing-dimorphic planthoppers that differed dramatically in their dispersal abilities stimulated an interest in the evolutionary consequences of this variation. Denno and his colleagues predicted that flightless insects should show much stronger population differentiation than their winged kin. Peterson & Denno (90) tested the hypothesis that levels of gene flow among planthopper populations were correlated with wing form (dispersal ability). They found that in both *P. marginata* and *P. dolus*, population-genetic subdivision was strongly correlated with dispersal ability, such that subdivision decreased as the level of macroptery increased. Peterson & Denno (91) expanded on this question by reviewing the literature on genetic isolation by distance for 43 species and host races of phytophagous insects. Somewhat surprisingly, they found that genetic isolation by distance was weak in sedentary as well as highly mobile species. They concluded that genetic isolation by distance was weak in strong dispersers because of the homogenizing effects of gene flow and that genetic isolation by distance was weak in sedentary insects because reduced gene flow resulted in population differentiation among virtually all populations. The strongest genetic isolation by distance was found in species with moderate dispersal abilities because modest dispersal creates genetic homogeneity at relatively small scales but significant divergence over longer distances. Peterson & Denno also used this data set to test and ultimately refute the long-standing hypothesis that genetic isolation by distance increases as diet specialization of phytophagous insects increases. Genetic isolation by distance did not differ among monophagous and polyphagous herbivores, casting serious doubt on the idea that diet specialization promotes diversification by influencing population-genetic subdivision.

EFFECTS OF PLANT STRESS ON HERBIVORE DYNAMICS

Entomologists have been interested in the effects of plant stress, especially drought stress, on herbivore population dynamics for nearly a century (8, 85). In many cases, stress from lack of water, altered nutrient availability, heat, shading, and flooding is correlated with insect outbreaks (112). By contrast, many herbivorous insects perform better on unstressed plants (67), and it has been a challenge for insect ecologists to reach a consensus about the effects of stress on the susceptibility of plants to insects (75, 85, 98). Within this context, Hanks & Denno (56) set out to determine the relative effect of drought stress on outbreaks of pests on urban trees. Hanks & Denno found that the distribution of the white peach scale, *Pseudaulacaspis pentagona*, on mulberry trees at urban sites and forested sites was determined by an interaction of water stress and natural enemies. Water potential is inversely related to water stress, and water potential of mulberry trees at forested sites was significantly higher compared to that of trees at urban sites. Water potential was positively correlated with the survival of scales, suggesting that scales should be most abundant on unstressed plants like those found at forested sites. Surprisingly, armored scales were actually more abundant at urban sites than at forested sites. Generalist natural enemies like phalangids, earwigs, and tree crickets were much more abundant at forested sites, and these generalist predators strongly reduced the survival and subsequent abundance of scales. Consequently, scales were largely relegated to the subset of mulberry trees at urban sites that were not water stressed and where generalist predators were rare.

In a similar study, Trumbule et al. (111) found that azalea lace bugs, *Stephanitis pyrioides*, were more abundant and caused more damage to azalea plants that were in stressful locations that received high light and low water. Subsequently, Trumbule & Denno (110) tested the hypothesis that azaleas grown under high light intensity and low water availability are stressed and promote lace bug outbreaks. In a series of greenhouse and field experiments, survival of caged azalea lace bugs was higher and female lace bugs were more fecund and preferred to feed and oviposit on shade-grown, well-watered azaleas rather than on azaleas grown in full sun with limited water, results all counter to the hypothesis. When they placed uncaged lace bugs on plants, however, lace bug survival was dramatically lower on plants in shaded woodlot habitats than on plants in open settings. Trumbule & Denno hypothesized that although shaded and watered azaleas were better host plants for lace bugs, generalist predators were more abundant in shaded habitats and were responsible for strong lace bug suppression. Subsequent work in this system confirmed this hypothesis (102).

Denno also investigated the effects of plant stress on herbivore dynamics in his beloved salt marsh system. Huberty & Denno (59) found that the closely related, phloem-feeding planthoppers *P. marginata* and *P. dolus* responded differently to plant stress. Huberty & Denno found that both planthoppers performed best when they were reared on luxuriant *Spartina* plants that were supplied with optimal levels of nitrogen and phosphorus. When plants were nitrogen stressed, the performance of *P. marginata* planthoppers was dramatically reduced, whereas the performance of *P. dolus* planthoppers was only slightly affected. *P. marginata* planthoppers were also negatively affected when plants were phosphorus stressed, although not to the same extent as when plants were nitrogen stressed. *P. dolus* planthoppers, however, were not affected at all when grown on phosphorus-stressed *Spartina*. Huberty & Denno (60) demonstrated in a companion study that *P. dolus* planthoppers avoid the effects of plant stress because they have significantly larger cibarial muscles that allow them to pump more phloem from their host plants than *P. marginata* planthoppers do. Huberty & Denno argued that these two species evolved divergent strategies for dealing with stressed host plants: The typically macropterous *P. marginata* invests in the flight musculature necessary to disperse from patches of stressed plants where it performs poorly, whereas the typically short-winged *P. dolus* invests in the cibarial musculature necessary for compensatory feeding.

In most of these cases, Denno and his students found that unstressed, vigorous plants were better hosts for herbivores, even though herbivore abundance was often not solely dictated by host plant quality, and some herbivores have strategies for dealing with stressed plants. Given Denno's penchant for synthesis, it is not surprising that Denno was driven to compare and integrate the results of work conducted by his laboratory with the work of other ecologists. Huberty & Denno (58) used data from over 80 published studies to test the hypothesis that drought stress increases herbivore performance and promotes outbreaks. They found that the performance and population growth of sap-feeding insects (those that feed on xylem, phloem, and mesophyll) and insects that induce galls were much lower on water-stressed plants than on unstressed plants. Furthermore, borers were the only insects that consistently benefited from feeding on water-stressed plants. The results of this review were surprising and profoundly challenged the long-held view that water stress promotes outbreaks of insect herbivores.

OMNIVORY

Denno had a long-standing interest in the evolution and ecological consequences of omnivory (feeding at more than one trophic level). In particular, Denno was interested in identifying factors that promote the evolution of omnivory and understanding the consequences of omnivory for omnivore-resource interactions. Denno & Fagan (19) took an extremely broad view of the evolution of omnivory and proposed that the mismatch in nitrogen stoichiometry between herbivores and their host plants and differences in nitrogen content between predators and their herbivorous prey promote omnivory. Basically, Denno & Fagan hypothesized that nitrogen-limited arthropods could enhance their nitrogen intake by broadening their diet to include nitrogen-rich prey. This extremely compelling paper strongly influenced how ecologists view (55, 114) and study nitrogen limitation in arthropods (84). As a result, there has been a greater integration of the nutritional requirements of secondary consumers in the study of ecological interactions (94) and critical evaluations of our assumptions about these nutritional requirements (113).

In a more detailed evaluation of the evolution of omnivory within a single group of insects, Eubanks et al. (44) focused on terrestrial lineages of the insect suborder Heteroptera and tested the idea that feeding on nitrogen-rich plant parts and polyphagy are correlated with the evolution of omnivory in this group. Feeding on high-nitrogen plant parts (seeds and pollen) and polyphagy were strongly associated with the evolution of omnivory within ancestrally herbivorous lineages and, likewise, omnivores that evolved within ancestrally predaceous lineages typically fed on nitrogen-rich plant parts of many different host plants. Results of this study and work on a wide range of other animals strongly suggest that omnivores that feed on both plants and prey represent a unique blend of adaptations found in their predaceous and herbivorous relatives (6, 7).

Denno's laboratory group asked questions about the ecological significance of omnivory in multiple study systems. Eubanks & Denno (42, 43) investigated the ecological consequences of omnivory in lima beans by the big-eyed bug, *Geocoris punctipes*. High-quality plant food, in this case lima bean pods, significantly increased the survival of big-eyed bug nymphs fed low-quality prey (pea aphids). Feeding on lima bean pods also allowed big-eyed bug nymphs to survive long periods without prey, and dispersal of big-eyed bugs was dramatically reduced by the presence of pods on plants. Not surprisingly, big-eyed bug populations were significantly larger in plots of lima beans with pods than in plots of lima beans without pods (42). Although the presence of pods acted as an alternative prey that reduced the per capita consumption of aphids and moth eggs by big-eyed bugs, the dramatic increase in the abundance of big-eyed bugs associated with pod feeding resulted in higher total prey consumption (43). Similarly, Frank et al. (51) found that adding grass seeds to cornfields could significantly decrease the incidence of cannibalism in omnivorous ground beetles and thus potentially lead to greater pest suppression. One of Denno's last students, Rachel Pearson, found that the salt marsh katydid, *Conocephalus spartinae*, was highly omnivorous and that in the field this katydid tracked variation in plant quality rather than changes in prey abundance, even though prey abundance affected its fitness to a greater extent than variation in plant quality (89).

HETEROGENEOUS HABITATS AND SPECIES INTERACTIONS

The role of habitat heterogeneity in mediating species interactions and shaping community structure was a recurring theme of Denno's work that was examined at a variety of dimensions and spatial scales. Despite the superficial appearance of a uniform habitat, salt marsh habitats are actually extremely heterogeneous due to the influence of tidal inundation on the accumulation (or lack of accumulation) of dead plant material, thatch, at the base of marsh plants. Denno (10) evaluated the impact of thatch on the diversity and composition of phytophagous insect communities by comparing the assemblage of sap-feeders inhabiting structurally complex *S. patens* with its dense matrix of associated thatch and sap-feeders found in the less complex vegetation of *S. alterniflora*. Species richness, diversity, and evenness were greater in *S. patens*. Denno (10) altered the structure of *S. patens* by removing its thick layer of thatch, thereby homogenizing the habitat. Habitat homogenization resulted in lower diversity and evenness of associated sap-feeders. In a related study, another species of salt marsh grass, *D. spicata*, yielded similar results. Owing to a dense accumulated layer of thatch and upright culms that support a tangle of leaf blades, *D. spicata* was much more architecturally complex than *S. alterniflora*. Correspondingly, the community of sap-feeders was richer and more diverse in *D. spicata* than in *S. alterniflora* (106).

A comprehensive study of spider communities in cordgrass habitats varying in thatch abundance and architectural complexity revealed a role for habitat heterogeneity in structuring communities at higher trophic levels as well (39). *S. patens* had a less diverse community of spiders than S. *alterniflora*, and *S. alterniflora* at high elevations housed a greater diversity of spiders than *S. alterniflora* along creeks, where inundation and seasonal destruction of the grass occurred. *S. alterniflora* provided architecture sufficient for colonization by web-builders that was lacking in *S. patens*, whereas the community of spiders in *S. patens* was dominated by hunting spiders (39). Langellotto & Denno (73) determined if this was a broad pattern in nature by performing a meta-analysis of the relevant literature. They found that the aggregation of predators in complex habitats is a generalized response across predatory taxa, including hunting and web-building spiders, hemipterans, mites, and parasitoids (73). The mechanisms underlying this response were poorly known in most cases, but further experiments on the marsh found that the presence of thatch provided refuge for spiders from cannibalism (74) and diminished the occurrence of intraguild predation (45, 46). Thus, complex-structured salt marsh habitats appear to promote spider abundance by both promoting their accumulation and enhancing their survival once they arrive.

Intraguild predation: predators consume other predators with which they compete for a shared prey resource

At a larger spatial scale, habitat heterogeneity in the form of host plant patches that vary in size and isolation was also found to influence herbivore communities and their interactions with natural enemies. Due to subtle differences in elevation on the marsh surface, *S. patens* often occurs in relatively large pure stands or as smaller patches or islands embedded in a matrix of *S. alterniflora*. Raupp & Denno (95) found no differences in species richness between small and large patches but discovered lower abundance of several species of sap-feeders on small habitat patches (12, 95). More recently, ecological traits such as mobility and fecundity of *S. patens* inhabitants were linked with critical habitat thresholds using 62 patches of *S. patens* spanning four orders of magnitude of patch size (83). Specialists and poor dispersers were highly influenced by spatial structure, but these factors had little effect on generalists and highly mobile species.

POSITIVE PREDATOR-PREDATOR INTERACTIONS

Although negative interactions among predators (intraguild predation) was a frequent focus of Denno's research (19, 45, 47, 49, 74), his laboratory also found that positive predator-predator interactions could play an important role in community dynamics. Losey & Denno (80) found that aphid suppression in alfalfa was dramatically increased when aphids were exposed to a combination of foliar-foraging predators (lady beetles) and ground-foraging predators (ground beetles) than when exposed to either predator alone. This synergistic suppression of aphids occurred because aphids dropped from plants in response to foliar-foraging predators (81), a behavior that made them accessible to ground-foraging predators

Top-down: natural enemy-mediated regulation of herbivore populations

Trophic cascade: an indirect effect of natural enemies on plants transmitted by the direct effect of enemies on herbivores

Bottom-up: plant-mediated regulation of herbivore populations

(79). This study was one of the first to document synergistic effects of invertebrate predators on their prey and was a forerunner of more recent studies that have shown positive effects of increasing predator diversity, especially functional group diversity, on herbivore suppression (50, 93).

FOOD WEB COMPLEXITY AND TROPHIC CASCADES

In the 1990s and early 2000s considerable dispute raged in the ecological literature on whether the top-down effects of natural enemies on herbivores were strong enough to cascade down through the food web to indirectly benefit primary producers (104). For terrestrial food webs dominated by phytophagous arthropods, there was accumulating support both for and against the occurrence of such trophic cascades (99), but a synthesis of the factors underlying these disparate findings was lacking. Finke & Denno (47, 49) approached the issue from a mechanistic perspective, attempting to explain why enemy effects cascade in some situations and not others. They conducted a series of field and greenhouse studies with the salt marsh community, examining the interactive effects of predator diversity, the presence of intraguild predators, and habitat complexity on the occurrence of trophic cascades. Increasing predator diversity dampened cascading effects of natural enemies on *Prokelisia* planthoppers and *S. alterniflora* cordgrass biomass (47, 48), but this effect was mediated by the presence of intraguild predators. As the proportion of intraguild predator species in the predator assemblage increased, the strength of the trophic cascade decreased. Therefore, a reticulate food web with high predator diversity and complex trophic interactions (e.g., intraguild predation) buffered the community against the occurrence of a trophic cascade. These studies were among the first to examine the cascading responses of herbivores and plants to manipulated diversity within the predator guild (rather than merely the presence/absence of predators), and they remain one of the most comprehensive mechanistic examinations of the effects of predator diversity loss on prey suppression and plant biomass (2, 100, 103).

TEMPORAL AND SPATIAL VARIATION IN TOP-DOWN AND BOTTOM-UP FACTORS

While some might find it tedious to work in the same system for over 30 years, Denno's in-depth studies of planthopper ecology gave him the unique perspective necessary to develop a mechanistic understanding of the complex, larger-scale processes governing the dynamics of the terrestrial salt marsh food web. As a result, he synthesized his extensive work on planthopper population dynamics; life-history strategies; species interactions such as competition, omnivory, plant-herbivore, predator-prey, and predator-predator interactions; and the mediating effects of habitat heterogeneity to explain temporal and spatial variation in the relative strengths of bottom-up (host plant resources) and top-down (predation) forces on phytophagous insects and their subsequent cascading effects on plant biomass. In a comprehensive approach that was a hallmark of Denno's experimental style, he integrated a series of manipulative studies from the laboratory and field with extensive sampling of field populations to develop a conceptual model of the processes at work on the salt marsh. What emerged was a general paradigm in this system of bottom-up primacy, whereby plants set the dynamic stage on which herbivorous insects and their natural enemies interact (21).

What Denno described was the existence of a landscape gradient across the salt marsh in the quality and structure of basal resources and the abundance of natural enemies that leads to spatial variation in the relative strength of top-down and bottom-up forces. Marsh habitats at lower elevations are characterized by nitrogen-rich *Spartina* that is free of thatch due to frequent tidal inundation that deposits nutrients and removes plant debris (13, 39). The high nitrogen content of the low-marsh *Spartina* promotes mass colonization, enhances survival and fecundity, and encourages rapid population growth of *Prokelisia* planthoppers (30, 88). At the same time, the paucity of thatch and greater tidal disturbance combine to reduce populations of many predators including *Pardosa* wolf spiders (39). The rarity of natural enemies, coupled with superior *Spartina* nutrition, promotes the largest planthopper populations and fosters outbreaks (30). Thus, Denno concluded that bottom-up forces generally prevail at lower elevations on the marsh.

In contrast, invertebrate predators play a greater role in planthopper suppression in certain high-elevation marsh habitats characterized by nitrogen-poor plants with abundant thatch. Impacts of natural enemies are more pronounced in these high-marsh habitats because the colonization-enhancing and growth-promoting effects of host plant nutrition on planthopper populations, and thus the opportunity for escape from control by predators, are not as strong. In addition, thatch encourages predator aggregation (38) and diminishes antagonistic predator-predator interactions such as cannibalism (74) and intraguild predation (45), increasing the overall abundance of predators. However, the top-down impact of natural enemies is mediated not only by host plants, but also by the identity of the herbivores themselves, with both defensive/escape behaviors and life-history strategies playing a role (20, 30). Predators play a more prominent role in the suppression of *Prokelisia* planthopper populations than other sap-feeders on the marsh, due in part to the ineffectiveness of the *Prokelisia* defense/escape response (20). Even between the two *Prokelisia* species, the relative strength of top-down effects also varies, mediated by differences in their life-history strategies. The highly mobile *P. marginata* planthoppers colonize nutritious host plants in very high numbers (13, 36), and in so doing, partially escape natural enemy impact (21). The less mobile *P. dolus* planthoppers are more susceptible to suppression by predators because their reduced dispersal ability delays their population response to nitrogen-rich plants.

But plant factors not only influence the strength of predator impacts on planthopper populations, they also dictate when such impacts are realized. Gratton & Denno (53) showed that *Prokelisia* planthoppers in habitats with high-nitrogen *Spartina* experience a temporal shift from population increases promoted by bottom-up processes during the summer to strong top-down suppression later in the season. The greater impact of natural enemies at the end of the season, despite the direct positive effects of nutrition on planthopper colonization and growth, is the result of a buildup of predators over the season. This buildup is the result of direct effects of habitat complexity on predator aggregation and retention and indirect effects on the availability of alternative prey.

Lewis & Denno (78) added an additional layer of complexity to this increasingly complicated picture by considering how spatial subsidies of predators interact with variation in host plants and resident natural enemies to influence populations of insect herbivores. They documented an annual spring/summer migration of *Pardosa* wolf spiders out of their overwintering habitat in upland *S. patens* into *S. alterniflora*, which results in a decline in *Pardosa* abundance with increasing distance from *S. patens* (25, 78). Because *Pardosa* is a voracious intraguild predator, this subsidy shifts the composition of the predator community, such that intraguild predators dominate the predator assemblage near *S. patens* and their impacts decline with increasing distance from *S. patens* patches. Thus, the opportunity for antagonistic interactions among predators also decreases with increasing distance from *S. patens*. However, the biomass of thatch is also greatest nearer *S. patens*, providing refuge from intraguild predation, promoting predator accumulation, and enhancing top-down control of planthoppers and their cascading effects on plants (38, 45, 74).

When Denno began his synthesis of factors influencing planthopper populations on the marsh, the historical controversy over the importance of top-down versus bottom-up impacts on phytophagous insect populations (41, 54) had already been supplanted by a more unified view (27, 61, 92). Therefore, the significance of Denno's contribution to this field comes not from his recognition that host plant factors can mediate the intensity of natural enemy impacts on herbivore populations, but from his identification of the larger-scale spatial and temporal patterns and his ability to provide a mechanistic explanation for why the patterns exist. Years of independent studies on the intricacies of individual species interactions provided the pieces of a puzzle that Denno, given his intimate knowledge of the system and knack for synthesis, was uniquely able to assemble into a complete picture.

A HOLISTIC VIEW OF DENNO AND HIS CAREER

Although Robert Denno loved butterflies, western plants, and diverse habitats, his research focused on a relatively nondiverse and simple system for almost 40 years. He kept his work fresh and relevant by incorporating the latest hot topics in ecology, be it top-down versus bottom-up regulation of populations, intraguild predation, or stoichiometry. Denno also kept his work fresh by having graduate students that worked outside the marsh, often focusing on plants of agricultural and landscape importance. This helped him intimately learn the insects and plants of many systems and explore diverse research questions without altering his own focus. Denno often expanded his research horizons by synergizing his interests with those of new students, postdocs, and visiting scientists (e.g., population geneticists). Denno made the most out of his travels and managed to conduct several significant research projects while leading field courses or visiting colleagues in a variety of places (e.g., Sweden, Costa Rica, Venezuela, Mexico, British Virgin Islands).

Denno's accomplishments as a scientist and a person were the direct result of his partnership with his wife of 42 years, Barbara Denno. They first met at a football game during their senior year of high school and were inseparable thereafter. Barb provided the intellectual and moral support that sustained Denno's career. But she also contributed directly to his research program by counting thousands (if not millions) of planthoppers, performing literature searches and tracking down citations, organizing his trips to meetings and other functions, and quelling his frustration over the disappearance of DOS-based word processing programs and minor computer glitches. The Dennos welcomed students, friends, and colleagues into their home. Parties at the Denno home were often fueled by rum and diet coke and full of thought-provoking discussions and tons of laughter and camaraderie. Bob was never known to take himself too seriously and he knew how to have a great time.

His synthetic approach to ecology and his ability to provide a complete story meant that Denno's talks at meetings were inevitably standing room only. This was despite (or perhaps because of) Denno's unconventional style of writing talks word-for-word beforehand and many, many, many practice sessions. This meticulous approach to public speaking resulted in Denno knowing exactly what he wanted to say, using precisely the right words, and packing more than seemingly possible into the allotted time.

Denno was an unmistakable force in insect ecology, not only because of the quality of his research, but also for his dedication to mentoring and the desire to make science a social endeavor. He was famous for his impromptu gatherings at scientific meetings where graduate students were welcomed and encouraged to interact with established scientists. When on departmental seminar visits, he devoted as much (if not more!) attention to discussions with students than he did with faculty members. His commitment to the training and promotion of his own students was unwavering, famously promising one future graduate student on his interview visit that "my job is to get you a job." In the end, Denno measured his own success not by the number of papers he published, but by his academic lineage of students and postdoctoral associates, of which there were many.

Denno's laboratory worked on an array of topics in insect ecology. Yet the majority of these studies included a deep understanding of the natural history of the system, a detailed, meticulous experimental approach, and a strong focus on synthesizing the results of research in that area. We hope that those who read this review, especially students, get a sense of the depth and breadth of Denno's research and the passion that he brought to science and to scientists. His deep commitment to digging deeper into his study system, to telling complete stories, and to developing future ecologists means that Denno's legacy will continue to shape the field of insect ecology for generations.

SUMMARY POINTS

1. Spatial and temporal heterogeneity of resources dictates the richness, diversity, and abundance of arthropod communities in a variety of habitats. Architecturally complex habitats and resource patches that vary spatially and temporally in quality provide opportunities for coexistence in diverse assemblages of herbivores and their natural enemies by relaxing competitive interactions and reducing effects of top-down forces. Specialization is often positively linked to the persistence of resources.
2. Herbivorous insects employ a variety of behavioral, physiological, and morphological adaptations to counter plant defenses, evaluate nutritional variation in their hosts, and mitigate attack from enemies. Strategies such as flightlessness, vein-cutting, and host and tissue specialization are widely distributed across insect taxa, whereas other strategies such as maternal care and the use of morphological shields against predators may be less widespread.
3. Habitat persistence selects for reduced dispersal capability in many insects because insects trade-off higher investment in reproduction with investment in the musculature associated with flight. Wing polymorphisms in which some individuals of a species do not produce fully functioning wings and flight muscles and some individuals do are a common form of reduced dispersal ability in insects.
4. Interspecific competition is an important ecological force that influences the population dynamics and community structure of phytophagous insects. Host plants mediate interspecific competition more frequently than natural enemies, physical factors, and intraspecific competition. The intensity of interspecific competition is not correlated with plant damage, can occur between distantly related species, is not dampened by temporal or spatial resource partitioning, and is often asymmetric.
5. Plant stress caused by lack of water often leads to reduced herbivore performance and smaller populations of herbivores. This occurs in a wide range of plants that inhabit diverse habitats and across many insect taxa. These results are opposite those predicted by the plant stress hypothesis (plant stress should promote outbreaks of phytophagous insects).
6. The bottom-up effects of plant nutrition and vegetation structure mediate the strength of top-down impacts on phytophagous insect populations. Predators are only able to exert significant suppressive effects on herbivore populations in habitats with low-quality host plants and complex vegetation structure where herbivores develop slowly and predators aggregate.

7. Robert Denno was an internationally respected figure in population and community ecology, known for his cutting-edge empirical studies and his penchant for synthesis. However, his lasting impact on the field of insect ecology is primarily as a devoted mentor who produced an extensive academic lineage of students and postdoctoral associates.

DISCLOSURE STATEMENT

The authors are not aware of any affiliations, memberships, funding, or financial holdings that might be perceived as affecting the objectivity of this review.

ACKNOWLEDGMENTS

We thank Merrill Peterson and Ian Kaplan for incredibly helpful comments on an earlier draft of this manuscript. M.D.E. acknowledges support from NSF DEB 0754462 and 0716983 and USDA NRI 2008-35302-04491. D.L.F. acknowledges support from USDA NIFA AFRI grant 2009-65104-05679. M.J.R. acknowledges support from USDA NRI grants 2005-35302-16269 and 2005-35302-16330.

LITERATURE CITED

1. Describes how herbivores consuming high-quality hosts reach invulnerable stages faster and suffer reduced levels of parasitism.

1. **Benrey B, Denno RF. 1997. The slow growth-high mortality hypothesis: a test using the cabbage butterfly. *Ecology* 78:987–99**
2. Bruno JF, O'Connor MI. 2005. Cascading effects of predator diversity and omnivory in a marine food web. *Ecol. Lett.* 8:1048–56
3. Campbell BC, Denno RF. 1976. The effect of the mosquito larvicides temefos and chlorophyrifos on the aquatic insect community of a New Jersey salt marsh. *Environ. Entomol.* 5:477–83
4. Campbell BC, Denno RF. 1978. The structure of the aquatic insect community associated with intertidal pools on a New Jersey salt marsh. *Ecol. Entomol.* 3:181–87
5. Connell JH. 1983. On the prevalence and relative importance of interspecific competition: evidence from field experiments. *Am. Nat.* 111:1119–44
6. Cooper W. 2002. Convergent evolution of plant chemical discrimination by omnivorous and herbivorous sceroglossan lizards. *J. Zool.* 257:53–66
7. Cooper WE, Vitt LJ. 2002. Distribution, extent, and evolution of plant consumption by lizards. *J. Zool.* 257:487–517
8. Craighead FC. 1925. Bark-beetle epidemics and rainfall deficiency. *J. Econ. Entomol.* 18:577
9. Denno RF. 1976. Ecological significance of wing-polymorphism in Fulgoroidea which inhabit salt marshes. *Ecol. Entomol.* 1:257–66
10. Denno RF. 1977. Comparisons of the assemblages of sap feeding insects (Homoptera-Hemiptera) inhabiting two structurally different salt marsh grasses of the genus *Spartina*. *Environ. Entomol.* 6:359–72
11. Denno RF. 1979. The relation between habitat stability and the migration tactics of planthoppers. *Misc. Publ. Entomol. Soc. Am.* 11:41–49
12. Denno RF. 1980. Ecotope differentiation in a guild of sap-feeding insects on the salt marsh grass, *Spartina patens*. *Ecology* 61:702–14
13. Denno RF. 1983. Tracking variable host plants in space and time. In *Variable Plants and Herbivores in Natural and Managed Systems*, ed. RF Denno, MS McClure, pp. 291–341. New York: Academic
14. Denno RF, Cothran WR. 1975. Niche relationships of a guild of necrophagous flies. *Ann. Entomol. Soc. Am.* 68:741–54
15. Denno RF, Cothran WR. 1976. Competitive interactions and ecological strategies of sarcophagid and calliphorid flies inhabiting rabbit carrion. *Ann. Entomol. Soc. Am.* 69:109–13

16. Denno RF, Donnely MA. 1981. Patterns of herbivory on *Passiflora* leaf tissues and species by generalized and specialized feeding insects. *Ecol. Entomol.* 6:11–16
17. Denno RF, Douglass LW, Jacobs D. 1985. Crowding and host plant nutrition: environmental determinants of wing-form in *Prokelisia marginata*. *Ecology* 66:1588–96
18. Denno RF, Douglass LW, Jacobs D. 1986. Effects of crowding and host plant nutrition on a wing-dimorphic planthopper. *Ecology* 67:116–23
19. Denno RF, Fagan WF. 2003. Might nitrogen limitation promote omnivory among carnivorous arthropods. *Ecology* 84:2522–31
20. Denno RF, Gratton C, Döbel HG, Finke DL. 2003. Predation risk affects relative strength of top-down and bottom-up impacts on insect herbivores. *Ecology* 84:1032–44
21. **Denno RF, Gratton C, Peterson MA, Langellotto GA, Finke DL, Huberty AF. 2002. Bottom-up forces mediate natural-enemy impact in a phytophagous insect community. *Ecology* 83:1443–58**
22. **Denno RF, Grissell EE. 1979. The adaptiveness of wing-dimorphism in the salt marsh-inhabiting planthopper, *Prokelisia marginata* (Homoptera: Delphacidae). *Ecology* 60:221–36**
23. Denno RF, Kaplan I. 1997. Plant-mediated interactions in herbivorous insects: mechanisms, symmetry, and challenging the paradigms of competition past. In *Ecological Communities: Plant Mediation in Indirect Interaction Webs*, ed. T Ohgushi, TP Craig, PW Price, pp. 19–50. Cambridge, UK: Cambridge Univ. Press
24. **Denno RF, Larsson S, Olmstead KL. 1990. Host plant selection in willow-feeding leaf beetles (Coleoptera: Chrysomelidae): role of enemy-free space and plant quality in host-plant selection by willow beetles. *Ecology* 71:124–37**
25. Denno RF, Lewis D, Gratton C. 2005. Spatial variation in the relative strength of top-down and bottom-up forces: causes and consequences for phytophagous insect populations. *Ann. Zool. Fenn.* 42:295–311
26. Denno RF, McCloud ES. 1985. Predicting fecundity from body size in the planthopper, *Prokelisia marginata* (Homoptera: Delphacidae). *Environ. Entomol.* 14:846–49
27. Denno RF, McClure MS. 1983. *Variable Host Plants and Herbivores in Natural and Managed Systems*. New York: Academic
28. **Denno RF, McClure MS, Ott JR. 1995. Interspecific interactions in phytophagous insects: competition reexamined and resurrected. *Annu. Rev. Entomol.* 40:297–331**
29. Denno RF, Olmstead KL, McCloud ES. 1989. Reproductive cost of flight capability: a comparison of life history traits in wing dimorphic planthoppers. *Ecol. Entomol.* 14:31–44
30. Denno RF, Peterson MA. 2000. Caught between the devil and the deep blue sea, mobile planthoppers elude natural enemies and deteriorating host plants. *Am. Entomol.* 46:95–109
31. Denno RF, Peterson MA, Gratton C, Cheng J, Langellotto GA, et al. 2000. Feeding-induced changes in plant quality mediate interspecific competition between sap-feeding herbivores. *Ecology* 81:1814–27
32. Denno RF, Raupp MJ, Tallamy DW, Reichelderfer CF. 1980. Migration in heterogeneous environments: differences in habitat selection between the wing forms of the dimorphic planthopper, *Prokelisia marginata* (Homoptera: Delphacidae). *Ecology* 61:859–67
33. **Denno RF, Roderick GK. 1990. Population biology of planthoppers. *Annu. Rev. Entomol.* 35:489–520**
34. Denno RF, Roderick GK. 1992. Density-related dispersal in planthoppers: effects of interspecific crowding. *Ecology* 73:1323–34
35. Denno RF, Roderick GK, Olmstead KL, Döbel HG. 1991. Density-related migration in planthoppers (Homoptera: Delphacidae): the role of habitat persistence. *Am. Nat.* 138:1513–41
36. Denno RF, Roderick GK, Peterson MA, Huberty AF, Döbel HG, et al. 1996. Habitat persistence underlies intraspecific variation in the dispersal strategies of planthoppers. *Ecol. Monogr.* 66:389–408
37. Denno RF, Schauff ME, Wilson SW, Olmstead KL. 1987. Practical diagnosis and natural history of two sibling salt marsh-inhabiting planthoppers in the genus *Prokelisia* (Homoptera: Delphacidae). *Proc. Entomol. Soc. Wash.* 89:687–700
38. Döbel HG, Denno RF. 1994. Predator-planthopper interactions. In *Planthoppers: Their Ecology and Management*, ed. RF Denno, TJ Perfect, pp. 325–99. New York: Chapman & Hall
39. Döbel HG, Denno RF, Coddington JA. 1990. Spider (Araneae) community structure in an intertidal salt marsh: effects of vegetation structure and tidal flooding. *Environ. Entomol.* 19:1356–70

21. Natural enemies most effectively suppress herbivore populations when plant quality is low and plant structural complexity is high.

22. Defines importance of wing dimorphism in planthoppers as an adaptation to temporally and spatially variable habitats.

24. A definitive test of enemy-free space as a determinant of dietary breadth in leaf beetles.

28. Interspecific competition is a widespread and important ecological force that influences population size and community composition of phytophagous insects.

33. Provides a synthetic worldwide review of population dynamics and tritrophic interactions of planthoppers.

40. Dussourd DE, Denno RF. 1991. Deactivation of plant defense: correspondence between insect behavior and secretory canal architecture. *Ecology* 72:1383–96
41. Ehrlich PR, Birch LC. 1967. The balance of nature and population control. *Am. Nat.* 150:554–67
42. Eubanks MD, Denno RF. 1999. The ecological consequences of variation in plants and prey for an omnivorous insect. *Ecology* 80:1253–66
43. **Eubanks MD, Denno RF. 2000. Host plants mediate omnivore-herbivore interactions and influence prey suppression. *Ecology* 81:865–75**
44. Eubanks MD, Styrsky JD, Denno RF. 2003. The evolution of omnivory in heteropteran insects. *Ecology* 84:2549–56
45. Finke DL, Denno RF. 2002. Intraguild predation diminished in complex-structured vegetation: implications for prey suppression. *Ecology* 83:643–52
46. Finke DL, Denno RF. 2003. Intra-guild predation relaxes natural enemy impacts on herbivore populations. *Ecol. Entomol.* 28:67–73
47. Finke DL, Denno RF. 2004. Predator diversity dampens trophic cascades. *Nature* 429:407–10
48. Finke DL, Denno RF. 2005. Predator diversity and the functioning of ecosystems: the role of intraguild predation in dampening trophic cascades. *Ecol. Lett.* 8:1299–306
49. Finke DL, Denno RF. 2006. Spatial refuge from intraguild predation: implications for prey suppression and trophic cascades. *Oecologia* 149:265–75
50. Finke DL, Snyder WE. 2008. Niche partitioning increases resource exploitation by diverse communities. *Science* 321:1488–90
51. Frank SD, Shrewsbury PM, Denno RF. 2010. Effects of alternative food on cannibalism and herbivore suppression by carabid larvae. *Ecol. Entomol.* 35:61–68
52. Gause GF. 1934. *The Struggle for Existence*. Baltimore, MD: Williams & Wilkins. 163 pp.
53. Gratton C, Denno RF. 2003. Seasonal shift from top-down to bottom-up impact in phytophagous insect populations. *Oecologia* 134:487–95
54. Hairston NG, Smith FE, Slobodkin LB. 1960. Community structure, population control, and competition. *Am. Nat.* 94:421–25
55. Hall SR. 2009. Stoichiometrically explicit food webs: feedbacks between resource supply, elemental constraints, and species diversity. *Annu. Rev. Ecol. Evol. Syst.* 40:503–28
56. Hanks LM, Denno RF. 1993. Natural enemies and plant water relations influence the distribution of an armored scale insect. *Ecology* 74:1081–91
57. Heady SE, Denno RF. 1991. Reproductive isolation in *Prokelisia* planthoppers (Homoptera: Delphacidae): acoustic differentiation and hybridization failure. *J. Insect Behav.* 4:367–90
58. Huberty AF, Denno RF. 2004. Plant water stress and its consequences for herbivorous insects: a new synthesis. *Ecology* 85:1383–98
59. Huberty AF, Denno RF. 2006. Consequences of nitrogen and phosphorus limitation for the performance of two phytophagous insects with divergent life-history strategies. *Oecologia* 149:444–55
60. Huberty AF, Denno RF. 2006. Trade-off in investment between dispersal and ingestion capability in phytophagous insects and its ecological implications. *Oecologia* 148:226–34
61. Hunter MD, Price PW. 1992. Playing chutes and ladders: heterogeneity and the relative roles of bottom-up and top-down forces in natural communities. *Ecology* 73:724–32
62. **Kaplan I, Denno RF. 2007. Interspecific interactions in phytophagous insects revisited: a quantitative assessment of competition theory. *Ecol. Lett.* 10:977–94**
63. Kaplan I, Halitschke R, Kessler A, Rehill B, Sardanelli S, Denno RF. 2008. Physiological integration of roots and shoots in plant defense strategies links above- and belowground herbivory. *Ecol. Lett.* 11:841–51
64. Kaplan I, Halitschke R, Kessler A, Sardanelli S, Denno RF. 2008. Constitutive and induced defenses to herbivory in above- and belowground plant tissues. *Ecology* 89:392–406
65. Kaplan I, Lynch ME, Dively GP, Denno RF. 2007. Leafhopper-induced plant resistance enhances predation risk in a phytophagous beetle. *Oecologia* 152:665–75
66. Kaplan I, Sardanelli S, Denno RF. 2009. Field evidence for indirect interactions between foliar-feeding insect and root-feeding nematode communities on *Nicotiana tabacum*. *Ecol. Entomol.* 34:262–70
67. Kennedy JS. 1958. Physiological condition of the host-plant and susceptibility to aphid attack. *Entomol. Exp. Appl.* 1:50–65

43. Plant feeding by omnivores results in larger and more stable omnivore populations and ultimately greater suppression of herbivorous prey by omnivores.

62. Insect herbivores commonly interact via nontraditional mechanisms, including plant-mediated and natural enemy–mediated indirect interactions and facilitation.

68. Kraft SK, Denno RF. 1982. Feeding responses of adapted and nonadapted insects to the defensive properties of *Baccharis halimifolia* (Compositae). *Oecologia* 52:156–63
69. Krischik VA, Denno RF. 1990. Differences in environmental response between the sexes of the dioecious shrub, *Baccharis halimifolia* (Compositae). *Oecologia* 83:176–81
70. Krischik VA, Denno RF. 1990. Patterns of growth, reproduction, defense, and herbivory in the dioecious shrub, *Baccharis halimifolia* (Compositae). *Oecologia* 83:182–90
71. Landis DA, Wratten SD, Gurr GM. 2000. Habitat management to conserve natural enemies of arthropod pests in agriculture. *Annu. Rev. Entomol.* 45:175–201
72. Langellotto GA, Denno RF. 2001. Benefits of dispersal in patchy environments: mate location by males of a wing-dimorphic insect. *Ecology* 82:1870–78
73. Langellotto GA, Denno RF. 2004. Responses of invertebrate natural enemies to complex-structured habitats: a meta-analytical synthesis. *Oecologia* 139:1–10
74. Langellotto GA, Denno RF. 2006. Refuge from cannibalism in complex-structured habitats: implications for the accumulation of invertebrate predators. *Ecol. Entomol.* 31:575–81
75. Larsson S. 1989. Stressful times for the plant stress-insect performance hypothesis. *Oikos* 56:277–83
76. Lawton JH. 1982. Vacant niches and unsaturated communities: a comparison of bracken herbivores at sites on two continents. *J. Anim. Ecol.* 51:573–95
77. Lawton JH, Strong DR. 1981. Community patterns and competition in folivorous insects. *Am. Nat.* 118:317–38
78. Lewis D, Denno RF. 2009. A seasonal shift in habitat suitability enhances an annual predator subsidy. *J. Anim. Ecol.* 78:752–60
79. Losey JE, Denno RF. 1998. Interspecific variation in the escape response of aphids: effect on risk of predation from foliar-foraging and ground-foraging predators. *Oecologia* 115:245–52
80. Losey JE, Denno RF. 1998. Positive predator-predator interactions: enhanced predation rates and synergistic suppression of aphid populations. *Ecology* 79:2143–52
81. Losey JE, Denno RF. 1998. The escape response of pea aphids to foliar-foraging predators: factors affecting dropping behavior. *Ecol. Entomol.* 23:53–61
82. Lynch ME, Kaplan I, Dively GP, Denno RF. 2006. Host plant-mediated competition via induced resistance: interactions between pest herbivores on potatoes. *Ecol. Appl.* 16:855–64
83. Martinson HM. 2010. *Critical patch sizes and the structure of salt marsh communities.* PhD diss. College Park: Univ. Maryland. 168 pp.
84. Matsumura M, Trafelet-Smith G, Gratton C, Finke DL, Fagan WF, Denno RF. 2004. Does intraguild predation enhance predator performance? A stoichiometric perspective. *Ecology* 85:2601–15
85. Mattson WJ, Haack RA. 1987. The role of drought in outbreaks of plant-eating insects. *BioScience* 37:110–18
86. Olmstead KL, Denno RF. 1992. Cost of shield defense for tortoise beetles (Coleoptera: Chrysomelidae). *Ecol. Entomol.* 17:237–43
87. Olmstead KL, Denno RF. 1993. Effectiveness of tortoise beetle larval shields against different predator species. *Ecology* 74:1394–405
88. Olmstead KL, Denno RF, Morton TC, Romeo JT. 1997. Influence of *Prokelisia* planthoppers on the amino composition and growth of *Spartina alterniflora*. *J. Chem. Ecol.* 23:303–21
89. Pearson RE. 2009. *Nutrient regulation by an omnivore and the effects on performance and distribution*. PhD diss. College Park: Univ. Maryland. 109 pp.
90. Peterson MA, Denno RF. 1997. The influence of intraspecific variation in dispersal strategies on the genetic structure of planthopper populations. *Evolution* 51:1189–206
91. Peterson MA, Denno RF. 1998. The influence of dispersal and diet breadth on patterns of genetic isolation by distance in phytophagous insects. *Am. Nat.* 152:428–46
92. Price PW, Bouton CE, Gross P, McPheron BA, Thompson ON, Weis AE. 1980. Interactions among three trophic levels: influence of plants on interactions between insect herbivores and natural enemies. *Annu. Rev. Ecol. Syst.* 11:41–65
93. Ramirez RA, Snyder WE. 2009. Scared sick? Predator-pathogen facilitation enhances the exploitation of a shared resource. *Ecology* 90:2832–39

80. A combination of predators with complimentary foraging tactics can synergistically suppress herbivore populations.

94. Raubenheimer D, Simpson SJ, Mayntz D. 2009. Nutrition, ecology and nutritional ecology: toward an integrated framework. *Func. Ecol.* 23:4–16
95. Raupp MJ, Denno RF. 1979. Influence of patch size on a guild of sap-feeding insects that inhabit the salt marsh grass *Spartina patens*. *Environ. Entomol.* 8:412–17
96. Raupp MJ, Denno RF. 1984. The suitability of damaged willow leaves as food for the imported willow leaf beetle, *Plagiodera versicolora* Laich. (Coleoptera: Chrysomelidae). *Ecol. Entomol.* 9:443–48
97. Raupp MJ, Sadof CS. 1989. Behavioral responses of a leaf beetle to injury-related changes in its salicaceous host. *Oecologia* 80:154–57
98. Raupp MJ, Shrewsbury PM, Herms DA. 2010. Arthropods in urban landscapes. *Annu. Rev. Entomol.* 55:19–38
99. Schmitz O, Hamback PA, Beckerman AP. 2000. Trophic cascades in terrestrial systems: a review of the effects of carnivore removals on plants. *Am. Nat.* 155:141–53
100. Schmitz OJ. 2007. Predator diversity and trophic interactions. *Ecology* 88:2415–26
101. Schoener TW. 1983. Field experiments on interspecific competition. *Am. Nat.* 122:155–74
102. Shrewsbury PM, Raupp MJ. 2006. Do top-down or bottom-up forces determine *Stephanitis pyroides* abundance in urban landscapes? *Ecol. Appl.* 16:262–72
103. Straub CS, Finke DL, Snyder WE. 2008. Are the conservation of natural enemy biodiversity and biological control compatible goals? *Biol. Control* 45:225–37
104. Strong DR. 1992. Are trophic cascades all wet? Differentiation and donor-control in speciose ecosystems. *Ecology* 73:747–54
105. Strong DR, Lawton JH, Southwood TRE, eds. 1984. *Ecological Communities*. Harvard, MA: Harvard Univ. Press. 313 pp.
106. Tallamy DW, Denno RF. 1979. Responses of sap-feeding insects (Homoptera: Hemiptera) to simplification of host plant structure. *Environ. Entomol.* 8:1021–28
107. Tallamy DW, Denno RF. 1981. Maternal care in *Gargaphia solani* (Hemiptera: Tingidae). *Anim. Behav.* 29:771–78
108. Tallamy DW, Denno RF. 1982. Life history trade-offs in *Gargaphia solani* (Hemiptera: Tingidae): the cost of reproduction. *Ecology* 63:616–20
109. Tallamy DW, Hansens EJ, Denno RF. 1976. A comparison of malaise trapping and aerial netting for sampling a horsefly and deerfly community. *Environ. Entomol.* 5:788–92
110. Trumbule RB, Denno RF. 1995. Light intensity, host plant irrigation, and habitat-related mortality as determinants of the abundance of azalea lace bug (Heteroptera: Tingidae). *Environ. Entomol.* 24:898–908
111. Trumbule RB, Denno RF, Raupp MJ. 1995. Management considerations for the azalea lace bug in landscape habitats. *J. Arboric.* 21:63–68
112. White TCR. 1969. An index to measure weather-induced stress of trees associated with outbreaks of psyllids in Australia. *Ecology* 50:905–9
113. Wilder SM, Eubanks MD. 2010. Might nitrogen limitation promote omnivory among carnivorous arthropods: comment. *Ecology*. 91:3114–17
114. Wilder SM, Rypstra AL, Elgar MA. 2009. The importance of ecological and phylogenetic conditions for the occurrence and frequency of sexual cannibalism. *Annu. Rev. Ecol. Evol. Syst.* 40:21–39
115. Wimp GM, Murphy SM, Finke DL, Huberty AF, Denno RF. 2010. Increased primary production shifts the structure and composition of a terrestrial arthropod community. *Ecology*. In press
116. Zera AJ, Denno RF. 1997. Physiology and ecology of dispersal polymorphism in insects. *Annu. Rev. Entomol.* 42:207–31

116. Habitat persistence selects for reduced dispersal capability and associated dispersal-fecundity trade-offs occur in virtually all insect taxa.

The Role of Resources and Risks in Regulating Wild Bee Populations

T'ai H. Roulston[1] and Karen Goodell[2]

[1]Department of Environmental Sciences, University of Virginia, Charlottesville, Virginia 22904-4123; email: tai.roulston@virginia.edu

[2]Department of Evolution, Ecology and Organismal Biology, The Ohio State University, Newark, Ohio 43055; email: goodell.18@osu.edu

Annu. Rev. Entomol. 2011. 56:293–312

First published online as a Review in Advance on September 3, 2010

The *Annual Review of Entomology* is online at ento.annualreviews.org

This article's doi:
10.1146/annurev-ento-120709-144802

0066-4170/11/0107-0293$20.00

Key Words

pollinators, direct effects, resource limitation, bees, Apiformes

Abstract

Recent declines of bee species have led to great interest in preserving and promoting bee populations for agricultural and wild plant pollination. Many correlational studies have examined the indirect effects of factors such as landscape context and land management practices and found great variation in bee response. We focus here on the evidence for effects of direct factors (i.e., food resources, nesting resources, and incidental risks) regulating bee populations and then interpret varied responses to indirect factors through their species-specific and habitat-specific effects on direct factors. We find strong evidence for food resource availability regulating bee populations, but little clear evidence that other direct factors are commonly limiting. We recommend manipulative experiments to illuminate the effects of these different factors. We contend that much of the variation in impact from indirect factors, such as grazing, can be explained by the relationships between indirect factors and floral resource availability based on environmental circumstances.

INTRODUCTION

Bees are a critical group of pollinators in many ecosystems, contributing to seed production in a wide diversity of wild plants (85). They are also the dominant pollinators for the portion of human food derived from animal-pollinated plants (60). While honey bees are the primary managed pollinator in agriculture, their decline in some countries (3, 86) has focused considerable attention on native, unmanaged bee populations as major contributors to agricultural pollination (41, 133). If honey bee populations decline further or cannot be increased to sustain the worldwide expansion of bee-pollinated crops (2), then it will be important to understand how to maintain pollination services of wild bees where they already predominate (104, 134) or to enhance their services where they make only modest contributions (62). Declines in populations of bee species other than honey bees, however, have been reported from several countries (6, 22). While maintenance of bee abundance assures the continuation of pollination services, bee diversity can improve reproduction in individual crop (41, 47, 119) and wild plant species (37, 105). Evidence also supports the positive role of pollinator diversity in maintaining plant community diversity (32).

Habitat loss is the most commonly cited factor affecting both pollination services and bee population and community declines (131). Recent work has modeled pollination as an ecosystem service dependent on landscape factors (61, 66), but studies of bee population responses rather than ecosystem services have produced complex or inconsistent results (13). A meta-analysis of the influence of disturbance factors on bee populations found that only extreme habitat loss produced statistically significant negative impacts on bee abundance and richness, and found great variation of bee population and community responses to different types of disturbance (131). We posit that factors such as disturbance have only indirect effects on bee populations and that improving our knowledge of direct effects (i.e., how they function and how they relate to indirect factors) will greatly improve our ability to augment wild bee population abundance and diversity. Here, we review what is known about factors hypothesized to have direct effects on bee populations and discuss whether varied responses to landscape factors may be due to their conflicting influence on these direct factors.

At a basic level, we expect that the factors that control bee populations are simple and direct. Individual species abundances should increase with food and nesting resources but be reduced by risk factors, such as predation. Niche differentiation of food and nesting resources should promote bee diversity, but individual risks and risks that affect the community as a whole, such as pesticides, should reduce it (55). **Figure 1** shows a conceptual model of the interactions between species traits and factors assumed to have direct effects on bee populations. Diet breadth determines which components of local resources a given species perceives as food (16, 71). Foraging range, because it correlates with body size (42), describes both the geographic area in which resources can be used and the amount of a resource that must be collected to provision offspring (76). Life-history traits include level of sociality and nesting habit (excavating belowground, excavating aboveground, or utilizing preexisting cavities), which influence bee population responses to disturbance (129). Factors with direct effects (food abundance and timing, nesting resources, and incidental risk) (**Figure 1**) should most clearly influence population size.

The indirect factors shown in our conceptual model, invasive species, habitat complexity, and land management, are a subset of potential categories. Others include habitat fragmentation, climate change, and anthropogenic disturbance, which may overlap in various ways. The key point is that indirect factors work by influencing the direct drivers of bee population growth. In order for indirect factors to produce consistent results, they must have consistent effects on the direct factors. First, we consider direct factors themselves and the extent to which their population effects are known. We point

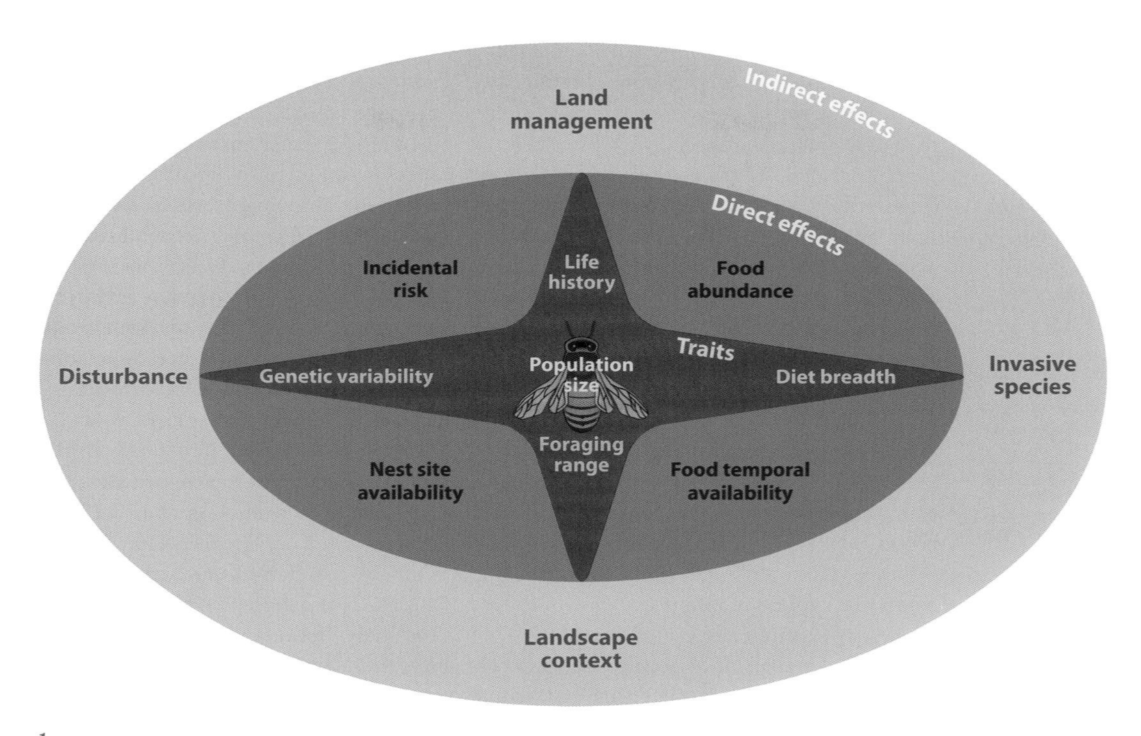

Figure 1

Direct and indirect effects. Bee populations are driven by direct effects and modulated by species traits. Indirect effects act by their influence on underlying direct factors.

out where important interactions between direct factors occur that modify their expected effects when acting individually, requiring joint consideration of these factors to understand their influence on bee populations. We argue that better understanding of direct effects and their interactions is needed for predictive models that focus on indirect factors. Ultimately, this information will be useful for creating prescriptive management plans to achieve specific population and community outcomes for bee community structure and function in pollination services.

DIRECT EFFECTS

Food Quantity and Temporal Distribution

Floral nectar and pollen are the primary energy source for most bee species, comprising both larval and adult diets (69). They are offered only by flowers, many of which present their rewards briefly. Pollen and nectar themselves are susceptible to abiotic conditions and consumption by other biotic communities, such as yeasts (45). Nonoverlapping phenologies of bees and flowers and lack of fit between bee food-gathering structures and flower parts can limit flower availability in a habitat. Widespread pollen specialization among bees, which can range from 15% to 60% of bee species in different biogeographic regions, further limits the potential bouquet of flower species available to a particular bee species in a particular habitat (71). Therefore, floral resource availability is hypothesized to be a major driver of population abundance and community diversity of bees.

Floral resource limitation of bee populations has been inferred by reproductive output of social colonies or solitary nests, positive correlations of either bee abundance or

Oligolectic: refers to bees that use a limited subset of available pollens, usually in a single plant family, to provision their offspring

bee diversity with floral resource abundance or diversity, and negative correlations between bee species thought to compete for food. Studies of oligolectic (pollen specialist) bees permit the simplest analyses of these relationships because resources are relatively easy to calculate and foragers are easy to find by observing their host plant. *Andrena hattorfiana* (Andrenidae) population size in southern Sweden correlates strongly with population sizes of their host plant, *Knautia arvensis* (65). The researchers calculated that each offspring requires the floral resources of two entire plants for complete development. A second specialist bee, *Calliopsis pugionis* (Andrenidae), was estimated to need a single plant to provide the pollen resources for 1,600 brood cells (122). Such estimates based on a resource economy offer tools for managing habitat to support viable populations of these species, as exemplified by Muller et al. (76).

Detailed studies of oligolectic bees also offer insights into the mechanisms and timing of resource limitation. *Dieunomia triangulifera* (Halictidae) foraging on its host, *Helianthus annuus*, took longer foraging trips and carried less pollen as resources declined seasonally and daily (72). Estimates of floral resource availability combined with foraging patterns and brood provision sizes predicted that individuals foraging early in the season when flowers were scarce would need a foraging radius of 8 km, but at peak bloom that radius declined to less than 1 km. Even with adequate floral resources present in the local landscape, the distance between them and bee nests can greatly influence bee productivity. When two specialist cavity-nesting species, *Hoplitis adunca* and *Chelostoma rapunculi* (Megachilidae), were placed at increasing distances from their host plants, provisioning rates dropped 23%–46% (140). While these studies support the influence of local floral abundance on reproductive success, it is notable that females of both *D. triangulifera* and *C. pugionis* mentioned above did not maximize their number of daily foraging bouts and did not forage every day (72, 122). Reduced foraging may reflect slower rates of oocyte maturation than of brood provisioning (95).

Resource limitation of reproductive output and population size of generalist species has also been tested but by using proxies for resource abundance because it is difficult to both specify and quantify all possible plant species that generalists might use. Brood cell production in *Osmia lignaria* (Megachilidae) in an agricultural landscape in California varied with the proximity to natural habitat at larger spatial scales (130). Similarly, snapshot estimates of local floral density and bee population abundance are often positively correlated (19, 88), and food supplementation of captive bumble bee colonies foraging in natural settings sometimes boosts reproductive success (84). There are exceptions to these findings (27), however, especially for bumble bees (20, 29, 40).

This correlative approach requires special attention to geographic and temporal scales of the study. Because current bee populations derive from resources present in the prior generation (weeks to more than a year previous) (23, 70), cross-site correlations based on snapshot estimates should be positive only if resource variation within sites across time is relatively low compared with cross-site variation. Areas under consistent farm management or with predictable rainfall are more likely to produce these conditions than highly variable communities, such as deserts, where different plant species have different rainfall requirements and often bloom intermittently and asynchronously (8). Evidence for temporal lags between resource levels in one year and bee abundances the following year is limited, partly because it is rarely considered (72, 88). The observation that founding bumble bee queen density correlates to floral resource density early in the spring in subalpine meadows, but not later in the summer (28), demonstrates that time lags can also result from short term fluctuations in floral resources. It seems likely that densities of eusocial bumble bees, which start colonies annually from a single foundress, track the availability of floral resources in the proximity of nest sites during the brief period of colony founding rather than later in the season when weather and flowers are less variable and foragers more plentiful. In general, short-term time lags should be more likely in population response to floral resources by multivoltine species, while univoltine species and first-generation multivoltine species should show evidence of interannual lags in population abundance.

Because bees are mobile, correlations between local floral density and forager density do not necessarily indicate resource limitation of population size but could reflect patterns of aggregation around pulses of rich resources, such as might be predicted by an ideal, free distribution of consumers relative to prey (91). Therefore, correlations between flower abundances relevant to a particular bee population may occur over a broader spatial or temporal scale than is typically studied. For example, local abundance and diversity of bumble bee species on grasslands in Iowa were better explained by floral resource abundance in the grasslands within a radius of 500–700 m of the site than by the same index of local floral resources (46). Large pulses of floral resources from oilseed rape crops strongly predicted bumble bee forager abundances on planted forage in German landscapes (126), and different bumble bee species responded to the proportion of these large patches of synchronously flowering crops at spatial scales ranging from 250 to 3000 m (127).

Timing and composition of bloom are also critical to bee species in ways that relate to species-specific traits such as diet breadth and flight season. Continual resource availability over the whole active season is needed by most social and multivoltine species. Although many studies find that the abundance of natural habitat is positively associated with bee populations, bee abundance and richness decline with increasing cover of the predominant natural vegetation type in the eastern United States—forest (132). Temperate deciduous forests can provide good springtime floral resources for bees, but the forest tends to lack both flowers and bees in the summertime (44). During one study of the impact of logging on bees in New York state, no flowers were seen in the control forest plot during the summer over two study years and only a single bee specimen was taken, in great contrast to logged plots, which had both bees and flowers (94). Therefore, the amount of natural habitat is not necessarily a blanket predictor for bee abundance and diversity.

The quality and composition of floral resources interact with species-specific traits to determine population abundances across sites. For example, bumble bee forager abundance in small plots within a 50 × 50 m swath of Norwegian grassland was determined largely by the presence of highly rewarding tubular flowers and not overall floral richness (43). Individual bee species responded to different measures of floral resource availability (88), suggesting that the diverse floral communities support diverse bee species by offering resources that benefit species differently.

Multivoltine: refers to species that produce more than one generation per year

Univoltine: refers to species that produce a single offspring generation per year

Nesting Resources

Bees use nests to protect adults and developing larvae from predators, parasites, environmental extremes, and incidental harm. The majority of species excavate simple tunnel systems in soil, but others require particular structures or resources that are potentially in limited supply, including mud, leaves, resin, pith, dead wood, narrow cavities, and large protected chambers (69, 96). Demonstrating that nesting resources are limiting is challenging because they are likely to correlate with local vegetation structure and thus floral resources as well. Ideally, nest site limitation should be demonstrated by associating an increase in nesting resources with a subsequent increase in population sizes of bees, without changing other important variables. We are unaware of any studies that have documented convincingly bee population response to augmented nesting resources independent of floral resources. The main evidence for nest site limitation is inference drawn from bee population sizes compared with estimates of nesting resources in unmanipulated (by the researcher) landscapes.

Stingless bees (Meliponini), which nest in tree cavities or hollow areas in the ground, use plant resins for nest construction (96). Because large cavities and resin-producing plants are more likely to occur in mature tropical forest than in agricultural areas, it is likely that the distance to forest constrains stingless bee populations. Ricketts (93) found that stingless bee abundance in coffee plantations declined with distance to forest, and suggested that nest site limitation was a potential mechanism. Eltz et al. (30) found that stingless bee nests were associated primarily with very large tropical forest trees, but that nest density was better explained by pollen resource availability than by nest availability in forests with different histories of disturbance.

Many bee species in the family Megachilidae (and some Colletidae) routinely nest in narrow cavities in wood, soil, or plant stems and separate their brood chambers with leaf pieces (*Megachile*), resin (*Heriades*), or mud (*Osmia*). Because they show great selectivity for cavity size (103) and leaf type (48), and because mud may be seasonally or locally rare, there is great potential for nest resource limitation. Numerous studies have used artificial cavities to record the presence of cavity-nesting bees (e.g., 63). Although trap nests provide evidence for the occurrence of species that will use them (125), inferring either population size or population change by this method is problematic. Trap nests compete for nesting female bees with natural cavities, and it is difficult to assess natural cavity availability (but see Reference 89) or the relative attractiveness of trap nests compared with natural cavities. In one study that captured numerous cavity-nesting bees by other methods, only a single bee was caught among 20 trap nests across 10 sites (36). One study used mark-recapture methodology with trap nests to estimate local population size (109) by comparing the number of marked bees released from and recovered in trap nests with those that colonized the trap independently. Given that bees show strong philopatry to natal nests (109), marked bees would have to be released away from nest sites, not in them, to permit equal sampling probability of marked and unmarked bees.

If narrow cavities are limiting resources, then providing trap nests at sites over time could increase population size. Ideally, this would be demonstrated by comparing estimates of bee abundance in trap nests with bee abundance sampled independently by other methods. Unfortunately, we are unaware of any studies that do this, but some find correlations consistent with nest site limitation. Moretti et al. (74) found that recent low-intensity fires in Switzerland had a differentially favorable impact on bee species that excavate dead wood or nest in preexisting cavities, consistent with the idea that a flush of nesting material resulting from fire releases particular guilds of bees from nest site limitation. One study inferred population increase through trap nest augmentation at study sites over five years by recording a 35-fold increase in the number of offspring reared from trap nests over time (110). Because bees reared in trap nests were returned to the trap nests for emergence each year, however, it is unclear if this represents population increase in general or just an increase in trap nest use driven by philopatry. In the sixth year of that study, bees were not returned to the trap nests, and trap nest occupancy, then composed of only nest-searching bees reared in the habitat, dropped by 80%–90%, to nearly the level at the beginning of the project. Thus, the 35-fold increase in brood cells over five years appears to be applicable only to the descendents of the first generation of trap-nesting females unless the trap nest population actually comprised most of the local population.

Cane et al. (15) inferred nest site limitation for cavity-nesting bees through field population estimates of the pollen specialist *Hoplitis biscutellae* (Megachilidae). They found that the bee was several times more abundant at its host plant, creosote bush, in urban desert fragments than it was in nearly pure stands of its host in nearby open desert and postulated that cavities were more available in urban areas than in natural desert.

Bumble bees nest in larger aerial or subterranean cavities, often vacant rodent burrows.

McFrederick & LeBuhn (68) found that bumble bee forager density was positively associated with rodent hole density. Similarly, Potts et al. (89) found an association between cavities >2 cm in diameter and the abundance of large cavity-nesting bees in the family Apidae, mainly *Apis mellifera*.

Observational studies of bee guilds or individual ground-nesting species have found correlates of local nesting density with environmental variables such as soil moisture, ground cover, slope aspect, and soil compaction (e.g., 57, 89). Fewer studies have correlated estimates of population density with substrate characteristics. Julier & Roulston (52) found irrigation associated with squash bee density on farms. Potts et al. (89) found a strong association between bare ground and ground-nesting bee populations. Although provocative, correlative studies such as these do not provide clear evidence that manipulating only nesting resources would result in greater bee populations. Farm irrigation could easily correlate with flowering resources and be a greater driver than preferred soil texture. Similarly, bare ground could also be associated with floral characteristics. In the study by Potts et al. (89), bare ground was positively associated with fire disturbance, which has been shown in some studies to correlate with floral resources for pollinators as well (12, 74, 87).

Overall, there is good reason to expect that nesting resources are potentially limiting, but there is little compelling evidence to show the scale, frequency, or severity with which nest site limitation occurs. Studies are very much needed to show the conditions under which manipulating nesting resources changes bee populations independent of changes in vegetation or other potentially direct effects.

Incidental Risks

Incidental risks to bees include sources of mortality that disrupt the reproduction of individuals and therefore potentially contribute to population regulation. They include a variety of biotic and abiotic factors that differ among bee species and communities and therefore must be considered separately for each species, guild, or habitat. Incidental risks can be devastating locally but are highly variable over time and space, so it is more difficult to predict their effect on population size outside of a specific context, for example, bees inhabiting a particular cropping system. Here, we consider types of incidental risk that have received substantial research attention for their effects on bees.

Tilling. Agricultural tilling involves turning over and mixing the top layer of soil, usually on an annual or semiannual basis. Many insects, including both pests and beneficials, may be affected by this activity. Tilling potentially crushes subterranean insects and exposes vulnerable stages to predators and disease. Numerous studies have examined the effects of tilling on invertebrates, and the overall conclusion is that macroinvertebrates, such as beetles and earthworms, are particularly sensitive to tilling (58). Tillage practices across farms vary greatly, in both depth and type of plow (e.g., moldboard plow, chisel plow, and rotary tiller), and different tilling methods have species-specific effects, such as shown for ground beetle populations (102). Thus, results may be complex.

Many ground-nesting bee species place their brood cells <30 cm from the surface (67, 79). The newly produced offspring remain in diapause in their nests from the end of the previous flight season until the beginning of the next. Because agricultural tillage commonly reaches to a depth of 15–30 cm, tilling is likely to destroy part or all of some subterranean bee nests. To date, no researcher has reported on an experiment that tills through a nesting aggregation to measure mortality, as has been done for ground beetles (102).

There are two published studies of the effects of tillage on the density of *Peponapis pruinosa*, a specialist bee pollinator of squashes, gourds, and pumpkins (52, 104). In some respects, this species should be among the most sensitive to tilling: It has no noncrop host plants in much of its range and it prefers to nest in agricultural fields directly below its host plant at a depth that places its brood cells within common tilling range (52). Despite this apparent susceptibility, the two studies produced conflicting results, with Shuler et al. (104) finding reduced populations in tilled fields and Julier & Roulston (52) finding no difference. There are various possible reasons for this discrepancy. First, a sufficient reservoir of bees on-site may survive by nesting beneath the till zone or outside the till zone. Despite the apparent preference for *P. pruinosa* to nest within the crop, the bees sometimes nest outside the crop area in aggregations that can persist at a site for years (56). Many ground-dwelling bee species that nest in aggregations show strong philopatry (137), which could lead to increased bee occupancy in safe sites, once established, regardless of the relative preference for a particular site.

Second, one study (52) focused exclusively on pumpkins while the other included all cultivated yellow-flowered *Cucurbita* available in regional agricultural systems. Because pumpkin is cultivated as a late-season holiday crop, it may not flower before its specialist bees emerge. If early bees disperse rather than wait, farms with safe nest sites but late flowers may serve as regional source populations and depend on late emergence or immigration to maintain a specialist bee population. Distinguishing among these scenarios (surviving tilling, avoiding tilling through philopatry at safe sites, and dispersing toward resources at regional scales) is important because each scenario provides different guidelines for promoting and preserving bee populations. If bees often survive tilling, then well-timed resources may be the most important factor for keeping wild bees in agricultural systems. If they do not survive tilling, then regional untilled land will act as sources and tilled land as sinks, as proposed by Kim et al. (57).

Parasites/Disease. Bees support numerous parasitic guilds, including insects (Diptera, Coleoptera, Hymenoptera, and Strepsiptera), arachnids (mites), and protozoans. These parasites can be classified into those that attack adult bees, usually while foraging, those that attack brood, those that attack the stored provisions of brood cells (cleptoparasites), and those that usurp the nests of eusocial bees and produce only sexual castes using their host's worker castes, such as cuckoo bumble bees. Together or individually, these natural enemies could influence individual bee or colony survival and reproduction and eventually population dynamics and community structure. A detailed account of the ecology and evolution of parasites of social bees has been compiled by Schmid-Hempel (99). In their review of cleptoparasites and natural enemies of bees, Wcislo & Cane (124) indicate that the evidence for demographic effects of natural enemies is scant and we would argue that rigorous studies demonstrating that natural enemies regulate bee populations are still lacking. Parasite and pathogen effects on bee populations are known to be problematic in managed bees such as honey bees (118) and other domesticated bees, such as the alfalfa leafcutting bee, in which unnaturally high densities and environmental stresses can increase exposure and susceptibility to parasites and pathogens (50). Research effort into the population effects of natural enemies (top-down effects) in wild bees is small relative to that in managed bees. Nevertheless, it is likely to be important, as top-down control of herbivorous insect density is not uncommon (e.g., 31). In fact, it is one of the central pillars of the theory behind biological control of pest outbreaks (116).

The best test for the importance of top-down factors in regulating the populations of insects involves experimental manipulations of predator or parasite density and measures of prey or host response in terms of density or demographic rates (31). Because demographic responses result from a difference between birth and death rates, experiments that include a manipulation of resources can determine the relative importance of these two direct factors. In bees, these experiments are tricky because manipulation of floral resources and parasite density often requires caging, which is likely to affect the normal foraging and nesting behavior of bees. Studies that assess the effect of natural or experimental variation in parasitism in the field have the advantage of retaining the natural context for behaviors, but the disadvantage of confounding factors, such as resource availability and exposure to multiple natural enemies (e.g., 49). When conducted in a comparative framework, for instance, across sites varying in environmental factors, they can help to isolate important correlates to parasite success. Replication of these studies over several years can provide important data on the magnitude of temporal variation in demographic effects of parasites.

Experimental manipulation of both floral resources and parasites within enclosures surrounding nesting populations of a twig-nesting solitary bee, *Osmia pumila*, indicated a greater influence of resources than of parasites on overall brood cell production; 60% more brood cells were produced when floral resources were doubled versus 12% mortality caused by parasites when present versus absent (38). Importantly, parasitism rates rose to 25% of brood cells under sparse floral resources, five times higher than under rich floral resource environments. Solitary bee females are expected to experience a special sensitivity to parasitism under conditions of sparse resources because increased foraging effort causes a trade-off with protecting the nest from brood parasites (e.g., 122). This trade-off has not yet been demonstrated in the field, but if it proves common it suggests exacerbated negative effects on overall reproductive output when floral resources become limiting. Confirmation of this interaction between direct factors under natural conditions is desirable.

A study of natural variation in resource availability and parasitism rates by conopid flies, common parasites of foraging bumble bees in Europe, showed that parasitism explained less of the variance in population-level reproductive output (male production and the number of males per worker) than did resource availability (100). Social parasitism of free-living bumble bees by their parasitic congeners (*Bombus* subgenus *Psithyrus)* directly reduces colony success by ovarian suppression of workers and the production of their own reproductives (120). In some studies, rates of *Psithyrus* attack of field-placed captive colonies reached 100% and higher rates of attack occurred under the most favorable floral resource conditions, presumably related to the higher densities of *Psithyrus* in resource-rich habitats (20, 40). This interaction between social parasitism and floral resources is potentially important because it indicates that parasitism limits a colony's maximum reproduction under the most favorable resource conditions.

Social parasitism: the exploitation of a social colony structure for the production of offspring by an individual unrelated to the colony, often a different species

Brood parasitism of solitary bees directly reduces reproductive output by causing offspring mortality after the female has fully invested in that individual. Surveys of parasitism in natural populations of solitary bees indicate that parasitism rates can be high but vary widely among years. Parasitism of the solitary bee *Osmia rufa* by six species of cleptoparasites and parasitoids accounted for 17% of overall brood cell production averaged over 30 field sites and five years but varied significantly across years (110). In this study, average attack rates of trap-nesting solitary bee nests by cleptoparasites and parasitoids were related to habitat age, but not to the diversity of trap nesters. An investigation of a trap-nesting population of *Osmia tricornis* found that parasitism rates of brood cells varied little (12%–16%) over three years, but the rates of parasitism by any one species and its rank importance varied dramatically over years (121). Because the stage and mode of attack of different parasite species differ, avoiding parasites requires various strategies and adaptation to particular parasites may be hindered. Minimizing parasitism in captive populations also requires multiple strategies (7).

Internal protozoan parasites of bumble bees are potential factors in declines of bumble bee species, particularly in association with the commercial rearing and importation of bumble bees for greenhouse pollination (81). *Crithidia bombi* and *Nosema bombi*, two protozoan gut parasites prevalent in field and commercial bumble bee colonies, show potential for large-scale negative effects on bumble bee colonies and populations. *C. bombi*, though not highly virulent under field conditions (49), can have strong negative effects on colony size and reproduction when infected colonies experience starvation or other stressful conditions (10, 11). Negative effects of *C. bombi* infection on foraging performance of bumble bee workers suggest the potential for additional negative consequences in field conditions (80).

The effects of *N. bombi*, generally thought to be the more virulent of the two, have provided similarly variable results. Natural infection of captive bumble bee colonies in the field with *N. bombi* showed greater production of reproductive individuals than uninfected colonies (49). A study of greenhouse colonies experimentally infected showed no significant effects of *N. bombi* infection on colony performance (128), yet laboratory studies indicate almost total loss of fitness of queens and males in colonies infected early in the colony development cycle (82). A field study of experimentally infected colonies indicated significant negative impacts of *N. bombi* on colony growth and reproduction (83). Field studies are likely to provide the most realistic picture of parasite impacts on later colony development, but their ability to discern effects on queen survival and establishment is more limited. Genetic differences in effects and local adaptation of parasites to bee hosts increase the chances that exposure of wild populations of bumble bees to parasites of commercially reared bumble bees could create harmful epidemics. Well-controlled field experiments that track infection and colony development and reproduction under a variety of field conditions and that include species of conservation concern are needed to assess the environmental conditions under which *N. bombi* threatens the persistence of local populations.

Empirical research on bumble bee communities suggests that both parasite diversity and parasite load increase with local abundance of a species (26), as predicted by theory (90). In competitively structured bumble bee communities, lower-ranking species may benefit if parasites suppress population growth of the most abundant species, potentially enhancing community diversity. To our knowledge, no studies have examined the effects of parasitism on community structure of bees other than bumble bees.

Predators. Despite the ability of females of most species to sting, bees still suffer predation by numerous types of predators. These include vertebrates that consume larvae and/or honey (111), but most predators eat adult bees. Abundance of the European bee-eater, *Merops apiaster*, a bird that predominantly consumes flying insects, was positively correlated with the abundance of honey bee colonies, its primary prey, suggesting the importance of prey density on predator density. The bee-eater, however, consumed less than 1% of honey bee workers, suggesting that the prey was more important to the predator than the predator to the prey (34). Other major predators include flies, wasps, ants, and spiders. Species in the robber fly genus *Mallophora* consume adult bees of a wide variety of species (21). Many crab spider species perch on flowers and pose sufficient risk to bees such that bees alter their foraging behavior for an extended period of time to avoid the area where an encounter previously took place (25). Species in the wasp genus *Philanthus* (the beewolves) are voracious consumers of a wide variety of bee and wasp species. In one study, Stubblefield et al. (113) sampled more than 4,000 prey from a nesting aggregation of *Philanthus sanborni* and found that nearly all observed flower-visiting bee and wasp species in the area had been taken as prey. A study conducted near an aggregation of *Philanthus bicinctus* found that the predator had significantly reduced the local forager density of bumble bees, which further led to a significant reduction in pollination of a bumble bee–pollinated plant, *Aconitum columbianum* (24). Predation by ants is so common in some systems that it has been implicated in favoring behavioral shifts to sociality as a defense mechanism in the facultatively social allodapine bee *Exoneura nigrescens* (Apidae) in Australia (139).

Although all these studies show that predation is common and can have local population impacts, there is still little known about demographic effects on a larger scale and which direct (e.g., food availability and nest site selection) and indirect (e.g., landscape features or land management regimes) factors influence predation on bee populations.

Pesticides. Many pesticides and some herbicides and fungicides are toxic to bees and clearly pose a substantial hazard wherever they are used extensively, including agricultural and residential areas as well as woodlands that are treated for forest pests. They have been shown to cause sudden death of honey bee colonies since the nineteenth century (51) and may be related to recent honey bee declines (117). They have also been shown to reduce alfalfa leafcutting bee (4) and bumble bee (35) productivity and have been implicated in ecosystem-wide reductions of pollinator services (54). Pesticides have many sublethal effects on bees, including the impairment of foraging behavior (73), reducing the likelihood of returning to the nest (138), slowing larval development (1), and impeding learning (5). Given the abundance and diversity of pesticides found in honey bee colonies (77), it appears that bees regularly come in contact with pesticides in anthropogenic landscapes. Gauging the risks of pesticides is greatly complicated, however, by their diverse chemistry, retention times, and formulations (35) as well as the likelihood of species-specific effects on bees (101). Some species-specific differences may be caused by differences in physiology and behavior, such as the potential for concentrating or diluting pesticides in social colonies or adding risk through the use of leaf material in nest building (101). Although the most common means of testing is through laboratory feeding trials, it is difficult to extrapolate these results to likely impacts without better knowledge of field exposure or potential combined effects of mortality and sublethal effects on population growth (106).

Various studies use organic and conventional farms as contrasting treatments presumed to correlate with pesticide exposure (104, 130). Pesticide use on conventional farms is seldom quantified, however. In addition, some pesticides approved in organic farming have both lethal and sublethal effects on bees (101, 114). Thus, a lack of significant differences in bee populations between conventional and organic farming could be generated by a lack of experimentally robust treatment categories to specify pesticide risk. For studies on farms, we recommend making comparisons between treatments on large acreages of single crops (such as orchards) where a particular pesticide is being used. For manipulative experiments, the approach by Gels et al. (35) of monitoring colony growth of bumble bees under restricted but free-flying conditions with realistic field exposure is a promising middle ground between overly controlled experiments that may not represent realistic exposure and natural experiments where documenting actual exposure may be difficult. We hope that future pesticide exposure research will work with multiple bee species representing different ecological groups (social and solitary, leafcutters, and ground-nesters) toward the creation of population models of risk, as outlined by Stark & Banks (106).

INDIRECT EFFECTS

We define indirect effects as those arising from factors of broad general impact that influence bee populations primarily through subsequent changes in factors that have more direct impacts. For example, logging in most habitats is more likely to have a bigger impact on bees through changes in vegetation (94) than through direct mortality, although direct mortality is possible and could be substantial in tropical ecosystems. Indirect effects include many categories explored by other researchers as potential drivers of bee populations, including proportion of natural habitat in an area, distance to natural habitat, fragmentation, farming method, habitat complexity, disturbance level, grazing, agriculture, and habitat loss. Many of these factors were included in the meta-analysis of Winfree et al. (131), who showed the outcome (positive or negative) of different disturbance types on bee populations. Here, we focus discussion on the extent to which factors that cause indirect effects can be examined more simply as drivers of direct effects. To the extent that direct effects are relatively simple to measure and have predictable effects on bee populations, land manipulations that alter direct factors should result in relatively simple responses by bee populations. For example, in their model of land management's effect on invertebrate trophic levels, Woodcock et al. (135) found that the key predictor of species richness for both phytophagous and predatory trophic levels was sward architecture.

Land Management

Of land management practices, grazing has been examined most often for its impact on bee populations. Most studies of grazing have considered grazing intensity on a relative scale from light to heavy. Generally, increased grazing intensity negatively affects bee populations (53, 64, 136). In these cases, increased grazing is associated with decreased abundance or diversity of floral resources. In other studies, grazing has a positive effect on bee populations (18, 123), and in these cases grazing has a positive effect on floral resources. Thus, the evidence on grazing to date points mainly to a fairly simple effect on floral resources, but one study found that bare ground (a surrogate for ground-nesting bee nesting substrate) provided information beyond that provided by flower response alone (123).

Disturbance

The only particular disturbance type to date examined in several studies has been fire. Fire may have direct effects by burning bees (78), but this has never been quantified directly or inferred as a major effect. Instead, the impact on bees has been most clearly related to floral resources, which tend to flush in the first years after fire and then decline, along with bees, several years later (75, 87). One study found evidence of species-specific postfire effects on bees most likely generated by shifts in the herbaceous plant community that favored large social bees over smaller solitary species (78). In a study of fire and fire surrogates as management practices, Campbell et al. (12) found that either type of disturbance, but especially the two in combination, decreased tree canopy, increased herbaceous vegetation, and increased bee abundance. Similarly, in a study of the effect of logging (2–3 years after the disturbance), bees were most abundant where floral abundance was greatest in the most disturbed habitats (94).

Although disturbance does not always cause local reduction in abundance and diversity of bees (α-diversity), presumably because local resources in disturbed habitats provide equal or better resource opportunities for bees, disturbance can still have negative impacts through reducing diversity at larger spatial scales (β-diversity). Winfree et al. (132) noted that, although cleared land had higher abundance and diversity of bees than native eastern pine forests, several bee species used only pine forest habitat. Quintero et al. (92) reported little difference in local-scale diversity of bee communities between paired disturbed and undisturbed patches across a 50-km-long elevational and precipitation gradient of Patagonian forest. Anthropogenic disturbance appeared to homogenize bee communities over broad spatial scales, however. There was less than expected turnover in bee species across the gradient in disturbed than in undisturbed patches. Similarly, paired restored and remnant riparian habitats supported similar bee abundances and diversities, but approximately half of the species were found in only one type of habitat (127). Species traits are likely an important filter in determining which species thrive in more disturbed communities.

Landscape Context

The abundance or diversity of bees estimated in a local area may be related to the type, amount, or connectivity of land surrounding the surveyed area. It is seldom possible to estimate floral resources or their temporal distribution, nesting substrate, and incidental risks over a large area, or to calibrate habitat connectivity for the bee populations sampled. Thus, several direct effects may be confounded when landscape context is examined. The most common variable used in studies of bees is the amount of seminatural habitat surrounding the sampling area. Although this is often a significant predictor variable for bee populations (62, 108, 130), it sometimes is not (17, 132). It may yield a positive statistical interaction with other indirect factors (98) or do so in a species-specific or habitat-specific manner (17, 59). This is to be expected because it is seldom possible to quantify the direct effects in the system (but see Reference 107) and because different species respond demographically at different geographic scales (108). Given that habitat loss had only modest effects on bee populations, and then only in extremely degraded habitats (131), landscapes that appear highly degraded to humans may still provide the necessary resources for bees. Even urban landscapes can be relatively rich in bees. Berlin, Germany, for instance, has 262 bee species, half the total for the entire country (14). Interestingly, the habitat that seems most hostile to bees, such that distance to another type of habitat is of critical importance, is agriculture. As with food abundance, the relationship between landscape context and direct effects, as well as independent effects of connectivity, is best understood for specialist bees with easily discernible host plants. Franzen & Nilssen (33) followed 63 habitat patches containing the plant host of the specialist bee *Andrena hattorfiana* (Andrenidae) over four years and were able to document the relationship between floral resources, bee population size, habitat connectivity, and bee extinction and colonization events. Generalist species and whole communities are much more complicated to track and predict, but this work should stand as a model for how these types of parameters may combine to regulate bee populations.

One often discussed form of landscape context is habitat fragmentation. Although some studies report negative effects of fragmentation overall (reviewed in Reference 13), species-specific responses are also common, with some species thriving, others indifferent, and others declining (9, 13, 15). Species-specific traits, therefore, are important. For example, larger bees may be less sensitive to fragmentation than smaller species that cannot fly between fragments (e.g., 9). Cane (13) addressed fragmentation as influencing a combination of direct effects calibrated for each species. The effects of fragmentation can be further influenced by the matrix habitat surrounding them. For instance, pine plantations increase connectivity for most bee species between isolated patches of South American forest, but they seem to act as a barrier for at least one bee species (115).

Invasive Species

The effects of invasive plant and insect species on native bees have been thoroughly and recently reviewed (39, 112).The primary impacts are through changes in resources [i.e., how strongly invasive insects compete for floral resources (97) and how the intrusion of novel plant species into the plant community changes resource availability]. These effects are highly species specific, both on the part of the invader and on the part of the natives.

Summary

Despite the many types of variables considered as predictors of bee population size or diversity, there seem to be very few demonstrated underlying causes for the expected and unexpected results from diverse studies. Floral resource abundance and diversity, the most clearly demonstrated limiting factor among direct effects, are the most frequently implicated factors in studies of indirect effects. Convincing evidence for the primacy of other factors will require holding floral resource abundance constant, preferably through a manipulative study. Undoubtedly, there are cases in which nesting substrate or parasites or other direct factors limit population size, but there is little evidence at this point that those cases are common. Thus, we suggest that knowledge of landscape effects and management effects on flowers used by bees will likely predict the outcome of those factors on bee populations.

We also suggest that manipulative experiments are very much needed in order to understand when factors with potential direct effects do in fact limit population size. How much bare soil do ground-nesting bees need? At what density of bees in a habitat do cavities or plant stems become limiting? How often do parasites have demographic effects on bees? How much effect does tilling have on bees nesting in agricultural lands? These are some of the direct effects that are expected to regulate bee populations and be influenced by the categorical variables currently being tested, yet we know very little about how they work. Knowledge of these areas will help connect land management programs more directly to bee populations and provide an informed means to maintain pollinators for both wild plants and agriculture.

SUMMARY POINTS

1. Floral resource availability is the primary direct factor influencing bee population abundance as supported by a wide variety of observational and experimental evidence.
2. While plausible, little evidence supports nest site limitation of bees.
3. Studying the influence of resource availability requires selection of appropriate spatial and temporal scales in which to assess correlations between bees and flowers.
4. While parasites, especially newly introduced pathogens, potentially limit bee populations, little evidence exists for wild bee populations.
5. Parasitism acting directly can interact in synergistic or antagonistic ways with floral resource availability to influence individual and population performance.
6. Teasing apart important influences on bee populations requires separating indirect effects, such as fragmentation, from direct effects, such as floral resources.
7. Bee susceptibility to factors acting indirectly depends on species-specific traits such as foraging range and diet breadth.

DISCLOSURE STATEMENT

The authors are not aware of any affiliations, memberships, funding, or financial holdings that might be perceived as affecting the objectivity of this review.

ACKNOWLEDGMENTS

We would like to thank the anonymous reviewers whose comments improved this manuscript.

LITERATURE CITED

1. Abbott VA, Nadeau JL, Higo HA, Winston ML. 2008. Lethal and sublethal effects of imidacloprid on *Osmia lignaria* and clothianidin on *Megachile rotundata* (Hymenoptera: Megachilidae). *J. Econ. Entomol.* 101:784–96
2. Aizen MA, Garibaldi LA, Cunningham SA, Klein AM. 2008. Long-term global trends in crop yield and production reveal no current pollination shortage but increasing pollinator dependency. *Curr. Biol.* 18:1572–75
3. Allen-Wardell G, Bernhardt P, Bitner R, Búrquez A, Buchmann S, et al. 1998. The potential consequences of pollinator declines on the conservation of biodiversity and stability of food crop yields. *Conserv. Biol.* 12:8–17

4. Alston DG, Tepedino VJ, Bradley BA, Toler TR, Griswold TL, Messinger SM. 2007. Effects of the insecticide phosmet on solitary bee foraging and nesting in orchards of Capitol Reef National Park, Utah. *Environ. Entomol.* 36:811–16
5. Bernadou A, Demares F, Couret-Fauvel T, Sandoz JC, Gauthier M. 2009. Effect of fipronil on side-specific antennal tactile learning in the honeybee. *J. Insect Physiol.* 55:1099–106
6. Biesmeijer JC, Roberts SPM, Reemer M, Ohlemuller R, Edwards M, et al. 2006. Parallel declines in pollinators and insect-pollinated plants in Britain and the Netherlands. *Science* 313:351–54
7. Bosch J, Kemp WP. 2002. Developing and establishing bee species as crop pollinators: the example of *Osmia* spp. (Hymenoptera: Megachilidae) and fruit trees. *Bull. Entomol. Res.* 92:3–16
8. Bowers JE, Dimmitt MA. 1994. Flowering phenology of six woody plants in the northern sonoran desert. *Bull. Torrey Bot. Club* 121:215–29
9. Brosi BJ. 2009. The effects of forest fragmentation on euglossine bee communities (Hymenoptera: Apidae: Euglossini). *Biol. Conserv.* 142:414–23
10. Brown MJF, Loosli R, Schmid-Hempel P. 2000. Condition-dependent expression of virulence in a trypanosome infecting bumblebees. *Oikos* 91:421–27
11. Brown MJF, Schmid-Hempel R, Schmid-Hempel P. 2003. Strong context-dependent virulence in a host-parasite system: reconciling genetic evidence with theory. *J. Anim. Ecol.* 72:994–1002
12. Campbell JW, Hanula JL, Waldrop TA. 2007. Effects of prescribed fire and fire surrogates on floral visiting insects of the blue ridge province in North Carolina. *Biol. Conserv.* 134:393–404
13. Cane JH. 2001. Habitat fragmentation and native bees: a premature verdict? *Conserv. Ecol.* 5(1):3
14. Cane JH. 2005. Bees, pollination and the challenges of sprawl. In *Nature in Fragments: The Legacy of Sprawl*, ed. EA Johnson, MW Klemens, pp. 109–24. New York: Columbia Univ. Press
15. Cane JH, Minckley RL, Kervin LJ, Roulston TH, Williams NM. 2006. Complex responses within a desert bee guild (Hymenoptera: Apiformes) to urban habitat fragmentation. *Ecol. Appl.* 16:632–44
16. Cane JH, Sipes SD. 2006. Floral specialization by bees: analytical methods and a revised lexicon for oligolecty. In *Plant-Pollinator Interactions: From Specialization to Generalization*, ed. NM Waser, J Ollerton, pp. 99–122. Chicago: Univ. Chicago Press
17. Carre G, Roche P, Chifflet R, Morison N, Bommarco R, et al. 2009. Landscape context and habitat type as drivers of bee diversity in European annual crops. *Agric. Ecosyst. Environ.* 133:40–47
18. Carvell C. 2002. Habitat use and conservation of bumblebees (*Bombus* spp.) under different grassland management regimes. *Biol. Conserv.* 103:33–49
19. Carvell C, Meek WR, Pywell RF, Nowakowski M. 2004. The response of foraging bumblebees to successional change in newly created arable field margins. *Biol. Conserv.* 118:327–39
20. Carvell C, Rothery P, Pywell RF, Heard MS. 2008. Effects of resource availability and social parasite invasion on field colonies of *Bombus terrestris*. *Ecol. Entomol.* 33:321–27
21. Cole FR, Pritchard AE. 1964. The genus *Mallophora* and related asilid genera in North America. *Univ. Calif. Publ. Entomol.* 36:43–100
22. Colla SR, Packer L. 2008. Evidence for decline in eastern North American bumblebees (Hymenoptera: Apidae), with special focus on *Bombus affinis* Cresson. *Biodivers. Conserv.* 17:1379–91
23. Danforth BN. 1999. Emergence dynamics and bet hedging in a desert bee, *Perdita portalis*. *Proc. R. Soc. Lond. B Biol. Sci.* 266:1985–94
24. Dukas R. 2005. Bumble bee predators reduce pollinator density and plant fitness. *Ecology* 86:1401–6
25. Dukas R, Morse DH. 2003. Crab spiders affect flower visitation by bees. *Oikos* 101:157–63
26. Durrer S, Schmid-Hempel P. 1995. Parasites and the regional distribution of bumble bees. *Ecography* 18:114–22
27. Ebeling A, Klein AM, Schumacher J, Weisser WW, Tscharntke T. 2008. How does plant richness affect pollinator richness and temporal stability of flower visits? *Oikos* 117:1808–15
28. Elliott SE. 2009. Subalpine bumble bee foraging distances and densities in relation to floral availability. *Environ. Entomol.* 38:748–56
29. Elliott SE. 2009. Surplus nectar available for subalpine bumble bee colony growth. *Environ. Entomol.* 38:1680–89

30. Eltz T, Bruhl CA, Imiyabir Z, Linsenmair KE. 2003. Nesting and nest trees of stingless bees (Apidae: Meliponini) in lowland dipterocarp forests in Sabah, Malaysia, with implications for forest management. *Forest Ecol. Manage.* 172:301–13
31. Floyd T. 1996. Top-down impacts on creosotebush herbivores in a spatially and temporally complex environment. *Ecology* 77:1544–55
32. Fontaine C, Dajoz I, Meriguet J, Loreau M. 2006. Functional diversity of plant-pollinator interaction webs enhances the persistence of plant communities. *PLoS Biol.* 4:129–35
33. Franzen M, Nilsson SG. 2010. Both population size and patch quality affect local extinctions and colonizations. *Proc. R. Soc. Lond. B Biol. Sci.* 277:79–85
34. Galeotti P, Inglisa M. 2001. Estimating predation impact on honeybees *Apis mellifera* L. by European bee-eaters *Merops apiaster* L. *Rev. D Ecol. La Terre Et La Vie* 56:373–88
35. Gels JA, Held DW, Potter DA. 2002. Hazards of insecticides to the bumble bees *Bombus impatiens* (Hymenoptera: Apidae) foraging on flowering white clover in turf. *J. Econ. Entomol.* 95:722–28
36. Giles V, Ascher JS. 2006. A survey of the bees of the Black Rock Forest preserve, New York (Hymenoptera: Apoidea). *J. Hymenopt. Res.* 15:208–31
37. Gomez JM, Bosch J, Perfectti F, Fernandez J, Abdelaziz M. 2007. Pollinator diversity affects plant reproduction and recruitment: the tradeoffs of generalization. *Oecologia* 153:597–605
38. Goodell K. 2003. Food availability affects *Osmia pumila* (Hymenoptera: Megachilidae) foraging, reproduction, and brood parasitism. *Oecologia* 134:518–27
39. Goodell K. 2008. Invasive exotic plant-bee interactions. In *Bee Pollination in Agroecosystems*, ed. RR James, TL Pitts-Singer, pp. 166–83. London: Oxford Univ. Press
40. Goulson D, Hughes WOH, Derwent LC, Stout JC. 2002. Colony growth of the bumblebee, *Bombus terrestris*, in improved and conventional agricultural and suburban habitats. *Oecologia* 130:267–73
41. Greenleaf SS, Kremen C. 2006. Wild bees enhance honey bees' pollination of hybrid sunflower. *Proc. Natl. Acad. Sci. USA* 103:13890–95
42. Greenleaf SS, Williams NM, Winfree R, Kremen C. 2007. Bee foraging ranges and their relationship to body size. *Oecologia* 153:589–96
43. Hegland SJ, Boeke L. 2006. Relationships between the density and diversity of floral resources and flower visitor activity in a temperate grassland community. *Ecol. Entomol.* 31:532–38
44. Heinrich B. 1976. Flowering phenologies: bog, woodland, and disturbed habitats. *Ecology* 57:890–99
45. Herrera CM, Garcia IM, Perez R. 2008. Invisible floral larcenies: microbial communities degrade floral nectar of bumble bee-pollinated plants. *Ecology* 89:2369–76
46. Hines HM, Hendrix SD. 2005. Bumble bee (Hymenoptera: Apidae) diversity and abundance in tallgrass prairie patches: effects of local and landscape floral resources. *Environ. Entomol.* 34:1477–84
47. Hoehn P, Tscharntke T, Tylianakis JM, Steffan-Dewenter I. 2008. Functional group diversity of bee pollinators increases crop yield. *Proc. R. Soc. B Biol. Sci.* 275:2283–91
48. Horne M. 1995. Leaf area and toughness: effects on nesting material preferences of *Megachile rotundata* (Hymenoptera: Megachilidae). *Ann. Entomol. Soc. Am.* 88:868–75
49. Imhoof B, Schmid-Hempel P. 1999. Colony success of the bumble bee, *Bombus terrestris* in relation to infections by two protozoan parasites, *Crithidia bombi* and *Nosema bombi*. *Insectes Soc.* 46:233–38
50. James RR. 2008. The problem of disease when domesticating bees. In *Bee Pollination in Agricultural Ecosystems*, ed. RR James, TL Pitts-Singer, pp. 124–41. Oxford, UK: Oxford Univ. Press
51. Johansen CA. 1977. Pesticides and pollinators. *Annu. Rev. Entomol.* 22:177–92
52. Julier HE, Roulston TH. 2009. Wild bee abundance and pollination service in cultivated pumpkins: farm management, nesting behavior and landscape effects. *J. Econ. Entomol.* 102:563–73
53. Kearns CA, Oliveras DM. 2009. Environmental factors affecting bee diversity in urban and remote grassland plots in Boulder, Colorado. *J. Insect Conserv.* 13:655–65
54. Kevan P. 1975. Forest application of the insecticide fenitrothion and its effects on wild bee populations (Hymenoptera: Apoidea) of lowbush blueberries (*Vaccinium*) in southern New Brunswick, Canada. *Biol. Conserv.* 7:301–9
55. Kevan PG, LaBerge WE. 1979. Demise and recovery of native pollinator populations through pesticide use and some economic implications. In *Proc. IVth Int. Symp. Pollination, Md. Agric. Exp. Sta. Spec. Misc. Publ.* 1:489–508

56. Kevan PG, Mohr NA, Offer MD, Kemp JR. 1988. The squash and gourd bee, *Peponapis pruinosa* (Hymenoptera: Anthophoridae), in Ontario, Canada. *Proc. Entomol. Soc. Ont.* 119:9–15
57. Kim J, Williams N, Kremen C. 2006. Effects of cultivation and proximity to natural habitat on ground-nesting native bees in California sunflower fields. *J. Kans. Entomol. Soc.* 79:309–20
58. Kladivko EJ. 2001. Tillage systems and soil ecology. *Soil Tillage Res.* 61:61–76
59. Kleijn D, van Langevelde F. 2006. Interacting effects of landscape context and habitat quality on flower visiting insects in agricultural landscapes. *Basic Appl. Ecol.* 7:201–14
60. Klein AM, Vaissiere BE, Cane JH, Steffan-Dewenter I, Cunningham SA, et al. 2007. Importance of pollinators in changing landscapes for world crops. *Proc. R. Soc. Lond. B Biol. Sci.* 274:303–13
61. Kremen C, Williams NM, Aizen MA, Gemmill-Herren B, LeBuhn G, et al. 2007. Pollination and other ecosystem services produced by mobile organisms: a conceptual framework for the effects of land-use change. *Ecol. Lett.* 10:299–314
62. Kremen C, Williams NM, Thorp RW. 2002. Crop pollination from native bees at risk from agricultural intensification. *Proc. Natl. Acad. Sci. USA* 99:16812–16
63. Krombein KV. 1967. *Trap-Nesting Wasps and Bees: Life Histories, Nests and Associates*. Washington, DC: Smithsonian Press. 570 pp.
64. Kruess A, Tscharntke T. 2002. Grazing intensity and the diversity of grasshoppers, butterflies, and trap-nesting bees and wasps. *Conserv. Biol.* 16:1570–80
65. **Larsson M, Franzen M. 2007. Critical resource levels of pollen for the declining bee *Andrena hattorfiana* (Hymenoptera, Andrenidae). *Biol. Conserv.* 134:405–14**
66. Lonsdorf E, Kremen C, Ricketts T, Winfree R, Williams N, Greenleaf S. 2009. Modelling pollination services across agricultural landscapes. *Ann. Bot.* 103:1589–600
67. Mathewson JA. 1968. Nest construction and life history of the eastern cucurbit bee, *Peponapis pruinosa* (Hymenoptera: Apoidea). *J. Kans. Entomol. Soc.* 41:255–61
68. McFrederick QS, LeBuhn G. 2006. Are urban parks refuges for bumble bees *Bombus* spp. (Hymenoptera: Apidae)? *Biol. Conserv.* 129:372–82
69. Michener CD. 2000. *The Bees of the World*. Baltimore: Johns Hopkins Univ. Press. 913 pp.
70. Minckley RL, Cane JH, Kervin LJ. 2000. Origins and ecological consequences of pollen specialization among desert bees. *Proc. R. Soc. Lond. B Biol. Sci.* 267:265–71
71. Minckley RL, Roulston TH. 2006. Incidental mutualisms and pollen specialization among bees. In *Plant-Pollinator Interactions: From Specialization to Generalization*, ed. NM Waser, J Ollerton, pp. 69–98. Chicago: Univ. Chicago Press
72. **Minckley RL, Wcislo WT, Yanega D, Buchmann SL. 1994. Behavior and phenology of a specialist bee (*Dieunomia*) and sunflower (*Helianthus*) pollen availability. *Ecology* 75:1406–19**
73. Mommaerts V, Reynders S, Boulet J, Besard L, Sterk G, Smagghe G. 2010. Risk assessment for side-effects of neonicotinoids against bumblebees with and without impairing foraging behavior. *Ecotoxicology* 19:207–15
74. Moretti M, de Bello F, Roberts SPM, Potts SG. 2009. Taxonomical versus functional responses of bee communities to fire in two contrasting climatic regions. *J. Anim. Ecol.* 78:98–108
75. Moretti M, Obrist MK, Duelli P. 2004. Arthropod biodiversity after forest fires: winners and losers in the winter fire regime of the southern Alps. *Ecography* 27:173–86
76. Muller A, Diener S, Schnyder S, Stutz K, Sedivy C, Dorn S. 2006. Quantitative pollen requirements of solitary bees: implications for bee conservation and the evolution of bee-flower relationships. *Biol. Conserv.* 130:604–15
77. Mullin CA, Frazier M, Frazier JL, Ashcraft S, Simonds R, et al. 2010. High levels of miticides and agrochemicals in North American apiaries: implications for honey bee health. *PLoS One* 5:1–19
78. Ne'eman G, Dafni A. 1999. Fire, bees, and seed production in a Mediterranean key species *Salvia fruticosa* Miller (Lamiaceae). *Isr. J. Plant Sci.* 47:157–63
79. Neff JL. 2003. Nest and provisioning biology of the bee *Panurginus polytrichus* Cockerell (Hymenoptera: Andrenidae), with a description of a new *Holeopasites* species (Hymenoptera: Apidae), its probable nest parasite. *J. Kans. Entomol. Soc.* 76:203–16
80. Otterstatter MC, Gegear RJ, Colla SR, Thomson JD. 2005. Effects of parasitic mites and protozoa on the flower constancy and foraging rate of bumble bees. *Behav. Ecol. Sociobiol.* 58:383–89

65. Directly relates bee population size to resource abundance for a system in which both are tractable.

72. Shows how seasonal changes in provisioning rates match seasonal changes in resource availability.

81. Otterstatter MC, Thomson JD. 2008. Does pathogen spillover from commercially reared bumble bees threaten wild pollinators? *PLoS One* 3:9
82. Otti O, Schmid-Hempel P. 2007. *Nosema bombi*: a pollinator parasite with detrimental effects. *J. Invertebr. Pathol.* 96:118–24
83. **Otti O, Schmid-Hempel P. 2008. A field experiment on the effect of *Nosema bombi* in colonies of the bumblebee *Bombus terrestris*. *Ecol. Entomol.* 33:577–82**

83. Investigates the effects of experimentally infected bumble bee colonies compared to controls in a field setting to test the effects of a widespread parasitic protozoan on colony performance.

84. Pelletier L, McNeil JN. 2003. The effect of food supplementation on reproductive success in bumblebee field colonies. *Oikos* 103:688–94
85. Petanidou T, Ellis WN. 1993. Pollinating fauna of a phryganic ecosystem: composition and diversity. *Biodivers. Lett.* 1:9–22
86. Potts SG, Biesmeijer JC, Kremen C, Neumann P, Schweiger O, Kuhn WE. 2010. Global pollinator declines: trends, impacts and drivers. *Trends Ecol. Evol.* 25:345–53
87. Potts SG, Vulliamy B, Dafni A, Ne'eman G, O'Toole C, et al. 2003. Response of plant-pollinator communities to fire: changes in diversity, abundance and floral reward structure. *Oikos* 101:103–12
88. Potts SG, Vulliamy B, Dafni A, Ne'eman G, Willmer P. 2003. Linking bees and flowers: How do floral communities structure pollinator communities? *Ecology* 84:2628–42
89. Potts SG, Vulliamy B, Roberts S, O'Toole C, Dafni A, et al. 2005. Role of nesting resources in organising diverse bee communities in a Mediterranean landscape. *Ecol. Entomol.* 30:78–85
90. Price PW, Westoby M, Rice B. 1988. Parasite-mediated competition: some predictions and tests. *Am. Nat.* 131:544–55
91. Pyke GH. 1984. Optimal foraging theory: a critical review. *Annu. Rev. Ecol. Syst.* 15:523–75
92. **Quintero C, Morales CL, Aizen MA. Effects of anthropogenic habitat disturbance on local pollinator diversity and species turnover across a precipitation gradient. *Biodivers. Conserv.* 19:257–74**

92. Shows that anthropogenic disturbance has led to a loss of β-diversity in bee communities across a large habitat gradient that is not evident by sampling diversity of paired disturbed and undisturbed sites within habitats.

93. Ricketts TH. 2004. Tropical forest fragments enhance pollinator activity in nearby coffee crops. *Conserv. Biol.* 18:1262–71
94. Romey WL, Ascher JS, Powell DA, Yanek M. 2007. Impacts of logging on midsummer diversity of native bees (Apoidea) in a northern hardwood forest. *J. Kans. Entomol. Soc.* 80:327–38
95. Rosenheim JA. 1996. An evolutionary argument for egg limitation. *Evolution* 50:2089–94
96. Roubik DW. 1989. *Ecology and Natural History of Tropical Bees*. Cambridge: Cambridge Univ. Press. 514 pp.
97. Roubik DW, Villanueva-Gutiérrez R. 2009. Invasive Africanized honey bee impact on native solitary bees: a pollen resource and trap nest analysis. *Biol. J. Linn. Soc.* 98:152–60
98. Rundlof M, Nilsson H, Smith HG. 2008. Interacting effects of farming practice and landscape context on bumblebees. *Biol. Conserv.* 141:417–26
99. Schmid-Hempel P. 1998. *Parasites in Social Insects*. Princeton, NJ: Princeton Univ. Press. 409 pp.
100. Schmid-Hempel P, Durrer S. 1991. Parasites, floral resources and reproduction in natural populations of bumblebees. *Oikos* 62:342–50
101. Scott-Dupree CD, Conroy L, Harris CR. 2009. Impact of currently used or potentially useful insecticides for canola agroecosystems on *Bombus impatiens* (Hymenoptera: Apidae), *Megachile rotundata* (Hymenoptera: Megachilidae), and *Osmia lignaria* (Hymenoptera: Megachilidae). *J. Econ. Entomol.* 102:177–82
102. Shearin AF, Reberg-Horton SC, Gallandt ER. 2007. Direct effects of tillage on the activity density of ground beetle (Coleoptera: Carabidae) weed seed predators. *Environ. Entomol.* 36:1140–46
103. Sheffield CS, Kevan PG, Westby SM, Smith RF. 2008. Diversity of cavity-nesting bees (Hymenoptera: Apoidea) within apple orchards and wild habitats in the Annapolis Valley, Nova Scotia, Canada. *Can. Entomol.* 140:235–49
104. Shuler RE, Roulston TH, Farris GE. 2005. Farming practices influence wild pollinator populations on squash and pumpkins. *J. Econ. Entomol.* 98:790–95
105. Slagle MW, Hendrix SD. 2009. Reproduction of *Amorpha canescens* (Fabaceae) and diversity of its bee community in a fragmented landscape. *Oecologia* 161:813–23
106. Stark JD, Banks JE. 2003. Population-level effects of pesticides and other toxicants on arthropods. *Annu. Rev. Entomol.* 48:505–19

107. Steffan-Dewenter I. 2002. Landscape context affects trap-nesting bees, wasps, and their natural enemies. *Ecol. Entomol.* 27:631–37
108. Steffan-Dewenter I, Munzenberg U, Burger C, Thies C, Tscharntke T. 2002. Scale-dependent effects of landscape context on three pollinator guilds. *Ecology* 83:1421–32
109. Steffan-Dewenter I, Schiele S. 2004. Nest-site fidelity, body weight and population size of the red mason bee, *Osmia rufa* (Hymenoptera: Megachilidae), evaluated by mark-recapture experiments. *Entomol. Gen.* 27:123–32
110. Steffan-Dewenter I, Schiele S. 2008. Do resources or natural enemies drive bee population dynamics in fragmented habitats? *Ecology* 89:1375–87
111. Storer TI, Vansell GH. 1935. Bee-eating proclivities of the striped skunk. *J. Mammal.* 16:118–21
112. Stout JC, Morales CL. 2009. Ecological impacts of invasive alien species on bees. *Apidologie* 40:388–409
113. Stubblefield JW, Seger J, Wenzel JW, Heisler MM. 1993. Temporal, spatial, sex-ratio and body-size heterogeneity of prey species taken by the beewolf *Philanthus sanbornii* (Hymenoptera: Sphecidae). *Philos. Trans. R. Soc. Lond. Ser. B* 339:397–423
114. Tasei JN, Carre S, Bosio PG, Debray P, Hariot J. 1987. Effects of the pyrethroid insecticide, WL85871 and phosalone on adults and progeny of the leaf-cutting bee *Megachile rotundata* F., pollinator of lucerne. *Pestic. Sci.* 21:119–28
115. Valdovinos FS, Chiappa E, Simonetti JA. 2009. Nestedness of bee assemblages in an endemic South American forest: the role of pine matrix and small fragments. *J. Insect Conserv.* 13:449–52
116. Van Dreische RG, Bellows TS. 1996. *Biological Control*. New York: Chapman & Hall
117. van Engelsdorp D, Evans JD, Saegerman C, Mullin C, Haubruge E, et al. 2009. Colony collapse disorder: a descriptive study. *PLoS One* 4:17
118. van Engelsdorp D, Meixner MD. 2010. A historical review of managed honey bee populations in Europe and the United States and the factors that may affect them. *J. Invertebr. Pathol.* 103:S80–S95
119. Vergara CH, Badano EI. 2009. Pollinator diversity increases fruit production in Mexican coffee plantations: the importance of rustic management systems. *Agric. Ecosyst. Environ.* 129:117–23
120. Vergara CH, Schroder S, Almanza MT, Wittmann D. 2003. Suppression of ovarian development of *Bombus terrestris* workers by *B. terrestris* queens, *Psithyrus vestalis* and *Psithyrus bohemicus* females. *Apidologie* 34:563–68
121. Vicens N, Bosch J, Blas M. 1994. Biology and population structure of *Osmia tricornis* Latreille (Hym., Megachilidae). *J. Appl. Entomol.* 117:300–6
122. Visscher PK, Danforth BN. 1993. Biology of *Calliopsis pugionis* (Hymenoptera: Andrenidae): nesting, foraging, and investment sex ratio. *Ann. Entomol. Soc. Am.* 86:822–32
123. Vulliamy B, Potts SG, Willmer PG. 2006. The effects of cattle grazing on plant-pollinator communities in a fragmented Mediterranean landscape. *Oikos* 114:529–43
124. Wcislo WT, Cane JH. 1996. Floral resource utilization by solitary bees (Hymenoptera: Apoidea) and exploitation of their stored foods by natural enemies. *Annu. Rev. Entomol.* 41:195–224
125. Westphal C, Bommarco R, Carre G, Lamborn E, Morison N, et al. 2008. Measuring bee diversity in different European habitats and biogeographical regions. *Ecol. Monogr.* 78:653–71
126. Westphal C, Steffan-Dewenter I, Tscharntke T. 2003. Mass flowering crops enhance pollinator densities at a landscape scale. *Ecol. Lett.* 6:961–65
127. Westphal C, Steffan-Dewenter I, Tscharntke T. 2006. Bumblebees experience landscapes at different spatial scales: possible implications for coexistence. *Oecologia* 149:289–300
128. Whittington R, Winston ML. 2003. Effects of *Nosema bombi* and its treatment fumagillin on bumble bee (*Bombus occidentalis*) colonies. *J. Invertebr. Pathol.* 84:54–58
129. Williams N, Crone EE, Roulston TH, Minckley RL, Packer L, Potts SG. 2010. Ecological and life history traits predict bee species responses to environmental disturbances. *Biol. Conserv.* In press
130. Williams NM, Kremen C. 2007. Resource distributions among habitats determine solitary bee offspring production in a mosaic landscape. *Ecol. Appl.* 17:910–21
131. Winfree R, Aguilar R, Vazquez DP, LeBuhn G, Aizen MA. 2009. A meta-analysis of bees' responses to anthropogenic disturbance. *Ecology* 90:2068–76
132. Winfree R, Griswold T, Kremen C. 2007. Effect of human disturbance on bee communities in a forested ecosystem. *Conserv. Biol.* 21:213–23

130. Investigates solitary bee reproductive performance as a function of the quality of its local habitat and the broader landscape that tracks habitat use by analyzing pollens bees gather from agricultural or surrounding natural habitats.

133. Winfree R, Williams NM, Dushoff J, Kremen C. 2007. Native bees provide insurance against ongoing honey bee losses. *Ecol. Lett.* 10:1105–13
134. Winfree R, Williams NM, Gaines H, Ascher JS, Kremen C. 2008. Wild bee pollinators provide the majority of crop visitation across land-use gradients in New Jersey and Pennsylvania, USA. *J. Appl. Ecol.* 45:793–802
135. Woodcock BA, Potts SG, Tscheulin T, Pilgrim E, Ramsey AJ, et al. 2009. Responses of invertebrate trophic level, feeding guild and body size to the management of improved grassland field margins. *J. Appl. Ecol.* 46:920–29
136. Xie ZH, Williams PH, Tang Y. 2008. The effect of grazing on bumblebees in the high rangelands of the Eastern Tibetan Plateau of Sichuan. *J. Insect Conserv.* 12:695–703
137. Yanega D. 1990. Philopatry and nest founding in a primitively social bee, *Halictus rubicundus*. *Behav. Ecol. Sociobiol.* 27:37–42
138. Yang EC, Chuang YC, Chen YL, Chang LH. 2008. Abnormal foraging behavior induced by sublethal dosage of imidacloprid in the honey bee (Hymenoptera: Apidae). *J. Econ. Entomol.* 101:1743–48
139. Zammit J, Hogendoorn K, Schwarz MP. 2008. Strong constraints to independent nesting in a facultatively social bee: quantifying the effects of enemies-at-the-nest. *Insectes Soc.* 55:74–78
140. Zurbuchen A, Cheesman S, Klaiber J, Müller A, Hein S, Dorn S. 2010. Long foraging distances impose high costs on offspring production in solitary bees. *J. Anim. Ecol.* 79:674–81

Venom Proteins from Endoparasitoid Wasps and Their Role in Host-Parasite Interactions

Sassan Asgari[1] and David B. Rivers[2]

[1]School of Biological Sciences, The University of Queensland, St. Lucia QLD 4072; email: s.asgari@uq.edu.au

[2]Department of Biology, Loyola University Maryland, Baltimore, Maryland 21210

Annu. Rev. Entomol. 2011. 56:313–35

First published online as a Review in Advance on September 3, 2010

The *Annual Review of Entomology* is online at ento.annualreviews.org

This article's doi:
10.1146/annurev-ento-120709-144849

0066-4170/11/0107-0313$20.00

Key Words

koinobiont, host manipulation, immunity, development

Abstract

Endoparasitoids introduce a variety of factors into their host during oviposition to ensure successful parasitism. These include ovarian and venom fluids that may be accompanied by viruses and virus-like particles. An overwhelming number of venom components are enzymes with similarities to insect metabolic enzymes, suggesting their recruitment for expression in venom glands with modified functions. Other components include protease inhibitors, paralytic factors, and constituents that facilitate/enhance entry and expression of genes from symbiotic viruses or virus-like particles. In addition, the venom gland may itself support replication/production of some viruses or virus-like entities. Overlapping functions and structural similarities of some venom, ovarian, and virus-encoded proteins suggest coevolution of molecules recruited by endoparasitoids to maintain their fitness relative to their host.

Idiobiont: an endoparasitoid that inhibits further development of its host following parasitization

Koinobiont: an endoparasitoid that allows further development of its host following parasitization

INTRODUCTION

Parasitoid wasps of the order Hymenoptera are one of the most fascinating groups of insects. From a practical point of view, they are invaluable in classical and augmentative biological control of various insect pests. From an evolutionary point of view, the mechanisms evolved by the wasps to ensure successful development of their progeny are phenomenal. Some of the components involved in these adaptations may be utilized, in the future, to improve biological control of insect pests by genetic modification of the parasitoids, their hosts, or crops. In addition, exploring the mechanisms involved in evasion/suppression of the host immune system has helped us tremendously to learn about how the immune system in insects works.

In endoparasitoids, various components are introduced into the host at oviposition that facilitate development of their progeny, including the venom fluid. Traditionally, venom has been defined as a toxic fluid that inflicts a sudden death or paralysis in the host/prey. However, more recent and modern definitions consider venom as secretions delivered through a wound that interferes with normal physiological processes to facilitate feeding or defense by the animal that produces the venom (36). Many venom proteins show structural and functional similarities to endogenous molecules that are involved in normal physiological and biochemical processes within animals. These proteins are found in multiple groups of venomous animals, indicating their convergent recruitment over the course of evolution into venoms, some of which are also recruited in the saliva of hematophagous insects (36). Venom from parasitoid wasps consists of a complex mixture of proteinaceous as well as nonproteinaceous components involved in host manipulation. In this review, we examine biochemical and functional properties of venom proteins from endoparasitoid wasps and explore the mechanisms by which they influence host-parasite interactions at the molecular level.

VENOM AND PARASITOID LIFE STRATEGIES: ENDOPARASITISM VERSUS ECTOPARASITISM

The mode of action of venom proteins from parasitoid wasps is dependent largely on their life strategy. Parasitoids are broadly divided into ectoparasitoids and endoparasitoids (6). Ectoparasitoids glue their eggs externally to the host; a larva hatches and continues to feed externally on the host until it pupates. This requires immobilization of the host to secure the food source for the wasp progeny. Therefore, ectoparasitoid venom causes mostly a long-term paralysis of the host and blocks its further development following parasitization (idiobionts). Nevertheless, there are exceptions in which paralysis is transient but developmental arrest is permanent in the host (20). In addition to paralysis and developmental arrest, accumulating evidence indicates that venom from ectoparasitic wasps also interferes with host immune responses (83, 87).

Endoparasitoids, on the other hand, deposit their eggs within the host hemocoel, in which they continue to grow as larvae, allowing development of their host (koinobionts) (6). Except for a few examples, a paralytic effect has not been reported for most endoparasitoids, although this might be due mainly to a mild and transient paralysis effect that has been overlooked. Most of the components present in endoparasitoid wasp venom fluids are involved in host regulation by directly or indirectly affecting immune functions, facilitating entry of symbiotic viruses, providing a suitable nutritional milieu, and delaying/arresting host development at the larval stage.

BIOCHEMICAL AND BIOPHYSICAL PROPERTIES OF ENDOPARASITOID VENOM PROTEINS

Venoms from parasitic Hymenoptera are rich sources of biomolecules containing small peptides, including neurotoxins, amines and mid- to high-molecular-weight enzymes involved in manipulation of host physiology. Similar to venom proteins from other venomous animals, endoparasitoid venom proteins are generally secretory; exist in protein families; show extensive similarity to body proteins, indicating their acquisition from those proteins; and have stable molecular structures. In general, the size of polypeptides from endoparasitoid wasp venoms is larger than that of predatory and social wasps and bees, and they are mostly acidic in nature (41, 47, 55, 75). Furthermore, glycosylation of venom proteins from endoparasitoids has been reported (5). Jones et al. (42) first observed unique features of signal peptides in venom proteins from hymenopterans, which were further analyzed by Moreau & Guillot (62), who used a larger number of venom proteins. The analysis showed that the signal peptides of endoparasitoid venoms are not, in general, unusual and are consistent with the definition of classical signal peptides. Major groups of polypeptides found in venom from endoparasitoids are enzymes, protease inhibitors, and paralytic factors and are discussed below.

Enzymes

Proteins with conserved enzymatic domains constitute a major proportion of venom components reported from endoparasitoid venoms (**Table 1**). The enzymes may be involved in host regulation or venom metabolism in the venom glands, but in most cases their exact role in parasitism is not clearly known. Below, we discuss the major enzyme groups reported so far; however, other enzymes that have been reported from endoparasitoid venoms but whose structure and function have not been characterized in detail are only listed in **Table 1**.

Hydrolases

A 94-kDa heterotetrameric protein, consisting of 30- and 18-kDa subunits, from *Asobara tabida* (Braconidae) that parasitizes *Drosophila* spp. has sequence similarities to several aspartylglucosaminidase subunits (63). In vitro functional analyses of the protein using standard assays failed to establish its enzymatic activity in venom extracts. However, it was suggested that the protein may be responsible for the transient paralytic effect observed in host larvae (see below) (11).

Acid phosphatases, which are associated mainly with lysosomes involved in degradation and cell death (13), have been reported from *Pteromalus puparum* (Pteromalidae) and *Pimpla hypochondriaca* (Ichneumonidae) (22, 115). Both enzymes are highly active under acidic conditions, with optimal activity at pH 4.8. However, they have little or no activity at neutral or alkaline conditions, raising the question about the activity of these enzymes under host hemolymph physiological conditions that are near neutral or slightly alkaline. The role of this protein in host-parasitoid interactions is still unknown, but it may be involved in the release of carbohydrates in the host, providing nutrients for the developing parasitoid, or it may interfere with host immunity by dephosphorylating immune related proteins (22). In *P. hypochondriaca*, the enzyme has no effect on hemocytes and their behavior (22). In addition to acid phosphatase, other hydrolase activities (β-glucosidase, esterase, β-galactosidase, esterase lipase, and lipase) were also detected in *P. hypochondriaca* but their functions in host-parasitoid interactions have not been investigated (22).

γ-Glutamyl Transpeptidase

A heterodimeric nonglycosylated protein of about 55 kDa in size, consisting of 36- and 18-kDa subunits, was isolated from the venom fluid of *Aphidius ervi* (Braconidae) (34). Sequence analysis of the protein indicated that it has significant similarities to γ-glutamyl transpeptidases (γ-GTs) from other insects. γ-GTs are enzymes that play a key role in glutathione metabolism by transferring the glutamyl moiety to a variety of acceptor molecules, thereby contributing to amino acid trafficking, promoting homeostasis, and protecting cells from oxidative stress (114). Like other γ-GTs, the

Table 1 Venom proteins reported from endoparasitoid wasps with known functions or structural similarities

Protein/peptide	Length (aa)[a]	Putative function	Species	Reference(s)
Enzymes				
Trehalase	585	Release of hydrocarbons	Ph	70
Laccase	680	Oxidation	Ph	74
Putative serine protease	248	Protease	Ph	70
Phenoloxidase I	699	Melanization	Ph	72
Phenoloxidase II	690	Melanization	Ph	72
Phenoloxidase III	708	Melanization	Ph	72
Reprolysin	539	Metalloprotease	Ph, Ma, Mh	21, 70
Chitinase	483	Chitinolytic	Cc, Ma, Mh	21, 47
Aspartylglucosaminidase-like	NA	Lysosomal	At	63
Acid phosphatase	405	Release of hydrocarbons?	Pp, Ph	22, 115
Phospholipase B	NA	Cytolytic?	Pt	102
γ-glutamyl transpeptidase	541	Induce apoptosis	Ae	34
Cathepsin	NA	Associated with lysozymes	Ma, Mh	21
Thiol reductase	222	Lysosomal thiol reductase	Ma, Mh	21
Protease inhibitors				
Cys-rich venom protein 1	85	Protease inhibitor	Ph	73
Cys-rich venom protein 2	77	Kunitz type protease inhibitor	Ph	73
Cys-rich venom protein 4	203	Pacifastin; protease inhibitor	Ph	73
Cys-rich venom protein 6	77	Protease inhibitor	Ph	73
$LbSPN_y$	411	Inhibitor of melanization	Lb	19
Neurotoxin-like/paralytic factors				
Pimplin	143	Paralytic factor	Ph	71
Cys-rich venom protein 5	115	Similar to conotoxins	Ph	73
Cys-rich venom protein 3	63	Similar to atracotoxins	Ph	73
Vn4.6	65	Similar to atracotoxins	Cr	4
VG3	232	Similar to allergen from fire ant	Ma, Mh	21
Icarapin	NA	Similar to *Apis mellifera* allergen	Ma, Mh	21
Other functions				
Vn50	388	Inhibitor of melanization	Cr	5
Vn1.5	14	Facilitate PDV gene expression	Cr	112
Calreticulin	403	Inhibit hemocyte spreading	Cr, Ma, Mh	21, 113
Virulence protein, P4	282	Rho-GAP protein affecting cell adhesion	Lb	51
Tetraspanin	NA	Involved in cell adhesion, motility	Ma, Mh	21
Ferritin	NA	An iron binding protein	Ma, Mh	21
TEGT	NA	Testis enhanced gene transcript, suppression of apoptosis	Ma, Mh	21
VG10	120	Ion transport-like protein	Ma, Mh	21
VG8	197	Heat-shock protein	Ma, Mh	21
VPr3	312	Antihemocyte aggregation	Ph	82
Vn.11	NA	Inhibit encapsulation	Pp	108

[a]If the protein has been reported from multiple species, amino acid length provided is from a representative.

Abbreviations: Ae, *Aphidius ervi*; At, *Asobara tabida*; Cc, *Chelonus* near *curvimaculatus*; Cr, *Cotesia rubecula*; Lb, *Leptopilina boulardi*; Ph, *Pimpla hypochondriaca*; Pp, *Pteromalus puparum*; Pt, *Pimpla turionellae*; Ma, *Microctonus aethiopoides*; Mh, *Microctonus hyperodae*. NA, not available.

precursor polypeptide is posttranslationally processed into the two smaller subunits that are noncovalently bound to each other. However, unlike other γ-GTs, which are transmembrane proteins, *A. ervi* γ-GT is a secreted protein. It was previously shown that venom from *A. ervi* castrates its host, *Acyrthosiphon pisum* (Aphididae), by inducing degradation of host ovarian tissues (27). Later, it was confirmed that γ-GT is responsible for inducing apoptosis in cells of the ovary, leading to host castration possibly by interfering with the delicate balance of glutathione, which causes oxidative stress in ovarian cells and triggers apoptosis (34).

Phenoloxidases

Tyrosinase phenoloxidase (PO) activity has been reported from the venom fluid of only one endoparasitoid, *P. hypochondriaca* (75). The enzyme became active in the absence of activators. Activation of prophenoloxidase (PPO) in the hemolymph occurs following activation of a cascade of serine proteases upon immune induction. The *P. hypochondriaca* PO enzyme is heat-sensitive and inhibited by phenylthiocarbamide, an inhibitor of PO. Sequencing of complementary DNA (cDNA) clones generated from *P. hypochondriaca* venom glands indicated the presence of three PPO-related proteins (PPOI–III; all ~80 kDa) with 60%–77% identity among the clones (72). A striking difference between *P. hypochondriaca* PPOs and others isolated from hemocytes is that *P. hypochondriaca* PPOs contain signal peptides, which destine them to the secretory pathway and eventually secretion into the venom sac. Their activity is regulated within the venom storage sac in association with inhibitors (see Protease Inhibitors, below). Hemocyte PPOs are apparently released into the hemolymph upon immune activation and lysis of hemocytes (67). In addition, the conserved proteolytic cleavage site, present in all insect PPOs, is absent in *P. hypochondriaca* PPOs (72). It has not been shown experimentally whether these PPOs are cleaved after injection into the host or upon activation in vitro, and their role in parasitism has not been determined.

PO activity has also been reported from venom of an ectoparasitoid, *Nasonia vitripennis* (Pteromalidae) (1). This study suggested that PO was directly related to cell death when tested in a cell culture (high five cells from *Trichoplusia ni*) and might be responsible for plasmatocyte cytolysis observed in *Sarcophaga bullata* pupae parasitized by *N. vitripennis* (87). Similarly, *P. hypochondriaca* POs may affect the host cellular immunity by binding to hemocytes and interfering with their function or by inducing cell death to reduce the number of hemocytes in circulation.

Apoptosis: the process of programmed cell death that involves a cascade of biochemical events leading to characteristic changes in cell morphology and ultrastructure prior to the onset of death

PO: phenoloxidase

PPO: prophenoloxidase

cDNA: complementary DNA

The putative amino acid sequence of a clone from a *P. hypochondriaca* venom cDNA library exhibited similarity to several laccases from fungi and *Drosophila* (74). In contrast to tyrosinase POs, which are active mostly in the hemolymph and involved in melanization and wound healing, laccases are involved in hardening and tanning of the cuticle (2). In addition, laccases contain signal peptides and are therefore secreted proteins, whereas tyrosinase POs are not secreted. In this respect, *P. hypochondriaca* PPOI–III are like laccases, but in terms of biochemical properties they are tyrosinase-type PPOs. The role of *P. hypochondriaca* laccase in host-parasite interaction remains to be determined.

Chitinases

A 52-kDa chitinase was found in the venom of the egg parasitoid *Chelonus* near *curvimaculatus* (Braconidae) containing all the conserved residues for chitinolytic activity (43, 47). Ironically, the enzyme exists in its active form in the chitin-lined venom storage sac. Despite significant sequence similarities to other known insect chitinases, *C. curvimaculatus* chitinase exhibits different structural and kinetic properties indicating its divergence from nonvenom arthropod chitinases (47). For example, the carboxyl terminus is highly diverged and lacks a Ser/Thr-rich domain after the signal peptide, and the catalytic domain has a higher specific activity and is less sensitive to substrate inhibition. As a consequence, the venom-recruited enzyme may have different substrate specificities contributing to its lack of activity against the chitin-lined venom reservoir. The significance of this venom enzyme in parasitism has not been explored, but it is likely involved in dissociation and degradation of cells from host tissues, facilitating endoparasitoid larval feeding.

cvp: cysteine-rich venom protein

Reprolysin

Members of the reprolysin family are metalloproteinases that require zinc for catalysis. They have been found mainly in snake venoms, but other members of the family have been reported from mammalians and invertebrates as well (91). The only known reprolysin-like molecule from endoparasitoid venoms is from *P. hypochondriaca* (rep1) (69). The size of the mature protein is 39.9 kDa, containing a conserved catalytic domain with two histidine residues involved in zinc binding and a conserved disintegrin domain. The latter domain is involved in binding to cell surface integrin receptors and key extracellular matrix proteins containing von Willebrand factor A domains (e.g., collagen) (92). Consequently, the binding results in the cleavage of the substrate, leading to destabilization of extracellular matrix or cell–extracellular matrix interaction. However, in vitro hemocyte aggregation assays demonstrated that venom fractions that contained rep1 did not affect host hemocyte aggregation (70). In addition, enzymatic activity of rep1 was not established using a common substrate, which may indicate substrate specificity of the metalloproteinase (69).

Recently, three reprolysin-like cDNAs (EpMP1–3) were identified from the venom of the ectoparasitic wasp *Eulophus pennicornis* (Eulophidae) (78). Although EpMP1–3 contain a metalloproteinase domain, they lack the disintegrin domain at the C terminus, which is present in rep1. Injection experiments using the recombinant EpMP3 indicated a possible role for the protein in the manipulation of host development by repressing molting and metamorphosis. *E. pennicornis* parasitizes the same host as *P. hypochondriaca*, *Lacanobia oleracea*. Although rep1 and EpMP1-3 do not belong to the same reprolysin group, rep1 may still be involved in manipulation of host development.

Protease Inhibitors

Several polypeptides with amino acid sequence similarities to protease inhibitors have been reported from endoparasitoid venoms. However, their protease inhibitory function is largely unexplored. In addition, because of a lack of information, we do not know whether polypeptide function is relevant to venom metabolism in the venom gland and storage sac, host manipulation, or both. The majority of these protease inhibitors have been reported from *P. hypochondriaca*. For example, PO inhibitors in the venom inhibit PO activity in the venom storage sac by binding to the protein possibly to avoid potential damages by the enzyme to venom sac tissue and other venom-stored components (75). Two venom fractions that showed PO inhibitory activity contained peptides of 12 and <3 kDa. The inhibitors are heat stable, whereas the venom PPO is heat sensitive.

Four cysteine-rich venom proteins (cvps) from *P. hypochondriaca* venom (cvp1, -2, -4, and -6) show significant sequence identity to known protease inhibitors (73). cvp1 is physically associated with PPO in the venom, suggesting that it may be involved in the inhibition of PO activity in venom. Dilution of venom injected into the host at parasitization may allow activation of venom PPO as shown in in vitro assays (75). cvp2 and cvp4 are similar to Kunitz- and pacifastin-type protease inhibitors. Their enzyme-inhibiting role has not been documented, but considering the presence of five enzymes other than PPOs in *P. hypochondriaca* venom, the inhibitors could well be involved in the inhibition of the other enzymes or host regulation.

A 4.6-kDa peptide from *Cotesia rubecula* (Braconidae) venom (Vn4.6) inhibits melanization in the hemolymph of its host, *Pieris rapae*

(4). The peptide sequence shows weak fold similarities to Kunitz-type protease inhibitors, but in trypsin and α-chymotrypsin enzyme assays using artificial substrates, Vn4.6 failed to inhibit the enzymes' activity.

A venom protein from *Leptopilina boulardi* (Figitidae), named $LbSPN_y$, belongs to the serine protease inhibitor (serpin) family (19). The reactive center loop of $LbSPN_y$ shows conserved residues with insect serpins involved in the melanization cascade. Experimental evidence suggests that $LbSPN_y$ inhibits activation of the PO cascade (see Effects on Humoral Responses, below).

Paralytic Factors

In contrast to the majority of idiobionts, paralysis is rare in koinobionts, except for a few of them that induce a transient paralysis in the host. This effect lasts from less than 5 min to about 2 h postinjection of venom. The exact reason for this temporary paralysis is not quite understood. Desneux et al. (26) tested two alternative hypotheses in their host-parasitoid system: (*a*) The transient paralysis interferes with the host defense behavior, facilitating oviposition success, and (*b*) transient paralysis is used by the parasitoid as a means to avoid self-superparasitism. Parasitism of aphids by two endoparasitoids, *Binodoxys communis* and *B. koreanus* (Braconidae), causes 4.5–8 min of paralysis in the host. Interestingly, evidence supported the second, but not the first, hypothesis. It would be appealing to test these hypotheses in other host-parasitoid systems to determine whether this is a more general evolutionary adaptation.

There are only a few reports indicating transient paralysis is caused by endoparasitoids. *Asobara tabida* venom injected into *D. melanogaster* larvae causes transient paralysis from 30 up to 120 min postinjection with similar symptoms occurring in natural parasitization (64). In addition to temporary paralysis, venom causes mortality in host larvae prior to or at pupation in the absence of the parasitoid. Interestingly, two different strains of *A. tabida* (A1 from France and WOPV from the Netherlands) have differential venom potency in causing paralysis and mortality in *D. melanogaster* larvae, with WOPV showing more severe effects compared to the A1 strain (64). Although the paralytic component has not been identified, it was suggested that the major venom protein component, an aspartylglucosaminidase-like protein, may be involved in the temporary paralysis by the production of aspartate, which is an excitatory neurotransmitter (63).

Serpins: serine protease inhibitors

In a closely related species, *Asobara japonica*, venom is injected into drosophilid host larvae prior to oviposition, causing a flaccid paralysis reminiscent of that elicited by scorpion toxins, further confirming that venom components are responsible for paralysis (37). In the absence of a developing endoparasitoid, venom induces mortality in habitual hosts (*D. simulans*, *D. lutescens*, and *D. auraria*), but not nonhabitual hosts (*D. bipectinata* and *D. ficusphila*), before pupariation, suggesting long-term effects of venom in addition to temporary paralysis immediately after parasitization. In the presence of a developing endoparasitoid, mortality was not observed prior to pupariation of the host, indicating that components secreted by the developing larva may inhibit lethal effects of venom on the host (37). Further, the paralytic effects of *A. japonica* venom can be reversed by injection of ovarian extracts if the treatment quickly (within 1 min) follows venom injection (58). These observations argue a close evolutionary relationship between this group of endoparasitoids and ectoparasitic braconids that produce paralytic venoms (62).

In addition, a venom fraction from *P. hypochondriaca* was shown to cause paralysis when injected into other developmental stages of the host, *Lacanobia oleracea*, and larvae of *Musca domestica* (75). This peptide, pimplin (71), is 22 kDa and is a heterodimeric polypeptide consisting of two polypeptides with molecular masses of 10.5 and 6.3 kDa that are linked through a disulfide bond. The peptide is unique and does not show sequence similarity to

VLP: virus-like particle

PDV: polydnavirus

THC: total hemocyte count

DHC: differential hemocyte count

existing sequences in GenBank. An interesting characteristic of this polypeptide is the presence of a stretch of a repeating dipeptide sequence, XP (P, proline), at the N terminus of the mature peptide. The authors suggested that this region may be subjected to proline-specific dipeptidyl peptidase activity in the host, further trimming the polypeptide (71).

In *Pimpla turionellae*, venom also induces paralysis in different developmental stages of the host, although venom was more effective against the pupal stage, which is the preferred host stage of the parasitoid in nature (32). Paralytic factors such as melittin, apamin, and noradrenaline have also been isolated from the venom (103).

A 4.2-kDa peptide (cvp3; 38 residues) from *P. hypochondriaca* venom shows similarity to ω-atracotoxins with exact alignment of six cysteine residues involved in intramolecular disulfide bonds in spider neurotoxins (73). However, two additional cysteine residues that are important in toxicity of spider atracotoxins are replaced with threonine and glutamate residues in cvp3. In addition, a 10.3-kDa peptide (cvp5) was reported from *P. hypochondriaca* with similarity to ω-conotoxins from the cone snail (*Conus geographus*) (73). The ω-conotoxins are 24- to 30-amino-acid peptides with three disulfide bonds that block calcium channels without interfering with cellular action potentials (56). Both cvp3 and cvp5 are fairly basic peptides with pI values of 8.33 and 12.14, respectively (73). Considering that pimplin has been reported as the main paralytic factor from this parasitoid, it is unlikely that cvp3 and cvp5 have any neurotoxicity; however, this needs to be investigated. Similarly, a Vn4.6 peptide from *C. rubecula* venom shows significant structural similarities to ω-atracotoxins from the Australian funnel-web spider (*Hadronyche versuta*) (4). However, no paralytic effect of this peptide on the host, *P. rapae*, has been established. Peptides such as cvp3, cvp5, and Vn4.6 could be reminiscent of ancestral neurotoxins that existed in the ecto-/endoparasitoid ancestor with modified functions.

VENOM AS A REGULATOR OF HOST IMMUNITY

The major immune response toward endoparasitoid eggs is the encapsulation response carried out by immunocytes (**Figure 1**). Capsule formation is accompanied mostly by melanin formation/melanization during which cytotoxic intermediate molecules are also produced (67). Therefore, it is essential for the endoparasitoid to inhibit encapsulation and melanization responses. Endoparasitoids have evolved myriad mechanisms to avoid/suppress host immune responses by passive and active means (104). Venom, viruses, or virus-like particles (VLPs) introduced by the female wasp at oviposition interact with host factors, leading to inactivation of host responses. However, combinations of passive and active mechanisms may be utilized by individual parasitoids.

In host-parasitoid systems associated with polydnaviruses (PDVs) and VLPs, venom may synergize their effects or may have overlapping functions. Considering the time it takes for PDV genes to be expressed and affect the immune system, venom usually provides immediate short-term protection for the parasitoid egg. In contrast, in those systems in which the wasp is devoid of any associated PDVs or VLPs, venom constituents are the key components that regulate host physiology and development and often have broader functions than venom from PDV-/VLP-associated wasps.

Effects on Cellular Immunity

Venom injected into the host at oviposition has variable effects on host hemocytes depending on the host-parasitoid system. These include reductions in total (THC) and differential hemocyte counts (DHC), induction of cell lysis or apoptosis, breakdown of cell cytoskeleton, and alterations of hemocyte behaviors such as spreading, attachment, and aggregation. Venom components from *Leptopilina* spp., which are devoid of PDVs, interfere with encapsulation in *Drosophila* species by inducing

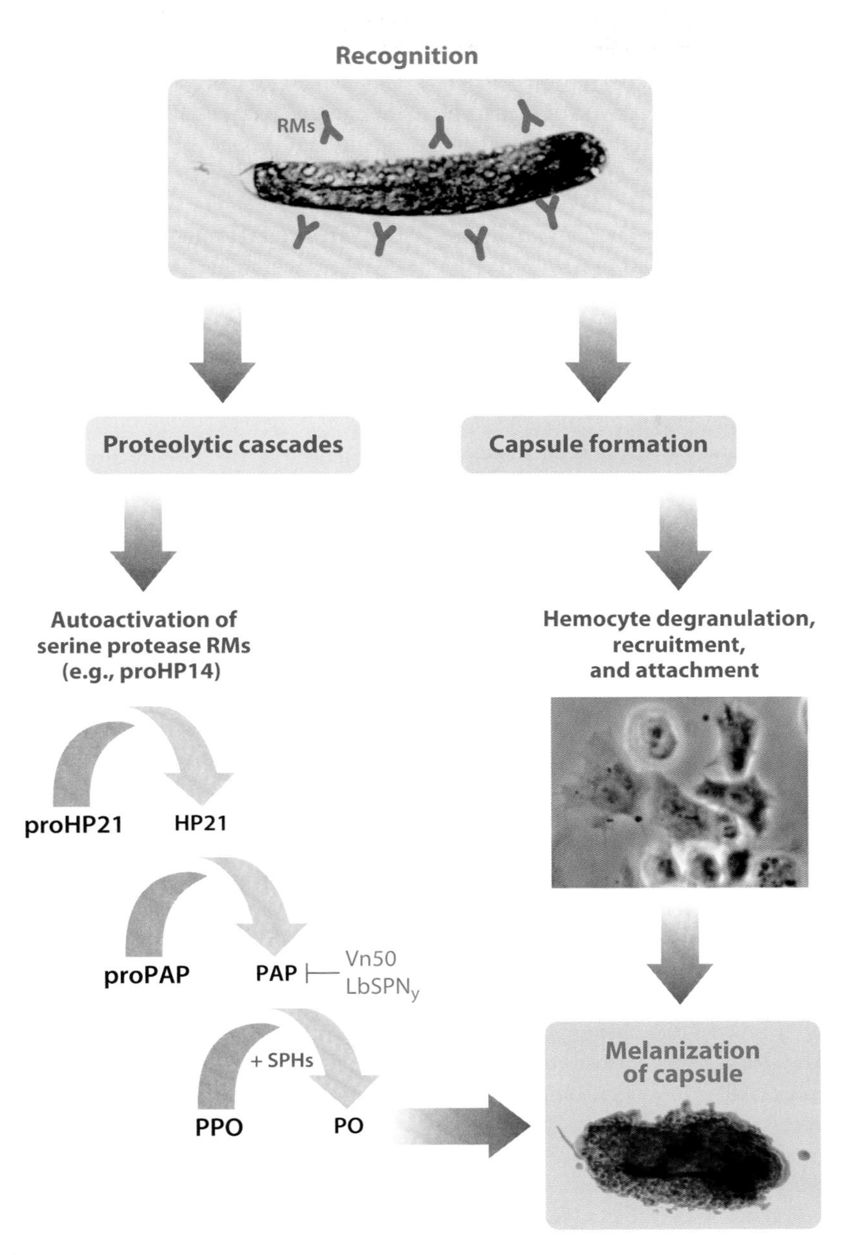

Figure 1

Encapsulation and melanization of endoparasitoid egg. Following recognition of the endoparasitoid egg surface as foreign by recognition molecules (RMs), proteolytic cascades are activated and the hemocyte encapsulation response is initiated, leading to the formation of a multicellular layer around the egg surface. Capsule formation is usually accompanied by melanin formation, although melanization does not seem to be necessary for the encapsulation response. Formation of melanin is the end result of a complex proteolytic cascade in which inactivated hemolymph proteins (proHP) are activated, leading to the activation of the prophenoloxidase (PPO)-activating proteinase (PAP), which in turn activates phenoloxidase (PO), the main enzyme involved in melanin formation (see text for more details). SPHs, serine protease homologs.

cell lysis in lamellocytes or by changing their morphology from discoidal to bipolar (49, 89). Some of these effects have been attributed to the VLPs that are produced in their venom glands (see below). Two strains of *Leptopilina boulardi*, IS_y and IS_m, have been identified as avirulent and virulent (suppress encapsulation) strains, respectively, toward *D. melanogaster*. However, their virulence is the opposite in another host, *D. yakuba* (29). In other words, IS_y successfully avoids encapsulation in *D. yakuba*, whereas IS_m eggs are heavily encapsulated in the host. Similar to natural parasitization, injection of venom from the *L. boulardi* IS_y strain reduces encapsulation of oil droplets injected into *D. yakuba* larvae, whereas venom from the IS_m strain has no effect on encapsulation rate (28). The immunosuppression effect 24 h after parasitization or venom injection is presumed to be due to the differential effect of parasitism by the two strains on THC and DHC rather than to lamellocyte modification. Experiments revealed that THC and the number of lamellocytes and plasmatocytes were significantly lower in *D. yakuba* larvae parasitized by the IS_y strain than in larvae parasitized by the IS_m strain, but the percentage of abnormal lamellocytes did not differ between the two strains (28). The immunosuppressive effect, however, was transient, which declined by 48 h postparasitization or postinjection of venom. Although the IS_m strain does not change the morphology of lamellocytes in *D. yakuba*, it induces the change in its habitual host, *D. melanogaster*, further confirming a species-specific effect of venom components on host hemocytes (29).

The main immunosuppressive factor in *L. boulardi* venom was determined to be a 30-kDa protein (P4) with a Ras domain homologous to the GTPase-activating protein (Rho-GAP) domain, which is absent from the avirulent strain (51). Proteins with a Rho-GAP domain usually function as inhibitors of Rho GTPase involved in various cellular pathways, including the regulation of the actin cytoskeleton, vesicular transport, and proliferation (12). Similar to natural parasitization, injection of total venom from the virulent strain significantly decreases the total lamellocyte number and increases the proportion of lamellocytes of the modified shape unable to perform encapsulation (50). Injection of purified P4 protein also mimics the effect on the lamellocyte morphology but not on numbers, suggesting that other venom components may be required to affect the numbers. The change in adhesion property of lamellocytes caused by P4 could be associated with Rho-GAP domain activity, leading to rearrangement/destabilization of filamentous actin cytoskeleton.

In *P. hypochondriaca*, injection of venom into pupae of the tomato moth, *L. oleracea*, leads to failure of hemocytes to form capsules around injected Sephadex beads (84). In addition, venom reduces THC by more than 50%, compared to control injected with a saline buffer, by inducing cell death. In addition to cell lysis, *P. hypochondriaca* venom affects spreading of plasmatocytes by destabilizing the actin cytoskeleton (81). Subsequently, a 33-kDa venom protein (VPr3) was isolated and characterized from the wasp venom and was shown to inhibit aggregation of hemocytes and the encapsulation response (23, 82). The venom protein does not show a significant sequence similarity to proteins with known functions. In addition to inhibition of the encapsulation response, *P. hypochondriaca* venom reduces phagocytosis of bacteria in vitro, which could well be related to the breakdown of cell cytoskeleton (84). As a consequence, venom-treated host pupae become more susceptible to microbial infections by *Bacillus cereus* or *Beauveria bassiana* (24).

Parasitization of *P. rapae* pupae by a non-PDV wasp, *Pteromalus puparum*, leads to a significant reduction in THC up to five days postparasitization, with the number of plasmatocytes significantly declining (14). Injection of venom alone significantly reduces the percentages of spreading plasmatocytes and encapsulated Sephadex beads in vitro. Although venom-treated hemocytes fail to spread or produce extended pseudopods, examination of actin cytoskeleton did not indicate any changes, suggesting that alterations in hemocyte behavior caused by *P. puparum* venom are not associated with destabilization of the actin cytoskeleton. Later, a 24.1-kDa venom protein (Vn.11) was characterized that mimics the same effects of the total venom (108).

In many host-parasitoid systems in which the wasp produces PDVs/VLPs, venom alone is not sufficient to protect the parasite from host immune responses. With few exceptions, in most studies, the combined effect of calyx fluid (containing the particles) and venom has been investigated and reported. Consequently, the exact effect of venom alone is unknown in most systems. In *Cotesia kariyai* and *C. glomeratus* (Braconidae) venom is essential for the eggs to avoid the encapsulation response, as removal of the venom apparatus leads to encapsulation of parasitoid eggs in *Pseudaletia separata* and *Pieris rapae*, respectively (45, 98). However, calyx fluid should accompany venom to avoid host immune responses. In *P. separata*, THC significantly drops shortly (within 1 h) after parasitization by *C. kariyai* (99). This drop was in response to venom, as PDVs alone did not cause the same effect. However, PDVs maintain the low THC 6 h postparasitization together with venom by inducing apoptosis in hemocytes and the hematopoietic organ. Similarly, both calyx and venom fluids are essential for protecting *C. glomeratus* eggs from the host encapsulation response, as neither alone inhibits the response (45). Further, in *Chelonus* near *curvimaculatus* (an egg parasitoid), effects on host immune suppression and development disruption require both venom and calyx fluids (54). In contrast, removal of venom glands from *Campoletis sonorensis* (Ichneumonidae) did not affect the success of parasitization, suggesting that venom is not essential for parasitism or ichnovirus function in this system (106).

Calreticulin has been implicated in insect cellular responses such as phagocytosis and encapsulation of foreign objects by playing a role in nonself recognition (3, 17). The protein is found on the surface of hemocytes and is enriched in early stage capsules. Calreticulin is a calcium (Ca^{2+})-binding protein with multifunctional properties (60). The protein has Ca^{2+}- and lectin-binding properties with chaperone functions (66). However, the role of calreticulin in many host-parasite interactions (e.g., 35, 66), cell spreading, and adhesion (35, 109) has been documented, and growing evidence suggests that the protein may mediate a broad array of cellular functions. A calreticulin from *C. rubecula* venom glands inhibits spreading of host hemocytes and consequently prevents capsule formation (113). However, the mechanism by which the venom calreticulin prevents encapsulation is unclear. Treatment of *P. hypochondriaca* venom with an anticalreticulin antibody prevented cell death in host hemocytes (86). Cell death is usually induced within 15 min after application of venom to *Trichoplusia ni* cells (BTI-TN-5B1-4) with an increase in intracellular calcium. Although the presence of calreticulin in *P. hypochondriaca* venom has not been confirmed, results suggest that a calreticulin-like protein could affect cellular calcium homeostasis in the cells. Calreticulin has also been reported from *Microctonus aethiopoides* and *M. hyperodae* venoms but its functional role in the venoms or host-parasitoid interactions has not been explored (21).

PAP: prophenoloxidase activating proteinase

Effects on Humoral Responses

Major humoral responses in insects include the production of antimicrobial peptides, wound healing, coagulation, and melanization (38). The humoral response most relevant to host-parasitoid interactions is melanization. Almost invariably, soon after the initiation of the encapsulation response a brown-black material (eumelanin) is deposited onto the foreign surface (**Figure 1**). Melanization involves the activation of a complex series of enzymatic reactions that lead to the activation of the zymogen (PPO) to active PO by PPO-activating proteinases (PAPs). For detailed biochemical pathways, readers are referred to other reviews (15, 18, 68).

Despite the co-occurrence of melanization and encapsulation, it is not clear whether melanization is essential for encapsulation and death of the developing egg. At least in one report, it was shown that melanization is not

SPH: serine protease homolog

required for encapsulation of *L. boulardi* parasitoid eggs by *D. melanogaster* larvae (88), but the effect on parasitism success was not determined. Nevertheless, the fact that most parasitoids inhibit the melanization response in their host hemolymph indicates that this inhibition is evolutionarily advantageous for the wasps. The inhibitory factors may be calyx fluid proteins (e.g., 8), PDV-expressed proteins (9, 57), or venom. In most host-parasitoid systems that have been explored, the inhibitory components seem to hinder activation of PPO either by inhibiting upstream reactions that lead to the activation of the zymogen or by directly affecting activation of PPO.

A 50-kDa glycoprotein (Vn50) from *C. rubecula* venom inhibits melanization (5). Ironically, the protein has significant similarity to hemolymph serine protease homologs (SPHs) that facilitate the activation of PPO by PAP after cleavage into a clip and a protease-like domain (40). Contrary to hemolymph SPHs, Vn50 is not cleaved into the two domains in the hemolymph and remains stable for 3 days postparasitization (111). In vitro binding assays revealed that both PPO and PAP bind to Vn50. In addition, in enzyme assays using purified PPO, PAP-1, and SPHs from *Manduca sexta*, Vn50 suppressed the activation of PPO to active PO in a concentration-dependent manner (111). This suggested that Vn50 may compete with host SPHs for binding to PPO and PAP. Therefore, the plausible model for inhibition of melanization by Vn50 is that because Vn50 is not cleaved into the two domains, and binds to PPO and PAP, it may not provide the required spatial position or conformational change in PPO for cleavage by PAP. Transformed *D. melanogaster* flies expressing Vn50 have significantly lower melanization activity and are more susceptible to fungal infection (100). In addition to Vn50, a 4.6-kDa peptide (Vn4.6) from *C. rubecula* venom inhibits melanization with an unknown mechanism (4).

Total venom from a virulent strain of *L. boulardi* (IS_y) also inhibits activation of PPO in the hemolymph of *D. yakuba* (19). A detailed analysis indicated that venom interferes with upstream steps, leading to the activation of PPO rather than affecting the PO activity itself. Similarly, venom extracts from *L. boulardi* inhibited melanization in *D. melanogaster* hemolymph by suppressing the oxidation of two diphenol eumelanin precursors, dopamine and 5,6-dihydroxyindole (46). A serpin ($LbSPN_y$; see Protease Inhibitors, above), which is expressed specifically in the venom glands, was identified to significantly inhibit activation of *D. yakuba* PPO. The underlying reaction(s) upstream of PPO activation affected by this serpin has not been identified.

COEVOLUTION OF MUTUALISTIC VIRUSES AND VIRUS-LIKE PARTICLES AND VENOM PROTEINS

Various types of viruses and VLPs are associated with the ovaries and venom glands of endoparasitoid wasps that are injected into the host upon oviposition. A major group of these virus-like entities are PDVs. Venom proteins and protein products of PDV genes expressed in the host have a fundamental similarity: They all seem to have derived from normal body proteins that are typically involved in regulatory processes. This can also be extended to the proteins that make up VLPs that are devoid of nucleic acids. The recruited genes, duplicated from the genome, are expressed in the venom glands or, in the case of PDVs, amplified in calyx cells for packaging as encapsidated DNA. Presumably, over the course of evolution, the number of genes increased by duplication/diversification events, leading to multigene families with new functions. Both venom and PDV protein products expressed in the host are involved in host regulation, often with overlapping functions. Antigenic similarities and hybridization of venom and PDV genes expressed in the host have also been reported (107), which suggests that they might have had common ancestral origins.

In many host-parasitoid systems, venom proteins are required for PDV function or synergistic effects. This ranges from complete independence of some ichneumonid PDVs (ichnoviruses) to variable dependency of braconid PDVs (bracoviruses) on venom. In most scenarios, the exact mechanism of synergism is not known; however, it seems that venom facilitates the entry and stability of PDV particles in the host. Venom from *Cotesia melanoscela* promotes the release of virions into the cytoplasm after uptake into host cells by facilitating the uncoating of PDVs at nuclear pores in host hemocytes (94). In *C. rubecula*, venom is not required for virus entry into host cells but is essential for expression of PDV genes (112). A small venom peptide comprising 14 amino acids (1.6 kDa; Vn1.5) was found to mediate expression of PDV genes. It is likely that the peptide may facilitate uncoating of the particles similar to *C. melanoscela* particles.

VLPs devoid of encapsidated nucleic acids produced in the venom glands have been reported from parasitoid wasps. *Meteorus pulchricornis* (Braconidae) ovaries lack the calyx region and PDVs but instead produce VLPs (MpVLPs) in the venom gland filaments (97). The particles induce apoptosis in *Pseudaletia separata* hemocytes both in vivo and in vitro (96, 97). While MpVLPs seem to be involved in the induction of apoptosis, the particles also affect spreading behavior of hemocytes (95). The effects, which include loss of focal adhesions and retraction of filopodia, are observed as early as 30 min following the exposure of hemocytes to MpVLPs. Therefore, the particles seem to provide immediate (first few minutes) as well as early (first few hours) protection for the developing parasitoid egg by affecting the adhesion and spreading properties of hemocytes and by inducing apoptosis in the cells.

Leptopilina spp. are specialist parasitoids of *Drosophila* spp. that produce VLPs in their venom (=long) glands. There are two strains of *L. boulardi* based on their ability to evade the host's immune system. The virulent strain is always able to suppress the host encapsulation response, whereas the eggs of the avirulent strain are often encapsulated by host hemocytes (49). Both strains produce VLPs in their venom glands; however, the particles are morphologically different (30, 49). Nevertheless, there is no direct experimental evidence to demonstrate that the VLPs are involved in immunosuppression. On the other hand, *L. heterotoma* VLPs bind and enter host lamellocytes and selectively destroy the cells via a nonapoptotic mechanism, thus preventing the encapsulation response (89). The particles do not seem to affect plasmatocytes. VLPs from *L. victoriae* morphologically resemble *L. heterotoma* VLPs and also share antigenic similarities with the particles (61). The venom fluid containing LvVLPs has immunosuppressive properties and causes lysis of lamellocytes.

The only conventional virus that has been identified and shown in association with venom glands from an endoparasitoid wasp is an entomopoxvirus (EPV). The virus was found replicating in the venom apparatus of *Diachasmimorpha longicaudata* (Braconidae) parasitizing the larvae of the Caribbean fruit fly, *Anastrepha suspensa* (Tephritidae) (52). DlEPV replicates both in the wasp and in the host infected hemocytes (53). Infected hemocytes exhibit typical apoptotic symptoms including blebbing and DNA concatenation (53). As a consequence, they lose their adhesive property and fail to perform the encapsulation response. These pathological effects do not seem to occur within the carrying wasps, and therefore DlEPV is considered a mutualistic virus that facilitates the parasite's survival in the host.

Teratocytes: cells derived from extraembryonic serosa tissue in parasitoid eggs

VENOM AS MODULATOR OF THE HOST NUTRITIONAL ENVIRONMENT

Numerous reports have documented changes in the hemolymph profile of lipids, proteins, and carbohydrates of host insects that have been altered by PDVs, VLPs, teratocytes, and venoms (reviewed in References 65 and 101). In nearly all cases involving endoparasitic wasps, the effects of venom alone have been difficult to ascertain because multiple agents are injected simultaneously into hosts. Despite this limitation, endoparasitoid venoms possess a wealth of proteins that influence, either directly or indirectly, the availability of nutrients present for feeding wasp larvae.

The braconid wasp *A. ervi* injects venom into the aphid *Acyrthosiphon pisum*, resulting in castration of the host (39). Host castration results from a protein in the venom, γ-glutamyl transpeptidase (γ-GT), that stimulates apoptotic pathways in cells of the germaria and ovariole sheath (34). The degeneration of host reproductive tissue is believed to relieve the developing parasitoids of competing with host tissues for available nutrients. Factors released from teratocytes facilitate digestion of the host reproductive tissue and transport liberated fatty acids to the wasp larvae (33). Presumably, the castration of larval testes in *Plutella xylostella* by *Cotesia vestalis* (Braconidae) and *Diadegma semiclausum* (Ichneumonidae) also decreases competition for host nutrients (7). It has yet to be determined whether venom from either wasp is capable of evoking testis degeneration in the absence of PDVs, but venom does appear to be required (7).

Venom from *P. hypochondriaca* contains six hydrolases: acid phosphatase, esterase lipase, β-glucosidase, esterase, lipase, and β-galactosidase (22). A function in venom has not been determined for any of these enzymes; however, the acid phosphatase has been predicted to be associated with liberating carbohydrates into the host hemolymph for consumption by wasp larvae (22), a function also predicted for venom trehalases (tre-1) present in this wasp.

Several endoparasitoids target host fat body to acquire nutrients to feed larvae. The action of PDV gene products alone or synergistically with venom has been reported to stimulate nutrient synthesis and release from host fat body (48, 65). Venom lipases or other hydrolases may function to digest fat body cells, allowing nutrients to flow into the hemolymph in a manner similar to teratocyte-induced release of proteins and lipids observed in *Meteorus pulchricornis* (Braconidae) and *Cotesia kariyai* (65, 97). *M. pulchricornis* relies on venom proteins to destroy actin and tubulin fibers in the cytoskeletal scaffolding of host fat body cells, which generates an inward collapse of the cytoskeleton (apoptosis) and thereby compromises the integrity of plasma membranes (65). Uçkan et al. (103) identified phospholipase B in venom from *Pimpla turionellae*, which may function to lyse fat body cells as has been reported with larval salivary enzymes of some ectoparasitoids (35).

VENOM AND ALTERATION OF HOST DEVELOPMENT

A delay or complete arrestment in host development is commonly associated with parasitism by endoparasitoids (59, 77, 79). In almost all host-parasitoid associations examined, host arrest appears to result from a hormone imbalance (i.e., ecdysteroids and juvenile hormone), through a disruption or deterioration of host endocrine tissue or possibly by inhibition of the ability of target tissues within the host insect to respond to hormonal signals. In recent years, research emphasis placed on examining host developmental changes has shifted more toward quantitative and qualitative analysis of viral gene expression in host tissues (10, 16) and viral mechanisms of immunosuppression (31). Consequently, the information presented in this section lacks mechanistic details on how endoparasitoid venoms trigger developmental delay or arrest in susceptible hosts. At the cellular level, virtually nothing is known about the mechanisms of host arrest and this represents an area of host-parasite biology in need of new investigations. What follows, then, is part speculation and part reported observations on how venom proteins from endoparasitoids manipulate the host condition to yield arrested host development.

Pruijssers et al. (79) observed that larvae of *Pseudoplusia includens* remained in a state of hyperglycemia if injected with bracovirus from *Microplitis demolitor* before reaching a critical larval weight. Host hyperglycemia was associated with reduced nutrient stores and corresponded with prothoracic glands remaining in a refractory state of ecdysteroid release. Consequently, host metamorphosis was inhibited and parasitized larvae remained in an arrested state of development until death (79). Although there is no direct evidence that any endoparasitoid venom triggers host arrest by a similar mechanism, numerous endo- and ectoparasitoids have been shown to evoke altered host metabolism, and in many instances, endoparasitoids have elicited either hyper- or hypoglycemia (reviewed in References 101 and 105). Such host responses could possibly account for the reduction in prothoracic gland activity resulting from *Toxoneuron nigriceps* attack of larval *Heliothis virescens* (76), conceivably leading to the arrestment of host development.

Reprolysin-like metalloproteinase is also a candidate for contributing to host arrestment because the enzyme is involved in the regulation of key developmental events in *D. melanogaster*, including stimulation of *Notch* signal transduction pathways during embryonic and imaginal disc pattern formation (93). Reprolysin has been reported from *P. hypochondriaca* (69).

CYTOTOXICITY OF VENOM PROTEINS

Cell death is a common feature of host-parasitoid relationships, frequently associated with immunosuppression of the host as well as the final demise of particular tissues. Depending on the maternal secretions injected into the host, and possibly on the dose of material and stage of development, cell death can be the result of stimulation of apoptotic and/or oncotic pathways (**Figure 2**). Apoptosis is generally initiated by either intrinsic pathways that involve mitochondrial release of caspase activators or extrinsic signaling that originates with cell surface receptors (i.e., Fas) and uses second messengers such as Ca^{2+}, IP_3, or cAMP to activate caspases (**Figure 2**). Both pathways culminate with the irreversible events of cell disassembly driven by caspase 3. In contrast, oncosis originates with altered plasma membrane integrity, yielding a net movement of ions and water into the cell, and ultimately results in cellular demise associated with disrupted intracellular homeostasis (**Figure 2**).

Unlike with ectoparasitic wasps, in which it is clear that venom is typically responsible for the induction of the host demise (80), the role of endoparasitoid venom in eliciting host or tissue-specific death is far less certain. *P. hypochondriaca* and *P. turionellae* produce venom that triggers host paralysis and appears to operate by cytotoxic and/or cytolytic mechanisms dependent on the susceptibility of the target cells (32, 44, 75). Parkinson et al. (69) identified a heterodimeric protein, pimplin, in venom from *P. hypochondriaca* that elicits paralysis when injected into the hemocoel of adult *M. domestica*. A 13-kDa protein isolated from this wasp venom demonstrates cytotoxicity toward *Spodoptera frugiperda* cells (Sf21) (75) and may account for the swelling and lysis evoked in cultured cells derived from *T. ni* (BTI-TN-5B1-4) when incubated with the venom (86). The cytotoxic action of *P. hypochondriaca* venom also extends to hemocytes collected from *L. oleracea* (84). Prior to cell death, Sf21 cells and hemocytes display degranulation and extensive cytoplasmic vacuolization (75, 84). The venom does not cause a loss in the integrity of host plasmatocyte or granular cell (granulocytes) plasma membranes, nor do hemocyte cytoskeletons disintegrate (rather they collapse around the nucleus) (81). The latter has also been observed with venom from *Pteromalus puparum* when host plasmatocytes and granular cells were stained with phallodin following an in vitro exposure to wasp venom (110). These cellular changes are features common to apoptotic mechanisms of cell death, although induction of apoptosis in host hemocytes by *P. hypochondriaca* venom is dose dependent (23, 81).

Analyses of a cDNA library constructed from venom glands extracted from *P. hypochondriaca* indicate that venom contains a laccase and also three putative PPOs (discussed above and see References 70 and 72). Proteins with PO activity, which includes laccases, have the potential to disrupt plasma membranes of susceptible cells, potentially evoking membrane blebbing, rounding, and swelling (1). Cells injured by venom from *P. hypochondriaca* display qualitative increases in $[Ca^{2+}]_i$ within 15 min after

Oncosis: form of cell death typically induced by injury or accident that may result in cellular swelling and lysis

exposure to venom, and the mitochondrial membrane potential ($\Delta\Psi_m$) drops to undetectable levels within 5 min posttreatment (86). Such cellular changes have been observed with *T. ni* cells (86), but not host hemocytes (81), exposed to venom. Treatment of wasp venom with a potent inhibitor of PO did not attenuate the toxic effects of venom on the BTI-TN-5B1-4 cells (86). In contrast, preincubation of venom with antibodies generated toward calreticulin reduced the cytotoxicity of the wasp venom, as evidenced by diminished cellular swelling and no increases in intracellular calcium (86). A role for calreticulin in eliciting apoptotic and oncotic cell death has also been suggested for venom from *N. vitripennis* (85).

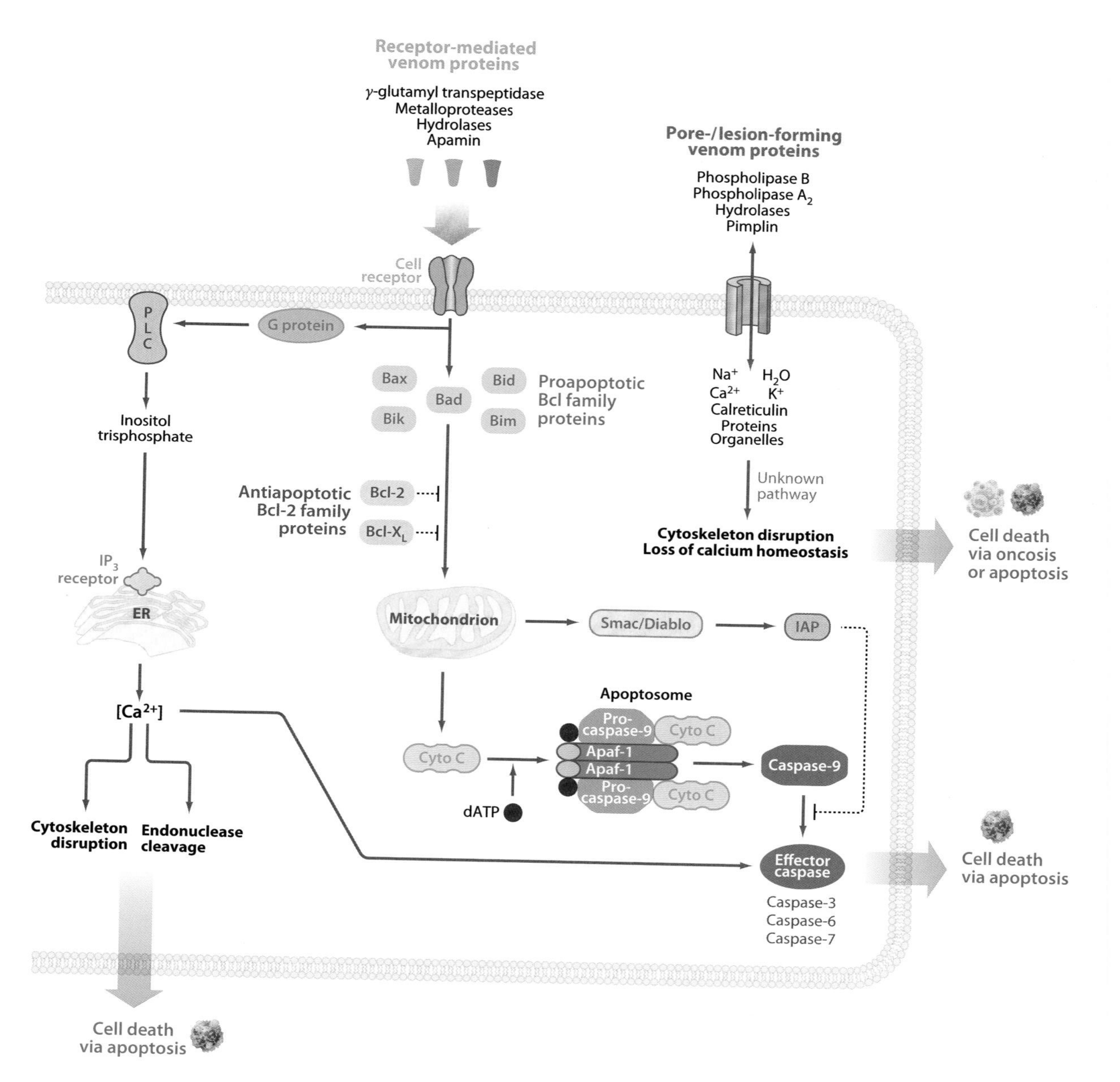

P. turionellae synthesizes venom that appears to be divergent in composition and function from that of *P. hypochondriaca*. *P. turionellae* venom is a complex mixture of proteins, polypeptides, catecholamines, and enzymes (103), and induces complete paralysis of host larvae (32). None of the venom constituents, however, has been shown to specifically evoke host paralysis or cell death. Morphological changes in cell structure similar to those triggered by *P. hypochondriaca* have been observed in cultured cells of *T. ni* (BTI-TN-5B1-4) and hemocytes from *Galleria mellonella* when incubated with cytotoxic doses of *P. turionellae* venom (32, 44). In addition, venom-induced increases in intracellular (cytosolic) calcium precede plasma membrane blebbing and cytoplasmic vacuolization. The source of the calcium elevations has yet to be discovered. In vitro assays using BTI-TN-5B1-4 cells incubated with venom indicate that a loss of membrane integrity occurs prior to the increase in $[Ca^{2+}]_i$ (44). This observation would argue that an influx of extracellular calcium is a key signal leading to cell death. However, although a variety of calcium channel blockers and osmotic protectants could slow the eventual demise of venom-treated cells, the rise in intracellular calcium was not attenuated (44). The exact mechanisms leading to cell death have not been determined, but the cellular changes and results from physiological assays are consistent with a necrotic/oncotic basis of death dependent on either pore/lesion formation or opening of existing ion channels (**Figure 2**).

Several species of parasitoids possess hydrolases in the venom (62). In social Hymenoptera, this class of enzymes can be both cytotoxic and cytolytic, often functioning in cell and tissue digestion to facilitate distribution or spreading of other agents in the venom (90). Surprisingly little information is available on the functionality of these enzymes in endoparasitoid venoms despite an apparently high level of the proteins in several venoms. Six different hydrolases have been detected in venom from *P. hypochondriaca*, with acid phosphatase the most abundant (22). Venom acid phosphatase, however, was found to have no influence on the cytotoxic effect of venom on hemocytes from *L. oleracea* (22). Other hydrolases in this venom have yet to be tested for toxicity, and cell types other than hemocytes also need to be examined.

Necrosis: deterioration of cell structures following cell death

γ-GT functions in pathways (transpeptidation) that protect cells from oxidative stress. Disruption of these pathways can lead to induction of apoptosis (**Figure 2**). Venom from the braconid *A. ervi* contains a γ-GT that triggers apoptosis in host cells of germaria and ovariole sheath, leading to castration of the aphid hosts (34). This enzyme may be a component of other endoparasitoid venoms as well and has been identified by genomic mining and proteomic analyses in venom from *N. vitripennis* (25), although no functional assays have been performed.

Figure 2

Potential mode of action of endoparasitoid venom proteins in eliciting death in host cells. Involvement of apoptotic and oncotic (necrotic) pathways is speculative based on biochemical and morphological observations described in the text. Biochemical activation of apoptosis generally involves two pathways: the intrinsic pathway, which is dependent on the release of cytochrome *c* (Cyto C) from mitochondria, which in turn leads to a cascade of events following the activation of caspase 9, and the extrinsic pathway(s), which begins with stimulation of cell surface receptors such as Fas and leads to the activation of one or more members of a family of cysteine-rich proteases called caspases. A third pathway originating with ER also can lead to the activation of multiple caspases. Far less is known about the pathways associated with necrotic/oncotic death. In general, following pore/lesion formation or opening of ion channels, ions and macromolecules move into the cell essentially unregulated, followed by water, and the result is loss of homeostasis of the intracellular environment, which can activate or suppress multiple pathways. Abbreviations: Apaf-1, apoptotic protease activating factor 1; Cyto C, cytochrome *c*; ER, endoplasmic reticulum; IAP, inhibitor of apoptosis; IP_3, inositol trisphosphate; PLC, phospholipase C; Smac (second mitochondria-derived activator of caspases)/Diablo, a mitochondrial protein that binds to IAP and relieves IAP inhibition of caspases.

CONCLUSION

Parasitoid wasps are remarkable for the level and diversity of peptides/proteins produced in their venom glands. Most of these proteins are involved in modification of the host physiological environment to facilitate development of the wasp progeny, most importantly compromising host immune responses and interrupting/inhibiting its development. These molecules may prove to be valuable in designing environment-friendly pest control strategies. Further, pharmaceutical properties of these components have not been explored. At present, high-end proteomics and molecular approaches are available to unravel the complexity of venomics from parasitic wasps. Recent attempts have been initiated in various laboratories to look further into the venomics of a range of endoparasitoids with different life strategies, which will clarify the evolution of these molecules and provide a springboard to elucidate the functional role of these bioactive molecules.

SUMMARY POINTS

1. In contrast to ectoparasitoids, the majority of endoparasitic wasps are koinobionts, which allow their hosts to develop after oviposition/parasitization.
2. Venom components from endoparasitoids are diverse and hence may have different functions based on the host-parasitoid system. These functions could be temporary paralysis, suppression of immune responses, modulation of nutritional environment, and alteration of development in the host.
3. Proteinaceous components in endoparasitoid venoms consist of enzymes, protease inhibitors, and paralytic and cytolytic factors.
4. Enzymes present in endoparasitoid venoms are structurally similar to insect metabolic enzymes, indicating their diverged evolution from cellular enzymes.
5. In some host-parasitoid systems, venom is required or enhances the effect of symbiotic viruses or VLPs introduced by endoparasitoids into the host at oviposition.

DISCLOSURE STATEMENT

The authors are not aware of any affiliations, memberships, funding, or financial holdings that might be perceived as affecting the objectivity of this review.

ACKNOWLEDGMENTS

We thank two anonymous reviewers for their constructive criticism.

LITERATURE CITED

1. Abt M, Rivers DB. 2007. Characterization of phenoloxidase activity in venom from the ectoparasitoid *Nasonia vitripennis* (Walker) (Hymenoptera: Pteromalidae). *J. Invertebr. Pathol.* 94:108–18
2. Andersen SO. 2005. Cuticular sclerotization and tanning. In *Comprehensive Molecular Insect Science*, ed. LI Gilbert, K Iatrou, S Gill, pp. 145–70. Oxford, UK: Elsevier
3. Asgari S, Schmidt O. 2003. Is cell surface calreticulin involved in phagocytosis by insect hemocytes? *J. Insect Physiol.* 49:545–50
4. Asgari S, Zareie R, Zhang G, Schmidt O. 2003. Isolation and characterization of a novel venom protein from an endoparasitoid, *Cotesia rubecula* (Hym: Braconidae). *Arch. Insect Biochem. Physiol.* 53:92–100

5. **Asgari S, Zhang G, Zareie R, Schmidt O. 2003. A serine proteinase homolog venom protein from an endoparasitoid wasp inhibits melanization of the host hemolymph. *Insect Biochem. Mol. Biol.* 33:1017–24**
6. Askew RR, Shaw MR. 1986. Parasitoid communities: their size, structure and development. In *Insect Parasitoids*, ed. J Waage, D Greathead, pp. 225–63. London: Academic
7. Bai S, Cai D, Li X. 2009. Parasitic castration of *Plutella xylostella* larvae induced by polydnaviruses and venom of *Cotesia vestalis* and *Diadegma semiclausum* polydnaviruses and venom of *Cotesia vestalis* and *Diadegma semiclausum*. *Arch. Insect Biochem. Physiol.* 70:30–43
8. Beck M, Theopold U, Schmidt O. 2000. Evidence for serine protease inhibitor activity in the ovarian calyx fluid of the endoparasitoid *Venturia canescens*. *J. Insect Physiol.* 46:1275–83
9. Beck MH, Strand MR. 2007. A novel polydnavirus protein inhibits the insect prophenoloxidase activation pathway. *Proc. Natl. Acad. Sci. USA* 104:19267–72
10. Beliveau C, Levasseur A, Stoltz D, Cusson M. 2003. Three related TrIV genes: comparative sequence analysis and expression in host larvae and Cf-124T cells. *J. Insect Physiol.* 49:501–11
11. Besson MT, Soustelle L, Birman S. 2000. Selective high-affinity transport of aspartate by a *Drosophila* homologue of the excitatory amino acid transporters. *Curr. Biol.* 10:207–10
12. Bourne HR, Sanders DA, McCormick F. 1991. The GTPase superfamily: conserved structure and molecular mechanisms. *Nature* 349:117–27
13. Bull H, Murray PG, Thomas D, Fraser AM, Nelson PN. 2002. Acid phosphatases. *Mol. Pathol.* 55:65–72
14. Cai J, Ye G-y, Hu C. 2004. Parasitism of *Pieris rapae* (Lepidoptera: Pieridae) by a pupal endoparasitoid, *Pteromalus puparum* (Hymenoptera: Pteromalidae): effects of parasitization and venom on host hemocytes. *J. Insect Physiol.* 50:315–22
15. **Cerenius L, Lee BL, Söderhäll K. 2008. The proPO-system: pros and cons for its role in invertebrate immunity. *Trends Immunol.* 29:263–71**
16. Chen YF, Shi M, Huang F, Chen X-x. 2007. Characterization of two genes of *Cotesia vestalis* polydnavirus and their expression patterns in the host *Plutella xylostella*. *J. Gen. Viol.* 88:3317–22
17. Choi JY, Whitten MMA, Cho MY, Lee KY, Kim MS, et al. 2002. Calreticulin enriched as an early-stage encapsulation protein in wax moth *Galleria mellonella* larvae. *Dev. Comp. Immunol.* 26:335–43
18. Christensen BM, Li JY, Chen CC, Nappi AJ. 2005. Melanization immune responses in mosquito vectors. *Trends Parasitol.* 21:192–99
19. Colinet D, Dubuffet A, Cazes D, Moreau S, Drezen JM, Poirié M. 2009. A serpin from the parasitoid wasp *Leptopilina boulardi* targets the *Drosophila* phenoloxidase cascade. *Dev. Comp. Immunol.* 33:681–89
20. Coudron TA, Kelly TJ, Puttler B. 1990. Developmental responses of *Trichoplusia ni* (Lepidoptera: Noctuidae) to parasitism by the ectoparasite *Euplectrus plathypenae* (Hymenoptera: Eulophidae). *Arch. Insect Biochem. Physiol.* 13:83–94
21. Crawford AM, Brauning R, Smolenski G, Ferguson C, Barton D, et al. 2008. The constituents of *Microctonus* sp. parasitoid venoms. *Insect Mol. Biol.* 17:313–24
22. Dani MP, Edwards JP, Richards EH. 2005. Hydrolase activity in the venom of the pupal endoparasitic wasp, *Pimpla hypochondriaca*. *Comp. Biochem. Physiol. B* 141:373–81
23. Dani MP, Richards EH. 2009. Cloning and expression of the gene for an insect haemocyte anti-aggregation protein (VPr3), from the venom of the endoparasitic wasp, *Pimpla hypochondriaca*. *Arch. Insect Biochem. Physiol.* 71:191–204
24. Dani MP, Richards EH, Edwards JP. 2004. Venom from the pupal endoparasitoid, *Pimpla hypochondriaca*, increases the susceptibility of larval *Lacanobia oleracea* to the entomopathogens *Bacillus cereus* and *Beauveria bassiana*. *J. Invertebr. Pathol.* 86:19–25
25. de Graaf DC, Brunain M, Scharlaken B, Peiren N, Devreese B, et al. 2010. Two novel proteins expressed by the venom glands of *Apis mellifera* and *Nasonia vitripennis* share an ancient C1q-like domain. *Insect Mol. Biol.* 19:11–26
26. Desneux N, Barta RJ, Delebecque CJ, Heimpel GE. 2009. Transient host paralysis as a means of reducing self-superparasitism in koinobiont endoparasitoids. *J. Insect Physiol.* 55:321–27
27. Digilio MC, Isidoro N, Tremblay E, Pennacchio F. 2000. Host castration by *Aphidius ervi* venom proteins. *J. Insect Physiol.* 46:1041–50

5. Identifies and characterizes a venom protein with similarity to mediators of PPO activation but instead inhibits activation of PPO.

15. A comprehensive review of the role of PO in invertebrate immunity.

28. Dubuffeta A, Douryb G, Labroussea C, Drezena JM, Cartonc Y, Poirié M. 2008. Variation of success of *Leptopilina boulardi* in *Drosophila yakuba*: the mechanisms explored. *Dev. Comp. Immunol.* 32:597–602
29. Dubuffeta A, Dupas S, Frey F, Drezena J-M, Poirié M, Cartonc Y. 2007. Genetic interactions between the parasitoid wasp *Leptopilina boulardi* and its *Drosophila* hosts. *Heredity* 98:21–27
30. Dupas S, Brehelin M, Frey F, Carton Y. 1996. Immune suppressive virus-like particles in a *Drosophila* parasitoid: significance of their intraspecific morphological variations. *Parasitology* 113:207–12
31. Dupas S, Gitau GW, Branca A, Le Ru BP, Silvain JF. 2008. Evolution of a polydnavirus gene in relation to parasitoid-host species immune resistance. *J. Hered.* 99:491–99
32. Ergin E, Uckan F, Rivers DB, Sak O. 2006. In vivo and in vitro activity of venom from the endoparasitic wasp *Pimpla turionellae* (L.) (Hymenoptera: Ichneumonidae). *Arch. Insect Biochem. Physiol.* 61:87–97
33. Falabella P, Perugino G, Caccialupi P, Riviello L, Varricchio P, et al. 2005. A novel fatty acid binding protein produced by teratocytes of the aphid parasitoid *Aphidius ervi*. *Insect Mol. Biol.* 14:195–205
34. Falabella P, Riviello L, Caccialupi P, Rossodivita T, Valente MT, et al. 2007. A g-glutamyl transpeptidase of *Aphidius ervi* venom induces apoptosis in the ovaries of host aphids. *Insect Biochem. Mol. Biol.* 37:453–65
35. Ferreira V, Molina MC, Valck C, Rojas Á, Aguilar L, et al. 2004. Role of calreticulin from parasites in its interaction with vertebrate hosts. *Mol. Immunol.* 40:1279–91
36. **Fry BG, Roelants J, Norman JA. 2009. Tentacles of venom: toxic protein convergence in the kingdom Animalia. *J. Mol. Evol.* 68:311–21**

36. A comprehensive review analyzing convergence of venom proteins across the Animal kingdom.

37. Furihata SX, Kimura MT. 2009. Effects of *Asobara japonica* venom on larval survival of host and nonhost *Drosophila* species. *Physiol. Entomol.* 34:292–95
38. Gillespie JP, Kanost MR, Trenczek T. 1997. Biological mediators of insect immunity. *Annu. Rev. Entomol.* 42:611–43
39. Giordana B, Milani A, Grimaldi A, Farneti R, Casartelli M, et al. 2003. Absorption of sugars and amino acids by the epidermis of *Aphidius ervi* larvae. *J. Insect Physiol.* 49:1115–24
40. Jiang H, Kanost MR. 2000. The clip-domain family of serine proteinases in arthropods. *Insect Biochem. Mol. Biol.* 30:95–105
41. Jones D. 1996. Biochemical interaction between chelonine wasps and their lepidopteran hosts: After a decade of research—the parasite is in control. *Insect Biochem. Mol. Biol.* 26:981–96
42. Jones D, Sawicki G, Wozniak M. 1992. Sequence, structure, and expression of a wasp venom protein with a negatively charged signal peptide and a novel repeating internal structure. *J. Biol. Chem.* 267:14871–78
43. Jones D, Wache S, Chhokar V. 1996. Toxins produced by arthropod parasites: salivary gland proteins of human body lice and venom proteins of chelonine wasps. *Toxicon* 34:1421–29
44. Keenan B, Uckan F, Ergin E, Rivers DB. 2007. Morphological and biochemical changes in cultured cells induced by venom from the endoparasitoid, *Pimpla turionellae*. See Ref. 87a, pp. 75–92
45. Kitano H. 1986. The role of *Apanteles glomeratus* venom in the defensive response of its host, *Pieris rapae crucivora*. *J. Insect Physiol.* 32:369–75
46. Kohler LJ, Carton Y, Mastore M, Nappi AJ. 2007. Parasite suppression of the oxidations of eumelanin precursors in *Drosophila melanogaster*. *Arch. Insect Biochem. Physiol.* 66:64–75
47. Krishnan A, Nair PN, Jones D. 1994. Isolation, cloning and characterization of new chitinase stored in active form in chitin-lined venom reservoir. *J. Biol. Chem.* 269:20971–76
48. Kroemer JA, Webb BA. 2004. Polydnavirus genes and genomes: emerging gene families and new insights into polydnavirus replication. *Annu. Rev. Entomol.* 49:431–56
49. Labrosse C, Carton Y, Dubuffet A, Drezen JM, Poirie M. 2003. Active suppression of *D. melanogaster* immune response by long gland products of the parasitic wasp *Leptopilina boulardi*. *J. Insect Physiol.* 49:513–22
50. Labrosse C, Eslin P, Doury G, Drezen JM, Poirie M. 2005. Haemocyte changes in *D. melanogaster* in response to long gland components of the parasitoid wasp *Leptopilina boulardi*: a Rho-GAP protein as an important factor. *J. Insect Physiol.* 51:161–70
51. Labrosse C, Stasiak K, Lesobre J, Grangeia A, Huguet E, et al. 2005. A RhoGAP protein as a main immune suppressive factor in the *Leptopilina boulardi* (Hymenoptera, Figitidae)-*Drosophila melanogaster* interaction. *Insect Biochem. Mol. Biol.* 35:93–103
52. Lawrence PO. 2002. Purification and partial characterization of an entomopoxvirus (DlEPV) from a parasitic wasp of tephritid fruit flies. *J. Insect Sci.* 2:10

53. Lawrence PO. 2005. Morphogenesis and cytopathic effects of the *Diachasmimorpha longicaudata* entomopoxvirus in host haemocytes. *J. Insect Physiol.* 51:221–33
54. Leluk J, Jones D. 1989. *Chelonus* sp. near *curvimaculatus* venom proteins: analysis of their potential role and processing during development of host *Trichoplusia ni*. *Arch. Insect Biochem. Physiol.* 10:1–12
55. Leluk J, Schmidt J, Jones D. 1989. Comparative studies on the protein composition of hymenopteran venom reservoirs. *Toxicon* 27:105–14
56. Lewis RJ. 2004. Conotoxins as selective inhibitors of neuronal ion channels, receptors and transporters. *IUBMB Life* 56:89–93
57. Lu Z, Beck MH, Wang Y, Jiang H, Strand MR. 2008. The viral protein Egf1.0 is a dual activity inhibitor of prophenoloxidase-activating proteinases 1 and 3 from *Manduca sexta*. *J. Biol. Chem.* 283:21325–33
58. Mabiala-Moundoungou ADN, Doury G, Eslin P, Cherqui A, Prevost G. 2010. Deadly venom of *Asobara japonica* parasitoid needs ovarian antidote to regulate host physiology. *J. Insect Physiol.* 56:35–41
59. Marris GC, Weaver RJ, Edwards JP. 2001. Endocrine interactions of the ectoparasitoid wasps with their hosts: an overview. In *Endocrine Interactions of Insects Parasites and Pathogens*, ed. JP Edwards, RJ Weaver, pp. 133–51. Oxford: Bios
60. Michalak M, Milner RE, Burns K, Opas M. 1992. Calreticulin. *Biochem. J.* 285:681–92
61. Morales J, Chiu H, Oo T, Plaza R, Hoskins S, Govind S. 2005. Biogenesis, structure, and immune-suppressive effects of virus-like particles of a *Drosophila* parasitoid, *Leptopilina victoriae*. *J. Insect Physiol.* 51:181–95
62. Moreau A, Guillot S. 2005. Advances and prospects on biosynthesis, structures and functions of venom proteins from parasitic wasps. *Insect Biochem. Mol. Biol.* 35:1209–23
63. Moreau SJM, Cherqui A, Doury G, Dubois F, Fourdrain Y, et al. 2004. Identification of an aspartylglucosaminidase-like protein in the venom of the parasitic wasp *Asobara tabida* (Hymenoptera: Braconidae). *Insect Biochem. Mol. Biol.* 34:485–92
64. Moreau SJM, Dingremont A, Doury G, Giordanengo P. 2002. Effects of parasitism by *Asobara tabida* (Hymenoptera: Braconidae) on the development, survival and activity of *Drosophila melanogaster* larvae. *J. Insect Physiol.* 48:337–47
65. Nakamatsu Y, Suzuki M, Harvey JA, Tanaka T. 2007. Regulation of the host nutritional milieu by ecto- and endoparasitoid venom. See Ref. 87a, pp. 37–56
66. Nakhasi HL, Pogue GP, Duncan RC, Joshi M, Atreya CD, et al. 1998. Implication of calreticulin function in parasite biology. *Parasitol. Today* 14:157–60
67. Nappi AJ, Christensen BM. 2005. Melanogenesis and associated cytotoxic reactions: applications to insect innate immunity. *Insect Biochem. Mol. Biol.* 35:443–59
68. Nappi AJ, Vass E. 1993. Melanogenesis and the generation of cytotoxic molecules during insect cellular immune reactions. *Pigment Cell Res.* 6:117–26
69. Parkinson N, Conyers C, Smith I. 2002. A venom protein from the endoparasitoid wasp *Pimpla hypochondriaca* is similar to snake venom reprolysin-type metalloproteases. *J. Invertebr. Pathol.* 79:129–31
70. **Parkinson N, Richards EH, Conyers C, Smith I, Edwards JP. 2002. Analysis of venom constituents from the parasitoid wasp *Pimpla hypochondriaca* and cloning of a cDNA encoding a venom protein. *Insect Biochem. Mol. Biol.* 32:729–35**

70. The first comprehensive analysis of venom components from an endoparasitoid wasp.

71. Parkinson N, Smith I, Audsley N, Edwards JP. 2002. Purification of pimplin, a paralytic heterodimeric polypeptide from venom of the parasitoid wasp *Pimpla hypochondriaca*, and cloning of the cDNA encoding one of the subunits. *Insect Biochem. Mol. Biol.* 32:1769–73
72. Parkinson N, Smith I, Weaver R, Edwards JP. 2001. A new form of arthropod phenoloxidase is abundant in venom of the parasitoid wasp *Pimpla hypochondriaca*. *Insect Biochem. Mol. Biol.* 31:57–63
73. Parkinson NM, Conyers C, Keen J, MacNicoll A, Smith I, et al. 2004. Towards a comprehensive view of the primary structure of venom proteins from the parasitoid wasp *Pimpla hypochondriaca*. *Insect Biochem. Mol. Biol.* 34:565–71
74. Parkinson NM, Conyers C, Keen JN, MacNicoll AD, Weaver ISR. 2003. cDNAs encoding large venom proteins from the parasitoid wasp *Pimpla hypochondriaca* identified by random sequence analysis. *Comp. Biochem. Physiol. C* 134:513–20
75. **Parkinson NM, Weaver RJ. 1999. Noxious components of venom from the pupa-specific parasitoid *Pimpla hypochondriaca*. *J. Invertebr. Pathol.* 73:74–83**

75. The first to demonstrate PO in venom from an endoparasitic wasp.

76. Pennacchio F, Strand MR. 2006. Evolution of developmental strategies in parasitic Hymenoptera. *Annu. Rev. Entomol.* 51:233–58
77. Pfister-Wilhelm R, Lanzrein B. 2009. Stage dependent influences of polydnaviruses and the parasitoid larvae on host ecdysteroids. *J. Insect Physiol.* 55:707–15
78. Price DRG, Bell HA, Hinchliffe G, Fitches E, Weaver R, Gatehouse JA. 2009. A venom metalloproteinase from the parasitic wasp *Eulophus pennicornis* is toxic towards its host, tomato moth (*Lacanobia oleracae*). *Insect Mol. Biol.* 18:195–202
79. Pruijssers AJ, Falabella P, Eum JH, Pennacchio F, Brown MR, Strand MR. 2009. Infection by a symbiotic polydnavirus induces wasting and inhibits metamorphosis of the moth *Pseudoplusia includens*. *J. Exp. Biol.* 212:2998–3006
80. Quicke DLJ. 1997. *Parasitic Wasps*. London: Chapman & Hall
81. Richards EH, Dani MP. 2007. Venom-induced apoptosis of insect hemocytes. See Ref. 87a, pp. 19–36
82. Richards EH, Dani MP. 2008. Biochemical isolation of an insect haemocyte anti-aggregation protein from the venom of the endoparasitic wasp, *Pimpla hypochondriaca*, and identification of its gene. *J. Insect Physiol.* 54:1041–19
83. Richards EH, Edwards JP. 2000. Parasitism of *Lacanobia oleracea* (Lepidoptera) by the ectoparasitoid, *Eulophus pennicornis*, is associated with a reduction in host haemolymph phenoloxidase activity. *Comp. Biochem. Physiol. B* 127:289–98
84. Richards EH, Parkinson NM. 2000. Venom from the endoparasitic wasp *Pimpla hypochondriaca* adversely affects the morphology, viability, and immune function of hemocytes from larvae of the tomato moth, *Lacanobia oleracea*. *J. Invertebr. Pathol.* 76:33–42
85. Rivers DB, Brogan A. 2008. Venom glands from the ectoparasitoid *Nasonia vitripennis* (Walker) (Hymenoptera: Pteromalidae) produce a calreticulin-like protein that functions in developmental arrest and cell death in the flesh fly host, *Sarcophaga bullata* Parker (Diptera: Sarcophagidae). In *Insect Physiology: New Research*, ed. RP Maes, pp. 259–78. New York: Nova Sci.
86. Rivers DB, Dani MP, Richards EH. 2009. The mode of action of venom from the endoparasitic wasp *Pimpla hypochondriaca* (Hymenoptera: Ichneumonidae) involves Ca^{+2}-dependent cell death pathways. *Arch. Insect Biochem. Physiol.* 71:173–90
87. **Rivers DB, Ruggiero L, Hayes M. 2002. The ectoparasitic wasp *Nasonia vitripennis* (Walker) (Hymenoptera: Pteromalidae) differentially affects cells mediating the immune response of its flesh fly host, *Sarcophaga bullata* Parker (Diptera: Sarcophagidae). *J. Insect Physiol.* 48:1053–64**

87. Demonstrates that venom from ectoparasitoids could also affect the host immune system.

87a. Rivers DB, Yoder JA, eds. 2007. *Recent Advances in the Biochemistry, Toxicity, and Mode of Action of Parasitic Wasp Venoms*. Kerala, India: Res. Signpost
88. **Rizki RM, Rizki TM. 1990. Encapsulation of parasitoid eggs in phenoloxidase-deficient mutants of *Drosophila melanogaster*. *J. Insect Physiol.* 36:523–29**

88. Demonstrates that PO is not required for the encapsulation response.

89. Rizki RM, Rizki TM. 1990. Parasitoid virus-like particles destroy *Drosophila* cellular immunity. *Proc. Natl. Acad. Sci. USA* 87:8388–92
90. Schmidt JO, Blum MS, Overall WL. 1986. Comparative enzymology of venoms from stinging hymenoptera. *Toxicon* 24:907–21
91. Seals DF, Courtneidge SA. 2003. The ADAMs family of metalloproteases: multidomain proteins with multiple functions. *Genes Dev.* 17:7–30
92. Serrano SMT, Kim J, Wang DY, Dragulev B, Shannon JD, et al. 2006. The cysteine-rich domain of snake venom metalloproteinases is a ligand for von Willebrand factor A domains—Role in substrate targeting. *J. Biol. Chem.* 281:39746–56
93. Sotillos S, Roch F, Campuzano S. 1997. The metalloprotease-disintegrin Kuzbanian participates in *Notch* activation during growth and patterning of *Drosophila* imaginal discs. *Development* 124:4769–79
94. **Stoltz DB, Guzo D, Belland ER, Lucarotti CJ, MacKinnon EA. 1988. Venom promotes uncoating in vitro and persistence in vivo of DNA from a braconid polydnavirus. *J. Gen. Virol.* 69:903–7**

94. First report showing the role of venom in polydnavirus infection of host cells at the cellular level.

95. Suzuki M, Miura K, Tanaka T. 2008. The virus-like particles of a braconid endoparasitoid wasp, *Meteorus pulchricornis*, inhibit hemocyte spreading in its noctuid host, *Pseudaletia separata*. *J. Insect Physiol.* 54:1015–22

96. Suzuki M, Miura K, Tanaka T. 2009. Effects of the virus-like particles of a braconid endoparasitoid, *Meteorus pulchricornis*, on hemocytes and hematopoietic organs of its noctuid host, *Pseudaletia separata*. *Appl. Entomol. Zool.* 44:115–25
97. Suzuki M, Tanaka T. 2006. Virus-like particles in venom of *Meteorus pulchricornis* induce host hemocyte apoptosis. *J. Insect Physiol.* 52:602–13
98. Tanaka T. 1987. Effect of the venom of the endoparasitoid *Apanteles kariyai* Watanabe, on the cellular defence reaction of the host, *Pseudaletia separata* Walker. *J. Insect Physiol.* 33:413–20
99. Teramoto T, Tanaka T. 2004. Mechanism of reduction in the number of the circulating hemocytes in the *Pseudaletia separata* host parasitized by *Cotesia kariyai*. *J. Insect Physiol.* 50:1103–11
100. Thomas P, Yamada R, Johnson KN, Asgari S. 2010. Ectopic expression of an endoparasitic wasp venom protein in *Drosophila melanogaster* affects immune function, larval development and oviposition. *Insect Mol. Biol.* 19:473–80
101. Thompson SN. 1993. Redirection of host metabolism and effects on parasites nutrition. In *Parasites and Pathogens of Insects*, ed. NE Beckage, SN Thompson, BA Federici, pp. 125–44. San Diego: Academic
102. Uçkan F, Ergin E, Rivers DB, Gençer N. 2006. Age and diet influence the composition of venom from the endoparasitic wasp *Pimpla turionellae* L. (Hymenoptera: Ichneumonidae). *Arch. Insect Biochem. Physiol.* 63:177–87
103. Uçkan F, Sinan S, Savasci S, Ergin E. 2004. Determination of venom components from the endoparasitoid wasp *Pimpla turionellae* L. (Hymenoptera: Ichneumonidae). *Ann. Entomol. Soc. Am.* 97:775–80
104. Vinson SB. 1990. How parasitoids deal with the immune system of their host: an overview. *Arch. Insect Biochem. Physiol.* 13:3–27
105. Vinson SB, Iwantsch GF. 1980. Host regulation by insect parasitoids. *Q. Rev. Biol.* 55:143–65
106. Webb BA, Luckhart S. 1994. Evidence for an early immunosuppressive role for related *Campoletis sonorensis* venom and ovarian proteins in *Heliothis virescens*. *Arch. Insect Biochem. Physiol.* 26:147–63
107. Webb BA, Summers MD. 1990. Venom and viral expression products of the endoparasitic wasp *Campoletis sonorensis* share epitopes and related sequences. *Proc. Natl. Acad. Sci. USA* 87:4961–65
108. Wu M-L, Ye G-y, Zhu Jy, Chen X-X, Hu C. 2008. Isolation and characterization of an immunosuppressive protein from venom of the pupa-specific endoparasitoid *Pteromalus puparum*. *J. Invertebr. Pathol.* 99:186–91
109. Yao L, Pike SE, Tosato G. 2002. Laminin binding to the calreticulin fragment vasostatin regulates endothelial cell function. *J. Leukoc. Biol.* 71:47–53
110. Ye G-Y, Zhu J, Zhang Z, Fang Q, Cai J, Hu C. 2007. Venom from the endoparasitoid *Pteromalus puparum* (Hymenoptera: Pteromalidae) adversely affects host hemocytes: differential toxicity and microstructural and ultrastructural changes in plasmatocytes and granular cells. See Ref. 87a, pp. 115–28
111. Zhang G, Lu Z-Q, Jiang H, Asgari S. 2004. Negative regulation of prophenoloxidase (proPO) activation by a clip-domain serine proteinase homolog (SPH) from endoparasitoid venom. *Insect Biochem. Mol. Biol.* 34:477–83
112. Zhang G, Schmidt O, Asgari S. 2004. A novel venom peptide from an endoparasitoid wasp is required for expression of polydnavirus genes in host hemocytes. *J. Biol. Chem.* 279:41580–85
113. Zhang G, Schmidt O, Asgari S. 2006. A calreticulin-like protein from endoparasitoid venom fluid is involved in host hemocyte inactivation. *Dev. Comp. Immunol.* 30:756–64
114. Zhang H, Forman HJ, Choi J. 2005. Gamma-glutamyl transpeptidase in glutathione biosynthesis. *Methods Enzymol.* 401:468–83
115. Zhu J-Y, Ye GY, Hu C. 2008. Molecular cloning and characterization of acid phosphatase in venom of the endoparasitoid wasp *Pteromalus puparum* (Hymenoptera: Pteromalidae). *Toxicon* 51:1391–99

Recent Insights from Radar Studies of Insect Flight

Jason W. Chapman,[1] V. Alistair Drake,[2,3] and Don R. Reynolds[1,4]

[1]Plant and Invertebrate Ecology Department, Rothamsted Research, Harpenden, Hertfordshire AL5 2JQ, United Kingdom; email: jason.chapman@bbsrc.ac.uk

[2]School of Physical, Environmental and Mathematical Sciences, The University of New South Wales at the Australian Defence Force Academy, Canberra, ACT 2600, Australia; email: a.drake@adfa.edu.au

[3]Institute of Applied Ecology, University of Canberra, ACT 2601, Australia

[4]Natural Resources Institute, University of Greenwich, Chatham, Kent ME4 4TB, United Kingdom; email: d.reynolds@greenwich.ac.uk

Annu. Rev. Entomol. 2011. 56:337–56

The *Annual Review of Entomology* is online at ento.annualreviews.org

This article's doi:
10.1146/annurev-ento-120709-144820

0066-4170/11/0107-0337$20.00

Key Words

foraging, migration, insect-monitoring radar, harmonic radar, orientation, Lévy flights

Abstract

Radar has been used to study insects in flight for over 40 years and has helped to establish the ubiquity of several migration phenomena: dawn, morning, and dusk takeoffs; approximate downwind transport; concentration at wind convergences; layers in stable nighttime atmospheres; and nocturnal common orientation. Two novel radar designs introduced in the late 1990s have significantly enhanced observing capabilities. Radar-based research now encompasses foraging as well as migration and is increasingly focused on flight behavior and the environmental cues influencing it. Migrant moths have been shown to employ sophisticated orientation and height-selection strategies that maximize displacements in seasonally appropriate directions; they appear to have an internal compass and to respond to turbulence features in the airflow. Tracks of foraging insects demonstrate compensation for wind drift and use of optimal search paths to locate resources. Further improvements to observing capabilities, and employment in operational as well as research roles, appear feasible.

INTRODUCTION

Flight boundary layer (FBL): a layer of the atmosphere, usually close to the ground, where the wind is light enough for the insect to make progress in any direction

VLR: vertical-looking radar

The predominant and most significant mode of transport among insects is powered flight in the adult stage (38). Although there are some notable exceptions to this generalization, e.g., ballooning Lepidoptera larvae (6) and marching bands of Mormon crickets, *Anabrus simplex* (106), the great majority of long-range insect movements are accomplished by flight. Insects fly for many reasons, but one useful and rather general distinction that can be made is between station-keeping movements, in which the individual remains within its current habitat patch, and movements that take it away, permanently or for an extended period, from this home range (27, 52). Probably, the most common form of station-keeping activity is foraging, in which movements are directed toward the acquisition and exploitation of resources (e.g., host plants, shelter, or mates). Such movements, e.g., odor-mediated anemotaxis in mate-searching male moths (10), cease when appetitive cues associated with the sought-after resource are encountered; they produce characteristically erratic flight trajectories made up of numerous short stages in different directions. Foraging insects need to maneuver precisely; this requires good control of their track, and thus foraging flights typically occur close to the vegetation canopy, within the insect's flight boundary layer (FBL) (107).

Movements that take an individual beyond its home range can be further divided into ranging and migration. Ranging is movement over an area in order to explore it and locate a new home range; like foraging, it stops when the sought after resource is found. Migration, in contrast, entails a temporary inhibition of responsiveness toward the appetitive cues that would otherwise arrest the individual's motion (27, 52). Migratory movement is sustained, characteristically undistracted, and produces rectilinear tracks; its function is to relocate the individual to a habitat that is (or will soon be) better endowed with resources than the current one. For many insect species, the most efficient way to achieve sustained (i.e., long-distance) movement is to utilize the fast winds above the insect's FBL (33). Consequently, migrating insects typically fly at high altitudes, sometimes as high as 2 or 3 km above the ground (11, 44). This strategy would seemingly often result in displacement in unfavorable directions, however, and indeed some large day-flying species (e.g., many migratory butterflies and dragonflies) migrate predominantly within their FBL (103) and maintain considerable control over their movement direction, though at the cost of significantly increased journey durations.

Movement plays a central role in all facets of the population dynamics, ecology, and evolution of insects, and thus knowledge of insect flight strategies should underpin many entomological disciplines. A recent theoretical framework (62) proposed that the primary challenge for movement ecologists is the characterization of the external factors, internal states, motion capabilities, and navigation capacities of their study organisms. These are difficult challenges indeed for small, flying organisms such as insects—even if studies are restricted to foraging flights—as all but the shortest flights quickly take the insect out of visual range. The scale of the problem increases dramatically in the case of long-range migratory flights, many of which take place at heights beyond the reach even of powerful optical devices.

The remote-sensing capability of radar provides a solution to the seemingly intractable difficulty of collecting ecologically relevant movement information from distant flying insects, and entomological radars have made a huge contribution to several distinct areas of movement research (16, 17, 31, 35, 84). During the 1970s and 1980s, the application of scanning radars to the study of high-altitude insect migration resulted in numerous insights into the migration ecology of a range of agricultural pests (81). The development of two new entomological radar techniques in the 1990s, namely vertical-looking radar (VLR) and harmonic radar, ushered in a new era of radar entomology in which descriptive observation gave way to hypothesis-driven analysis and even some experimentation. A synthesis of the important findings emerging from the last 10 years of radar entomology studies is thus timely. Developments in radar technology continue, and the new opportunities that these developments provide for entomologists are also outlined in this review.

EARLY ENTOMOLOGICAL RADARS AND THEIR CONTRIBUTION TO MIGRATION STUDIES

Insects were first positively identified as a source of radar echoes in 1949 (24), and the detection of a locust swarm in 1954 alerted entomologists to the technology's potential for their discipline (68). However, it was not until 1968, when G.W. Schaefer deployed a radar designed specifically for locust observation to the Sahara, that findings of real entomological value were generated (97). Most early entomological radars were of the scanning pencil-beam type, in which the beam is directed upward at a series of elevation angles and turned steadily in azimuth. Targets appear as bright marks (paints) on a display known as a plan-position indicator (PPI), and by following the paints over a series of scans, it is possible to estimate target speed and direction (97). Observations have been made mostly with units transmitting at a wavelength of 3.2 cm using components from marine radars, but versions operating at 9 mm have been produced for the purpose of studying smaller species such as planthoppers (85).

A number of other radar designs were trialed in the 1970s and 1980s (81), but only an aircraft-mounted type with a beam that pointed directly downward saw much use (45, 116). The speed of the aircraft enabled transects to be obtained, allowing the lateral extent of a migration to be determined. A study in the southern United States, for example, showed that a cloud of moths originating from a cropping region remained extant for over 400 km as it drifted downwind (116). Despite such successes, no airborne entomological radar survives; and since the early 1990s, use of scanning radars has also declined. The main reason is that radar entomologists have increasingly recognized the need for long-term datasets and have sought methods that are less labor-intensive in order to acquire these at practicable cost. The new VLRs also provide more precise information about target identity and this has further encouraged the switch away from scanning systems. However, scanning observations have continued in China (42), where the radars are now partially automated and PPI images are acquired and analyzed digitally (22).

PPI: plan position indicator

The early scanning radars were used mainly to study the migrations of pest species, including grasshoppers and locusts; noctuid, pyralid, and tortricid moths; and planthoppers. The principal findings have been

- that migratory flight occurs by day and by night, with the former commencing around midmorning (69) and the latter with a mass takeoff at dusk (32, 89);
- that migrants regularly attain heights of a few hundred meters and flight at 1 km is not unusual (32, 83, 89);
- that movement is almost always approximately downwind (32, 37);
- that during daytime, when the atmosphere is typically convective (i.e., mixed by thermally driven updrafts and downdrafts), insects become concentrated in plumes of rising air (69);
- that in the stable atmosphere, which typically forms at night over land, insects are distributed nearly uniformly in the horizontal but frequently concentrate around particular heights (i.e., form layers) in response to features of the temperature or wind profile (30, 37, 80);
- that in a stable atmosphere most larger insects exhibit some degree of mutual alignment, often in a direction different to that in which they are moving (29, 83, 89);
- and that the steady nighttime migrations in stable conditions are sometimes disrupted by atmospheric disturbances, especially sea breezes, thunderstorm outflows, and flows down sloping terrain, that not only change the direction of a

movement but also gather the migrants into line concentrations that propagate across the landscape (28, 90).

These features of insect migratory flight have continued to be observed as scanning-radar studies have been undertaken on further species and in additional regions (5, 42), and appear universal.

The contribution of radar to the study of a particular species has often been to establish that high-altitude migrations occur, that they do so frequently, and that a significant proportion of the population is involved (e.g., 19, 32, 40, 45, 81). Radar observations provide various measures of the intensity of a migration, such as the cumulative number of migrants passing per unit length of a line drawn in whatever direction the investigator chooses (32, 80). Comparison of this number with infestation levels in the destination region allows the ecological significance of the movement to be determined. Multidisciplinary programs with radar observations occupying a central role have contributed to the recognition of the migratory circuits of several pest species, both in temperate regions (42) and in the semiarid subtropics (26, 32, 80).

NEW-GENERATION ENTOMOLOGICAL RADARS

Vertical-Looking Entomological Radar

Since the late 1990s, most radar observations of insect migration have been made with a design in which echoes are recorded electronically as targets pass through a stationary upward-pointing beam. Each echo shows a gradual rise and then fall, over a period of a few seconds, as the insect causing it traverses the beam. Echoes are partitioned by range, which corresponds to height in this configuration, so profiles of insect numbers are accumulated. The number of traverses in each range bin is readily determined by computer analysis of the digitized recordings, and a migration rate for the bin altitude can be estimated from this. Year-long datasets were first achieved with a radar of this type operating in Texas in 1990 (4). The speed of the target can be determined from the rate of the rise and the fall, and for larger species the wing-beat frequency can be extracted by Fourier analysis (36).

The most sophisticated variant of this design (16, 21) incorporates target interrogation, in which the polarization of the transmitted wave is rotated continuously and, synchronously, the beam is nutated (wobbled slightly) to produce an additional time variation in the echo. The now rather complicated form of the echo contains information about the target's movement, heading direction, its size and its shape, and analyses have been developed to extract all these parameters (48, 100). The ability to discriminate between targets of different size, shape, and wing-beat frequency, and thus to accumulate samples of track directions, headings, and speeds for specific target classes, is unique to this radar design and enables significantly more precise biological analyses.

VLRs operate automatically, under microcomputer control, so it is practicable to install them in locations—perhaps far from the researcher's home laboratory—where insect activity of particular interest is most likely to occur (**Figure 1*a***). Datasets extending over several years, and with millions of individual traverses recorded, are now available for sites in the United Kingdom and Australia. These have provided unprecedented opportunities for empirical, large-sample studies of insect migratory behavior (see below). The Australian radars are also developing an operational role, as their outputs are routinely drawn upon by the Australian Plague Locust Commission in support of its operations aimed at forecasting and managing locust outbreaks.

Harmonic Radar

Harmonic radars rely on a rectifier circuit (which in biological applications is incorporated into a tag worn by an individual animal that is to be tracked) that generates an echo with exactly half the wavelength of the transmitted wave. As there are few naturally occurring

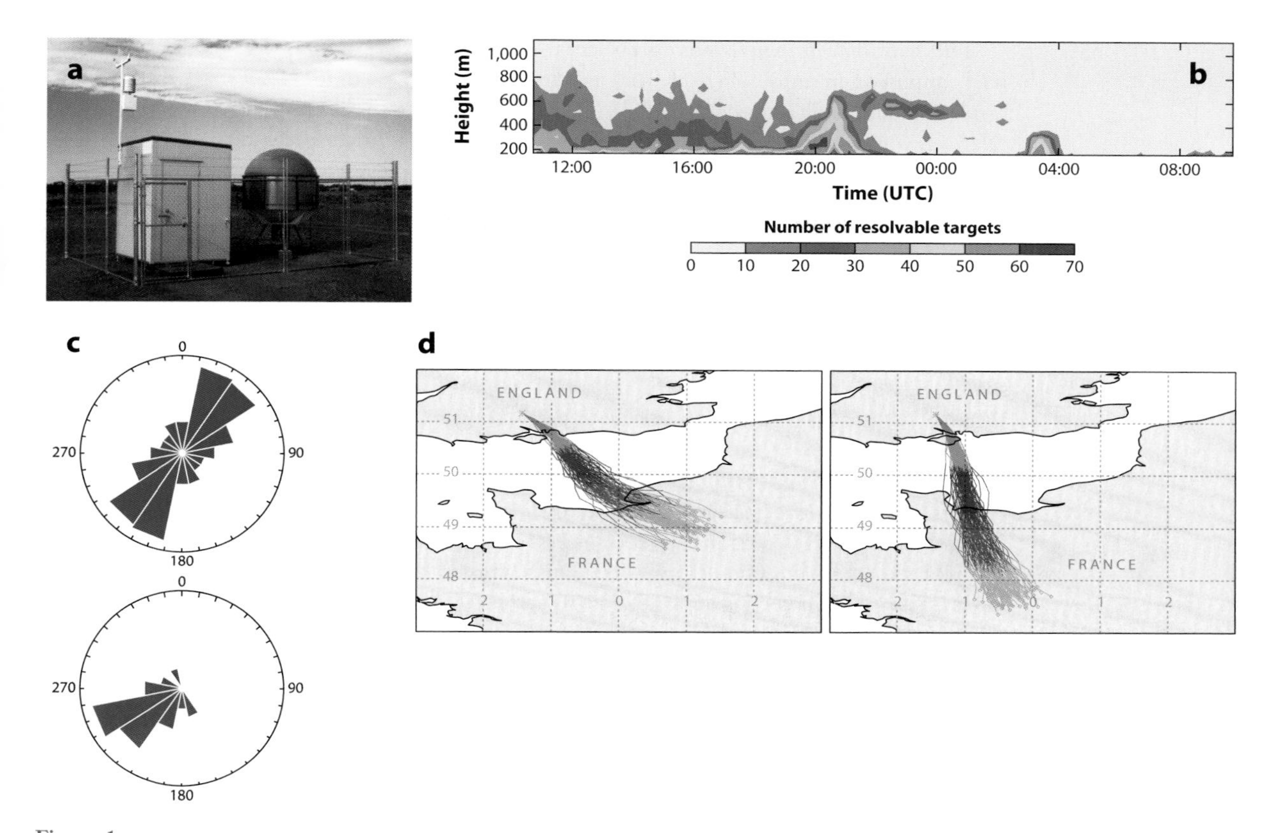

Figure 1

(*a*) A vertical-looking entomological radar (VLR) for automatic monitoring of high-altitude insect migration. This unit is situated at Thargomindah, in the remote inland of Queensland, Australia (photo courtesy of V.A. Drake). (*b*) Time/height plot of insect numbers recorded by a VLR at Malvern, England, on June, 25–26, 2003. The color scale bar refers to the number of resolvable insect targets detected at each sampling height during each 5-min sampling period. The figure shows many of the typical features in the vertical profile, namely, the extended daytime flight period and the activity peaks at dusk and dawn. On this occasion, an elevated layer formed after 2200 h and persisted until dawn. Modified with permission from Reference 82. (*c*) Circular distributions of (*top*) body alignment and (*bottom*) displacement direction of insects recorded by a VLR during a single night's migratory flight of *Autographa gamma* moths over the United Kingdom. Although there is a 180° ambiguity in the body alignment directions, other evidence shows that the migrants were orienting themselves toward the southwest, close to the direction of their windborne displacement. (*d*) Simulated 8-h migration trajectories for (*left*) 100 inert particles and (*right*) 100 *A. gamma* moths released from a VLR site in southern England on August 10, 2006. The different colors represent successive 2-h sections of trajectory from 2000 to 0400 coordinated universal time (UTC). The coastlines of southern England and northern France and lines of latitude and longitude are shown. The effect of flight heading and altitude selection by the moths produces improvements in both distance covered and direction (i.e., moth trajectories were 24° closer to the seasonally preferred direction of 180°). Modified with permission from Reference 12.

rectifiers, a receiver operating at the harmonic wavelength detects the echo from the transponder without being swamped by echoes from terrain features and vegetation. The particular advantage of harmonic tags for entomological applications is that they operate passively and therefore do not require a battery; this means that, with modern microelectronic components, they can be made light enough to be carried by medium-sized insects such as honey bees (*Apis mellifera*) and noctuid moths without any obvious effects on behavior. Their main limitation is that the returned signal contains no identification information. Another constraint is that terrain and vegetation, although not sources of echo, still act as barriers and create shadow areas within which tags are undetectable.

The harmonic principle was first successfully employed in entomological work in the form of handheld direction finders (54). These relatively low-cost devices allow tagged insects to be relocated once or twice a day; individuals are then identified by reading a number on the tag. Harmonic direction finders continue to be used in studies of the lifetime tracks of individuals through their habitat (115). A harmonic radar, in contrast, measures ranges as well as direction. The unit developed for entomological applications in the early 1990s (92) (**Figure 2*a***) is similar in many respects to the conventional scanning entomological radars then in use: It transmits with a wavelength of 32 mm, has a narrow beam (and so locates the target quite precisely), completes a scan in 3 s, and presents its output on a PPI display. This radar is capable of detecting transponders to a range of around 900 m; its spatial resolution is ~5 m. The signals returned by the tags have a wavelength of 16 mm, and to match this optimally the transponder antennas, which are constructed from fine wire with the diode and an inductive loop at the midpoint, are 16 mm long. They weigh 1–12 mg and are mounted dorsally, pointing upward (**Figure 2*b***); this provides reasonable performance and minimizes obstruction of the insect's normal activities.

Development of a true harmonic radar is technically demanding and until recently only a single unit has been available. In addition, the lack of identification means that only one or two insects can be followed at a time, and because of the shadowing problem observations have been possible only in flat and relatively open landscapes. Nevertheless, a range of species have been studied, both by observing natural behaviors and through manipulative experimentation.

RADAR STUDIES OF WINDBORNE INSECT MIGRANTS

Long-range windborne migration, the process by which migrants ascend above their FBL to be transported downwind, occurs in all the major orders of insects and has evidently evolved many times (11, 44). Migration allows species to exploit temporary breeding habitats and escape adverse seasonal conditions, and provides the potential to breed continuously throughout the year. Utilization of fast, high-altitude air currents enables individuals to cover enormous distances (hundreds or even thousands of kilometers) with just a few days or nights of flight. These strategies often benefit migrants relative to less mobile species, and consequently many migrant species have become hugely abundant and often these are now important pests of crops (34) or vectors of animal or plant diseases (79). Windborne insect migration occurs on a colossal scale, far exceeding (at least in numerical terms) the migratory flux of birds (12, 16, 46, 50). Yet in comparison to birds we know relatively little about the behavioral adaptations of insects that undertake long-range movements (1, 2, 50). The new VLRs, however, have opened a window onto this previously inaccessible realm and have already produced some important findings. Concurrent work with scanning radar in China has also contributed and is also summarized here.

Diurnal and Seasonal Patterns of Migratory Activity

Over ten years of continuous VLR observations in southern England, in conjunction with a dedicated aerial-sampling program (20), have revealed that even in the comparatively cool Atlantic climate of the United Kingdom the magnitude of the migratory flux is immense. A bioflow of 3 billion insects crossing any 1-km stretch of the southern English countryside has been estimated for a typical summer month (16). This figure exceeds the total number of migratory passerine birds involved in the entire Palaearctic-African migration system (46)! The radars have shown that numbers aloft increase dramatically as temperature rises, and have not indicated any reduction in numbers at the highest temperatures experienced in a typical British summer (21, 99, 118).

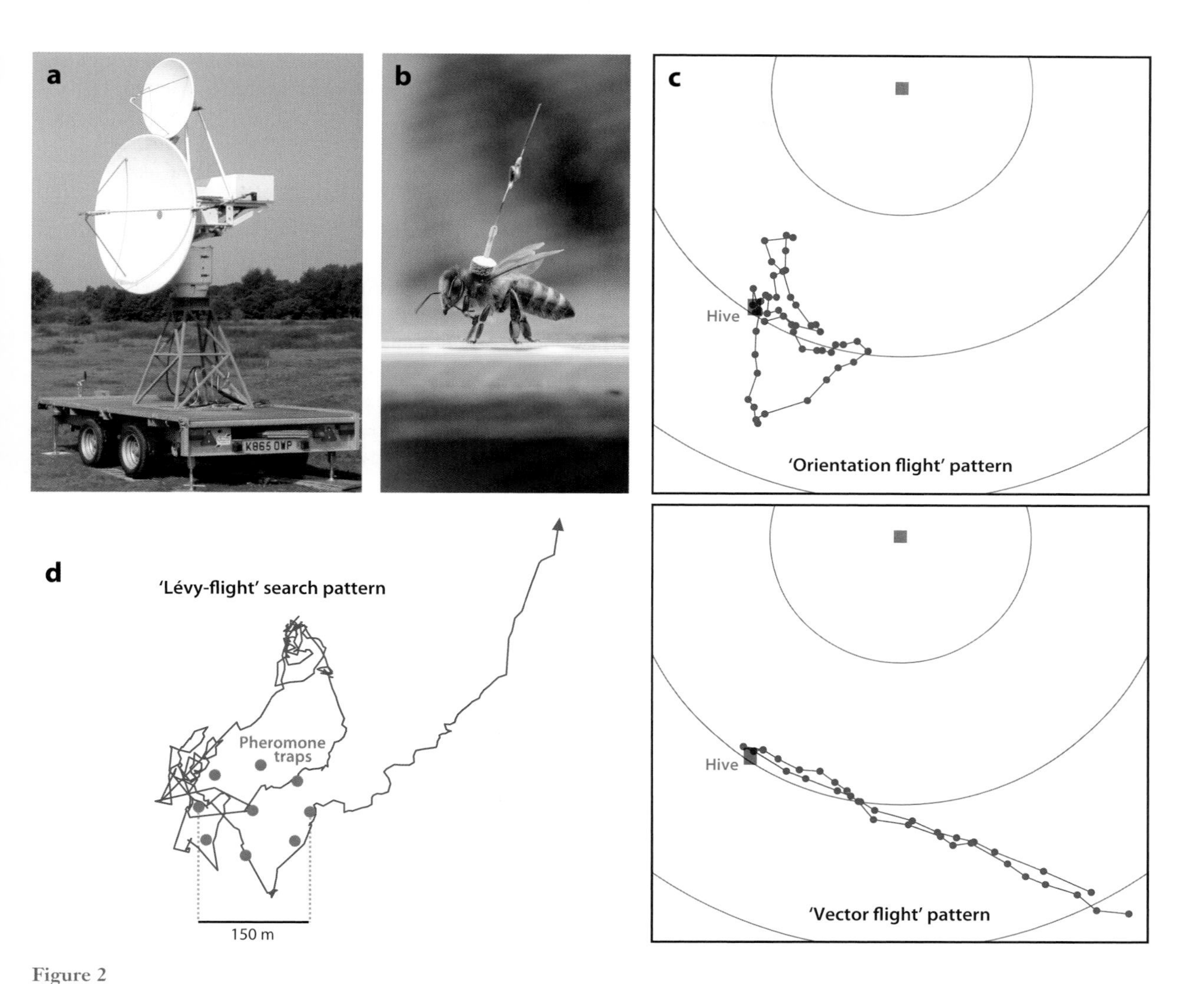

Figure 2

(*a*) The scanning harmonic radar for following individual low-flying insects tagged with transponders (photo courtesy of A.D. Smith). (*b*) Honey bee wearing a radar transponder (photo courtesy of M. Garcia-Alonso). (*c*) Two examples of honey bee flight tracks: (*top*) an orientation flight by a young (6-day-old) worker bee; (*bottom*) part of a vector flight to a forage patch from which the bee returned with a pollen load. The latter flight goes out of range before returning along the same path. The hive is marked by a red square; the radar range rings are at 116-m intervals. Modified with permission from Reference 9. (*d*) A Lévy-flight search pattern recorded from a male *Agrotis segetum* moth during a pheromone-confusion trial. The ring of pheromone traps (*dark yellow dots*) is 150 m across. Modified with permission from Reference 73.

The diurnal pattern of insect flight activity over England is highly consistent (16, 21, 82, 99, 120). There is generally a discrete take-off of crepuscular species during the dawn twilight period (**Figure 1*b***); this flight tends to be short-lived, although occasionally it continues for some time (82). Day-flying migrants take off from mid-morning onward, as atmospheric convection develops, and generally descend in the late afternoon (**Figure 1*b***). Occasionally, small numbers of day-flying species continue their migration into the night (20). However, nocturnal migration is typically initiated by a mass takeoff at dusk (**Figure 1*b***), stimulated by falling light levels, and the nocturnal migrants usually fly throughout the night, sometimes concentrated into narrow altitudinal layers (16, 78, 82, 117, 120). There is almost no overlap between these three phases of migratory activity, which, at least in the United Kingdom, involve different taxa. By day, microinsects such as cereal aphids and their hymenopteran parasitoids—which are all too small to be detected by the VLRs—predominate but large species including carabid beetles and hoverflies are also abundant (19, 20). The dusk peak is composed partly of crepuscular insects such as lacewings and small Diptera that fly for a relatively short period of an hour or two around sunset (13, 20), and partly of the nocturnal migrant fauna that flies throughout the night. The dominant large nocturnal migrants are generally noctuid moths (12, 14, 15, 30, 40, 42, 43, 120), with grasshoppers and locusts also abundant in the semiarid subtropics (32, 35, 80, 89, 90).

Atmospheric boundary layer (ABL): the bottommost 1–2 km of the atmosphere, within which the air is slowed by surface friction; usually it is mixed by thermal updrafts and downdrafts by day but is stable (though turbulent) by night

Migration is associated with spatiotemporal habitat variation, which in temperate regions is driven primarily by the (relatively) regular and predictable seasonal pattern of temperatures, whereas in warmer semiarid and arid regions it is usually rainfall that is limiting and this is often erratic. In the former case, insect migrations in spring and early summer are toward higher latitudes, with populations tracking the poleward extension of the growing season of their plant hosts; a reverse movement occurs in autumn. Radar has helped to document migration circuits of this type and to identify behavioral adaptations that facilitate the migrations for several moth species that are serious agricultural pests in North America and China, including *Helicoverpa zea* (112), *H. armigera* (42), *Spodoptera exigua* (40), and *Mythimna separata* (43). In Australia, VLR observations revealed that northward movements of Australian plague locusts, *Chortoicetes terminifera*, were more frequent and intense than had previously been realized, leading to recognition of a semiregular seasonal to-and-fro movement between winter and summer rainfall regions at temperate and subtropical latitudes, respectively (26, 35).

Selection of Flight Altitude and Layer Concentrations

An hour or so after the dusk emigration, the vertical profile of insect aerial density usually assumes a relatively stable state where (in the absence of disturbances) changes occur only gradually. These midevening profiles, when the aerial population has come into equilibrium with the atmospheric boundary layer (ABL), may assume various forms, but one of the most common is the concentration of the migrants into layers of shallow depth (~100–150 m) but broad horizontal extent (30, 33, 44, 71) (**Figure 1*b***). Early radar observations established that nocturnal insect layers are a frequent phenomenon in subtropical regions, and more recently it has been demonstrated that they also often occur in temperate regions (40, 41, 42, 78, 117, 118, 120). The vertical stratification of insect aerial density is often paralleled (and presumably influenced) by the stratified nature of the nocturnal ABL. In fine weather, a nocturnal temperature inversion usually forms in the first few hundred meters above the ground, and insect layers frequently occur near the inversion top where warm, fast-moving airflows provide optimal conditions for long-distance transport. In the relatively cool nighttime conditions of temperate regions, layers are often confined to the warmest zone of the atmosphere because the cooler airstreams above or below are unsuitable for insect flight (78, 117, 118). When air temperatures are warmer, however, migrants can select their flight altitude, and moths in particular are often found at the altitude where the air is moving fastest, thereby maximizing their nightly displacement (12, 14, 15, 117).

The cues and sensory mechanisms involved in layer formation may be similar to those used to achieve orientation and are discussed further below. One recent suggestion that is specific to layering is that a turbophoretic mechanism tends to advect migrants into regions with low turbulence, which typically coincide with zones of maximum air temperature and/or wind speed (71). Accumulation in layers through passive turbophoresis would be very slow, but if the migrants can amplify the effect by actively initiating ascents or descents in response to turbulent vertical air motions, then concentration into layers that match those regularly observed appears plausible (71, 72).

In fair weather over land, thermal convection is usually well developed by midmorning, and the resulting up- and downdrafts often become organized into systems of cells. Insects tend to be concentrated into regions of convective uplift typically ~0.5–1 km across (69, 97) that move along with the large-scale airflow. Daytime migration is thus highly heterogeneous, at least on the small scale, and time-height profiles from VLRs show this process well (120) (**Figure 1*b***). In contrast to the nighttime situation, daytime layers are reported only occasionally (21, 51, 99). They may originate from the dawn takeoff, are usually found in the stable air above the convecting region of the atmosphere, and are liable to be disrupted later in the day as the convective plumes extend upward (82).

Common Orientation in Relation to the Wind

Perhaps the most surprising finding to emerge from the earliest radar entomology observing campaigns was that the body alignments (i.e., flight headings) of large nocturnal migrants, such as grasshoppers and noctuid moths, often showed a high degree of common orientation (83, 97) (**Figure 1*c***). Prior to this discovery, it had been assumed that nocturnal insects flying at high altitudes would orient themselves at random due to the presumed lack of environmental cues to which they could respond (72). Radar studies have since documented the ubiquity of common orientation among large nocturnal migrants and have shown that they often maintain headings that are relatively close to the downwind direction (29, 41, 72, 80, 81, 90, 97, 117). Downwind orientation is potentially highly adaptive for large insects, as addition of their air speed to the wind speed significantly increases their displacement distance in a given time.

The question of how nocturnal migrants flying at altitudes of several hundreds of meters, and under low illumination levels, can select wind-related flight headings has proved difficult to resolve. Determination of the wind direction by visual perception of their movement direction relative to ground features (a type of optomotor reaction) seems an exceptionally challenging task given the low rates of optic flow and the dim nighttime illumination (88), although some nocturnal insects have extremely sensitive visual capacities (88, 108, 109). It seems unlikely that a visually mediated mechanism can explain all the observed patterns of nocturnal common orientation, but a wind-mediated mechanism similar to that already considered for layer formation (see above) may be able to do so (72), and a common mechanism for these two phenomena might account for the frequency with which they occur together.

How an actively flying insect could distinguish turbulent accelerations from those produced by its wing-beating is unclear. There is growing evidence that insects' antennae, via their action on Johnston's organs, play important roles in the stabilization of flight (96), regulation of migratory activity (58), and detection of extremely subtle environmental cues (110, 121); that the antennae might be capable of detecting faint turbulent flows is therefore at least plausible. An intriguing prediction from the proposed model is that nocturnal migrants attempting to orient themselves downwind will be misled by the change of wind direction with altitude (the Ekman spiral) that occurs in the boundary layer and will show a consistent offset to the right of the wind direction in the Northern Hemisphere and to the left in the Southern Hemisphere. Radar observations of medium-sized insects migrating over England at night have demonstrated a highly significant bias in their offsets that is indeed to the right of the downwind direction (72); an obvious test of the theory is therefore to determine whether similar migrants in the Southern

Hemisphere exhibit a corresponding bias to the left.

Selection of Seasonally Beneficial Tailwinds and Optimal Migration Trajectories

Many species of insect undertake seasonal migrations to higher latitudes to take advantage of temporary breeding habitats, and then escape the onset of harsh winter climates by returning to lower latitudes in the fall (44). To cover the huge distances required within the short developmental window (between assuming the fully flight-capable adult form and sexual maturation) available to most species, insects predominantly make use of the fast airstreams at higher altitudes. Simple reliance on windborne transport would seemingly often lead to displacement in highly disadvantageous directions and could cause the progeny of spring migrants to become trapped at high latitudes at summer's end—the so-called Pied Piper phenomenon (55). However, large-scale return migrations of windborne migrants are now well documented (12, 14, 40, 42, 59, 98). As wind speeds aloft are typically faster than insect airspeeds, a mechanism that ensures regular transport in seasonally beneficial directions would be highly adaptive. Initiating migratory flights only on occasions when the wind is blowing in a broadly favorable direction, and then letting the wind do the rest, appears a good strategy, especially for weak flyers, but it is dependent on a means for identifying suitable winds. In some species, return migrations in the fall are achieved by responding to particular meteorological conditions associated with winds blowing toward the equator, such as cold air temperatures (59), though this appears to be a high-risk strategy given the characteristically unpredictable weather of the temperate zone.

Recent radar observations indicate that large moths employ a sophisticated behavioral repertoire to exploit wind transport to move in the seasonally appropriate direction. The orientation strategies of the silver Y moth, *Autographa gamma*, which migrates annually between north Africa and northern Europe, have received particular attention (12, 14, 15). *A. gamma* evidently has a compass, and it uses it not only to select nights with favorably directed winds, but also to align its flight heading so that it partially compensates for crosswind drift while maximizing its displacement speed (14, 15). Thus, these moths are clearly not at the mercy of the wind, and analysis of their flight trajectories indicates that selection of appropriate orientation and height of flight can increase displacement distances in favorable directions by up to 40% compared to passive movement downwind (12) (**Figure 1*d***). These findings imply that this species is able to complete its migrations between summer-breeding and winter-breeding regions in as little as three or four nights. Similar behavioral mechanisms have been found in other large insects that engage in seasonal windborne migration (12). The compass utilized by nocturnal insects remains to be elucidated, although a magnetic mechanism (67), perhaps calibrated by the twilight pattern of polarized light (61), appears most likely. Day-flying migrants such as butterflies use a sun compass to maintain seasonally beneficial flight directions when migrating within their FBL (58, 60, 63, 103), and they possibly use it also to select favorable winds when migrating at high altitudes (12, 104).

BEHAVIORAL RESEARCH WITH HARMONIC RADAR

The relatively short working ranges of scanning harmonic radars have led to these units being used exclusively for the characterization of local movements, especially foraging flights. Even with this limitation, harmonic radar has almost as many potential applications as there are species capable of carrying the transponders and that inhabit reasonably open landscapes (92, 93). Harmonic radar has already contributed to studies on bee neuroethology (9, 56, 57, 87), pollinator ecology (8, 65), odor-mediated anemotaxis (73, 77, 94, 105), optimal searching strategies (73, 75–77), and short-range dispersal (64, 66).

Contribution of Harmonic Radar to Studies of Bee Behavior

Honey bees provide convenient subjects for manipulative behavioral experiments (101), and researchers studying them quickly adopted harmonic-radar technology when it became available. A second unit, recently commissioned in Germany, is also being applied to the study of bee flight (56).

Orientation flights by naïve honey bee foragers. The ability to learn the precise location of sites that a foraging individual will return to, such as a nest or reliable forage patch, is an essential component of the navigational capacity of central-place foragers such as honey bees. Beekeepers have known for many years that honey bees make repeated orientation flights that allow inexperienced worker bees to learn the hive location with respect to landscape features, and such flights are a prerequisite for successful homing of foragers. However, little was known about the nature of these exploratory movements beyond the immediate vicinity of the hive until a pioneering harmonic radar study recorded the paths of inexperienced workers throughout their first flights (9) (**Figure 2*c***). The radar tracks revealed the ontogeny of the orientation flights of naïve bees: The area covered, and the maximal range from the hive, progressively increased on successive trips, but this was achieved by faster flight on later trips rather than an increase in flight duration (9). Furthermore, the progressive changes in flight behavior were related to the number of trips an individual engaged in rather than to the worker's age. This flexible timeframe for completing the series of exploratory flights allows bees to cope with variable weather conditions and forage availability (9).

Vector flights and drift compensation in social bees. Honey bee and bumble bee workers have to solve the challenging problem of returning swiftly and efficiently to their nest site after long and convoluted foraging excursions into often unfamiliar territory. Social bees were believed to achieve this by a mechanism known as path integration, which involves the continual updating of the flight vector that would take them home by integrating all the distances covered and angles steered during each component of the outgoing trip (23). Direction of travel is ascertained from the sun's azimuth (or, alternatively, from patterns of polarized sky light), and distance traveled is gauged by the passage of movement-induced optic flow over the bee's retina—the so-called odometer (102). Whereas the use of path integration for returning directly to a nest was established many years ago in pedestrian foragers such as desert ants, *Cataglyphis fortis* (111), the practical difficulties of quantifying the return trajectories of flying insects prevented investigation of whether social bees used the same procedure. A study by Riley et al. (87) provided the first direct quantification of goal-oriented homeward flights in foraging honey bees and established that they do indeed rely on path integration to make a bee-line back to their hive (**Figure 2*c***). Bees captured at an established feeding station, and then displaced before release, embarked on the vector flight that would have taken them directly to their nest if they had not been artificially displaced. During these vector flights, they ignored unfamiliar landscape features along the new route, and so in essence they were flying in automatic pilot mode (87).

In contrast to pedestrian central-place foragers, social bees have to compensate for the effects of crosswind drift if they are to maintain direct routes between foraging sites and their nest. This appears to be a difficult task, as fine-scale variation in wind speed and direction may be considerable at the heights at which social bees fly. However, visualization of the complete vector tracks of returning bumble bees, *Bombus terrestris* (91), and honey bees (86, 87), with harmonic radar demonstrated that both species can compensate for lateral wind drift, even when flying in strong crosswinds, by heading partly into the wind and moving on an oblique course over the ground. They most likely gauge the degree of compensation necessary by adjusting their headings until the direction of the optic flow they experience occurs at the angle to the sun's azimuth corresponding to their intended track (91).

Lévy flight: a power-law step-length distribution characterized by frequent short steps and a small proportion of much longer ones

The honey bee waggle dance and forager recruitment. The astonishing discovery by Karl von Frisch that worker honey bees engaging in the waggle dance transmit a coded, abstract message to their hivemates communicating the location of a new food source was one of the great breakthroughs of twentieth-century biology. Although there was some initial skepticism, recent experimental studies provide convincing evidence that the dance language contains information on the distance to the food source (as measured by the amount of optic flow experienced en route) as well as its direction (39, 102). What was missing was direct quantification of the actual flight paths of bees recruited by the waggle dance, and so harmonic radar was used to provide this description and thus to test how effectively recruited bees translate the encoded information (86). In this experiment, bees were trained to visit an artificial feeder lacking odor and visual cues, and then harmonic radar was used to track the flights of naïve bees recruited by the waggle dance performed by the trained bees. The flight paths of naïve control bees took them from the hive directly into the vicinity of the feeder, but, crucially, experimental naïve bees that were released from a displaced location undertook vector flights directly to the vicinity of where the feeder would have been if the bees had not been displaced (86), thus confirming von Frisch's claim that the dance communicates both distance and direction information.

Contribution of Harmonic Radar to an Understanding of Movement Ecology

The ability to track individuals over distances of several hundred meters that harmonic radar provides has revealed features of insect foraging and dispersal flights that can be understood as adaptations to the particular environment or as consequences of the particular population history.

Optimal searching patterns in honey bees and moths. One of the most exciting recent developments in animal movement research has been the development of biologging technologies that allow high-resolution quantification of the entire movement trajectories of individual animals, and the concomitant development of sophisticated analytical techniques for interpreting the movement behaviors captured by these techniques (62, 74, 95, 113). Size and weight constraints generally preclude the use of most tracking devices for studies of long-range migration in insects (but see Reference 114 for an exception), but harmonic radar has made characterization of insect foraging flights possible and a significant development of our conceptual understanding of insect searching strategies has ensued.

There are several situations in which foraging honey bees have to conduct searches when they have no prior information of where the resource will be found, for example, searching for a new forage patch or locating their hive after becoming disoriented due to the accumulation of errors in their path integration system. Theoretically, the optimal strategy for locating patchily distributed resources in an uncertain environment is a type of scale-free, random searching pattern known as a Lévy flight (74).

A reanalysis of radar-derived flight trajectories by A.M. Reynolds demonstrated that honey bees employ Lévy-flight searches when trying to locate profitable food sources (76, 77) or find their hive after artificial displacement (75). When they have some limited knowledge of where the resource is expected to be, they deploy a looping search strategy, with randomly directed movements from a central point, but if this fails they eventually switch to a freely roaming search to cover more ground (75).

Many insects have to systematically search for randomly distributed resources that advertise their presence with attractive odors, e.g., male moths searching for conspecific females that release sex pheromones. There have been numerous studies on the upwind anemotaxis used to follow an odor plume to its source (10), but few investigations of the behaviors employed to find the plume in the first place. Landscape-scale harmonic radar studies of male *Agrotis segetum* moths indicated that when odor filaments of sufficient concentration are encountered, males that are flying downwind change course and start flying crosswind, while crosswind fliers start moving upwind (73); the effectiveness of such a plume-locating strategy could be demonstrated using realistic modeling of plume structure, pheromone-detection thresholds, and wind conditions. More complex search patterns were also noted, including a type of Lévy flight (73) (**Figure 2*d***). Investigation of further flight-capable insect species for Lévy flights and other optimal searching strategies would appear worthwhile.

Dispersal rates in butterfly metapopulations. Dispersal capability is a key factor in the survival of insect populations, particularly for those species that exist as metapopulations in spatially fragmented landscapes where there is a distinct probability of local extinction (47). Dispersal is achieved through a suite of morphological, physiological, genetic, and environmental factors, and individual variation in dispersal rates can be affected by any and all of these. These complex interactions have been extensively studied in metapopulations of the Glanville fritillary butterfly, *Melitaea cinxia*, in the Åland Islands (Finland) by Hanski and colleagues (47). In an experiment in the United Kingdom, harmonic radar was used to track free-flying *M. cinxia* females in order to record their tracks over periods of several hours. The results confirmed that variation in dispersal was associated with population history: Females from newly established (1-year-old) isolated Finnish populations were more dispersive than those from older (>5 years) isolated populations (66). Previous work had also indicated that variation in the gene encoding for the enzyme phosphoglucose isomerase (*Pgi*) was associated with individual variation in flight metabolism in this butterfly (64). The radar tracking demonstrated that the higher flight metabolic rate translated into higher dispersal rates in the field, with heterozygotic *Pgi* individuals moving longer distances and flying at lower ambient temperatures than homozygous individuals (64). The radar studies thus helped to establish connections between organizational levels ranging from molecular variation in a single gene, via flight physiology, to landscape-scale patterns of butterfly movement.

FMCW: frequency-modulated continuous wave

FUTURE PROSPECTS

The increasing availability, affordability, and reliability of radar, signal processing, and microcomputing hardware, and the rapidly increasing power of the last two, will allow further development and wider deployment of entomological radar technology. In the case of harmonic radar, modification of the receive antenna to produce a pair of vertically overlapping beams would allow the target's height of flight to be inferred. For vertical beam radars, separate transmit and receive antennas, and perhaps use of frequency-modulated continuous wave (FMCW) rather than pulse transmission, would largely eliminate the ~150-m blind zone just above the ground of current designs. Incorporation of Doppler processing would allow observation of target ascent and descent; this appears to be the most promising technique for studying the rather cryptic and poorly characterized termination phase of migratory flight.

The size, shape, and wing-beat frequency values determined for each individual insect by a VLR with full target interrogation capability are already allowing inferences to be drawn about the identity of the detected targets (13, 18, 25). With a priori information on the local migrant insect fauna, assignment of target numbers, speeds, and directions to broad taxonomic classes (e.g., large grasshoppers, large moths, small moths)—and sometimes even to a particular numerically dominant species—is possible. Fully automated monitoring of locusts, and perhaps also of some pest moth species, can thus be envisaged. Extending the rather small current dataset of measurements of the radar properties of known species (49), and investigating intraspecific variance, would help to develop identification confidence.

As millimeter-wavelength radar technology becomes more affordable, deployment of 9- and 3-mm-wavelength VLRs alongside current 32-mm units at each observing site can be contemplated. Flight activity could then be monitored for insects ranging in size from sphingid moths and locusts, through planthoppers (~1–2 mg; 85), to aphids, small Diptera, and minute parasitic wasps (~0.05–0.5 mg). Radars operating at 9 and 3 mm have been developed for atmospheric observations and on warm cloudless days detect insects in large numbers (119). The great profusion of small taxa will present a target identification challenge that will be best tackled by sampling, at least occasionally, with a balloon-borne net (20).

An alternative way of exploiting the falling hardware costs would be to install additional 32-mm-wavelength VLRs in a network to explore how migratory activity varies over a region. A chain of these units across a prominent migration route, for example, the Mississippi flyway in North America, would reveal the timing and lateral extent of migration events and how these vary with season, wind, and weather; additional units at higher or lower latitudes would strengthen knowledge of distances moved and the seasonal advance and retreat of populations. Similar networks can be envisaged for Europe and eastern China, while in Australia an additional five or six VLRs would provide comprehensive coverage of the inland region where locust outbreaks most frequently develop. On a smaller scale, operating a mobile unit for a few weeks at varying distances from a long-term observing site would establish the scale over which a VLR's observations are representative.

Finally, modern weather surveillance radars, especially those with dual-polarization capability, routinely detect insects in fine warm weather and are being deployed in networks covering entire countries. Dual polarization can discriminate insects, with their elongated bodies held approximately horizontal, from both hydrometeors and birds (3); it also reveals patterns of common orientation (70) (a phenomenon long familiar to radar entomologists, see above), which helps to establish the biological identity of the targets. Weather-radar observations of insect activity are beginning to contribute to forecasts and "now casts" of migratory pest incidence (53).

Turning from technology to biology, the orientation behavior observed in nocturnal windborne migrants (12, 14, 15) clearly warrants laboratory investigation to determine its neurophysiological basis; identification and characterization of magnetoreceptors (67) and of sensory organs for detecting turbulent air motions are obvious priorities. The association of layer concentrations and common orientation, and the variation of orientation with height and time, can be examined with archival VLR datasets, but FMCW radars, with range resolutions of only 1–2 m, and simultaneous meteorological profiling of wind, turbulence, and temperature may be needed to fully resolve the causes of these phenomena. Observations from a VLR network (see above) would establish whether seasonally advantageous flight headings occur throughout a species' range and how these headings vary with longitude and latitude or according to local geography. A radar on the Falsterbo Peninsula in southwest Sweden, for example, might resolve whether high-flying insects take the shortest sea crossing to Denmark by flying west—as low-flying red admiral butterflies, *Vanessa atalanta*, have been observed visually to do (7)—or steer a southward course more in accordance with their presumed long-range goal. VLR network data might also elucidate how some seasonal migrants are able to move differentially on winds from certain directions, avoiding displacement in highly disadvantageous directions (e.g., poleward in the fall). Is this achieved by sensing a cue (e.g., the relative humidity of the airstream) while still on the ground, or do potential migrants have to take off and then rapidly terminate their flight if they sense that they are being displaced in the wrong direction?

CONCLUSIONS

Radar has now been used by entomologists for over 40 years. After a brief but exciting discovery phase, the focus of its application became careful studies of economically significant pests, usually undertaken in conjunction with simultaneous trapping, survey, and visual observation campaigns. The aims of this research were primarily to establish whether these pests were migratory and to determine the scale, direction, timing, and frequency of movements and the conditions in which they occur. Almost in passing, behaviors modulating the incidence of migratory flight and the displacements achieved were recognized. More recently, technical developments and a new radar design allowed more precise observations to be accumulated over much longer periods, while the introduction of harmonic detection extended radar work into a new field—the study of foraging and searching flight. As the entomological use of radar has matured, behavioral questions have increasingly been targeted, and descriptive and quantitative analyses of observations have been supplemented by formal hypothesis testing and, in the case of harmonic radar, manipulative experimentation. However, observations with a more ecological focus continue in regions where migrant insect pests have a significant economic or welfare impact, and in this more applied context the new automated units may be able to fulfill an operational role—most likely for locust management.

Radar continues, as it began, to reveal remarkable insect behaviors that appear to represent sophisticated evolved adaptations. It is now clear, thanks in good part to radar work, that insects are highly effective migrants and foragers. Over the past decade, radar-led research has refined our knowledge of the strategies and sensory capabilities employed by insects in flight. Some specific and bold hypotheses have been developed to account for the observed phenomena, and these have been tested with both new and archival radar datasets. Much of this research is at an early stage and it is reasonable to suppose that there is still much to learn. The potential of current entomological radar types is surely far from exhausted, and further technological advances can be expected to provide additional opportunities for those willing to grasp them.

SUMMARY POINTS

1. Entomological radars have made a huge contribution to understanding windborne insect migration.
2. Two new designs, VLR and harmonic radar, have significantly enhanced the observational capabilities of researchers studying high-altitude migration and low-altitude foraging flight, respectively.
3. VLRs have shown that windborne migrant moths use environmental cues to select flight headings and flight altitudes that optimize their migratory trajectories.
4. Harmonic radar has demonstrated that social bees navigate to their goal by path integration (with wind-drift compensation) and progressive learning of landmarks, and confirmed that honey bees communicate both direction and distance information through the waggle dance. Harmonic radar has also shown that insect foragers employ Lévy-flight optimal search strategies.
5. Further technological improvements will enable radar entomology to remain in the forefront of research into insect flight behavior.

DISCLOSURE STATEMENT

The authors are not aware of any affiliations, memberships, funding, or financial holdings that might be perceived as affecting the objectivity of this review.

ACKNOWLEDGMENTS

We thank Laura Burgin, Lynda Castle, Mònica Garcia-Alonso, Juliet Osborne, Andy Reynolds, Alan Smith, and Eric Warrant for provision of images and constructive comments on the manuscript. Rothamsted Research receives grant-aided support from the UK Biotechnology and Biological Sciences Research Council.

LITERATURE CITED

1. Åkesson S, Hedenström A. 2007. How migrants get there: migratory performance and orientation. *BioScience* 57:123–33
2. Alerstam T. 2006. Conflicting evidence about long-distance animal navigation. *Science* 313:791–94
3. Bachmann S, Zrnic D. 2007. Spectral density of polarimetric variables separating biological scatterers in the VAD display. *J. Atmos. Ocean. Technol.* 24:1186–98
4. Beerwinkle KR, Lopez JD, Schleider PG, Lingren PD. 1995. Annual patterns of aerial insect densities at altitudes from 500 to 2400 meters in east-central Texas indicated by continuously-operating vertically-oriented radar. *Southwest. Entomol. Suppl.* 18:63–79
5. Beerwinkle KR, Lopez JD, Witz JA, Schleider PG, Eyster RS, Lingren PD. 1994. Seasonal radar and meteorological observations associated with nocturnal insect flight at altitudes to 900 meters. *Environ. Entomol.* 23:676–83
6. Bell JR, Bohan DA, Shaw EM, Weyman GS. 2005. Ballooning dispersal using silk: world fauna, phylogenies, genetics and models. *Bull. Entomol. Res.* 95:69–114
7. Brattström O, Kjellén N, Alerstam T, Åkesson S. 2008. Effects of wind and weather on red admiral, *Vanessa atalanta*, migration at a coastal site in southern Sweden. *Anim. Behav.* 76:335–44
8. Cant ET, Smith AD, Reynolds DR, Osborne JL. 2005. Tracking butterfly flight paths across the landscape with harmonic radar. *Proc. R. Soc. Lond. B* 272:785–90
9. Capaldi EA, Smith AD, Osborne JL, Fahrbach SE, Farris SM, et al. 2000. Ontogeny of orientation flight in the honeybee revealed by harmonic radar. *Nature* 403:537–40
10. Cardé RT, Willis MA. 2008. Navigational strategies used by insects to find distant, wind-borne sources of odor. *J. Chem. Ecol.* 34:854–66
11. Chapman JW, Drake VA. 2010. Insect migration. In *Encyclopedia of Animal Behaviour*, ed. MD Breed, J Moore, 2:161–66. Oxford, UK: Academic
12. **Chapman JW, Nesbit RL, Burgin LE, Reynolds DR, Smith AD, et al. 2010. Flight orientation behaviors promote optimal migration trajectories in high-flying insects. *Science* 327:682–85**

12. The first demonstration that the compass-mediated selection of optimal flight headings is widespread among large windborne insect migrants.

13. Chapman JW, Reynolds DR, Brooks SJ, Smith AD, Woiwod IP. 2006. Seasonal variation in the migration strategies of the green lacewing *Chrysoperla carnea* species complex. *Ecol. Entomol.* 31:378–88
14. Chapman JW, Reynolds DR, Hill JK, Sivell D, Smith AD, Woiwod IP. 2008. A seasonal switch in compass orientation in a high-flying migratory moth. *Curr. Biol.* 18:R908–9
15. Chapman JW, Reynolds DR, Mouritsen H, Hill JK, Riley JR, et al. 2008. Wind selection and drift compensation optimize migratory pathways in a high-flying moth. *Curr. Biol.* 18:514–18
16. Chapman JW, Reynolds DR, Smith AD. 2003. Vertical-looking radar: a new tool for monitoring high-altitude insect migration. *BioScience* 53:503–11
17. Chapman JW, Reynolds DR, Smith AD. 2004. Migratory and foraging movements in beneficial insects: a review of radar monitoring and tracking methods. *Int. J. Pest Manag.* 50:225–32
18. Chapman JW, Reynolds DR, Smith AD, Riley JR, Pedgley DE, Woiwod IP. 2002. High-altitude migration of the diamondback moth, *Plutella xylostella*, to the UK: a study using radar, aerial netting and ground trapping. *Ecol. Entomol.* 27:641–50

19. Chapman JW, Reynolds DR, Smith AD, Riley JR, Telfer MG, Woiwod IP. 2005. Mass aerial migration in the carabid beetle *Notiophilus biguttatus*. *Ecol. Entomol.* 30:264–72
20. Chapman JW, Reynolds DR, Smith AD, Smith ET, Woiwod IP. 2004. An aerial netting study of insects migrating at high-altitude over England. *Bull. Entomol. Res.* 94:123–36
21. Chapman JW, Smith AD, Woiwod IP, Reynolds DR, Riley JR. 2002. Development of vertical-looking radar technology for monitoring insect migration. *Comput. Electron. Agric.* 35:95–110
22. Cheng DF, Wu KM, Tian Z, Wen LP, Shen ZR. 2002. Acquisition and analysis of migration data from the digitized display of a scanning entomological radar. *Comput. Electron. Agric.* 35:63–75
23. Collett M, Collett TS. 2000. How do insects use path integration for their navigation? *Biol. Cybern.* 83:245–59
24. Crawford AB. 1949. Radar reflections in the lower atmosphere. *Proc. Inst. Radio Eng.* 37:404–5
25. Dean TJ, Drake VA. 2005. Monitoring insect migration with radar: the ventral-aspect polarization pattern and its potential for target identification. *Int. J. Remote Sens.* 26:3957–74
26. Deveson ED, Drake VA, Hunter DM, Walker PW, Wang HK. 2005. Evidence from traditional and new technologies for northward migrations of Australian plague locusts (*Chortoicetes terminifera*) (Walker) (Orthoptera: Acrididae) to western Queensland. *Aust. Ecol.* 30:928–43
27. Dingle H, Drake VA. 2007. What is migration? *BioScience* 57:113–21

27. Part of a special section on animal migration that provides a useful introduction to the study of migration (see also Reference 1).

28. Drake VA. 1982. Insects in the sea-breeze front at Canberra: a radar study. *Weather* 37:134–43
29. Drake VA. 1983. Collective orientation by nocturnally migrating Australian plague locusts, *Chortoicetes terminifera* (Walker) (Orthoptera: Acrididae): a radar study. *Bull. Entomol. Res.* 73:679–92
30. Drake VA. 1984. The vertical distribution of macroinsects migrating in the nocturnal boundary layer: a radar study. *Bound.-Layer Meteorol.* 28:353–74
31. Drake VA. 2002. Automatically operating radars for monitoring insect pest migrations. *Entomol. Sin.* 9:27–39
32. Drake VA, Farrow RA. 1983. The nocturnal migration of the Australian plague locust, *Chortoicetes terminifera* (Walker) (Orthoptera: Acrididae): quantitative radar observations of a series of northward flights. *Bull. Entomol. Res.* 73:567–85
33. Drake VA, Farrow RA. 1988. The influence of atmospheric structure and motions on insect migration. *Annu. Rev. Entomol.* 33:183–210
34. Drake VA, Gatehouse AG, eds. 1995. *Insect Migration: Tracking Resources Through Space and Time*. Cambridge, UK: Cambridge Univ. Press. 478 pp.

34. Provides a snapshot of insect migration research at the start of the period with which this review is primarily concerned.

35. Drake VA, Gregg PC, Harman IT, Wang HK, Deveson ED, et al. 2001. Characterizing insect migration systems in inland Australia with novel and traditional methodologies. In *Insect Movement: Mechanisms and Consequences*, ed. IP Woiwod, DR Reynolds, CD Thomas, pp. 207–33. Wallingford, UK: CAB Int. 458 pp.
36. Drake VA, Harman IT, Wang HK. 2002. Insect monitoring radar: stationary-beam operating mode. *Comput. Electron. Agric.* 35:111–37
37. Drake VA, Helm KF, Readshaw JL, Reid DG. 1981. Insect migration across Bass Strait during spring: a radar study. *Bull. Entomol. Res.* 71:449–66
38. Dudley R. 2002. *The Biomechanics of Insect Flight: Form, Function, Evolution*. Princeton, NJ: Princeton Univ. Press. 536 pp.
39. Esch HE, Zhang S, Srinivasan MV, Tautz J. 2001. Honeybee dances communicate distances measured by optic flow. *Nature* 411:581–83
40. Feng HQ, Wu KM, Cheng DF, Guo YY. 2003. Radar observations of the beet armyworm *Spodoptera exigua* (Lepidoptera: Noctuidae) and other moths in northern China. *Bull. Entomol. Res.* 93:115–24
41. Feng HQ, Wu KM, Ni YX, Cheng DF, Guo YY. 2006. Nocturnal migration of dragonflies over the Bohai Sea in northern China. *Ecol. Entomol.* 31:511–20
42. Feng HQ, Wu XF, Wu B, Wu KM. 2009. Seasonal migration of *Helicoverpa armigera* (Lepidoptera: Noctuidae) over the Bohai Sea. *J. Econ. Entomol.* 102:95–104
43. Feng HQ, Zhao XC, Wu XF, Wu B, Wu KM, et al. 2008. Autumn migration of *Mythimna separata* (Lepidoptera: Noctuidae) over the Bohai Sea in northern China. *Environ. Entomol.* 37:774–81
44. Gatehouse AG. 1997. Behavior and ecological genetics of wind-borne migration by insects. *Annu. Rev. Entomol.* 42:475–502

45. Greenbank DO, Schaefer GW, Rainey RC. 1980. Spruce budworm (Lepidoptera: Tortricidae) moth flight and dispersal: new understanding from canopy observations, radar, and aircraft. *Mem. Entomol. Soc. Can.* 110:1–49
46. Hahn S, Bauer S, Liechti F. 2009. The natural link between Europe and Africa—2.1 billion birds on migration. *Oikos* 118:624–26
47. Hanski I. 1999. *Metapopulation Ecology*. Oxford: Oxford Univ. Press. 328 pp.
48. Harman IT, Drake VA. 2004. Insect monitoring radar: analytical time-domain algorithm for retrieving trajectory and target parameters. *Comput. Electron. Agric.* 43:23–41
49. Hobbs SE, Aldhous AC. 2006. Insect ventral radar cross-section polarization dependence measurements for radar entomology. *IEE Proc.-Radar Sonar Navig.* 153:502–8
50. Holland RA, Wikelski M, Wilcove DS. 2006. How and why do insects migrate? *Science* 313:794–96
51. Irwin ME, Thresh JM. 1988. Long range aerial dispersal of cereal aphids as virus vectors in North America. *Philos. Trans. R. Soc. Lond. B* 321:421–46
52. Kennedy JS. 1985. Migration: behavioral and ecological. In *Migration: Mechanisms and Adaptive Significance*, ed. MA Rankin, *Contrib. Mar. Sci.* 27(Suppl.):5–26. Port Aransas: Mar. Sci. Inst., Univ. Tex., Austin
53. Leskinen M, Markkula I, Koistinen J, Pylkkö P, Ooperi S, et al. 2010. Pest immigration warning by an atmospheric dispersion model, weather radars and traps. *J. Appl. Entomol.* In press
54. Mascanzoni D, Wallin H. 1986. The harmonic radar: a new method of tracing insects in the field. *Ecol. Entomol.* 11:387–90
55. McNeil JN. 1987. The true armyworm, *Pseudoletia unipuncta*: a victim of the pied piper or a seasonal migrant? *Insect Sci. Appl.* 8:591–97
56. Menzel R, Fuchs J, Nadler L, Weiss B, Kumbischinski N, et al. 2010. Dominance of the odometer over serial landmark learning in honeybee navigation. *Naturwissenschaften* 97:763–67
57. Menzel R, Greggers U, Smith A, Berger S, Brandt R, et al. 2005. Honey bees navigate according to a map-like spatial memory. *Proc. Natl. Acad. Sci. USA* 102:3040–45
58. Merlin C, Gegear RJ, Reppert SM. 2009. Antennal circadian clocks coordinate sun compass orientation in migratory monarch butterflies. *Science* 325:1700–4
59. Mikkola K. 2003. Red admirals *Vanessa atalanta* (Lepidoptera: Nymphalidae) select northern winds on southward migration. *Entomol. Fenn.* 14:15–24
60. Mouritsen H, Frost BJ. 2002. Virtual migration in tethered flying monarch butterflies reveals their orientation mechanisms. *Proc. Natl. Acad. Sci. USA* 99:10162–66
61. Muheim R, Phillips JB, Åkesson S. 2006. Polarized light cues underlie compass calibration in migratory songbirds. *Science* 313:837–39
62. Nathan R, Getz WM, Revilla E, Holyoak M, Kadmon R, et al. 2008. A movement ecology paradigm for unifying organismal movement research. *Proc. Natl. Acad. Sci. USA* 105:19052–59
63. Nesbit RL, Hill JK, Woiwod IP, Sivell D, Bensusan KJ, Chapman JW. 2009. Seasonally-adaptive migration headings mediated by a sun compass in the painted lady butterfly (*Vanessa cardui*). *Anim. Behav.* 78:1119–25
64. Niitepõld K, Smith AD, Osborne JL, Reynolds DR, Carreck NL, et al. 2009. Flight metabolic rate and *Pgi* genotype influence butterfly dispersal rate in the field. *Ecology* 90:2223–32
65. Osborne JL, Clark SJ, Morris RJ, Williams IH, Riley JR, et al. 1999. A landscape-scale study of bumble bee foraging range and constancy, using harmonic radar. *J. Appl. Ecol.* 36:519–33
66. Ovaskainen O, Smith AD, Osborne JL, Reynolds DR, Carreck NL, et al. 2008. Tracking butterfly movements with harmonic radar reveals an effect of population age on movement distance. *Proc. Natl. Acad. Sci. USA* 105:19090–95
67. Phillips JB, Jorge PE, Muheim R. 2010. Light-dependent magnetic compass orientation in amphibians and insects: candidate receptors and candidate molecular mechanisms. *J. R. Soc. Interface* 7:S241–56
68. Rainey RC. 1955. Observation of desert locust swarms by radar. *Nature* 175:77–78
69. Reid DG, Wardhaugh KG, Roffey J. 1979. Radar studies of insect flight at Benalla, Victoria, in February 1974. *CSIRO Aust. Div. Entomol. Tech. Pap. 16.* 21 pp.
70. Rennie SJ, Illingworth AJ, Dance SL, Ballard SP. 2010. The accuracy of Doppler radar wind retrievals using insects as targets. *Meteorol. Appl.* 17:In press

71. Reynolds AM, Reynolds DR, Riley JR. 2009. Does a 'turbophoretic' effect account for layer concentrations of insects migrating in the stable night-time atmosphere? *J. R. Soc. Interface* 6:87–95
72. Reynolds AM, Reynolds DR, Smith AD, Chapman JW. 2010. A single wind-mediated mechanism explains high-altitude 'nongoal oriented' headings and layering of nocturnally-migrating insects. *Proc. R. Soc. Lond. B* 277:765–72
73. Reynolds AM, Reynolds DR, Smith AD, Svensson GP, Löfstedt. 2007. Appetitive flight patterns of male *Agrotis segetum* moths over landscape scales. *J. Theor. Biol.* 245:141–49
74. Reynolds AM, Rhodes CJ. 2009. The Lévy flight paradigm: random search patterns and mechanisms. *Ecology* 90:877–87
75. Reynolds AM, Smith AD, Menzel R, Greggers U, Reynolds DR, Riley JR. 2007. Displaced honey bees perform optimal scale-free search flights. *Ecology* 88:1955–61
76. Reynolds AM, Smith AD, Reynolds DR, Carreck NL, Osborne JL. 2007. Honeybees perform optimal scale-free searching flights when attempting to locate a food source. *J. Exp. Biol.* 210:3763–70
77. Reynolds AM, Swain JL, Smith AD, Martin AP, Osborne JL. 2009. Honeybees use a Lévy flight search strategy and odor-mediated anemotaxis to relocate food sources. *Behav. Ecol. Sociobiol.* 64:115–23
78. Reynolds DR, Chapman JW, Edwards AS, Smith AD, Wood CR, et al. 2005. Radar studies of the vertical distribution of insects migrating over southern Britain: the influence of temperature inversions on nocturnal layer concentrations. *Bull. Entomol. Res.* 95:259–74
79. Reynolds DR, Chapman JW, Harrington R. 2006. The migration of insect vectors of plant and animal viruses. *Adv. Virus Res.* 67:453–517
80. Reynolds DR, Riley JR. 1988. A migration of grasshoppers, particularly *Diabolocatantops axillaris* (Thunberg) (Orthoptera: Acrididae), in the West African Sahel. *Bull. Entomol. Res.* 97:1974–83
81. Reynolds DR, Riley JR. 1997. *The Flight Behaviour and Migration of Insect Pests: Radar Studies in Developing Countries*. NRI Bull. 71. Chatham, UK: Nat. Resour. Inst. 114 pp.
82. Reynolds DR, Smith AD, Chapman JW. 2008. A radar study of emigratory flight and layer formation at dawn over southern Britain. *Bull. Entomol. Res.* 98:35–52
83. Riley JR. 1975. Collective orientation in night-flying insects. *Nature* 253:113–14
84. Riley JR. 1989. Remote sensing in entomology. *Annu. Rev. Entomol.* 34:247–71
85. Riley JR. 1992. A millimetric radar to study the flight of small insects. *Electron. Commun. Eng. J.* 4:43–48
86. **Riley JR, Greggers U, Smith AD, Reynolds DR, Menzel R. 2005. The flight paths of honeybees recruited by the waggle dance. *Nature* 435:205–7**

86. Confirms that von Frisch's hypothesis was correct all along.

87. Riley JR, Greggers U, Smith AD, Stach S, Reynolds DR, et al. 2003. The automatic pilot of honeybees. *Proc. R. Soc. Lond. B* 270:2421–24
88. Riley JR, Kreuger U, Addison CM, Gewecke M. 1988. Visual detection of wind-drift by high-flying insects at night: a laboratory study. *J. Comp. Physiol. A* 162:793–98
89. Riley JR, Reynolds DR. 1979. Radar-based studies of the migratory flight of grasshoppers in the middle Niger area of Mali. *Proc. R. Soc. Lond. B* 204:67–82
90. Riley JR, Reynolds DR. 1983. A long-range migration of grasshoppers observed in the Sahelian zone of Mali by two radars. *J. Anim. Ecol.* 52:167–83
91. Riley JR, Reynolds DR, Smith AD, Edwards AS, Osborne JL, et al. 1999. Compensation for wind drift by bumble-bees. *Nature* 400:126
92. Riley JR, Smith AD. 2002. Design considerations for an harmonic radar to investigate the flight of insects at low altitude. *Comput. Electron. Agric.* 35:151–69
93. Riley JR, Smith AD, Reynolds DR, Edwards AS, Osborne JL, et al. 1996. Tracking bees with harmonic radar. *Nature* 379:29–30
94. Riley JR, Valeur P, Smith AD, Reynolds DR, Poppy GM, Löfstedt C. 1998. Harmonic radar as a means of tracking the pheromone-finding and pheromone-following flight of male moths. *J. Insect Behav.* 11:287–96
95. Rutz C, Hays GC. 2009. New frontiers in biologging science. *Biol. Lett.* 5:289–92
96. Sane SP, Dieudonne A, Willis MA, Daniel TL. 2007. Antennal mechanosensors mediate flight control in moths. *Science* 315:863–86
97. **Schaefer GW. 1976. Radar observations of insect flight. In *Insect Flight. Symp. R. Entomol. Soc. No.* 7, ed. RC Rainey, pp. 157–97. Oxford: Blackwell Sci. 287 pp.**

97. The classic foundation paper of radar entomology.

98. Showers WB. 1997. Migratory ecology of the black cutworm. *Annu. Rev. Entomol.* 42:393–425
99. Smith AD, Reynolds DR, Riley JR. 2000. The use of vertical-looking radar to continuously monitor the insect fauna flying at altitude over southern England. *Bull. Entomol. Res.* 90:265–77
100. Smith AD, Riley JR, Gregory RD. 1993. A method for routine monitoring of the aerial migration of insects by using a vertical-looking radar. *Philos. Trans. R. Soc. B* 340:393–404
101. Srinivasan MV. 2010. Honey bees as a model for vision, perception, and cognition. *Annu. Rev. Entomol.* 55:267–84
102. Srinivasan MV, Zhang S, Altwein M, Tautz J. 2000. Honeybee navigation: nature and calibration of the 'odometer'. *Science* 287:851–53
103. Srygley RB, Dudley R. 2008. Optimal strategies for insects migrating in the flight boundary layer: mechanisms and consequences. *Integr. Comp. Biol.* 48:119–33
104. Stefanescu C, Alarcón M, Àvila A. 2007. Migration of the painted lady butterfly, *Vanessa cardui*, to north-eastern Spain is aided by African wind currents. *J. Anim. Ecol.* 76:888–98
105. Svensson GP, Valeur PG, Reynolds DR, Smith AD, Riley JR, et al. 2001. Mating disruption in *Agrotis segetum* monitored by harmonic radar. *Entomol. Exp. Appl.* 101:111–21
106. Sword GA, Lorch PD, Gwynne DT. 2005. Migratory bands give crickets protection. *Nature* 433:703
107. Taylor LR. 1974. Insect migration, flight periodicity and the boundary layer. *J. Anim. Ecol.* 43:225–38
108. Theobald JC, Warrant EJ, O'Carroll DC. 2010. Wide-field motion tuning in nocturnal hawkmoths. *Proc. R. Soc. Lond. B* 277:853–60
109. Warrant EJ, Kelber A, Gislén A, Greiner B, Ribi W, Wcislo WT. 2004. Nocturnal vision and landmark orientation in a tropical halictid bee. *Curr. Biol.* 14:1309–18
110. Warren B, Gibson G, Russell IJ. 2009. Sex recognition through midflight mating duets in *Culex* mosquitoes is mediated by acoustic distortion. *Curr. Biol.* 19:485–91
111. Wehner W, Srinivasan MV. 1981. Searching behavior of desert ants, genus *Cataglyphis* (Formicidae, Hymenoptera). *J. Comp. Physiol.* 142:315–38
112. Westbrook JK. 2008. Noctuid migration in Texas within the nocturnal aeroecological boundary layer. *Integr. Comp. Biol.* 48:99–106
113. Wikelski M, Kays RW, Kasdin NJ, Thorup K, Smith JA, Swenson GW Jr. 2007. Going wild: what a global small-animal tracking system could do for experimental biologists. *J. Exp. Biol.* 210:181–86
114. Wikelski M, Moskowitz D, Adelman JS, Cochran J, Wilcove DS, May ML. 2006. Simple rules guide dragonfly migration. *Biol. Lett.* 2:325–29
115. Williams DW, Li G, Gao R. 2004. Tracking movements of individual *Anoplophora glabripennis* (Coleoptera: Cerambycidae) adults: application of harmonic radar. *Environ. Entomol.* 33:644–49
116. Wolf WW, Westbrook JK, Raulston J, Pair SD, Hobbs SE. 1990. Recent airborne radar observations of migrant pests in the United States. *Philos. Trans. R. Soc. Lond. B* 328:619–30
117. Wood CR, Chapman JW, Reynolds DR, Barlow JF, Smith AD, Woiwod IP. 2006. The influence of the atmospheric boundary layer on nocturnal layers of moths migrating over southern Britain. *Int. J. Biometeorol.* 50:193–204
118. Wood CR, Clark SJ, Barlow JF, Chapman JW. 2010. Insect migration at high-altitudes: a systematic study of the meteorological conditions correlated with nocturnal layers in the UK. *Agric. For. Entomol.* 12:113–21
119. Wood CR, O'Connor EJ, Hurley RA, Reynolds DR, Illingworth AJ. 2009. Cloud-radar observations of insects in the UK convective boundary layer. *Meteorol. Appl.* 16:491–500
120. Wood CR, Reynolds DR, Wells PM, Barlow JF, Woiwod IP, Chapman JW. 2009. Flight periodicity and the vertical distribution of high-altitude moth migration over southern Britain. *Bull. Entomol. Res.* 99:525–35
121. Yorozu S, Wong A, Fischer BJ, Dankert H, Kernan MJ, et al. 2009. Distinct sensory representations of wind and near-field sound in the *Drosophila* brain. *Nature* 458:201–5

Arthropod-Borne Diseases Associated with Political and Social Disorder

Philippe Brouqui

Faculté de Médecine, Unité de Recherche sur les Maladies Infectieuses et Tropicales Emergentes, CNRS-IRD UMR 6236/198, 13385 Marseille cedex 5, France; email: philippe.brouqui@univmed.fr

Annu. Rev. Entomol. 2011. 56:357–74

First published online as a Review in Advance on September 3, 2010

The *Annual Review of Entomology* is online at ento.annualreviews.org

This article's doi:
10.1146/annurev-ento-120709-144739

0066-4170/11/0107-0357$20.00

Key Words

homeless, typhus, *Bartonella quintana*, louse, bed bug, tungiasis, myiasis

Abstract

The living conditions and the crowded situations of the homeless, war refugees, or victims of a natural disaster provide ideal conditions for the spread of lice, fleas, ticks, flies and mites. The consequence of arthropod infestation in these situations is underestimated. Along with louse-borne infections such as typhus, trench fever, and relapsing fever, the relationship between *Acinetobacter* spp.–infected lice and bacteremia in the homeless is not clear. Murine typhus, tungiasis, and myiasis are likely underestimated, and there has been a reemergence of bed bugs. Attempted eradication of the body louse, despite specific measures, has been disappointing, and infections with *Bartonella quintana* continue to be reported. The efficacy of ivermectin in eradicating the human body louse, although the effect is not sustained, might provide new therapeutic approaches. Arthropod-borne diseases continue to emerge within the deprived population. Public health programs should be engaged rapidly to control these pests and reduce the incidence of these transmissible diseases.

INTRODUCTION

In recent decades, the movements of political, economic, and environmental refugees due to conflict and warfare or associated with the increase of population density and size, urbanization, and persistent poverty (especially in expanding peri-urban slums) lead the poorer segments of populations to live in overcrowded and unhygienic conditions, providing ideal situations for the spread of ectoparasites such as lice, fleas, ticks, mites, bed bugs, and flies. The living conditions found in refugee camps during war (84) or in jails (87) are also commonly observed for migrants and asylum seekers in developed countries (38).

SOCIAL DETERMINANTS OF POVERTY

In 2008, it was estimated that 26 million persons worldwide had been forced to flee their homes without crossing national borders, with most of them living in refugee camps (51). Others cross the borders to enter neighboring developing countries illegally. Those countries host 70% of the global refugee population, estimated at 30 to 40 million illegal migrants worldwide, 5 to 8 million entering Europe, and 11 million entering the United States (52). In Marseille, France, migrants account for up to 60% of the homeless, defined as a person that lacks customary and regular access to a conventional dwelling or residence, and migrants account for up to 15% of the homeless in eastern European countries (15, 16). This is a growing social and public health problem in wealthy, western developed countries. The number of homeless people living in the United States, the United Kingdom, and France has been estimated to be at least 500,000, 120,000, and 400,000, respectively (7). The homeless population is estimated to be over 1,500 in Marseilles, France (81), and greater than 5,500 in Tokyo (97). Only 3%–10% of homeless people become sedentary, the vast majority traveling across the countries and enhancing the risk of transmitting infectious disease such as epidemic typhus (104).

Homeless adults, as well as children, are prime candidates for exposure to ectoparasites (30, 81). Although typically prevalent in rural communities in upland areas of countries close to the equator, body lice are increasingly encountered in developed countries. The most commonly encountered fleas that parasitize humans are the cat flea, *Ctenocephalides felis*, the rat flea, *Xenopsylla cheopis*, and the human flea, *Pulex irritans*, but several recent reports have identified outbreaks of tungiasis caused by the sand flea, *Tunga penetrans*, specifically in underprivileged communities of Brazil (20). Ticks belonging to the Ixodidae, including the genera *Dermacentor*, *Rhipicephalus*, and *Ixodes*, are frequent parasites of humans. Among mites, *Sarcoptes scabiei* var. *hominis* is an obligate parasite of human skin responsible for scabies. *Liponyssoides sanguineus* is a hematophagous rodent mite that can also bite humans. More recently, the bed bug *Cimex lectularius* and the fly *Dermatobia hominis* have been reported to bite humans. Finally, the threat posed by the ectoparasite is often not from the ectoparasites themselves but from the associated infectious diseases that they may transmit to humans and the local complications and immunologic disease they can induce.

LOUSE-BORNE DISEASES

There are more than three thousand species of lice, but only three affect humans: *Pediculus humanus capitis* (head louse), *Pediculus humanus humanus* (body louse), and *Phthirus pubis* (pubic louse). *P. humanus capitis* affects all levels of society, *P. pubis* is transmitted sexually, and *P. humanus humanus* is associated more commonly with poor hygiene and extreme poverty. Head lice infestations have been prevalent among humans for thousands of years (1), and the body louse was found in textiles dating from the Jewish revolt against the Romans in AD 66–73. The human body louse potentially first arose from a common ancestor of the head louse at a time when humans began to wear clothes (55). The divergence of the head louse and body louse does not appear to result from a single event (114), and they can be differentiated into three divergent mitochondrial clades, A, B, and C (85). Body lice belong to the unique phylotype A, which is distributed worldwide, whereas head lice are distributed among all three phylotypes, each with a specific geographic distribution (65). Recently, a new hypothesis for the emergence of body lice has been proposed, suggesting that humans with both poor hygiene and head louse infestations provided an opportunity for head louse variants able to ingest a larger blood meal (a characteristic of body lice) to colonize clothing (**Figure 1**) (64).

The Head Louse

Although head lice are not yet recognized to transmit infectious agents, *B. quintana* DNA has been detected recently in lice collected on the heads of homeless persons in San Francisco, California (12). During an epidemiological investigation in a public school in Buffalo, New York, the distribution of head lice was associated with sex, age, socioeconomic status, crowding, methods of storing garments, and family size, but not with hair length (100). Poverty and ignorance appeared to contribute to the persistence of the disease. A recent study in Nepal reported that the prevalence of simultaneous infestation with both head and body lice might be as high as 59% in street children (76).

The Body Louse

Severe infestation with the body louse, *P. humanus humanus*, also known as vagabond's disease, is an issue of concern among the homeless. Close body-to-body contact is strongly associated with louse transmission. For that reason, infestation occurs more frequently in crowded environments such as homeless shelters, refugee camps, and jails, especially when hygienic standards are lacking. Lice are host specific and live in clothes, and they feed exclusively on humans several times a day (21, 86). (The **Supplemental Movie** shows body lice on homeless clothes. Lice are moving quickly and reach their preferred location: the folds of the inner belts of underwear, trousers,

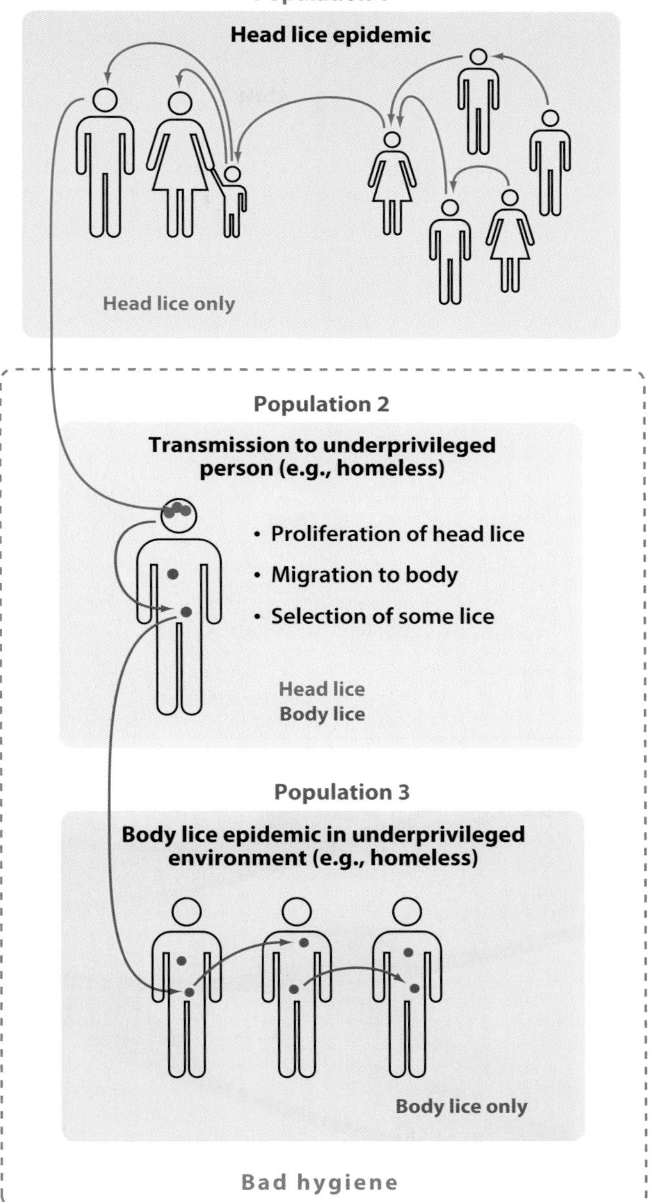

Figure 1

The hypothetical origin of body louse outbreak. Head lice outbreak in deprived populations leads to infestation of clothes and selection of "large blood feeder variant" at the origin of body louse outbreak. From Reference 64. Copyright © 2010 Li et al.

or skirts. Follow the **Supplemental Material link** from the Annual Reviews home page at **http://www.annualreviews.org**.)

Body lice are defenseless, and their only natural enemy is their host (17). A body louse's life cycle begins as an egg, laid in the folds of clothing. Because the body louse is highly susceptible to cold, the eggs are usually attached to inner clothing, close to the skin, where the body temperature reaches 29°C–32°C. When seeking lice or their eggs, the inner belts of underwear, trousers, or skirts are therefore the best places to look. Louse eggs are held in place by an adhesive produced by the female's accessory gland (67). When held at a constant temperature (i.e., when clothes are not removed), the eggs hatch 6–9 days after being laid. The emerging louse immediately moves onto the skin to feed before returning to the clothing, where it remains until feeding again. A louse typically feeds five times a day (**Figure 2**). The growing louse molts three times, usually at days 3, 5, and 10, after hatching. After the final molt, the mature louse typically lives for another 20 days. Digestion of the blood meal is rapid. Erythrocytes are quickly hemolyzed and remain liquefied. At maturity, lice can mate immediately, and both the male and the female

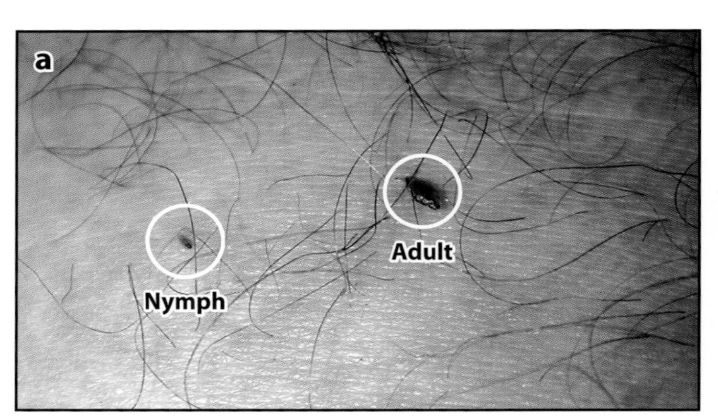

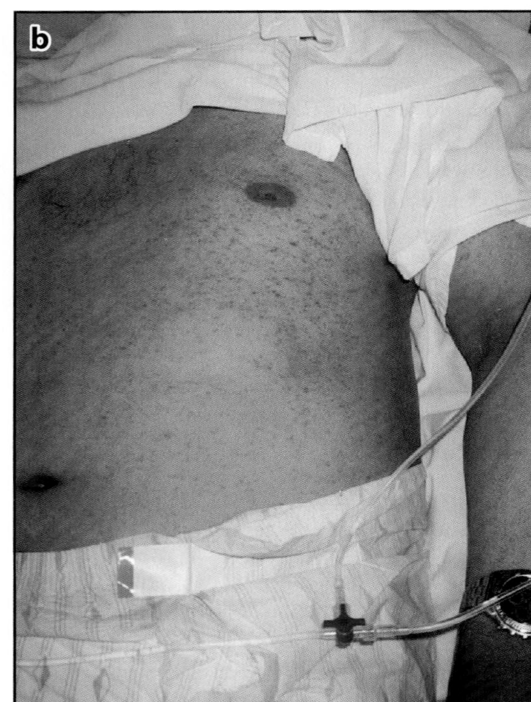

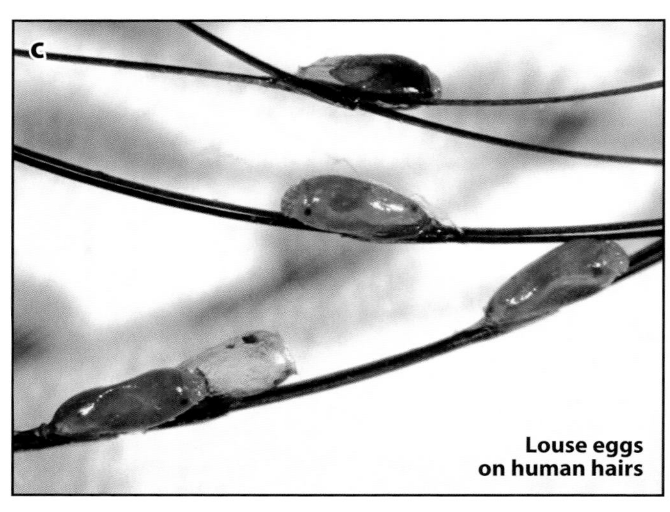

Figure 2

(*a*) Nymph and an adult body louse feeding on a homeless person. (*b*) Epidemic typhus in a homeless person returning from Algeria. Reported in Reference 71 (Niang M, Brouqui P, Raoult D. Epidemic typhus imported from Algeria. 1999. *Emerg. Infect. Dis.* 5:716–18). Note the extensive maculopapular rash on the trunk and arms. (*c*) Head lice eggs on the hair of an adult homeless person.

continue to feed throughout the prolonged mating process (67). Females lay about eight eggs a day. Daily mating is necessary because they lack a spermatheca for sperm storage, and females must mate before laying eggs. Population density is variable; usually only a few lice are observed on the same host, although we have observed people with more than 300 lice (34). Theoretically, a pair of mating lice can generate 200 lice during their 1-month life span. A population can increase by as much as 11% per day, but this rate is rarely observed (31). Although merely theoretical, this calculation shows how rapidly a louse infestation could develop. Humidity is a critical factor for lice, which are susceptible to rapid dehydration (17). The optimal humidity for survival is in the range of 70%–90% (67); they cannot survive when this value falls below 40%. Conversely, under conditions of extremely high humidity, louse feces become sticky and can fatally stick the lice to clothing. The louse's only method of rehydration is to feed on blood. The small diameter of the proboscis prevents the rapid uptake of blood, and frequent, small meals are necessary (17).

Temperature is also highly influential on the louse's physiology. Laboratory lice prefer a temperature between 29°C and 32°C (67). In the wild, lice maintain this temperature range by nestling in clothing. However, if a host becomes too hot because of fever or heavy exercise, infesting lice will leave. Body lice die at 50°C, and this temperature is critical when washing clothes, as water or soap alone will not kill lice. Eggs can survive at lower temperatures than adults, but their life span never exceeds 16 days.

Louse-Borne Diseases

Although the body louse has been experimentally shown to harbor both *Acinetobacter* spp. and *Yersinia pestis*, the agent of plague, at this time only three louse-borne diseases are recognized: trench fever, first described during World War I and caused by *B. quintana*; epidemic typhus caused by *Rickettsia prowazekii*; and relapsing fever caused by the spirochete *Borrelia recurrentis* (86). All these diseases are associated with louse infestation and poverty. Louse-borne infections have recently reemerged in jails of Rwanda and in refugee camps in Burundi (84), in a rural community in the Andes of Peru (86), and in rural louse-infested populations of Russia (109), but this reemergence has also occurred in large and modern cities of developed countries, especially in the homeless population.

Trench fever. Trench fever is an old disease, much older than previously believed. *B. quintana* DNA was detected in the dental pulp of a 4,000-year-old man (29), and trench fever has recently been identified as a disease of Napoleon's soldiers, as *B. quintana* DNA was detected in lice found in a mass grave in Lithuania (80). The name "trench fever" was chosen because the disease was associated with both Allied and German troops during World War I. At this time, the causative bacteria and a potential curative treatment were unknown. As a result, the environmental measures of sanitary discipline, improving trench construction, regular bathing, and treating lice infestation, were instituted. These measures had an unexpected success, underlying the importance of environmental control in body lice–related diseases (3). It has been estimated that trench fever affected 1,000,000 people during World War I (68). Epidemics of the disease were most frequently reported in Russia and on the eastern, central, and western European fronts during World War I and World War II. The disease was supposedly imported from the eastern front by German soldiers in 1914, and British troops were responsible for its spread to Mesopotamia (68). After the war, the incidence of trench fever fell dramatically. During World War II, trench fever reemerged and large-scale epidemics of the disease were again reported. Confusion is apparent in early articles on *B. quintana*, because it was named *Rickettsia quintana*, *Rickettsia weigli*, *Rickettsia da Rochalimae*, or *Rickettsia pediculi* (86). A rickettsia-like organism, named *Rickettsia quintana*, was proposed as the

Endocarditis: inflammation of the inner layer of the heart, the endocardium

etiologic agent of trench fever (18). Vinson & Fuller (112) reported the first successful axenic cultivation of the agent, which had been reclassified as *Rochalimaea quintana*. *R. quintana*, which has been subject to taxonomic reclassification, is now named *B. quintana*. The earliest studies for *B. quintana* were carried out on human volunteers (113) followed by macaques (18). McNee et al. (69) were the first to suggest that lice had a role in the transmission of trench fever. Experimental transmission of trench fever to human volunteers by infected lice demonstrated their role as early as 1918 (106). The ubiquity of trench fever has been further demonstrated, with cases reported in Ireland, China, Mexico, and Burundi (66, 68, 111). Recent investigations indicate the reemergence of *B. quintana* as an organism of medical importance. Evidence of *B. quintana* in the homeless has been reported in France (13, 14, 28, 42), the United States (53), Japan (97), and Russia (96). In this population *B. quintana* causes trench fever (36), chronic bacteremia (14, 94, 103), endocarditis (28, 37), bacillary angiomatosis (56), and undifferentiated fever in HIV-infected persons (57).

Trench fever results from a primary infection with *B. quintana*. It is an acute disease, with sudden onset of high grade fever, headaches, dizziness, and a characteristic shin pain as the most frequently observed signs. Dizziness and headaches are sometimes so sudden that they were reported to cause soldiers to fall into the trenches. There is usually no rash (49). The first episode of fever may last from 2 to 4 days and sometimes be followed by relapses every 4–6 days in the more prolonged form, giving the name of Quintane fever, the origin of the bacterial name (*Rickettsia*) *B. quintana*. The incubation period typically varies from 15 to 25 days but may be reduced to 6 days in experimental infections. Although trench fever often results in prolonged disability, no deaths have been reported. Other clinical manifestations of *B. quintana* infection are chronic bacteremia (33), lymphadenopathy (79), endocarditis (82), and bacillary angiomatosis in immunocompromised hosts (56).

Persistent and chronic bacteremia has long been recognized. Among 104 louse feeders enrolled for typhus vaccine production, all but 14 became ill. Of these 14, 5 had prolonged bacteremia. Among the symptomatic patients, two-thirds experienced several episodes of trench fever; these were not relapses, as they were several months apart. Asymptomatic carriers do not frequently have antibodies (58). In our experience, chronic bacteremia can last up to 78 weeks (33). A definitive link between chronic bacteremia and endocarditis, although likely, has not yet been proven (37).

Infectious endocarditis due to *B. quintana* was first reported in three non-HIV-infected homeless men in France (28). All three patients required valve replacements because of extensive valvular damage, and pathological investigation confirmed the diagnosis of endocarditis. *B. quintana* endocarditis is most often observed in homeless people with chronic alcoholism and exposure to body lice and in patients without previously known valvulopathy (37). *Bartonella* endocarditis is usually indolent and culture negative. As a result, diagnosis is often delayed, resulting in a higher mortality rate compared with endocarditis caused by other microorganisms (83).

Bacillary angiomatosis is a vascular proliferative disease most often involving the skin, but it may involve other organs such as the spleen or liver. The disease was first described in HIV-infected patients (105) and organ transplant recipients (54), but it can also rarely affect immunocompetent patients (108). Bacillary angiomatosis may be caused by both *B. quintana* and *B. henselae*, the agent of cat scratch disease (see Flea-Borne Diseases, below).

Epidemic typhus. Epidemic typhus is caused by *R. prowazekii*, an obligate intracellular bacterium from the $\alpha2$ group of proteobacteria. The main reservoir of the bacterium is humans, but the flying squirrel has been suggested as a possible reservoir in the United States. It causes a life-threatening, acute exanthematic febrile illness. The mortality rate varies from 0.7% to 60% for untreated cases depending

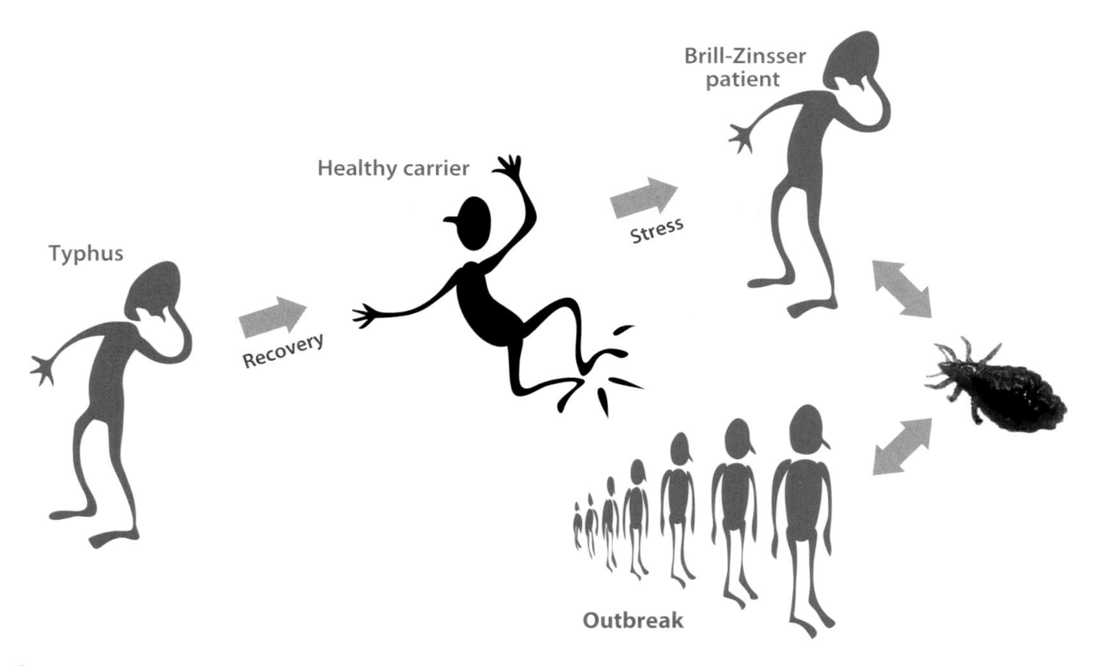

Figure 3

Hypothetical reemergence of epidemic typhus through a bacteremic Brill-Zinsser patient. The *Rickettsia prowazekii*–infected louse appears reddish brown. (Source Dr. Hervé Tissot Dupont).

on the age of patient, with a case fatality ratio lower than 5% in patients less than 30 years old. In a report of 5,747 consecutive cases occurring in Warsaw, Poland, in the Jewish quarter between January and July 1917, none of the 44 infants under 18 months died (41). This suggests that in some cases the mildness of the disease may lead to an underestimate of the importance of the outbreak. In self-resolving cases, the bacteria can persist for life in humans, and under stressful conditions recrudescence may occur as a milder form of Brill-Zinsser disease (86). Because *R. prowazekii* bacteremia occurs in Brill-Zinsser disease, it can initiate an outbreak of epidemic typhus when body louse infestations are prevalent in the population (4) (**Figure 3**). Outbreaks of epidemic typhus have always been associated with war, famine, refugee camps, cold weather, poverty, or gaps in public health management. Epidemics have been reported in recent decades in Burundi and Russia, and sporadic cases have been reported in Algeria (84, 109). In developed countries, similar poor living conditions predispose homeless populations to a high prevalence of body lice infestation (13). In a recent study in Marseilles of homeless people living in shelters, we demonstrate that significant antibody titers to *R. prowazekii* were present in 0.75% of sera (16). Moreover, in Marseilles, a sporadic case of imported typhus from Algeria in a homeless patient, a case of Brill-Zinsser, and an acute autochthonous case of epidemic typhus have been reported (4, 71, 104). Similarly, 2 of 176 homeless persons in Houston, Texas, had significant antibody titers to *R. prowazekii*, raising the possible threat of typhus in this population (90). Although no outbreaks of typhus have been identified yet in the homeless population, this disease is likely to reemerge anytime in such a situation. An extensive review of epidemic typhus has recently been published (9).

Epidemic relapsing fever. Epidemic relapsing fever is caused by the spirochete *Borrelia recurrentis* (63). Although the disease has

Meningismus: the triad of nuchal rigidity (neck stiffness), photophobia (sensitivity to bright light), and headache when present without actual infection or inflammation

Jarisch-Herxheimer reaction: occurs when large quantities of toxins are released into the body as bacteria (typically spirochetal bacteria) die due to antibiotic treatment

disappeared over large regions of the world, it is still an important endemic disease in northeastern Africa, especially in the highlands of Ethiopia (70), where it is thought to be among the top ten reasons for up to 25% of hospital admission and is associated with significant morbidity and mortality (26). Outbreaks are ongoing in Sudan and antibodies to *B. recurrentis* have been detected in rural Andean communities in Peru (78). As with epidemic typhus, relapsing fever affects mostly military and civilian populations disrupted by war and other disasters. The disease is commonly reported among slum dwellers, prisoners, and other impoverished and overcrowded segments of the population. Few data are available regarding the occurrence of relapsing fever in the homeless population. In Marseilles, we found a significantly higher seroprevalence of antibodies to *B. recurrentis* in this population. That seroprevalence increased in 2002, suggesting that an unnoticed small outbreak had occurred (16).

The illness begins abruptly with chills, headache, and fever. Most of these symptoms, which are associated with myalgia, arthralgia, abdominal pains, anorexia, dry cough, and fatigue, are mild for the first few days. Fever ranges between 39.5°C and 40°C. A cough is frequently prominent and could be associated with both epistasis and hemoptysis. Neurological involvement is usual (19). The most commonly reported neurological symptom is meningismus, characterized by neck stiffness, headache, and photophobia, which is generally not severe unless associated with subarachnoid hemorrhage. Encephalitis and encephalopathy occur occasionally, manifesting as seizures and somnolence. Physical signs may be observed, such as conjunctivitis, petechial skin rash on the trunk, splenomegaly that is often tender, and hepatomegaly. Jaundice is possible and is a diagnostic clue in louse-associated diseases.

One of the complications of louse-borne relapsing fever is bleeding, purpura and epistaxis being the more common findings. Other hemorrhagic phenomena include hemoptysis, hematemesis, hematuria, cerebral hemorrhages, bloody diarrhea, retinal hemorrhage, and spleen rupture. Clinical characteristics of relapsing fever are an initial febrile episode terminating in the crisis phenomenon, followed by an interval of apyrexia of variable length, which is followed by relapse, with return of fever and other clinical manifestations (102). Periods of relapse are less severe and shorter than the first febrile attack, with each relapse being less severe. Occasionally, no relapses are observed. The duration of the primary febrile attack averages 5.5 days. The duration of apyretic intervals averages 9.25 days (range of 3–27 days). Most patients have only one relapse, although a few have two. The duration of relapse averages 1.9 days. Peak temperatures are lower during relapses.

The most common method for detection of *Borrelia* is standard Giemsa staining of blood films, but protocols used for malaria diagnosis that use water to lyse erythrocytes usually also lyse spirochetes and are therefore useless for relapsing fever diagnosis. Although the relapsing fever–specific protein GlpQ has been used in epidemiological studies (16), serology is of little use. Polymerase chain reaction (PCR) is useful, although not accessible to many laboratories in endemic countries (88). Finally, without treatment, the death rate varies from 10% to 40%; antibiotic therapy decreases it to 2% to 4% (102). The treatment itself can be problematic, as the Jarisch-Herxheimer reaction may occur in up to 75% of treated patients, with a threefold greater risk in patients over 14 years old (70).

Other louse-associated bacteria. *Acinetobacter baumannii* is a gram-negative bacterium implicated mainly in hospital-acquired infection and is often resistant to antibiotics. It is also a prevalent cause of severe community-acquired pneumonia, endocarditis, and meningitis particularly in alcoholics. Two clones of *A. baumannii* have been characterized in lice from over the world, one of which is sensitive to antibiotics. *A. baumannii* DNA was later detected in 21% of 622 lice collected worldwide, demonstrating that *A. baumannii* is common in human body lice. Whether lice are infected through *A. baumannii* bacteremic alcoholic patients or transmit the agent to humans is still debated (60, 61).

Yersinia pestis. *Y. pestis* is the etiological agent of plague, one of the most deadly arthropod-borne diseases. Because of its role in the plague pandemic, *Xenopsylla cheopis*, the rat flea, is considered the classic vector of plague (see below). Several other arthropods have been suggested in the transmission of plague (27). Transmission of plague by clothes was first observed in 1665 and reported by M. Baltazard (8). Later, infected *P. humanus humanus* was collected from septicemic patients during familial plague in south Morocco in the 1940s (27). Although no direct louse-bite transmission was demonstrated, the role of lice in plague transmission was strongly suspected, leading to a suggested alternative transmission cycle in the plague pandemic (27). Such transmission has recently been achieved in our laboratory by adaptation of a rabbit experimental model previously used to study the louse-borne transmission of several pathogens including *B. quintana* (47).

FLEA-BORNE DISEASES

Fleas are widespread, and because they are not adapted to specific hosts, they often bite humans (101). The most common fleas that parasitize humans are the cat flea, *C. felis*, the rat flea, *X. cheopis*, and the human flea, *P. irritans*. Fleas transmit plague (*X. cheopis*, *P. irritans*) (27), murine typhus (*X. cheopis*), flea-borne spotted rickettsiosis caused by *R. felis* (*C. felis*, *P. irritans*), and cat-scratch disease due to *Bartonella henselae* (*C. felis*). Antibodies to *Bartonella* spp. have been retrieved with a high prevalence in intravenous drug users in inner-city Baltimore (24). Moreover, *B. quintana* has been recovered from *C. felis* and *P. irritans*, suggesting a potential role of fleas in the transmission (92, 95).

Murine typhus is transmitted mainly by the fleas of rodents. It is associated with cities and ports where urban rats (*Rattus rattus* and *Rattus norvegicus*) are abundant. It is also found in warm humid climates where the abundance of food and shelter supports a large population of rat reservoirs. Murine typhus represent 13.8% of febrile illness etiologies in Bedouin children in Israel (98). In the United States, cases are concentrated in suburban areas of Texas and California, and the most important reservoirs of infection in these areas are opossums and cats. The cat flea, *C. felis*, has been identified as the principal vector. Murine typhus is a febrile illness in which headaches, rash, and arthralgia are present in more than half of the patients (22). Current information on murine typhus in the southern United States is complicated by the overlapping distribution and the cross-reactivity of *R. typhi* with *R. felis*, the agent of cat flea typhus. In their study of 176 homeless in Houston, Reeves et al. (90) set up an antibody cutoff of 1/256 to avoid possible serum cross-reactivity with the spotted fever group (SFG) *R. felis*. With this cutoff, 9.6% of tested homeless people had antibody to *R. typhi*.

Cat flea typhus is caused by *R. felis*, an emerging pathogen reported first in 1990 in *C. felis* and named the ELB agent (El Labs, Soquel, CA) and then reclassified as a spotted fever group rickettsia (59). It was later detected by PCR in other fleas, including *P. irritans*, *C. felis*, and *C. canis*, worldwide. Cases of human infections have yet to be identified using molecular tools. The first case was detected in the blood of a patient from Texas. Since then, cases were reported in France, Brazil, Mexico, Germany, the Canary Islands, and Tunisia (91, 115). *R. felis* infects as many as 15% of *C. felis*, but the incidence of *R. felis* infection in patients is unknown. Clinical features of cat flea typhus are poorly described and are probably mistaken for murine typhus for the reasons cited above. In our study on 930 homeless people from Marseilles, the seroprevalence of both *R. felis* and *R. typhi* was not significantly different from that of the control population (16). However, homeless people as well as others living under poor environmental conditions such as urban shantytowns are likely to be exposed. Unlike plague, where the role of lice and *P. irritans* plays an important role in human-to-human transmission and consequent outbreaks (8), it is

Tick-borne encephalitis: a viral infection of the central nervous system affecting humans caused by the tick-borne encephalitis virus transmitted by *Ixodes* ticks

likely that flea-transmitted diseases occur more as an endemic disease and require more medical surveillance to be discovered in the exposed population.

Tungiasis is caused by the smallest flea, *Tunga penetrans*, the only species of the genus that affects human. Both males and females feed on blood but no potentially transmissible agent has been yet detected. *T. penetrans* is found on dogs, cats, and rats at a prevalence rate comparable to humans in Brazil. Tungiasis is still a neglected health problem of poor communities living in the tropics (45). It occurs in underdeveloped communities and in slums of urban centers in the dry season in tropical climates. Although reported mostly in the Caribbean, Latin America, Africa, and the Indian subcontinent, it has been reported in some areas in the United States. Homeless children are more affected than adults (30). There is no report of tungiasis in homeless adults of developed countries. Tungiasis causes only superficial skin disease, but bacterial secondary infections are relatively frequent and may develop to a severe disease with debilitating sequelae (32).

TICK-BORNE DISEASES

Ticks belonging to the Ixodidae, particularly the genera *Dermacentor*, *Rhipicephalus*, and *Ixodes*, are frequent parasites of humans. The 30-fold increase in the prevalence of viral tick-borne encephalitis in central and eastern European countries from 1992 to 1993 has been suggested to be associated with unemployment, poverty, and environmental changes, but virtually no data are available on tick-borne transmission in this specific population (107). Tick bites usually go unnoticed and the tick may remain attached to the host without any local symptoms for several hours or days, the time usually necessary for disease transmission.

Mediterranean spotted fever is caused by *Rickettsia conorii*, which is transmitted to humans by the brown dog tick, *Rhipicephalus sanguineus*. In Marseille, during summer months many dogs harbor *Rh. sanguineus*, which has occasionally been found on humans (73). Although the seroprevalence of antibodies to tick-borne (spotted fever group) rickettsia is not different between the homeless general populations of Marseille (16), we have reported a severe case of Mediterranean spotted fever in these patients, one of whom was bitten by 22 ticks (43).

MITE-BORNE DISEASES

Rickettsial Pox and Scabies

Liponyssoides sanguineus is a hematophagous biting mite of the rat, mouse, and other domestic rodents. It also bites humans and is responsible for the transmission of *Rickettsia akari*, the etiologic agent of rickettsial pox (48). This disease has been reported mainly in New York (23), the former Soviet Union, Slovenia, Ukraine, the Republic of Korea, and the People's Republic of Korea. Rickettsial pox was persistent in New York City, with 34 new cases diagnosed in an 18-month survey (72). Nine percent of intravenous drug users in Harlem, New York City, and 16% of those in inner-city Baltimore had antibodies to *R. akari* (23, 25). In our population of homeless people in Marseilles, the seroprevalence of *R. akari* was 0.2%, not different from that of the control group (16).

Scabies is caused by the mite *Sarcoptes scabiei* var. *hominis* (Arachnida) (21). It is an obligate parasite of human skin. Human-to-human transmission usually occurs after prolonged skin contact, as scabies is highly contagious. Scabies is a widespread disease occurring irrespective of race or socioeconomic condition, but epidemics occur more frequently under unsanitary conditions. Poverty, poor nutrition, homelessness, dementia, and poor hygiene are associated risk factors (44). Increased levels of scabies have been reported during periods of war. In addition, scabies occurs more frequently in winter owing both to the biology of the mites and to overcrowding. A global epidemiologic assessment is not available, but global estimates account for 300 million case of scabies (5% of the world population) (44). Scabies is present in

8.8% of the slum population in Fortaleza, Brazil (46). In the homeless, scabies was reported with a prevalence varying from 3.8% to 56.5%, depending on the population tested (2, 6). A diagnosis of scabies is based on clinical presentation and microscopic examination of the mite after skin scraping. A PCR-based detection using a highly conserved region of *S. scabiei* microsatellite 15 (Sarms 15) is effective in the diagnosis of scabies in paraffin-embedded skin biopsy (11), but this method is not routinely used for diagnosis.

FLY-BORNE DISEASES

Myiasis is the infestation of the body with dipterous larvae, also known as maggots. Soiled clothing attracts flies, especially in areas where sanitation is poor, such as landfills and slums, where fly prevalence is high (30). In the United States one-third of reported cases of myiasis occur in homeless people (99). Although *D. hominis* is the most prevalent species of fly in tropical areas, most cases of human myiasis in the United States are caused by noninvasive blowflies, such as *Lucilia sericata*, laying eggs in preexisting wounds (99).

BED BUGS

Bed bugs, *Cimex lectularius*, have the ability to disperse great distances. Local spreading, called active dispersal, refers to bed bugs walking short distances, such as when they attempt to reach hosts from their dark resting places. Active dispersal is the main method by which bed bugs spread from room to room, or floor to floor via ventilation ducts, in hotels, trains, cruise ships, or long-term care facilities. Bed bugs can also travel longer distances via passive dispersal; i.e., they are transported by humans in clothing, luggage, or furniture. The rapid turnover of residents in certain locations, e.g., backpackers' hostels and immigrant and guest-worker shelters, is consequently a risk factor for spreading bed bug infestations. Furthermore, overcrowded accommodations (dormitories, military barracks), along with poor hygiene and deprived conditions (homeless shelters, refugee camps), are factors facilitating the bed bug burden.

The incidence of skin disease secondary to infestation with the human bed bug has recently increased in the United States, the United Kingdom, and Canada (39, 110). The Toronto Public Health documented complaints of bed bug infestations from 46 locations in 2003, most commonly apartments (63%), shelters (15%), and rooming houses (11%). Pest control operators in Toronto ($n = 34$) reported treating bed bug infestations at 847 locations in 2003, most commonly single-family dwellings (70%), apartments (18%), and shelters (8%). Bed bug infestations were reported at 20 (31%) of 65 homeless shelters. In one of the affected shelters, 4% of the residents reported having bed bug bites (50).

The typical skin lesion consists of a pruritic erythematous maculopapule, 5 mm to 2 cm in diameter, with a central hemorrhagic crust or vesicle at the bite site, similar to arthropod bites. Classically, pruritus is exacerbated in the morning, temporarily resolving by the evening. Atypical forms vary from asymptomatic or paucisymptomatic to purpuric vesicular and bullous lesions. The lesions are usually numerous (5–21), depending on the intensity of infestation, and are preferentially on unclothed areas. The typical bed bug–bite distribution follows a line or curve (**Figure 4**). Bacteria such as *Wolbachia* have been detected in the bed bug (89). The transmission of more than 40 human diseases has been attributed to bed bugs, but there has yet to be a reported case of bed bug–borne infection. Bed bug infestations can have an adverse effect on health and on the quality of life in the general population, particularly among homeless persons living in shelters, and in some cases cause severe anemia (40, 77).

DIAGNOSIS OF ARTHROPOD-BORNE DISEASES

Except for relapsing fever, in which the blood smear is still the gold standard to demonstrate the presence of the spirochete (see above),

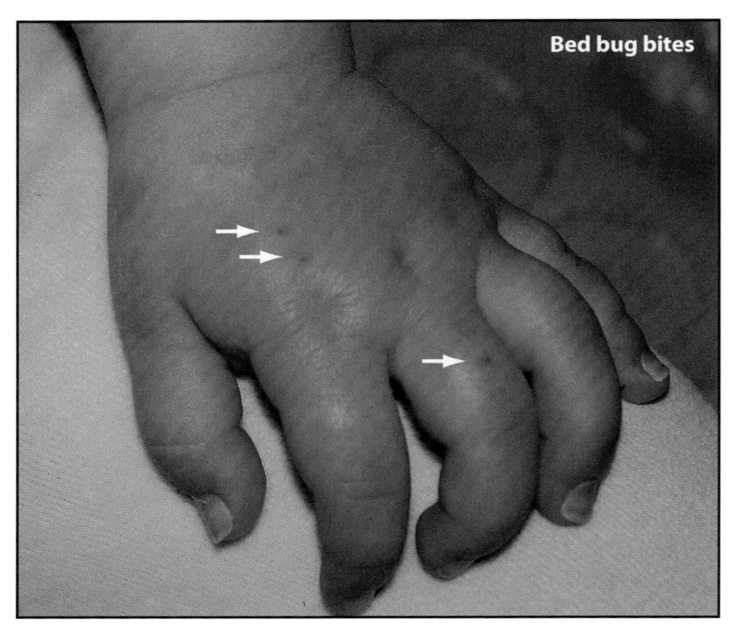

Figure 4

Typical bed bug bite distribution on a children's hand. Image courtesy of Dr. Pascal Delaunay.

Rickettsia and *Bartonella* diagnostics are based on the detection of a specific antibody by indirect fluorescent antibody testing. Serological cross-reaction occurs between closely related bacterial species. Consequently, a cross-adsorption assay allowing removal of the cross-reacting antibody is mandatory to confirm the species involved (62). This diagnostic is now available for *B. recurrentis*, because this spirochete is cultivable in vitro (75). Molecular detection and cultivation of the infectious agents for *Rickettsia*, *Bartonella*, and *Borrelia* are available, but they are restricted to specialized laboratories.

PREVENTION AND TREATMENT

Delousing and Pest Control

Over the long term, control of lice has largely been a failure. Although the simplest method for delousing is a complete change of clothing, this is not always practical or even acceptable. Other simple measures, such as washing clothes in water over 60°C or higher, can be effective (86). Powder dusting of all clothing with 10% DDT, 1% malathion, or 1% permethrin dust is another alternative (86). There is no need to disinfect other belongings, with the exception of recently used blankets or clothing. Ivermectin, a macrocyclin lactone used to treat onchocerciasis, causes paralysis in many nematodes and arthropods. It has been used to treat lice-infected swine and cattle and has been used successfully in the treatment of human head lice (10). Oral ivermectin dramatically reduces the prevalence of body lice infestation and pruritus in the homeless, but the effect is transient (5, 35). Thus, any additional measure(s) able to decrease the effect of ectoparasite-based pruritus on homeless people may be useful.

For bed bugs, fumigants, which are frequently used by nonprofessionals, do not provide any residual protection and can pose an immediate health risk to the user. Aerosolized insecticides are quick-killing agents that can be accurately applied to an area, e.g., mattress or cracks and crevices in furniture, but it is always best to vacuum first. The best option is a residual insecticide, spread by a professional in all hiding areas identified during the inspection process. However, pest control is often a challenge. Regular inspection, hygiene procedures, ongoing maintenance, and general education of the population optimize prevention.

Treatment of Specific Infections

Treatment of paucisymptomatic, persistent *B. quintana* bacteremia may be important for the prevention of endocarditis in these patients (33). We recommend that patients with *B. quintana* bacteremia be treated with gentamycin (3 mg per kilogram of body weight intravenously once daily for 14 days) in combination with doxycycline (200 mg per os daily) for 28 days, or 6 weeks in the case of endocarditis (93). The recommended treatment for epidemic typhus is doxycycline. A single dose of 200 mg is as effective as the conventional therapy for epidemic typhus (84). Doxycycline prophylaxis may be used in epidemic situations in conjunction with delousing measures.

In patients with louse-borne relapsing fever, a single dose of 100 mg of doxycycline is adequate treatment for most patients and is effective in clearing the spirochetes from blood smears (74). Tick-borne rickettsioses should be treated with 200 mg of doxycycline per day for 1–5 days.

CONCLUSION

Infestations with scabies, tungiasis, myiasis, and lice are responsible for a non-negligible morbidity in poor communities worldwide (44). Murine typhus and other flea-borne infections are likely to be underreported. However, among arthropod-borne diseases in these deprived populations, louse-borne infection is particularly worrisome. Despite effective treatment of *B. quintana* bacteremia and the efforts made to delouse at-risk populations, *B. quintana* remains endemic (16). Unnoticed outbreaks of epidemic typhus and relapsing fever have been reported (16, 90). The uncontrolled louse infestation in deprived populations should alert the community to the possibility of severe reemerging louse-borne infections.

SUMMARY POINTS

1. Poverty, lack of hygiene, cold weather, and overcrowding are conditions commonly found in refugee camps, in homeless shelters, or in slums of developed countries, and lead to proliferation of lice and emergence of louse-borne diseases.
2. Eradication of lice is a challenge in such situations and new therapeutic approaches are needed.
3. Consequently, *B. quintana* infection is still uncontrolled, and epidemic typhus can reemerge at any time.
4. Flea-borne diseases, including murine typhus, are underreported in these populations.
5. Bed bug outbreaks are more often reported in developed countries and their medical importance is still under investigation.

DISCLOSURE STATEMENT

The author is not aware of any affiliations, memberships, funding, or financial holdings that might be perceived as affecting the objectivity of this review.

ACKNOWLEDGMENTS

I especially thank Professor Didier Raoult for critically reading the manuscript and for providing important insights into the evolutionary and phylogenetic history of lice.

LITERATURE CITED

1. Araujo A, Ferreira LF, Guidon N, Maues Da Serra FN, Reinhard KJ, Dittmar K. 2000. Ten thousand years of head lice infection *Parasitol. Today* 16:269
2. Arfi C, Dehen L, Benassaia E, Faure P, Farge D, et al. 1999. Dermatologic consultation in a precarious situation: a prospective medical and social study at the Hopital Saint-Louis in Paris. *Ann. Dermatol. Venereol.* 126:682–86
3. Atenstaedt RL. 2007. The response to the trench diseases in World War I: a triumph of public health science. *Public Health* 121:634–9

4. Badiaga S, Brouqui P, Raoult D. 2005. Autochthonous epidemic typhus associated with *Bartonella quintana* bacteremia in a homeless person. *Am. J. Trop. Med. Hyg.* 72:638–39

5. Badiaga S, Foucault C, Rogier C, Doudier B, Rovery C, et al. 2008. The effect of a single dose of oral ivermectin on pruritus in the homeless. *J. Antimicrob. Chemother.* 62:404–9

6. Badiaga S, Menard A, Tissot DH, Ravaux I, Chouquet D, et al. 2005. Prevalence of skin infections in sheltered homeless. *Eur. J. Dermatol.* 15:382–86

7. Badiaga S, Raoult D, Brouqui P. 2008. Preventing and controlling emerging and reemerging transmissible diseases in the homeless. *Emerg. Infect. Dis.* 14(9):1353–59

8. Baltazard M, Bahmanyar M, Mostachfi P, Eftekhari M, Mofidi C. 1960. Research on plague in Iran. *Bull. World Health Organ.* 23:141–55

9. Bechah Y, Capo C, Mege JL, Raoult D. 2008. Epidemic typhus. *Lancet Infect. Dis.* 8:417–26

10. Bell TA. 1998. Treatment of *Pediculus humanus* var. *capitis* infestation in Cowlitz County, Washington, with ivermectin and the LiceMeister comb. *Pediatr. Infect. Dis. J.* 17:923–24

11. Bezold G, Lange M, Schiener R, Palmedo G, Sander CA, et al. 2001. Hidden scabies: diagnosis by polymerase chain reaction. *Br. J. Dermatol.* 144:614–18

12. Bonilla DL, Kabeya H, Henn J, Kramer VL, Kosoy MY. 2009. *Bartonella quintana* in body lice and head lice from homeless persons, San Francisco, California, USA. *Emerg. Infect. Dis.* 15:912–15

13. Brouqui P, Houpikian P, Dupont HT, Toubiana P, Obadia Y, et al. 1996. Survey of the seroprevalence of *Bartonella quintana* in homeless people. *Clin. Infect. Dis.* 23:756–59

14. Brouqui P, Lascola B, Roux V, Raoult D. 1999. Chronic *Bartonella quintana* bacteremia in homeless patients. *N. Engl. J. Med.* 340:184–89

15. Brouqui P, Raoult D. 2006. Arthropod-borne diseases in homeless. *Ann. N. Y. Acad. Sci.* 1078:223–35

16. Brouqui P, Stein A, Dupont HT, Gallian P, Badiaga S, et al. 2005. Ectoparasitism and vector-borne diseases in 930 homeless people from Marseilles. *Medicine* 84:61–68

17. Burgess IF. 1995. Human lice and their management. *Adv. Parasitol.* 36:271–342

18. Byam W, Lloyd LL. 1920. Trench fever: its epidemiology and endemiology. *Proc. R. Soc. Med.* 13:1–27

19. Cadavid D, Barbour AG. 1998. Neuroborreliosis during relapsing fever: review of the clinical manifestations, pathology, and treatment of infections in humans and experimental animals. *Clin. Infect. Dis.* 26:151–64

20. Cestari TF, Pessato S, Ramos-e-Silva. 2007. Tungiasis and myiasis. *Clin. Dermatol.* 25:158–64

21. Chosidow O. 2000. Scabies and pediculosis. *Lancet* 355:819–26

22. Civen R, Ngo V. 2008. Murine typhus: an unrecognized suburban vectorborne disease. *Clin. Infect. Dis.* 46:913–18

23. Comer JA, Diaz T, Vlahov D, Monterroso E, Childs JE. 2001. Evidence of rodent-associated *Bartonella* and *Rickettsia* infections among intravenous drug users from Central and East Harlem, New York City. *Am. J. Trop. Med. Hyg.* 65:855–60

24. Comer JA, Flynn C, Regnery RL, Vlahov D, Childs JE. 1996. Antibodies to *Bartonella* species in inner-city intravenous drug users in Baltimore, Md. *Arch. Intern. Med.* 156:2491–95

25. Comer JA, Tzianabos T, Flynn C, Vlahov D, Childs JE. 1999. Serologic evidence of rickettsialpox (*Rickettsia akari*) infection among intravenous drug users in inner-city Baltimore, Maryland. *Am. J. Trop. Med. Hyg.* 60:894–98

26. Cutler SJ, Abdissa A, Trape JF. 2009. New concepts for the old challenge of African relapsing fever borreliosis. *Clin. Microbiol. Infect.* 15:400–6

27. Drancourt M, Houhamdi L, Raoult D. 2006. *Yersinia pestis* as a telluric, human ectoparasite-borne organism. *Lancet Infect. Dis.* 6:234–41

28. Drancourt M, Mainardi JL, Brouqui P, Vandenesch F, Carta A, et al. 1995. *Bartonella* (*Rochalimaea*) *quintana* endocarditis in three homeless men. *N. Engl. J. Med.* 332:419–23

29. Drancourt M, Tran-Hung L, Courtin J, Lumley H, Raoult D. 2005. *Bartonella quintana* in a 4000-year-old human tooth. *J. Infect. Dis.* 191:607–11

30. Estrada B. 2003. Ectoparasitic infestations in homeless children. *Semin. Pediatr. Infect. Dis.* 14:20–24

31. Evans FC, Smith FE. 1952. The intrinsic rate of natural increase for the human louse *Pediculus humanus*. *Am. Nat.* 86:299–310

5. A double-blind randomized study on the effect of ivermectin on homeless people with pruritus.

14. Discovery of the relationship between homelessness and severe infection due to *B. quintana*.

16. The largest study on arthropod-borne diseases and homelessness demonstrating the medical importance and challenge of eradicating lice in this population.

27. The most recent review on plague and ectoparasites.

32. Feldmeier H, Eisele M, Saboia-Moura RC, Heukelbach J. 2003. Severe tungiasis in underprivileged communities: case series from Brazil. *Emerg. Infect. Dis.* 9:949–55
33. Foucault C, Barrau K, Brouqui P, Raoult D. 2002. *Bartonella quintana* bacteremia among homeless people. *Clin. Infect. Dis.* 35:684–89
34. Foucault C, Brouqui P, Raoult D. 2006. *Bartonella quintana* characteristics and clinical management. *Emerg. Infect. Dis.* 12:217–23
35. Foucault C, Ranque S, Badiaga S, Rovery C, Raoult D, Brouqui P. 2006. Oral ivermectin in the treatment of body lice. *J. Infect. Dis.* 193:474–76
36. Foucault C, Rolain JM, Raoult D, Brouqui P. 2004. Detection of *Bartonella quintana* by direct immunofluorescence examination of blood smears of a patient with acute trench fever. *J. Clin. Microbiol.* 42:4904–6
37. Fournier PE, Lelievre H, Eykyn SJ, Mainardi JL, Marrie TJ, et al. 2001. Epidemiologic and clinical characteristics of *Bartonella quintana* and *Bartonella henselae* endocarditis: a study of 48 patients. *Medicine* 80:245–51
38. Global Health Watch 2. 2008. *An alternative world health report. Access to health care for migrants and asylum-seekers.* **http://www.ghwatch.org/ghw2**
39. Goddard J, de Shazo R. 2008. Rapid rise in bed bug populations: the need to include them in the differential diagnosis of mysterious skin rashes. *South. Med. J.* 101:854–55
40. Goddard J, deShazo R. 2009. Bed bugs (*Cimex lectularius*) and clinical consequences of their bites. *JAMA* 301:1358–66
41. Goodall EW. 1920. Typhus fever in Poland, 1916 to 1919. *Proc. R. Soc. Med.* 13:261–76
42. Guibal F, de La Salmonière P, Rybojad M, Hadjrabia S, Dehen L, Arlet G. 2001. High seroprevalence to *Bartonella quintana* in homeless patients with cutaneous parasitic infestations in downtown Paris. *J. Am. Acad. Dermatol.* 44:219–23
43. Hemmersbach-Miller M, Parola P, Raoult D, Brouqui P. 2004. A homeless man with maculopapular rash who died in Marseille, France. *Clin. Infect. Dis.* 38:1412, 1493–94
44. Hengge UR, Currie BJ, Jager G, Lupi O, Schwartz RA. 2006. Scabies: a ubiquitous neglected skin disease. *Lancet Infect. Dis.* 6:769–79
45. Heukelbach J, de Oliveira FA, Hesse G, Feldmeier H. 2001. Tungiasis: a neglected health problem of poor communities. *Trop. Med. Int. Health* 6:267–72
46. Heukelbach J, Wilcke T, Winter B, Feldmeier H. 2005. Epidemiology and morbidity of scabies and pediculosis capitis in resource-poor communities in Brazil. *Br. J. Dermatol.* 153:150–56
47. **Houhamdi L, Lepidi H, Drancourt M, Raoult D. 2006. Experimental model to evaluate the human body louse as a vector of plague. *J. Infect. Dis.* 194:1589–96**

47. Demonstrates the hypothetical role of lice as a vector of plague in an experimental model.

48. Huebner RJ, Jellison WL, Pomerantz C. 1946. Rickettsialpox—a newly recognized rickettsial disease. IV. Isolation of a rickettsia apparently identical with the causative agent of rickettsialpox from *Allodermanyssus sanguineus*, a rodent mite. *Public Health Rep.* 61:1677–82
49. Hurst A. 1942. Trench fever. *Br. Med. J.* 12:318–21
50. Hwang SW, Svoboda TJ, De Jong IH, Kabasele KJ, Gogosis E. 2005. Bed bug infestations in an urban environment. *Emerg. Infect. Dis.* 11:533–38
51. IDMC. 2007. *Internal displacement: global overview of trends and developments in 2006.* Geneva: Internal Displacement Monitoring Centre. **http://www.internal-displacement.org/**
52. IOM (International Organization for Migration). *Global estimates and trends.* **http://www.iom.ch/jahia/Jahia/lang/en/pid/1**
53. Jackson LA, Spach DH. 1996. Emergence of *Bartonella quintana* infection among homeless persons. *Emerg. Infect. Dis.* 2:141–44
54. Kemper CA, Lombard CM, Deresinski SC, Tompkins LS. 1990. Visceral bacillary epithelioid angiomatosis: possible manifestations of disseminated cat scratch disease in the immunocompromised host: a report of two cases. *Am. J. Med.* 89:216–22
55. Kittler R, Kayser M, Stoneking M. 2003. Molecular evolution of *Pediculus humanus* and the origin of clothing. *Curr. Biol.* 13:1414–17
56. Koehler JE, Quinn FD, Berger TG, LeBoit PE, Tappero JW. 1992. Isolation of *Rochalimaea* species from cutaneous and osseous lesions of bacillary angiomatosis. *N. Engl. J. Med.* 327:1625–31

57. Koehler JE, Sanchez MA, Tye S, Garrido-Rowland CS, Chen FM, et al. 2003. Prevalence of *Bartonella* infection among human immunodeficiency virus-infected patients with fever. *Clin. Infect. Dis.* 37:559–66
58. Kostrzewski J. 1950. Epidemiology of trench fever. *Med. Dosw. Mikrobiol.* 2:19–51
59. La Scola B, Meconi S, Fenollar F, Rolain JM, Roux V, Raoult D. 2002. Emended description of *Rickettsia felis* (Bouyer et al. 2001), a temperature-dependent cultured bacterium. *Int. J. Syst. Evol. Microbiol.* 52(Pt. 6):2035–41
60. La Scola B, Raoult D. 2004. *Acinetobacter baumannii* in human body louse. *Emerg. Infect. Dis.* 10:1671–73
61. La Scola B, Fournier PE, Brouqui P, Raoult D. 2001. Detection and culture of *Bartonella quintana*, *Serratia marcescens*, and *Acinetobacter* spp. from decontaminated human body lice. *J. Clin. Microbiol.* 39:1707–9
62. La Scola B, Raoult D. 1997. Laboratory diagnosis of rickettsioses: current approaches to diagnosis of old and new rickettsial diseases. *J. Clin. Microbiol.* 35:2715–27
63. Larsson C, Andersson M, Bergstrom S. 2009. Current issues in relapsing fever. *Curr. Opin. Infect. Dis.* 22:443–49
64. Li W, Ortiz G, Fournier PE, Gimenez G, Reed DL, et al. 2010. Genotyping of human lice suggests multiple emergences of body lice from local head louse populations. *PLoS Negl. Trop. Dis.* 4(3):e641
65. Light JE, Allen JM, Long LM, Carter TE, Barrow L, et al. 2008. Geographic distributions and origins of human head lice (*Pediculus humanus capitis*) based on mitochondrial data. *J. Parasitol.* 94:1275–81
66. Logan JS. 1989. Trench fever in Belfast, and the nature of the 'relapsing fevers' in the United Kingdom in the nineteenth century. *Ulster Med. J.* 58:83–88
67. Maunder JW. 1983. The appreciation of lice. *Proc. R. Inst. G. B.* 55:1–31
68. Maurin M, Raoult D. 1996. *Bartonella* (*Rochalimaea*) *quintana* infections. *Clin. Microbiol. Rev.* 9:273–92
69. McNee JW, Renshaw A, Brunt EH. 1916. "Trench fever": a relapsing fever occurring with the British forces in France. *Br. Med. J.* 12:225–34
70. Mitiku K, Mengistu G. 2002. Relapsing fever in Gondar, Ethiopia. *East Afr. Med. J.* 79:85–87
71. Niang M, Brouqui P, Raoult D. 1999. Epidemic typhus imported from Algeria. *Emerg. Infect. Dis.* 5:716–18
72. Paddock CD, Zaki SR, Koss T, Singleton J Jr, Sumner JW, et al. 2003. Rickettsialpox in New York City: a persistent urban zoonosis. *Ann. N. Y. Acad. Sci.* 990:36–44
73. Parola P, Raoult D. 2001. Ticks and tickborne bacterial diseases in humans: an emerging infectious threat. *Clin. Infect. Dis.* 32:897–928
74. Perine PL, Krause DW, Awoke S, McDade JE. 1974. Single-dose doxycycline treatment of louse-borne relapsing fever and epidemic typhus. *Lancet* 2:742–44
75. Porcella SF, Raffel SJ, Schrumpf ME, Schriefer ME, Dennis DT, Schwan TG. 2000. Serodiagnosis of Louse-Borne relapsing fever with glycerophosphodiester phosphodiesterase (GlpQ) from *Borrelia recurrentis*. *J. Clin. Microbiol.* 38:3561–71
76. Poudel SK, Barker SC. 2004. Infestation of people with lice in Kathmandu and Pokhara, Nepal. *Med. Vet. Entomol.* 18:212–13
77. Pritchard MJ, Hwang SW. 2009. Cases: severe anemia from bedbugs. *Can. Med. Assoc. J.* 181:287–88
78. Raoult D, Birtles RJ, Montoya M, Perez E, Tissot-Dupont H, et al. 1999. Survey of three bacterial louse-associated diseases among rural Andean communities in Peru: prevalence of epidemic typhus, trench fever, and relapsing fever. *Clin. Infect. Dis.* 29:434–36
79. Raoult D, Drancourt M, Carta A, Gastaut JA. 1994. *Bartonella* (*Rochalimaea*) *quintana* isolation in patients with chronic adenopathy, lymphopenia, and a cat. *Lancet* 343:977
80. Raoult D, Dutour O, Houhamdi L, Jankauskas R, Fournier PE, et al. 2006. Evidence for louse-transmitted diseases in soldiers of Napoleon's Grand Army in Vilnius. *J. Infect. Dis.* 193:112–20
81. Raoult D, Foucault C, Brouqui P. 2001. Infections in the homeless. *Lancet Infect. Dis.* 1:77–84
82. Raoult D, Fournier PE, Drancourt M, Marrie TJ, Etienne J, et al. 1996. Diagnosis of 22 new cases of *Bartonella endocarditis*. *Ann. Intern. Med.* 125:646–52
83. Raoult D, Fournier PE, Vandenesch F, Mainardi JL, Eykyn SJ, et al. 2003. Outcome and treatment of *Bartonella endocarditis*. *Arch. Intern. Med.* 163:226–30
84. Raoult D, Ndihokubwayo JB, Tissot-Dupont H, Roux V, Faugere B, et al. 1998. Outbreak of epidemic typhus associated with trench fever in Burundi. *Lancet* 352:353–58
85. Raoult D, Reed DL, Dittmar K, Kirchman JJ, Rolain JM, et al. 2008. Molecular identification of lice from pre-Columbian mummies. *J. Infect. Dis.* 197:535–43

58. Describes the first experimental human cases of trench fever.

81. Highlights the risk for homeless people to acquire infectious diseases, including those not arthropod borne.

84. Reports the largest recent outbreak of epidemic typhus in refugee camps of Burundi.

86. Raoult D, Roux V. 1999. The body louse as a vector of reemerging human diseases. *Clin. Infect. Dis.* 29:888–911

87. Raoult D, Roux V, Ndihokubwayo JB, Bise G, Baudon D, et al. 1997. Jail fever (epidemic typhus) outbreak in Burundi. *Emerg. Infect. Dis.* 3:357–60
88. Ras NM, Lascola B, Postic D, Cutler SJ, Rodhain F, et al. 1996. Phylogenesis of relapsing fever *Borrelia* spp. *Int. J. Syst. Bacteriol.* 46:859–65
89. Rasgon JL, Scott TW. 2004. Phylogenetic characterization of *Wolbachia* symbionts infecting *Cimex lectularius* L. and *Oeciacus vicarius* Horvath (Hemiptera: Cimicidae). *J. Med. Entomol.* 41:1175–78
90. Reeves WK, Murray KO, Meyer TE, Bull LM, Pascua RF, et al. 2008. Serological evidence of typhus group rickettsia in a homeless population in Houston, Texas. *J. Vector. Ecol.* 33:205–7
91. Richter J, Fournier PE, Petridou J, Haussinger D, Raoult D. 2002. *Rickettsia felis* infection acquired in Europe and documented by polymerase chain reaction. *Emerg. Infect. Dis.* 8:207–8
92. Rolain JM, Bourry O, Davoust B, Raoult D. 2005. *Bartonella quintana* and *Rickettsia felis* in Gabon. *Emerg. Infect. Dis.* 11:1742–44
93. Rolain JM, Brouqui P, Koehler JE, Maguina C, Dolan MJ, Raoult D. 2004. Recommendations for treatment of human infections caused by *Bartonella* species. *Antimicrob. Agents Chemother.* 48:1921–33
94. Rolain JM, Foucault C, Guieu R, La Scola B, Brouqui P, Raoult D. 2002. *Bartonella quintana* in human erythrocytes. *Lancet* 360:226–28
95. Rolain JM, Franc M, Davoust B, Raoult D. 2003. Molecular detection of *Bartonella quintana*, *B. koehlerae*, *B. henselae*, *B. clarridgeiae*, *Rickettsia felis*, and *Wolbachia pipientis* in cat fleas, France. *Emerg. Infect. Dis.* 9:338–42
96. Rydkina EB, Roux V, Gagua EM, Predtechenski AB, Tarasevich IV, Raoult D. 1999. *Bartonella quintana* in body lice collected from homeless persons in Russia. *Emerg. Infect. Dis.* 5:176–78
97. Sasaki T, Kobayashi M, Agui N. 2002. Detection of *Bartonella quintana* from body lice (Anoplura: Pediculidae) infesting homeless people in Tokyo by molecular technique. *J. Med. Entomol.* 39:427–29
98. Shalev H, Raissa R, Evgenia Z, Yagupsky P. 2006. Murine typhus is a common cause of febrile illness in Bedouin children in Israel. *Scand. J. Infect. Dis.* 38:451–55
99. Sherman RA. 2000. Wound myiasis in urban and suburban United States. *Arch. Intern. Med.* 160:2004–14
100. Slonka GF, Fleissner ML, Berlin J, Puleo J, Harrod EK, Schultz MG. 1977. An epidemic of pediculosis capitis. *J. Parasitol.* 63:377–83
101. Sousa CA. 1997. Fleas, flea allergy, and flea control: a review. *Dermatol. Online J.* 3:7
102. Southern PM Jr, Sanford JP. 1969. Relapsing fever. A clinical and microbiological review. *Medicine* 48:129–49
103. Spach DH, Kanter AS, Dougherty MJ, Larson AM, Coyle MB, et al. 1995. *Bartonella* (*Rochalimaea*) *quintana* bacteremia in inner-city patients with chronic alcoholism. *N. Engl. J. Med.* 332:424–28
104. Stein A, Purgus R, Olmer M, Raoult D. 1999. Brill-Zinsser disease in France. *Lancet* 353:1936
105. Stoler MH, Bonfiglio TA, Steigbigel RT, Pereira M. 1983. An atypical subcutaneous infection associated with acquired immune deficiency syndrome. *Am. J. Clin. Pathol.* 80:714–18
106. Strong RP, Swift HF, Opie EL, MacNeal WJ, Baetjer W, et al. 1918. Report on progress of trench fever investigation. *JAMA* 70:1597–99
107. Sumilo D, Bormane A, Asokliene L, Vasilenko V, Golovljova I, et al. 2008. Socio-economic factors in the differential upsurge of tick-borne encephalitis in Central and Eastern Europe. *Rev. Med. Virol.* 18:81–95
108. Tappero JW, Koehler JE, Berger TG, Cockerell CJ, Lee TH, et al. 1993. Bacillary angiomatosis and bacillary splenitis in immunocompetent adults. *Ann. Intern. Med.* 118:363–65
109. Tarasevich I, Rydkina E, Raoult D. 1998. Outbreak of epidemic typhus in Russia. *Lancet* 352:1151
110. Ter Poorten MC, Prose NS. 2005. The return of the common bedbug. *Pediatr. Dermatol.* 22:183–87
111. Vinson JW. 1964. Etiology of trench fever in Mexico. *Ind. Trop. Health* 5:109–14
112. Vinson JW, Fuller HS. 1961. Studies on trench fever. I. Propagation of Rickettsia-like microorganisms from a patient's blood. *Pathol. Microbiol.* 24(Suppl.):152–66

86. Reviews the role of lice in the reemergence of typhus and other louse-borne diseases.

113. Vinson JW, Varela G, Molina-Pasquel C. 1969. Trench fever. 3. Induction of clinical disease in volunteers inoculated with *Rickettsia quintana* propagated on blood agar. *Am. J. Trop. Med. Hyg.* 18:713–22
114. Yong Z, Fournier PE, Rydkina E, Raoult D. 2003. The geographical segregation of human lice preceded that of *Pediculus humanus* capitis and *Pediculus humanus* humanus. *C. R. Biol.* 326:565–74
115. Zavala-Velazquez JE, Ruiz-Sosa JA, Sanchez-Elias RA, Becerra-Carmona G, Walker DH. 2000. *Rickettsia felis* rickettsiosis in Yucatan. *Lancet* 356:1079–80

Ecology and Management of the Soybean Aphid in North America

David W. Ragsdale,[1] Douglas A. Landis,[2] Jacques Brodeur,[3] George E. Heimpel,[1] and Nicolas Desneux[4]

[1]Department of Entomology, University of Minnesota, St. Paul, Minnesota 55108; email: ragsd001@umn.edu, heimp001@umn.edu

[2]Department of Entomology, Michigan State University, East Lansing, Michigan 48824; email: landisd@msu.edu

[3]Institut de Recherche en Biologie Végétale, Département des Sciences Biologiques, Université de Montréal, Québec H1X 2B2, Canada; email: jacques.brodeur@umontreal.ca

[4]INRA (French National Institute for Agricultural Research), Sophia Antipolis 06903, France; email: nicolas.desneux@sophia.inra.fr

Annu. Rev. Entomol. 2011. 56:375–99

First published online as a Review in Advance on September 20, 2010

The *Annual Review of Entomology* is online at ento.annualreviews.org

This article's doi:
10.1146/annurev-ento-120709-144755

0066-4170/11/0107-0375$20.00

Key Words

invasive species, biological control, economic threshold, IPM, *Aphis glycines*

Abstract

The soybean aphid, *Aphis glycines* Matsumura, has become the single most important arthropod pest of soybeans in North America. Native to Asia, this invasive species was first discovered in North America in July 2000 and has rapidly spread throughout the northcentral United States, much of southeastern Canada, and the northeastern United States. In response, important elements of the ecology of the soybean aphid in North America have been elucidated, with economic thresholds, sampling plans, and chemical control recommendations widely adopted. Aphid-resistant soybean varieties were available to growers in 2010. The preexisting community of aphid natural enemies has been highly effective in suppressing aphid populations in many situations, and classical biological control efforts have focused on the addition of parasitoids of Asian origin. The keys to sustainable management of this pest include understanding linkages between the soybean aphid and other introduced and native species in a landscape context along with continued development of aphid-resistant varieties.

INTRODUCTION

Invasive species: a nonindigenous species that has undergone substantial spread and population growth, often to the point of achieving pest or weed status

Integrated pest management (IPM): an ecosystem-based strategy that relies on a combination of techniques (e.g., biological control, cultural practices, pesticides) to prevent or control pest populations

SBA: soybean aphid

USDA-APHIS: U.S. Department of Agriculture - Animal and Plant Health Inspection Service

Invasive species represent a serious global threat to natural and managed systems (24, 119). In agriculture, invasive species can reduce yields, increase control costs, and result in increased reliance on pesticides (42, 54, 140), which can disrupt existing integrated pest management (IPM) systems (124, 142). Moreover, the establishment of an invasive exotic species has the potential to cause cascading ecological impacts that may extend into natural systems as well (153). The soybean aphid, *Aphis glycines* Matsumura, in North America has emerged as a classic case of an invasive, exotic species that has become a major source of economic loss in soybean production and whose presence has aided other invasive species (59). Here, we review the invasion history, ecology, economic impacts, and management of the soybean aphid (SBA) in North America. A review of these topics is timely considering that the rate of insect invasions is anticipated to increase in the future because of globalization and climate change (92, 151), and because major soybean-producing regions of the world, in particular South America (Brazil, Argentina, Paraguay, and Bolivia) and central India (United Nations, FAO data, **http://faostat.fao.org/default.aspx**), are at risk of being invaded by this pest.

INVASION HISTORY

The invasion of SBA into the Americas from Asia was anticipated by Kogan & Turnipseed (84). This prediction was realized in July 2000 when SBA was first detected in Wisconsin, and by the end of summer 2000, SBA was found in 10 northcentral U.S. states (**Figure 1**). Although no prior records occur, SBA was likely present in the United States prior to its detection but went unnoticed until populations reached damaging levels (126). Alternatively, its capability for rapid expansion suggests that even recent infestations could become extensive. The aphid was first detected in late summer 2002 (111) in southwestern Québec, and by 2003, all soybean-growing regions throughout the province were infested, with 51 of 54 sampled fields colonized by SBA. By 2009, the aphid had colonized 30 states and three Canadian provinces (**Figure 1**). Aphids can enter low-level jet streams and migrate long distances, for instance, between primary and secondary hosts or between geographic areas. SBA also has a great capacity to disperse within and between fields (127). Tethered alates engage in up to 11 h of active flight, covering an average estimated distance of 6.7 km during a single flight period (162). Interestingly, SBA was also discovered infesting soybean in Australia in 1999 (148) but has not become a major pest there, probably because of a lack of suitable primary hosts (**http://www.ars-grin.gov/**).

The origins of the U.S. infestation are not known. Analysis of USDA-APHIS (U.S. Department of Agriculture - Animal and Plant Health Inspection Service) aphid detections coupled with climate-matching of SBA spread in the United States suggested that Japan could have been the source of the infestation (148). More recently, the genetic diversity and differentiation among 2 South Korean and 10 North American populations have been compared (109, 110). These data indicate that South Korea was not the source of the North American invasion and suggest a pattern of a small colonizing population, followed by rapid clonal amplification and subsequent large-scale differentiation.

SOYBEAN APHID ECOLOGY IN NORTH AMERICA

SBA is a heteroecious (host-alternating) aphid that utilizes a woody primary host for overwintering and an herbaceous secondary host in the growing season. In North America, neither of the known primary hosts in Asia, *Rhamnus davurica* or *R. japonica*, is present to any large extent outside botanical gardens. In contrast, an invasive buckthorn species from Europe, *R. cathartica* (common buckthorn), is widespread (149) and is used by SBA as its principal overwintering host (150). In one study, early-season SBA density in soybean

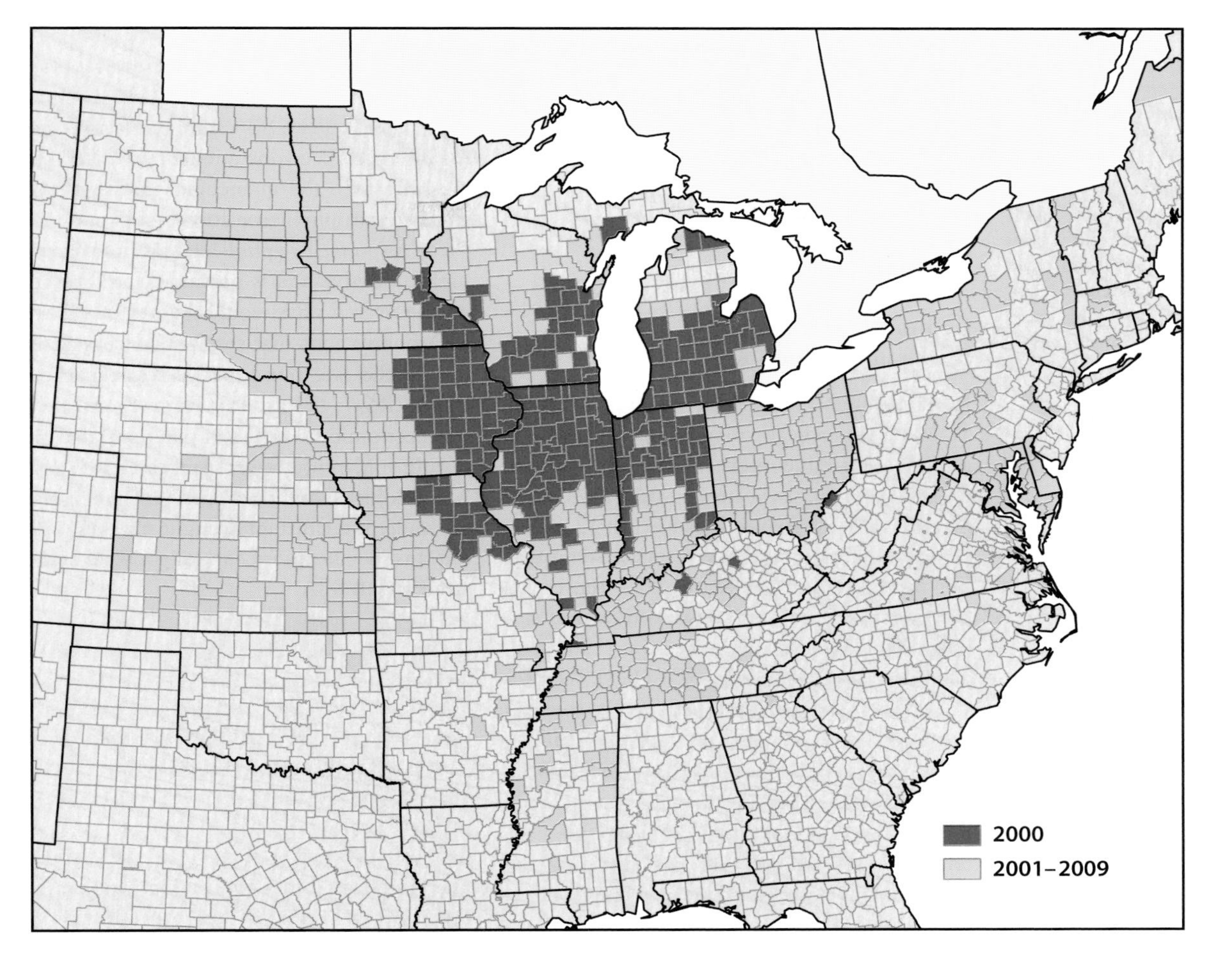

Figure 1

Distribution of the soybean aphid (SBA) in North America (125, 148). Red represents the initial 10 states (and counties within those states) that reported SBA by the end of summer 2000. Yellow represents the current known distribution (summer 2009) of SBA one decade after its initial discovery, with data recorded by county for the United States. For Canada, fine detail is not available, but pale yellow indicates provinces that reported SBA by 2009.

was best explained by the ratio of buckthorn density to field area (2). U.S. native *Rhamnus* spp., particularly *R. alnifolia* and *R. lanceolata*, are also potential primary hosts (150), and *Frangula alnus* (=*R. frangula*) is a less preferred but nonetheless acceptable primary host (65). *Glycine max* (soybean) is the principal secondary host of SBA in North America, although studies suggest reproduction is possible on Carolina horsenettle, *Solanum carolinense* (23), and red clover, *Trifolium pratense* (126). This exotic aphid from Asia survives in North America by using a primary host (common buckthorn) introduced from northern Europe and a secondary host (soybean) from Asia.

Several studies have examined SBA growth rates in relation to temperature, and as a result, the optimal and upper lethal temperatures are well understood. SBA has an optimal development temperature of 28°C and exhibits a rapid decline in reproductive output as temperatures approach 35°C (102, 105). This constraint, along with reduced abundance of primary hosts, likely explains why SBA's range has not yet extended to the southern soybean-producing regions in North America (**Figure 1**). Temperature-dependent growth rates have also been incorporated into models of SBA population dynamics (29).

Nonpersistent virus: an insect-vectored virus acquired by feeding on infected plants and immediately able to transmit virions to susceptible hosts via contaminated mouthparts without the need for a latent period between acquisition and transmission

Economic threshold (ET): the pest population level at which a control action must be taken to prevent economic damage to a crop

Economic injury level (EIL): the population level of a pest that will cause economic damage to a crop

ECONOMIC IMPORTANCE IN NORTH AMERICA

Economic Impact on Soybean Production

In the United States, there are approximately 32 M ha of soybeans planted each year, with a production value in excess of US$27 billion. Over 80% of the soybeans grown in the United States (26 M ha) are produced in just 12 northcentral states (**http://www.nass.usda.gov/**). For producers in these states, SBA represents the first insect pest to consistently cause significant yield losses over wide areas (126), with yield decreases as high as 40% (125). As a result, pest management practices for soybean producers have changed dramatically. Prior to the arrival of SBA, less than 2% of the soybean acreage in the northcentral states was scouted for arthropod pests, and IPM programs centered almost exclusively on defoliating insects. Currently, 77% of the soybean acreage is routinely scouted for insects and mites, which represents a 40-fold increase in scouting activity since the arrival of SBA (138). In a survey of IPM practices adopted by soybean producers, 84%–94% considered scouting reports crucial for making management decisions for SBA (118). SBA has resulted in a tremendous economic impact on soybean production in the northcentral United States as producers must now budget for scouting and insecticidal control of aphids to remain profitable (138).

Injury to Soybean

SBA-feeding injury manifests itself as a reduction in plant height, resulting in reduced pod set and fewer seeds within pods at maturity (5, 125, 128). At high aphid densities, seed size can also be reduced and protein content increases with a concomitant decrease in oil content (5). Aphid injury consists of the removal of photosynthate as vast quantities of plant sap are removed (147). With SBA a unique form of feeding injury has been documented in which aphids interfere with the quenching process that restores chlorophyll to a low-energy state (99). Plant response to aphid-feeding injury can be highly variable and is dependent in large part on the overall physiological status of the crop. SBA populations that reach their peak density during the late reproductive soybean growth stages (R6–7; 63) are less likely to cause serious economic loss than populations that peak during mid-reproductive stages (R3–5) (125). Aphid injury could be exacerbated when plants are under intermittent drought stress (75, 130), but this specific injury has not yet been demonstrated with SBA.

Transmission of Viruses

SBA is a competent vector of many plant viruses (22, 30, 31, 70, 76), and in this section we briefly review its ability to transmit plant-pathogenic viruses to crops other than soybean. SBA was, in part, responsible for widespread virus epidemics observed in snap bean (113) and has been implicated in a continuing problem with *Potato virus Y* in seed potato production (30, 31) and in squash production, demonstrating that the economic impact of this invasive aphid reaches well beyond soybean production. In soybean, outbreaks of *Soybean mosaic virus* have not occurred, although SBA is a competent vector and outbreaks were anticipated. Moreover, insecticide applications have not been effective in preventing spread of *Soybean mosaic virus* (122), so host plant resistance may be the most effective tool available to control spread of nonpersistent viruses in soybean.

MANAGEMENT OF THE SOYBEAN APHID

Development of Economic Thresholds

Economic thresholds (ETs) and economic injury levels (EILs) for SBA were developed using whole-plant enumerative counts of aphids per plant. To count aphids on 30 plants per field is time-consuming and might lead to lower adoption rates for SBA IPM (125). A binomial sampling plan was developed that allows rapid assessment of aphid density and leads to a recommendation to either treat or resample (71). This binomial sampling protocol was validated in multiple states, and the correct management decision was attained 79% of the time. When the incorrect decision was reached, the binomial sampling plan was always more conservative, meaning a treatment decision was reached when aphid density was, in fact, below the ET (72). Other sampling plans based on aphid density at specific plant nodes (103) are available but have not yet been incorporated into any IPM decision tool.

The ET and EIL were developed using a common experimental protocol in six northcentral states over a three-year period to produce a dataset of 19 location-years (125). The ET for SBA is 250 aphids per plant. This ET has proven applicable over a wide range of yield, price expectations, and control costs. The current ET is valid through growth stage R5 (full-size pods, immature seeds) and was developed on aphid-susceptible soybean varieties. This ET provides producers with a seven-day treatment window before aphid populations are projected to exceed the EIL of 674 aphids per plant. The current threshold is a balance between preventing catastrophic losses and conserving natural enemies with the goal of applying a single foliar spray per season to prevent aphid-induced yield loss. In a three-year comparison of the calculated ET versus treatment at a specified plant growth stage or use of seed treatments alone to control aphids, the best return on investment was the 250 aphid per plant ET (78). Valid ETs and EILs for aphid populations that reach peak density later in the growing season at plant growth stage R6 (full-size green seed in pods) are not yet available (125). The ET for SBA will need refinement as yield expectations, control costs, natural enemy abundance, aphid-resistant soybean plants, and other factors affecting aphid population growth are better understood.

Chemical Management

Prior to the discovery of SBA in North America, only 2 of the 12 northcentral U.S. states (Illinois and Ohio) reported any insecticide use in soybean, and the area treated in those 2 states was <1% of that state's soybean crop. Thus, in 2000, less than 0.1% of soybean acreage in the northcentral United States was treated for insect pests [United States Department of Agriculture–National Agricultural Statistical Service (USDA-NASS), **http://www.nass.usda.gov/**]. In contrast, by 2006, over 13% of the soybeans grown in the northcentral United States were treated with insecticides (USDA-NASS), suggesting that SBA has been responsible for a 130-fold increase in the use of insecticides on soybean fields. A similar situation has been observed in Canada, where, except for rare infestations of the fall armyworm, *Pseudaletia unipuncta*, insecticides were not used in soybean fields prior to the invasion of SBA. However, following a severe SBA infestation in 2007, 57% of the soybeans grown in Québec that were insured were treated with insecticides (Financière agricole du Québec; **http://www.fadq.qc.ca/**).

Most producers rely on foliar application of pyrethroids and organophosphate insecticides to control SBA, and during SBA outbreak years, up to 57% of a given state's soybean acreage had been treated with insecticides (USDA-NASS). Pyrethroids provide good efficacy at low use rates, and formulation allows for longer persistence than that typically found with organophosphate insecticides. Organophosphate insecticides continue to be used because they are less expensive and, unlike pyrethroids, most have some activity against the two-spotted spider mite, *Tetranychus urticae* (120), an occasional pest of soybean in dry years. Overall, insecticide use has increased production costs by US$16–$33 per ha (125). In addition, these insecticides can negatively affect

Natural enemy: an organism that kills or otherwise reduces the numbers of another organism

USDA-NASS: U.S. Department of Agriculture-National Agricultural Statistical Service

natural enemies because of lethal and multiple sublethal effects (34, 77). The use of neonicotinoid insecticides applied as seed treatments has become common (100). However, these compounds do not persist long enough within plants to provide economic control of SBA (78).

Conservation biological control: a general approach that refers to the provision of necessary requirements for biological control agents and the avoidance of practices that interfere with their beneficial activities

Host Plant Resistance

Several soybean varieties and plant introductions (PIs) have been identified as resistant to SBA and present antibiosis, antixenosis, and tolerance as mechanisms of resistance (44, 64, 67, 94). The Germplasm Resources Information Network lists 16 plant introductions as resistant and 6 as mostly resistant (USDA-ARS, **http://www.ars.usda.gov/main/main.htm**). All these accessions were initially identified in laboratory or greenhouse studies. Hill et al. (67) were the first to identify the varieties Jackson and Dowling as resistant to SBA via antibiosis and PI 71506 as having an antixenosis form of aphid resistance. Resistance in Dowling, a maturity group VIII soybean variety, has been further characterized as having a single dominant gene designated *Rag1* (Resistance to *Aphis glycines* gene 1) (69). Aphid resistance in the variety Jackson is also a single dominant gene and maps to the same linkage group (M) as Dowling and may simply be an allele of *Rag1* (65, 68, 69). Another gene showing antibiosis from PI 243540 is *Rag2* and is associated with linkage group F (107). Additional antibiosis was identified in PI 567541B and PI 567598B (106), and in these cases resistance appears to be controlled by two unnamed recessive genes. On aphid-resistant lines K1639, Pioneer 951397, Jackson, and Dowling, probing aphids required significantly longer time to reach the phloem and ingested from phloem tissues for only 2–7 min compared to >60 min on a susceptible control variety (43). In 2010, growers in the United States had SBA-resistant varieties (*Rag1*) available for the first time, and in field tests it appears that, although the primary mechanism of resistance is antibiosis, these resistant varieties also show some degree of tolerance to aphid feeding (21).

As aphid-resistant plants become available, aphids that perform equally well on resistant and susceptible varieties have been identified (81). The biotype that was found infesting lines bearing the *Rag1* gene was initially called the Ohio biotype because it was first recognized from a 2006 Ohio study. Subsequently, this biotype has been designated biotype2, with biotype1 referring to the aphid that commonly colonizes soybeans lacking any resistance genes (66). In 2007, SBA found colonizing *F. alnus* in Indiana were placed on soybean plants containing either *Rag1* or *Rag2* and readily colonized *Rag2* soybeans. These field-collected aphids have been designated biotype3 (65). Several aphid species have overcome single-gene resistance in many crops (147); however, the existence of aphid biotypes is often only the result of laboratory observations (143). Nonetheless, releases of single-gene resistance may be a poor way to deploy resistant soybean germplasm, and a combination of antibiosis and antixenosis may result in more stable resistance (146). At present, we know little about the geographic distribution of SBA biotypes or if this pattern could change over the course of a growing season as summer migrants colonize additional soybean fields.

Natural and Conservation Biological Control

The arrival of SBA in North America prompted intense interest in the factors that may control it in its introduced range. Much of this work has focused on understanding the roles of existing natural enemies in North America and the potential to conserve them to achieve biological control of SBA.

Natural enemies in Asia. In China, SBA is attacked by over 55 taxa of natural enemies, including predators, parasitoids, and pathogens (154), and seminal works in Asia suggested that natural enemies play a major role in controlling this pest (18, 93). Working in Indonesia, Van den Berg et al. (145) reported that the coccinellid *Harmonia arcuata* was the major predator reducing SBA infestations and recommended conservation of early-season natural enemies. Following the North American invasion of SBA, several studies were initiated in China to determine the relative impact of natural enemies on SBA population growth. Liu et al. (96) found that a combination of natural enemies dominated by the parasitoids (*Lysiphlebus* sp.), the coccinellid predators *Propylea japonica* and *Scymnus* (*Neopullus*) *babai*, and the dipteran predator *Paragus tibialis* reduced SBA populations by as much as 12-fold over uncaged plants. In a similar study, Miao et al. (108) found 16 species of natural enemies (3 parasitoids and 13 predators) that suppressed SBA below economic levels in a two-year study. These researchers explored the relative importance of natural enemy taxa by using cages with different mesh sizes. They concluded that, although parasitoids provided detectable aphid suppression, the combination of parasitoids and predators (dominated by *P. japonica*) was most effective in suppressing SBA.

Natural enemies in North America. Immediately after the discovery of SBA in North America, studies were initiated to determine the impact of natural enemies on SBA in the new invaded range. A diverse community of generalist natural enemies, both indigenous and naturalized, started to exploit SBA following invasion in North America (**Table 1**). Rutledge et al. (132) reported 43 predator taxa associated with SBA in soybean, including 30 species of Carabidae, 5 Coccinellidae, and 1 each of Anthocoridae, Cantharidae, Chamaemyiidae, Chrysopidae, Forficulidae, Hemerobiidae, Lampyridae, Nabidae, and Syrphidae. In subsequent no-choice trials, 15 of these taxa fed on SBA, with *Forficula auricularia*, *Coccinella septempunctata*, *Harmonia axyridis*, *Hippodamia convergens*, *Nabis* spp., and *Chrysopa* spp. consuming the highest numbers of aphids. That seminal survey (conducted in Michigan and Indiana in 2001–2002) hinted that natural enemy communities can differ widely from state to state. In Québec, Coccinellidae, Anthocoridae, Chrysopidae, Syrphidae, and Nabidae were detected in SBA-infested soybean, with Coccinellidae the most abundant predator group observed (111). Among the seven coccinellid species found, *Propylea quatuordecimpunctata*, *H. axyridis*, *Coleomegilla maculata lengi*, and *C. septempunctata* were the most common. Nine species of predatory flies were found attacking SBA in Michigan, including seven Syrphidae and one each of Cecidomyiidae and Chamaemyiidae (79). In New York, among 59 species of carabids collected in soybean fields, the exotic *Agonum muelleri* was dominant and confirmed to eat SBA in no-choice tests (55). This species spent considerable time climbing on soybean, where it reduced the population of apterous adult SBA and indirectly the numbers of nymphs these adults produce (55, 56). Finally, both harvestmen (1) and the coccinellid *Scymnus louisianae* (10) consume SBA.

Parasitoid communities of SBA in North America have been documented. In New York, three species of braconid parasitoids (*Aphidius* and *Praon* spp.) were incidentally detected in studies focusing largely on pathogens (114). Subsequent targeted studies in Michigan revealed six hymenopteran parasitoids attacking SBA: *Lysiphlebus testaceipes*, *Aphidius colemani*, *Binodoxys kelloggensis* (123), a *Praon* sp. (braconids), *Aphelinus asychis*, and a member of the *Aphelinus varipes* complex (aphelinids) (79, 115). In addition, *Aphelinus certus*, a species presumed to have been introduced accidentally, has been found attacking SBA both in soybean fields and in buckthorn in the eastern United States and Canada (59). Despite the occurrence of multiple species, parasitoids have been so far only a minor component of SBA natural enemies in North America compared with Asia. However, parasitism rates, particularly by *A. certus* and the presumably native *L. testaceipes*, have increased over the course of the soybean invasion (59, 115).

Nielsen & Hajek (114) reported seven species of pathogenic fungi recovered from SBA in New York, including *Pandora neoaphidis*, *Conidiobolus thromboides*, *Entomophthora chromaphidis*, *Pandora* sp., *Zoophthora occidentalis*, *Neozygites fresenii*, and

Table 1 Arthropod predators, parasitoids, and entomopathogenic fungi confirmed to attack soybean aphid in North America

Order	Family	Scientific name	Reference(s)
Predators			
Arachnida	Opiliones	*Phalangium opilio* Linnaeus	1
Coleoptera	Carabidae	*Agonum muelleri* Herbst	55, 56
		Anisodactylus sanctaecrusis Fabricius	132
		Bembidion quadrimaculatum Say	46, 132
		Clivina impressefrons LeConte	46, 132
		Elaphropus anceps LeConte	46, 132
		Harpalus herbivigus Say	132
		Poecilus chalcites Say	132
		Poecilus lucublandus Say	132
		Pterostichus melanarius Illiger	132
		Scymnus louisianae J. Chapin	10
	Coccinellidae	*Coccinella septempunctata* Linnaeus	17, 26, 27, 47, 111, 116, 132, 159
		Coleomegilla maculata DeG.	17, 111, 116, 132
		Coleomegilla maculata lengi Timberlake	111
		Cycloneda munda Say	26, 116
		Harmonia axyridis Pallas	17, 26, 27, 47, 50, 111, 116, 132, 159
		Hippodamia convergens Guerin-Meneville	16, 26, 27, 116, 132
		Hippodamia tredecimpunctata Linnaeus	116
		Hippodamia variegata Goeze	52, 116
		Propylea quatuordecimpunctata Linnaeus	111
	Staphylinidae	*Philonthus thoracicus* Gravenhorst	132
Dermaptera	Forficulidae	*Forficula auricularia* Linnaeus	132
Diptera	Cecidomyiidae	*Aphidoletes aphidimyza* Rondani	79, 116
	Chamaemyiidae	*Leucopis glyphinivora* Tanasijtshuk	79
	Syrphidae	*Allograpta obliqua* Say	79, 116
		Eupeodes americanus Wiedemann	116
		Eupeodes volucris Osten Sacken	79, 116
		Paragus hemorrhous Meigen	79
		Sphaerophoria contigua Macquart	79, 116
		Syrphus rectus Osten Sacken	79
		Toxomerus geminatus Say	116
		Toxomerus marginatus Say	79, 116
Heteroptera	Anthocoridae	*Orius insidiosus* Say	17, 26, 27, 37, 47, 57, 58, 132
	Miridae	*Chlamydatus associatus* Uhler	26
		Plagiognathus sp. Fieber	26, 27
	Nabidae	*Nabis* sp.	25, 27, 132
Neuroptera	Chrysopidae	*Chrysopa* sp.	27, 132
		Chrysoperla carnea Stephens	17
Parasitoids			
Hymenoptera	Aphelinidae	*Aphelinus asychis* Walker	79, 116
		Aphelinus certus Yasnosh	59

(*Continued*)

Table 1 (*Continued*)

Order	Family	Scientific name	Reference(s)
		Aphelinus varipes complex	79, 116
	Braconidae	*Aphidius colemani* Viereck	17, 79, 116
		Aphidius ervi Haliday	116
		Binodoxys communis Gahan	16, 59
		Binodoxys kelloggensis Pike, Starý & Brewer	79, 123
		Diaeretiella rapae McIntosh	116
		Ephedrus sp.	116
		Lysiphlebus testaceipes Cresson	25, 79, 116
		Praon sp.	79, 114
Entomopathogens			
Entomophthorales	Ancylistaceae	*Conidiobolus thromboides* Drechsler	114
	Entomophthoraceae	*Entomophthora chromaphidis* Burger & Swain	114
		Pandora neoaphidis Humber	114
		Pandora sp.	114
		Zoophthora occidentalis Batko	114
		Zoophthora radicans Batko	82
	Neozygitaceae	*Neozygites fresenii* Remaud et Keller	114
Hypocreales		*Lecanicillium lecanii* Gams et Zare	114

Lecanicillium lecanii. Of these, *P. neoaphidis* was the most abundant species, responsible for an 84% infection rate and subsequent collapse of the SBA population in 2003. In Minnesota, the dominant entomopathogen was *P. neoaphidis* (90%) (82), with *C. thromboides* (9%) and *Zoophthora radicans* (1%) comprising the remainder of the aphid infections in a two-year field study.

Impact of North American endemic natural enemy communities. Fox et al. (47) studied the potential impact of natural enemies on newly establishing SBA females in early-spring field trials in Michigan. They found that ground and foliar predators reduced aphids over a 24-h period in 50% of trials. The same authors demonstrated that, when protected from predation (predator communities dominated by *H. axyridis*, *Orius insidiosus*, and *Leucopus* spp.), SBA populations increased rapidly but declined when cages were subsequently opened (46). Using a similar exclusion cage design in Indiana, Desneux et al. (37) showed that a single generalist predator species, *O. insidiosus*, can decrease SBA population growth early in the season. Predators exerted top-down control of SBA populations across a wide range of agronomic treatments, whereas bottom-up impacts through host plant quality were minimal (25), and the impact of natural enemy communities results in a trophic cascade; i.e., predator control of SBA results in increased soybean yields (28, 128).

Biology of important natural enemies. Among natural enemies of SBA identified in the early years of the invasion into North America, coccinellids and the insidious flower bug, *O. insidiosus*, have emerged as particularly important (132).

The exotic lady beetles *H. axyridis* and *C. septempunctata* attack SBA in the majority of the northcentral U.S. field studies, and *C. septempunctata* is attracted to SBA-induced plant volatiles in the laboratory (163). Both coccinellids exhibit a type II functional response resulting in high rates of predation on SBA (159). When fed a diet of SBA, *H. axyridis* has a higher intrinsic rate of increase (0.238 per day) than either *P. quatuordecimpunctata* (0.215 per day) or *C. maculata lengi* (0.134 per day) (111).

Frequently associated with forested habitats (55), *H. axyridis* is commonly found preying on SBA both on soybean and on its overwintering host *Rhamnus cathartica* and is attracted to olfactory cues from naturally infested buckthorn leaves as well as visual cues from leaf silhouettes (3).

O. insidiosus can oviposit and feed on several host plants including soybean (97, 98) and can use alternative prey to build up populations in early soybean cropping season (160) prior to SBA arrival (132). Both immature and adult stages feed on SBA even when the aphids are present at low densities in the field (57, 58), and *O. insidiosus* can successfully survive, develop, and reproduce on a diet of SBA (11). SBA defensive behaviors reduce efficiency of *O. insidiosus* (13, 36); although 50% of attacks usually fail after initial probing by the predator (36), all aphids probed invariably die (13). *O. insidiosus* also feeds on soybean thrips, *Neohydatothrips variabilis*, a prey species on which it has higher fitness (12), and shows a preference for thrips over SBA (14, 36). The presence of soybean thrips may disrupt control of SBA by *O. insidiosus* in the short term through a dilution effect or preference for thrips (apparent commensalism; 36), but longer-term effects seem to favor increased suppression of SBA in the presence of thrips through apparent competition or amensalism because *O. insidiosus* reproduction is increased in the presence of soybean thrips (**Figure 2**) (12, 160).

Relative effectiveness of natural enemies. Several studies have attempted to characterize the relative effectiveness of SBA natural enemy species and guilds. A factorial field study in Michigan tested the separate and combined effects of SBA predators and parasitoids by releasing *L. testaceipes* into field cages with and without ambient levels of predators (28). Predators alone (primarily Coccinellidae) produced strong and season-long aphid suppression. However, the generalist parasitoid alone provided only minor aphid suppression, and only when predators were excluded. In a subsequent experiment, strong aphid suppression (36- to 86-fold reduction) was observed in coccinellids exposed to treatments, but only minor reduction due to small predators (primarily *O. insidiosus*) and parasitoids was observed, with aphids rapidly reaching EILs when coccinellids were excluded (27). At the plant level, coccinellid impacts resulted in a trophic cascade that restored soybean biomass and yield, whereas small natural enemies provided only minor protection against yield loss.

The importance of predation of SBA by coccinellids varies geographically in the northcentral United States, complicating their incorporation into IPM decision rules. Despite a relatively minor role of *O. insidiosus* on SBA population growth in Michigan (26), in states where coccinellids are sparse during the cropping season (or present only late in the season), as in Indiana (37, 160) and Nebraska (9), *O. insidiosus* appears to be important in SBA limitation. *O. insidiosus* suppressed SBA populations in laboratory trials (131) and was identified as a key predator of SBA in early cropping soybean fields in Indiana fields (37), but here aphid colonization dynamics are likely different because of the paucity of overwintering hosts (buckthorn) compared to that of northern states. Because *O. insidiosus* establishes in fields before SBA arrival, it can help to delay or prevent aphids from establishing in soybean fields (9, 37). However, once aphid density reaches a certain threshold (estimated in laboratory conditions at 32–64 aphids per plant depending on plant complexity; 36, 131); *O. insidiosus* is no longer able to significantly limit SBA population growth in the field (37).

Conservation of SBA natural enemies. Given the important role that existing natural enemy communities can play in regulating SBA populations, there has been considerable interest in conservation and enhancement of their populations and impacts. Several authors have examined the impacts of conventional (48) and reduced-risk insecticides (86, 117) on SBA and its natural enemies. Others have explored the impacts of alternative production systems including living-mulches (134), showing that an

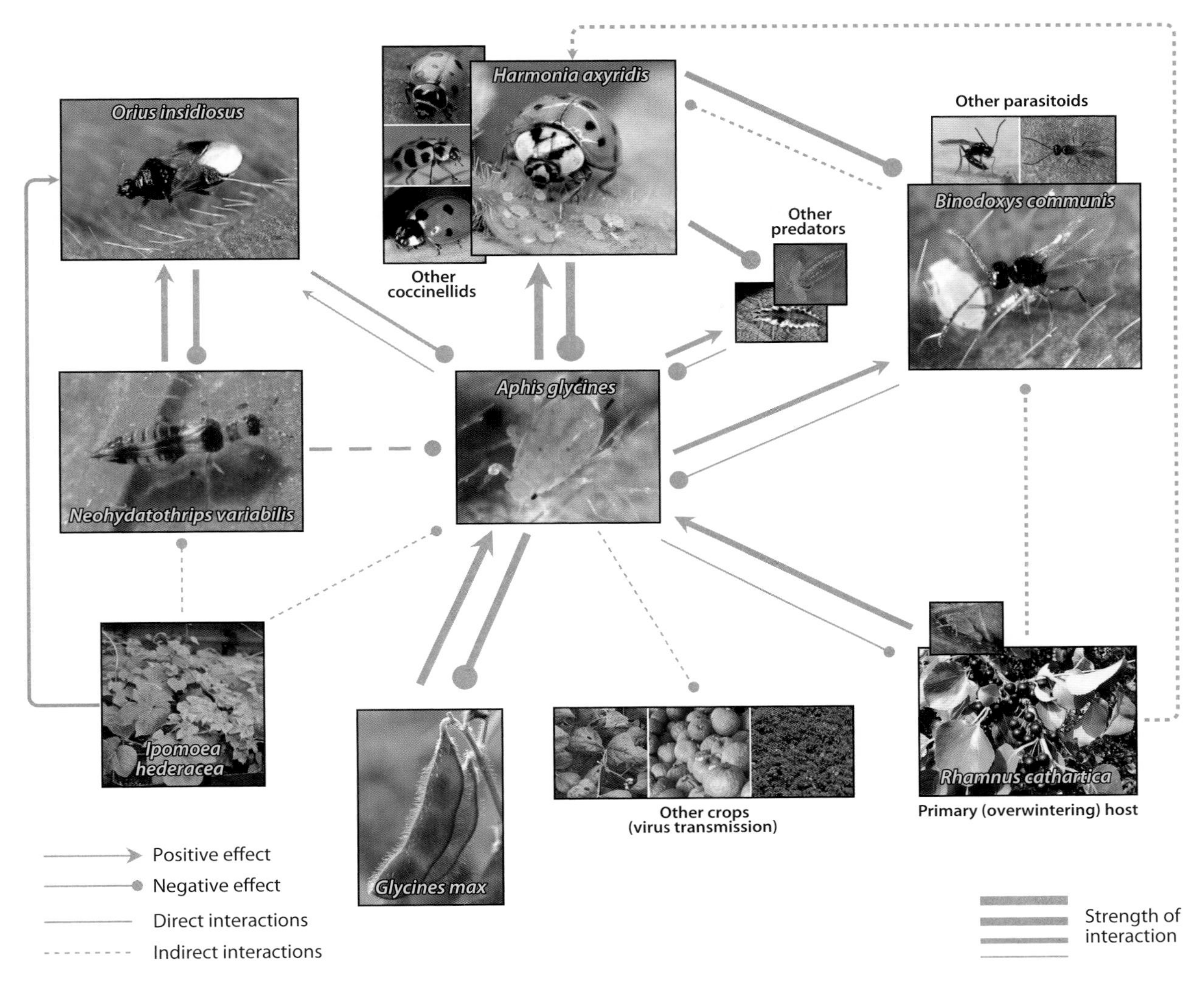

Figure 2

Multitrophic interactions and potential linkages between the soybean aphid and other biotic components of North American ecosystems. Lines with arrows indicate a positive effect in the direction of the arrows, and lines with circles indicate negative effects in the direction of the circles. Solid lines indicate direct interactions, and dashed lines indicate indirect interactions (mediated by another component of the system). Thickness of lines is roughly proportional to the known or suspected strength of the interaction. Signs and magnitudes of arrows were derived from studies cited and discussed in the review. Photos from D. Cappaert: *Coccinella septempunctata*, *Coleomegilla maculata*, and *Hippodamia convergens* (i.e., other coccinellids), *Chrysoperla* sp. and hoverfly larvae (i.e., other predators), and *Aphidius colemani*; from C.D. Difonzo: snap beans and pumpkins; from D. Hansen: *Binodoxys communis*; from M. Kogan: *Neohydatothrips variabilis*; from J.G. Lundgren: *Ipomoea hederacea*; from J.C. Malausa: *Lysiphlebus testaceipes*; from R.J. O'Neil: *Glycines max*; from J. Samanek: *Rhamnus cathartica*; from Z.S. Wu: *Aphis glycines*; and from H.J.S. Yoo: *Orius insidiosus*.

alfalfa living-mulch treatment increased predator abundance and delayed SBA population increase. Vegetation diversity can also be used as a tool for conserving natural enemies in agricultural landscapes and enhancing their potential as biological control agents (91). Lundgren et al. (98) demonstrated that *O. insidiosus* is more abundant in weedy than in weed-free soybean plots, that females prefer to oviposit onto weeds rather than into soybean,

SOYBEAN APHID AND BIOFUEL PRODUCTION

Biofuel production systems are likely to change agricultural landscapes and affect SBA management systems. In the northcentral United States, biofuel production has focused on the use of corn grain to produce ethanol. In 2007, corn acreage reached its recent peak, reducing local landscape diversity and resulting in a loss of biocontrol services in soybean estimated at US$58 million (89). Further expansion of biofuel production could negatively affect biocontrol of SBA if biofuel crops are unsuitable for natural enemies or replace habitats that are critical for their persistence (90). For example, wheat and alfalfa may provide an early-season source of aphids for reproduction of key SBA natural enemies such as coccinellids. If these crops are replaced by ones that do not support robust natural enemy communities, SBA outbreaks may become more frequent or intense. Alternatively, some biofuel crops may be more suitable for natural enemies than the crops they replace. Recent works suggest that floristically diverse biofuel crops such as mixed prairie or diverse switchgrass stands support higher abundance and diversity of beneficial insects than do monoculture crops (53). Future policy regarding the production of biofuel feedstocks should consider the ecosystem services they provide to agricultural landscapes.

and that nymphs live longest on these weeds (notably on morning glory, *Ipomoea hederacea*) (97, 98).

The influence of landscape structure on SBA–natural enemy interactions has been investigated in four midwestern U.S. states (Michigan, Iowa, Minnesota, and Wisconsin) (50). Predators, principally coccinellid beetles, dominated the natural enemy community of soybean and significantly reduced SBA populations. *H. axyridis* was the dominant exotic coccinellid in all states, comprising 45%–62% of the total coccinellid community, followed by *C. septempunctata* (13%–30%) (51). The level of biological control provided by natural enemies varied with landscape diversity. Landscapes dominated by corn and soybean fields provided less biological control to soybean than did landscapes with an abundance of crop and noncrop habitats. Landscape diversity and composition at a scale of 1.5 km surrounding the focal field explained the greatest proportion of the variation in both coccinellid abundance and biocontrol service. Landscape structure also significantly influenced the composition of coccinellid communities, with native coccinellids more abundant in low-diversity landscapes with an abundance of grassland habitat while exotic coccinellids were associated with abundance of forested habitats (51). Biocontrol services to soybean are worth at least US$239 million per year to the producers in these four states (89), and biofuel-driven growth in corn planting in 2007 resulted in lower landscape diversity, altering the supply of aphid natural enemies to soybean fields and reducing biological control by 24% (89).

Classical biological control: pest-control strategy in which natural enemies are imported and released against an invasive pest from that pest's native range

Contribution of natural enemies to SBA cycling. Major outbreaks of SBA were observed in 2001, 2003, 2005, and 2007. This pattern of outbreaks in odd-number years was noted by many researchers (129) and led to investigations of its origin. In Indiana, an eight-year study suggests that a combination of intrinsic aphid dynamics amplified by late-season predation by coccinellids best explains the phenomenon, with *H. axyridis*, *C. maculata lengi*, and *H. convergens* the most abundant species. A long-term study of the population dynamics of coccinellids in Michigan suggests that *H. axyridis* is particularly responsive to SBA, with its abundance increasing sharply in years following SBA outbreaks, and its average abundance has more than doubled since 2000 (59). Increased *H. axyridis* survival may lead to increased SBA predation in the spring following outbreak years, thus resulting in the cycle (59, 129). Detailed modeling studies have suggested a similar relationship for wheat aphids and coccinellid predators in Europe (6).

Classical Biological Control

Classical biological control of SBA is seen as a promising management option for a number of reasons. First, SBA rarely attains pest status in its native Asia (154) and a number of studies in Asia indicated that both insect predators and parasitoids were instrumental in keeping densities below economically important levels (96, 108, 145, 154). Second, the natural enemy fauna of aphid parasitoids in North America apparently largely lacks effective species (8, 25, 37, 79, 95, 115, 116, 132). While parasitism levels of SBA in China often exceed 10%, they are typically far below 1% in North America (59, 96). These observations suggest that release from Asian parasitoids may be an important contributing factor to the pest status of SBA in North America, and consequently that classical biological control using Asian parasitoids has the potential to suppress SBA below economically important levels (60, 73).

Foreign exploration and taxa recovered. Exploration for parasitoids and other specialized natural enemies of SBA began in 2001 and has been conducted in China, South Korea, and Japan. Over 40 populations of parasitoids and one population of a chamaemyiid predator were collected during these explorations and brought into quarantine laboratories in the United States for further study. The parasitoids include *Aphelinus* spp. as well as members of the aphidiine Braconidae, and various populations of both of these groups are cryptic species, making a complete taxonomic characterization of the collections difficult (41, 61, 155). At a minimum, aphidiine braconids recovered in Asia include *Binodoxys communis*, *Binodoxys koreanus*, *Lipolexis gracilis*, *Lysiphlebia japonica*, and *Lysiphlebus orientalis*, as well as *Aphelinus* species belonging to three species complexes: *A. varipes*, *A. asychis*, and *A. mali*. Of these, four represent species new to science: *B. koreanus* (32, 41), *L. orientalis* (139), and two as yet unnamed *Aphelinus* species, *A.* sp. nr. *gossypii* and *A.* sp. nr. *engaeus*.

***Binodoxys communis* as a parasitoid of SBA.** Of the species recovered in Asia, *B. communis* is the only one, as of this writing, for which a permit (from USDA-APHIS) allowing field release in North America has been granted (155). The permit was granted on the basis of laboratory studies demonstrating specificity of *B. communis* to a subset of aphids in the genus *Aphis* and of field studies of native aphids suggesting the presence of strong ecological filters that would limit the exposure of native aphids to attack by *B. communis*.

Laboratory host specificity studies on *B. communis* were done on 20 aphid species within 11 genera and two tribes, all from the aphid subfamily Aphidinae (33). Reproduction in microcosm assays was highest on SBA (58% successful parasitism), and nil or <1% on 10 aphid species. For the remaining nine species, parasitism levels ranged between 3% and 50%, with *B. communis* host use appearing to be related to the phylogenetic proximity of aphid species to *A. glycines* (33). Three native species of *Aphis* exhibited some levels of suitability for *B. communis*, *A. monardae*, *A. oestlundi*, and *A. asclepiadis*, with *A. monardae* second only to SBA in suitability to *B. communis*. However, further laboratory and field studies provided evidence of two ecological filters that should greatly limit the ability of *B. communis* to exploit native aphids. Eighteen species of ants were found tending the three native aphids in surveys of various native prairie sites in Minnesota, with more than half of the colonies tended by ants (156, 157). The most commonly encountered ant species was *Lasius neoniger*, which strongly interferes with parasitism of the native nontarget aphid *A. monardae* by *B. communis* (156). The second ecological filter is a physical refuge for native aphids on their host plants. *A. monardae* is a specialist on the native North American prairie plant *Monarda fistulosa* and feeds mainly within internal spaces of the inflorescences, which are apparently difficult for *B. communis* to access (155, 156). These ecological filters are likely to restrict parasitism of native nontarget aphid species by *B. communis*.

Release of *Binodoxys communis*. Releases of *B. communis* into North America began in 2007, with permission to do laboratory studies in Canada following in 2009. As of this writing, there has been no documentation of successful overwintering despite the fact that this

IGP: intraguild predation

parasitoid was collected in an area of China that provided a good climate match to the northcentral United States. Potential reasons for nonestablishment include absence of a suitable overwintering host and difficulty for *B. communis* to overwinter on SBA in *Rhamnus* stands (59). For example, while it is conceivable that *B. communis* could be transported to *Rhamnus* stands as eggs or larvae within alate aphids, alate aphids are parasitized at a lower rate than apterae in laboratory studies (59, 158). Another possibility is that strong predation of parasitized aphids by coccinellids and other intraguild predators severely depresses *B. communis* populations in soybean during the summer, reducing the numbers that survive to enter the overwintering season (17). Intraguild predation (IGP) has been evaluated in the field by using counts of *B. communis* mummies chewed by predators, and the fraction of chewed mummies tends to increase over the season (59) and as the SBA density increases and more predators are attracted to aphid aggregations (16). Other possible reasons for poor establishment include factors related to genetic bottlenecks in *B. communis*, and rapid emigration of *B. communis* from release sites coupled with Allee effects that increase with the distance from the release site (74). These hypotheses are under investigation.

Exotic parasitoids other than *Binodoxys communis*. A number of other Asian parasitoids merit discussion. A population of *Aphelinus atriplicis* (previously released against the Russian wheat aphid, *Diuraphis noxia*, in the early 1990s) was found to attack and develop on SBA in soybean fields (60, 154). This strain was released again in Minnesota and Wisconsin in 2002 (60), but establishment remains undocumented. *Aphelinus certus* has been found attacking SBA in eastern North America since 2005 (59). This species has not been intentionally introduced as far as we are aware and thus was likely cointroduced with SBA. If this is the case, it supports the hypothesis of at least two introductions of the SBA into North America: one in or near Wisconsin and one in eastern North America. Beyond these species are a number of *Aphelinus* spp. and aphidiine braconids that have been alluded to above. Prominent among these are two unnamed *Aphelinus* species with a narrow host range and *Lysiphlebus orientalis*, which is thelytokous in addition to exhibiting a narrow host range (139).

Intraguild Predation and Interactions with Ants

IGP may limit the effectiveness of natural enemies of SBA in North America (16, 17). In laboratory microcosms, when *H. axyridis* adults were present with *Aphidoletes aphidimyza* or *Chrysoperla carnea* larvae, the lady beetle acted as an intraguild predator (**Figure 2**) (49). However, intraguild feeding did not result in a release of aphid populations compared with microcosms containing only the intraguild and aphid prey. A similar result was found in a field cage experiment in which *H. axyridis* reduced numbers of *A. aphidimyza* and *C. carnea* larvae but also resulted in significantly fewer aphids. Thus, in both laboratory and field studies the direct impact of *H. axyridis* on SBA overcame its negative impact as an intraguild predator. Taken as a whole, the studies contrasting natural enemy impacts on SBA suggest that when present *H. axyridis* and *C. septempunctata* are keystone predators in the SBA system.

Herbert & Horn (62) recorded the ant species *Monomorium minimum*, *Formica subsericea*, and *Lasius neoniger* tending SBA populations in Ohio. In laboratory trials they demonstrated that *M. minimum* harassed or killed *O. insidiosus* and *H. axyridis* and reduced parasitism by *A. colemani*; however, in field trials exclusion of ants had no effect on SBA control (27). Tending by *L. neoniger* can strongly interfere with parasitism of nontarget aphids by *B. communis* and presumably for SBA as well. Whereas relatively high levels of tending were reported at low SBA densities in Ohio (62), virtually no tending was reported in Minnesota fields with relatively high SBA densities (157).

IMPLICATIONS FOR BASIC ECOLOGY AND INVASION BIOLOGY

While the SBA has been studied primarily as an agricultural pest in North America, insights from its invasion are relevant to population biology, food web ecology, and invasion biology theory. Studies of SBA have contributed to our understanding of top-down versus bottom-up effects, IGP, trophic cascades in food webs, and landscape control on herbivore–natural enemy interactions (16, 25, 27, 28, 49–51). Population modelers have used the SBA system to elucidate a novel formulation of exponential growth based on cumulative density-dependent feedback (29, 101). Finally, the SBA system has been proposed as an example of invasional meltdown (sensu 135, 136). Researchers have documented interconnection of SBA with multiple Eurasian species (several represented in **Figure 2**) including the earthworm *Lumbricus terrestris*; *R. cathartica*; oat crown rust, *Puccinia coronata*; European starling, *Sturnus vulgaris*; an Asian predatory flatworm, *Bipalium adventitium*; *H. axyridis*; *C. septempunctata*; *Agonum muelleri*; and *Aphelinus certus* (59).

The presence of a new nonsaturated niche of soybean-related resources (i.e., SBA) has promoted the population-level enrichment of a number of coccinellid species, in particular the exotic *H. axyridis*. Since its first recorded establishment in 1988 in southeastern Louisiana (19), *H. axyridis* has gradually spread throughout much of the United States and southern Canada (83). This species was well established in the northcentral United States by the late 1990s (thus before the arrival of SBA), but the invasion of SBA has likely facilitated its population buildup within North America (59). A 15-year survey of predators in soybean fields in Michigan (1994–2008) showed that populations of *H. axyridis* generally remained low from 1994 to 2000 but doubled following the arrival of SBA (59). Although the beneficial impact of *H. axyridis* on top-down suppression of SBA populations in soybean and of other pest aphids in other crops is clear, the coccinellid has multiple negative impacts as well (59, 83). For example, IGP by *H. axyridis* on other natural enemies may be problematic both for the establishment of new biological control agents against the SBA and the potential natural colonization of soybean fields by endemic aphid parasitoids (16, 17, 49).

The presence of the aphid has also induced a drastic change in arthropod community composition in soybean, in particular the relative proportions of aphidophagous species. Prior to SBA arrival, the coccinellids *Coleomegilla maculata*, *Coccinella septempunctata*, and *Hippodamia convergens* represented less than 5% of predators sampled in soybean fields in Iowa (4, 88). Following SBA arrival, populations of coccinellids increased and accounted for 39% (2004) and for 30% (2005) of the predator fauna (133). Conversely, *O. insidiosus* has decreased as a proportion of predators in soybean agroecosystems since the arrival of SBA in North America, more specifically in states where coccinellid populations have expanded. Schmidt et al. (133) reported that the percentage of predators that were *O. insidiosus* in Iowa soybean fields dropped from over 40% in 1977–1978 to less than 17% in 2004–2005.

INTEGRATION OF MANAGEMENT STRATEGIES

Management tactics for SBA were developed soon after its arrival in North America. For example, the ET is widely adopted and is set high enough to allow for maximum response by natural enemies. In 2010, several seed companies released soybean varieties resistant to SBA, providing producers with another tactic for SBA management. Classical biological control programs are also being developed, and one parasitoid, *B. communis*, has been released with additional species under evaluation.

As thresholds were developed, a key consideration was to understand the effectiveness of resident natural enemies in keeping aphid populations below the EIL. Under ideal conditions, the laboratory SBA population doubling time is as little as 1.5 days (105), whereas field population doubling times average 6.8 days (125). This difference in population dynamics can be attributed to density-independent and density-dependent mortality factors. ETs could conceivably be adjusted to explicitly incorporate natural enemy densities (161), but to date, effective sampling strategies for predators have not been developed. When aphid populations need to be controlled with insecticidal sprays, all registered chemical products are broad-spectrum materials that negatively affect natural enemies (34), and several authors have warned that the overuse of insecticides can result in resurgence of aphid populations (77, 85). One of the most commonly used insecticides against SBA is the pyrethroid lambda-cyhalothrin, which is lethal to natural enemies at field doses (39, 48) and can disrupt olfactory orientation toward host-infested plants and oviposition behavior in aphid parasitoids (38). Systemic insecticides such as the neonicotinoids, most commonly used at planting as a seed treatment, also have lethal and sublethal effects on aphid natural enemies (34). In laboratory studies, thiamethoxam reduced the emergence of the aphid parasitoid *Aphelinus gossypii* (144) and induced trembling, paralysis, and loss of coordination in exposed *H. axyridis* (112). The neonicotinoid imidacloprid negatively affected larval development, mobility, and adult fecundity in coccinellids (121, 137), and adverse effects on parasitoids have been documented as well (34, 87). Thus, if ETs are to be refined to incorporate natural enemy density, then negative effects of insecticides on natural enemies will need to be more thoroughly understood. This will require investigation of both lethal and sublethal (behavioral and physiological) effects of insecticides used against the soybean aphid (34, 35, 40).

The development of soybean varieties that are resistant to SBA also prompts questions about the interaction of host plant resistance and biological control. Interactions can be negative if resistant varieties interfere with biological control agents (7), or they can be positive if the two sources of mortality are complementary or synergistic (e.g., 15, 45, 80). Experiments aimed at determining these relationships for the SBA are underway in North America.

As aphid-resistant soybean varieties become more widely used, it is clear that a modified ET will need to be developed. The current ET assumes a population doubling time of 6.8 days, yet population doubling times on aphid-resistant varieties are typically in the range of 10–14 days (20). Aphid-resistant soybean varieties are not immune to colonization by SBA. Indeed, SBA biotypes have been discovered that overcome known resistance genes, and these biotypes reproduce at the same rate as SBA on susceptible soybean varieties (65). It is unclear if these biotypes will present a problem to growers, and work needs to be done on the geographic distribution of these biotypes and their ability to overwinter.

To date, all aphid-resistant soybean varieties employ a single gene for resistance, which has led to the rapid development of resistant biotypes (143). In addition, aphid-resistant varieties are often sold with an insecticidal seed treatment, typically one of several neonicotinoids. Published results demonstrate that in most years there is no yield advantage of a seed treatment, and during SBA outbreaks, seed treatments lose efficacy within 35 days following planting, well before SBA populations typically reach the ET (104). Coupling seed treatments with aphid resistance is thus a serious misuse of available IPM strategies. Although it is unclear how much of the soybean crop is treated with seed treatments, one early estimate was that 20% of soybean and 71% of corn is seed treated with one or more active ingredients (152), and industry reports indicate that seed treatments are likely to increase across all commodities (141).

CONCLUSIONS AND FUTURE OUTLOOK

The arrival of SBA in North America has linked a growing set of exotic organisms of Eurasian origin. Together, these invaders are having major impacts on both agricultural and natural ecosystems. Much remains to be discovered regarding this remarkable insect, including such basic information as the country(ies) of origin for the North American invasion, patterns of movement of the insect at landscape and regional scales, and overwintering ecology. Development of resistant varieties and effective classical biological control are likely to reduce but not eliminate SBA as a pest of soybean. In addition, SBA has already adapted to resistant varieties in laboratory studies, which presents researchers with challenges such as discovering biotypes and determining whether biotypes persist in the field. Opportunities also exist to explore the potential for regional or even continental control of the aphid via management of its primary hosts, and offer a unique opportunity to link the interests of agricultural producers and natural area land managers, both of whom have incentive to manage invasive buckthorn species. Active programs exploring classical biological control of buckthorn offer one alternative, as do physical and herbicidal control programs. Finally, molecular tools have just begun to be used to explore the biology and ecology of SBA. Use of these tools is likely to further enhance our understanding of this important pest and our ability to manage its multiple impacts.

SUMMARY POINTS

1. SBA was first discovered in North America in 2000 in Wisconsin and has subsequently spread throughout the continent. It is a major constraint to profitable soybean production in North America. SBA feeding injury reduces yields, and SBA is an efficient vector of several viruses to soybean and other crops.
2. Reliable sampling methods and ETs have been developed and are widely adopted by producers, and control strategies rely on insecticides.
3. A large guild of existing aphidophagous enemies are capable of controlling SBA but are affected by landscape structure and within-field management practices. A classical biological control effort is underway with release of one parasitoid species from Asia, *Binodoxys communis*, but establishment has not been documented as of 2009.
4. SBA appears to be involved in a broad invasional meltdown that includes a web of interactions between a number of invasive species, including SBA's main overwintering host, the common buckthorn, which is facilitated by exotic earthworms and birds and in turn serves as the overwintering host for oat rust. SBA itself facilitates populations of the multicolored Asian lady beetle (*Harmonia axyridis*), which has negative impacts on native lady beetle communities.

DISCLOSURE STATEMENT

The authors are not aware of any affiliations, memberships, funding, or financial holdings that might be perceived as affecting the objectivity of this review.

ACKNOWLEDGMENTS

One of the key laboratories studying the impact of natural enemies on SBA was that of Dr. Robert (Bob) J. O'Neil, who unfortunately passed away in February 2008. Bob challenged us to think broadly about biological and natural control, and this legacy lives on as we strive to unravel the complex relationships between invasive species and the response of native and introduced natural enemies. We also wish to thank Rob Venette for developing the distribution map of SBA in North America. Funding from the North Central Soybean Research Program, USDA-RAMP, USDA-CAR, USDA-NRI, NSF, and the generous support of state-based soybean checkoff programs and our respective institutions is gratefully acknowledged.

LITERATURE CITED

1. Allard CM, Yeargan KV. 2005. Effect of diet on development and reproduction of the harvestman *Phalangium opilio* (Opiliones: Phalangiidae). *Environ. Entomol.* 34:6–13
2. Bahlai CA, Sikkema S, Hallett RH, Newman J, Schaafsma AW. 2010. Modelling distribution and abundance of soybean aphid in soybean fields using measurements from the surrounding landscape. *Environ. Entomol.* 39:50–56
3. Bahlai CA, Welsman JA, Macleod EC, Schaafsma AV, Hallett RH, Sears MK. 2008. Role of visual and olfactory cues from agricultural hedgerows in the orientation behavior of multicolored Asian lady beetle (Coleoptera: Coccinellidae). *Environ. Entomol.* 37:973–79
4. Bechinski EJ, Pedigo LP. 1981. Ecology of predaceous arthropods in Iowa soybean agroecosystems. *Environ. Entomol.* 10:771–78
5. Beckendorf EA, Catangui MA, Riedell WE. 2008. Soybean aphid feeding injury and soybean yield, yield components, and seed composition. *Agron. J.* 100:237–46
6. Bianchi FJJA, Van der Werf W. 2004. Model evaluation of the function of prey in noncrop habitats for biological control by ladybeetles in agricultural landscapes. *Ecol. Model.* 171:177–93
7. Birch ANE, Geoghegan IE, Majerus MEN, McNicol JW, Hackett CA, et al. 1999. Tri-trophic interactions involving pest aphids, predatory 2-spot ladybirds and transgenic potatoes expressing snowdrop lectin for aphid resistance. *Mol. Breed.* 5:75–83
8. Brewer MJ, Noma T. 2010. Habitat affinity of resident natural enemies of the invasive *Aphis glycines* (Hemiptera: Aphididae), on soybean, with comments on biological control. *J. Econ. Entomol.* 103:583–96
9. Brosius TR, Higley LG, Hunt TE. 2007. Population dynamics of soybean aphid and biotic mortality at the edge of its range. *J. Econ. Entomol.* 100:1268–75
10. Brown GC, Sharkey MJ, Johnson DW. 2003. Bionomics of *Scymnus* (*Pullus*) *louisianae* J. Chapin (Coleoptera: Coccinellidae) as a predator of the soybean aphid, *Aphis glycines* Matsumura (Homoptera: Aphididae). *J. Econ. Entomol.* 96:21–24
11. Butler CD, O'Neil RJ. 2007. Life history characteristics of *Orius insidiosus* (Say) fed *Aphis glycines* Matsumura. *Biol. Control* 40:333–38
12. Butler CD, O'Neil RJ. 2007. Life history characteristics of *Orius insidiosus* (Say) fed diets of soybean aphid, *Aphis glycines* Matsumura and soybean thrips, *Neohydatothrips variabilis* (Beach). *Biol. Control* 40:339–46
13. Butler CD, O'Neil RJ. 2006. Defensive response of soybean aphid (Hemiptera: Aphididae) to predation by insidious flower bug (Hemiptera: Anthocoridae). *Ann. Entomol. Soc. Am.* 99:317–20
14. Butler CD, O'Neil RJ. 2008. Voracity and prey preference of insidious flower bug (Hemiptera: Anthocoridae) for immature stages of soybean aphid (Hemiptera: Aphididae) and soybean thrips (Thysanoptera: Thripidae). *Environ. Entomol.* 37:964–72
15. Cai QN, Ma XM, Zhao X, Cao YZ, Yang XQ. 2009. Effects of host plant resistance on insect pests and its parasitoid: a case study of wheat-aphid-parasitoid system. *Biol. Control* 49:134–38
16. Chacón JM, Heimpel GE. 2010. Density-dependent intraguild predation of an aphid parasitoid. *Oecologia* 164:213–20
17. Chacón JM, Landis DA, Heimpel GE. 2008. Potential for biotic interference of a classical biological control agent of the soybean aphid. *Biol. Control* 46:216–25
18. Chang YD, Lee JY, Youn YN. 1994. Primary parasitoids and hyperparasitoids of the soybean aphid, *Aphis glycines* (Homoptera: Aphididae). *Korean J. Appl. Entomol.* 33:51–55
19. Chapin JB, Brou VA. 1991. *Harmonia Axyridis* (Pallas), the third species of the genus to be found in the United States (Coleoptera: Coccinellidae). *Proc. Entomol. Soc. Wash.* 93:630–35
20. Chiozza MV, O'Neal ME, MacIntosh GC. 2010. Constitutive and induced differential accumulation of amino acid in leaves of susceptible and resistant soybean plants in response to the soybean aphid (Hemiptera: Aphididae). *Environ. Entomol.* 39:856–64

21. Chiozza MV, O'Neal ME, MacIntosh GC, Chandrasena DI, Tinsley NA, et al. 2010. Host plant resistance for soybean aphid management: a multi-environment study. *Crop Sci.* In press
22. Clark AJ, Perry KL. 2002. Transmissibility of field isolates of soybean viruses by *Aphis glycines*. *Plant Dis.* 86:1219–22
23. Clark TL, Puttler B, Bailey WC. 2006. Is horsenettle, *Solanum carolinense* L. (Solanaceae), an alternate host for soybean aphid, *Aphis glycines* Matsumura (Hemiptera: Aphididae)? *J. Kans. Entomol. Soc.* 79:380–83
24. Clavero M, García-Berthou E. 2005. Invasive species are a leading cause of animal extinctions. *Trends Ecol. Evol.* 20:110
25. **Costamagna AC, Landis DA. 2006. Predators exert top-down control of soybean aphid across a gradient of agricultural management systems. *Ecol. Appl.* 16:1619–28**
26. Costamagna AC, Landis DA. 2007. Quantifying predation on soybean aphid through direct field observations. *Biol. Control* 42:16–24
27. Costamagna AC, Landis DA, Brewer MJ. 2008. The role of natural enemy guilds in *Aphis glycines* suppression. *Biol. Control* 45:368–79
28. Costamagna AC, Landis DA, Difonzo CD. 2007. Suppression of soybean aphid by generalist predators results in a trophic cascade in soybeans. *Ecol. Appl.* 17:441–51
29. Costamagna AC, van der Werf W, Bianchi FJJA, Landis DA. 2007. An exponential growth model with decreasing *r* captures bottom-up effects on the population growth of *Aphis glycines* Matsumura (Hemiptera: Aphididae). *Agric. For. Entomol.* 9:297–305
30. Davis JA, Radcliffe EB. 2008. The importance of an invasive aphid species in vectoring a persistently transmitted potato virus: *Aphis glycines* is a vector of potato leafroll virus. *Plant Dis.* 92:1515–23
31. Davis JA, Radcliffe EB, Ragsdale DW. 2005. Soybean aphid, *Aphis glycines* Matsumura, a new vector of Potato virus Y in potato. *Am. J. Potato Res.* 82:197–201
32. Desneux N, Barta RJ, Delebecque CJ, Heimpel GE. 2009. Transient host paralysis as a means of reducing self-superparasitism in koinobiont endoparasitoids. *J. Insect Physiol.* 55:321–27
33. **Desneux N, Barta RJ, Hoelmer KA, Hopper KR, Heimpel GE. 2009. Multifaceted determinants of host specificity in an aphid parasitoid. *Oecologia* 160:387–98**
34. Desneux N, Decourtye A, Delpuech JM. 2007. The sublethal effects of pesticides on beneficial arthropods. *Annu. Rev. Entomol.* 52:81–106
35. Desneux N, Denoyelle R, Kaiser L. 2006. A multi-step bioassay to assess the effect of the deltamethrin on the parasitic wasp *Aphidius ervi*. *Chemosphere* 65:1697–706
36. Desneux N, O'Neil RJ. 2008. Potential of an alternative prey to disrupt predation of the generalist predator, *Orius insidiosus*, on the pest aphid, *Aphis glycines*, via short-term indirect interactions. *Bull. Entomol. Res.* 98:631–39
37. **Desneux N, O'Neil RJ, Yoo HJS. 2006. Suppression of population growth of the soybean aphid, *Aphis glycines* Matsumura, by predators: the identification of a key predator and the effects of prey dispersion, predator abundance, and temperature. *Environ. Entomol.* 35:1342–49**
38. Desneux N, Pham-Delègue MH, Kaiser L. 2004. Effect of a sublethal and lethal dose of lambda-cyhalothrin on oviposition experience and host searching behaviour of a parasitic wasp, *Aphidius ervi*. *Pest Manag. Sci.* 60:381–89
39. Desneux N, Rafalimanana H, Kaiser L. 2004. Dose-response relationship in lethal and behavioural effects of different insecticides on the parasitic wasp *Aphidius ervi*. *Chemosphere* 54:619–27
40. Desneux N, Ramirez-Romero R, Kaiser L. 2006. Multi step bioassay to predict recolonization potential of emerging parasitoids after a pesticide treatment. *Environ. Toxicol. Chem.* 25:2675–82
41. Desneux N, Starý P, Delebecque CJ, Gariepy TD, Barta RJ, et al. 2009. Cryptic species of parasitoids attacking the soybean aphid, *Aphis glycines* Matsumura (Hemiptera: Aphididae), in Asia: *Binodoxys communis* Gahan and *Binodoxyx koreanus* Stary sp. n. (Hymenoptera: Braconidae: Aphidiinae). *Ann. Entomol. Soc. Am.* 102:925–36
42. Desneux N, Wajnberg E, Wyckhuys KAG, Burgio G, Arpaia S, et al. 2010. Biological invasion of European tomato crops by *Tuta absoluta*: ecology, history of invasion and prospects for biological control. *J. Pest Sci.* 83:197–215

25. Showed that in the early invasion period, soybean aphid was primarily controlled via top-down influences of generalist predators, and suggested complementary roles of different predator species.

33. Assesses the host range of the exotic parasitoid *Binodoxys communis* and highlights the multifaceted nature of factors determining host specificity in parasitoids.

37. Demonstrates that a single generalist predator species, *O. insidiosus*, can be responsible for decreasing soybean aphid population growth in soybean field early in the season.

43. Diaz-Montano J, Reese JC, Louis J, Campbell LR, Schapaugh WT. 2007. Feeding behavior by the soybean aphid (Hemiptera: Aphididae) on resistant and susceptible soybean genotypes. *J. Econ. Entomol.* 100:984–89

44. Diaz-Montano J, Reese JC, Schapaugh WT, Campbell LR. 2006. Characterization of antibiosis and antixenosis to the soybean aphid (Hemiptera: Aphididae) in several soybean genotypes. *J. Econ. Entomol.* 99:1884–89

45. Farid A, Johnson JB, Shafii B, Quisenberry SS. 1998. Tritrophic studies of Russian wheat aphid, a parasitoid, and resistant and susceptible wheat over three parasitoid generations. *Biol. Control* 12:1–6

46. Fox TB, Landis DA, Cardoso FF, Difonzo CD. 2004. Predators suppress *Aphis glycines* Matsumura population growth in soybean. *Environ. Entomol.* 33:608–18

47. Fox TB, Landis DA, Cardoso FF, Difonzo CD. 2005. Impact of predation on establishment of the soybean aphid, *Aphis glycines* in soybean, *Glycine max*. *Biocontrol* 50:545–63

48. Galvan TL, Koch RL, Hutchison WD. 2005. Toxicity of commonly used insecticides in sweet corn and soybean to multicolored Asian lady beetle (Coleoptera: Coccinellidae). *J. Econ. Entomol.* 98:780–89

49. Gardiner MM, Landis DA. 2007. Impact of intraguild predation by adult *Harmonia axyridis* (Coleoptera: Coccinellidae) on *Aphis glycines* (Hemiptera: Aphididae) biological control in cage studies. *Biol. Control* 40:386–95

50. Gardiner MM, Landis DA, Gratton C, DiFonzo CD, O'Neal M, et al. 2009. Landscape diversity enhances biological control of an introduced crop pest in the north-central USA. *Ecol. Appl.* 19:143–54

50. Demonstrates that landscape structure within a 1.5-km radius of a focal soybean field is the key determinant of the level of soybean aphid biocontrol achieved by the natural enemy community.

51. Gardiner MM, Landis DA, Gratton C, Schmidt NP, O'Neal M, et al. 2009. Landscape composition influences patterns of native and exotic lady beetle abundance. *Divers. Distrib.* 15:554–64

52. Gardiner MM, Parsons GL. 2005. *Hippodamia variegata* (Goeze) (Coleoptera: Coccinellidae) detected in Michigan soybean fields. *Great Lakes Entomol.* 38:164–69

53. Gardiner MM, Tuell J, Isaacs R, Gibbs J, Ascher J, Landis DA. 2010. Implications of three model biofuel crops for beneficial arthropods in agricultural landscapes. *BioEnergy Res.* 3:6–19

54. Haack RA, Herard F, Sun JH, Turgeon JJ. 2010. Managing invasive populations of Asian longhorned beetle and citrus longhorned beetle: a worldwide perspective. *Annu. Rev. Entomol.* 55:521–46

55. Hajek AE, Hannam JJ, Nielsen C, Bell AJ, Liebherr JK. 2007. Distribution and abundance of Carabidae (Coleoptera) associated with soybean aphid (Hemiptera: Aphididae) populations in central New York. *Ann. Entomol. Soc. Am.* 100:876–86

56. Hannam JJ, Liebherr JK, Hajek AE. 2008. Climbing behaviour and aphid predation by *Agonum muelleri* (Coleoptera: Carabidae). *Can. Entomol.* 140:203–7

57. Harwood JD, Desneux N, Yoo HJS, Rowley DL, Greenstone MH, et al. 2007. Tracking the role of alternative prey in soybean aphid predation by *Orius insidiosus*: a molecular approach. *Mol. Ecol.* 16:4390–400

58. Harwood JD, Yoo HJS, Greenstone MH, Rowley DL, O'Neil RJ. 2009. Differential impact of adults and nymphs of a generalist predator on an exotic invasive pest demonstrated by molecular gut-content analysis. *Biol. Invasions* 11:895–903

59. Heimpel GE, Frelich LE, Landis DA, Hopper KR, Hoelmer KA, et al. 2010. European buckthorn and Asian soybean aphid as components of an extensive invasional meltdown in North America. *Biol. Invasions* 12:2913–31

60. Heimpel GE, Ragsdale DW, Venette R, Hopper KR, O'Neil RJ, Rutledge CE, Wu Z. 2004. Prospects for importation biological control of the soybean aphid: anticipating potential costs and benefits. *Ann. Entomol. Soc. Am.* 97:249–58

61. Heraty JM, Woolley JB, Hopper KR, Hawks DL, Kim JW, Buffington M. 2007. Molecular phylogenetics and reproductive incompatibility in a complex of cryptic species of aphid parasitoids. *Mol. Phylogenet. Evol.* 45:480–93

62. Herbert JJ, Horn DJ. 2008. Effect of ant attendance by *Monomorium minimum* (Buckley) (Hymenoptera: Formicidae) on predation and parasitism of the soybean aphid *Aphis glycines* Matsumura (Hemiptera: Aphididae). *Environ. Entomol.* 37:1258–63

63. Herman JC. 1988. How a soybean plant develops. *Spec. Rep. Iowa State Coop. Ext. No. 53*

64. Hesler LS, Dashiell KE, Lundgren JG. 2007. Characterization of resistance to *Aphis glycines* in soybean accessions. *Euphytica* 154:91–99
65. Hill CB, Crull L, Herman T, Voegtlin DJ, Hartman GL. 2010. A new soybean aphid (Hemiptera: Aphididae) biotype identified. *J. Econ. Entomol.* 103:509–15
66. Hill CB, Kim KS, Crull L, Diers BW, Hartman GL. 2009. Inheritance of resistance to the soybean aphid in soybean PI 200538. *Crop Sci.* 49:1193–200
67. Hill CB, Li Y, Hartman GL. 2004. Resistance to the soybean aphid in soybean germplasm. *Crop Sci.* 44:98–106
68. Hill CB, Li Y, Hartman GL. 2006. Soybean aphid resistance in soybean Jackson is controlled by a single dominant gene. *Crop Sci.* 46:1606–8
69. Hill CB, Li Y, Hartman GL. 2006. A single dominant gene for resistance to the soybean aphid in the soybean cultivar Dowling. *Crop Sci.* 46:1601–5
70. Hill JH, Alleman R, Hogg DB, Grau CR. 2001. First report of transmission of Soybean mosaic virus and Alfalfa mosaic virus by *Aphis glycines* in the New World. *Plant Dis.* 85:561
71. **Hodgson EW, Burkness EC, Hutchison WD, Ragsdale DW. 2004. Enumerative and binomial sequential sampling, plans for soybean aphid (Homoptera: Aphididae) in soybean. *J. Econ. Entomol.* 97:2127–36**
72. Hodgson EW, McCornack BP, Koch KA, Ragsdale DW, Johnson KD, et al. 2007. Field validation of speed scouting for soybean aphid. *Crop Manag.* doi:10.1094/CM-2007-0511-01-RS
73. Hoelmer KA, Kirk AA. 2005. Selecting arthropod biological control agents against arthropod pests: Can the science be improved to decrease the risk of releasing ineffective agents? *Biol. Control* 34:255–64
74. Hopper KR, Roush RT. 1993. Mate finding, dispersal, number released, and the success of biological control introductions. *Ecol. Entomol.* 18:321–31
75. Huberty AF, Denno RF. 2004. Plant water stress and its consequences for herbivorous insects: a new synthesis. *Ecology* 85:1383–98
76. Iwaki M, Roechan M, Hibino H, Tochihara H, Tantera DM. 1980. A persistent aphid borne virus of soybean, Indonesian soybean dwarf virus. *Plant Dis.* 64:1027–30
77. Johnson KD, O'Neal ME, Bradshaw JD, Rice ME. 2008. Is preventative, concurrent management of the soybean aphid (Hemiptera: Aphididae) and bean leaf beetle (Coleoptera: Chrysomelidae) possible? *J. Econ. Entomol.* 101:801–9
78. Johnson KD, O'Neal ME, Ragsdale DW, Difonzo CD, Swinton SM, et al. 2009. Probability of cost-effective management of soybean aphid (Hemiptera: Aphididae) in North America. *J. Econ. Entomol.* 102:2101–8
79. Kaiser ME, Noma T, Brewer MJ, Pike KS, Vockeroth JR, Gaimari SD. 2007. Hymenopteran parasitoids and dipteran predators found using soybean aphid after its midwestern United States invasion. *Ann. Entomol. Soc. Am.* 100:196–205
80. Kalule T, Wright DJ. 2002. Tritrophic interactions between cabbage cultivars with different resistance and fertilizer levels, cruciferous aphids and parasitoids under field conditions. *Bull. Entomol. Res.* 92:61–69
81. Kim KS, Hill CB, Hartman GL, Mian MAR, Diers BW. 2008. Discovery of soybean aphid biotypes. *Crop Sci.* 48:923–28
82. Koch KA, Potter BD, Ragsdale DW. 2010. Non-target impacts of soybean rust fungicides on the fungal entomopathogens of soybean aphid. *J. Invert. Pathol.* 103:156–64
83. Koch RL. 2003. The multicolored Asian lady beetle, *Harmonia axyridis*: a review of its biology, uses in biological control, and non-target impacts. *J. Insect Sci.* 3:1–16
84. Kogan M, Turnipseed SG. 1987. Ecology and management of soybean arthropods. *Annu. Rev. Entomol.* 32:507–38
85. Kraiss H, Cullen EM. 2008. Efficacy and nontarget effects of reduced-risk insecticides on *Aphis glycines* (Hemiptera: Aphididae) and its biological control agent *Harmonia axyridis* (Coleoptera: Coccinellidae). *J. Econ. Entomol.* 101:391–98
86. Kraiss H, Cullen EM. 2008. Insect growth regulator effects of azadirachtin and neem oil on survivorship, development and fecundity of *Aphis glycines* (Homoptera: Aphididae) and its predator, *Harmonia axyridis* (Coleoptera: Coccinellidae). *Pest Manag. Sci.* 64:660–68

71. The ET is based on a sample of 30 plants that takes 1 h to complete. The binomial sampling plan shortens sampling to 15 min and facilitates adoption of IPM by growers.

87. Krischik VA, Landmark AL, Heimpel GE. 2007. Soil-applied imidacloprid is translocated to nectar and kills nectar-feeding *Anagyrus pseudococci* (Girault) (Hymenoptera: Encyrtidae). *Environ. Entomol.* 36:1238–45
88. Lam WKF, Pedigo LP. 1998. Response of soybean insect communities to row width under crop-residue management systems. *Environ. Entomol.* 27:1069–79
89. Landis DA, Gardiner MM, van der Werf W, Swinton SM. 2008. Increasing corn for biofuel production reduces biocontrol services in agricultural landscapes. *Proc. Natl. Acad. Sci. USA* 105:20552–57
90. Landis DA, Werling BP. 2010. Arthropods and biofuel production systems in North America. *Insect Sci.* 17:1–17
91. Landis DA, Wratten SD, Gurr GM. 2000. Habitat management to conserve natural enemies of arthropod pest in agriculture. *Annu. Rev. Entomol.* 45:173–201
92. Levine JM, D'Antonio CM. 2003. Forecasting biological invasions with increasing international trade. *Conserv. Biol.* 17:322–26
93. Li CS, Luo RW, Yang CL, Shang YF, Zhao JH, Xin XQ. 2000. Studies on the biology and control of *Aphis glycines*. *Soybean Sci.* 19:337–40
94. Li Y, Hill CB, Carlson SR, Diers BW, Hartman GL. 2007. Soybean aphid resistance genes in the soybean cultivars Dowling and Jackson map to linkage group M. *Mol. Breed.* 19:25–34
95. Lin LA, Ives AR. 2003. The effect of parasitoid host-size preference on host population growth rates: an example of *Aphidius colemani* and *Aphis glycines*. *Ecol. Entomol.* 28:542–50
96. Liu J, Wu KM, Hopper KR, Zhao KJ. 2004. Population dynamics of *Aphis glycines* (Homoptera: Aphididae) and its natural enemies in soybean in northern China. *Ann. Entomol. Soc. Am.* 97:235–39
97. Lundgren JG, Fergen JK, Riedell WE. 2008. The influence of plant anatomy on oviposition and reproductive success of the omnivorous bug *Orius insidiosus*. *Anim. Behav.* 75:1495–502
98. Lundgren JG, Wyckhuys KAG, Desneux N. 2009. Population responses by *Orius insidiosus* to vegetational diversity. *Biocontrol* 54:135–42
99. Macedo TB, Bastos CS, Higley LG, Ostlie KR, Madhavan S. 2003. Photosynthetic responses of soybean to soybean aphid (Homoptera: Aphididae) injury. *J. Econ. Entomol.* 96:188–93
100. Magalhaes LC, Hunt TE, Siegfried BD. 2008. Development of methods to evaluate susceptibility of soybean aphid to imidacloprid and thiamethoxam at lethal and sublethal concentrations. *Entomol. Exp. Appl.* 128:330–36
101. Matis JH, Kiffe TR, van der Werf W, Costamagna AC, Matis TI, Grant WE. 2009. Population dynamics models based on cumulative density dependent feedback: a link to the logistic growth curve and a test for symmetry using aphid data. *Ecol. Model.* 220:1745–51
102. McCornack BP, Carrillo MA, Venette RC, Ragsdale DW. 2005. Physiological constraints on the overwintering potential of the soybean aphid (Homoptera: Aphididae). *Environ. Entomol.* 34:235–40
103. McCornack BP, Costamagna AC, Ragsdale DW. 2008. Within-plant distribution of soybean aphid (Hemiptera: Aphididae) and development of node-based sample units for estimating whole-plant densities in soybean. *J. Econ. Entomol.* 101:1488–500
104. McCornack BP, Ragsdale DW. 2006. Efficacy of thiamethoxam to suppress soybean aphid populations in Minnesota soybean. *Crop Manag.* doi:10.1094/CM-2006-0915-01-RS
105. McCornack BP, Ragsdale DW, Venette RC. 2004. Demography of soybean aphid (Homoptera: Aphididae) at summer temperatures. *J. Econ. Entomol.* 97:854–61
106. Mensah C, DiFonzo C, Wang DC. 2008. Inheritance of soybean aphid resistance in PI567541B and PI567598B. *Crop Sci.* 48:1759–63
107. Mian MAR, Kang ST, Beil SE, Hammond RB. 2008. Genetic linkage mapping of the soybean aphid resistance gene in PI 243540. *Theor. Appl. Genet.* 117:955–62
108. Miao J, Wu KM, Hopper KR, Li GX. 2007. Population dynamics of *Aphis glycines* (Homoptera: Aphididae) and impact of natural enemies in Northern China. *Environ. Entomol.* 36:840–48
109. Michel AP, Zhang W, Jung JK, Kang S, Mian MAR. 2009. Population genetic structure of the soybean aphid, *Aphis glycines*. *Environ. Entomol.* 38:1301–11
110. Michel AP, Zhang W, Jung JK, Kang S, Mian MAR. 2009. Cross-species amplification and polymorphism of microsatellite loci in the soybean aphid, *Aphis glycines*. *J. Econ. Entomol.* 102:1389–92

111. Mignault MP, Roy M, Brodeur J. 2006. Soybean aphid predators in Quebec and the suitability of *Aphis glycines* as prey for three Coccinellidae. *Biocontrol* 51:89–106
112. Moser SE, Obrycki JJ. 2009. Non-target effects of neonicotinoid seed treatments; mortality of coccinellid larvae related to zoophytophagy. *Biol. Control* 51:487–92
113. Nault BA, Shah DA, Straight KE, Bachmann AC, Sackett WM, et al. 2009. Modeling temporal trends in aphid vector dispersal and cucumber mosaic virus epidemics in snap bean. *Environ. Entomol.* 38:1347–59
114. Nielsen C, Hajek AE. 2005. Control of invasive soybean aphid, *Aphis glycines* (Hemiptera: Aphididae), populations by existing natural enemies in New York State, with emphasis on entomopathogenic fungi. *Environ. Entomol.* 34:1036–47
115. Noma T, Brewer MJ. 2008. Seasonal abundance of resident parasitoids and predatory flies and corresponding soybean aphid densities, with comments on classical biological control of soybean aphid in the Midwest. *J. Econ. Entomol.* 101:278–87
116. Noma T, Gratton C, Colunga-Garcia M, Brewer MJ, Mueller EE, et al. 2010. Relationship of soybean aphid (Hemiptera: Aphididae) to soybean plant nutrients, landscape structure, and natural enemies. *Environ. Entomol.* 39:31–41
117. Ohnesorg WJ, Johnson KD, O'Neal ME. 2009. Impact of reduced-risk insecticides on soybean aphid and associated natural enemies. *J. Econ. Entomol.* 102:1816–26
118. Olson KD, Badibanga T, DiFonzo CD. 2008. Farmers' awareness and use of IPM for soybean aphid control: report of survey results for the 2004, 2005, 2006, and 2007 crop years. *Staff Pap. P08-12. Dep. Appl. Econ., Univ. Minn., St. Paul.* **http://ageconsearch.umn.edu/bitstream/45803/2/p08-12.pdf**
119. Olson LJ. 2006. The economics of terrestrial invasive species: a review of the literature. *Agric. Res. Econ. Rev.* 35:178–94
120. Ostlie K, Potter B. 2009. Managing two-spotted spider mites on soybeans in Minnesota. *Univ. Minn. Ext.* **http://www.soybeans.umn.edu/pdfs/2009/2009ManagingSpiderMitesinMinnesotaSoybean.pdf**
121. Papachristos DP, Milonas PG. 2008. Adverse effects of soil applied insecticides on the predatory coccinellid *Hippodamia undecimnotata* (Coleoptera: Coccinellidae). *Biol. Control* 47:77–81
122. Pedersen P, Gran C, Cullen E, Hill JH. 2007. Potential for integrated management of soybean virus disease. *Plant Dis.* 91:1255–59
123. Pike KS, Starý P, Brewer MJ, Noma T, Langley S, Kaiser M. 2007. A new species of *Binodoxys* (Hymenoptera: Braconidae: Aphidiinae), parasitoid of the soybean aphid, *Aphis glycines* Matsumura, with comments on biocontrol. *Proc. Entomol. Soc. Wash.* 109:359–65
124. Pimentel D, Lach L, Zuniga R, Morrison D. 2000. Environmental and economic costs of nonindigenous species in the United States. *BioScience* 50:53–65
125. Ragsdale DW, McCornack BP, Venette RC, Potter DB, Macrae IV, et al. 2007. Economic threshold for soybean aphid (Hemiptera: Aphididae). *J. Econ. Entomol.* 100:1258–67
126. Ragsdale DW, Voegtlin DJ, O'Neil RJ. 2004. Soybean aphid biology in North America. *Ann. Entomol. Soc. Am.* 97:204–8
127. Rhainds M, Brodeur J, Borcard D, Legendre P. 2008. Toward management guidelines for soybean aphid, *Aphis glycines*, in Quebec. II. Spatial distribution of aphid populations in commercial soybean fields. *Can. Entomol.* 140:219–34
128. Rhainds M, Roy M, Daigle G, Brodeur J. 2007. Toward management guidelines for the soybean aphid in Quebec. I. Feeding damage in relationship to seasonality of infestation and incidence of native predators. *Can. Entomol.* 139:728–41
129. Rhainds M, Yoo HJS, Kindlmann P, Voegtlin D, Castillo D, et al. 2010. Two-year oscillation cycle in abundance of soybean aphid in Indiana. *Agric. For. Entomol.* 12:251–57
130. Riedell WE. 1989. Effects of Russian wheat aphid infestation on barley plant response to drought stress. *Physiol. Plant.* 77:587–92
131. Rutledge CE, O'Neil RJ. 2005. *Orius insidiosus* (Say) as a predator of the soybean aphid, *Aphis glycines* Matsumura. *Biol. Control* 33:56–64
132. Rutledge CE, O'Neil RJ, Fox TB, Landis DA. 2004. Soybean aphid predators and their use in integrated pest management. *Ann. Entomol. Soc. Am.* 97:240–48
133. Schmidt NP, O'Neal ME, Dixon PM. 2008. Aphidophagous predators in Iowa soybean: a community comparison across multiple years and sampling methods. *Ann. Entomol. Soc. Am.* 101:341–50

111. Characterizes the predatory fauna associated with the soybean aphid in commercial fields in the northern range of the pest.

125. The first paper to describe the development of the EIL and the ET for soybean aphid in North America.

134. Schmidt NP, O'Neal ME, Singer JW. 2007. Alfalfa living mulch advances biological control of soybean aphid. *Environ. Entomol.* 36:416–24
135. Simberloff D. 2006. Invasional meltdown 6 years later: important phenomenon, unfortunate metaphor, or both? *Ecol. Lett.* 9:912–19
136. Simberloff D, Von Holle B. 1999. Positive interactions of nonindigenous species: invasional meltdown? *Biol. Invasions* 1:21–32
137. Smith SF, Krischik VA. 1999. Effects of systemic imidacloprid on *Coleomegilla maculata* (Coleoptera: Coccinellidae). *Environ. Entomol.* 28:1189–95
138. Song F, Swinton SM. 2009. Returns to integrated pest management research and outreach for soybean aphid. *J. Econ. Entomol.* 102:2116–25
139. Starý P, Rakhshani E, Tomanović Ž, Hoelmer K, Kavallieratos NG, et al. 2010. A new species of *Lysiphlebus* Förster 1862 (Hymenoptera: Braconidae, Aphidiinae) attacking soybean aphid, *Aphis glycines* Matsumura (Hemiptera: Aphididae) from China. *J. Hymenoptera Res.* 19:184–91
140. Suckling DM, Brockerhoff EG. 2010. Invasion biology, ecology, and management of the light brown apple moth (Tortricidae). *Annu. Rev. Entomol.* 55:285–306
141. Taylor S. 2009. New and upcoming advancements in seed treatments. *Indiana CCA Conf. Notes.* 2 pp.
142. Thomas MB. 1999. Ecological approaches and the development of 'truly integrated' pest management. *Proc. Natl. Acad. Sci. USA* 96:5944–51
143. Tonet GL, DaSilva RFP. 1995. Resistance of wheat genotypes of C biotype of *Schizaphis graminum* (Rondani) (Homoptera, Aphididae). *Pesqui. Agropecu. Bras.* 30:1283–87
144. Torres JB, Silva-Torres CSA, de Oliveira JV. 2003. Toxicity of pymetrozine and thiamethoxam to *Aphelinus gossypii* and *Delphastus pusillus*. *Pesqui. Agropecu. Bras.* 38:459–66
145. Van den Berg H, Ankasah D, Muhammad A, Rusli R, Widayanto HA, et al. 1997. Evaluating the role of predation in population fluctuations of the soybean aphid *Aphis glycines* in farmers' fields in Indonesia. *J. Appl. Ecol.* 34:971–84
146. van Emden HF. 1997. Host-plant resistance to insect pests. In *Techniques for Reducing Pesticide Use*, ed. D Pimentel, pp. 129–152. Chichester, UK: Wiley
147. van Emden HF, Harrington R. 2007. *Aphids as Crop Pests*. Cambridge, MA: CABI North Am. Off. 800 pp.
148. Venette RC, Ragsdale DW. 2004. Assessing the invasion by soybean aphid (Homoptera: Aphididae): Where will it end? *Ann. Entomol. Soc. Am.* 97:219–26
149. Voegtlin DJ, O'Neil RJ, Graves WR. 2004. Tests of suitability of overwintering hosts of *Aphis glycines*: identification of a new host association with *Rhamnus alnifolia* L'Heritier. *Ann. Entomol. Soc. Am.* 97:233–34
150. Voegtlin DJ, O'Neil RJ, Graves WR, Lagos D, Yoo HJS. 2005. Potential winter hosts of soybean aphid. *Ann. Entomol. Soc. Am.* 98:690–93
151. Walther GR, Roques A, Hulme PE, Sykes MT, Pysek P, et al. 2009. Alien species in a warmer world: risks and opportunities. *Trends Ecol. Evol.* 24:686–93
152. White KE, Hoppin JA. 2004. Seed treatment and its implication for fungicide exposure assessment. *J. Expo. Anal. Environ. Epidemiol.* 14:195–203
153. Williamson M. 1996. *Biological Invasions*. London: Chapman & Hall. 256 pp.
154. Wu ZS, Schenk-Hamlin D, Zhan WY, Ragsdale DW, Heimpel GE. 2004. The soybean aphid in China: a historical review. *Ann. Entomol. Soc. Am.* 97:209–18
155. Wyckhuys KAG, Hopper KR, Wu KM, Straub C, Cratton C, Heimpel GE. 2007. Predicting potential ecological impact of soybean aphid biological control introductions. *Biocontrol. News Inf.* 28:N30–34
156. Wyckhuys KAG, Koch RL, Heimpel GE. 2007. Physical and ant-mediated refuges from parasitism: implications for non-target effects in biological control. *Biol. Control* 40:306–13
157. **Wyckhuys KAG, Koch RL, Kula RR, Heimpel GE. 2009. Potential exposure of a classical biological control agent of the soybean aphid, *Aphis glycines*, on non-target aphids in North America. *Biol. Invasions* 11:857–71**
158. Wyckhuys KAG, Stone L, Desneux N, Hoelmer KA, Hopper KR, Heimpel GE. 2008. Parasitism of the soybean aphid *Aphis glycines* by *Binodoxys communis*: the role of aphid defensive behavior and parasitoid reproductive performance. *Bull. Entomol. Res.* 98:361–70

157. A risk assessment of a classical biological control agent of soybean aphid that takes into account both exposure and effects of a biological control agent on nontarget species.

159. Xue Y, Bahlai CA, Frewin A, Sears MK, Schaafsma AV, Hallett RH. 2009. Predation by *Coccinella septempunctata* and *Harmonia axyridis* (Coleoptera: Coccinellidae) on *Aphis glycines* (Homoptera: Aphididae). *Environ. Entomol.* 38:708–14
160. Yoo HJS, O'Neil RJ. 2009. Temporal relationships between the generalist predator, *Orius insidiosus*, and its two major prey in soybean. *Biol. Control* 48:168–80
161. Zhang W, Swinton SM. 2009. Incorporating natural enemies in an economic threshold for dynamically optimal pest management. *Ecol. Model.* 220:1315–24
162. Zhang Y, Wang L, Wu KM, Wyckhuys KAG, Heimpel GE. 2008. Flight performance of the soybean aphid, *Aphis glycines* (Hemiptera: Aphididae) under different temperature and humidity regimens. *Environ. Entomol.* 37:301–06
163. Zhu JW, Park KC. 2005. Methyl salicylate, a soybean aphid-induced plant volatile attractive to the predator *Coccinella septempunctata*. *J. Chem. Ecol.* 31:1733–46

A Roadmap for Bridging Basic and Applied Research in Forensic Entomology

J.K. Tomberlin,[1] R. Mohr,[1] M.E. Benbow,[2] A.M. Tarone,[1] and S. VanLaerhoven[3]

[1]Department of Entomology, Texas A&M University, College Station, Texas 77843; email: jktomberlin@ag.tamu.edu

[2]Department of Biology, University of Dayton, Dayton, Ohio 45469-2320

[3]Department of Biology, University of Windsor, Windsor, Ontario, N9B 3P4 Canada

Annu. Rev. Entomol. 2011. 56:401–21

First published online as a Review in Advance on September 7, 2010

The *Annual Review of Entomology* is online at ento.annualreviews.org

This article's doi:
10.1146/annurev-ento-051710-103143

0066-4170/11/0107-0401$20.00

Key Words

conceptual framework, succession, community assembly, quantitative genetics, functional genomics, *Daubert* standard

Abstract

The National Research Council issued a report in 2009 that heavily criticized the forensic sciences. The report made several recommendations that if addressed would allow the forensic sciences to develop a stronger scientific foundation. We suggest a roadmap for decomposition ecology and forensic entomology hinging on a framework built on basic research concepts in ecology, evolution, and genetics. Unifying both basic and applied research fields under a common umbrella of terminology and structure would facilitate communication in the field and the production of scientific results. It would also help to identify novel research areas leading to a better understanding of principal underpinnings governing ecosystem structure, function, and evolution while increasing the accuracy of and ability to interpret entomological evidence collected from crime scenes. By following the proposed roadmap, a bridge can be built between basic and applied decomposition ecology research, culminating in science that could withstand the rigors of emerging legal and cultural expectations.

INTRODUCTION

Daubert **standard:** the court standard used to evaluate scientific information used in court either as evidence or the interpretation of evidence

Population genetics: the study of the flow of genetic material within and among populations of a species

Postmortem interval (PMI): the time from death of an individual to discovery of their remains

The criteria established by *Daubert v. Merrell Dow Pharmaceuticals, Inc.* are used to evaluate scientific evidence prior to its admission in court (124). The *Daubert* decision mandated scientific evidence (*a*) be testable, (*b*) have a known error rate, (*c*) be peer-reviewed, and (*d*) be accepted by the specific scientific community employing the technique (37). This ruling profoundly altered the landscape of the forensic sciences and continues to affect them today.

A 2009 National Research Council (NRC) report (93) indicated a need for major improvements in many forensic science disciplines in order to increase accuracy and meet the *Daubert* standard. In hindsight, this report was inevitable. Calls have been made for at least ten years to restructure the forensic sciences to fit the model of self-criticism and review used by pure sciences (114). Increased media exposure of unconscious and/or fraudulent data analysis, interpretation, and presentation by forensic experts has added to highly visible exonerations (79), raising the awareness of the general public. Now that this issue of objective reliability has been brought to the attention of the forensic science community, it can no longer be ignored (78, 116). Questions such as "What are the limitations and error rates of evidence interpretation?" and "What can be done to reduce these limitations?" need to be answered by all the forensic sciences through basic research in each discipline. Answering such questions gives forensics a rigorous scientific foundation, which provides more objective and reliable evidence interpretation in a legal setting.

The NRC report (93) outlined specific areas where forensic disciplines could improve by addressing basic research questions using methods and standard practices common to basic science research. In this paper, we discuss the field of forensic entomology and propose a roadmap to guide research and practice to address the concerns of the NRC report. We propose a framework for forensic entomology (**Figure 1**) applications with a common language to streamline research questions, techniques, and data output that focuses on basic science. We believe that resulting research will eventually produce guidelines that meet the *Daubert* criteria and provide greater insight into the function of natural ecological systems, thereby developing a stronger link between basic and applied sciences. To demonstrate our concept, we provide examples of research systems that conduct basic scientific studies with direct, practical applications to the field of forensic entomology. These systems can be bridged with the wider forensic sciences, just as the principles of population genetics have been applied to forensic DNA analysis (92).

The Present State of Forensic Entomology

Forensic entomologists are frequently asked to examine arthropod evidence recovered from human remains and determine how long the arthropods were present. This time period has often been interpreted as the postmortem interval (PMI), or time since death (24). Assessment of the PMI has been grounded in arthropod development rates and community succession of arthropods. Following seminal casework by Beregeret (13) and Mégnin (87), the assumption that forensic entomologists provide the actual PMI was widely accepted. Case studies demonstrating entomologists' ability to accurately estimate the PMI have been published in books (49), research articles (10, 69), and national forensic science conference proceedings (64), and have been reinforced by popular media (120). However, arthropod-based PMI predictions are acknowledged to be associated with a number of assumptions, which can lead to severe deviations from the true PMI if violated (25). The European Forensic Entomology Association recognizes that the onset of arthropod colonization does not always coincide with the actual time of death, and in some instances can occur without death (e.g., myiasis) (4). Consequently, Amendt et al. (4) proposed that forensic entomologists reconsider their conclusions

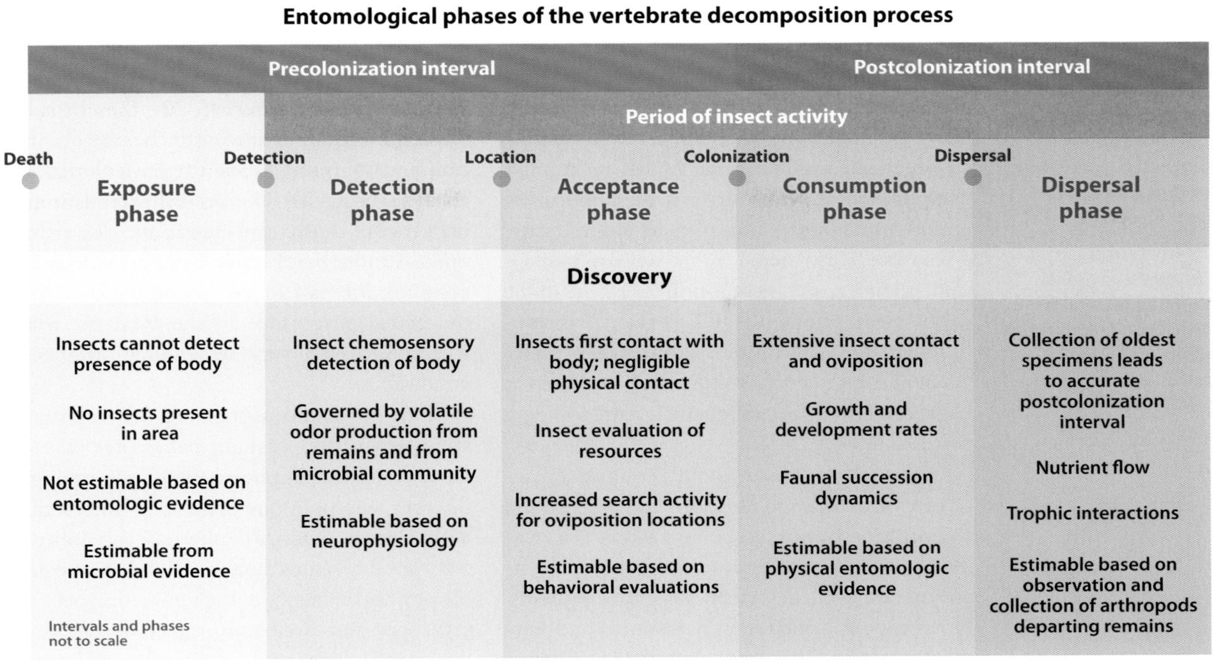

Entomological phases of the vertebrate decomposition process

Precolonization interval			Postcolonization interval	
	Period of insect activity			
Death	Detection	Location	Colonization	Dispersal
Exposure phase	**Detection phase**	**Acceptance phase**	**Consumption phase**	**Dispersal phase**
		Discovery		
Insects cannot detect presence of body	Insect chemosensory detection of body	Insects first contact with body; negligible physical contact	Extensive insect contact and oviposition	Collection of oldest specimens leads to accurate postcolonization interval
No insects present in area	Governed by volatile odor production from remains and from microbial community	Insect evaluation of resources	Growth and development rates	Nutrient flow
Not estimable based on entomologic evidence	Estimable based on neurophysiology	Increased search activity for oviposition locations	Faunal succession dynamics	Trophic interactions
Estimable from microbial evidence		Estimable based on behavioral evaluations	Estimable based on physical entomologic evidence	Estimable based on observation and collection of arthropods departing remains

Intervals and phases not to scale

Figure 1

Framework proposed for entomological phases of the decomposition process for vertebrate remains (phases not to scale).

in terms of the period of insect activity (PIA), defined as the time from arthropod colonization until discovery of the remains.

A more explicit recognition that PMI and PIA are usually, but not always, strongly related will promote better interpretation of results and systematic evaluation of potential sources of error, a key feature of *Daubert* standards and the NRC recommendations. Consequently, those implementing current methods need to be mindful of data limitations. The proposed framework will guide research that builds understanding of the sources of error in forensic entomology by applying basic science.

Conceptual Framework for Forensic Entomology

This proposed framework is a generality, meant primarily to guide future research. Forensic entomology should be framed in terms of multidisciplinary ecological concepts to advance understanding of the carrion decomposition process and to explain observed error and variation. The framework is based on these concepts to make it objective and to provide a roadmap for the application of principles of ecology and molecular biology to forensic entomology. Furthermore, we advocate the use of terms that reflect the basic dynamics of decomposition that transcend forensics and reflect concepts that can be used to guide basic research. Undoubtedly, the temporal scale of the given intervals and phases will change depending on the arthropod species and specific environmental factors examined.

Background and rationale. There are important elements composing the behavioral ecology of arthropods that use carrion (e.g., human remains). These include, but are not limited to, (*a*) evolutionary underpinnings of effective foraging, (*b*) carrion signaling

Period of insect activity (PIA): the time interval encompassing arthropod association with decomposing remains

Precolonization interval (pre-CI): the time interval from death of an individual to their colonization by arthropod(s)

Postcolonization interval (post-CI): the time interval from arthropod colonization of decomposing remains to dispersal

characteristics, (*c*) control modes of arthropod behavioral cascades, and (*d*) mechanisms of host location and selection.

According to optimal foraging theory (OFT) as defined by MacArthur & Pianka (85), an organism maximizes its fitness by capitalizing on necessary resources while minimizing energy expenditure; however, its ability to use a resource is restricted by its sensory perception, memory, and locomotion (61). One of the major assumptions of OFT is that, over evolutionary time, the fitness advantage of more efficient foraging drives individual behaviors to converge into a species' characteristic foraging patterns (105).

Although questions about the mechanisms of carrion arthropod location, acceptance, and colonization of new resource patches are relatively new in forensic entomology, these mechanisms are well described for other systems. Host-finding behavior in parasitoids and herbivorous insects has been described as a series of decision steps with important neural events (145). It is likely that arthropods make decisions when locating carrion. We hypothesize that adult carrion arthropods such as blow flies (Diptera: Calliphoridae) emerge under one of three scenarios. In the first scenario, arthropods emerge into a habitat where both appropriate carrion resources and mates are present, requiring limited searching behavior. Alternatively, they may find a habitat that contains appropriate carrion but no mates, or vice versa. In all three cases, arthropods must detect and exploit resources to maximize their reproductive fitness. For the last two cases, they must also disperse from their natal habitat. In the proposed framework, different neurobiological events and the ensuing arthropod choices divide the PMI continuum into five phases: exposure, detection, acceptance, consumption, and dispersal (**Figure 1**). These discrete ecological phases can in turn be used to accurately and more precisely describe the phases of the PMI. It is important to note that each of the phases, particularly the consumption phase, may be prematurely ended by the recovery or destruction of the remains. In these instances, the time of discovery/recovery would function as the endpoint.

Precolonization interval. At the broadest level, the PMI is divided into the precolonization interval (pre-CI) and the postcolonization interval (post-CI). The pre-CI extends immediately after death until colonization by arthropods. Although in **Figure 1** the pre-CI visually accounts for half of the decomposition time, the actual percentage of the PMI for which it accounts will vary depending on specific conditions.

The mechanisms/motivations of neural stimulations and ensuing behavioral cascades have significant implications for the timing of discrete necrophilous arthropod interactions with carrion. Due to inherent variability in arthropod foraging behaviors and the negligible physical evidence of the interaction between arthropod and carrion prior to colonization, estimation of the pre-CI is currently problematic. The pre-CI phase is generally overlooked in the literature (87, 90, 97, 99), indicating a large void in understanding decomposition ecology. Research on this topic will help determine how PMI and PIA relate to each other, leading to a better understanding of error associated with entomologically based predictions.

The exposure phase is the time between death and exposure of remains to initial artnropod detection. In most instances, it is of negligible duration, as the body is instantly exposed to the environment. In some cases, however, remains are artificially contained, preserved, or protected from the arthropod activity, including such treatments as cold storage in a morgue, embalming, or deep burial. In criminal cases, wrapping (48), burning (8), and placement of remains in contained facilities (e.g., the trunk of a car) can substantially delay, if not outright prevent, natural arthropod succession. Because arthropods have no interaction with the remains, the duration of the exposure phase cannot currently be estimated using entomological evidence. In the future, it may become measurable, perhaps through microbial and/or biochemical assays. Nevertheless, this phase should not be ignored as part of the total PMI.

The detection phase is made up of two stages: activation and searching. The activation stage begins when arthropods first detect decomposition cues. Arthropod resource-finding behavior is regulated by two control systems: allothetic, or the processing of external stimuli, and idiothetic, the processing of endogenous stimuli and memory (147). These control systems work in tandem to determine if neuronal stimulation will result in a behavioral cascade (89). As a result, arthropod response to carrion is shaped by external factors such as temperature, precipitation, wind speed, time of day (19), and internal factors such as mating status (39) and ovarian development.

At long distances, cues are likely to take the form of volatile chemicals produced by the carrion itself (141), the endogenous bacterial community (16, 66, 72), or semiochemicals produced by other organisms using the remains (33, 80, 124, 150). Arthropods must differentiate relevant cues from the complex suite of background odors (119), so their sensory systems may be specifically sensitive to particular odors or odor blends (15). This sensitivity to chemical properties of a resource allows individuals to discriminate among available patches and selectively forage (65). Such a mechanism can partially explain characteristic arthropod colonization and succession on carrion. As the chemical profile of the carrion changes over the decomposition process (141), species within later succession waves are activated to seek out the carrion, while the earlier colonizers are no longer activated (148). With identification of chemical cues that activate primary colonizers, and the taphonomic and/or microbiological carrion conditions that produce these cues, the duration of this phase may become measurable.

The searching stage is the time between sensory activation of arthropods and their physical contact with the carrion and is similar to the classic host-finding stage of parasitoid, hematophagous, or herbivorous arthropods (104). Among parasitoids, host microhabitat cues play as important a role in host-finding as cues presented by the hosts themselves (142). Habitat location and searching, therefore, are frequently a preliminary step to host searching. Long-distance cues are often allelochemicals released by host plants, from volatile compounds in frass, and/or by pheromones released by herbivores (146). Other long-distance cues include auditory cues from feeding or mating hosts, visual cues of plant damage, and/or the actual host itself (106). For necrophilous arthropods, long-distance cues take the form of chemical products of decomposition, body fluids, and/or the visual image of the resource. Once the carrion resource is detected, it is likely that shorter-distance cues are used by arthropods, just as is found with parasitoids. This switch is due to higher reliability and informational quality of short-range signals compared to longer-distance cues, the classic reliability-detectability trade-off of parasitoid host-finding (142). Short-distance cues of carrion take the form of low-volatility decomposition products, allelochemicals released by microbial colonists, or kairomones released by conspecific or heterospecific blow flies or other saprophages.

Arthropods may use a variety of related searching strategies and behaviors (147) to identify and track the exact location of carrion. These search strategies and their efficiency may be affected by environmental conditions such as darkness (5), humidity, and/or wind patterns (19). Like the activation stage, the mechanisms of searching behavior have an idiothetic component, which can greatly affect the speed and efficiency of foraging (147). Physiological state and learning can significantly accelerate host-finding (138). Searching behavior of a species is also limited by its sensory performance, memory capability, and structure and locomotion ability (61). Once activated, searching tends to proceed directly regardless of conditions (5), so long as the stimulus remains above the arthropod's activation threshold.

Three components are important for estimating the duration of the searching stage for a species: (*a*) appropriate identification of the activating cues, (*b*) identification of external and internal factors that modify activation, and (*c*) characterization of spatial distribution and rate of taxis. Once these components have been identified, modeling the system to determine the relationship between cues and attraction of the targeted arthropod to the carrion source can provide an understanding of the nature and variation of insect response to cues, as well as error associated with predictions based on such models.

The first two components could be achieved through neurophysiological analysis of arthropod response to different carrion- and microbe-derived odor blends. The third component could be evaluated through both laboratory behavioral assays, such as locomotion trials and olfactometry, and field trial and observation, including mark-release-recapture studies. Accurately identifying the chemical activators and attractors of carrion is a critical gap in our understanding of necrophilous arthropod behavior and has direct relevance to PIA estimates.

The acceptance phase is the period of time from physical contact of an arthropod with carrion until the arthropod begins to establish residence on that resource. As the searching stage is similar to parasitoid host-finding (145), the acceptance phase of the necrophilous arthropod is similar to the host acceptance phase of parasitoid and phytophagous arthropods. In parasitoids, this takes the form of antennation to identify a host, followed by testing the host for suitability. During this stage, arthropods use close-range cues including color, shape, size, movement, sound, and taste to evaluate the resource (144). Like parasitoids, carrion arthropods must positively identify a resource and then determine its suitability for oviposition, which could include chemotactile contact (26). Blow flies likely use a similar combination of chemosensory taste receptors on their tarsi and labellum.

Solid acceptance criteria are critical to fitness, particularly for females evaluating oviposition sites (96). Reproductive strategies may be dictated by exposure to resources. The number of eggs deposited may vary depending on the female physiological state and the size, nutritional quality, or age of the resource (96). Experienced parasitoid females evaluate a potential site more quickly than naïve females (138); however, older females and females with a high egg load accept lower-quality oviposition sites more readily and are less likely to leave a suboptimal patch (43). Aggregative semiochemicals may play a role in influencing host acceptance, particularly for species that oviposit gregariously, such as *Cochliomyia macellaria* (Fabricius) (32) (Diptera: Calliphoridae); alternatively, semiochemicals from competitors, predators, or prey items may also shape acceptance phase behaviors (45). For species that merely feed on carrion, rather than using it to rear offspring, the acceptance behavior may be under much less selection and is a balance of phagostimulatory and deterrent inputs either as volatile or contact chemicals (26).

Accurate estimation of the activation phase requires understanding inter- and intraspecies variability in innate behavior of carrion arthropods. The behavior of primary colonizers is particularly important, as acceptance marks the onset of direct arthropod contact and extensive physical colonization of carrion.

Postcolonization interval. The second of the broad divisions of the PMI, the post-CI, is initiated at colonization (i.e., oviposition) and lasts until the departure of arthropods, either following complete decomposition or upon discovery and removal of remains in the case of forensics. It commences when arthropods begin to leave discrete evidence of their presence on remains, either as feeding damage or through oviposition. The post-CI currently represents a minimum postmortem interval (4, 128, 134).

The consumption phase is the time between the onset of colonization and arthropod departure from the remains when they no longer provide appropriate resources for sustenance or development. The consumption stage is characterized by successive waves of arthropods extensively using the carrion as a food source for themselves and/or their offspring. It is currently the best understood of the proposed phases of decomposition relevant for PIA estimates, including studies on principles of larval development rates (24), seasonality and carcass size (31), taxon structure of successional waves (118), presence of antemortem toxins (70), and effects of intraguild predation (112). This phase is most often estimated using species-specific known larval and pupal development rates of Diptera (24) using mass (154), length (152), or stage of development (24). However, increasing evidence of biogeographic variation in forensically important species makes use of locally derived development data critical for accurate assessment of this phase (101, 128).

The dispersal phase includes the movement of arthropods previously feeding on the remains to their departure. Dispersal can occur due to the need to pupate (58, 131); to disturbance of the remains; to lack of resources (51, 53); or to interactions with abiotic factors, such as temperature, rain, or sunlight (52). Although not initially thought to affect the development cycle of dispersing individuals, dispersal prior to completion of the consumption phase could result in extended development times (11).

This proposed framework provides a flexible list of terms to describe ecologically relevant phases of decomposition, allowing researchers to describe and communicate the temporal and physical aspects of studies. It can apply to the use of carrion by an individual arthropod, by a single population, or by the entire necrophilous arthropod community.

Universal application of this framework in future research would allow for a more concrete understanding of ecology and evolution within the practice of forensic entomology. We suggest that an overall model for characterizing each phase be taken from Tinbergen's (132) "four questions" of animal behavior: causation, ontogeny, phylogeny, and adaptation. The proximate causes of "How does the behavior occur?" and "How does it change over the organism's life?" are significant primarily from a forensic perspective, and the ultimate causes of "How did it develop?" and "How does it affect reproductive fitness?" are important to understanding the natural variation associated with decomposition. Within this framework most of the underlying ecological and genetic mechanisms remain unknown or understudied. Not until the mechanisms driving carrion location, acceptance, and colonization in arthropods are better understood will we have the foundational science required to make predictions that better meet the *Daubert* criteria.

ECOLOGY OF DECOMPOSITION

As an example of how this framework can be used to facilitate basic and applied research, we consider an important issue for forensic entomology: variability in carrion succession under natural conditions. **Figure 2** illustrates the visual differences in decomposition among six swine carcasses that were euthanized at the same time and photographed on the same day during decomposition.

A fundamental question of ecology asks how and why communities of organisms assemble under various environmental conditions. Living organisms act as discrete patches of space, nutrients, and energy that are made available soon after death. In many ecosystems, these remains are food-falls that act as resource subsidies to the local habitat, often with a functional impact on the surrounding ecosystem (16). The process by which these resources become available to the ecosystem is limited by the natural succession of organisms that occupy and modify the patch over time. This succession is defined by the species that assemble from the wider regional species pool (139). As species modify resources in ways limiting or facilitating additional use by other species, they affect the rate and permanence of species assembly. The biological interactions that occur during colonization are complex and often habitat dependent (16).

Abiotic variables affect networks of local community patches (i.e., metacommunities) by habitat modification and production of conditions that limit species distribution, competitive ability, and persistence in the landscape. Understanding how abiotic conditions interact with biotic communities can be important for

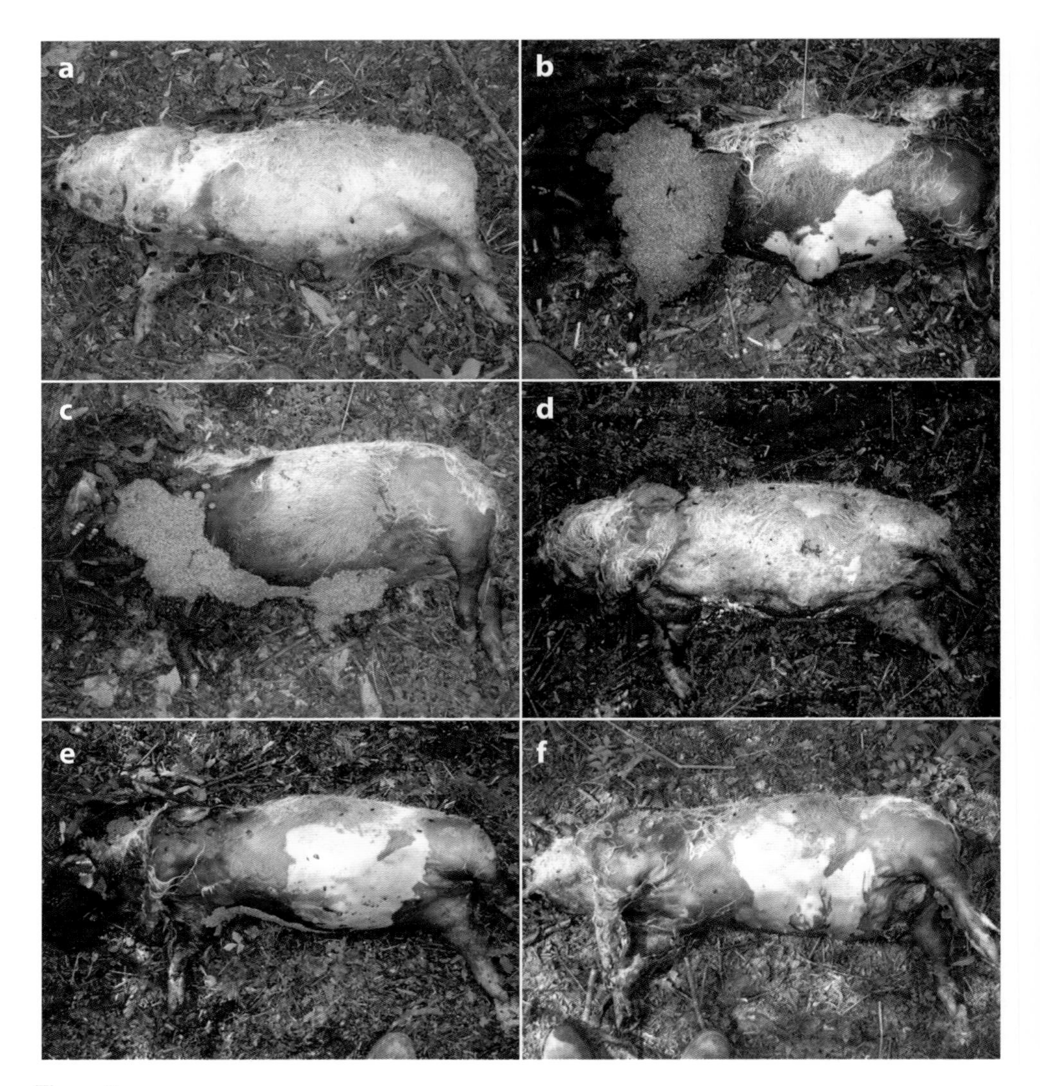

Figure 2

Six replicate swine carcasses all euthanized and placed in the forest on the same day and then photographed here on the same day during decomposition. Note that two of the six (panels *b* and *c*) have large larval blow fly masses, indicating differences occurring during decomposition.

Metacommunity: several ecological communities that are linked together through the dispersal and gene flow of multiple, often interacting, species in a local geographic area

predicting metacommunity assembly (16, 103), with implications for identifying and describing variation in arthropod assembly on human remains under different conditions (139). By further expanding our understanding of how these mechanisms operate over space and time, we will be able to predict more accurately how and in what way a carrion resource will be colonized.

ECOLOGICAL APPLICATIONS TO FORENSIC ENTOMOLOGY

In the criminal justice system, forensic entomology uses data derived from arthropods that have evolved to colonize and consume decomposing animal remains (19, 59). Understanding the timing of arthropod colonization of a body is useful in estimating a post-CI. As previously mentioned, many abiotic factors can affect entomologically based post-CI estimates (23). Substantial variation in the arrival and succession of arthropods on remains can reduce the effectiveness of using entomological succession data in criminal cases (**Figure 2**).

Because arthropods have predictable life histories, habitats, known distributions, behaviors, and/or developmental rates, the presence/absence and developmental stage of certain species at a crime scene can provide important information about when, where, and how a particular death occurred (19). Arthropods play a natural role in the decomposition of carrion by consuming organic material and recycling energy and nutrients. They are thought to follow predictable rules of community assembly, but this idea has not been robustly tested in replicated field and/or laboratory studies. When an organism dies, bacteria that were once held in equilibrium by the immune system immediately begin to digest proteins, lipids and carbohydrates as energy sources, creating both gaseous and liquid by-products that act as olfactory cues for colonization by arthropods (16, 72). In most instances initial colonizers are adult blow flies that feed and lay eggs or live larvae on the remains (6, 19, 88).

Decomposing remains act as a food resource patch for newly hatched larvae that develop at temperature-dependent rates (**Figure 2**). The presence of blow fly larvae attracts predators and parasites such as beetles (113), mites (99), ants (50), wasps, and spiders (99) that parasitize or feed on the eggs, larvae, and/or pupae of the flies. This suite of species is followed by other species that come to feed on previously eaten or conditioned (e.g., dry skin) remains in a succession of arthropod species that colonize and ultimately decompose the carrion to dry bones and hair (19).

In most studies of forensic entomology, swine carcasses have been used as models for human decomposition (99). Many studies have evaluated arthropod colonization of swine remains during the post-CI (**Figure 1**). Few have examined the time interval from death to initial insect contact (140) or colonization (56, 87, 90, 97). Based on the few studies examining initial contact, the temporal variation can span from 30 s (140) to several hours (149) or days (121, 133) after death. Lacking information on this phase limits our understanding of the ecological variation of the entire decomposition process.

Quantitative genetics: the study of the inheritance of complex traits

QUANTITATIVE GENETICS AS A MEANS OF DECREASING ERROR IN FORENSIC ENTOMOLOGY

The analysis of DNA has set the standard by which other forensic sciences are measured (115). Molecular research is well established in forensic entomology species identification (153), but an understanding of the role of genetics in development and behavior of necrophilous arthropods will help decrease error in forensic entomology. Although this section focuses on blow fly biology, the principles discussed in this section can apply to other forensically informative arthropods.

Quantitative genetics attempts to identify and understand variation in continuously variable phenotypes (28, 38, 84, 86). The basic premise underlying quantitative genetic research is that phenotypes can be affected by genetic differences among individuals, environment, and/or by interactions between the two. This basic concept is demonstrated with a reaction norm (**Figure 3**), which can be used in part to identify the contributions of each component and interaction to a trait. This concept is shown by the equation

$$P = G + E + G \times E$$

where P is the phenotype, G is the genotype, E is the environment, and $G \times E$ is the interaction between genotype and environment.

Forensic entomologists use two quantitative traits, body size and development time, to estimate the post-CI using blow fly evidence (127). Accordingly, it stands to reason that forensic entomologists should approach the study of these phenotypes and their use in the forensic setting within a quantitative genetic framework. Doing so will enable the discipline

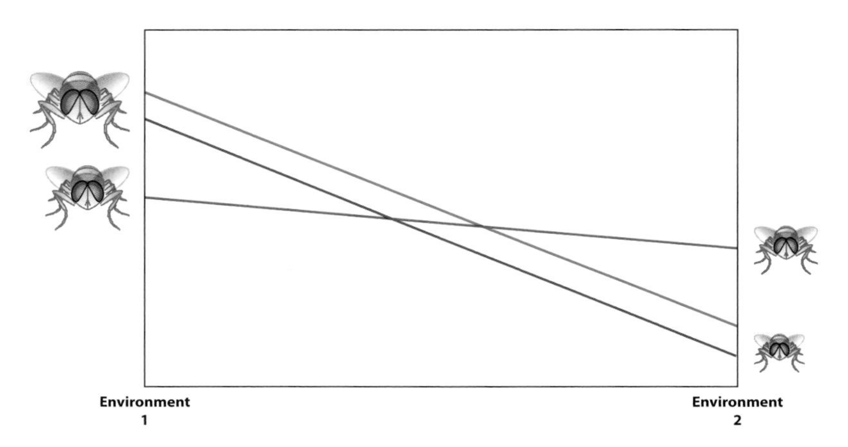

Figure 3

A theoretical reaction norm for blow fly body size. E1 and E2 represent different environments. The colored lines connect average phenotype scores for different genotypes. A significant difference in the phenotypes (the larger flies in E1) between environments means that the trait is plastic with respect to the environments tested. A significant difference between genotypes (*green and blue lines*) means there is a genetic component to body size variation between these two groups. A significant difference in the slopes of the lines for each genotype (*red line versus the green and blue lines*) is a genotype by environment interaction. Genotype by environment interactions observed among populations are indicative of local adaptation, when the phenotype affects the fitness of an organism.

to reduce error in estimates with blow fly evidence by helping practitioners understand how deviations from published developmental data may occur and subsequently allow them to account for these factors in analyses. To date, the discipline has addressed some aspects of the basic quantitative genetic equation but overall has not considered blow fly development in a quantitative genetic context.

The concept of plasticity, an environmental response, in blow fly developmental phenotypes is widely appreciated, and such studies will likely continue. The effect of temperature on development rates has resulted in numerous studies of species-specific developmental times under laboratory-controlled treatments (17, 18, 91). Such studies aid investigators in predicting development rates under field conditions experienced in real-world casework (57). However, other environmental and biological factors, including larval density and food moisture (54, 127), can influence fly development (27, 75).

The genetic side of the quantitative genetic equation has been less fully appreciated in forensic entomology. Because each species has its own unique developmental profile, it is important to correctly identify the species collected as evidence. Accordingly, there is a body of research designed to enable the use of gene sequence as a means of species identification (153). However, it may also be important to determine population-specific development. Grassberger & Reiter (57) made reference to zoogeographic differences among *Lucilia sericata* (Meigen) (Diptera: Calliphoridae) populations to explain differences between development rates of a local Viennese population and that reported by other studies. Likewise, Greenberg (58) found intraspecific differences in development rates of *L. sericata* between Russian and midwestern United States populations. A more recent comparison of *L. sericata* populations from California, Michigan, and West Virginia found significant differences in pupal length, pupal weight, and minimum development times (126) and further found that body size differences among populations could be a result of intraspecific variation in the point at which larvae physiologically commit to pupation (128). These observations are also supported by evidence that *L. sericata* populations from Sacramento and San Diego, California, and Easton, Massachusetts, develop at different rates (44). Although the field of forensic entomology is beginning to conduct studies of intraspecific variation in development, it is clear that, even within relatively small geographic areas, there are differences between populations of forensically informative flies. This variation means that, just as it is important to know which species of blow fly was collected, it may also be important to know the originating source populations.

These observations of blow fly population differentiation in developmental rates are supported by a wealth of quantitative genetic observations in other Diptera. There are repeated occurrences among *Drosophila* species of population-based differences in development times and body size (71, 73, 95, 98, 136). Some of the most striking observations are chromosomal inversions exhibiting population frequencies that strongly correlate with latitudinal clines of body size. These inversion clines occur in both the ancestral and newly inhabited continents occupied by *Drosophila melanogaster* (Meigen) (22, 77, 108, 109) and *Drosophila subobscura* (Collin) (21, 47), indicating the presence of selective pressure maintaining latitudinal variation in these traits. Among nondrosophilids, different populations of *Scathophaga stercoraria* L. (Diptera: Scathophagidae) exhibit heritably variable body sizes and development times (14, 34, 110), and populations of *Rhagoletis pomonella* (Walsh) (Diptera: Tephritidae) exhibit a developmental duration cline from Mexico to Michigan (40).

Genetic variation in plasticity, resulting in a genotype by environment interaction, has also been demonstrated for fly populations. The leaf miner *Liriomyza sativae* (Diptera: Agromyzidae) exhibits genotype-dependent shifts in development time on different plant hosts (143). Similarly, the goldenrod gall midge, *Dasineura folliculi* (Felt) (Diptera: Cecidomyiidae), shows different host preferences between genotypes (35). These host races also have differences in wing size, abdominal segment allometry, and ovipositor length. Clearly, populations of nonforensically informative flies can have variable development times, body sizes, and/or morphology, warranting more detailed investigation of the quantitative genetics of life-history traits in forensically informative flies.

POPULATION GENETICS IN FORENSIC ENTOMOLOGY

As evidence mounts for intraspecific development rate differences among forensically informative populations, it becomes necessary to account for such variation, just as interspecific developmental differences are presently considered. Doing so will lead to greater acceptance of entomological evidence under the *Daubert* criteria, allowing for more confident and accurate estimations of the PIA.

Fortunately, there is a wealth of research outlining how distinct populations within a species are identified by population genetic analysis. It is routinely done in conservation genetics (reviewed in Reference 12), in many gene-mapping experiments when population structure correlates with a phenotype (156), and within the forensic sciences, as human populations have different frequencies of the same alleles, requiring the use of population-specific databases to more accurately identify individuals that belong to those groups (92). Tests of deviation from Hardy-Weinberg (HW) equilibrium (76), inbreeding coefficients/fixation indices such as Wright's F_{st} (68, 155), and isolation by distance (3, 42, 111) can all be used to identify differentiated populations. F_{st} can also be compared to quantitative trait divergence among populations (Q_{st}) to identify the effects of selection on a trait (83).

Often inbreeding coefficients are assessed with a small panel (tens of loci) of neutral markers such as microsatellites, describing the general level of differentiation among groups. It is possible for populations to diverge genetically without diverging in phenotype (i.e., F_{st} indicates different populations, but Q_{st} does not). Similarly, populations may phenotypically

Functional genomics: the genome-wide study of gene function

diverge (i.e., F_{st} indicates no population structure, but Q_{st} differs) if gene flow is high and the trait of interest experiences differential selection among the environments encountered by specific populations (as in the *Drosophila* examples). In these cases, more detailed evaluations such as functional genomic studies may be necessary to find regions of fly genomes associated with causal variation in phenotypic differences. When such variations are observed, it is critical that the forensic science community makes genetic observations in forensically informative fly species, as this will identify error in PIA estimates due to population-level differentiation.

The presence of rare or unique combinations of genetic material will thus aid in identifying distinct populations of a species and potentially reveal portions of the genome that correspond to phenotypic divergence. Genomic tools have already been used to identify regions of the genome that have differentiated between specific canine breeds (151), human populations (94), and reproductively isolated forms of *Anopheles gambiae* (Giles) (Diptera: Culicidae) (137). Such data could be used as markers of phenotypic differences between populations and will enable functional genomics studies (9) that are useful for predicting blow fly age (129, 130). Currently, genomic data are publically available for only one potentially forensically informative species, *Sarcophaga crassipalpis* (Macquart) (Diptera: Sarcophagidae) (62); unfortunately, this species is rarely used as a forensic indicator (2).

Clearly, at least some of the population genetic analyses described here must be conducted on fly species of forensic importance if unique/divergent populations are to be identified. Unfortunately, these types of analyses have not been routine practice in forensic entomology. To date, there have been a number of analyses of conserved mitochondrial and nuclear DNA sequence variation (153), but these particular analyses have not reliably identified specific populations of forensically informative flies. Cuticular hydrocarbon profiles have been evaluated in *Phormia regina* (Meigen) (Diptera: Calliphoridae). These traits vary extensively across *Drosophila* populations (30, 36, 135) and could be used to differentiate between sexes and populations (20), but initial work has not been pursued. There have been several amplified fragment length polymorphism (AFLP) and microsatellite studies of *P. regina*, *L. sericata*, *Lucilia illustris* (Meigen), and *Cochliomyia hominivorax* (Coquerel) (41, 60, 101, 102). Results in the United States indicated that nonselected markers are unlikely to change significantly across populations and that developmental data collected from flies caught at the same time may result in genotype-biased results (101, 102). In Sweden (41) and Brazil (60), a large degree of inbreeding has been found, but genetic variation was maintained across the populations, likely due to large effective population sizes. Additional studies such as these will be necessary to gain an understanding of how the flow of genes within and among populations of blow flies might affect variation in traits of interest.

A NEED FOR MOLECULAR TOOLS IN FORENSIC ENTOMOLOGY

One limitation to population and functional genetic analyses of forensically informative species is the lack of sufficient genetic sequence data. Frequently, genetic analyses of blow flies are restricted to mitochondrial DNA or ribosomal sequences, which are typically similar enough in sequence among species to enable comparison but different enough to distinguish close relatives. However, genetic tests that are useful for identifying populations rely either on genotyping a panel of neutral loci, such as microsatellites (41) and AFLPs (101), or on a dense array of genomic data to evaluate genome-wide patterns of relatedness (94, 151). As these genomic resources are generally lacking in the forensic entomology community, there should be a focus on the identification and evaluation of genetic markers (41, 101), which can be used to identify population structure and regions of the genome that have undergone selection.

Once genetic and genomic tools are available, there are some steps for the next generation of forensic entomology research. There should be efforts to characterize the degree of phenotypic divergence among fly populations. In cases in which genetic variation for forensically informative traits is demonstrated, studies should be conducted to determine if population identity correlates with variation in phenotype and to characterize the population structure present in the species. If no correlation between trait variation and population membership exists, then finding the genetic variation that correlates with phenotypic divergence must be attempted. If populations develop differently and markers for these divergent phenotypes can be found, evidentiary flies should be considered only locally informative. As appropriate, flies should be assigned to their proper population and/or phenotypic class before their ages are estimated with developmental data. This practice should result in lower error rates for PIA estimates, due to a better fit of predicted development rates to true development rates. In the cases in which genetic variation in forensically informative traits is demonstrated but marker loci are not found, efforts must be made to explore the full range of expected variation, which should then be used to calculate confidence intervals on predictions with arthropod evidence. Although this will likely result in larger confidence intervals for entomologically derived predictions, they will be more realistic and solidly supported by basic science.

FORENSIC ENTOMOLOGY IS AN APPLICATION OF ECOLOGICAL GENETICS

Ecological genetics/genomics is devoted to understanding the inheritance of ecologically relevant phenotypes. Accordingly, the field relies heavily on studying the quantitative and population genetics of adaptations (7, 46, 67, 74). Natural variation in developmental rates, life-history traits, stress tolerance, oviposition preferences, mating behaviors, disease resistance, and predator avoidance behaviors (to name a few) affect the survival and/or reproductive output of an organism in its natural environment. Furthermore, this variation affects the duration of each phase or stage of our proposed framework (**Figure 1**). Understanding the underlying causes of natural variation would lead to a greater appreciation of the variation associated with both pre-CI and post-CI.

Ecological genetics: the study of the inheritance of ecologically important phenotypes

Ecologically relevant phenotypes are also part of the forensic entomology roadmap presented in this paper. Like many organisms, necrophilous flies must find and attract mates. They need to detect, locate, and evaluate a resource to colonize and ensure that their offspring gain access to that resource. Once the flies have colonized a resource, they must compete with conspecifics and heterospecifics for an ephemeral food source. All the while, they must avoid predation, parasitism, disease, and more successful competitors in a variety of environmental settings. These challenges are not unique to forensically informative flies but are encountered in a wide variety of organisms employing a range of adaptations to meet those challenges. Many of those adaptations (as discussed above) have been studied within the context of ecological genetics with great success. A detailed understanding of the sources and consequences of variation in fly development could provide the discipline of forensic entomology with needed valid error rates in blow fly development time. Forensic entomologists should not consider their research just as a problem in applied developmental biology, but rather a problem in applied ecological genetics.

CONCLUSION

In the late 1970s and early 1980s, evolutionary and ecological scientists evaluated the state of their fields and resolved to develop a more hypothesis-driven approach to research (1, 29, 55, 63, 81, 82, 100, 107, 122, 125). This movement provided the foundation for improved science that led both to notable advances in evolutionary ecology and to a maturation of the field. This change paved the way for rapid advances in technological and theoretical developments, as seen in the successful application of sophisticated ecological genomics research discussed above. We propose that the state of forensic entomology has reached a similar point of development. The field has made great strides in documenting and characterizing many aspects of terrestrial carrion decomposition through observational studies. Although there is always a need for better descriptions, it is equally important to ask quantitative and mechanistic questions that address why carrion decay proceeds as it does and why the participants behave as they do. These are not questions that can be answered under the current observational paradigm and are addressed in this review.

The publication of the NRC report has brought the field to a crossroads. It can either continue down the current path of conducting research that limits the application of data to the traditional assumptions of forensics, or it can embrace a research agenda devoted to identifying and understanding the underlying mechanisms and sources of error associated with arthropod-based predictions. Pursuing the course suggested here will align the field with standards and requirements imposed by the *Daubert* standard, by the 2009 NRC report, and by basic science. We have provided a roadmap for the field to make choices that will continue to strengthen the science at a more efficient and rapid pace with a unified language and scientific philosophy.

> *"Although obstacles exist both inside and outside forensic science, the time is ripe for the traditional forensic sciences to replace antiquated assumptions of uniqueness and perfection with a more defensible empirical and probabilistic foundation."*
>
> *Saks & Koehler* (115)

SUMMARY POINTS

1. The NRC released a report in 2009 calling for forensic-related research built on a stronger scientific foundation.
2. We propose a new roadmap and framework to unify basic and applied decomposition research.
3. The roadmap is intended to advocate the use of terms that reflect the basic dynamics of decomposition that transcend forensics and reflect concepts that can be used to guide basic research.
4. The framework is grounded in understanding the ecological, evolutionary, and genetic mechanisms happening during carrion decomposition.
5. The framework advocates a standard language for describing the ecological activities occurring during decomposition and identifies major intervals and phases that characterize ecologically important transitions of the process.
6. We demonstrate that the pre-CI is largely understudied and could hold the key to the basic parameters regulating community assembly.
7. Forensic entomology can be aligned with basic biology research by studying the phenotypes inherent to the proposed framework in an ecological genetic context.

FUTURE ISSUES

1. The role of microbes associated with decomposing remains as mediators of colonization by arthropods needs to be further explored.

2. The ecological role of microbes on decomposing remains and how they mediate trophic interactions, nutrient processing and ecosystem services should be described and quantified.
3. Common mechanisms of arthropod detection, attraction and use of organic living and nonliving resources should be identified.
4. Error rates for predicting the duration of carrion decomposition by understanding sources of abiotic and biotic variation during the process should be developed.
5. Estimates of the PIA as it relates to the PMI of a corpse should be validated.
6. Genetic and genomic tools for necrophilous arthropod species need to be developed.
7. Ecological genetic studies of phenotypes used to make PMI estimates and that are important for colonization of, and survival on, carrion should be conducted.

DISCLOSURE STATEMENT

The authors are not aware of any affiliations, memberships, funding, or financial holdings that might be perceived as affecting the objectivity of this review.

ACKNOWLEDGMENTS

The Department of Entomology and Agrilife Research at Texas A&M University provided financial support to A.M.T. and J.K.T. The University of Dayton provided financial support to M.E.B. We would like to thank J. Conner, J. Wells, and C. Picard for comments on earlier versions of this manuscript.

LITERATURE CITED

1. Abbott I, Abbott LK, Grant PR. 1977. Comparative ecology of Galapagos ground finches (*Geospiza*-Gould)—evaluation of importance of floristic diversity and inter-specific competition. *Ecol. Monogr.* 47:151–84
2. Alessandrini F, Mazzanti M, Onofri V, Turchi C, Tagliabracci A. 2008. MtDNA analysis for genetic identification of forensically important insects. *Forensic Sci. Int. Genet. Suppl. Ser.* 1:584–85
3. Alexandersson R, Agren J. 2000. Genetic structure in the nonrewarding, bumblebee-pollinated orchid *Calypso bulbosa*. *Heredity* 85:401–9
4. Amendt J, Campobasso CP, Gaudry E, Reiter C, LeBlanc HN, Hall MJR. 2007. Best practice in forensic entomology—standards and guidelines. *Int. J. Legal Med.* 121:90–104
5. Amendt J, Zehner R, Reckel F. 2007. The nocturnal oviposition behavior of blowflies (Diptera: Calliphoridae) in Central Europe and its forensic implications. *Forensic Sci. Int.* 175:61–64
6. Ames C, Turner B. 2003. Low temperature episodes in development of blowflies: implications for postmortem interval estimation. *Med. Vet. Entomol.* 17:178–86
7. Anderson AR, Hoffmann AA, McKechnie SW, Umina PA, Weeks AR. 2005. The latitudinal cline in the In(3R)Payne inversion polymorphism has shifted in the last 20 years in Australian *Drosophila melanogaster* populations. *Mol. Ecol.* 14:851–58
8. Anderson GS. 2005. Effects of arson on forensic entomology evidence. *Can. Soc. Forensic Sci. J.* 38:49–67
9. Arbeitman MN, Furlong EE, Imam F, Johnson E, Null BH, et al. 2002. Gene expression during the life cycle of *Drosophila melanogaster*. *Science* 297:2270–75
10. Arnaldos MI, Garcia MD, Romera E, Presa JJ, Luna A. 2005. Estimation of postmortem interval in real cases based on experimentally obtained entomological evidence. *Forensic Sci. Int.* 149:57–65

11. Arnott S, Turner B. 2008. Post-feeding larval behaviour in the blowfly, *Calliphora vicina*: effects on postmortem interval estimates. *Forensic Sci. Int.* 177:162–67
12. Avise JC. 2004. *Molecular Markers, Natural History, and Evolution*. Sunderland, MA: Sinauer Assoc. 684 pp.
13. Bergeret M. 1855. Infanticide, momification du cadaver. Decouverte du cadaver d'un enfant nouveau-ne dans une dheminee ou il setait momifie. Determination de l'epoque de la naissance par la presence de numphes et de larves d'insectes dans le cadaver et par l'etude de leurs metamorphoses. *Ann. Hyg. Legal Med.* 4:442–52
14. Blanckenhorn WU. 2002. The consistency of quantitative genetic estimates in field and laboratory in the yellow dung fly. *Genetica* 114:171–82
15. Bruce TJA, Wadhams LJ, Woodcock CM. 2005. Insect host location: a volatile situation. *Trends Plant Sci.* 10:269–74
16. **Burkepile DE, Parker JD, Woodson CB, Mills HJ, Kubanek J, et al. 2006. Chemically mediated competition between microbes and animals: microbes as consumers in food webs. *Ecology* 87:2821–31**

16. Demonstrated that microbes associated with decomposing tissue regulate consumption by animals in higher trophic levels.

17. Byrd JH, Butler JF. 1996. Effects of temperature on *Cochliomyia macellaria* (Diptera: Calliphoridae) development. *J. Med. Entomol.* 33:901–5
18. Byrd JH, Butler JF. 1997. Effects of temperature on *Chrysomya rufifacies* (Diptera: Calliphoridae) development. *J. Med. Entomol.* 34:353–58
19. **Byrd JH, Castner JL, eds. 2010. *Forensic Entomology: The Utility of Arthropods in Legal Investigations*. Boca Raton, FL: CRC. 681 pp. 2nd ed.**

19. A summary of the use of entomological evidence in criminal investigations.

20. Byrne AL, Camann MA, Cyr TL, Catts EP, Espelie KE. 1995. Forensic implications of biochemical differences among geographic populations of the black blow fly, *Phormia regina* (Meigen). *J. Forensic Sci.* 40:372–77
21. Calboli FCF, Gilchrist GW, Partridge L. 2003. Different cell size and cell number contribution in two newly established and one ancient body size cline of *Drosophila subobscura*. *Evolution* 57:566–73
22. Calboli FCF, Kennington WJ, Partridge L. 2003. QTL mapping reveals a striking coincidence in the positions of genomic regions associated with adaptive variation in body size in parallel clines of *Drosophila melanogaster* on different continents. *Evolution* 57:2653–58
23. Catts EP. 1992. Problems in estimating the postmortem interval in death investigations. *J. Agric. Entomol.* 9:245–55
24. Catts EP, Goff ML. 1992. Forensic entomology in criminal investigations. *Annu. Rev. Entomol.* 37:253–72
25. Catts EP, Haskell NH, eds. 1990. *Entomology and Death: A Procedural Guide*. Clemson, SC: Joyce's Print Shop, Inc. 182 pp.
26. Chapman RF. 2003. Contact chemoreception in feeding by phytophagous insects. *Annu. Rev. Entomol.* 48:455–84
27. Clark K, Evans L, Wall R. 2006. Growth rates of the blowfly, *Lucilia sericata*, on different body tissues. *Forensic Sci. Int.* 156:145–49
28. Conner JK, Hartl DL. 2004. *A Primer of Ecological Genetics*. Sunderland, MA: Sinauer Assoc. 304 pp.
29. Connor EF, Simberloff D. 1978. Species number and compositional similarity of 635 Galapagos flora and avifauna. *Ecol. Monogr.* 48:219–48
30. Coyne JA, Wicker-Thomas C, Jallon JM. 1999. A gene responsible for a cuticular hydrocarbon polymorphism in *Drosophila melanogaster*. *Genet. Res.* 73:189–203
31. Davies L. 1999. Seasonal and spatial changes in blowfly production from small and large carcasses at Durham in lowland northeast England. *Med. Vet. Entomol.* 13:245–51
32. De'ath G, Fabricius KE. 2000. Classification and regression trees: a powerful yet simple technique for ecological data analysis. *Ecology* 81:3178–92
33. Dethier VG. 1947. *Chemical Insect Attractants and Repellents*. Philadelphia, PA: The Blakiston Co.
34. Demont M, Blanckenhorn WU, Hosken DJ, Garner TWJ. 2008. Molecular and quantitative genetic differentiation across Europe in yellow dung flies. *J. Evol. Biol.* 21:1492–503
35. Dorchin N, Scott ER, Clarkin CE, Luongo MP, Jordan S, Abrahamson WG. 2009. Behavioural, ecological and genetic evidence confirm the occurrence of host-associated differentiation in goldenrod gall-midges. *J. Evol. Biol.* 22:729–39

36. Etges WJ, de Oliveira CC, Ritchie MG, Noor MAF. 2009. Genetics of incipient speciation in *Drosophila mojavensis*: II. Host plants and mating status influence cuticular hydrocarbon QTL expression and G × E interactions. *Evolution* 63:1712–30
37. Faigman DL. 2002. Science and the law: Is science different for lawyers? *Science* 297:339–40
38. Falconer DS. 1989. *Introduction to Quantitative Genetics*. New York: Longman Wiley. 438 pp.
39. Fauvergue X, Lo Genco A, Lo Pinto M. 2008. Virgins in the wild: Mating status affects the behavior of a parasitoid foraging in the field. *Oecologia* 156:913–20
40. Feder JL, Berlocher SH, Roethele JB, Dambroski H, Smith JJ, et al. 2003. Allopatric genetic origins for sympatric host-plant shifts and race formation in *Rhagoletis*. *Proc. Natl. Acad. Sci. USA* 100:10314–19
41. Florin AB, Gyllenstrand N. 2002. Isolation and characterization of polymorphic microsatellite markers in the blowflies *Lucilia illustris* and *Lucilia sericata*. *Mol. Ecol. Notes* 2:113–16
42. Forrest AD, Hollingsworth ML, Hollingsworth PM, Sydes C, Bateman RM. 2004. Population genetic structure in European populations of *Spiranthes romanzoffiana* set in the context of other genetic studies on orchids. *Heredity* 92:218–27
43. Frechette B, Dixon AFG, Claude A, Jean L. 2004. Age and experience influence patch assessment for oviposition by an insect predator. *Ecol. Entomol.* 29:578–83
44. **Gallagher MB, Sandhu S, Kimsey R. 2010. Variation in development time for geographically distinct populations of the common green bottle fly, *Lucilia sericata* (Meigen). *J. Forensic Sci.* 55:438–42**
45. Gião J, Godoy W. 2007. Ovipositional behavior in predator and prey blowflies. *J. Insect Behav.* 20:77–86
46. Gilbert GS. 2002. Evolutionary ecology of plant diseases in natural ecosystems. *Annu. Rev. Phytopathol.* 40:13–43
47. Gilchrist GW, Huey RB, Serra L. 2001. Rapid evolution of wing size clines in *Drosophila subobscura*. *Genetica* 112:273–86
48. Goff ML. 1992. Problems in estimation of postmortem interval resulting from wrapping of the corpse: a case study from Hawaii. *J. Agric. Entomol.* 9:237–43
49. Goff ML. 2000. *A Fly for the Prosecution: How Insect Evidence Helps Solve Crimes*. Cambridge, MA: Harvard Univ. Press. 225 pp.
50. Goff ML, Win BH. 1997. Estimation of postmortem interval based on colony development time for *Anoplolepsis longipes* (Hymenoptera: Formicidae). *J. Forensic Sci.* 42:1176–79
51. Gomes L, Godoy WA, Von Zuben CJ. 2006. A review of postfeeding larval dispersal in blowflies: implications for forensic entomology. *Naturwissenschaften* 93:207–15
52. Gomes L, Gomes G, Von Zuben CJ. 2009. The influence of temperature on the behavior of burrowing in larvae of the blowflies, Chrysomya albiceps and Lucilia cuprina, under controlled conditions. *J. Insect. Sci.* 9:14
53. Gomes L, Von Zuben CJ. 2005. Postfeeding radial dispersal in larvae of Chrysomya albiceps (Diptera: Calliphoridae): implications for forensic entomology. *Forensic Sci. Int.* 155:61–64
54. Goodbrod JR, Goff ML. 1990. Effects of larval population density on rates of development and interactions between two species of *Chrysomya* (Diptera: Calliphoridae) in laboratory culture. *J. Med. Entomol.* 27:338–43
55. Grant PR, Abbott I. 1980. Inter-specific competition, island biogeography and null hypotheses. *Evolution* 34:332–41
56. Grassberger M, Frank C. 2004. Initial study of arthropod succession on pig carrion in a central European urban habitat. *J. Med. Entomol.* 41:511–23
57. Grassberger M, Reiter C. 2001. Effect of temperature on *Lucilia sericata* (Diptera: Calliphoridae) development with special reference to the isomegalen- and isomorphen-diagram. *Forensic Sci. Int.* 120:32–36
58. Greenberg B. 1990. Behavior of postfeeding larvae of some Calliphoridae and a muscid (Diptera). *Ann. Entomol. Soc. Am.* 83:1210–14
59. Greenberg B. 1991. Flies as forensic indicators. *J. Med. Entomol.* 20:565–77
60. Griffiths AM, Evans LM, Stevens JR. 2009. Characterization and utilization of microsatellite loci in the New World screwworm fly, *Cochliomyia hominivorax*. *Med Vet. Entomol.* 23(Suppl.)1: 8–13
61. Grünbaum D. 1998. Using spatially explicit models to characterize foraging performance in heterogeneous landscapes. *Am. Nat.* 151:97–113

44. A comparison of development rates of different blow fly populations raised in the same environment, demonstrating population-specific growth rates.

62. Hahn DA, Ragland GJ, Shoemaker DD, Denlinger DL. 2009. Gene discovery using massively parallel pyrosequencing to develop ESTs for the flesh fly *Sarcophaga crassipalpis*. *BMC Genomics* 10:234
63. Harvey PH, Colwell RK, Silvertown JW, May RM. 1983. Null models in ecology. *Annu. Rev. Ecol. Syst.* 14:189–211
64. Haskell NH. 2007. Insect evidence distribution: tabulation of primary indicator species, the life stage, and the season of year used in final analysis from 100 random North American cases. *Proc. Am. Acad. Forensic Sci., San Antonio, Tex., 2007*, 13:220. Colorado Springs: Am. Acad. Forensic Sci.
65. Hengeveld GM, van Langevelde F, Groen TA, de Knegt HJ. 2009. Optimal foraging for multiple resources in several food species. *Am. Nat.* 174:102–10
66. Hobson RP. 1936. Sheep blow fly investigations. III. Obsrevations on the chemotropism of *Lucilia sericata*. *Ann. Appl. Biol.* 23:845–51
67. Hoffmann AA, Watson M. 1993. Geographical variation in the acclimation responses of *Drosophila* to temperature extremes. *Am. Nat.* 142:S93–113
68. Holsinger KE, Weir BS. 2009. Genetics in geographically structured populations: defining, estimating and interpreting F(ST). *Nat. Rev. Genet.* 10:639–50
69. Huntington TE, Higley LG, Baxendale FP. 2007. Maggot development during morgue storage and its effect on estimating the post-mortem interval. *J. Forensic Sci.* 52:453–58
70. Introna F, Campobasso CP, Goff ML. 2001. Entomotoxicology. *Forensic Sci. Int.* 120:42–47
71. James AC, Azevedo RB, Partridge L. 1997. Genetic and environmental responses to temperature of *Drosophila melanogaster* from a latitudinal cline. *Genetics* 146:881–90
72. **Janzen DH. 1977. Why fruits rot, seeds mold, and meat spoils. *Am. Nat.* 111:691–713**

72. Recognized that microbes have ecological roles in nature other than nutrient recyclers.

73. Johnson FM, Schaffer HE. 1973. Isozyme variability in species of the genus *Drosophila*. VII. Genotype-environment relationships in populations of *D. melanogaster* from the Eastern United States. *Biochem. Genet.* 10:149–63
74. Jones CD. 2005. The genetics of adaptation in *Drosophila sechellia*. *Genetica* 123:137–45
75. Kaneshrajah G, Turner B. 2004. Calliphora vicina larvae grow at different rates on different body tissues. *Int. J. Legal Med.* 118:242–44
76. Karlsson S, Mork J. 2005. Deviation from Hardy-Weinberg equilibrium, and temporal instability in allele frequencies at microsatellite loci in a local population of Atlantic cod. *ICES J. Mar. Sci.* 62:1588–96
77. Kennington WJ, Hoffmann AA, Partridge L. 2007. Mapping regions within cosmopolitan inversion In(3R)Payne associated with natural variation in body size in *Drosophila melanogaster*. *Genetics* 177:549–56
78. Koppl R. 2005. How to improve forensic science. *Eur. J. Law Econ.* 20:255–86
79. Koppl R. 2007. CSI for real: how to improve forensic science. *Reason Found. Policy Study* 364
80. Lam K, Babor D, Duthie B, Babor EM, Moore M, Gries G. 2007. Proliferating bacterial symbionts on house fly eggs affect oviposition behaviour of adult flies. *Anim. Behav.* 74:81–92
81. Levins R, Lewontin R. 1980. Dialectics and reductionism in ecology. *Synthese* 43:47–78
82. Loehle C. 1983. Evaluation of theories and calculation tools in ecology. *Ecol. Model.* 19:239–47
83. Lopez-Fanjul C, Fernandez A, Toro MA. 2003. The effect of neutral nonadditive gene action on the quantitative index of population divergence. *Genetics* 164:1627–33
84. Lynch M, Walsh B. 1998. *Genetics and Analysis of Quantitative Traits*. Sunderland, MA: Sinauer Assoc. 980 pp.
85. MacArthur RH, Pianka ER. 1966. On the optimal use of a patchy environment. *Am. Nat.* 100:603–9
86. Mackay TF. 2001. The genetic architecture of quantitative traits. *Annu. Rev. Genet.* 35:303–39
87. Mégnin P. 1894. La faune des cadavres, application de l'entomologie a la médecine légale. In *Volume 101B. Encyclopédie Scientifique des Aide-Mémoire*. Paris: Masson, Francen
88. Merritt RW, Benbow ME. 2009. Forensic entomology. In *Encyclopedia of Forensic Science*, ed. A Jamieson, A Moenssens, pp. 1–12. Hoboken, NJ: Wiley
89. Mittelstaedt H. 1962. Control systems of orientation in insects. *Annu. Rev. Entomol.* 7:177–98
90. Motter MG. 1898. A contribution to the study of the fauna of the grave. A study of one hundred and fifty disinterments, with some additional experimental observations. *J. N.Y. Entomol. Soc.* 6:201–31

91. Nabity PD, Higley LG, Heng-Moss TM. 2006. Effects of temperature on development of Phormia regina (Diptera: Calliphoridae) and use of developmental data in determining time intervals in forensic entomology. *J. Med. Entomol.* 43:1276–86
92. Natl. Res. Counc. (U.S.). Comm. DNA Forensic Sci.: An Update. 1996. *The Evaluation of Forensic DNA Evidence*. Washington, DC: Natl. Acad. Press. 254 pp.
93. Natl. Res. Counc. (U.S.). Comm. Identifying Needs Forensic Sci. Community; Comm. Sci. Law Policy Glob. Aff.; Comm. Appl. Theor. Stat. Div. Eng. Phys. Sci. 2009. *Strengthening Forensic Science in the United States: A Path Forward*. 352 pp. Washington, DC: Natl. Acad. Press
94. Olshen AB, Gold B, Lohmueller KE, Struewing JP, Satagopan J, et al. 2008. Analysis of genetic variation in Ashkenazi Jews by high density SNP genotyping. *BMC Genet.* 9:14
95. Oudman L, Van Delden W, Kamping A, Bijlsma R. 1991. Polymorphism at the Adh and alpha Gpdh loci in *Drosophila melanogaster*: effects of rearing temperature on developmental rate, body weight, and some biochemical parameters. *Heredity* 67(Pt. 1):103–15
96. Papaj DR. 2000. Ovarian dynamics and host use. *Annu. Rev. Entomol.* 45:423–48
97. Parmenter RR, MacMahon JA. 2009. Carrion decomposition and nutrient cycling in a semiarid shrub–steppe ecosystem. *Ecol. Monogr.* 79:637–61

97. Offers a comprehensive review of carrion decomposition ecology.

98. Parsch J, Russell JA, Beerman I, Hartl DL, Stephan W. 2000. Deletion of a conserved regulatory element in the *Drosophila* Adh gene leads to increased alcohol dehydrogenase activity but also delays development. *Genetics* 156:219–27
99. Payne JA. 1965. A summer carrion study of the baby pig *Sus scrofa* Linnaeus. *Ecology* 46:592–602

99. A seminal work in decomposition ecology that serves as part of the foundation of forensic entomology.

100. Peters RH. 1976. Tautology in evolution and ecology. *Am. Nat.* 110:1–12
101. Picard CJ, Wells JD. 2009. Survey of the genetic diversity of Phormia regina (Diptera: Calliphoridae) using amplified fragment length polymorphisms. *J. Med. Entomol.* 46:664–70

101. A population genetic analysis of U.S. populations of a forensically informative blow fly species, demonstrating genetic relatedness of flies collected at the same time, but little spatial genetic structure.

102. Picard CJ, Wells JD. 2010. The population genetic structure of North American *Lucilia sericata* (Diptera: Calliphoridae), and the utility of genetic assignment methods for reconstruction of postmortem corpse relocation. *Forensic. Sci. Int.* 195:63–67
103. Piechnik DA, Lawler SP, Martinez ND. 2008. Food-web assembly during a classic biogeographic study: Species' "trophic breadth" corresponds to colonization order. *Oikos* 117:665–74
104. Prokopy RJ, Owens ED. 1983. Visual detection of plants by herbivorous insects. *Annu. Rev. Entomol.* 28:337–64
105. Pyke GH. 1984. Optimal foraging theory: a critical review. *Annu. Rev. Ecol. Syst.* 15:523–75
106. Quicke DLJ. 1997. *Parasitic Wasps*. London: Chapman & Hall. 470 pp.
107. Quinn JF, Dunham AE. 1983. On hypothesis-testing in ecology and evolution. *Am. Nat.* 122:602–17
108. Rako L, Blacket MJ, McKechnie SW, Hoffmann AA. 2007. Candidate genes and thermal phenotypes: identifying ecologically important genetic variation for thermotolerance in the Australian *Drosophila melanogaster* cline. *Mol. Ecol.* 16:2948–57
109. Rako L, Hoffmann AA. 2006. Complexity of the cold acclimation response in *Drosophila melanogaster*. *J. Insect Physiol.* 52:94–104
110. Reim C, Teuschl Y, Blanckenhorn WU. 2006. Size-dependent effects of larval and adult food availability on reproductive energy allocation in the Yellow Dung Fly. *Funct. Ecol.* 20:1012–21
111. Riginos C, Nachman MW. 2001. Population subdivision in marine environments: the contributions of biogeography, geographical distance and discontinuous habitat to genetic differentiation in a blennioid fish, *Axoclinus nigricaudus*. *Mol. Ecol.* 10:1439–53
112. Rosa GS, de Carvalho LR, dos Reis SF, Godoy WAC. 2006. The dynamics of intraguild predation in Chrysomya albiceps Wied. (Diptera: Calliphoridae): interactions between instars and species under different abundances of food. *Neotropical Entomol.* 35:775–80
113. Rozen DE, Engelmoer DJP, Smiseth PT. 2008. Antimicrobial strategies in burying beetles breeding on carrion. *Proc. Natl. Acad. Sci. USA* 105:17890–95
114. Saks MJ. 2001. Model prevention and remedy of erroneous convictions act. *Ariz. State Law J.* 33:665–718
115. Saks MJ, Koehler JJ. 2005. The coming paradigm shift in forensic identification science. *Science* 309:892–95
116. Saks MJ, Koehler JJ. 2008. The individualization fallacy in forensic science evidence. *Vanderbilt Law Rev.* 61:199–219

117. Deleted in proof
118. Schoenly K. 1992. A statistical analysis of successional patterns in carrion-arthropod assemblages: implications for forensic entomology and determination of the postmortem interval. *J. Forensic Sci.* 37:1489–513
119. Schröder R, Hilker M. 2008. The relevance of background odor in resource location by insects: a behavioral approach. *BioScience* 58:308–16
120. Schweitzer NJ, Saks MJ. 2007. The CSI effect: Popular fiction about forensic science affects the public's expectations about real forensic science. *Jurimetrics* 47:357–64
121. Shalaby OA, deCarvalho LM, Goff ML. 2000. Comparison of patterns of decomposition in a hanging carcass and a carcass in contact with soil in a xerophytic habitat on the Island of Oahu, Hawaii. *J. Forensic Sci.* 45:1267–73
122. Simberloff D. 1980. Succession of paradigms in ecology: essentialism to materialism and probabilism. *Synthese* 43:3–39
123. Solomon SM, Hackett EJ. 1996. Setting boundaries between science and law: lessons from Daubert v. Merrell Dow Pharmaceuticals, Inc. *Sci. Technol. Hum. Values* 21:131–56
124. Statheropoulos M, Spiliopoulou C, Agapiou A. 2005. A study of volatile organic compounds evolved from the decaying human body. *Forensic Sci. Int.* 153:147–55
125. Strong DR, Szyska LA, Simberloff DS. 1979. Tests of community-wide character displacement against null hypotheses. *Evolution* 33:897–913
126. Tarone AM. 2007. *Lucilia sericata Development: Plasticity, Population Differences, and Gene Expression*. East Lansing: Mich. State Univ. 248 pp.
127. Tarone AM, Foran DR. 2006. Components of developmental plasticity in a Michigan population of *Lucilia sericata* (Diptera: Calliphoridae). *J. Med. Entomol.* 43:1023–33
128. Tarone AM, Foran DR. 2008. Generalized additive models and *Lucilia sericata* growth: assessing confidence intervals and error rates in forensic entomology. *J. Forensic Sci.* 53:942–49
129. Tarone AM, Foran DR. 2010. Gene expression during blow fly development: improving the precision of age estimates in forensic entomology. *J. Forensic Sci.* In press
130. Tarone AM, Jennings KC, Foran DR. 2007. Aging blow fly eggs using gene expression: a feasibility study. *J. Forensic Sci.* 52:1350–54
131. Tessmer JW, Meek CL. 1996. Dispersal and distribution of Calliphoridae (Diptera) immatures from animal carcasses in southern Louisiana. *J. Med. Entomol.* 33:665–69
132. Tinbergen N. 1963. On aims and methods of ethology. *Z. Tierpsychol.* 20:410–33
133. Tomberlin JK, Sheppard DC, Joyce JA. 2005. Black soldier fly (Diptera: Stratiomyidae) colonization of pig carrion in south Georgia. *J. Forensic Sci.* 50:152–53
134. Tomberlin JK, Wallace JR, Byrd JH. 2006. Forensic entomology: myths busted! *Forensic Mag.* 3:10–14
135. Toolson EC, Kupersimbron R. 1989. Laboratory evolution of epicuticular hydrocarbon composition and cuticular permeability in *Drosophila pseudoobscura*: effects on sexual dimorphism and thermal-acclimation ability. *Evolution* 43:468–73
136. Trotta V, Calboli FC, Ziosi M, Guerra D, Pezzoli MC, et al. 2006. Thermal plasticity in *Drosophila melanogaster*: a comparison of geographic populations. *BMC Evol. Biol.* 6:67
137. Turner TL, Hahn MW, Nuzhdin SV. 2005. Genomic islands of speciation in *Anopheles gambiae*. *PloS Biol.* 3:e285
138. Ueno K, Ueno T. 2005. Effect of wasp size, physiological state, and prior host experience on host-searching behavior in a parasitoid wasp (Hymenoptera: Ichneumonidae). *J. Ethol.* 23:43–49
139. VanLaerhoven SL. 2010. Ecological theory and its application in forensic entomology. See Ref. 19, pp. 493–518
140. VanLaerhoven SL, Anderson GS. 1999. Insect succession on buried carrion in two biogeoclimatic zones of British Columbia. *J. Forensic Sci.* 44:32–43
141. Vass AA, Barshick SA, Sega G, Caton J, Skeen JT, et al. 2002. Decomposition chemistry of human remains: a new methodology for determining the postmortem interval. *J. Forensic Sci.* 47:542–53
142. Vet LEM, Dicke M. 1992. Ecology of infochemical use by natural enemies in a tritrophic context. *Annu. Rev. Entomol.* 37:141–72

143. Via S. 1984. The quantitative genetics of polyphagy in an insect herbivore. 1. Genotype-environment interaction in larval performance on different host plant-species. *Evolution* 38:881–95
144. Vinson SB. 1976. Host selection by insect parasitoids. *Annu. Rev. Entomol.* 21:109–33
145. Vinson SB. 1985. The behavior of parasitoids. In *Comprehensive Insect Physiology, Biochemistry, and Pharmacology: Nervous System*, ed. GA Kerkut, LI Gilbert, pp. 417–69. New York: Pergamon
146. Vinson SB. 1991. Chemical signals used by insect parasitoids. *Redia, Geornale Zool.* 124:15–42
147. Visser JH. 1986. Host odor perception in phytophagous insects. *Annu. Rev. Entomol.* 31:121–44
148. Voss SC, Spafford H, Dadour IR. 2009. Annual and seasonal patterns of insect succession on decomposing remains at two locations in Western Australia. *Forensic Sci. Int.* 193:26–36
149. Watson EJ, Carlton CE. 2003. Spring succession of necrophilous insects on wildlife carcasses in Louisiana. *J. Med. Entomol.* 40:338–47
150. Watts JE, Merritt GC, Goodrich BS. 1981. The ovipositional response of the Australian sheep blowfly, *Lucilia cuprina*, to fleece-rot odours. *Aust. Vet. J.* 57:450–54
151. Wayne RK, Ostrander EA. 2007. Lessons learned from the dog genome. *Trends Genet.* 23:557–67
152. Wells JD, Lamotte LR. 2010. Estimating the postmortem interval. See Ref. 19, pp. 367–88
153. Wells JD, Stevens JR. 2008. Application of DNA-based methods in forensic entomology. *Annu. Rev. Entomol.* 53:103–20
154. Williams H. 1984. A model for the aging of fly larvae in forensic entomology. *Forensic Sci. Int.* 25:191–99
155. Wright S. 1965. The interpretation of population-structure by F-statistics with special regard to systems of mating. *Evolution* 19:395–420
156. Zhao K, Aranzana MJ, Kim S, Lister C, Shindo C, et al. 2007. An *Arabidopsis* example of association mapping in structured samples. *PLoS Genet.* 3:e4

Visual Cognition in Social Insects

Aurore Avarguès-Weber,[1,2] Nina Deisig,[1,2] and Martin Giurfa[1,2]

[1]Centre de Recherches sur la Cognition Animale, Université de Toulouse, F-31062 Toulouse Cedex 9, France

[2]Centre de Recherches sur la Cognition Animale, CNRS, F-31062 Toulouse Cedex 9, France; email: giurfa@cict.fr

Annu. Rev. Entomol. 2011. 56:423–43

First published online as a Review in Advance on September 24, 2010

The *Annual Review of Entomology* is online at ento.annualreviews.org

This article's doi:
10.1146/annurev-ento-120709-144855

0066-4170/11/0107-0423$20.00

Key Words

learning, nonelemental learning, Hymenoptera, vision

Abstract

Visual learning admits different levels of complexity, from the formation of a simple associative link between a visual stimulus and its outcome, to more sophisticated performances, such as object categorization or rules learning, that allow flexible responses beyond simple forms of learning. Not surprisingly, higher-order forms of visual learning have been studied primarily in vertebrates with larger brains, while simple visual learning has been the focus in animals with small brains such as insects. This dichotomy has recently changed as studies on visual learning in social insects have shown that these animals can master extremely sophisticated tasks. Here we review a spectrum of visual learning forms in social insects, from color and pattern learning, visual attention, and top-down image recognition, to interindividual recognition, conditional discrimination, category learning, and rule extraction. We analyze the necessity and sufficiency of simple associations to account for complex visual learning in Hymenoptera and discuss possible neural mechanisms underlying these visual performances.

INTRODUCTION

Visual learning refers to an individual's capacity of acquiring experience-based information pertaining to visual stimuli so that adaptive responses can be produced when viewing such stimuli again. This capacity admits different levels of complexity, from the establishment of a simple associative link connecting a visual stimulus (e.g., a specific color) and its outcome (e.g., a reward or a punishment) to more sophisticated performances such as learning to categorize distinct objects (e.g., animal versus nonanimal) or apprehending abstract rules applicable to unknown visual objects (e.g., "larger than," "on top of," or "inside of").

The first situation, the establishment of univocal, unambiguous links between a visual target and its outcome, constitutes a case of elemental learning. For instance, what is learned for a color is valid only for that color and not for different ones. In contrast, learning about categories or rules constitutes a case of nonelemental learning, as appropriate responses can be transferred to unknown stimuli for which the subject has no personal experience, as long as the stimuli satisfy the learned category or rule. In these cases, the subject's response is flexible and relatively independent of the physical nature of the stimuli considered.

Social Hymenoptera, particularly bees (*Apis* spp. and *Bombus* spp.), ants, and wasps (several genera), which are at the center of this article, are interesting models for the study of visual learning because in their natural context they have to solve a diversity of visual problems of varying complexity. For instance, these insects learn and memorize the local cues characterizing the places of interest, which are essentially the hive and the food sources (27, 28, 68, 71, 85, 112). Honey bees (*Apis mellifera*), and to a minor extent bumble bees (*Bombus terrestris*), are flower constant, which means that they forage on a unique floral species as long as it offers a profitable nectar and/or pollen reward (14, 40, 46). This capacity is based partly on visual cues provided by flowers such as colors or patterns. Learning and memorizing the visual cues of the exploited flower through their association with a nectar and/or pollen reward are what allow a bee forager to track a particular flower species in the field (68). Similarly, learning abilities for landmark constellations, complex natural scenes, and celestial cues used in navigation (e.g., azimuthal position of the sun, polarized light pattern of the blue sky) ensure a safe return to the nest and enhance foraging efficiency (15, 16, 24, 74, 103).

Visual capacities are highly developed in social Hymenoptera. Bees, wasps, and some ant species see the world in color (3, 8, 9, 12, 21, 60, 61, 63, 70), perceive shapes and patterns (23, 33, 60, 61, 84, 102), and resolve movements with a high temporal resolution (86). One of the reasons why bees, ants, and wasps constitute an attractive model for the study of visual learning resides precisely in the existence of controlled experimental methods for the study of these capacities in the laboratory.

VISUAL CONDITIONING OF BEES

Visual conditioning of honey bees (98) has uncovered the perceptual capabilities of these insects and has been used to this end for nearly a century. This protocol exploits the fact that free-flying honey bees learn visual cues such as colors, shapes, patterns, depth, and motion contrast, among others (33, 34, 59, 98, 102), when these cues are presented together with a reward of sucrose solution. Each bee is individually marked by means of a color spot on the thorax or the abdomen so that individual performances can be recorded. The marked bee is generally displaced by the experimenter toward the experimental site, where it is rewarded with sucrose solution to promote its regular return. Such pretraining is performed without presenting the training stimuli in order to avoid uncontrolled learning. When the bee starts visiting the experimental place actively (i.e., without being displaced by the experimenter), the training stimuli are presented and the choice of the appropriate visual target rewarded with sucrose. This basic protocol has been used to study visual learning in other bee species such as bumble bees (24, 63, 82), leaf-cutter bees (9), stingless bees (6, 72), and wasps (3, 60).

In studies on pattern vision, the plane of stimulus presentation is extremely important. Early works (47–49, 98) presented stimuli on the horizontal plane of an experimental table, and the position of the stimuli were varied to verify that bees were indeed choosing a rewarded pattern rather than using position information. In this situation it is difficult to determine which specific information contained in the patterns is used by the insects to make their choices because they can approach the patterns from any possible direction. Later, vertical presentation was preferred because it constrained the approach direction to the patterns (99), thereby forcing a frontal approach and perception. In this way, it was possible to determine whether bees are sensitive to different pattern cues such as orientation, bilateral and radial symmetry, and center of gravity. In all cases, insects have to be trained and tested individually to achieve a precise control of the experience of each subject. It is also important to control the distance at which a choice is made because visual orientation and choice are mediated by different visual cues at different distances or angles subtended by the target (33, 34, 37). The time between visits to the experimental place also has to be recorded, as it reflects the appetitive motivation of the bee (75) and thus its motivation to learn. The associations built in this context can be classical, operant, or both; i.e., they may link a visual stimulus (conditioned stimulus, or CS) and a sucrose reward (unconditioned stimulus, or US), the response of the animal (e.g., landing) and the US, or both, respectively. The experimental framework is nevertheless mainly operant, because the bee's behavior is a determinant for obtaining or not obtaining the sucrose reinforcement.

Visual conditioning of freely flying insects does not allow visual learning to be studied at the cellular level. Because bees freely fly during the experiment, studying neural activity in visual centers in the brain simultaneously remains impossible thus far. Recently, however, a protocol for visual conditioning of harnessed bees has been developed (50, 51). This protocol, based on pioneer studies by Kuwabara (56), consists of training a harnessed bee to extend its proboscis to colors (50) or motion cues (51) paired with sucrose solution. Hungry bees reflexively extend the proboscis when their antennae are touched with sucrose solution, the equivalent of a nectar reward. In this protocol, colors or patterns are paired with a sucrose reward to create a Pavlovian association in which the visual stimuli are the CS and sucrose is the US. Learning, however, is poor in this protocol. It takes 2 days to reach an acquisition level that is approximately 40%, and this is possible only if bees have their antennae previously cut. The reasons for the apparent interference of the antennae on visual learning remain unknown. Cutting the antennae may affect the general motivation of the bee so that the sucrose reward is not as attractive as expected by the experimenter (19). Improving this protocol is a priority for future research on visual learning as it will allow combining behavioral quantification with access to the nervous system.

VISUAL CONDITIONING OF ANTS

Visual learning in ants has been studied mostly in the context of insect navigation. Experiments with the Sahara desert ant, *Cataglyphis bicolor*, have increased our understanding of navigation strategies based on celestial cues and landmarks (see Reference 103 for a review). In other ant species such as, *Melophorus bagoti* (10, 74), *Gigantiops destructor* (64), and *Formica rufa* (39, 45), similar questions have been investigated, thus placing importance on determining how ants use visual cues to negotiate their spatial environment. These works are not examined here because they rarely focus on the learning process itself, which is the main framework of our review. Although learning of visual cues underlies navigation processes studied in these ants, learning curves or memory retention tests are generally absent from these works, thus making difficult any analysis in terms of the associative links established during spatial learning in ants. It is therefore difficult to determine what is and what is not elemental in these performances.

Nevertheless, a recent work on *Cataglyphis aenescens* and *Formica cunicularia* has used an experimental design that reproduces basic features of visual training in bees (8), as it focuses on color learning and discrimination. These ants were trained in a Y-maze to choose and discriminate monochromatic lights of constant intensity associated with a food reward. Using this kind of design could help researchers dissect in a more controlled way the nature of associative learning in ants.

ATTENTIONAL AND EXPERIENCE-DEPENDENT MODULATION OF VISUAL LEARNING

The first study on bee learning and memory that used controlled protocols for characterizing individual learning and memory employed colors as rewarding stimuli (66). Free-flying honey bees were trained to choose a rewarded monochromatic light and were then presented in dual-choice situations with the rewarded light versus an alternative color on a horizontal plane. This study reported learning curves for different wavelengths and showed that, under these experimental conditions, bees learned all wavelengths after few learning trials. Some wavelengths, particularly 413 nm, were learned faster than others, requiring only one to three acquisition trials (66; but see below). This result argued in favor of innate biases in color learning, probably reflecting the intrinsic biological relevance of the color signals that are learned faster (66). Indeed, color-naïve honey bees in their first foraging flight prefer those colors that experienced bees learn faster (35), and those colors seem to correspond to floral colors highly associated with profitable nectar reward (35).

Visual learning, as studied in these color-conditioning experiments, is elemental, as bees are just presented with a single color target paired with sucrose solution. It was supposed to be a fast form of learning (66; see above), compared, for instance, to learning of visual patterns, which usually takes longer (20 or more trials). Recent studies on bumble bee and honey bee color learning (21, 29) have nevertheless introduced a new twist to these conclusions. It was long thought that what an animal sees and visually learns is constrained by its perceptual machinery with no or less place for experience-dependent modulations of perception. Studies on honey bees (22, 29) and bumble bees (21) have shown that this idea is wrong: In some cases, learning one and the same color may occur after few trials but in other cases it may take more than 20 trials (**Figure 1*a***). The critical feature is how the bees learn the task. Absolute conditioning, in which a subject is trained with a single color rewarded with sugar water, yields generally fast learning. Differential conditioning, in which the same subject learns to discriminate a rewarded from a nonrewarded color, takes more trials, even if the rewarded color is the same as in absolute conditioning. When these animals are asked to discriminate colors in a test, their performance differs dramatically. Whereas bees trained in differential conditioning discriminate colors that are very similar (**Figure 1*c***), bees trained in absolute conditioning do not discriminate the same pair of colors (**Figure 1*b***) (21, 29). Similar results were obtained in ants trained to discriminate colors in a Y-maze (8).

Comparable results were obtained in a study on pattern learning and discrimination by honey bees (32, 89). Whereas differential conditioning results in a visual recognition strategy that uses the cues present in the whole pattern, absolute conditioning results in a recognition strategy that restricts cue sampling mainly to the lower half of the pattern. In other words, bees recognize a pattern differently, depending on the kind of learning implicit to the conditioning task. In both cases (color and pattern learning), however, differential conditioning increases the demands imposed on the perceptual system of the bees, which must not only go where a rewarded stimulus is presented (absolute conditioning) but also discriminate

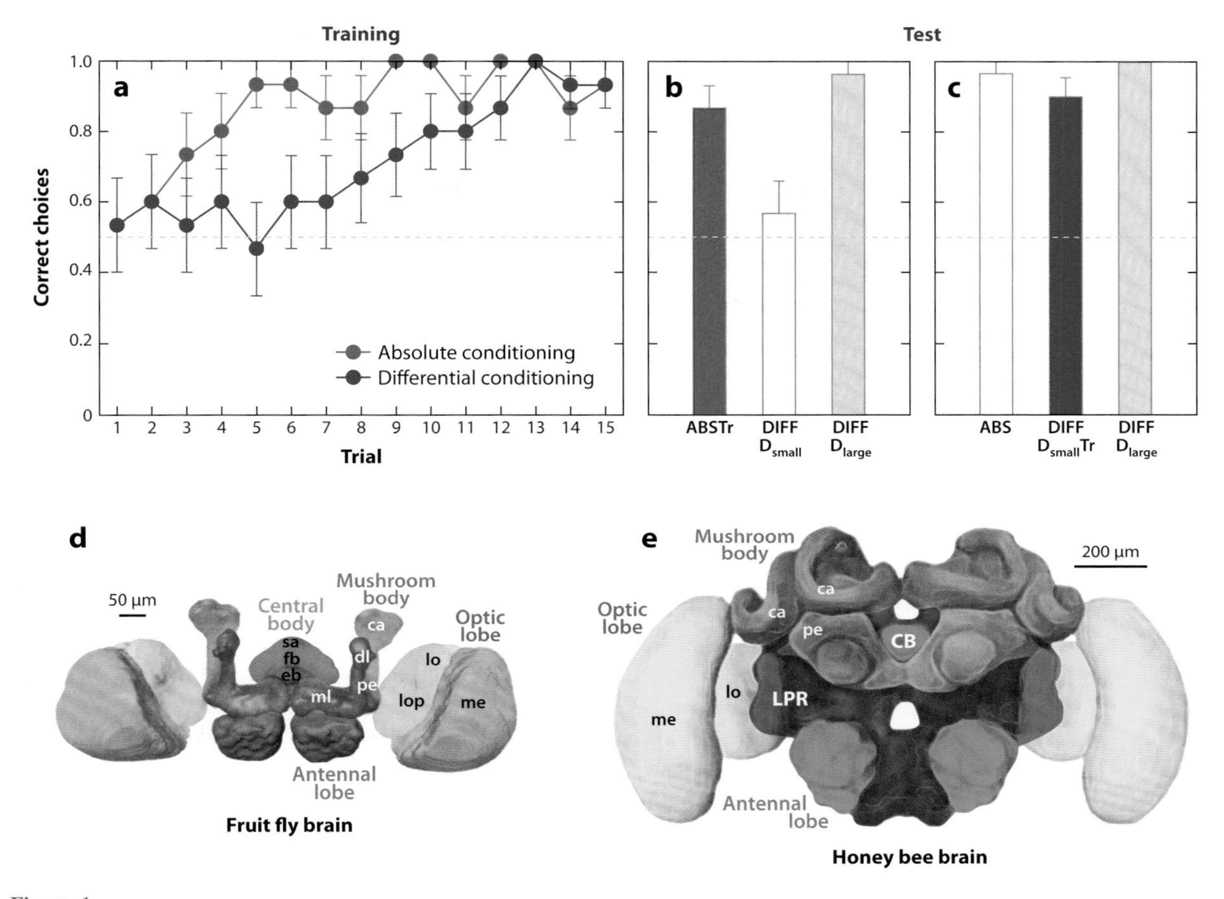

Figure 1

(*a, b, c*) Attention-like processes in honey bees. Performance of the free-flying bees trained with colors under absolute and differential conditioning. (*a*) Acquisition along 15 trials (mean ± standard error; $n = 15$ bees for each curve). (*b*) Tests of the group trained with absolute conditioning. The red bar depicts the test presenting the trained situation (ABSTr), i.e., the single color that was previously rewarded. The white bar depicts the test presenting a novel differential situation (DIFF D_{small}), i.e., the color that was previously rewarded versus a new color that was similar to the trained one. The gray bar depicts the test presenting a novel differential situation (DIFF D_{large}), i.e., the color that was previously rewarded versus a new color that was different from the trained one. (*c*) Tests of the group trained with differential conditioning. The blue bar depicts the test presenting the trained situation (DIFF D_{small}Tr), i.e., the previously rewarded and the nonrewarded colors that were similar. The white bar depicts the test presenting just the previously rewarded color (ABS). The gray bar depicts the test presenting the previously rewarded color versus a novel color different from the rewarded one (DIFF D_{large}). (*d*) Three-dimensional reconstruction of the fruit fly (*Drosophila melanogaster*) brain (courtesy of Armin Jenett). (*e*) Three-dimensional reconstruction of a honey bee brain. Abbreviations: me, medulla; lo, lobula; lop, lobula plate; ca, calyx; dl, dorsal lobe; ml, medial lobe; pe, peduncle; CB, central body; eb, ellipsoid body; fb, fan-shaped body; sa, superior arch; LPR, lateral protocerebrum. Adapted from Reference 29.

it from a nonrewarding alternative (differential conditioning). The difference in performance therefore suggests that attentional processes are involved because in differential conditioning the bee has to focus on the difference and not on the mere presence of a visual target, thus making learning slower. In any case, the result goes against the idea that the difference between two colors is an immutable property constrained by the visual machinery.

When Menzel characterized color learning (66, 67; see above), studies on pattern perception were simultaneously performed by Wehner (99–101) and others (e.g., 1), continuing the tradition started by von Frisch's students (47–49, 105–107). In contrast to Menzel's work, these studies focused not on learning but on the perceptual capabilities of bees confronted with pattern discrimination tasks. Visual conditioning was also used in these and later works on pattern perception (for reviews, see References 59, 84, and 102), but a quantification of acquisition curves and/or a characterization of pattern memory was absent from these works. This tradition was continued in the 1970s, 1980s, and even 1990s, as visual learning was used mainly as a tool to answer questions on visual perception and discrimination. Yet, some experiments showed that in pattern vision, as in color vision, what a bee perceives depends on its previous visual experience and of possible attentional processes. Zhang & Srinivasan (115) showed, for instance, that the previous visual experience of a bee can speed up the analysis of the retinal image when a familiar object or scene is encountered. They first attempted to train bees to distinguish between a ring and a disk when each shape was presented as a textured figure placed a few centimeters in front of a similarly textured background. The figures are, in principle, detectable through the relative motion that occurs at the figure borders, which are at a different distance than the background when bees fly toward the targets. Despite intensive training, the bees were incapable of learning the difference between a ring and a disk, a discrimination that usually poses no problems when the bees experience these stimuli as plain (nontextured) shapes. Zhang & Srinivasan then trained a group of bees to this "easy" problem, presenting a plain black disk and ring a few centimeters in front of a white background. The bees could, as expected, easily learn the task. They were then confronted with the difficult problem of learning the textured disk versus the ring, and this time they immediately solved the discrimination. Thus, pretraining with plain stimuli primed the pattern recognition system in such a way that it was able to detect shapes that otherwise could not be distinguished. It may be that such pretraining triggers attentional processes that allow better focusing on the targets that have to be discriminated.

Uncovering how attentional processes and learning modulate visual perception constitutes an unexplored and promising research field. The existence of attentional processes in insect brains is not far-fetched, and recent research has located such processes in precise structures of the insect brain. In the fruit fly, *Drosophila melanogaster*, attention can be demonstrated and characterized at the physiological level (96). A fruit fly fixed stationary within a circular arena, and tracking a visual object (a vertical black bar) moving at a constant frequency around it, exhibits anticipatory behavior consistent with attention for the bar tracked. Such an anticipatory tracking has a neural correlate in the form of a transient increase in a 20–30 Hz local-field potential recorded in a region of the brain called the medial protocerebrum (**Figure 1*d***) (96). In other words, the 20–30 Hz response in the fly brain is correlated with transitions to behavioral tracking. This response is not only anticipatory, but also selective to the stimulus presented, increased by novelty and salience and reduced when the fly is in a sleep-like state (96). Moreover, the use of mutants showed that a subset of neurons of the mushroom bodies, which are a higher-order structure of the insect brain (**Figure 1*d,e***), is required for both the tracking response and the 20–30 Hz response (96). This result is consistent with the finding that mushroom bodies are required for choice behavior of *D. melanogaster* facing contradictory visual cues (90). In this case, individually tethered flies flying stationary are trained in a circular arena in which one kind of visual stimulus (say, a T pattern) represents a permitted flight direction, while another kind of visual stimulus (say, an inverted T pattern) represents a forbidden flight direction associated with a displeasing heat beam on the thorax. Tang & Guo (90) conditioned flies to choose one of two directions in response to color and shape cues; after the training, flies were tested with contradictory cues. Wild-type flies made a discrete choice that switched from one alternative to the other as the relative salience of color and shape cues gradually changed, but this ability was greatly diminished in mutant flies with miniature mushroom bodies or with chemically ablated mushroom bodies. In other words, mushroom bodies mediate the assessment of the relative saliency of conflicting visual cues (90, 110) and are also involved in improving the extraction of visual cues after pretraining in *D. melanogaster* (77). The mushroom bodies of hymenopterans may play similar roles (**Figure 1*e***), thus favoring attention and better problem solving and discrimination.

Yet, visual learning and the neural circuits mediating it are still poorly understood in the fruit fly. The mushroom bodies, which are the main site for olfactory memories, are not directly involved in visual learning because in *D. melanogaster*, contrary to hymenopterans, there is no direct input from the visual areas of the brain to these structures (108). Recent studies have succeeded in identifying the precise neuronal substrates of two forms of visual memory in the *D. melanogaster* brain outside the mushroom bodies (62). Memories for pattern elevation and orientation were retraced to different neuronal layouts of the central complex, a median structure of the insect brain (**Figure 1*d***). Liu et al. (62) showed that two neuronal layers of the central complex are required to achieve visual discriminations based on pattern elevation or orientation. In all cases, visual short-term memory was studied, thus leaving open the question of the localization of visual long-term memory. In bees and wasps, the localization of visual memories may differ from that of *D. melanogaster*. In contrast to the fruit fly, visual areas of the hymenopteran brain provide direct input to the mushroom bodies (26), thus making these structures a candidate for the localization of visual memories in addition to the central complex (**Figure 1*d***).

COMPLEX FORMS OF VISUAL LEARNING (THAT MAY NOT BE SO COMPLEX)

Only in the 1990s did researchers become interested in the existence of cognitive processing in insects, and the honey bee was the model chosen to address most of the works performed in that direction. Such a delay with respect to the cognitive revolution, which flourished from the late 1970s to the early 1980s (73), can only be explained by the reluctance to view invertebrates, and therefore insects, as organisms capable of nonelemental, higher-order forms of learning. For instance, the main idea with respect to visual pattern learning, which is still sometimes defended, was that insects can only view isolated spots, blobs, and bars without the capacity of integrating them into a given configuration (52–55). Even a basic capacity of recognition systems such as generalization, the ability of individuals to respond to stimuli that, despite being different from a trained target, are nevertheless perceptually similar to it (81, 83), was and is considered by some researchers as too high-level for a honey bee (54, 55). Yet, dozens of works had already shown that honey bees generalize their choice of visual patterns to novel figures that have some similarity to those that have been trained (e.g., 1, 100). This refusal of generalization capacities is consistent with the preconception that insects have limited plasticity and should be viewed rather as reflex machines reacting to specific features in the environment to which they are tuned.

In the past decade, however, researchers have found evidence showing that bees are not reflex machines and that they exhibit visual learning capabilities that were only suspected in some vertebrates. Some of these capacities are surprising and may be viewed as nonelemental. However, an alternative view could argue that it is possible to explain them based on simple, elemental associations. These experiments, reviewed in the next section, were not conceived to address these opposite views, so that we are currently unable to determine whether these performances are forms of elemental or higher-order learning.

Visually Based Individual Recognition in Wasps

The capacity of individuals to recognize their distinctive identity has long been dismissed in social insects due to the cognitive requirements that such performance may impose on colonies made of thousand of individuals. For instance, Wilson (104) stated that "... insect societies are, for the most part, impersonal ... The sheer size of the colonies and the short life of the members make it inefficient, if not impossible, to establish individual bonds." However, not all social insects live in huge, overcrowded societies. Small, relatively primitive colonies of bumble bees, wasp, and some ant species are based on dominance hierarchies where individual recognition may be crucial for responding appropriately to a conspecific. Indeed, recent studies have shown that *Pachycondyla villosa* ant queens recognize each other using olfactory, cuticular cues (20). In the visual domain, studies on the paper wasp, *Polistes fuscatus*, have shown that individual recognition is achieved through learning the yellow-black patterns of the wasp faces and/or abdomens (92). In another species, *Polistes dominulus*, more variable patterns with larger black components were carried by individuals ranking higher in the nest hierarchy. Altering these facial and/or abdominal color patterns induced aggression against such animals, whether or not their patterns were made to signal higher or lower ranking. These results, however, were challenged by another study (9a), which could not find evidence supporting the hypothesis that the facial patterns of *P. dominulus* act as hierarchy or quality signals. Size, on the contrary, was highly correlated with social dominance in this wasp (9a). Although these results question the validity of the hypothesis that posits that visual facial patterns that contain more black areas or more black spots are associated with dominant wasps, they do not exclude the possibility that visual patterns are used as individual identity markers rather than as status markers. In this scenario, wasps would recognize each other on the basis of their facial features, and each individual mask, regardless of its amount of black areas or spots, would have an unambiguous outcome in terms of its ranking in the social structure (i.e., mask A $\rightarrow$ α individual and mask B $\rightarrow$ β individual). Wasps would learn a series of elemental associations between mask patterns and social ranking. Given the small size of colonies, in which 5 to 10 individuals can coexist, storing several memories, one for each individual, seems plausible. If this were the case, a fundamental goal would be to characterize the storage capacity of the visual memory as related to colony size.

Observatory Learning in Bumble Bees

Recent studies on bumble bees (57, 58, 109) have shown that these insects copy other bees' learned foraging preferences by observing their choices of visual, rewarded targets. Bumble bees, *B. terrestris*, are influenced by other conspecifics when sampling unfamiliar flowers, such that they land on unknown flowers where other bees have been (57). This occurs even when naïve bees are separated from experienced foragers by a transparent screen so that they can neither sample the flowers by themselves nor interact with their foraging conspecifics (109). Similarly, naïve bees abandon an unrewarding species and switch to a more rewarding alternative more quickly if accompanied by experienced foragers (58).

As surprising as this performance may appear, it can be accounted for by an elemental form of associative learning called second-order conditioning (76), which involves two connected associations. In this scenario, an animal first learns an association between a CS and an US (CS1 + US) and then experiences a pairing between a new conditioned stimulus, CS2 and CS1 (CS1 + CS2). In this way, CS2 becomes meaningful, directly through its association with CS1 and indirectly with the US. How would this apply to the observational learning of bumble bees? One could propose that naïve bumble bees would first associate the presence of a conspecific with reward (CS1 + US) simply by foraging close to experienced foragers. Afterward, observing a conspecific landing on a given color may allow an association between color and conspecific (CS2 + CS1) to be established (58). These connected elemental links may thus underlie the observational learning of bees. This hypothesis is supported by the fact that honey bees can learn second-order associations while searching for food. They learn to connect both two odors (odor 1 + sucrose reward; Odor 2 + Odor 1) (7) and one odor and one color (odor + sucrose reward; color + odor) (42).

Symbolic Matching to Sample and Other Forms of Conditional Discrimination in Honey Bees and Bumble Bees

Symbolic matching to sample is a term used to describe an experimental situation in which the correct response to a problem depends on a specific background or condition. In other words, animals have to learn, for instance, that given condition A, response C is correct, while for condition B, response D is correct. Symbolic matching to sample is a form of conditional discrimination because a given stimulus, the sample (also called the occasion setter), sets the condition for the next choice. Using this design, Zhang et al. (117) trained honey bees to fly though a compound Y-maze consisting of a series of interconnected cylinders. The first cylinder carried the sample stimulus (e.g., a vertical or a horizontal black-and-white grating). The second and third cylinders had two exits apiece. Each exit presented a visual stimulus so that the bee had to choose between them. In the second cylinder, bees had to choose between a blue and a green square. In the third cylinder, they had to choose between a radial sectored pattern and a ring pattern. Correct sequences of choices were "Vertical–Green–Ring" and "Horizontal–Blue–Radial." Only after making a succession of correct choices (i.e., in both the second and the third cylinders) could a bee reach a feeder with sucrose solution. The bees learned to master these successive associations between different kinds of visual cues (117). This finding was also extended to other sensory modalities, as the same principle applied when visual cues were combined with odors in a similar protocol (87).

Conditional learning allows other variants that, depending on the number of occasion setters and discriminations involved, have received different names. For instance, another form of conditional discrimination involving two occasion setters is the so-called transwitching problem. In this problem, an animal is trained differentially with two stimuli, A and B, and with two different occasion setters, C and D. When C is available, stimulus A is rewarded, whereas stimulus B is not (A+ versus B-); the opposite occurs (A- versus B+) when D is available. The transwitching problem is also considered a form of contextual learning because the occasion setters C1 and C2 can be viewed as contexts determining the appropriateness of each choice. Bumble bees have been trained in a transwitching problem to choose a 45° grating and to avoid a 135° grating to reach a feeder, and to do the opposite to reach their nest (28). Here, the nest and the feeder provide the appropriate contexts defining what has to be chosen. Bumble bees can also learn that an annular or a radial disc must be chosen, depending on the disc's association with a 45° or a 135° grating either at the feeder or the nest entrance: At the nest, access was allowed by the combinations 45° + radial disc and 135° + annular disc, but not by the combinations 45° + annular disc and 135° + radial disc; at the feeder, the opposite applied (27). In both cases, the potentially competing visuomotor associations were insulated from each other because they were set in different contexts.

Solving this kind of problem can be viewed as a form of nonelemental learning and thus as a sophisticated form of cognitive visual processing. Indeed, as for other forms of conditional discrimination, one could describe this protocol as CA+, CB- (if C then A, but not B), and DA-, DB+ (if D then B, but not A). Each stimulus, A, B, C, and D, therefore is rewarded as often as it is nonrewarded, such that solutions cannot be based on the mere outcome of A, B, C or D. A higher-order solution then would be to learn the outcome of each particular configuration, CA, CB, DA, or DB. However, an alternative explanation could argue that what the insects do is establish hierarchical simple associations as the ones underlying second-order conditioning (see above). Indeed, one could imagine that bees learn to associate a radial disc with sucrose reward and that they then learn to associate a 45° grating with the radial disc. This is a relatively simple strategy that is probably used by bees for navigational purposes (116, 117) when they are confronted with successions of different landmarks en route to the goal.

A critical factor determining one strategy or the other therefore may be the temporal order of stimulus presentation. If these are presented serially, learning chains of simple associations could be primed, while simultaneous presentation of stimuli may prime learning of configurations and their specific outcome. An example of the latter is the case of honey bees trained to solve a biconditional discrimination, AC+, BD+, AD-, BC-, in which all four stimuli were presented simultaneously and were rewarded as often as they were nonrewarded (80). Four different gratings combining one color (yellow or violet = A or B) with one orientation (horizontal or vertical = C or D) were used in such a way that bees had to learn that, for instance, yellow-horizontal (AC) and violet-vertical (BD) were rewarded while yellow-vertical (AD) and violet-horizontal (BC) were nonrewarded. Bees learned to choose the rewarded stimuli, although the colors and orientations were ambiguous when considered alone. Thus, they learned the configurations and not the specific outcome of the elements (80). Again, cumulative experience is a critical factor promoting configural learning (36). Although few learning trials promote processing of a compound color stimulus made of A and B elements as the sum of A and B, increasing number of trials results in bees treating the compound AB as a unique entity that is different from its composing elements (36).

NONELEMENTAL VISUAL LEARNING

The visual performances of the previous section could be accounted for by elemental associations despite their sophistication. A higher level of complexity, however, is reached when animals respond in an adaptive manner to novel stimuli that they have never encountered before and that do not predict a specific outcome per se based on the animals' past experience. Such a positive transfer of learning (78) is therefore different from elemental forms of learning, which link known stimuli or actions to specific rewards (or punishments). In the cases considered in this section, the insects' responses have in common the transfer to novel stimuli, which cannot be explained based on the previous knowledge that the animal has of these stimuli.

Categorization of Visual Stimuli in Honey Bees

Visual categorization refers to the classification of visual stimuli into defined functional groups (44). It can be defined as the ability to group distinguishable objects or events on the basis of a common feature or set of features and therefore to respond similarly to them (94, 113). A typical categorization experiment trains an animal to extract the basic attributes of a category and then tests it with novel stimuli that were never encountered before and that may or may not present the attributes of the category learned. If the animal chooses the novel stimuli on the basis of these attributes, it classifies them as belonging to the category and exhibits therefore positive transfer of learning.

Several studies have shown the ability of visual categorization in free-flying honey bees trained to discriminate different patterns and shapes. For instance, van Hateren et al. (95) trained bees to discriminate two given gratings presented vertically with different orientations (e.g., 45° versus 135°) by rewarding one of these gratings with sucrose solution. Each bee was trained with a changing succession of pairs of different gratings, one of which was always rewarded and the other not. Despite the difference in pattern quality, all the rewarded patterns had the same edge orientation and all the nonrewarded patterns also had a common orientation, which was perpendicular to the rewarded one. Under these circumstances, the bees had to extract and learn the orientation that was common to all rewarded patterns to solve the task. This was the only cue predicting reward delivery. In the tests, bees were presented with novel patterns, all of which were all nonrewarded, and exhibited the same stripe orientations as the rewarding and nonrewarding patterns employed during the training. In such transfer tests, bees chose the appropriate orientation despite the novelty of the structural details of the stimuli. Thus, bees could categorize visual stimuli on the basis of their global orientation.

They can also categorize visual patterns based on their bilateral symmetry. When trained with a succession of changing patterns to discriminate bilateral symmetry from asymmetry, they learn to extract this information from different figures and to transfer it to novel symmetrical and asymmetrical patterns (31). Similar conclusions apply to other visual features such as radial symmetry, concentric pattern organization and pattern disruption (see Reference 5 for a review), and even photographs belonging to a given class (e.g., radial flower, landscape, plant stem) (118).

How could bees appropriately classify different photographs of radial flowers if they vary in color, size, outline? An explanation was provided by Stach et al. (88), who showed that different coexisting orientations can be considered at one time and that they can be integrated into a global stimulus representation that is the basis for the category (88). Thus, a radial flower would be, in fact, the conjunction of five or more radiating edges. In addition to focusing on a single orientation, honey bees assembled different features to build a generic pattern representation, which could be used to respond appropriately to novel stimuli sharing such a basic layout. Honey bees trained with a series of complex patterns sharing a common layout comprising four edges oriented differently remembered these orientations simultaneously in their appropriate positions and transferred their response to novel stimuli that preserved the trained layout. These results show that honey bees extract regularities in their visual environment and establish correspondences among correlated features. Therefore, they may generate a large set of object descriptions from a finite set of elements.

This capacity can explain recent findings showing that honey bees learn to recognize human faces if trained to do so (23). In this case, bees were rewarded with sugar water to choose and distinguish people's faces from photographs. Bees chose the appropriate photograph, thus showing a capacity to discriminate this particular kind of stimuli. Does this mean that bees realize that two different persons are behind two discriminated photographs? Not really. To the bees, the photographs were merely strange flowers. The question then would be what information from the photographs was used to recognize the right stimulus. This question was tackled by a work that studied whether bees can bind the features of a face-like stimulus (two dots in the upper part representing the eyes, a vertical line below as the nose, and a horizontal line in the lower part as the mouth) and recognize faces using this basic configuration (2). Bees did indeed distinguish between different variants of the face-like stimuli, thus showing that they discriminate between these options, but grouped together and reacted, therefore, similarly to faces if trained to do so. Stimuli made of the same elements (two dots, a vertical line, and a horizontal line) but not preserving the configuration of a face were not recognized as positive, thus showing that bees learn that the rewarded stimulus consists of a series of elements arranged in the specific spatial configuration of a face (2). Furthermore, if trained with real faces, bees can learn to recognize novel views of a face by interpolating between or averaging views they have experienced (25).

In any case, honey bees show positive transfer of learning from a trained to a novel set of stimuli, and their performances are consistent with the definition of categorization. Visual stimulus categorization therefore is not a prerogative of certain vertebrates. However, this result might not be surprising as it allows an elemental learning interpretation. To understand this elemental interpretation, the possible neural mechanisms underlying categorization should be considered. If we allow that visual stimuli are categorized on the basis of specific features such as orientation, the neural implementation of category recognition could be relatively simple. The feature(s) allowing stimulus classification would activate specific neuronal detectors in the optic lobes, the visual areas of the bee brain. Examples of such feature detectors are the orientation detectors whose orientation and tuning have already been characterized by means of electrophysiological recordings in the honey bee optic lobes (111). Thus, responding to different gratings having a common orientation of, say, 60° is simple, because all these gratings elicit the same neural activation in the same set of orientation detectors despite their different structural quality.

In the case of category learning, the activation of an additional neural element is needed. Such an element would be a reward neuron whose activity would substitute for sucrose reward. Such a neuron has been identified in the honey bee brain, VUM_{mx1} (ventral unpaired median neuron located in the maxillar neuromere 1) (43). VUM_{mx1} mediates olfactory learning in the honey bee because it contacts the olfactory circuit at its key processing stages in the brain. In other words, when an odor activates the olfactory circuit, concomitant sucrose stimulation activates VUM_{mx1}, thus providing the basis for neural coincidence between odor and reward. The branching of VUM_{mx1} makes it specific for the olfactory circuit and thus for olfactory learning (43). Other VUM neurons whose function is still unknown are present in the bee brain (79). Some of them could provide the neural basis of reward in associative visual learning.

Category learning could thus be reduced to the progressive reinforcement of an associative neural circuit relating visual-coding and reward-coding neurons, similar to that underlying simple associative (e.g., Pavlovian) conditioning. From this perspective, even if categorization is viewed as a nonelemental learning form because it involves positive transfer of learning, it may simply rely on elemental links between conditioned and unconditioned stimuli.

An even simpler alternative may account for this performance. The mechanism explained above could be viewed as a form of supervised learning, in which a visual network is instructed by the external signal of the reinforcement neuron to respond to the right combination of features. Recent modeling work on the vertebrate visual system has shown that visual networks can learn to extract the distinctive features of a category without any kind of supervision (65). The model relies on spike-timing-dependent plasticity (STDP), which is a learning rule that modifies synaptic strength as a function of the relative timing of pre- and postsynaptic spikes. When a neuron is repeatedly presented with similar inputs, STDP has the effect of concentrating high synaptic weights on afferents that systematically fire early, while postsynaptic spike latencies decrease. Masquelier et al. (65) showed that a network that exhibits STDP and is repeatedly presented with natural images of a given category becomes progressively tuned to respond better to the features that correspond to prototypical patterns of the category. Those features that are both salient and consistently present in the images are highly informative and enable robust object recognition. Testing whether similar neural mechanisms underlie object categorization in the insect visual system would be a fascinating endeavor.

Rule Learning in Honey Bees

In rule learning, the animal learns relations between objects, not the objects themselves. Typical examples are the so-called rules of sameness and of difference. Sameness and difference rules are demonstrated through the protocols of delayed matching to sample (DMTS) and delayed nonmatching to sample (DNMTS), respectively. In DMTS, animals are presented first with a sample and then with a set of stimuli, one of which is identical to the sample and is reinforced. Because the sample is changed regularly, animals must learn the sameness rule, i.e., "always choose what is shown to you (the sample), independent of what else is shown to you." In DNMTS, the animal must learn the opposite, i.e., "always choose the opposite of what is shown to you (the sample)." Honey bees foraging in a Y-maze learn both rules (38). Bees were trained in a DMTS problem in which they were presented with a changing nonrewarded sample (i.e., one of two different colored disks or one of two different black-and-white gratings, vertical or horizontal) at the entrance of a maze (**Figure 2**). The bees were rewarded only if they chose the stimulus identical to the sample once within the maze. Bees trained with colors and presented in transfer tests with unknown black-and-white gratings solved the problem and chose the grating identical to the sample at the entrance of the maze. Similarly, bees trained with the gratings and tested with colors in transfer tests also solved the problem and chose the novel color corresponding to that of the sample grating at the maze entrance. Transfer was not limited to modalities within the visual domain (pattern versus color) but could also operate between different domains such as olfaction and vision (38). Furthermore, bees also mastered a DNMTS task, thus showing that they learn a rule of difference between stimuli (38). The capacity of honey bees to solve a DMTS task has recently been analyzed with respect to the working memory underlying it (114). It was found that the sample is stored for approximately 5 s (114), a period that coincides with the duration of other visual and olfactory short-term memories characterized in simpler forms of associative learning in honey bees (69; see above). Moreover, bees trained in a DMTS task can learn to pay attention to one of two different samples presented successively in a flight tunnel and can transfer the learning of this sequence weight to novel samples (114).

Despite the honey bees' evident capacity to solve relational problems such as the DMTS or the DNMTS tasks, such capacities are not unlimited. In some cases, biological constraints may impede the bee from solving a particular problem for which rule extraction is necessary. It is therefore interesting to focus on a different example of rule learning that bees could not master, the transitive inference problem (4). In this problem, animals must learn a transitive rule, i.e., A > B, B > C, then A > C. Preference for A over C in this context can be explained by two strategies: (*a*) deductive reasoning (97) in which the experimental subjects construct and manipulate a unitary and linear representation of the implicit hierarchy A > B > C, and (*b*) responding as a function of reinforced and non-reinforced experiences (91), in which animals choose among stimuli on the basis of the effective number of reinforced and nonreinforced experiences (A is always reinforced whereas C is always nonreinforced).

To determine whether bees can learn a transitive rule, they were trained with five different visual stimuli (A, B, C, D, and E) in a multiple discrimination task: A+ versus B-, B+ versus C-, C+ versus D-, D+ versus E- (4). Training involved overlapping adjacent premise pairs (A > B, B > C, C > D, D > E), which underlie a linear hierarchy A > B > C > D > E. After training, bees were tested with B versus D, a nonadjacent pair of stimuli that were never explicitly trained together. In theory, B and D have equivalent associative strengths because they are, in principle, equally associated with reinforcement or no reinforcement during training. Thus, if bees were guided by the stimulus' associative strength, they should choose randomly between B and D. If, however, bees used a transitive rule, they should prefer B to D.

Honey bees learned the premise pairs as long as these were trained as uninterrupted, consecutive blocks of trials (4). But if shorter and interspersed blocks of trials were used, such that bees had to master all pairs practically simultaneously, performance collapsed and bees did not learn the premise pairs. The bees' choice

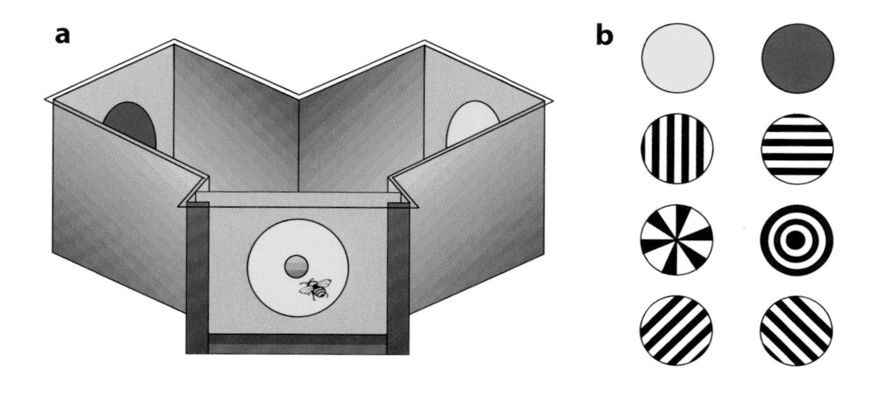

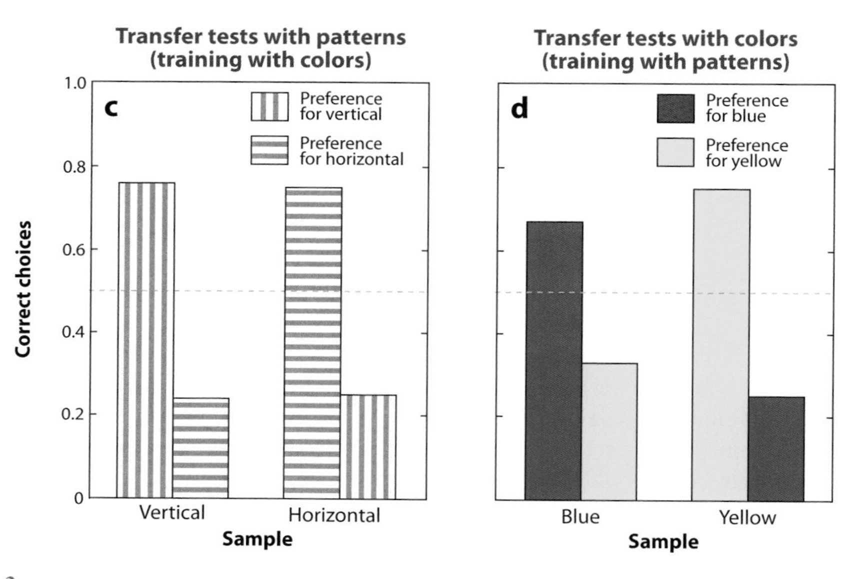

Figure 2

Rule learning in honey bees. Honey bees trained in a delayed matching-to-sample task to collect sugar solution in (*a*) a Y-maze on a series of (*b*) patterns or colors learn a rule of sameness. (*c*, *d*) Transfer tests with novel stimuli. (*c*) In Experiment 1, bees trained on the colors were tested on the patterns. (*d*) In Experiment 2, bees trained on the patterns were tested on the colors. In both cases, bees chose the novel stimuli corresponding to the sample, although they had no experience with such test stimuli. Adapted from Reference 38.

was significantly influenced by their experience with the last pair of stimuli (D+ versus E-) such that they preferred D and avoided E. In the tests, no preference for B to D was found. Although this result agrees with an evaluation of stimuli in terms of their associative strength (see above), during training bees more often visited B when it was rewarding (B+ vs. C-) than D when it was rewarding (D+ vs. E-), such that a preference for B should have been expected if only the associative strength were guiding the bees' choices. It was then concluded that bees do not establish transitive inferences between stimuli but rather guide their choices by the joint action of a recency effect (preference of the last rewarded stimulus, D) and by an evaluation of the associative strength of the stimuli (in which case preference for B should be evident). As the former supports choice of D while the latter supports choice of B, equal choice of B and D in the tests could be explained (4). In any case, memory constraints (in this case the fact that simultaneous mastering of the different premise pairs was not possible and the fact that the last excitatory memory seems to predominate over previous memories) impeded learning the transitive rule. Recently, Cheng & Wignall (11) demonstrated that failure to master several consecutive visual discriminations is due to response competition occurring when animals are tested. This may explain why bees in the transitive inference protocol were unable to master the successive short blocks of training with different premise pairs.

Counting

Counting could be useful in navigation tasks where the number of landmarks encountered during a foraging trip or near the hive may contribute to efficient orientation of free-flying bees. Furthermore, it could also improve foraging through evaluation of food source profitability (e.g., number of flowers in a patch). Whether or not honey bees estimate numerosity has been addressed recently by two different works that reached similar conclusions (18, 41).

Dacke & Srinivasan (18) were inspired by Chittka & Geiger's (13) pioneering work suggesting that bees may count landmarks en route to the feeder. In Dacke & Srinivasan's protocol, bees were trained to fly into a tunnel to find a food reward after a given number of landmarks. The shape, size, and positions of the landmarks were changed in the different testing conditions in order to avoid any confounding factor(s). Bees showed a stronger preference to land after the correct number of landmark in nonrewarded tests. This behavior was observed when bees were trained to collect reward after one, two, three, or four landmarks but not more, thus indicating a limit in their counting capacity.

A similar limit was found in a DMTS protocol (41; see above) in which bees had to choose the stimulus containing the same number of items as a sample. The authors controlled for low-level cues such as cumulated area and edge length, configuration identity, and illusionary shape similarity formed by the elements. Their results showed that honey bees have the capacity to match visual stimuli in a DMTS task as long as the number of items does not exceed four. Together with Dacke & Srinivasan's work (18), this result indicates that the bee brain can deal with a real numerosity concept, even if it is limited to a number of four. Interestingly, the same limit was found in humans when time exposure of items to be counted is limited (17).

CONCLUSION

Almost one hundred years of research on visual learning in bees and other social Hymenoptera, starting with Karl von Frisch's (98) first demonstrations on color and pattern learning in bees, have yielded an impressive amount of information about how honey bees, bumble bees, and wasps see the world and learn about visual cues in their environment. New discoveries in this field have shown that, in addition to simple forms of visual learning, social Hymenoptera also master complex forms of visual learning, from conditional discriminations and observational learning to rule learning. Although the cognitive capabilities of bees and wasps may surprise owing to their sophistication, limitations related to natural life seem inescapable. For instance, in the case of wasps learning facial mask patterns of conspecifics, one could imagine that interindividual recognition is certainly possible but probably has limitations in terms of the number of individuals that can be learned and remembered. Similarly, mastering several different associations simultaneously would be facilitated if these are organized serially or hierarchically in chains of associations that can mediate successful navigation in a complex environment. But if these associations have to be mastered simultaneously at the same place, learning them would probably be difficult given the bees' biological specialization as a serial forager. In this case, learning configurations of stimuli may be more adaptive than learning each component separately.

If bees and wasps exhibit such a high degree of complex forms of visual learning, which kind of limitation do they present as a model for unraveling the mechanisms of these phenomena? The main limitation resides so far in the impossibility of addressing questions related to the cellular and molecular mechanisms underlying these learning forms. Learning protocols have exploited the advantage of not restraining the animals' movements so that the behaviors recorded express all the potential of the insect brain. However, they are limiting because no access to the brain is so far possible in a flying bee. As mentioned above, new protocols in which bees learn color-reward and motion cues–reward associations under restrained conditions (50, 51) are promising because they allow the neural circuits involved in these learning forms to be accessed (30). The critical question would be then to what extent do restraining conditions limit the expression of more complex forms of visual learning?

Why should bees and wasps continue to be attractive for research on visual cognition despite this technical limitation? The answer is simple: They exhibit sophisticated visual performances that cannot at this time be uncovered in a fruit fly. Future research in social insects should benefit from a comparative analysis between the visual performances and mechanisms of bees and flies and overcome the historic burden of not having a window open to the neural and molecular basis of visual learning, irrespective of the level of complexity considered.

SUMMARY POINTS

1. Visual learning admits different levels of complexity, from the formation of a simple associative link between a visual stimulus and its outcome (elemental learning), to more sophisticated performances such as object categorization or rules learning, which allow flexible responses beyond simple forms of learning (nonelemental learning). Social insects excel at visual learning in a foraging and navigation context.
2. Honey bees, ants, and bumble bees learn to associate different kinds of visual cues such as colors or patterns with food reward. Even if these associations remain elemental, performance can be modulated by the complexity of the task, thus suggesting attentional processes in these insects. Attention- and experience-dependent changes in visual discrimination can be traced to the neural level. Studies in the fruit fly suggest that these processes can be located at the level of mushroom bodies, a structure of the insect brain involved in learning and memory.
3. Social insects exhibit sophisticated visual abilities. Wasps recognize each other on the basis of facial marks, bumble bees learn to choose profitable food sources by observing the choice of other bees, and honey bees and bumble bees learn to solve different kinds of conditional discriminations. These performances admit both elemental and nonelemental explanations.
4. Depending on their past experience, honey bees respond in an adaptive manner to novel stimuli that they have never encountered before. Such a positive transfer of learning, characteristic of nonelemental forms of learning, has been shown in studies demonstrating categorization, learning of rules such as sameness or difference, and basic counting abilities in bees.

DISCLOSURE STATEMENT

The authors are not aware of any affiliations, memberships, funding, or financial holdings that might be perceived as affecting the objectivity of this review.

ACKNOWLEDGMENTS

We thank the support of the French National Research Agency (ANR; Project APICOLOR), the French Research Council (CNRS), and the University Paul Sabatier (Project APIGENE). Aurore Avarguès-Weber and Nina Deisig contributed equally to this review. This review is dedicated to Prof. Dr. Randolf Menzel on his seventieth anniversary.

LITERATURE CITED

1. Anderson AM. 1972. The ability of honey bees to generalize visual stimuli. In *Information Processing in the Visual Systems of Arthropods*, ed. R Wehner, pp. 207–12. Berlin: Springer
2. Avarguès-Weber A, Portelli G, Benard J, Dyer A, Giurfa M. 2010. Configural processing enables discrimination and categorization of face-like stimuli in honeybees. *J. Exp. Biol.* 213:593–601
3. Beier W, Menzel R. 1972. Untersuchungen über den Farbensinn der deutschen Wespe (*Paravespula germanica* F., Hymenoptera, Vespidae): verhaltensphysiologischer Nachweis des Farbensehens. *Zool. Jb. Physiol.* 76:441–54
4. Benard J, Giurfa M. 2004. A test of transitive inferences in free-flying honeybees: unsuccessful performance due to memory constraints. *Learn. Mem.* 11:328–36
5. Benard J, Stach S, Giurfa M. 2006. Categorization of visual stimuli in the honeybee, *Apis mellifera*. *Anim. Cogn.* 9:257–70
6. Biesmejer JC, Giurfa M, Koedam D, Potts SG, Joel DM, Dafni A. 2005. Convergent evolution: floral guides, stingless bee next entrances, and insectivorous pitchers. *Naturwissenchaften* 92:444–50
7. Bitterman ME, Menzel R, Fietz A, Schäfer S. 1983. Classical conditioning of proboscis extension in honeybees (*Apis mellifera*). *J. Comp. Psychol.* 97:107–19
8. Camlitepe Y, Aksoy V. 2010. First evidence of fine colour discrimination ability in ants (Hymenoptera, Formicidae). *J. Exp. Biol.* 213:72–77
9. Campan R, Lehrer M. 2002. Discrimination of closed shapes by two species of bee, *Apis mellifera* and *Megachile rotundata*. *J. Exp. Biol.* 205:559–72

9a. Cervo R, Dapporto L, Beani L, Strassmann JE, Turillazzi S. 2008. On status badges and quality signals in the paper wasp *Polistes dominulus*: body size, facial colour patterns and hierarchical rank. *Proc. R. Soc. B* 275:1189–96

10. Cheng K, Narendra A, Sommer S, Wehner R. 2009. Traveling in clutter: navigation in the central Australian desert ant *Melophorus bagoti*. *Behav. Process.* 80:261–68
11. Cheng K, Wignall AE. 2006. Honeybees (*Apis mellifera*) holding on to memories: Response competition causes retroactive interference effects. *Anim. Cogn.* 9:141–50
12. Chittka L, Beier W, Hertel H, Steinmann E, Menzel R. 1992. Opponent colour coding is a universal strategy to evaluate the photoreceptor inputs in Hymenoptera. *J. Comp. Physiol. A* 170:545–63
13. Chittka L, Geiger K. 1995. Can honey bees count landmarks? *Anim. Behav.* 49:159–64
14. Chittka L, Thomson JD, Waser NM. 1999. Flower constancy, insect psychology, and plant evolution. *Naturwissenschaften* 86:361–77
15. Collett TS, Collett M. 2002. Memory use in insect visual navigation. *Nat. Rev. Neurosci.* 3:542–52
16. Collett TS, Graham P, Durier V. 2003. Route learning by insects. *Curr. Opin. Neurobiol.* 13:718–25
17. Cowan N. 2001. The magical number 4 in short-term memory: a reconsideration of mental storage capacity. *Behav. Brain Sci.* 24:87–114
18. Dacke M, Srinivasan MV. 2008. Evidence for counting in insects. *Anim. Cogn.* 11:683–89
19. de Brito Sanchez M, Chen C, Li J, Liu F, Gauthier M, Giurfa M. 2008. Behavioral studies on tarsal gustation in honeybees: sucrose responsiveness and sucrose-mediated olfactory conditioning. *J. Comp. Physiol. A* 194:861–69

20. D'Ettorre P, Heinze J. 2005. Individual recognition in ant queens. *Curr. Biol.* 15:2170–74
21. Dyer AG, Chittka L. 2004. Fine colour discrimination requires differential conditioning in bumblebees. *Naturwissenschaften* 91:224–27
22. Dyer AG, Neumeyer C. 2005. Simultaneous and successive colour discrimination in the honeybee (*Apis mellifera*). *J. Comp. Physiol. A* 191:547–57
23. Dyer AG, Neumeyer C, Chittka L. 2005. Honeybee (*Apis mellifera*) vision can discriminate between and recognise images of human faces. *J. Exp. Biol.* 208:4709–14
24. Dyer AG, Rosa MGP, Reser D. 2008. Honeybees can recognise images of complex natural scenes for use as potential landmarks. *J. Exp. Biol.* 211:1180–86
25. Dyer AG, Vuong QC. 2008. Insect brains use image interpolation mechanisms to recognize rotated objects. *PLoS ONE* 3:e4086
26. Ehmer B, Gronenberg W. 2002. Segregation of visual input to the mushroom bodies in the honeybee (*Apis mellifera*). *J. Comp. Neurol.* 451:362–73
27. Fauria K, Colborn M, Collett TS. 2000. The binding of visual patterns in bumblebees. *Curr. Biol.* 10:935–38
28. Fauria K, Dale K, Colborn M, Collett TS. 2002. Learning speed and contextual isolation in bumblebees. *J. Exp. Biol.* 205:1009–18
29. Giurfa M. 2004. Conditioning procedure and color discrimination in the honeybee *Apis mellifera*. *Naturwissenschaften* 91:228–31
30. Giurfa M. 2007. Behavioral and neural analysis of associative learning in the honeybee: a taste from the magic well. *J. Comp. Physiol. A* 193:801–24
31. Giurfa M, Eichmann B, Menzel R. 1996. Symmetry perception in an insect. *Nature* 382:458–61
32. Giurfa M, Hammer M, Stach S, Stollhoff N, Müller-Deisig N, Mizyrycki C. 1999. Pattern learning by honeybees: conditioning procedure and recognition strategy. *Anim. Behav.* 57:315–24
33. Giurfa M, Lehrer M. 2001. Honeybee vision and floral displays: from detection to close-up recognition. In *Cognitive Ecology of Pollination*, ed. L Chittka, JD Thomson, pp. 61–82. Cambridge, UK: Cambridge University Press
34. Giurfa M, Menzel R. 1997. Insect visual perception: complex abilities of simple nervous systems. *Curr. Opin. Neurobiol.* 7:505–13
35. Giurfa M, Núñez JA, Chittka L, Menzel R. 1995. Colour preferences of flower-naive honeybees. *J. Comp. Physiol. A* 177:247–59
36. Giurfa M, Schubert M, Reisenman C, Gerber B, Lachnit H. 2003. The effect of cumulative experience on the use of elemental and configural visual discrimination strategies in honeybees. *Behav. Brain Res.* 145:161–69
37. Giurfa M, Vorobyev M, Kevan P, Menzel R. 1996. Detection of coloured stimuli by honeybees: minimum visual angles and receptor specific contrasts. *J. Comp. Physiol. A* 178:699–709
38. Giurfa M, Zhang S, Jenett A, Menzel R, Srinivasan MV. 2001. The concepts of 'sameness' and 'difference' in an insect. *Nature* 410:930–33
39. Graham P, Collett TS. 2002. View-based navigation in insects: how wood ants (*Formica rufa* L.) look at and are guided by extended landmarks. *J. Exp. Biol.* 205:2499–509
40. Grant V. 1951. The fertilization of flowers. *Sci. Am.* 12:1–6
41. Gross HJ, Pahl M, Si A, Zhu H, Tautz J, Zhang S. 2009. Number-based visual generalisation in the honeybee. *PLoS ONE* 4:e4263
42. Grossmann KE. 1971. Belohnungsverzögerung beim Erlernen einer Farbe an einer künstlichen Futterstelle durch Honigbienen. *Z. Tierpsychol.* 29:28–41
43. Hammer M. 1993. An identified neuron mediates the unconditioned stimulus in associative olfactory learning in honeybees. *Nature* 366:59–63
44. Harnard S. 1987. *Categorical Perception. The Groundwork of Cognition*. Cambridge, UK: Cambridge University Press
45. Harris RA, Hempel de Ibarra N, Graham P, Collett TS. 2005. Ant navigation: priming of visual route memories. *Nature* 438:302
46. Heinrich B. 1979. "Majoring" and "Minoring" by foraging bumblebees, *Bombus vagans*: an experimental analysis. *Ecology* 60:245–55

47. Hertz M. 1929. Die Organisation des optischen Feldes bei der Biene I. *Z. Vergl. Physiol.* 8:693–748
48. Hertz M. 1929. Die Organisation des optischen Feldes bei der Biene II. *Z. Vergl. Physiol.* 11:107–45
49. Hertz M. 1933. Über figurale Intensitäten und Qualitäten in der optischen Wahrnehmung der Biene. *Biol. Zb.* 53:11–40
50. Hori S, Takeuchi H, Arikawa K, Kinoshita M, Ichikawa N, et al. 2006. Associative visual learning, color discrimination, and chromatic adaptation in the harnessed honeybee *Apis mellifera* L. *J. Comp. Physiol. A* 192:691–700
51. Hori S, Takeuchi H, Kubo T. 2007. Associative learning and discrimination of motion cues in the harnessed honeybee *Apis mellifera* L. *J. Comp. Physiol. A* 193:825–33
52. Horridge A. 2000. Seven experiments on pattern vision of the honeybee, with a model. *Vis. Res.* 40:2589–603
53. Horridge A. 2003. The effect of complexity on the discrimination of oriented bars by the honeybee (*Apis mellifera*). *J. Comp. Physiol. A* 189:703–14
54. Horridge A. 2006. Visual discriminations of spokes, sectors, and circles by the honeybee (*Apis mellifera*). *J. Insect Physiol.* 52:984–1003
55. Horridge A. 2009. Generalization in visual recognition by the honeybee (*Apis mellifera*): a review and explanation. *J. Insect Physiol.* 55:499–511
56. Kuwabara M. 1957. Bildung des bedingten Reflexes von Pavlovs Typus bei der Honigbiene, *Apis mellifica*. *J. Fac. Sci. Hokkaido Univ. Ser. VI Zool.* 13:458–64
57. Leadbeater E, Chittka L. 2005. A new mode of information transfer in foraging bumblebees? *Curr. Biol.* 15:R447–48
58. Leadbeater E, Chittka L. 2007. The dynamics of social learning in an insect model, the bumblebee (*Bombus terrestris*). *Behav. Ecol. Sociobiol.* 61:1789–96
59. Lehrer M. 1997. Honeybee's visual orientation at the feeding site. In *Orientation and Communication in Arthropods*, ed. M Lehrer, pp. 115–44. Basel: Birkhäuser
60. Lehrer M, Campan R. 2004. Shape discrimination by wasps (*Paravespula germanica*) at the food source: generalization among various types of contrast. *J. Comp. Physiol. A* 190:651–63
61. Lehrer M, Campan R. 2005. Generalization of convex shapes by bees: What are shapes made of? *J. Exp. Biol.* 208:3233–47
62. Liu G, Seiler H, Wen A, Zars T, Ito K, et al. 2006. Distinct memory traces for two visual features in the *Drosophila* brain. *Nature* 439:551–56
63. Lotto RB, Chittka L. 2005. Seeing the light: illumination as a contextual cue to color choice behavior in bumblebees. *Proc. Natl. Acad. Sci. USA* 102:3852–56
64. Macquart D, Latil G, Beugnon G. 2008. Sensorimotor sequence learning in the ant *Gigantiops destructor*. *Anim. Behav.* 75:1693–701
65. Masquelier T, Thorpe SJ. 2007. Unsupervised learning of visual features through spike timing dependent plasticity. *PLoS Comput. Biol.* 3:e31
66. Menzel R. 1967. Untersuchungen zum Erlernen von Spektralfarben durch die Honigbiene (*Apis mellifica*). *Z. Vergl. Physiol.* 56:22–62
67. Menzel R. 1968. Das Gedächtnis der Honigbiene für Spektralfarben. I. Kurzzeitiges und langzeitiges Behalten. *Z. Vergl. Physiol.* 60:82–102
68. Menzel R. 1985. Learning in honey bees in an ecological and behavioral context. In *Experimental Behavioral Ecology and Sociobiology*, ed. B Hölldobler, M Lindauer, pp. 55–74. Stuttgart, Ger.: Fischer
69. Menzel R. 1999. Memory dynamics in the honeybee. *J. Comp. Physiol. A* 185:323–40
70. Menzel R, Backhaus W. 1991. Colour vision in insects. In *Vision and Visual Dysfunction. The Perception of Colour*, ed. P Gouras, pp. 262–88. London: MacMillan
71. Menzel R, Greggers U, Hammer M. 1993. Functional organization of appetitive learning and memory in a generalist pollinator, the honey bee. In *Insect Learning: Ecological and Evolutionary Perspectives*, ed. D Papaj, AC Lewis, pp. 79–125. New York: Chapman & Hall
72. Menzel R, Steinmann E, de Souza J, Backhaus W. 1988. Spectral sensitivity of photoreceptors and colour vision in the solitary bee, *Osmia rufa*. *J. Exp. Biol.* 136:35–52
73. Miller GA. 2003. The cognitive revolution: a historical perspective. *Trends Cognit. Sci.* 7:141–44

74. Narendra A, Si A, Sulikowski D, Cheng K. 2007. Learning, retention and coding of nest-associated visual cues by the Australian desert ant, *Melophorus bagoti*. *Behav. Ecol. Sociol.* 61:1543–53
75. Núñez JA. 1982. Honeybee foraging strategies at a food source in relation to its distance from the hive and the rate of sugar flow. *J. Apicult. Res.* 21:139–50
76. Pavlov IP. 1927. *Lectures on Conditioned Reflexes*. New York: International Publishers
77. Peng Y, Xi W, Zhang W, Zhang K, Guo A. 2007. Experience improves feature extraction in *Drosophila*. *J. Neurosci.* 27:5139–45
78. Robertson I. 2001. *Problem Solving*. Hove, UK: Psychology
79. Schröter U, Malun D, Menzel R. 2007. Innervation pattern of suboesophageal ventral unpaired median neurones in the honeybee brain. *Cell Tissue Res.* 327:647–67
80. Schubert M, Giurfa M, Francucci S, Lachnit H. 2002. Nonelemental visual learning in honeybees. *Anim. Behav.* 64:175–84
81. Shepard RN. 1987. Towards a universal law of generalisation for psychological science. *Science* 237:1317–23
82. Spaethe J, Tautz J, Chittka L. 2001. Visual constraints in foraging bumblebees: Flower size and color affect search time and flight behavior. *Proc. Natl. Acad. Sci. USA* 98:3898–903
83. Spence KW. 1937. The differential response to stimuli varying within a single dimension. *Psychol. Rev.* 44:430–44
84. Srinivasan MV. 1994. Pattern recognition in the honeybee: recent progress. *J. Insect Physiol.* 40:183–94
85. Srinivasan MV. 2010. Honey bees as a model for vision, perception, and cognition. *Annu. Rev. Entomol.* 55:267–84
86. Srinivasan MV, Poteser M, Kral K. 1999. Motion detection in insect orientation and navigation. *Vis. Res.* 39:2749–66
87. Srinivasan MV, Zhang SW, Zhu H. 1998. Honeybees link sights to smells. *Nature* 396:637–38
88. Stach S, Benard J, Giurfa M. 2004. Local-feature assembling in visual pattern recognition and generalization in honeybees. *Nature* 429:758–61
89. Stach S, Giurfa M. 2005. The influence of training length on generalization of visual feature assemblies in honeybees. *Behav. Brain Res.* 161:8–17
90. Tang S, Guo A. 2001. Choice behavior of *Drosophila* facing contradictory visual cues. *Science* 294:1543–47
91. Terrace HS, McGonigle B. 1994. Memory and representation of serial order by children, monkeys and pigeons. *Curr. Dir. Psychol. Sci.* 3:180–85
92. Tibbets EA. 2002. Visual signals of individual identity in the wasp *Polistes fuscatus*. *Proc. Biol. Sci.* 269:1423–28
93. Tibbets EA, Dale J. 2004. A socially enforced signal of quality in a paper wasp. *Nature* 432:218–22
94. Troje F, Huber L, Loidolt M, Aust U, Fieder M. 1999. Categorical learning in pigeons: the role of texture and shape in complex static stimuli. *Vis. Res.* 39:353–66
95. Van Hateren JH, Srinivasan MV, Wait PB. 1990. Pattern recognition in bees: orientation discrimination. *J. Comp. Physiol. A* 197:649–54
96. Van Swinderen B, Greenspan RJ. 2003. Salience modulates 20–30 Hz brain activity in *Drosophila*. *Nat. Neurosci.* 6:579–86
97. Von Fersen L, Wynne CDL, Delius JD. 1990. Deductive reasoning in pigeons. *Naturwissenschaften* 77:548–49
98. Von Frisch K. 1914. Der Farbensinn und Formensinn der Biene. *Zool. Jb. Physiol.* 37:1–238
99. Wehner R. 1967. Pattern recognition in bees. *Nature* 215:1244–48
100. Wehner R. 1971. The generalization of directional visual stimuli in the honeybee, *Apis mellifera*. *J. Insect Physiol.* 17:1579–91
101. Wehner R. 1972. Dorsoventral asymmetry in the visual field of the bee, *Apis mellifera*. *J. Comp. Physiol. A* 77:256–77
102. Wehner R. 1981. Spatial vision in arthropods. In *Invertebrate Visual Centers and Behavior. Handbook of Sensory Physiology*, ed. H Autrum, VII/6C:287–616. Berlin: Springer
103. Wehner R. 2003. Desert ant navigation: how miniature brains solve complex tasks. *J. Comp. Physiol. A* 189:579–88

104. Wilson EO. 1971. *The Insect Societies*. Cambridge, MA: Belknap Press
105. Wolf E. 1933. Critical frequency of flicker as a function of intensity of illumination for the eye of the bee. *J. Gen. Physiol.* 17:7–19
106. Wolf E. 1934. Das Verhalten der Biene gegenüber flimmernden Feldern und bewegten Objekten. *Z. Vergl. Physiol.* 20:151–61
107. Wolf E, Zerrahn-Wolf G. 1935. The effect of light intensity, area, and flicker frequency on the visual reactions of the honey bee. *J. Gen. Physiol.* 18:853–63
108. Wolf R, Wittig T, Liu L, Wustmann G, Eyding D, Heisenberg M. 1998. *Drosophila* mushroom bodies are dispensable for visual, tactile, and motor learning. *Learn. Mem.* 5:166–78
109. Worden BD, Papaj DR. 2005. Flower choice copying in bumblebees. *Biol. Lett.* 1:504–7
110. Xi W, Peng Y, Guo J, Ye Y, Zhang K, et al. 2008. Mushroom bodies modulate salience-based selective fixation behavior in *Drosophila*. *Eur. J. Neurosci.* 27:1441–51
111. Yang EC, Maddess T. 1997. Orientation-sensitive neurons in the brain of the honey bee (*Apis mellifera*). *J. Insect Physiol.* 43:329–36
112. Zeil J, Kelber A, Voss R. 1996. Structure and function of learning flights in ground-nesting bees and wasps. *J. Exp. Biol.* 199:245–52
113. Zentall TR, Galizio M, Critchfield TS. 2002. Categorization, concept learning and behavior analysis: an introduction. *J. Exp. Anal. Behav.* 78:237–48
114. Zhang S, Bock F, Si A, Tautz J, Srinivasan M. 2005. Visual working memory in decision making by honey bees. *Proc. Natl. Acad. Sci. USA* 102:5250–55
115. Zhang S, Srinivasan MV. 1994. Prior experience enhances pattern discrimination in insect vision. *Nature* 368:330–33
116. Zhang SW, Bartsch K, Srinivasan MV. 1996. Maze learning by honeybees. *Neurobiol. Learn. Mem.* 66:267–82
117. Zhang SW, Lehrer M, Srinivasan MV. 1999. Honeybee memory: navigation by associative grouping and recall of visual stimuli. *Neurobiol. Learn. Mem.* 72:180–201
118. Zhang SW, Srinivasan MV, Zhu H, Wong J. 2004. Grouping of visual objects by honeybees. *J. Exp. Biol.* 207 3289–98

Evolution of Sexual Dimorphism in the Lepidoptera

Cerisse E. Allen,[1,2] Bas J. Zwaan,[2] and Paul M. Brakefield[2]

[1]Division of Biological Sciences, University of Montana, Missoula, Montana 59812; email: cerisse.allen@mso.umt.edu

[2]Institute of Biology, Leiden University, 2300 RA Leiden, The Netherlands; email: b.j.zwaan@biology.leidenuniv.nl, p.m.brakefield@biology.leidenuniv.nl

Annu. Rev. Entomol. 2011. 56:445–64

First published online as a Review in Advance on September 7, 2010

The *Annual Review of Entomology* is online at ento.annualreviews.org

This article's doi:
10.1146/annurev-ento-120709-144828

0066-4170/11/0107-0445$20.00

Key Words

sexual selection, natural selection, fecundity selection, life-history evolution, secondary sexual traits

Abstract

Among the animals, the Lepidoptera (moths and butterflies) are second only to beetles in number of described species and are known for their striking intra- and interspecific diversity. Within species, sexual dimorphism is a source of variation in life history (e.g., sexual size dimorphism and protandry), morphology (e.g., wing shape and color pattern), and behavior (e.g., chemical and visual signaling). Sexual selection and mating systems have been considered the primary forces driving the evolution of sexual dimorphism in the Lepidoptera, and alternative hypotheses have been neglected. Here, we examine opportunities for sexual selection, natural selection, and the interplay between the two forces in the evolution of sexual differences in the moths and butterflies. Our primary goal is to identify mechanisms that either facilitate or constrain the evolution of sexual dimorphism, rather than to resolve any perceived controversy between hypotheses that may not be mutually exclusive.

INTRODUCTION

Opportunity for selection: the variance in relative fitness in a population

It is impossible to describe the butterflies and moths without making reference to their beauty and diversity. The approximately 137,000 named species of Lepidoptera (7) exhibit a bewildering array of shapes, sizes, colors, life histories, and habits. Our ability to understand the processes that have led to such diversity is complicated by extensive variation among individuals within species. Sexual dimorphism—differences in behavioral, morphological, physiological, and/or life-history traits between males and females—is an especially ubiquitous form of intraspecific variation in the butterflies and moths. Although iconic examples of elaborate male ornaments and exaggerated traits used in courtship and mate attraction are found in the Lepidoptera (e.g., UV-reflective wing patterns; 59), sexually dimorphic traits can have a mix of sexual and nonsexual functions (e.g., arctiid moth auditory apparatus; 143) or function exclusively in nonreproductive contexts (e.g., female-limited mimetic color patterns; 70).

Within the Lepidoptera, there is rampant sexual dimorphism as well as extensive within-sex polymorphism (84, 140). This presents a unique opportunity to examine dimorphism in both sexual and nonsexual contexts. Existing reviews have focused on those aspects of sexual dimorphism that can be attributed to sexual selection or mating systems (110, 144). Here, our goal is to focus on the recent literature and examine the opportunities for sexual and natural selection, and the interplay between the two forces, in the evolution of behavioral, morphological, physiological, and life-history differences between males and females in the butterflies and moths. We do not aim to catalog all the sex differences exhibited by 137,000 species, but we hope to highlight mechanisms that may facilitate or constrain the evolution of sexual dimorphisms in this diverse insect group.

SEXUAL VERSUS NATURAL SELECTION AND THE EVOLUTION OF SEXUAL DIMORPHISM

The selective forces driving the evolution of sexual dimorphism have been subject to intense and ongoing debate. Darwin and Wallace discussed the causation of sexually dimorphic mimicry in *Papilio* butterflies and plumage dimorphism in passerine birds over the course of several years and ultimately disagreed about the mechanisms involved (66). Although both initially acknowledged that sexual and natural selection were important in the evolution of sex differences (4, 66, 68), they eventually advocated different positions: Darwin arguing for sex-specific sexual selection (female choice of bright male coloration) and Wallace for sex-specific natural selection (intense predation on females favoring cryptic or mimetic coloration). Despite the fact that the two hypotheses are not necessarily exclusive (113), research has generally focused on sexual selection as the primary cause of dimorphisms. The practice (unwarranted; 114) whereby the degree of sexual dimorphism is employed as a proxy estimate for the strength of sexual selection (e.g., 47) reflects this bias. Moreover, there has been a tendency to treat natural selection primarily as a constraint on the degree of sexual dimorphism that evolves due to sexual selection (6, 72).

Competition for mates is at the heart of sexual selection. Potential mates or potential rivals are the agents of selection on traits that influence mating or fertilization success through competitive ability or mate choice (4, 6). The strength of sexual selection, and thus its ability to drive dimorphic evolution, requires that individuals of the same sex have unequal access to mates, and furthermore that the sexes differ in the opportunity for selection (114). For example, if large body size gives males a competitive advantage so that only the largest individuals mate, but size has no effect on a female's ability to access mates, the sex difference in the opportunity for selection results in selection for sexual size dimorphism. In contrast, selection for sexual size dimorphism is extremely weak if sexual selection acts similarly in males and females, for example, if there is mutual mate choice and large size is similarly advantageous to both sexes.

The evolution of sexual dimorphism likely involves complex interactions between sexual and natural selection (26, 113). Theoretical models typically treat the two selective forces as antagonistic (e.g., 48). Constraints due to natural selection can result from the costs of mate choice itself, which can expose individuals to increased predation or decreased mating opportunities (6, 19). Moreover, natural selection can oppose sexual selection if dimorphic traits are costly to produce, or reduce survival or fecundity. For example, if a prolonged growth period in one sex also increases larval vulnerability to predators (11, 16), overall selection for sexual size dimorphism is reduced. Alternatively, natural selection could enhance sex differences evolving under sexual selection (114). For example, if larger males are better able to compete for mates (sexual selection to increase male body size) and smaller females are better protected from predators by virtue of improved flight performance (natural selection to decrease female body size), both forces working together can drive the evolution of extreme sexual size dimorphism (144). Note that in both of these examples, sexual selection generates the driving force for the evolution of dimorphism, and natural selection limits or enhances the degree of difference between the sexes.

Natural selection alone may be as important a force as sexual selection in driving the evolution of sexual dimorphism. Although this is well supported by theory (e.g., 18, 117), natural selection has received much less attention in this context (51), and consequently, the nonsexual origins of dimorphic traits are often undocumented (132). Whereas sexual selection operates on variation in mating success, natural selection acts on all other sources of variation in fecundity and survival (6). Sources of sex-specific natural selection fall into two broad categories. First, divergent selection pressures on males and females can result from ecological interactions, such as resource competition (reviewed in Reference 113). Sexually dimorphic niche partitioning eliminates disruptive selection that might otherwise drive speciation, in a process analogous to other mechanisms of partitioning such as ontogenetic niche shifts, feeding polymorphisms, and dispersal (18, 74). Second, fecundity selection can drive males and females toward sex-specific optima for traits that maximize reproduction, or reduce the costs associated with mating (4, 71, 105, 113).

Sexual size dimorphism: the difference in adult body size between males and females in a population; one of the most easily recognized forms of sexual dimorphism

Fecundity selection: a form of natural selection for increased fertility or egg production

In an initially monomorphic population, ecological competition alone can drive the evolution of sexual dimorphism (117). However, different fitness interests of the sexes can also set the stage for ecological divergence. For example, sexually dimorphic energetic or nutritional requirements may result in selection for dimorphisms in feeding morphology (e.g., 132) or lead to sex differences in vulnerability to predators (e.g., 93). Alternatively, sexual selection might frequently co-opt dimorphic traits that initially evolved under natural selection, if those traits indicate general genetic quality or general defensive ability (13, 143).

CONTEXTS FOR SEXUALLY DIMORPHIC EVOLUTION IN THE LEPIDOPTERA

The diversity of life histories, habitats, body forms, and behaviors exhibited by butterflies and moths makes any attempt at generalization fraught with difficulty. However, there are several common characteristics with broad implications for the generation and maintenance of sexual dimorphism within populations: (*a*) The majority of female butterflies and moths mate multiply, although populations may contain a mix of monandrous and polyandrous individuals (80, 137). This may lead to sperm competition and/or sexual conflict over female remating and open the door for complex interactions between sexual and natural selection (115, 142).

W and Z chromosomes: females are the heterogametic sex in Lepidoptera, typically carrying WZ sex chromosomes; males are homogametic, typically ZZ

(*b*) Both sexes often make costly investments in reproduction—females to eggs, and males to apyrene (infertile) sperm (31) and to spermatophores that frequently contain nutrients and protective, antiaphrodisiac, or other compounds (15, 142). Consequently, fecundity selection might maximize quality over quantity in males as well as females. (*c*) Adults often rely on visual, chemical, or auditory communication. The signals themselves (visual, auditory) may be easily intercepted by predators (22) or expose the receiver to predators when actively orienting to the signal (e.g., males orienting to female chemical signals; 128). Thus, sex-specific predation could drive the evolution of dimorphic signals or act on signals that evolve via sexual selection. (*d*) Many traits involved in social interactions and reproductive isolation between species exhibit sex-linked or sex-limited expression (119). This can provide a mechanism to limit the genetic correlation between the sexes.

To understand the ultimate causes of sexual dimorphism, it is critical to determine the strength of selection and whether selection is likely to favor the evolution of monomorphism or dimorphism. This task is complicated because selection is often poorly quantified in natural populations (but see, e.g., 63, 64). Below, we discuss how selective forces and underlying genetic mechanisms may influence the evolution of monomorphism, dimorphism, and within-sex polymorphism in butterfly and moth populations (some examples of sexually dimorphic traits are shown in **Figure 1**).

SEXUAL DIMORPHISM IN LIFE-HISTORY TRAITS

Typical life-history traits are age and size at hatching, size and number of offspring, and reproductive life span (123). Although life-history traits together are expected to evolve to an optimal life-history configuration in a given environment, individual life-history traits may not be maximized in the direction of increased fitness because of trade-offs between traits closely related to fitness (35). The evolution of reproductive division of labor (i.e., separate sexes), and subsequent natural selection toward sex-specific reproductive optima, can readily lead to the evolution of sexually dimorphic life histories visible at the behavioral, morphological, and physiological levels (34). Sexual selection can also shape the evolution of life-history traits when the reproductive potential of males and females differs. Below, we discuss patterns of variation in sexually dimorphic life-history traits and highlight causal factors, including natural selection, sexual selection, and sexually antagonistic coevolution.

Gene Regulation and Life-History Evolution

For sexual dimorphism to evolve, genetic variation must have independent effects on both sexes (the between-sex genetic correlation must be <1 or >-1; 71). Many traits related to reproduction and life history are sex-linked in the Lepidoptera (119, 135), even though the W and Z chromosomes compose less than 5% of the genome (76). In addition, the absence of dosage compensation (135) may facilitate the evolution of Z-linked modifier loci, allowing sex-limited expression of autosomal genes (10, 135). Although this probably facilitates the evolution of sexual dimorphism, the existence of any between-sex genetic correlation can limit the evolution of dimorphic life histories or limit the extent to which selection can maximize fitness in one sex (e.g., 36). Natural and sexual selection can act on variation in gene expression, so that sex-specific gene regulation occurs as a result of selection driving males and females toward different reproductive optima. For example, evolutionary changes in the sex determination pathway in the dipteran *Drosophila melanogaster* appear to involve selection primarily for better males, and/or better females (102). Such changes in gene regulation and expression may have further consequences for the sexual fitness of one or both sexes, and antagonistic fitness effects in one sex as a result of selection for increased fitness in the other sex are expected to play a role.

Figure 1

Bilateral gynandromorphs illustrate sexual dimorphism in wing size, wing shape, wing color pattern, the size and shape of antennae, color of body hairs, and abdominal characteristics in butterfly and moth species. Complete bilateral gynandromorphs can result from nondisjunction of ZZ chromosomes in early embryonic development. Females are typically ZW, but Z0 individuals develop as female also. Panel *a* courtesy of USDA APHIS PPQ Archive, USDA APHIS PPQ, **http://bugwood.org/**. Panel *b* courtesy of D.J. Emlen. Panels *c* and *d* courtesy of N.H. Patel.

Body Size and Sexual Size Dimorphism

Recent analyses demonstrate that many lepidopteran species exhibit female-biased sexual size dimorphism (73% of 48 species in Reference 124). Size differences are established during the larval period (37, 131) by developmental and physiological mechanisms (e.g., number of larval instars and hormonal regulation; 124). Because females of many species are capital breeders (i.e., they allocate larval resources to reproduction), and large size is related directly to fecundity (12, 42, 145), selection for large female body size appears driven by natural selection for increased fecundity (105). Increased female size may influence the level of polyandry (e.g., *Lobesia botrana*; 134), further increasing fecundity because polyandrous females may have higher lifetime fecundity than monandrous females

Protandry: in a single age cohort, the earlier emergence (pupal eclosion) of males compared to females

(e.g., *Pieris napi*; 137). Thus, there are two critical questions regarding the evolution of SSD: What limits the evolution of even larger size in females, and what limits male size?

Trade-offs between the fitness gains due to increased fecundity and the costs due to large juvenile size (increased predation risk; 11) appear to limit the adult size of both sexes. Furthermore, in seasonal environments, time limits on egg deposition and maturation due to temperature may further limit female lifetime fecundity (12). Thus, selection on females for increased size will be limited by increased predation risk and time and temperature constraints. Males also suffer costs of increased juvenile size, but they are generally expected to have less to gain than females from increased adult size (but see below). As a consequence, selection favors small male size and female-biased sexual size dimorphism (see Reference 49 for a comprehensive review of the adaptive significance of growth decisions in butterflies). In lycaenid butterflies, the fact that female body size is much more robust to changes in developmental temperature, whereas males developing at high temperatures mature at smaller size to gain benefits of early emergence, is consistent with such sex-specific fitness optima (42–44).

Six percent of surveyed lepidopteran species (124) exhibit male-biased sexual size dimorphism. Such patterns are typically explained by the benefits of large male size in competition for access to mates (i.e., intrasexual selection; 40). However, male-biased sexual size dimorphism may also arise when males invest significantly in offspring and male size is directly correlated with increased reproductive output. For instance, in *Pieris napi*, spermatophores provide nitrogen equivalent to 70 eggs (57), representing a costly investment in offspring. In species with small spermatophores, such gifts are less costly (41). Larval food limitation (or unpredictable resources) may favor multiple mating by females (to increase reproductive nutritional gains), so that males are selected to provide competitive nuptial gifts (reviewed in Reference 144). Sexually antagonistic coevolution can occur, even when providing a costly spermatophore benefits both males and females (24). Females that mate multiply may forage for the resources contained in nuptial gifts; for example, female arctiid moths provision eggs with pyrrolizidine alkaloids obtained from multiple spermatophores, although they may fertilize those eggs using sperm from select males only (14; see also 21, for an example in heliconiid butterflies). Puddling behavior (feeding on dung, carrion, or mud) by males is associated with increased abdominal sodium stores and the degree of polyandry in some nymphalid butterflies (but not others; 85), suggesting that females could also forage for these resources by mating multiply.

Growth rate and development time are intimately related to sexual size dimorphism and protandry; however, each represents a life-history trait that may be targeted independently by selection. Butterflies and moths achieve increased body size primarily by increasing developmental time (131), either by adding larval instars (typical of many moth species) or through cumulative weight gain over several larval instars in species with fixed instar number (typical of butterflies; 43, 131). Sexual dimorphism in instar number is associated with female-biased sexual size dimorphism (38). The costs of predation or parasitism associated with prolonged growth and time limits on activity in seasonal environments may constrain the evolution of development time and, as a consequence, adult size. Growth rate within a particular larval instar may be subject to developmental constraints (131), and males may pay a particularly high cost for rapid development in the pupal period (49). Moreover, if males trade off size and development time under selection for protandry (below), between-sex genetic correlations for size may limit the extent of sexual dimorphism in these traits. Evidence suggests, however, that selection may break apparent constraints between traits including egg size, development time, and pupal size (45, 91).

Protandry

Protandry is widespread among butterflies and moths, with the notable exception of bagworms

(Psychidae) (106). Protandry can be achieved if males decrease their development time (148) or increase their growth rate (146) relative to that of females. Direct selection on protandry resulting from selection for increased mating success of early-emerging males accompanied by pre-emergence mortality of females and postemergence mortality of males may determine the degree of protandry (the adaptive explanation, cf 149). Alternatively, protandry may evolve as an indirect consequence of selection for increased adult size, in which females benefit more from increased size (see above) than males do (the incidental explanation; 91, 149).

Empirical evidence generally supports the adaptive explanation in butterflies. Under direct selection for protandry, the existence of protandry should be correlated with nonoverlapping generations (i.e., seasonal environments), whereas the degree of protandry should be insensitive to environmental conditions. This appears to be the case among populations along a latitudinal gradient in the butterfly *Pararge aegeria* (91), supporting the adaptive explanation. Among *Heliconius* species, adult males guard female pupae and mate with the females before eclosion is complete (pupal mating; 9). Direct female benefits are suggested in *Heliconius charithonia*, in which female pupae produce volatile compounds that allow males to locate and guard them (39). The cost to females of "no mate choice" in this system may be outweighed by the benefits of mating with high-quality males capable of guarding pupae. In the butterflies *Lycaena tityrus* and *Bicyclus anynana*, however, the degree of protandry is related to temperature (increases with decreasing developmental temperature; 42, 148), suggesting that the strength of selection acting on protandry may depend on specific environmental conditions and/or physiological constraints due to the effects of development time on weight loss and body composition at low versus high developmental temperatures.

Evidence concerning genetic variation underlying sexual dimorphism in life-history traits is scarce. Although sex-specific genetic variation exists for pupal weight and development time in *B. anynana* (148), the response to direct selection on the degree of protandry was extremely limited and quickly reached a plateau (149). The estimated genetic correlation between males and females for development time was 0.98, leaving very little scope for sex-specific responses to selection. In addition, decreased fertility among extreme combinations of early- versus late-emerging males and females may have strongly opposed selection for increased protandry (148). The lack of response to selection in such experiments is consistent with a legacy of strong positive selection acting directly on protandry in this species, leading to strong genetic correlations that prevent further evolutionary change, at least in the short term. However, variation in body size and/or development time might not always reflect underlying genetic variation. In some cases, the lack of response to selection may be due to environmental variation affecting traits such as the number of larval instars, total development time, and final adult size (see, e.g., Reference 131). More studies are needed to elucidate both the nature of past selection and the potential for evolutionary change in individual life-history traits.

Sexual Dimorphism in Physiology

Sex differences in metabolic rate may influence the evolution of sex differences in reproductive behavior, thermal ecology, and associated flight morphology, as shown in the butterfly *P. aegeria* (e.g., 83, 139). In *Melitaea cinxia*, flight metabolic rate is related to female dispersal but appears to influence intrasexual competition and mate location in males (90). In the butterfly *B. anynana*, males and females differ in their mass-specific resting metabolic rate and in the plasticity of resting metabolic rate in response to larval and adult thermal environments (98). Selection for increased adult starvation resistance (adults must survive to reproduce at the end of a long dry season) in *B. anyana* reveals the existence of sex-specific life-history trade-offs: Increased starvation resistance involves increased resource acquisition and altered reproductive allocation by females but decreased resting metabolic rate by males (99). Additional sex-specific trade-offs, such as trade-offs between melanic wing coloration and immunity (e.g., 125) could also span the larval-adult developmental boundary (cf 17) and influence sex-specific life-history evolution.

Sexually dimorphic life-history traits appear to be associated primarily with reproductive division of labor between the sexes and with subsequent optimization of resource allocation to reproduction in each sex. Understanding how such dimorphisms evolve will be facilitated greatly by a genetic analysis of underlying mechanisms. Gender is by far the most important correlate of gene expression in *D. melanogaster* (104); establishing how much of this variation contributes to key life-history traits is a challenge for the study of sexual dimorphism in the Lepidoptera as well.

SEXUAL DIMORPHISM IN MORPHOLOGICAL TRAITS

Sexual dimorphism occurs in a wide variety of morphological traits, including wing size and shape; size and color of pigment patches and UV-reflective regions on the wings; body size and body composition (e.g., relative thorax or abdomen size); color and density of body hairs; size and shape of sensory structures such as eyes, auditory organs, and antennae; structures associated with production and release of chemical attractants and size and shape of genitalia (reviewed for butterflies in Reference 110; see also, e.g., 106, 107, 143). These traits are intimately connected to the same key tasks of feeding, mate location, dispersal, and oviposition central to life-history evolution. Selection acting on performance—locomotion, communication, or reproduction—can drive morphological evolution (e.g., 64, 122). While sexual selection often affects butterfly and moth morphology, natural selection (discussed in Reference 69) and complex interactions between the two forces (discussed in Reference 144) are also important. Below, we discuss the opportunity for sexual and natural selection to drive the evolution of sex differences in morphology and highlight specific traits where complex interactions between the selective mechanisms can be identified.

Opportunities for Sexual Selection on Butterfly and Moth Morphology

Larger sex differences are typically expected to evolve under sexual selection rather than natural selection because of directional selection resulting from mate choice and intrasexual competition (4). However, the strength of sexual selection in natural populations is often unmeasured. Male-biased operational sex ratios in many lepidopteran populations (e.g., 129) suggest the existence of intense male mate competition and thus strong sexual selection. However, because the operational sex ratio does not measure the sex difference in the opportunity for selection, it is a poor estimator for the strength of sexual selection (114). Moreover, populations with female-biased operational sex ratios (e.g., *Hypolimnas bolina*; 25) can experience greater than expected intersexual conflict or male-male competition, highlighting the need to quantify contemporary selection acting on traits of interest rather than assume that strong directional selection exists when the sexes differ in morphology (discussed in Reference 68).

Demonstrating sexual selection via female choice requires evidence for a preference and evidence that sexually selected male signals are costly and condition dependent (13). Visual, auditory, and chemical signals are frequently involved in courtship and mate signaling in the Lepidoptera. Consequently, female mate choice may be influenced by male signaling behavior and male body condition, by the female's condition, or by chemicals transferred by males in copula (reviewed in Reference 142). Female wax moths (Pyralidae) prefer males with exaggerated ultrasonic calls, but the adaptive nature of the preference is unclear because the benefits of choice vary with female body condition (54). Although sexually dimorphic auditory calls used by noctuid, arctiid, geometrid, and crambid moths (involving dimorphic sound production organs in at least some species, e.g., 107) are often critical to male and/or female mating success, costs or condition dependency of calls has not been established (112).

Strong evidence for female choice based on bright male wing coloration is limited to a handful of butterfly systems in which preference exists for UV-reflective patches (structural colors; 32, 58, 59). The function of UV-reflective patches on moth wings is not well understood (77). Structural colors may be more likely to reflect male condition than pigment-based color patches because color reflection requires fine precision and regularity in the reflecting tissues (61). However, both UV-reflective patches (60) and melanic wing patches (e.g., 138) can exhibit condition dependency in both sexes. Here again, quantified selection estimates under multiple environmental conditions are necessary to understand the dynamics and benefits of mate choice.

Lepidopteran mating systems commonly feature both pre- and postcopulatory competitive interactions between males. Divergence in male genital morphology is associated with the degree of polyandry across butterfly and moth species (5), suggesting strong directional selection that results from sperm competition. However, the existence of environmental variation in male and female genital shape (e.g., 87) cautions against making broad assumptions concerning the presence or strength of sexual selection in the absence of quantified estimates.

The evolutionary effects of male mate preferences and mutual mate choice may be equally difficult to predict. Male mate choice for indicators of female fecundity may be more likely to reinforce differences that evolve via fecundity selection than to drive the evolution of exaggerated traits (19). Because male butterflies and moths also make costly investments in reproduction, female choice of male body size (e.g., an indicator of male defensive chemical content; 14) or condition-dependent colors might also be viewed as reinforcing the effects of fecundity selection. Alternatively, sperm competition can lead to postcopulatory mate choice, whereby males differentially allocate fertile sperm and/or larger spermatophores to larger females (e.g., 117).

Recent studies that account for all (directional and nondirectional) selection that arises from mating interactions (e.g., 133) show that sexual selection may be strongly stabilizing as well as directional. Chemical mate signaling by female moths and butterflies can generate strong directional selection on male competitive ability (and traits that improve detection and response) and simultaneous, weakly stabilizing selection on female signals (and traits related to signal production or dissemination; 10). Male use of female visual signals appears quite common in the day-flying butterflies (e.g., 36, 67) and may have similar effects. In addition, sexual selection arising from mutual mate choice (discussed in Lepidoptera in Reference 19) may reinforce sexual monomorphism as opposed to dimorphism.

Opportunities for Natural Selection

The existence of sex-specific reproductive optima has several critical consequences for the evolution of sexually dimorphic morphology. (*a*) Differences in reproductive allocation affect both body size (and sexual size dimorphism) and allocation to (and therefore size and shape of) body parts such as the thorax and abdomen (e.g., 121). (*b*) Different resource needs (e.g., perching sites versus oviposition host plants), or sex differences in the value of a common resource such as nectar or basking sites (e.g., 82, 86, 136), can result in sex-specific predation risk (sex-limited predation might also drive habitat segregation; 33, 56). (*c*). Fecundity or predator-mediated selection might target different traits in males and females. For example, in species with a nuptial flight, selection to minimize energetic costs or predation risk during mating can affect allocation to flight muscle in the carrying sex (2). Fecundity selection to minimize the costs of mate searching is also expected to target male characters (including wing size, wing shape, and flight muscle mass) associated with performance in flight-based mating tactics (reviewed in Reference 144).

Sex-limited mimicry: a form of sexual dimorphism in which only one sex mimics a noxious or distasteful species and the other sex is nonmimetic

Opposing sexual and natural selection can either limit or facilitate the evolution of sexual dimorphism. For example, male heliconiinine butterflies may choose mates on the basis of wing coloration (67), but natural selection favoring Müllerian mimicry might enforce sexual monomorphism in these species. Alternatively, male preferences for female wing coloration can oppose natural selection acting on thermoregulatory properties of female wing colors (36) and maintain female polymorphism (89). Lack of correspondence between the degree of sexual dimorphism in traits such as eye size, critical for signal detection, and mating systems (i.e., with presumed sexual selection) may result from the conflicting demands of natural selection on visual acuity (111).

Signal partitioning (cf 94) can be an evolutionary response to multiple conflicting selection pressures that result from the defensive and sexual function of traits such as wing color patches or eyespots. Signals that are attractive to mates may become restricted to wing surfaces that are easily covered (e.g., to the dorsal surface in species that rest with closed wings; 94), whereas sexually monomorphic defensive signals may be displayed on visible wing surfaces. Sexual dimorphism is also a form of signal partitioning—limiting trait expression to the sex with the greatest benefits and minimizing costs of signal expression to the other sex.

Widespread lepidopteran defensive traits, including alkaloid sequestration, ultrasound perception and production, predator-oriented flight, crypsis, aposematism, and mimicry, highlight the role predator-mediated selection plays in morphological evolution. Males typically exhibit greater flight activity (3, 62), but females are considered more vulnerable to predators (23, 53, 92, 93) because they must compensate for reproductive allocation in flight (e.g., 120). Consequently, the intensity of predation can be sex specific and/or drive males and females to different optimum values (93) for traits including wing size, shape, color, and pattern. Behavioral dimorphism, such as increased male flight height (56), might also result from strong predator-mediated selection.

The form of selection acting on defensive traits depends on the dynamics of predator-prey interactions (e.g., the effects of predator learning; Reference 1). Selection may vary from strongly directional to strongly stabilizing and exhibit positive or negative frequency dependency. Thus, it is unlikely that the role of natural selection is merely a negative limitation on strong sexual selection. In cryptic and Batesian mimicry systems, negative frequency-dependent selection (55) may strongly favor within-sex polymorphism and sexual dimorphism (discussed in Reference 69). In contrast, selection on aposematism and Müllerian mimicry is strongly stabilizing, is positively frequency-dependent, and is expected to drive local monomorphism (78). Directional selection on escape performance (e.g., 126) may be balanced by selection to minimize the costs of lost feeding or mating opportunities resulting from defense (e.g., 128). However, the balance between opposing forces of selection on performance can favor the evolution of plastic, dimorphic, or polymorphic traits that match escape tactics to condition. The existence of environmental variation in performance-related traits makes the evolution of these traits extremely difficult to predict in the absence of precise measurements of selection under a number of relevant environmental conditions (65).

Evolution of Sexually Dimorphic Mimicry Systems

Although mimicry is widespread among insects, sex-limited mimicry is especially prevalent in the Lepidoptera. Two key features of sex-limited mimicry are relevant to discussion of the selective forces that generate sexual dimorphism. (*a*) Sex-limited mimicry in the Lepidoptera is associated primarily with Batesian mimicry systems. Palatable mimics benefit when their population size is low relative to the distasteful model species (strong, negative frequency-dependent selection; 55). (*b*) Female-limited mimicry occurs in butterflies, in which both sexes are active in daylight, whereas male-limited mimicry occurs only in moths and only in species in which the male is active in daylight and the female is active at night (e.g., 141). Batesian mimics can copy the wing size, wing shape, and flight behavior of their distasteful model (28, 121, 140), but wing patterns are the best characterized of the traits involved in mimicry.

Sexual selection was typically assumed to be the major force driving the evolution of sex-limited mimicry (for a review of hypotheses, see Reference 69). Thus, the bright, nonmimetic male wing patterns were expected to evolve as a consequence of female mating preferences, and the ancestral form was assumed to have been sexually monomorphic and female-like (68). Recent analyses in *Papilio* butterflies demonstrate the opposite: Mimetic female wing patterns are largely derived, and the ancestral form was probably sexually monomorphic and male-like (68). Thus, the evolution of color pattern sexual dimorphism in this group appears associated with the evolution of a novel defensive trait in females (27, 68). The major alleles coding for female mimetic forms in *Papilio* show a similarly ancient origin consistent with the evolutionary association between female-limited mimicry and sexual dimorphism (27). Whether this holds generally true for all groups of female-limited mimics remains to be seen. However, a similar evolutionary trend was demonstrated in *Morpho* butterflies, which are nonmimetic but several species do exhibit sexually dimorphic wing patterns. In this group, the ancestral state appears to have been sexually monomorphic, with both sexes exhibiting bright iridescent wing patterns (96). Sexually dimorphic wing patterns are largely derived, and again, the origin of the sex difference is associated with acquisition of a defensive trait (cryptic wing coloration) by females (96).

The reverse situation, male-limited mimicry, is known only from moths. In several species of saturniid moths (*Callosamia promethea*, *C. securifera*, and *Eupackardia calleta*), the day-flying males are color pattern mimics of the distasteful butterfly *Battus philenor* (141; also see discussion in Reference 55). Females have nonmimetic color patterns and fly at night, although they release mate-attracting pheromones during the day (86). Precopulatory preferences exhibited by females do not appear to be based on visual signals (86). Although further study is needed, the association between male-limited mimicry and sexually dimorphic flight activity suggests that natural selection acting on the sex more vulnerable to visual predators drove the evolution of sexually dimorphic wing patterns in these species of saturniid moths.

Extreme within-sex polymorphism also occurs in some sex-limited mimicry systems. In the same population, multiple female forms can mimic different species, and nonmimetic female forms may also occur (140). Polymorphism can maintain low apparent frequencies of the palatable mimic relative to its model and would be favored by the negative frequency-dependent selection characteristic of Batesian mimicry (78). Thus, Batesian mimicry systems likely experience a combination of (*a*) frequency-dependent fitness benefits of Batesian mimicry and (*b*) female-biased predation (53, 62, 92, 93) due to reproductive constraints on the evolution of female body design and flight performance (79). This suggests that natural selection favors both the evolution of sex-limited mimicry and the maintenance of mimetic female polymorphism (68–70). Even in Müllerian systems, where sexes are typically monomorphic, sexually dimorphic flight activity, mate-searching behaviors, and/or flight height may facilitate mimicry evolution by enhancing sex-specific, naturally selected benefits of mimicry (56).

Sexual selection may also act on wing coloration in sex-limited mimicry systems. Male-male competitive interactions that strongly disfavor novel male wing patterns may maintain low genetic and phenotypic variability in *Papilio* (e.g., 52, 73). However, it is not clear why traits that potentially advertise male competitive ability are under strong stabilizing selection rather than directional selection (see Reference 13), and the costs and benefits to males of investing in melanic or other pigmented patches are poorly characterized. Signal partitioning in mimetic *Papilio polyxenes* suggests the potential for complex interactions between sexual and natural selection in Batesian mimicry systems. In this species, the ventral wing surface is sexually monomorphic and mimetic, whereas the dorsal wing surface is sexually dimorphic, and only females are effective dorsal mimics (29, 52).

Sex pheromones: chemical signals released in the environment to attract mates and or to communicate with the opposite sex

Evolution of Female and Male Sex Pheromones

Like visual and acoustic signals, lepidopteran sex pheromones have several potential functions: to prevent heterospecific mating by signaling species identity, to directly influence mate choice by acting as honest signals of mate quality, to mediate receptivity or acceptance of courtship, and to influence intrasexual competition by signaling competitor presence or competitive ability. Complex interactions can occur between these differing modes of communication, and as a result, there can be complex interactions between sexual and natural selection acting on chemical signals and their related morphological structures (e.g., 81). Together, these forces have led to the elaboration of the structures, chemical components, and signaling behaviors associated with pheromone communication. Reversals in the identity of the signaling versus receiving sex between night-flying (female signaling) and day-flying (male signaling) species highlight the roles that both sexual and natural selection play in shaping these sexually dimorphic traits in butterflies and moths. Moreover, the analysis of pheromone signaling systems provides us with detailed examples of the mechanisms targeted by selection.

The majority of females of the higher Lepidoptera (Ditrysia) release blends of long-chain unsaturated alcohols, acetates, or aldehydes to attract potential mates over long distances (75, 108, 130). The use of unique blends can allow closely related species to communicate over exclusive channels. Selection to minimize the costs of hybrid mating is strongly stabilizing but also strongly directional where closely related species overlap (e.g., 50). The result is low variability of pheromone blend within species but large differences between species (50). In moths, female pheromone blends appear to be a key component in the evolution of reproductive isolation and species diversification (109, 147).

Female moths typically release sex pheromones at night (but see, e.g., Reference 86) to attract males over long distances. Female pheromone-releasing structures (located between the terminal abdominal segments) and calling behavior are often highly elaborated (97). In contrast, the antennae, arrayed with chemosensory and mechanosensory organs, are frequently much more elaborate in males. Sexual dimorphism in these sensory structures appears to be shaped by predator-mediated natural selection, in addition to sexual selection for efficient pheromone detection and response. *Plodia interpunctellata* (Pyralidae), *Agrotis segetum*, and *Spodoptera littoralis* (Noctuidae) males appear to trade off mating success and survival, adjusting their bat-avoidance behavior according to the quality (i.e., chemical blend) and/or dosage (concentration or proximity) of the chemical signal produced by calling females (116, 127).

It is becoming clear that chemical signals produced by males are crucial for later courtship and mating success, even in the day-flying butterflies for which visual communication appears to dominate early phases of mate recognition (20). Pheromone-releasing structures of males include elaborate, modified wing scales (androconia) and eversible abdominal structures (coremata), and they are particularly diverse in many groups including Arctiidae (e.g., 143), Hesperiidae (e.g., 100), Danaidae (e.g., 20, 101), and Nymphalidae (e.g., 30). Pheromones are transferred to the female antennae during courtship, for example, during the hovering flight of male danaids (20, 101). Females assess both visual and chemical signals presented by males during courtship (e.g., 32, 46, 95). Female preference for one type of signal over the other may depend on which signal is a better indicator of male quality or male age (95). When males sequester defensive chemicals for use in pheromone biosynthesis, the pheromone blend itself may be an honest indicator of male quality that can be used by females in mate choice (e.g., 14, 20, 21) and consequently subject to directional sexual selection. Because defensive compounds may be derived from larval food plants, female mate choice for diet-derived male pheromones may be involved in the evolution of host plant shifts (103).

The relationship between male sex pheromones, sexual selection, and speciation is also illustrated in the nymphalid genus *Bicyclus*. The number, color, morphology, and position of the male androconia are discriminating traits in the phylogeny (30), suggesting an important role for pheromone biology in speciation. Indeed, *B. anynana* males release short-range courtship pheromones using rapid wing flicks. The female's antennae are sensitive to each major component of the male blend, and the sex pheromone indicates male quality and age and is involved in mate choice by females (32, 88).

CAUSES AND CONSEQUENCES OF SEXUAL DIMORPHISM

The existence of male choosiness and mutual mate choice, as well as strong stabilizing selection resulting from intra- or intersexual interactions, challenges our understanding of the mechanisms that drive the evolution of dimorphic behaviors, morphological traits, and life histories in the Lepidoptera. For life-history traits, natural selection driving males and females toward independent reproductive optima appears to be one of the major forces behind the evolution of sexual dimorphism in body size, emergence time, and physiological traits such as metabolic rate. Genetic evidence from model insect species, such as *D. melanogaster*, is consistent with natural selection playing a primary role in life-history evolution, and newly developed genomic resources (8) will make similar analyses possible in many more lepidopteran species. Sex-specific reproductive optima can set the stage for sexual selection and sexual conflict. However, sex differences in allocation to reproduction can expose males and females to a host of sex-specific selection pressures that act on morphological traits and result in the evolution of sexual dimorphisms. Sexually dimorphic traits, particularly signaling traits, are frequently associated with speciation, but it is not clear whether this is generally due to sexual selection (reproductive character displacement, cf 4) or natural selection against the costs of interspecific mating (e.g., 50). Determining the roles of sexual and natural selection in the origin and maintenance of such sexually dimorphic traits will require an effort to measure sexual and natural selection in more natural populations, to quantify the costs and condition dependency of sexual dimorphic traits, and to analyze the genetic mechanisms that limit the genetic correlation between the sexes. The Lepidoptera have long been central to discussions about the ultimate causes of sexual dimorphism, and we hope this review has highlighted the potential for complex interactions between the two forces in the evolution of dramatic sex differences exhibited in the butterflies and moths.

SUMMARY POINTS

1. The approximately 137,000 known species of butterflies and moths are known for their extreme diversity both between and within species. Much of the within-species variation is related to sexual differences in life history, behavior, and morphology. Differences between males and females in body size, time to adult emergence, wing color patterns related to antipredator defense or mate attraction, and the structures associated with the production and detection of chemical signals are widespread forms of sexual dimorphism in the Lepidoptera.

2. Debate over the ultimate causes of sexual dimorphism can be traced back to discussions between Darwin and Wallace, who disagreed about whether sexual selection or natural selection was responsible for differences in color pattern between male and female butterflies. Since then, it has been generally assumed that sexual selection is the major driving force in the evolution of sexual dimorphism in general and in the Lepidoptera in particular. A growing body of theory and empirical work demonstrates that natural selection alone can drive dimorphism, and that the two forces often interact in complex ways to affect the evolution of sexually dimorphic traits.
3. The evolution of sexually dimorphic life-history traits in butterflies and moths appears to have been driven largely by the evolution of different reproductive roles, accompanied by natural selection to optimize resource allocation to reproduction separately in each sex. Because both males and females can invest heavily in reproduction, sex-specific optima can set the stage for mate competition and sexual selection on life-history traits.
4. Sex differences in reproductive roles may be a source of sex-specific predation or of other forms of natural selection that affect males and females differently. Recent analyses show how sex-limited mimicry and sex differences in cryptic coloration may be associated with natural selection acting on females, rather than sexual selection acting on males.
5. Although sexual selection affects the evolution of many dimorphic traits, particularly those involved in chemical or visual signaling, we have only limited information concerning the strength or direction of sexual selection acting on these traits in the Lepidoptera.

DISCLOSURE STATEMENT

The authors are not aware of any affiliations, memberships, funding, or financial holdings that might be perceived as affecting the objectivity of this review.

ACKNOWLEDGMENTS

We thank D.J. Emlen, J. Rolff, and two anonymous reviewers for insightful comments on earlier versions of this manuscript, D.J. Emlen for help with preparation of **Figure 1**, and the Dutch Scientific Organization (NWO) for funding to C.E.A. (ALW 813.04.002) in developing this project. We apologize to all authors whose works could not be cited in this review because of space limitations. Address listed for C.E. Allen is her current affiliation. Current affiliation for B.J. Zwaan is Genetics, Plant Sciences Group, Wageningen University and Research Centre, 6708 PB Wageningen, The Netherlands. Current affiliation of P.M. Brakefield is University Museum of Zoology, Cambridge, Cambridge CB2 3EJ United Kingdom.

LITERATURE CITED

1. Allen JA. 1988. Frequency-dependent selection by predators. *Philos. Trans. R. Soc. London Ser. B* 319:485–503
2. Almbro M, Kullberg C. 2009. The downfall of mating: the effect of mate-carrying and flight muscle ratio on the escape ability of a pierid butterfly. *Behav. Ecol. Sociobiol.* 63:413–20
3. Altermatt F, Baumeyer A, Ebert D. 2009. Experimental evidence for male-biased flight-to-light behavior in two moth species. *Entomol. Exp. Appl.* 130:259–65
4. Andersson M. 1984. *Sexual Selection*. Princeton, NJ: Princeton Univ. Press

5. Arnqvist G. 1998. Comparative evidence for the evolution of genitalia by sexual selection. *Nature* 393:784–86
6. Arnqvist G, Rowe L. 2005. *Sexual Conflict*. Princeton, NJ: Princeton Univ. Press
7. Beccaloni GW, Scoble MJ, Robinson GS, Pitkin B, eds. 2003. *The Global Lepidoptera Names Index* (LepIndex). **http://www.nhm.ac.uk/entomology/lepindex** [Accessed April 14, 2010]
8. Beldade P, McMillan WO, Papanicolaou A. 2008. Butterfly genomics eclosing. *Heredity* 100:150–57
9. Beltran M, Jiggins CD, Brower AVZ, Bermingham E, Mallet J. 2007. Do pollen feeding, pupal-mating and larval gregariousness have a single origin in *Heliconius* butterflies? Inferences from multilocus DNA sequence data. *Biol. J. Linn. Soc.* 92:221–39
10. Bengtsson BO, Lofstedt C. 2007. Direct and indirect selection in moth pheromone evolution: population genetical simulations of asymmetric sexual interactions. *Biol. J. Linn. Soc.* 90:117–23
11. Berger D, Walters R, Gotthard K. 2006. What keeps insects small?—Size dependent predation on two species of butterfly larvae. *Evol. Ecol.* 20:575–89
12. Berger D, Walters R, Gotthard K. 2008. What limits insect fecundity? Body size- and temperature-dependent egg maturation and oviposition in a butterfly. *Funct. Ecol.* 22:523–29
13. Berglund A, Bisazza A, Pilastro A. 1996. Armaments and ornaments: an evolutionary explanation of traits of dual utility. *Biol. J. Linn. Soc.* 58:385–99
14. Bezzerides AL, Iyengar VK, Eisner T. 2008. Female promiscuity does not lead to increased fertility or fecundity in an Arctiid moth (*Utetheisa ornatrix*). *J. Insect. Behav.* 21:213–21
15. Bissoondath CJ, Wiklund C. 1996. Male butterfly investment in successive ejaculates in relation to mating system. *Behav. Ecol. Sociobiol.* 39:285–92
16. Blanckenhorn WU, Dixon AFG, Fairbairn DJ, Foellmer MW, Gibert P, et al. 2007. Proximate causes of Rensch's rule: Does sexual size dimorphism in arthropods result from sex differences in development time? *Am. Nat.* 169:245–57
17. Boggs CL, Freeman KD. 2005. Larval food limitation in butterflies: effects on adult resource allocation and fitness. *Oecologia* 144:353–61
18. Bolnick DI, Doebeli M. 2003. Sexual dimorphism and adaptive speciation: two sides of the same ecological coin. *Evolution* 57:2433–49
19. Bonduriansky R. 2001. The evolution of mate choice in insects: a synthesis of ideas and evidence. *Biol. Rev.* 76:305–39
20. Boppre M. 1984. Chemically mediated interactions between butterflies. In *The Biology of Butterflies*, ed. R Vane-Wright, PR Ackery, pp. 259–75. Symp. R. Entomol. Soc. London, No. 11. London: Academic
21. Cardoso MZ, Roper JJ, Gilbert LE. 2009. Prenuptial agreements: Mating frequency predicts gift-giving in *Heliconius* species. *Entomol. Exp. Appl.* 131:109–14
22. Casey TM, Joos BA. 1983. Morphometrics, conductance, thoracic temperature, and flight energetics of noctuid and geometrid moths. *Physiol. Zool.* 56:160–73
23. Chai P, Srygley RB. 1990. Predation and the flight, morphology, and temperature of Neotropical rain-forest butterflies. *Am. Nat.* 135:748–65
24. Chapman T. 2006. Evolutionary conflicts of interest between males and females. *Curr. Biol.* 16:R744–54
25. Charlat S, Reuter M, Dyson EA, Hornett EA, Duplouy A, et al. 2007. Male-killing bacteria trigger a cycle of increasing male fatigue and female promiscuity. *Curr. Biol.* 17:273–77
26. Chenoweth SF, Rundle HD, Blows MW. 2008. Genetic constraints and the evolution of display trait sexual dimorphism by natural and sexual selection. *Am. Nat.* 171:22–34
27. Clark R, Vogler AP. 2009. A phylogenetic framework for wing pattern evolution in the mimetic mocker swallowtail *Papilio dardanus*. *Mol. Ecol.* 18:3872–84
28. Clarke CA, Sheppard PM. 1962. Disruptive selection and its effect on a metrical character in butterfly *Papilio dardanus*. *Evolution* 16:214–26
29. Codella SG Jr, Lederhouse RC. 1989. Intersexual comparison of mimetic protection in the black swallowtail butterfly, *Papilio polyxenes*: experiments with captive blue jay predators. *Evolution* 43:410–20
30. Condamin M. 1973. *Monographie du genre Bicyclus (Lepidoptera Satyridae)*. Dakar: Ifan-Dakar
31. Cook PA, Wedell N. 1996. Ejaculate dynamics in butterflies: a strategy for maximizing fertilization success? *Proc. R. Soc. Lond. B* 263:1047–51

32. Costanzo K, Monteiro A. 2007. The use of chemical and visual cues in female choice in the butterfly *Bicyclus anynana*. *Proc. R. Soc. Lond. B* 274:845–51
33. Croft DP, Morrell LJ, Wade AS, Piyapong C, Ioannou CC, et al. 2006. Predation risk as a driving force for sexual segregation: a cross-population comparison. *Am. Nat.* 167:867–78
34. Crowley PH. 2000. Sexual dimorphism with female demographic dominance: age, size, and sex ratio at maturation. *Ecology* 81:2592–605
35. de Jong G, van Noordwijk AJ. 1992. Acquisition and allocation of resources: genetic (co)variances, selection, and life histories. *Am. Nat.* 139:749–70
36. **Ellers J, Boggs CL. 2003. The evolution of wing color: Male mate choice opposes adaptive wing color divergence in *Colias* butterflies. *Evolution* 57:1100–6**
37. Esperk T, Tammaru T. 2006. Determination of female-biased sexual size dimorphism in moths with a variable instar number: the role of additional instars. *Eur. J. Entomol.* 103:575–86
38. Esperk T, Tammaru T, Nylin S, Teder T. 2007. Achieving high sexual size dimorphism in insects: Females add instars. *Ecol. Entomol.* 32:243–56
39. Estrada C, Yildizhan S, Schulz S, Gilbert LE. 2010. Sex-specific chemical cues from immatures facilitate the evolution of mate guarding in Heliconius butterflies. *Proc. R. Soc. Lond. B* 277:407–13
40. Fairbairn DJ. 1997. Allometry for sexual size dimorphism: pattern and process in the coevolution of body size in males and females. *Annu. Rev. Ecol. Syst.* 28:659–87
41. Ferkau C, Fischer K. 2006. Costs of reproduction in male *Bicyclus anynana* and Pieris napi butterflies: effects of mating history and food limitation. *Ethology* 112:1117–27
42. Fischer K, Fiedler K. 2000. Sex-related differences in reaction norms in the butterfly *Lycaena tityrus* (Lepidoptera: Lycaenidae). *Oikos* 90:372–80
43. Fischer K, Fiedler K. 2001. Dimorphic growth patterns and sex-specific reaction norms in the butterfly Lycaena hippothoe sumadiensis. *J. Evol. Biol.* 14:210–18
44. Fischer K, Fiedler K. 2001. Sexual differences in life-history traits in the butterfly *Lycaena tityrus*: a comparison between direct and diapause development. *Entomol. Exp. Appl.* 100:325–30
45. Fischer K, Zwaan BJ, Brakefield PM. 2007. Realized correlated responses to artificial selection on preadult life-history traits in a butterfly. *Heredity* 98:157–64
46. Friberg M, Vongvanich N, Borg-Karlson A-K, Kemp DJ, Merilaita S, Wiklund C. 2008. Female mate choice determines reproductive isolation between sympatric butterflies. *Behav. Ecol. Sociobiol.* 62:873–86
47. Gage MJG, Parker GA, Nylin S, Wiklund C. 2002. Sexual selection and speciation in mammals, butterflies and spiders. *Proc. R. Soc. Lond. B* 269:2309–16
48. Gavrilets S, Arnqvist G, Friberg U. 2001. The evolution of female mate choice by sexual conflict. *Proc. R. Soc. Lond. B* 268:531–39
49. Gotthard K. 2008. Adaptive growth decisions in butterflies. *BioScience* 58:222–30
50. Groot AT, Horovitz JL, Hamilton J, Santangelo RG, Schal C, Gould F. 2006. Experimental evidence for interspecific directional selection on moth pheromone communication. *Proc. Natl. Acad. Sci. USA* 103:5858–63
51. Hedrick AV, Temeles EJ. 1989. The evolution of sexual dimorphism in animals: hypotheses and tests. *Trends Ecol. Evol.* 4:136–38
52. Herrell J, Hazel W. 1995. Female-limited variability in mimicry in the swallowtail butterfly *Papilio polyxenes* Fabr. *Heredity* 75:106–10
53. Ide JY. 2006. Sexual and seasonal differences in the frequency of beak marks on the wings of two *Lethe* butterflies. *Ecol. Res.* 21:453–59
54. Jia FY, Greenfield MD. 1997. When are good genes good? Variable outcomes of female choice in wax moths. *Proc. R. Soc. Lond. B* 264:1057–63
55. Joron M. 2003. Mimicry. In *Encyclopedia of Insects*, ed. RT Carde, VH Resh, pp. 714–26. New York: Academic
56. Joron M. 2005. Polymorphic mimicry, microhabitat use, and sex-specific behaviour. *J. Evol. Biol.* 18:547–66
57. Karlsson B. 1998. Nuptial gifts, resource budgets, and reproductive output in a polyandrous butterfly. *Ecology* 79:2931–40

36. Demonstrates how male mate preferences limit the evolution of female wing color in response to natural selection, thereby maintaining polymorphism in females.

58. Kemp DJ. 2007. Female butterflies prefer males bearing bright iridescent ornamentation. *Proc. R. Soc. Lond. B* 274:1043–47

59. Kemp DJ. 2007. Female mating biases for bright UV iridescence in a butterfly *Eurema hecabe* (Pieridae). *Behav. Ecol.* 19:1–8

60. Kemp DJ. 2008. Resource-mediated condition dependence in a sexually dichromatic butterfly. *Evolution* 62:2346–58

61. Kemp DJ, Rutowski RL. 2007. Condition dependence, quantitative genetics, and the potential signal content of iridescent UV butterfly coloration. *Evolution* 61:168–83

62. Kingsolver JG. 1995. Fitness consequences of seasonal polyphenism in Western White butterflies. *Evolution* 49:942–54

63. Kingsolver JG. 1995. Viability selection on seasonally polyphenic traits: wing melanin pattern in Western White butterflies. *Evolution* 49:932–41

64. Kingsolver JG. 1999. Experimental analyses of wing size, flight, and survival in the Western White Butterfly. *Evolution* 53:1479–90

65. Kingsolver JG, Gomulkiewicz R. 2003. Environmental variation and selection on performance curves. *Integr. Comp. Biol.* 43:470–77

66. Kottler MJ. 1980. Darwin, Wallace, and the origin of sexual dimorphism. *P. Am. Philos. Soc.* 124:203–26

67. Kronforst MR, Young LG, Kapan DD, McNeely C, O'Neill RJ, Gilbert LE. 2006. Linkage of butterfly mate preference and wing color preference cue at the genomic location of wingless. *Proc. Natl. Acad. Sci. USA* 103:6575–80

68. Kunte K. 2008. Mimetic butterflies support Wallace's model of sexual dimorphism. *Proc. R. Soc. Lond. B* 275:1617–24

69. Kunte K. 2009. Female-limited mimetic polymorphism: a review of theories and a critique of sexual selection as balancing selection. *Anim. Behav.* 78:1029–36

70. Kunte K. 2009. The diversity and evolution of Batesian mimicry in *Papilio* swallowtail butterflies. *Evolution* 63:2707–16

71. Lande R. 1980. Sexual dimorphism, sexual selection, and adaptation in polygenic characters. *Evolution* 34:292–305

72. Lande R. 1981. Models of speciation by selection on polygenic traits. *Proc. Natl. Acad. Sci. USA* 78:3721–25

73. Lederhouse RC, Scriber JM. 1996. Intrasexual selection constrains the evolution of the dorsal color pattern of male black swallowtail butterflies, *Papilio polyxenes*. *Evolution* 50:717–22

74. Leimar O. 2005. The evolution of phenotypic polymorphism: randomized strategies versus evolutionary branching. *Am. Nat.* 165:669–81

75. Liénard MA, Strandh M, Hedenström E, Johansson T, Löfstedt C. 2008. Key biosynthetic gene subfamily recruited for pheromone production prior to the extensive radiation of Lepidoptera. *BMC Evol. Biol.* 8:270

76. Löfstedt C. 1993. Moth pheromone genetics and evolution. *Philos. Trans. R. Soc. B* 340:167–77

77. Lyytinen A, Lindström L, Mappes J. 2004. Ultraviolet reflection and predation risk in diurnal and nocturnal Lepidoptera. *Behav. Ecol.* 15:982–87

78. Mallet J, Joron M. 1999. Evolution of diversity in warning color and mimicry: polymorphisms, shifting balance, and speciation. *Annu. Rev. Ecol. Syst.* 30:201–33

79. Marden JH, Chai P. 1991. Aerial predation and butterfly design: how palatability, mimicry, and the need for evasive flight constrain mass allocation. *Am. Nat.* 138:15–36

80. McNamara KB, Elgar MA, Jones TM. 2009. Large spermatophores reduce female receptivity and increase male paternity success in the almond moth, *Cadra cautella*. *Anim. Behav.* 77:931–36

81. Mellström HL, Wiklund C. 2009. Males use sex pheromone assessment to tailor ejaculates to risk of sperm competition in a butterfly. *Behav. Ecol.* 20:1147–51

82. Mendoza-Cuenca L, Macías-Ordóñez R. 2005. Foraging polymorphism in *Heliconius charitonia* (Lepidoptera: Nymphalidae): morphological constraints and behavioural compensation. *J. Trop. Ecol.* 21:407–15

59. Discusses how to demonstrate conclusively that sexual selection drives exaggerated male traits, and tests this empirically in a butterfly where females prefer bright iridescent wing patterns of males.

66. Details the long discussion between Darwin and Wallace about the origins of sexual dimorphism and highlights their views on the evolution of female-limited mimicry and crypsis in butterflies.

68. Uses *Papilio* phylogeny to directly test whether sexual or natural selection drives the evolution of female-limited mimicry.

83. Merckx T, Karlsson B, Van Dyck H. 2006. Sex- and landscape-related differences in flight ability under suboptimal temperatures in a woodland butterfly. *Funct. Ecol.* 20:436–41
84. Miller JC, Janzen DH, Hasswachs W. 2007. *100 Butterflies and Moths*. Cambridge, MA: Belknap Press of Harvard Univ. Press
85. Molleman F, Grunsven RHA, Liefting M, Zwaan BJ, Brakefield PM. 2005. Is male puddling behavior of tropical butterflies targeted at sodium for nuptial gifts or activity? *Biol. J. Linn. Soc.* 86:345–61
86. Morton ES. 2009. The function of multiple mating by female promethea moths, *Callosamia promethea* (Drury) (Lepidoptera: Saturniidae). *Am. Midl. Nat.* 162:7–18
87. Mutanen M, Kaitala A. 2006. Genital variation in a dimorphic moth *Selenia tetralunaria* (Lepidoptera, Geometridae). *Biol. J. Linn. Soc.* 87:297–307
88. Nieberding CM, de Vos H, Schneider MV, Lassance JM, Estramil N, et al. 2008. The male sex pheromone of the butterfly *Bicyclus anynana*: towards an evolutionary analysis. *PLOS One* 3:e2751
89. Nielsen MG, Watt WB. 2000. Interference competition and sexual selection promote polymorphism in Colias (Lepidoptera, Pieridae). *Funct. Ecol.* 14:718–30
90. Niitepõld K. 2009. *Flight Metabolic Rate in the Glanville Fritillary Butterfly*. Helsinki: Univ. Helsinki
91. **Nylin S, Wiklund C, Wickman PO, Garcia-Barros E. 1993. Absence of trade-offs between sexual size dimorphism and early male emergence in a butterfly. *Ecology* 74:1414–27**

91. A rare, direct test of the role of sexual versus natural selection in the evolution of protandry, using populations of *Pararge aegeria* along a latitudinal cline.

92. Ohsaki N. 1995. Preferential predation on female butterflies and the evolution of Batesian mimicry. *Nature* 378:173–75
93. Ohsaki N. 2005. A common mechanism explaining the evolution of female-limited and both-sex Batesian mimicry in butterflies. *J. Anim. Ecol.* 74:728–34
94. Oliver JC, Robertson KA, Monteiro A. 2009. Accommodating natural and sexual selection in butterfly wing pattern evolution. *Proc. R. Soc. Lond. B* 276:2369–75
95. Papke RS, Kemp DJ, Rutowski RL. 2007. Multimodal signaling: Structural UV reflectance predicts male mating success better than pheromones in the butterfly *Colias eurytheme* L. (Pieridae). *Anim. Behav.* 73:47–54
96. Penz CM, DeVries PJ. 2002. Phylogenetic analysis of *Morpho* butterflies (Nymphalidae, Morphinae): implications for classification and natural history. *Am. Mus. Novit.* 3374:1–33
97. Percy-Cunningham JE, MacDonald JA. 1987. Biology and ultrastructure of sex pheromone producing glands. In *Pheromone Biochemistry*, ed. GD Prestwich, GJ Blomquist, pp. 27–75. New York: Academic
98. Pijpe J, Brakefield PM, Zwaan BJ. 2007. Phenotypic plasticity of starvation resistance in the butterfly *Bicyclus anynana*. *Evol. Ecol.* 21:589–600
99. **Pijpe J, Brakefield PM, Zwaan BJ. 2008. Increased life span in a polyphenic butterfly artificially selected for starvation resistance. *Am. Nat.* 171:81–90**

99. Demonstrates that the evolution of different sex roles is reflected in physiological mechanisms that underlie the response to selection for a key life-history trait (starvation resistance).

100. Pivnick KA, Lavoie-Dornik J, McNeil JN. 1992. The role of the androconia in the mating-behavior of the European Skipper, *Thymelicus-Lineola*, and evidence for a male sex-pheromone. *Physiol. Entomol.* 17:260–68
101. Pliske TE. 1975. Courtship behavior of monarch butterfly, *Danaus plexippus* L. *Ann. Entomol. Soc. Am.* 68:143–51
102. Pomiankowski A, Nothiger R, Wilkins A. 2004. The evolution of the *Drosophila* sex-determination pathway. *Genetics* 166:1761–73
103. Quental TB, Patten MM, Pierce NE. 2007. Host plant specialization driven by sexual selection. *Am. Nat.* 169:830–36
104. Ranz JM, Machado CA. 2006. Uncovering evolutionary patterns of gene expression using microarrays. *Trends Ecol. Evol.* 21:29–37
105. Reeve JP, Fairbairn DJ. 1999. Change in sexual size dimorphism as a correlated response to selection on fecundity. *Heredity* 83:697–706
106. Rhainds M, Davis DR, Price PW. 2009. Bionomics of bagworms (Lepidoptera: Psychidae). *Annu. Rev. Entomol.* 54:209–26
107. Rodríguez-Loeches L, Barro A, Pérez M, Coro F. 2009. Anatomic and acoustic sexual dimorphism in the sound emission system of *Phoenicoprocta capistrata* (Lepidoptera: Arctiidae). *Naturwissenschaften* 96:531–36
108. Roelofs WL. 1995. Chemistry of sex attraction. *Proc. Natl. Acad. Sci. USA* 92:44–49

109. Roelofs WL, Rooney AP. 2003. Molecular genetics and evolution of pheromone biosynthesis in Lepidoptera. *Proc. Natl. Acad. Sci. USA* 100(Suppl. 2):14599
110. Rutowski RL. 1997. Sexual dimorphism, mating systems and ecology in butterflies. In *Mating Systems in Insects and Arachnids*, ed. JC Choe, BJ Crespi, pp. 257–72. Cambridge: Cambridge Univ. Press
111. Rutowski RL. 2000. Variation of eye size in butterflies: inter- and intraspecific patterns. *J. Zool.* 252:187–95
112. Sanderford MV. 2009. Acoustic courtship in the Arctiidae. In *Tiger Moths and Woolly Bears: Behavior, Ecology, and Evolution of the Arctiidae*, pp. 193–206. Oxford: Oxford Univ. Press
113. **Shine R. 1989. Ecological causes for the evolution of sexual dimorphism: a review of the evidence. *Q. Rev. Biol.* 64:419–41**
114. Shuster SM, Wade MJ. 2003. *Mating Systems and Strategies*. Princeton, NJ: Princeton Univ. Press
115. Simmons LW. 2001. *Sperm Competition and Its Evolutionary Consequences in the Insects*. Princeton, NJ: Princeton Univ. Press
116. Skals N, Anderson P, Kanneworff M, Lofstedt C, Surlykke A. 2005. Her odours make him deaf: cross-modal modulation of olfaction and hearing in a male moth. *J. Exp. Biol.* 208:595–601
117. Slatkin M. 1984. Ecological causes of sexual dimorphism. *Evolution* 38:622–30
118. Solensky MJ, Oberhauser KS. 2009. Male monarch butterflies, *Danaus plexippus*, adjust ejaculates in response to intensity of sperm competition. *Anim. Behav.* 77:465–72
119. Sperling FAH. 1994. Sex-linked genes and species-differences in Lepidoptera. *Can. Entomol.* 126:807–18
120. Srygley RB. 2001. Sexual differences in tailwind drift compensation in *Phoebis sennae* butterflies (Lepidoptera: Pieridae) migrating over seas. *Behav. Ecol.* 12:607–11
121. Srygley RB, Chai P. 1990. Flight morphology of Neotropical butterflies: palatability and distribution of mass to the thorax and abdomen. *Oecologia* 84:491–99
122. Srygley RB, Chai P. 1990. Predation and the elevation of thoracic temperature in brightly colored Neotropical butterflies. *Am. Nat.* 135:766–87
123. Stearns SC. 1992. *The Evolution of Life Histories*. New York: Oxford Univ. Press
124. Stillwell RC, Blanckenhorn WU, Teder T, Davidowitz G, Fox CW. 2010. Sex differences in phenotypic plasticity affect variation in sexual size dimorphism in insects: from physiology to evolution. *Annu. Rev. Entomol.* 55:227–45
125. Stoehr AM. 2010. Responses of disparate phenotypically-plastic, melanin-based traits to common cues: limits to the benefits of adaptive plasticity? *Evol. Ecol.* 24:287–98
126. Strobbe F, McPeek MA, DeBlock M, DeMeester L, Stoks R. 2009. Survival selection on escape performance and its underlying phenotypic traits: a case of many-to-one mapping. *J. Evol. Biol.* 22:1172–82
127. Svensson GP, Lofstedt C, Skals N. 2004. The odour makes the difference: Male moths attracted by sex pheromones ignore the threat by predatory bats. *Oikos* 104:91–97
128. Svensson GP, Löfstedt C, Skals N. 2007. Listening in pheromone plumes: disruption of olfactory-guided mate attraction in a moth by a bat-like ultrasound. *J. Insect. Sci.* 7:59
129. Svensson MGE, Marling E, Lofqvist J. 1998. Mating behavior and reproductive potential in the turnip moth *Agrotis segetum* (Lepidoptera: Noctuidae). *J. Insect. Behav.* 11:343–59
130. Symonds MRE, Elgar MA. 2008. The evolution of pheromone diversity. *Trends Ecol. Evol.* 23:220–28
131. Tammaru T, Esperk T, Ivanov V, Teder T. 2010. Proximate sources of sexual size dimorphism in insects: locating constraints on larval growth schedules. *Evol. Ecol.* 24:161–75
132. Temeles EJ, Pan IL, Brennan JL, Horwitt JN. 2000. Evidence for ecological causation of sexual dimorphism in a hummingbird. *Science* 289:441–43
133. Thomas ML, Simmons LW. 2009. Sexual selection on cuticular hydrocarbons in the Australian field cricket, Teleogryllus oceanicus. *BMC Evol. Biol.* 9:162
134. Torres-Vila LM, Rodríguez-Molina MC, McMinn M, Rodríguez-Molina A. 2005. Larval food source promotes cyclic seasonal variation in polyandry in the moth *Lobesia botrana*. *Behav. Ecol.* 16:114–22
135. Traut W, Sahara K, Marec F. 2008. Sex chromosomes and sex determination in Lepidoptera. *Sex Dev.* 1:332–46
136. Turlure C, Van Dyck H. 2009. On the consequences of aggressive male mate-locating behavior and microclimate for female host plant use in the butterfly Lycaena hippothoe. *Behav. Ecol. Sociobiol.* 63:1581–91

113. Remains the best synthesis of ideas about the role that natural selection plays in generating and maintaining sexual dimorphism.

137. Välimäki P, Kaitala A. 2007. Life history tradeoffs in relation to the degree of polyandry and developmental pathway in *Pieris napi* (Lepidoptera, Pieridae). *Oikos* 116:1569–80
138. Van Dyck H, Matthysen E, Wiklund C. 1998. Phenotypic variation in adult morphology and pupal color within and among families of the speckled wood butterfly *Pararge aegeria*. *Ecol. Entomol.* 23:465–72
139. Van Dyck H, Wiklund C. 2002. Seasonal butterfly design: morphological plasticity among three developmental pathways relative to sex, flight and thermoregulation. *J. Evol. Biol.* 15:216–25
140. Vane-Wright RI. 1975. An integrated classification for polymorphism and sexual dimorphism in butterflies. *J. Zool.* 177:329–37
141. Waldbauer GP, Sternburg JG. 1975. Saturniid moths as mimics: an alternative interpretation of attempts to demonstrate mimetic advantage in nature. *Evolution* 29:650–58
142. Wedell N. 2005. Female receptivity in butterflies and moths. *J. Exp. Biol.* 208:3433–40
143. Weller SJ, Jacobson NL, Conner WE. 1999. The evolution of chemical defences and mating systems in tiger moths (Lepidoptera: Arctiidae). *Biol. J. Linn. Soc.* 68:557–78
144. Wiklund C. 2003. Sexual selection and evolution of butterfly mating systems. In *Butterflies: Ecology and Evolution Taking Flight*, ed. CL Boggs, WB Watt, PR Ehrlich, pp. 67–90. Chicago: Univ. Chicago Press
145. Wiklund C, Karlsson B, Leimar O. 2001. Sexual conflict and cooperation in butterfly reproduction: a comparative study of polyandry and female fitness. *Proc. R. Soc. Lond. B* 268:1661–67
146. Wiklund C, Nylin S, Forsberg J. 1991. Sex-related variation in growth-rate as a result of selection for large size and protandry in a bivoltine butterfly, *Pieris napi*. *Oikos* 60:241–50
147. Wyatt TD. 2003. *Pheromones and Animal Behaviour*. Cambridge, UK: Cambridge Univ. Press
148. Zijlstra WG, Kesbeke F, Zwaan BJ, Brakefield PM. 2002. Protandry in the butterfly *Bicyclus anynana*. *Evol. Ecol. Res.* 4:1229–40
149. Zwaan BJ, Zijlstra WG, Keller M, Pijpe J, Brakefield PM. 2008. Potential constraints on evolution: sexual dimorphism and the problem of protandry in the butterfly *Bicyclus anynana*. *J. Genet.* 87:395–405

Forest Habitat Conservation in Africa Using Commercially Important Insects

Suresh Kumar Raina,[1] Esther Kioko,[2] Ole Zethner,[3] and Susie Wren[1]

[1]Commercial Insects Programme, International Center of Insect Physiology and Ecology, 00100 Nairobi, Kenya; email: sraina@icipe.org; swren@icipe.org

[2]National Museums of Kenya, 00100 Nairobi, Kenya; email: ekioko@museums.or.ke

[3]Independent Environmental Services Professional, Kongevej 4, 4450 Jyderup, Denmark; email: zet-kou@vip.cybercity.dk

Annu. Rev. Entomol. 2011. 56:465–85

First published online as a Review in Advance on September 7, 2010

The *Annual Review of Entomology* is online at ento.annualreviews.org

This article's doi:
10.1146/annurev-ento-120709-144805

0066-4170/11/0107-0465$20.00

Key Words

biodiversity, forest rejuvenation, pollination services, behavioral biology, value chain approach

Abstract

African forests, which host some of the world's richest biodiversity, are rapidly diminishing. The loss of flora and fauna includes economically and socially important insects. Honey bees and silk moths, grouped under commercial insects, are the source for insect-based enterprises that provide income to forest-edge communities to manage the ecosystem. However, to date, research output does not adequately quantify the impact of such enterprises on buffering forest ecosystems and communities from climate change effects. Although diseases/pests of honey bees and silk moths in Africa have risen to epidemic levels, there is a dearth of practical research that can be utilized in developing effective control mechanisms that support the proliferation of these commercial insects as pollinators of agricultural and forest ecosystems. This review highlights the critical role of commercial insects within the environmental complexity of African forest ecosystems, in modern agroindustry, and with respect to its potential contribution to poverty alleviation and pollination services. It identifies significant research gaps that exist in understanding how insects can be utilized as ecosystem health indicators and nurtured as integral tools for important socioeconomic and industrial gains.

INTRODUCTION

Forest habitat conservation influences the life cycle of many insect species and is critical to sustaining beneficial insects such as pollinators, which in turn contribute to forest rejuvenation (2, 66, 121). Threats to insect biodiversity are escalating with the loss of habitats due to increasing pressures from encroachment, overuse, and burning and felling of trees (96, 109, 119). Although insects are a critical natural resource in ecosystems, particularly those of forests, general understanding of their activities and their interactions with animal and plant species within the ecosystem is poor. In addition to their roles as efficient pollinators and natural/biological pest control agents, some insect species are important indicators in ecosystem management (18). Beneficial insects, often termed commercial insects (107, 109), include silk moths and honey bees. These can generate income for forest-dwelling and forest-adjacent human communities and can offer important forest conservation incentives at the community level.

This review is based on a thorough study of the literature produced from 1995 to 2010 on the characteristics and application of commercial insects to ecosystem management and enterprise development. Although the volume of this material is significant, it is apparent that there are research gaps. These gaps are filled partially by considerable research that has been conducted, but not yet officially published, on the pivotal and integral role of commercial insects in forest conservation and poverty alleviation in Africa. This review places specific emphasis on using improved understanding of the classification and behavioral biology of commercial insects in optimizing their productivity and their utilization as a driving force for forest conservation. It further emphasizes the utilization of pollination services in modern agriculture systems and livelihood diversification strategies that use the integrated value chain approach as a tool to assist African governments to conserve valuable forest ecosystems (3, 5, 42, 109, 114). Several authors have supported conservation strategies for protected forests through the direct involvement of local communities that have traditionally depended on forest resources (73, 99, 128). These strategies have helped to stimulate community-driven forest management efforts, thus reversing tropical forest degradation through provision of incentives to local communities (13). To this effect, this review verifies the direct links between commercial insects and forest conservation and identifies the main drivers that are integral to future forest conservation. Significant gaps exist in the body of published research material on the impact of climate change on pollinator species in African forest systems, and the extent to which increasing habitat stress affects the diversity and health of these populations, particularly the relationship between overall health and resilience of forest systems and climate fluctuations. There is a lack of suitable impact assessment methodologies for measuring these relationships and for evaluating the extent to which they buffer forest ecosystems and forest-edge communities from the effects of climate change.

Participatory approaches are effective components of sustainable forest co-management systems that lead to environmental and socioeconomic gains. In Africa, research approaches that mirror the two pillars of community-based development have resulted in equitable benefit sharing and improved conservation of degraded African forests. These are (*a*) identification and selection of the best potential commercial insect species and sources of food plants for pollination services, and (*b*) a value chain approach that transforms research into goods and services through training and development of sustainable and ethical community-driven enterprises based on forest resources (13, 109).

IDENTIFICATION AND CATEGORIZATION OF COMMERCIAL INSECTS

Commercial insects are divided into various categories on the basis of their morphologies and modes of development. Two specific insect

Table 1 Potential commercial insect species (honey bees and silk moths) in Africa

Family	Scientific Name	Status of commercial use	Country(ies)
Silk moths (Lepidoptera)			
Lasiocampidae	*Gonometa regia* Aurivillius	Commercially used	Uganda, Kenya
	Gonometa nysa Druce	Commercially used	Cameroon, Nigeria, Congo, Ivory Coast
	Gonometa rufobrunnea Aurivillius	Commercially used	South Africa, Namibia
	Gonometa nigrottoi	Commercially used	Kenya
	Gonometa podocarpi Aurivillius	Commercially used	Uganda, Kenya
	Gonometa titan Holland	Commercially used	Nigeria
	Leichriolepsis leucostigma Humpner	Commercially used	Uganda, Kenya
	Pachymeta stigma Strand	Commercially used	Uganda
	Pachypasa subfascia Walker	Commercially used	Uganda, Kenya, Tanzania
	Borocera cajani Vinson	Commercially used	Madagascar
Saturniidae	*Argema mimosae* Boisduval	Commercially used	South Africa, Malawi, Mozambique
	Argema besanti Rebel	Has potential for commercial use	Kenya
	Epiphora mythmnia Westwood	Has potential for commercial use	Malawi, South Africa, Mozambique, Kenya
	Epiphora bauhiniae Guérin-Méneville	Commercially used	Ghana, Gambia, Nigeria, Sudan
Thaumetopoeidae	*Anaphe carteri* Walsm	Commercially used	Uganda, Nigeria
	Anaphe moloneyi Druce	Commercially used	Uganda, Nigeria
	Anaphe panda Boisduval	Commercially used	Uganda, Tanzania, Kenya, Nigeria
	Anaphe reticulata Walker	Commercially used	Nigeria, Uganda, Kenya
	Anaphe venata Butler	Commercially used	Uganda
	Anaphe vuillet De Jouan	Commercially used	Congo, Sudan, Nigeria
Honey bees (Hymenoptera)			
Apidae	*Apis mellifera scutellata* L.	Commercially used	Most of Africa
	Apis mellifera monticola L.	Commercially used	Kenya, Sudan, Uganda
	Apis mellifera litorea L.	Commercially used	Kenya, Mozambique
	Apis mellifera capensis L.	Commercially used	South Africa
	Apis mellifera lamarckii L.	Commercially used	Egypt
	Hypotrigona gribodoi (Magretti)	Commercially used	Kenya, Uganda
	Meliponula bocandei Spinola	Commercially used	East Africa
	Meliponula ferruginea Lepeletier	Commercially used	East Africa
	Meliponula nebulata (Smith)	–	–
	Pleibeina hildebrandti (Friese)	Has potential for commercial use	East Africa
	Dactylurina Cockerell	Has potential for commercial use	East Africa

orders have, since ancient times, played highly significant roles in a wide bionetwork (53, 72) of activities. The important groups of commercial insects, honey bees, stingless bees, and mulberry and wild silk moths, are holometabolous (**Table 1**). Bees are hymenopterans, often with some degree of social behavior; moths are lepidopterans and are often large and attractive insects (54, 129). Besides supporting conservation of the ecosystem, these species are commercially valued today as pollinators of forest, horticultural, and agricultural crops and for their marketable products such as honey, wax, silk fibers, and moth pigment (dyes) (28, 60, 109).

Honey Bees

Morphometric and molecular techniques have been used to select and identify potential species for developing income-generating enterprises to support forest habitat conservation in Africa (34, 69, 75). Honey bee populations that span the continent from South Africa to Ethiopia are classified as *Apis mellifera scutellata* Lepeletier (Apidae) and have been identified through factor analysis as one primary cluster (105). Morphocluster formation and inclusivity are highly sensitive to sampling distance intervals, and within the *A. m. scutellata* region there are subtle physical variations. Larger bees associated with high altitudes have been traditionally classified as *A. m. monticola* Smith, although recently it was revealed that the group is not homogeneous. In Kenya, three races of the honey bee, *A. mellifera*, i.e., *A. m. scutellata*, *A. m. monticola*, and *A. m. litorea*, are differentiated by size, cubital index, and abdominal color banding pattern. These differences were used to assess the extent of interbreeding and hybridization between the races under seasonal pressure (108). Using multivariate analysis, Amssalu et al. (3) determined that five statistically separable morphoclusters were established in Ethiopia occupying ecologically different areas: (*a*) *A. m. jemenitica* Ruttner in the northwest and eastern arid and semiarid lowlands; (*b*) *A. m. scutellata* L. in the west, south, and southwest humid midlands; (*c*) *A. m. bandasii* Radloff & Hepburn, in the central moist highlands; (*d*) *A. m. monticola* Smith in the northern mountainous highlands; and (*e*) *A. m. woyi-gambell* Amssalu, Nuru, Radloff & Hepburn in southwestern semiarid to subhumid lowland parts of Ethiopia. Moreover, some areas with high inter- and intracolonial variances suggest introgression among these defined honey bee populations.

Using the *Dra*I RFLP (restriction fragment length polymorphism) of the COI-COII mitochondrial region, three highly divergent lineages in the mitochondrial DNA of African honey bees have been recorded. These lineages were analyzed with 738 colonies from 64 African localities (33). The African lineage A was found in all the localities except those from northeastern Africa. In this region, two newly described lineages (O and Y), possibly originating from the Near East, were observed in high proportions. This finding suggests an important differentiation of Ethiopian and Egyptian honey bees from those of other African areas. The A lineage is also present in high proportion in populations on the Iberian Peninsula and Sicily. Furthermore, eight populations from Morocco, Guinea, Malawi, and South Africa have been assayed with six microsatellite loci and compared to a set of eight additional populations from Europe and the Middle East. The African populations displayed higher genetic variability than European populations at all microsatellite loci studied, suggesting that African populations have larger effective sizes than European populations. According to their microsatellite allele frequencies, the eight African populations cluster together but are divided into two subgroups: the populations from Morocco and those from the other African countries. The populations from southern Europe show very low levels of Africanization (Africanized honey bees are hybrids of the African honey bee and various European honey bees) at nuclear microsatellite loci.

Nuclear and mitochondrial DNA often displayed discordant patterns of differentiation in African honey bees (126). Polyandry in honey bees is beneficial for colony performance as queen bees mate with many males, creating numerous patrilines within colonies that are genetically distinct. Accumulated differences in foraging rates, food storage, and population growth led to impressive boosts in the fitness of genetically diverse colonies (57, 74). The superiority of Africanized over European honey bees in tropical and subtropical regions of the New World is well documented but poorly understood (116). The Africanized bees develop significantly higher protein levels in the hemolymph than Carniolan bees (*A. m. carnica* Pollman), with approximately 10 times the levels in bees fed with sucrose alone. The superior food conversion efficiency of Africanized bees may be one of the reasons for their superiority both in the wild and for beekeeping in tropical and subtropical regions (22).

Stingless Bees

Honey bees coexist in the forest and other ecosystems with many competitors, among them the highly social stingless bees of the genera *Melipona* (Illiger) and *Trigona* (Jurine) (Apidae) (57, 109). These are tropical bees of African origin that have dispersed to other tropical and subtropical parts of the world. They form a large group of bees, composing the tribe Meliponini, with nonfunctional stings. About 500 species of stingless bees exist worldwide, and they are the only highly social bees besides the common honey bee, *A. mellifera* (111). Currently in Africa, meliponiculture is practiced only in Kenya, Tanzania, Uganda, and Angola on a small to medium scale (20, 71). These stingless bees constitute the principal pollinators of forest, agricultural, and horticultural plant species, ensuring their survival and contributing to food security for innumerable rural households (46). Stingless bees usually live in permanent colonies that consist of the queen, which lays the eggs, drones, and thousands of workers, and they are found in many forms of nests but are commonly located in tree cavities. A few species build underground nests and some build exposed nests surrounded by hard brittle layers hanging over tree branches (26, 133). These bees show strong hygienic behavior and remove dead brood to avoid the spread of disease in the colony (79). Stingless bees use a wide range of materials including resin, sand particles, and their excrement in building their nests, which are waterproof and highly resistant to predators. In the nest, stingless bees use propolis with antimicrobial properties (87) to mummify intruders.

African Wild Silk Moths

Another group of commercial insects that inhabit the same forest ecosystem as honey bees are African wild silk moths. Potential species suitable for enterprise development have been selected globally for commercial production using morphometrics and molecular markers (41, 95). These wild silk moths include some of the largest and most spectacular of all Lepidoptera, of the family Saturniidae, such as the moon or lunar moths (*Actias luna* L. and *Argema mimosae* Boisduval), atlas moths (*Attacus atlas* L.), and emperor moths (*Saturnia pavonia* L.). There are approximately 1,861 species in 162 saturniid genera and nine subfamilies including the Cercophaninae and Oxyteninae. Saturniidae, Lasiocampidae, Thaumetopoeidae, and other wild silk moths are important sources of wild silk and/or human food in a number of cultures in Africa (62) and other parts of the world. To improve the phylogenetic framework for such studies, the relationship across all nine subfamilies plus all five tribes of the Saturniidae has been examined through four protein-coding gene regions (113). Recent studies (7) provided for the first time WildSilkbase, a catalogue of expressed sequence tags (ESTs) generated from several tissues at different developmental stages of three economically important saturniid silk moths. Currently, the database comprises 57,113 ESTs, which are clustered and assembled into 4,019 contigs and 10,019 singletons. Wild silk moth development depends on the heterogeneity of plant quality and consequently affects the value of the produced silk (30). The diversified feeding pattern encourages the coexistence of several species of silk moths with different pollination abilities in forests. A survey of *Gonometa postica* Walker (Lepidoptera: Lasiocampidae) and its host plants revealed significant differences in abundance of larvae and pupae in the Imba and Mumoni forests of eastern Kenya (30). Host plant species richness did not differ between the two forests, but their evenness was significantly higher in Imba than in Mumoni. Therefore, a study was undertaken to assess the development time and quality of cocoons of *G. postica* larvae to support conservation and commercial production of silk. Minor host plants harbor significantly more pupae of *G. postica* than the major host plants in the two forests (32). This wild silk moth species produces high-quality wild silk but is greatly affected by six species of larval parasitoids in eastern Kenya (31).

Prospects for developing a wild silk industry in Africa would be brighter if silkworm survival under mass production conditions could be improved (141). A study of the survival of *Anaphe panda* (Lepidoptera: Thaumetopoeidae) (**Figure 1**) with and without protection by net sleeves as a means to prevent access of hymenopteran parasitoids in two different forest habitats, a natural and a modified one, in the Kakamega Forest of western Kenya showed that overall cohort survival was significantly higher in the natural forest than in the modified forest, but that larval survival increased more than threefold by protection with net sleeves in both habitat types. In the modified forest, only 16.8% of unprotected larvae survived to the pupal stage and formed cocoons, whereas 62.3% survived in the same environment when protected with net sleeves (77).

A few other moths such as noctuids and sphingids are important pollinators of many tree species in various agroforestry systems. Moth pollination is particularly prevalent during warm evenings and in proximity to the moth-rich forestland. The pollination efficiency of moths was examined during four seasons by counting the number of pollen grains that individual moths deposited onto the stigma of virgin female flowers of *Silene vulgaris* (Moench) Garcke (maidenstears). Pollen receipt indicated pollination by a guild of noctuid and sphingid moths (98, 109).

Figure 1

Anaphe panda cocoon nest in the Kakamega Forest.

Mulberry Silk Moths

In several favorable geographical locations in Africa, the globally domesticated bombycid *Bombyx mori* L. (Lepidoptera: Bombycidae) has been introduced as a component of a series of rural livelihood improvement programs linked to forest conservation (107, 109, 141). It feeds exclusively on leaves of the white mulberry tree, *Morus alba* Rosales (Moraceae), and related species. Various silk moth races, mostly bivoltines, have been reared in captivity across the world for over three thousand years. Silk cocoons from these domesticated species are the source of 95% of commercial silk yarn produced in the world. *Bombyx mandarina* Moore (Lepidoptera: Bombycidae) is the only wild relative of *B. mori* found in northern India and Southeast Asia (54). The genetic potential of silkworms and mulberry cultivars germplasm from a wide sampling of commercial locations in Asia was evaluated (11, 110, 140). Several mulberry cultivars were screened in Africa to select the best varieties for different agroecological zones across the continent. The performance of the *B. mori* bivoltine hybrid Shaanshi BV-333 was evaluated on six mulberry cultivars from India, based on economically important characters related to rearing and mulberry leaf quality. Cultivars Kanva-2/M5, Thailand, Thika, and S-36 were superior in rearing performance over the other tested cultivars (1). Similarly, the economic and field performance of various bivoltine races of *B. mori* was evaluated in Africa to select for high-yielding best-suited silkworm races (93). Based on the evaluation of the raw silk produced in several locations during the long and short rainy seasons, the ICIPE I race was found to be the most economical strain for conditions in East Africa. The cleanliness and neatness percentages of raw silk were 97% and 96%, respectively (92). (Raw silk cleanliness, neatness, and evenness are the characteristics that determine its suitability for weaving.) Rheology, the science of the deformation and flow of matter, can provide a powerful tool in the quest to learn from natural polymer fiber technology (50).

The sequencing and analysis of the *B. mori* genome (81) have been important in comparative and functional genomics studies (65). It has an estimated haploid nuclear genome size of 530 Mb broken into 28 chromosomes (42). Knowledge of the secondary structure of the ribosomal RNA (rRNA) molecules is important in phylogenetic analyses and identification of evolutionarily conserved structural elements in the silkworm races (94). The organization of the silkworm races' genes in the genome and their clustering and regulatory patterns in common functions or pathways have been studied, resulting into a genome fingerprinting of the silkworm, *B. mori* (89). Thus, the biological sequences (amino-acid-coding triplet) for potential species of the silkworm help to explain the function of gene homologs and make possible several studies such as insect domestication and control of diseases and pests (1, 77).

Diseases of Bees and Silk Moths

In Africa, the spread of parasites and predators threatens the productivity of commercial insects. *Varroa* mites (Parasitiformes: Varroidae), a parasite of *A. mellifera*, and viral and bacterial diseases have caused devastating harm to honey bee populations in Africa and beyond (35, 86, 117). However, stingless bees have not yet been found to be affected by *Varroa* mites and common honey bee diseases (59). On the other hand, wild and domesticated populations of silk moths suffer from many polyhedrosis viruses, bacterial and fungal diseases, and hymenopteran and dipteran parasitoids, threatening their commercial utilization (9, 67). The mechanisms underlying the suppression of insect immunity and the impacts of different immune responses to parasites and pathogens have been elucidated (52, 123, 136). The honey bee and silk moth genomes have been sequenced to address the problems in bee and moth health and their use in pollination services and commercialization (24, 81).

BEHAVIORAL BIOLOGY FOR OPTIMIZING THE PRODUCTIVITY OF COMMERCIAL INSECTS

To assess the combined ecological and economic benefits that arise from commercial insect rearing, specific parameters of their population dynamics and behavioral biology have been studied (14, 63, 134). For instance, production of eggs in the wild silk moth *G. postica* can be optimized through the use of net-sleeved cages in an indoor environment, compared with natural outdoor rearing (91). On the other hand, egg-laying is accelerated in domesticated *B. mori* through artificial insemination and the presence of matured and fertile eupyrene spermatozoa in the vestibulum (58). Moths can detect changes in environmental carbon dioxide (CO_2) with extremely high sensitivity, leading to changes in their oviposition behavior (44). Oviposition patterns in the silk moths *Samia cynthia ricini* (Drury), *S. c. pryeri* Jordan (Lepidoptera: Saturniidae), and their reciprocal F2 hybrids were investigated under different photoperiods, and 75% of eggs were laid during scotophase (49). The photoperiod and temperature experienced during the embryonic and larval stages also determine the development of bivoltine *B. mori* (Lepidoptera Bombycidae) (40). Possibly the development of seasonal morphs in bivoltine *B. mori* is regulated by a novel function of the diapause hormone orchestrated by the ecdysteroids and juvenile hormone, the presence of which precludes metamorphic changes (132, 64, 130, 139).

Most species of social insects have singly mated queens, but in some species each queen mates with numerous males to create a colony with a genetically diverse worker force. For instance, in *A. mellifera*, polyandry by queens is an adaptation to counter diseases within their colonies (124). The age at which worker honey bees begin foraging varies under different colony conditions. Juvenile hormone mediates this behavioral plasticity, and worker interactions influence both juvenile hormone titers and age at first foraging (51). Apart from honey bees, the stingless bee *Plebeia* spp. Schwarz (Hymenoptera: Apidae: Meliponini) displays a wealth of behavioral variation, especially with respect to the oviposition process. Brood production is seasonal and adult workers occur as summer and winter bees. Division of labor is similar to other stingless bees, but cell building activity is carried out by a specialized group of workers in this species (133). In general, improved understanding of the biology and behavior of commercial insects such as bees and silk moths has greatly supported the development of adapted production systems in the forests of Africa (107).

MONITORING AND MANAGEMENT TOOLS

To date, geographic information systems (GIS) and global positioning systems (GPS) are used extensively in agroecological management and research. For instance, use of GPS and GIS in the rainforest of Kibale National Park, in Uganda, and the methodology necessary to acquire high-accuracy spatial measurements allowed swift data collection on the spatial dispersion of individual rainforest trees useful for the food of commercial insects (27). The application of GPS and associated spatial tools at various stages of planning and implementation of area-wide programs linking commercial insects with forest conservation has been used to identify the spatial distribution and genetic diversity of species in the forest. Several variations were recorded in wild silk moth populations, especially with regard to silk cocoon abundance on host and nonhost plants (76) (**Figure 2**). Two important conclusions of this study are that native forests should be managed in a sustainable way and that indigenous tree species should be used predominantly in reforestation campaigns to maintain the habitat of important insect species.

Upon application of GIS technology, the spatial distribution and nesting biology were examined for naturally occurring colonies of *A. m. scutellata* in the Okavango River Delta of Botswana. Colonies had a density of 4.2 per km^2 but exhibited considerable spatial clumping (78). GIS is also used as a visualization tool for honey characterization and as a management tool for the regulation of a protected designation of origin, enabling documentation that the entire product is traditionally and entirely manufactured within the specific region and thus acquires unique properties. This application is called SMHGIS (Sierra Morena Honey Geographic Information System) (125).

Corridors for Biodiversity Conservation

Studies in Africa showed that the biodiversity of commercial insects could be enhanced by developing corridors between major forests (23, 90). These forest corridors are highly valuable refuges for biodiversity and act as buffers during the dearth period, which is of particular importance in the context of increasing climate variations. Corridors are commonly used to connect fragments of wildlife habitat, yet the identification of conservation corridors typically neglects processes of habitat selection and movement for target organisms. Instead, corridor designs are often based on binary patterns of habitat suitability. GIS is useful in tracing and mapping these corridors. Spatial cost-benefit

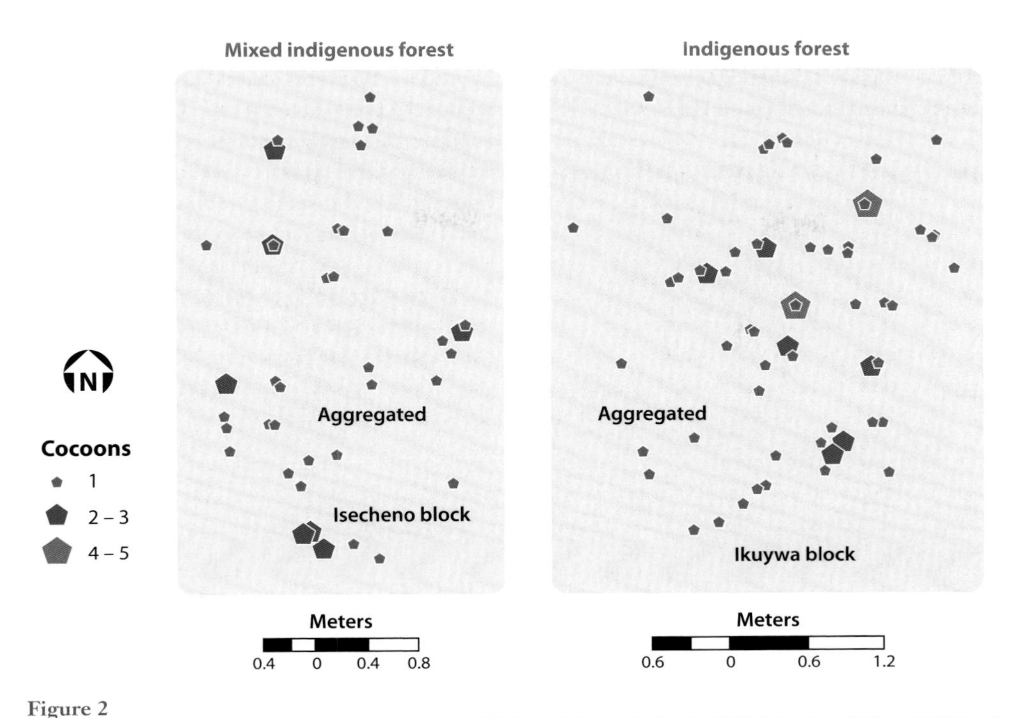

Figure 2

Aggregated distribution of the Boisduval wild silk moth *Anaphe panda* cocoon nests in Isecheno block (mixed indigenous forest) and Ikuywa block (indigenous forest) of Kakamega Forest in western Kenya (76).

analysis is a powerful mechanism for conservation planning in demonstrating the synergies between biodiversity conservation and economic development (90). Tools such as GIS offer new ways to design, implement, and study corridors as landscape linkages more objectively and holistically (23).

To improve the survival potential of existing biodiversity by increasing its resilience against mounting land-use pressures and climate fluctuations, it is essential that the full spectrum of biotopes and landscapes is conserved. Preserved areas must maintain typical species and communities, as well as provide endemic sinks and maintain movement and gene flow where possible (118). Although there has been much debate on the role of landscape corridors or linkages for animal movement and residency, there has been little exploration on their role in maintaining ecological associations. These corridors, even narrow and deep ones, are of greatest value when there is minimal anthropogenic impact. Corridors have been proposed to reduce isolation and increase population persistence in fragmented landscapes (19), yet little research has been conducted on the relationship between landscape types and corridor designs. To select the most effective designs in relationship to the landscape types supportive to commercial insects, a hypothesis was tested that assumed that corridors increase patch colonization by the buckeye butterfly, *Junonia coenia* Hübner (Lepidoptera: Nymphalidae), regardless of the butterfly's initial distance from a patch. Colonization did not change with distance in the corridor, and at long distances (128–192 m) the butterflies released in corridors were twice as likely to colonize open patches as those released in the forest (45). This work on interpatch distances, referred to as stepping stones, may be relevant for commercial insects, determining the relative effectiveness of corridors in reducing isolation in different types of fragmented landscapes. The conservation value of corridors is highest relative to other habitat configurations (23) when longer distances separate patches in fragmented landscapes.

Tracking Tools and Threat Reduction

Forest management in Africa has transitioned from reliance on traditional tree-based systems to wider approaches that protect and recognize the forest's integral role in biodiversity. To this effect, scientists working alongside government, development, and commercial sectors have collaborated in the development of ecosystem-based management approaches that maintain system complexity and function (97). Management effectiveness tracking tools (METT) have been designed to monitor progress in improving management effectiveness within forest systems. METT uses a simple questionnaire approach to monitoring and research management planning. Three protected Kenyan forest reserves in Arabuko, Kakamega, and Mwingi were subjected to METT monitoring and evaluation processes, conducted on both forest trees and biodiversity. To evaluate the effectiveness of protected areas over time, a framework was developed to integrate conservation management steps into an operational system. Moreover, effective measures were developed to raise living standards and the socioeconomic status of local communities in an environmentally compatible way through community-based commercial insect enterprise development (70). In addition to METT, researchers have applied threat reduction assessments (TRA) to provide a means of tracking the effectiveness of any project in reducing biodiversity threats. By utilizing the TRA methodology (109), it was revealed that many protected forests in eastern Africa are under serious threat from forest degradation, encroachment, fire, charcoal burning, illegal grazing, and a general overexploitation of forest resources. To enhance the relevance and impact of TRA, participatory monitoring systems have been incorporated to assist forest-edge communities to take an active role in conserving their forest resources, increase their awareness of the types and causes of risks and threats, and to mitigate factors that lead to an improvement in the welfare of these ecosystems. This approach, coupled with the development of commercial insect enterprises, has already shown positive impact on forest conservation and on the socioeconomic status of the communities living in the study locations (109).

COMMERCIAL INSECTS AS DRIVERS OF FOREST CONSERVATION

Africa's loss of forest cover and associated biodiversity has led to serious environmental deterioration, the consequences of which are, among others, changing weather patterns leading to extensive droughts that in turn result in sharply decreased food productivity and rising levels of rural poverty found across these previously rich and abundant lands (109). Empirical studies demonstrate that habitat loss has a wholly negative effect on biodiversity (12, 83). Protected areas are a cornerstone of local, regional, and global strategies for the conservation of biodiversity. However, the ecological performance of these areas, both in terms of the representation and the maintenance of key biodiversity features, remains poorly understood. The ecological performance of protected areas has been reviewed (37). Closed canopy forests in Kenya today cover less than 1.8% and woodlands cover less than 15% of the total land area, and every year these forests further decrease in size and regeneration capacity (109). Across the African continent, forests are under increasing pressure, as the number of people living within five kilometers of their boundaries, most of whom directly benefit from a whole range of forest resources, is rapidly growing. Improving forest resource management, which includes a strong biodiversity conservation component, requires a strategic mix of law enforcement and local capacity building with community participation based on incentives through the diversification of livelihood options. Agroforestry practices, such as snail farming, beekeeping, fish farming, and vegetable production, need to be considered relevant components of practical strategies aimed at addressing deforestation, particularly in highly populated areas (6). However, one of the greatest barriers to stimulating awareness and action concerning forest conservation is the lack of linkage and cooperation between forest managers, development and private sectors, and civil society (38).

LINKAGES

Several authors have demonstrated effective methods of achieving the integral goals of conserving biodiverse rich forests and improving the livelihoods of forest-adjacent communities. Research on apiculture and sericulture has developed technologies suitable to promote efficient mechanisms that support the poor in Africa, linking the communities with biodiversity resource management and providing markets for their products (61, 103, 106). Case studies from Kenya, Ghana, and Madagascar have demonstrated how commercial insect enterprises linked to innovative forest conservation practices have reduced pressures on forests and their productive natural resource base as well as their biodiversity. To ensure the economic success of the enterprises and maximize the returns to the participating communities, fully equipped market centers for silk/honey products have been developed. Training has been given to guarantee that the correct value-addition methods are used. As a result, sound market linkages have been achieved for the sale of these products (6, 29, 109, 112).

Linking Forest Rejuvenation and Pollination Services

Deforestation is caused mostly by rising human pressure, resulting in the severe loss of pollinators (47, 116), and has reduced the level and effectiveness of carbon sinks and nitrogen cycles across the globe. Research conducted over the past decade has increasingly highlighted and emphasized the need to estimate the level of pollinator decline that has resulted from the loss of forest ecosystem in Africa (55, 102). Conservation efforts in Africa now acknowledge the importance of pollinator services (28, 66) as recorded in the United States, for example, interaction of bumble bees in the preservation of threatened plant species in fragmented landscapes such as the tallgrass prairie biome in the western United States (48). Apart from insects, birds equally contribute to forest rejuvenation (127). Between 60% and 90% of plant species require an animal pollinator. The sensitivity of mobile organisms to ecological factors that operate across spatial scales makes the services provided by a given community of mobile agents, such as pollinator insects, highly contextual (66). The alkali bee *Nomia melanderi* Cockerell (Hymenoptera: Aphididae) has been used as an effective pollinator of commercial alfalfa (*Medicago sativa* L.) (Fabaceae) seed production (21) in intensively agricultural landscapes in the western United States.

There is growing evidence that natural pollination services conducted by specific species of silk moths, bees, and butterflies rejuvenate forests and give rise to new and diverse seeds and seedlings by pollen dispersal between flowers of different tree species (25). The apiculture and sericulture industries in Africa and elsewhere have strong interest in maintaining sustainable forest resources. They provide key breeding environments and nectar resources to enable a critical biological insurance for adaptation of future pollinator species and to maintain integral ecoservices as well as commercial insect productivity. The honey bee is currently undergoing extensive die-offs (mostly in the United States) because of colony collapse disorder. Researchers predict that in most regions of the world there will be a decline in agricultural production due to honey bee losses (137, 138). Tropical crops pollinated primarily by social bees are most susceptible to pollination failure from habitat loss (115). Quantifying these general relationships can help to predict the effects of land-use change on pollinator communities and crop productivity and can inform landscape conservation efforts of the type of balance that is needed between native species and people. Designing agricultural areas that integrate land use and ecosystem functions is a practical approach for promoting sustainable agricultural practices (85).

The importance of insect pollination to the forest buffer zone and agricultural fields is unequivocal. However, whether this service is provided largely by wild pollinators (genuine ecosystem service) or by managed pollinators (commercial service), and which of these requires immediate action amid reports of pollinator decline, remains contested (17). This situation has led to demands for a response by land managers, conservationists, and political decision makers to the impending global pollinator crisis. It becomes apparent that perceptions are driven mainly by reported declines of crop-pollinating honey bees in North America, and bumble bees and butterflies in Europe, whereas native pollinator communities elsewhere including Africa show mixed responses to environmental change (39).

Linking Forest Habitat Conservation to Livelihoods

Based on the information gathered on commercial insects and their role in ecosystem management, for example, in the Kakamega, Arabuko-Sokoke, and Mwingi forests of Kenya, it has been demonstrated that commercial insects are important conservation tools. By encouraging and assisting people dwelling across forest edges to develop viable enterprises based on commercial insect products, such as honey and silk, these rural communities are more motivated to conserve these highly valuable natural resources that are under significant pressure due to encroachment, overuse, and tree felling (109). Researchers have shown that proximity to a forest ensures better honey yields. For instance, honey yield is doubled in hives less than 1 km from the forest compared to those farther than 3 km (120). This factor has helped resident communities as well as researchers and conservationists to understand that forest management begins at the forest edge—the buffer zone (21, 48).

The value chain approach to commercial insect enterprise development conducted as action research initiatives in Africa is an effective and efficient method of linking commercial insects and forest conservation to achieve incentive-led, community-driven conservation, while tangibly improving the livelihoods of the participating communities. Several authors have proposed the use of the value chain approach in forest conservation (104, 122). It implies the breaking down of the series of research and business functions into strategically relevant operational activities through which utility is added to products and services (15, 68). The methodology and mechanisms used to scale up existing conservation-compatible livelihood activities in forest-adjacent villages were specifically designed to provide communities the ability to achieve lucrative and sustainable income as a direct result of activities that are essentially supportive to forest conservation, as well as to the commercial enterprises themselves (82). Direct payments show potential as an innovative tool for engaging local communities or resource users in conservation and as a mechanism for channeling global investments in biodiversity conservation services to site-based initiatives (80). Private philanthropy, premium pricing for biodiversity-related goods via certification schemes, such as the organic EU-ISO 65 and the US-NOP, and development of entirely new markets for environmental services through the value chain approach can bear the costs of often underfunded tropical conservation activities (4, 131). There are two components of value chain analysis: the industry value chain and the organization's internal value chain (101, 104). The value chain approach utilized for African forest conservation has linked commercial insect–based silk and honey production systems to the consumer market by using technology transfer, training, and product and marketplace development (107, 109) as the main tools (**Figure 3**). The economics of these enterprises has also been evaluated in the African context (106). Beneficial linkage between livelihoods and conservation has formed the basis of multidisciplinary approaches to motivate local communities to become practically involved in participatory forest management and conservation. Forest-adjacent communities have

Figure 3

Illustration of the value chain approach linking commercial insect–based products to the consumer market using technology transfer, training, products, and marketplace development in Africa.

understood this logic in their efforts to save their natural resources, demonstrated by action research initiatives conducted in eastern and western Africa over the past decade (5, 120).

INCENTIVE-LED, COMMUNITY-DRIVEN BIODIVERSITY AND FOREST HABITAT CONSERVATION

The efforts to protect ecosystems have expanded from pure biodiversity conservation to encompass measures to improve human welfare. The result is a policy shift toward allowing local resource users to benefit from sustainable use of natural resources in protected areas. Evaluation methods are being developed for economic models that incorporate site characteristics and location to predict economic returns for a variety of potential land uses (100). These studies pioneered a complementary set of practical forest activities that promote climate change mitigation and are in alignment with the recent United Nations Climate Change Conference (CO15), which took place in Copenhagen in 2009. The studies have specifically stressed the need to reduce emissions and enhance carbon sequestration from forests and buffer zones through the creation of tangible incentives that stimulate and incentivize adoption by rural communities (16, 88). Although cost-benefit analyses are common in many areas of policy, they are not typically used in conservation planning (90).

The management of tropical protected areas is a contentious issue and often leads to an unproductive polarization of viewpoints supporting either protectionist or sustainable development paradigms (56). Effective management requires a mix of private ownership, common property management, and central government involvement to maximize benefits to local people and ensure long-term protection of biodiversity (8). Case studies from Madagascar, Kenya, and Ghana provide insight into strengthening local management capacities and cooperation (36). For instance, bee diversity in Madagascar is important for both agriculture and indigenous ecosystem management (29). Several African institutions have collaborated to solve a number of biodiversity conservation issues related to the beneficial exploitation of commercial insects for forest habitat conservation and to generating productive landscape and livelihood improvements (112). Overexploitation of forest resources, driven by extreme poverty of these neighboring communities, has been responsible for the degradation of the forest.

Over the past 14 years, conservation initiatives conducted in African forests have focused on an integrated approach to forest conservation and have included biodiversity assessment, management planning, habitat restoration, and community-driven commercial insect enterprise development. Habitat monitoring in the buffer and core zones has shown that there has been no loss of habitat in the focal project forest area (43, 109). Incomes of the participating members increased by 16%–20% as a result of their commercial insect enterprise activities and pollination services.

Further work using interdisciplinary research and methodology conducted with pastoral populations in Kenya, Namibia, and Tanzania has aimed at integrating ecosystem services and their drivers (8, 84). As for European Union and United States organic certification standards, the tools developed for the International FairWild Standard now provide accredited certification for sustainable wild harvesting systems that provide market recognition through the product certification label. Compliance with these standards enables natural resource users to better manage their operations and, once certification is gained, to harness the opportunity to access attractive global markets (135).

SUMMARY POINTS

1. This review highlights some of the practical mechanisms and strategies that are in place that were proposed and initiated by several African institutions to conserve forest biodiversity and its ecosystem through commercial insect enterprises.
2. Illegal encroachment, burning, overuse, and tree felling by forest-edge communities pose threats to forest and biodiversity conservation in Africa.
3. This review provides a new way of looking at the development of strategies for sustainable management of forest habitat and its biodiversity to improve livelihood by developing enterprises using commercial insects.
4. Many key parameters in the selection of commercial insects in Africa have been evaluated.
5. The value chain and biodiversity approaches have been highlighted to establish strong linkages between commercial insects and forest conservation.

FUTURE ISSUES

1. Increasing the number and impact of commercial insect–based enterprises in Africa will provide tangible incentives to communities living around these forests to positively manage and protect their ecosystems and the pollinator species.
2. Greater research emphasis is called for in areas such as the application of insects as indicators of ecosystem health to measure the impact of climate change and the level of effectiveness of mitigation efforts.
3. Further research focus is required on the development of suitable methodologies that measure how, and to what level, commercial insect enterprises buffer forest ecosystems and forest-edge communities from the effects of climate change.
4. The rising incidences of death of honey bees and other pollinators in the ecosystems, possibly due to the use of pesticides for agricultural productivity, coupled with attacks by pests and predators, require critical thinking about the mutually beneficial integration of agricultural and ecosystem services.
5. Applied research on the breeding and management of pollinators in modern agricultural systems should be concentrated and developed to a meaningful scale where conclusive results can provide answers to this international issue.
6. To enable the rural poor to derive a livelihood from active participation in forest conservation, biologists need to understand that long-term solutions to biodiversity loss must be built around social programs that enable local people to thrive at the micro- and macroscales (10).

DISCLOSURE STATEMENT

The authors are not aware of any affiliations, memberships, funding, or financial holdings that might be perceived as affecting the objectivity of this review.

ACKNOWLEDGMENTS

We wish to thank Christian Borgemeister, Director General of *icipe*, for his critical, editorial comments. We also acknowledge the assistance received from Dolorosa Osogo, Elliud Muli, Everlyn Nguku, Sospeter Makau, and Irene Ogendo. Most of the research on commercial insects in Africa was initiated and supported by IFAD and UNDP-GEF grants.

LITERATURE CITED

1. Adolkar VV, Raina SK, Kimbu DM. 2007. Evaluation of various mulberry *Morus* spp. (Moraceae) cultivars for the rearing of the bivoltine hybrid race Shaanshi BV-333 of the silkworm *Bombyx mori* (Lepidoptera: Bombycidae). *Int. J. Trop. Insect Sci.* 27:6–14
2. Allsopp MH, de Lange WJ, Veldtman R. 2008. Valuing insect pollination services with cost of replacement. *PLoS ONE* 3(9):e3128
3. Amssalu B, Nuru A, Radloff SE, Hepburn HR. 2004. Multivariate morphometric analysis of honeybees (*Apis mellifera*) in the Ethiopian region. *Apid* 35:71–81
4. Andrew B, Whitten T. 2003. Who should pay for tropical conservation, and how could the costs be met? *Oryx* 37:238–50

5. Appiah M. 2001. Co-partnership in forest management: the Gwira-Banso joint forest management project in Ghana. *Environ. Dev. Sustain.* 3:343–60
6. Appiah M, Blay D, Damnyag L, Dwomoh FK, Pappinen A, Luukkanen O. 2009. Dependence on forest resources and tropical deforestation in Ghana. *Environ. Dev. Sustain.* 11:471–87
7. Arunkumar KP, Tomar A, Daimon T, Shimada T, Nagaraju J. 2008. An EST database of wild silk moths. *BMC Genomics* 9:338
8. Ashley C. 2000. *Applying Livelihood Approaches to Natural Resource Management Initiatives: Experiences in Namibia and Kenya*. Hong Kong: Chameleon. 30 pp.
9. Babu SM, Gopalaswamy G, Chandramohan N. 2005. Identification of an antiviral principle in *Spirulina platensis* against *Bombyx mori* nuclear polyhedrosis virus (BmNPV). *Indian J. Biotechnol.* 4:384–88
10. Baird IG, Dearden P. 2003. Biodiversity conservation and resource tenure regimes: a case study from Northeast Cambodia. *Environ. Manag.* 32:541–50
11. Banerjee R, Roychowdhuri S, Sau H, Das BK, Ghosh P, Saratchandra B. 2007. Genetic diversity and interrelationship among mulberry genotypes. *J. Genet. Genomics* 34:691–97
12. Bedward M, Ellis MV, Simpson CC. 2009. Simple modelling to assess if offsets schemes can prevent biodiversity loss, using examples from Australian woodlands. *Biol. Conserv.* 142:2732–42
13. Blay D, Appiah M, Damnyag L, Dwomoh FK, Luukkanen O, Pappinen A. 2008. Involving local farmers in rehabilitation of degraded tropical forests: some lessons from Ghana. *Environ. Dev. Sustain.* 10:503–18
14. Breed MD, Guzmán-Novoa E, Hunt GJ. 2003. Defensive behavior of honey bees: organization, genetics, and comparisons with other bees. *Annu. Rev. Entomol.* 49:271–98
15. Breite R, Vanharanta H. 2004. Value chain methodology for dynamic business environments. *Evol. Supply Chain Manag.* 3:249–64
16. Brown ME. 2009. Markets, climate change, and food security in West Africa. *Environ. Sci. Technol.* 43:8016–20
17. Buchmann SL, Nabhan GP. 1995. *The Forgotten Pollinators*. Washington, DC: Island. 292 pp.
18. Buchs W. 2003. Biotic indicators for biodiversity and sustainable agriculture. *J. Agric. Ecosyst. Environ.* 98:1–3
19. Bullock WL, Samways MJ. 2005. Conservation of flower-arthropod associations in remnant African grassland corridors in an afforested pine mosaic. *Biodivers. Conserv.* 14:3093–103
20. Byarugaba D. 2004. Stingless bees (Hymenoptera: Apidae) of Bwindi impenetrable forest, Uganda and Abayanda indigenous knowledge. *Int. J. Trop. Insect Sci.* 24:117–21
21. Cane JH. 2008. A native ground-nesting bee (*Nomia melanderi*) sustainably managed to pollinate alfalfa across an intensively agricultural landscape. *Apidologie* 39:315–23
22. Cappelari FA, Turcatto AP, Morais MM, De Jong D. 2009. Africanized honey bees more efficiently convert protein diets into hemolymph protein than do Carniolan bees (*Apis mellifera carnica*). *Genet. Mol. Res.* 8:1245–49
23. Chetkiewicz CB, St. Clair CC, Boyce MS. 2006. Corridors for conservation: integrating pattern and process. *Annu. Rev. Ecol.* 37:317–42
24. Consortium HGS. 2006. Insights into social insects from the genome of the honeybee *Apis mellifera*. *Nature* 443:931–49
25. Degen B, Roubik DW. 2004. Effects of animal pollination on pollen dispersal, selfing, and effective population size of tropical trees: a simulation study. *Biotropica* 36:165–79
26. Dollin AE, Dollin LJ, Sakagami SF. 1997. Australian stingless bees of the genus *Trigona* (Hymenoptera: Apidae). *Invertebr. Taxon.* 11:861–96
27. Dominy NJ, Duncan BW. 2001. GPS and GIS methods in an African rainforest: applications to tropical ecology and conservation. *Conserv. Ecol.* 5:537–49
28. **Eardley C, Roth D, Clarke J, Buchmann S, Gemmill B, eds. 2006. *Pollinators and Pollination: A Resource Book for Policy and Practice*. Pretoria: Afr. Pollinator Initiat. (API). 77 pp.**
29. Eardley CD, Gikungu M, Schwarz MP. 2009. Bee conservation in sub-Saharan Africa and Madagascar: diversity, status and threats. *Apidologie* 40:355–66
30. Fening KO, Kioko EN, Raina SK, Mueke JM. 2008. Monitoring wild silkmoth, *Gonometa postica* Walker, abundance, host plant diversity and distribution in Imba and Mumoni woodlands in Mwingi, Kenya. *Int. J. Biodivers. Sci. Manag.* 4:104–11

28. A key report on the importance of pollination services and policy issues on the conservation of pollinators.

31. Fening KO, Kioko EN, Raina SK, Mueke JM. 2009. Parasitoids of the African wild silkmoth, *Gonometa postica* (Lepidoptera: Lasiocampidae) in the Mwingi forests, Kenya. *J. Appl. Entomol.* 133:411–15
32. Fening KO, Kioko EN, Raina SK, Mueke JM. 2010. Effect of seasons and larval food plants on the quality of *Gonometa postica* cocoons. *Phytoparasitica* 38:111–19
33. Franck P, Garnery L, Loiseau A, Oldroyd BP, Hepburn HR, et al. 2001. Genetic diversity of the honeybee in Africa: microsatellite and mitochondrial data. *Heredity* 86:420–30
34. Francoy TM, Wittmann D, Drauschke M, Müller S, Steinhage V, et al. 2008. Identification of Africanized honey bees through wing morphometrics: two fast and efficient procedures. *Apidologie* 39:488–94
35. Frazier M, Muli E, Conklin T, Schmehl D, Torto B, et al. 2010. A scientific note on *Varroa destructor* found in East Africa; threat or opportunity? *Apidologie* 41:463–65
36. Fritz-Vietta NVM, Röttger C, Stoll-Kleemann S. 2009. Community-based management in two biosphere reserves in Madagascar—distinctions and similarities: What can be learned from different approaches? *Madag. Conserv. Dev.* 4:86–97
37. Gaston KJ, Jackson SF, Cantú-Salazar L, Cruz-Piñón G. 2008. The ecological performance of protected areas. *Annu. Rev. Ecol. Evol. Syst.* 39:93–113
38. Ghazoul J. 2001. Barriers to biodiversity conservation in forest certification. *Conserv. Biol.* 15:315–31
39. Ghazoul J. 2005. Buzziness as usual? Questioning the global pollination crisis. *Trends Ecol. Evol.* 20:367–73
40. Gianotti L, Rolla M, Arvat E, Belliti D, Valetto MR, et al. 1999. Effects of photoperiod and temperature on seasonal morph development and diapause egg oviposition in a bivoltine race (Daizo) of the silkmoth, *Bombyx mori* L. *J. Insect Physiol.* 45:101–6
41. Gogoi B, Goswami BC. 2000. Morphometric studies on muga silkworm, *Antheraea assama* Westwood. *Int. J. Wild Silk Moths Silk* 5:213–15
42. Goldsmith MR, Shimada T, Abe H. 2005. The genetics and genomics of the silkworm, *Bombyx mori*. *Annu. Rev. Entomol.* 50:71–100
43. Gordon I. 2003. Harnessing butterfly biodiversity for improving livelihoods and forest conservation: the Kipepeo Project. *J. Environ. Dev.* 12:82–98
44. Guerenstein PG, Yepez EA, van Haren J. 2004. Floral CO_2 emission may indicate food abundance to nectar-feeding moths. *Naturwissenschaften* 91:329–33
45. Haddad N. 2000. Corridor length and patch colonization by a butterfly, *Junonia coenia*. *Conserv. Biol.* 14:738–45
46. Heard TA. 1999. The role of stingless bees in crop pollination. *Annu. Rev. Entomol.* 44:183–206
47. Hillers KJ. 2004. Pollinator perils. *Curr. Biol.* 14:R1035–36
48. Hines HM, Hendrix SD. 2005. Bumble bee (Hymenoptera: Apidae) diversity and abundance in tallgrass prairie patches: effects of local and landscape floral resources. *Environ. Entomol.* 34:1477–84
49. Hitoshi S. 1993. Oviposition pattern in *Samia* silk moths (Lepidoptera: Saturniidae). *Jpn. J. Appl. Entomol. Zool.* 37:163–67
50. Holland C, Terry EA, Porter D, Vollrath F. 2007. Natural and unnatural silks. *Polymer* 48:3388–92
51. Huang ZY, Robinson GE. 1996. Regulation of honey bee division of labor by colony age demography. *Behav. Ecol. Sociobiol.* 39:147–58
52. Hurd H. 2009. Evolutionary drivers of parasite-induced changes in insect life-history traits: from theory to underlying mechanisms. *Adv. Parasitol.* 68:85–110
53. Illgner PM, Nel E, Robertson MP. 1998. Beekeeping and local self-reliance in rural southern Africa. *Geogr. Rev.* 88:349–62
54. Imms AD. 1957. *A General Textbook of Entomology: Including the Anatomy, Physiology, Development and Classification of Insects*. London: Methuen. 886 pp.
55. Ingram JC, Dawson TP. 2005. Inter-annual analysis of deforestation hotspots in Madagascar from high temporal resolution satellite observations. *Int. J. Remote Sens.* 26:1447–61
56. Jayanath A, Herath G. 2009. A critical review of multi-criteria decision making methods with special reference to forest management and planning. *Ecol. Econ.* 68:2535–48
57. Jha S, Vandermeer JH. 2009. Contrasting foraging patterns for Africanized honeybees, native bees and native wasps in a tropical agroforestry landscape. *J. Trop. Ecol.* 25:13–22

58. Karube F, Kobayashi M. 1999. Presence of eupyrene spermatozoa in vestibulum accelerates oviposition in the silkworm moth, *Bombyx mori*. *J. Insect Physiol.* 45:947–57
59. Kasina JM. 2007. *Bee pollinators and economic importance of pollination in crop production: case of Kakamega, western Kenya*. PhD thesis. Hochsch. ULB Bonn. 140 pp.
60. Kasina JM, Mburu J, Kraemer M, Holm-Mueller K. 2009. Economic benefit of crop pollination by bees: a case of Kakamega small-holder farming in western Kenya. *J. Econ. Entomol.* 102:467–73
61. Kioko EN, Raina SK, Mueke JM. 1999. Conservation of the African wild silk moths for economic incentives to rural communities of the Kakamega forest in Kenya. *Int. J. Wild Silk Moths Silk* 4:1–5
62. Kioko EN, Raina SK, Mueke JM. 2000. Survey on the diversity of wild silkmoth species in east Africa. *East Afr. J. Sci.* 2:1–6
63. Kocher SD, Tarpy DR, Grozinger CM. 2009. The effects of mating and instrumental insemination on queen honey bee flight behaviour and gene expression. *Insect Mol. Biol.* 19:153–62
64. Konopova B, Jindra M. 2008. Broad-complex acts downstream of Met in juvenile hormone signaling to coordinate primitive holometabolan metamorphosis. *Development* 135:559–68
65. Koshy N, Ponnuvel KM, Sinha RK, Qadri SMH. 2008. Silkworm nucleotide databases—current trends and future prospects. *Bioinformation* 2:308–10
66. Kremen C, Williams NM, Aizen MA, Gemmill-Herren B, Iebuhn G, et al. 2007. Pollination and other ecosystem services produced by mobile organisms: a conceptual framework for the effects of land-use change. *Ecol. Lett.* 10:299–314
67. Kumar SR, Ramanathan G, Subhakaran M, Inbaneson SJ. 2009. Antimicrobial compounds from marine halophytes for silkworm disease treatment. *Int. J. Med. Med. Sci.* 1:184–91
68. Lee H, Billington C. 1995. The evolution of supply-chain-management models and practice at Hewlett-Packard. *Interfaces* 25:42–63
69. Lobo JA. 2008. Morphometric isozymic and mitochondrial variability of Africanized honeybees in Costa Rica. *Escuela Apidol.* 39:488–94
70. Lü Y, Chen L, Fu B, Liu S. 2002. A framework for evaluating the effectiveness of protected areas: the case of Wolong Biosphere Reserve. *Landsc. Urban Plan.* 63:213–23
71. Macharia J, Raina S, Muli E. 2007. Stingless bees in Kenya. *Bees Dev. J.* 83:9
72. Makoto N, Ichizen M, Toshio M. 2001. Relations between scale of sericulture farming and scale merit worked on the farming and/or its profitability. *J. Sericult. Sci. Jpn.* 70:145–54
73. Mamo G, Sjaastad E, Vedeld P. 2007. Economic dependence on forest resources: a case from Dendi District, Ethiopia. *For. Policy Econ.* 9:916–27
74. Mattila HR, Seeley TD. 2007. Genetic diversity in honey bee colonies enhances productivity and fitness. *Science* 317:362–64
75. Mayo M, Watson AT. 2007. Automatic species identification of live moths. *Knowl.-Based Syst.* 20:195–202
76. **Mbahin N, Raina SK, Kioko EN, Mueke JM. 2007. Spatial distribution of cocoon nests and egg clusters of the silkmoth *Anaphe panda* (Lepidoptera: Thaumetopoeidae) and its host plant *Bridelia micrantha* (Euphorbiaceae) in the Kakamega Forest of western Kenya. *Int. J. Trop. Insect Sci.* 27:138–44**

76. Demonstrates the importance of GIS in biodiversity studies in reference to wild silk moths in indigenous forests.

77. Mbahin N, Raina SK, Kioko EN, Mueke JM. 2010. Use of sleeve nets to improve survival of the Boisduval silkworm, *Anaphe panda*, in the Kakamega Forest of western Kenya. *J. Insect Sci.* 10:Article 6
78. McNally LC, Schneider SS. 1996. Spatial distribution and nesting biology of colonies of the African honey bee *Apis mellifera scutellata* (Hymenoptera: Apidae) in Botswana, Africa. *Environ. Entomol.* 25:643–52
79. Medina LM, Hart AG, Ratnieks FLW. 2009. Hygienic behavior in the stingless bees *Melipona beecheii* and *Scaptotrigona pectoralis* (Hymenoptera: Meliponini). *Genet. Mol. Res.* 8:571–76
80. Milne S, Niesten E. 2009. Direct payments for biodiversity conservation in developing countries: practical insights for design and implementation. *Oryx* 43:530–41
81. Mita K, Kasahara M, Sasaki S, Nagayasu Y, Yamada T, et al. 2004. The genome sequence of silkworm, *Bombyx mori*. *DNA Res.* 11:27–35
82. **Mitchell J, Coles C, Keane J. 2009. Trading up: how a value chain approach can benefit the rural poor. *COPLA Glob. – Overseas Dev. Inst., London.* 94 pp.**

82. Thorough analysis of how the value chain approach can practically help the rural poor to gainfully participate in trade in efforts to alleviate poverty.

83. Moinde-Fockler NN, Oguge NO, Karere GM, Otina D, Suleman MA. 2007. Human and natural impacts on forests along lower Tana River, Kenya: implications towards conservation and management of endemic primate species and their habitat. *Biodivers. Conserv.* 16:1161–73
84. Monela GC, Kajembe GC, Kaoneka ARS, Kowero G. 2001. Household livelihood strategies in the miombo woodlands of Tanzania: emerging trends. *J. For. Nat. Conserv.* 73:17–33
85. Morandin LA, Winston ML, Abbott VA, Franklin MT. 2007. Can pastureland increase wild bee abundance in agriculturally intense areas? *Basic Appl. Ecol.* 8:117–24
86. Morse RA, Flottum K. 1997. *Honey Bee Pests, Predators and Diseases*. Medina, OH: A. I. Root. 575 pp.
87. Muli EM, Maingi JM, Macharia J. 2008. Antimicrobial properties of propolis and honey from the Kenyan stingless bee, *Dactylutina schimidti*. *Apiacta* 43:49–61
88. Murcia C, Kattan G. 2009. Application of science to protected area management: overcoming the barriers. *Ann. Mo. Bot. Gard.* 96:508–20
89. Nagaraja GM, Nagaraju J. 1995. Genome fingerprinting of the silkworm, *Bombyx mori*, using random arbitrary primers. *Electrophoresis* 16:1633–38
90. Naidoo R, Ricketts TH. 2006. Mapping the economic costs and benefits of conservation. *PLoS Biol.* 4:e360
91. Ngoka BM, Kioko EN, Raina SK, Mueke JM, Kimbu DM. 2007. Semi-captive rearing of the African wild silkmoth *Gonometa postica* (Lepidoptera: Lasiocampidae) on an indigenous and a non-indigenous host plant in Kenya. *Int. J. Trop. Insect Sci.* 27:183–90
92. Nguku EK, Adolkar VV, Raina SK. 2007. Evaluation of raw silk produced by bivoltine silkworm *Bombyx mori* L. (Lepidoptera: Bombycidae) races in Kenya. *J. Text. Appar. Technol. Manag.* 5:1–9
93. Nguku EK, Adolkar VV, Raina SK, Mburugu KG, Mugenda OM, Kimbu DM. Performance of six bivoltine *Bombyx mori* (Lepidoptera: Bombycidae) silkworm strains in Kenya. *Open Entomol. J.* 3:1–6
94. Niehuis O, Naumann CM, Misof B. 2005. Identification of evolutionary conserved structural elements in the mt SSU rRNA of Zygaenoidea (Lepidoptera): a comparative sequence analysis. *Org. Divers. Evol.* 6:17–32
95. Ojha NG, Reddy RM, Hansda G, Sinha MK, Suryanarayana N, Prakash V. 2009. Status and potential of jata, a new race of Indian tropical tasar silkworm (*Antheraea mylitta* Drury). *Acad. J. Entomol.* 2:80–84
96. Ostfeld RS, Logiudice K. 2003. Community disassembly, biodiversity loss, and the erosion of an ecosystem service. *Ecology* 84:1421–27
97. Perry DA. 1998. The scientific basis of forestry. *Annu. Rev. Ecol. Syst.* 29:435–66
98. Pettersson MW. 1991. Pollination by a guild of fluctuating moth populations: option for unspecialization in *Silene vulgaris*. *J. Ecol.* 79:591–604
99. Polansky C. 2003. Participatory forest management in Africa: lessons not learned. *Int. J. Sustain. Dev. World Ecol.* 10:109–18
100. Polasky S, Nelson E, Camm J, Csuti B, Fackler P, et al. 2008. Where to put things? Spatial land management to sustain biodiversity and economic returns. *Biol. Conserv.* 141:1505–24
101. Porter ME. 1985. *Competitive Advantage*. New York: Free Press
102. Potts SG, Biesmeijer JC, Kremen C, Neumann P, Schweiger O, Kunin WE. 2010. Global pollinator declines: trends, impacts and drivers. *Trends Ecol. Evol.* 25:345–53
103. Poulton C, Kydd J, Dorward A. 2006. Overcoming market constraints on pro-poor agricultural growth in sub-Saharan Africa. *Dev. Policy Rev.* 24:243–77
104. Purnomo H, Guizol P, Muhtaman DR. 2009. Governing the teak furniture business: a global value chain system dynamic modelling approach. *Environ. Model. Softw.* 24:1391–401
105. Radloff S, Hepburn R. 2000. Population structure and morphometric variance of the *Apis mellifera scutellata* group of honeybees in Africa. *Genet. Mol. Biol.* 23:305–16
106. Raina SK. 2000. *The Economics of Apiculture and Sericulture Modules for Income Generation in Africa*. Cardiff, UK: IBRA. 86 pp.
107. Raina SK. 2004. *A Practical Guide for Raising and Utilizing Silkworms and Honey Bees in Africa*. Cardiff, UK: IBRA. 164 pp.

108. Raina SK, Kimbu DM. 2005. Variations in races of the honeybee *Apis mellifera* (Hymenoptera: Apidae) in Kenya. *Int. J. Trop. Insect Sci.* 25:281–91
109. Raina SK, Kioko EN, Gordon I, Nyandiga C. 2009. *Improving Forest Conservation and Community Livelihoods through Income Generation from Commercial Insects in Three Kenyan Forests.* Nairobi, Kenya: Icipe Sci. Press. 87 pp.

109. Shows the practicality of using commercial insects as tangible incentives for forest-adjacent communities to participate in conservation.

110. Rao CGP, Seshagiri SV, Ramesh C, Ibrahim BK, Nagaraju H, Chandrashekaraiah. 2006. Evaluation of genetic potential of the polyvoltine silkworm (*Bombyx mori* L.) germplasm and identification of parents for breeding programme. *J. Zhejiang Univ. Sci.* 7:215–20
111. Rasmussen C, Cameron SA. 2006. A molecular phylogeny of the Old World stingless bees (Hymenoptera: Apidae: Meliponini) and the non-monophyly of the large genus *Trigona*. *Syst. Entomol.* 32:26–39
112. Razafimanantosoa T, Ravoahangimalala OR, Craig CL. 2006. Indigenous silk moth farming as a means to support Ranomafana National Park. *Madag. Conserv. Dev.* 1:34–39
113. Regier JC, Grant MC, Mitter C, Cook CP, Peigler RS, Rougerie R. 2008. Phylogenetic relationships of wild silk moths (Lepidoptera: Saturniidae) inferred from four protein-coding nuclear genes. *Syst. Entomol.* 33:219–22
114. Ribot JC. 1999. Decentralisation, participation and accountability in Sahelian forestry: legal instruments of political-administrative control. *Africa* 69:23–65
115. Ricketts TH, Regetz J, Steffan-Dewenter I, Cunningham SA, Kremen C, et al. 2008. Landscape effects on crop pollination services: Are there general patterns? *Ecol. Lett.* 11:499–515
116. Rúa P, Jaffé R, Olio RD, Muñoz I, Serrano J. 2009. Biodiversity, conservation and current threats to European honeybees. *Apidologie* 40:263–84
117. Sammataro D, Gerson U, Needham G. 2000. Parasitic mites of honey bees: life history, implications and impact. *Annu. Rev. Entomol.* 45:519–48
118. Samways MJ. 2004. Insects in biodiversity conservation: some perspectives and directives. *Biodivers. Conserv.* 2:258–82
119. Samways MJ. 2007. Insect conservation: a synthetic management approach. *Annu. Rev. Entomol.* 52:465–87
120. Sande SO, Crewe RM, Raina SK, Nicolson SW, Gordon I. 2009. Proximity to a forest leads to higher honey yield: another reason to conserve. *Biol. Conserv.* 142:2703–9
121. Schabel HG. 2006. *Forest Entomology in East Africa: Forest Insects of Tanzania.* Dordrecht, The Netherlands: Springer. 328 pp.
122. Scherr SJ, White A, Kaimowitz D. 2003. *A New Agenda for Forest Conservation and Poverty Reduction: Making Markets Work for Low-Income Producers.* Washington, DC: Forest Trends/CIFOR. 90 pp.
123. Schmid O, Theopold U. 2005. Immune defense and suppression in insects. *BioEssays* 13:343–46
124. Seeley TD, Tarpy DR. 2007. Queen promiscuity lowers disease within honeybee colonies. *Proc. R. Soc. Ser. B* 274:67–72
125. Serrano S, Jiménez-Hornero FJ, Gutiérrez de Ravé E, Jodral ML. 2008. GIS design application for "Sierra Morena Honey" designation of origin. *Comput. Electron. Agric.* 64:307–17
126. Shaibi T, Muñoz I, Dall'Olio R, Lodesani M, De la Rúa P, Moritz RFA. 2009. *Apis mellifera* evolutionary lineages in Northern Africa: Libya, where orient meets occident. *Insect. Soc.* 56:293–300
127. Sharam GJ, Sinclair ARE, Turkington R. 2009. Serengeti birds maintain forests by inhibiting seed predators. *Science* 325:51
128. Singh PP. 2008. Exploring biodiversity and climate change benefits of community-based forest management. *Glob. Environ. Change* 18:468–78
129. Snodgrass RE. 1993. *Principles of Insect Morphology*. Ithaca, NY: Cornell Univ. Press. 768 pp.
130. Suzuki R, Tase A, Fujimoto Z, Shiotsuki T, Yamazaki T. 2009. NMR assignments of juvenile hormone binding protein in complex with JH III. *Biomol. NMR Assign.* 3:73–76
131. Taylor PL. 2005. A fair trade approach to community forest certification? A framework for discussion. *J. Rural Stud.* 2:433–47
132. Truman JW, Riddiford LM. 2002. Endocrine insights into the evolution of metamorphosis in insects. *Annu. Rev. Entomol.* 47:467–500
133. Van Benthem DJ, Imperatriz-Fonseca VL, Velthuis HHW. 1995. Biology of the stingless bee *Plebeia remota* (Holmberg): observations and evolutionary implications. *Insect. Soc.* 42:71–87

134. Veldtman R, McGeoch MA, Scholtz CH. 2007. Fine-scale abundance and distribution of wild silk moth pupae. *Bull. Entomol. Res.* 97:15–27
135. Welford L, Le Breton G. 2008. Bridging the gap: phytotrade Africa's experience of the certification of natural products. *For. Trees Livelihoods* 18:69–79
136. Wilson-Rich N, Spivak M, Fefferman NH, Starks PT. 2009. Genetic, individual and group facilitation of disease resistance in insect societies. *Annu. Rev. Entomol.* 54:405–23
137. Winfree R, Williams NM, Dushoff J, Kremen C. 2007. Native bees provide insurance against ongoing honey bee losses. *Ecol. Lett.* 10:1105–13
138. Winfree R, Williams N, Gains H, Ascher J, Kremen C. 2008. Wild bee pollinators provide the majority of crop visitation across land use gradients in New Jersey and Pennsylvania, USA. *J. Appl. Ecol.* 45:793–802
139. Yamanak A, Tsurumaki J, Endo K. 2000. Neuroendocrine regulation of seasonal morph development in a bivoltine race (Daizo) of the silkmoth, *Bombyx mori* L. *J. Insect Physiol.* 46:803–8
140. Zanatta DB, Bravo JP, Barbosa JF, Munhoz REF, Fernandez MA. 2009. Evaluation of economically important traits from sixteen parental strains of the silkworm *Bombyx mori* l (Lepidoptera: Bombycidae). *Neotropical Entomol.* 38:327–31
141. Zethner O, Koustrup R, Raina SK. 2008. *African Ways of Silk. Ancient Threads—New Possibilities*. CASAS No. 57. 88 pp. Cape Town, South Africa

Systematics and Evolution of Heteroptera: 25 Years of Progress

Christiane Weirauch[1] and Randall T. Schuh[2]

[1]Department of Entomology, University of California, Riverside, California; email: christiane.weirauch@ucr.edu

[2]Division of Invertebrate Zoology, American Museum of Natural History, New York, New York 10024; email: schuh@amnh.org

Annu. Rev. Entomol. 2011. 56:487–510

First published online as a Review in Advance on September 7, 2010

The *Annual Review of Entomology* is online at ento.annualreviews.org

This article's doi:
10.1146/annurev-ento-120709-144833

0066-4170/11/0107-0487$20.00

Key Words

true bugs, Hemiptera, cladistics, morphology, molecular data

Abstract

Heteroptera, or true bugs, are part of the most successful radiation of nonholometabolous insects. Twenty-five years after the first review on the influence of cladistics on systematic research in Heteroptera, we summarize progress, problems, and future directions in the field. The few hypotheses on infraordinal relationships conflict on crucial points. Understanding relationships within Gerromorpha, Nepomorpha, Leptopodomorpha, Cimicomorpha, and Pentatomomorpha is improving, but progress within Enicocephalomorpha and Dipsocoromorpha is lagging behind. Nonetheless, the classifications of several superfamily-level taxa within the Pentatomomorpha, such as Aradoidea, Coreoidea, and Pyrrhocoroidea, are still unaffected by cladistic studies. Progress in comparative morphology is slow and drastically impedes our understanding of the evolution of major clades. Molecular systematics has dramatically contributed to accelerating the generation and testing of hypotheses. Given the fascinating natural history of true bugs and their status as model organisms for evolutionary studies, integration of cladistic analyses in a broader biogeographic and evolutionary context deserves increased attention.

INTRODUCTION

Heteroptera: the true bugs, a suborder of Hemiptera comprising seven infraorders and ~40,000 species

Hemiptera: an order of paurometabolous insects distinguished by their unique sucking mouthparts and composed of the suborders Sternorrhyncha, Auchenorrhyncha, Coleorrhyncha, and Heteroptera

Cladistics: grouping by synapomorphy based on the parsimony criterion and testing for congruence among characters in an automated and repeatable procedure

With more than 40,000 described species, Heteroptera (Hemiptera) are part of the most successful radiation of nonholometabolous insects. Systematic research on Heteroptera has made significant progress since the first review on the influence of cladistics on Heteropteran classification more than two decades ago (108). Since then, phylogenetic analyses on all taxonomic levels and the onset of molecular systematics have increasingly influenced the understanding of relationships within Heteroptera. Explicit cladistic hypotheses now exist for relationships among the seven infraorders (153, 158) and for relationships within five infraorders (40, 54, 61, 70, 112, 118). The use of cladistic methods has provided rigorous tests and rejections of hypotheses, such as those on the monophyly of Anthocoridae (116), Lygaeidae (66), and Triatominae within Reduviidae (149). Cladistics is also crucial in evaluating the placement of new taxa, such as the family Curaliidae in 2008 (117), and for proposing testable hypotheses on a wide array of topics such as historical biogeography, parental care, and prey capture strategies. Finally, cladistics has allowed for the generation of detailed species-level schemes of critical groups such as *Lygus* (120) and other biologically compelling taxa (48) and for the testing of a broad range of evolutionary scenarios related to habitat choice and adaptations (8). Schuh & Slater (114) offered a comprehensive treatment of heteropteran classification and natural history down to the subfamily level. More recently, Forero (49) reviewed systematic concepts in the order Hemiptera but treated individual infraorders only briefly. This review assesses the development of heteropteran systematics during the past 25 years while providing perspectives on where the field might develop.

MONOPHYLY OF HETEROPTERA AND POSITION WITHIN HEMIPTERA

The monophyly of Heteroptera is usually assumed to be well documented on the basis of both morphological and molecular evidence. Surprisingly, however, there are few uncontradicted morphological synapomorphies for the group. A widely used diagnostic character, the subdivision of the forewing into a proximal coriaceous and distal membranous portion, the so-called hemelytron, is derived within Heteroptera (153). Based on evidence provided by Carver et al. (29) and Wheeler et al. (153), three diagnostic characters of Heteroptera are synapomorphies of the group: labium inserted anteriorly on the head as opposed to posteriorly as in Sternorrhyncha, Auchenorrhyncha, and Coleorrhyncha (**Figure 3*a***); presence of metathoracic glands in adults (**Figure 3*d***); and immatures with dorsal abdominal glands. Fischer et al. (47) proposed an open rhabdom of the ommatidium as an additional synapomorphy for Heteroptera. This type of rhabdom is unique among Hemiptera, and even though this character was not tested in a cladistic framework, it may well hold up to tests of congruence. Finally, Zrzavy (160) mentioned a four-segmented antenna with two intersegmental sclerites as a character uniting the Heteroptera.

The sister-group relationship of Heteroptera with Coleorrhyncha is relatively uncontroversial and supported by morphological and molecular evidence. Originally proposed by Schlee (104), this theory has been expanded by Popov & Sherbakov (98, 99). The clade including the two hemipteran suborders is usually referred to as Heteropterodea (104), although the name Prosorrhyncha (27) is occasionally used. Wheeler et al. (153) incorporated characters treated as synapomorphies by Schlee in their combined morphological and molecular analysis and found that, in addition to several molecular synapomorphies, the reduction of the number of antennal flagellomeres and the wings being held flat over the abdomen are indeed synapomorphies for the two clades. Campbell et al. (27) analyzed 18S rDNA data for a set of taxa including one peloridiid, three Heteroptera, and a broad but small sample of nonheteropteran Hemiptera and found support for the Coleorrhyncha + Heteroptera clade. A contrasting hypothesis was proposed

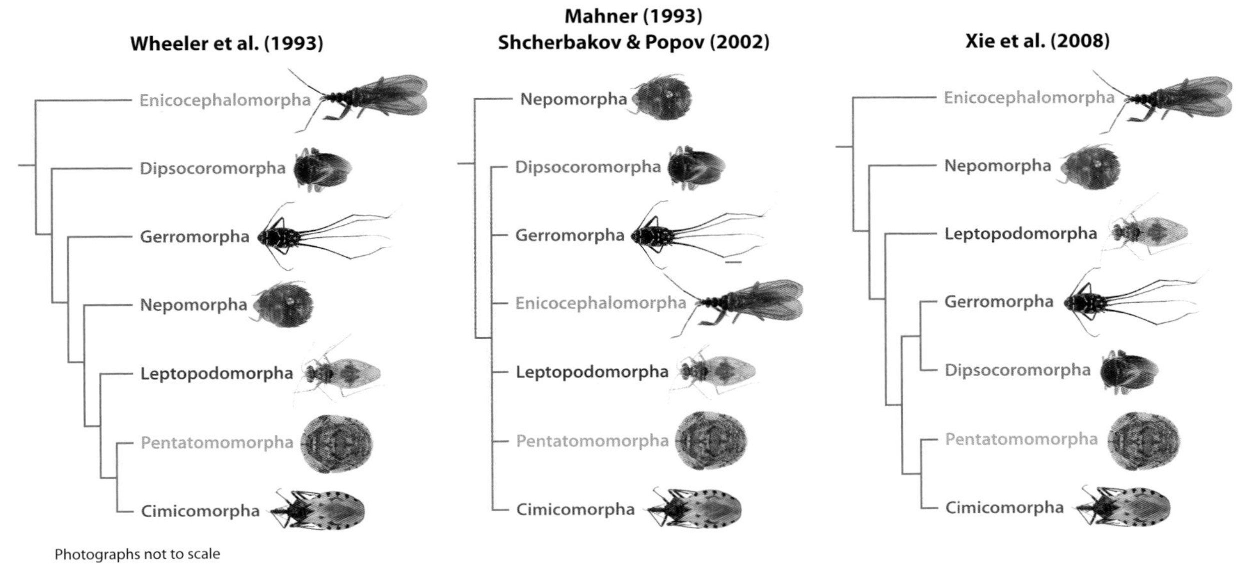

Figure 1

Alternative hypotheses on infraordinal relationships of Heteroptera. Representative species are color coded.

by Yang (159) based on male genitalic characters but without conducting a formal cladistic analysis; Coleorrhyncha was treated as more closely related to only part of the Heteroptera, i.e., Dipsocoromorpha, Nepomorpha, and Gerromorpha. All other analyses subsequent to the work by Schlee, i.e., Sorensen et al. (128), Ouvrard et al. (92), and Bourgoin & Campbell (23), recovered Heteropterodea as a clade.

HIGHER-LEVEL SYSTEMATICS OF HETEROPTERA

Fewer than a handful of cladistic analyses have addressed relationships among the seven infraorders of Heteroptera during the past 25 years. Wheeler et al. (153) analyzed ~669 bp of 18S rDNA and 31 morphological characters for 29 taxa (**Figure 1**); this analysis remains the most comprehensive analysis of infraordinal relationships to date with respect to number of taxa and characters included. The Wheeler et al. hypothesis treats Enicocephalomorpha as the sister group to the remaining Heteroptera, the Euheteroptera; Dipsocoromorpha as sister group to the Neoheteroptera; Gerromorpha as sister group to the Panheteroptera; and Nepomorpha as sister taxon to the Leptopodomorpha + Geocorisae. Geocorisae comprise the Cimicomorpha and Pentatomomorpha, the land bugs. Our discussions of infraordinal level taxa below follow a sequence concordant with the conclusions of Wheeler et al. (153) in treating Enicocephalomorpha and Dipsocoromorpha first, then progressing into semiaquatic, aquatic, and shore bugs, and concluding with the land bugs (**Figure 2**).

Mahner (88) proposed a competing hypothesis using morphological characters that placed Nepomorpha as the sister taxon to the remaining Heteroptera (**Figure 1**). Although not based on a formal cladistic analysis, Mahner evaluated several morphological character complexes that previously had not been considered in a phylogenetic context. He noted, among other points, that a tentorium as found in nonheteropteran Hemiptera is restricted to the Nepomorpha within Heteroptera and may be plesiomorphic in that infraorder. He concluded that the reduction of the tentorium in the remaining Heteroptera might thus be synapomorphic. However, tentorium data for

Heteroptera infraorders

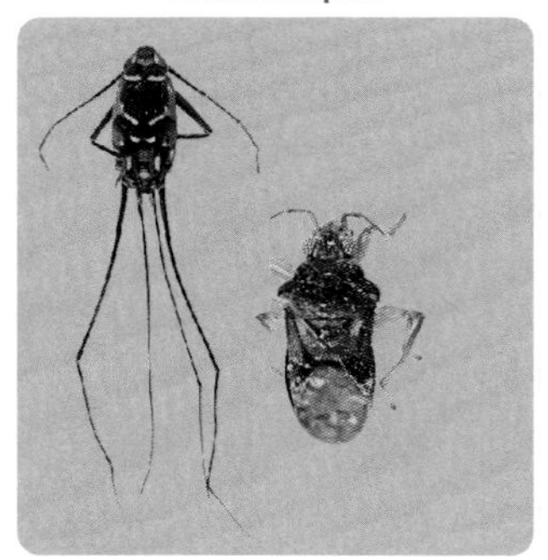

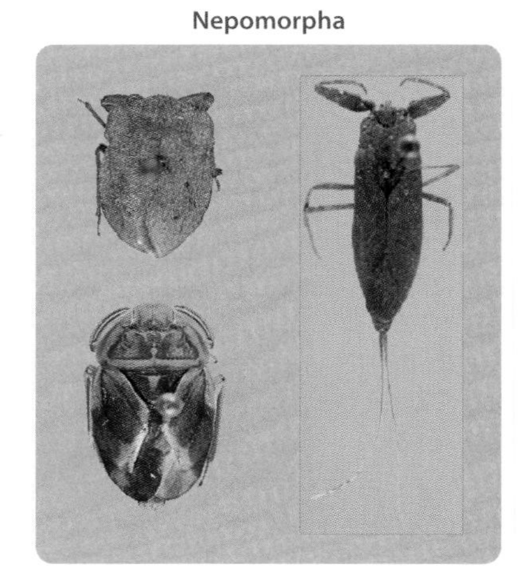

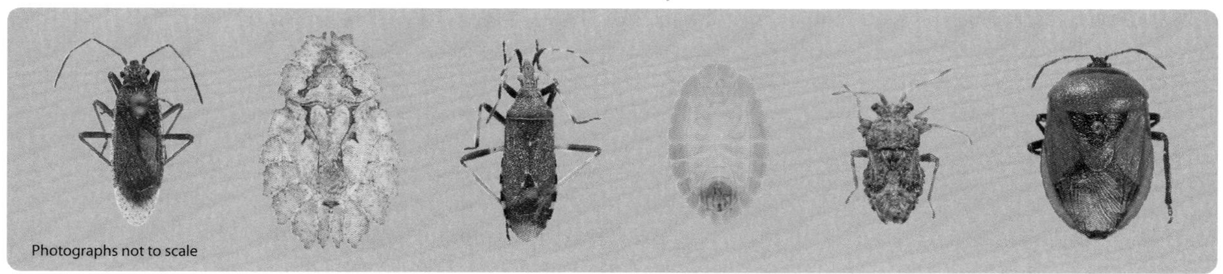

Figure 2

Morphological diversity in the seven infraorders of Heteroptera.

the infraorders Enicocephalomorpha and Dipsocoromorpha are still lacking, which precludes better informed tests of this hypothesis.

Sherbakov & Popov (121), although not strictly cladistic, presented a scheme of relationships for the infraorders of Heteroptera based on fossil and Recent taxa and morphological evidence that is congruent with Mahner's hypotheses (**Figure 1**). In their hypothesis, Nepomorpha is the sister taxon to the remaining Heteroptera, the latter group united by the presence of elongate antennae. Relationships among the remaining infraorders are not resolved.

Xie et al. (158) tripled the amount of 18S rDNA used by Wheeler et al. (153) for 26 representatives of Heteroptera and 11 outgroups (**Figure 1**). All infraorders were monophyletic in their analysis, and infraordinal relationships were recovered as Enicocephalomorpha (Nepomorpha (Leptopodomorpha ((Gerromorpha + Dipsocoromorpha) + (Cimicomorpha + Pentatomomorpha)))), thus providing a novel hypothesis of infraordinal relationships.

The only sister-group relationship that is not contradicted by any of these hypotheses, although not unequivocally supported, is Cimicomorpha + Pentatomomorpha, the Geocorisae. This clade was also supported in combined morphological and molecular analyses by Schuh et al. (118) and in morphological analyses focusing on single morphological character complexes (47). The position of Aradoidea is still under dispute as a sister group to the Cimicomorpha (77), to the Leptopodomorpha + Geocorisae (135), or as sister taxon to the remaining Pentatomomorpha, the Trichophora (30, 70, 153). Support for the sister-group relationship to Trichophora is derived from morphological, combined morphological and ribosomal molecular data, and whole mitochondrial genome analyses combined with sound analytical approaches. It seems the currently best argued hypothesis.

THE BASAL CLADES: ENICOCEPHALOMORPHA AND DIPSOCOROMORPHA

Enicocephalomorpha, or unique-headed bugs, are the putative sister group to all remaining Heteroptera (114, 131, 134, 153). They comprise two families, Enicocephalidae (five subfamilies) and Aenictopecheidae (four subfamilies), the latter possibly paraphyletic, and approximately 325 described species (114). The monophyly of the Enicocephalomorpha is uniquely supported by the novel structure of the head, the specialized raptorial forelegs, the distinctive forewings, and specialized gland morphology. Although their monophyly has not been disputed, neither has it been tested in a rigorous phylogenetic framework. Published sequence data are lacking except for the data of Wheeler et al. (153). Wygodzinsky & Schmidt (156) revised the New World Enicocephalomorpha and provided a provisional scheme of relationships based on a list of 14 morphological characters, but without formal analysis or the inclusion of outgroups in their discussion. The large (~250 described species) Enicocephalinae is paraphyletic in this scheme, but Aenictopecheidae and Alienatinae are monophyletic. Azar et al. (18) analyzed this dataset to discuss the position of their new Lebanon amber fossil as the sister taxon to the Enicocephalinae + Alienatinae but also failed to include outgroups. Grimaldi et al. (55) conducted a cladistic analysis of Recent and Dominican amber species of *Enicocephalus* Westwood and concluded that the fossil species are nested in various positions within the genus. To date, this study remains the only cladistic analysis of any genus-level taxon within Enicocephalomorpha. Tests on the monophyly and relationships of all suprageneric taxa are past due.

The monophyly of the infraorder Dipsocoromorpha is the most controversial within the Heteroptera and there is little agreement on synapomorphies (114, 130). Five morphologically divergent families comprise ~300 described species (114). Most of these tiny, cryptic bugs have grossly asymmetrical and uniquely modified male genitalia, which provide an excellent character system for cladistic analysis that surprisingly remains unexploited. The current family- and subfamily-level classification is untested and molecular data are unavailable except for those published in Wheeler et al. (153) and Schuh et al. (118). Systematic activity in recent years has focused largely on descriptive and revisionary work (67, 68), as well as on morphological studies on spermathecae (96), ovaries (133), wing venation (101), and chromosomes (56).

Most morphological studies in the Heteroptera (34, 35, 100) have neglected the Enicocephalomorpha and Dipsocoromorpha, leaving questions of homology unanswered. This lack of information impedes our ability to trace the evolution of entire character complexes in the basal groups, including male and female genitalia and wing venation. Depending on the hypothesis of relationships within Enicocephalomorpha, the male phallus in Heteroptera has evolved once or at least twice. The latter scenario by Stys (132) treated the median intermittent organ in Aenictopecheidae as nonhomologous to the phallus found in Euheteroptera.

SEMIAQUATIC, AQUATIC, AND SHORE BUGS: GERROMORPHA, NEPOMORPHA, AND LEPTOPODOMORPHA

Gerromorpha, or semiaquatic bugs, have achieved model status in heteropteran systematics thanks to the early (from 1980s onward) and rigorous efforts of Nils Moeller Andersen. He and his students have incorporated cladistics into revisionary studies, used cladistic hypotheses to test biogeographic and evolutionary scenarios, and incorporated fossils into analyses of Recent taxa (see Fossils and Hypothesis Testing, below). Family-level relationships are now extensively tested and show strong congruence with hypotheses originally proposed by Andersen (3, 9) that treated Mesoveliidae as the sister group to the remaining Gerromorpha and Hydrometroidea as sister taxon to the Gerroidea (40). This last study, a combined morphological and molecular analysis, found Veliidae to be paraphyletic, a result that is begging for additional analyses with a more comprehensive sample of veliids.

Nepomorpha, or true water bugs, have received much less attention of late than Gerromorpha. Rieger's (102) comparative morphological study focused on the position of Ochteridae within Nepomorpha but also provided a hypothesis on family- and superfamily-level relationships within the infraorder. Mahner (88) assembled and analyzed morphological data on Nepomorpha and, together with some original observations, he interpreted these character data in a cladistic framework (7). More recently, Hebsgaard et al. (61) combined morphological characters derived from Mahner's study with molecular data. The three hypotheses agree in placing the Nepoidea as the sister group to the remaining Nepomorpha and in treating Helotrephidae, Notonectidae, and Pleidae as a clade. The relationships of Aphelocheiridae, Potamocoridae, Naucoridae, and Corixidae vary among the hypotheses. The monophyly of Nepomorpha has only recently been disputed based on the analysis of nine nepomorphan mitochondrial genomes (71). Pleidae are recovered as the sister group to Geocorisae + Leptopodomorpha + remaining Nepomorpha, and the authors therefore suggested elevating this group to the infraordinal level, i.e., Plemorpha. Given the small taxon sample of this analysis and the lack of a combined effort including morphology, we perceive this change of classification as premature.

Cladistic analyses at the family and genus levels in Nepomorpha have seen some activity in recent years. We discuss Keffer's (72) analysis of Nepidae below in the comparative morphology section. In addition, genus groups were analyzed within Micronectidae (139) and the toad bug genus *Nerthra* has been the subject of two analyses (32, 33).

Hypotheses of family-level relationships within Leptopodomorpha (~335 species), or shore bugs, have not changed since the seminal publication by Schuh & Polhemus (112) that proposed Saldidae + Aepophilidae to represent the sister group to Omaniidae + Leptopodidae based on morphological character data. Analyses below the family rank are restricted to one evaluating the generic concept of *Saldula* Van Duzee from a cladistic perspective (81) and more recently a comprehensive analysis of the genus *Pseudosaldula* Cobben (113).

CIMICOMORPHA: PLANT FEEDERS, PREDATORS, BLOOD SUCKERS

Cimicomorpha comprise >20,000 species currently placed in 17 families. Members of this group show a wide range of adaptations to diverse habitats and life-history strategies (114, 117), including predation and blood feeding in the Reduviidae, mostly plant feeding in the Miroidea, and ectoparasitism in the Cimicidae and Polyctenidae. Cimicomorphan relationships were first analyzed in a cladistic framework by Schuh & Stys (116), who tested earlier theories of Kerzhner (74) using a ground plan coding approach. Among the major conclusions of Schuh & Stys were the sister-group relationship of Reduvioidea (Reduviidae + Pachynomidae) to the remaining Cimicomorpha, evidence for the paraphyly of Anthocoridae sensu lato, resulting in Lasiochilidae and Lyctocoridae elevated to family rank and Miroidea to comprise Thaumastocoridae + (Tingidae + Miridae). Tian et al. (138) analyzed a molecular dataset comprising 46 cimicomorphan taxa and ~3,300 bp of DNA. They found Miroidea to be polyphyletic, with Tingidae repeatedly recovered as the sister taxon to all remaining Cimicomorpha; Reduviidae to be monophyletic but never recovered in a basal position; Cimiciformes (Naboidea + Cimicoidea) to be paraphyletic; and Cimicoidea to be monophyletic. The Cimicomorpha have subsequently been the focus of a combined morphological and molecular analysis (118), the results of which are congruent with earlier proposed hypotheses (116). This combined analysis, using an exemplar taxon approach and comprising 92 taxa and ~3,500 bp of DNA, found a monophyletic Reduvioidea to be the sister group to the remaining Cimicomorpha. Miroidea was recovered as monophyletic, comprising monophyletic Miridae and Tingidae, and was nested within Cimicomorpha. Cimiciformes was monophyletic under an expanded concept including Cimicoidea, Naboidea, Joppeicidae, and Microphysidae. The position of the single included representative of Velocipedidae that fell outside the Geocorisae was viewed as an artifact. Thaumastocoridae was polyphyletic in this analysis, with Xylastodorinae as the sister taxon to Cimiciformes + Miroidea and Thaumastocorinae recovered as the sister taxon to the Pentatomomorpha. In addition, Nabidae was paraphyletic with respect to Medocostidae and Joppeicidae. Within Cimicoidea, the recently described family Curaliidae Schuh, Weirauch and Henry was nested within a clade that also contains Cimicidae, Polyctenidae, and Plokiophilidae.

Cimicomorpha have also undergone cladistic analyses at the family level and below. For Reduviidae, separate analyses based on morphological (149) and molecular data (151) have tested hypotheses of subfamily-level relationships within assassin bugs that corroborate the nesting of phymatine or ambush bugs within Reduviidae, establish a sister-group relationship between Ectrichodiinae + Tribelocephalinae, show Reduviinae to be polyphyletic, and recover Triatominae as monophyletic. The blood-feeding Triatominae, or kissing bugs, have attracted considerable attention and are discussed in more detail below (see Molecular Markers: Present and Future). Cladistic analyses at a lower systematic level have focused on Peiratinae, among a few other subfamilies (36, 90). The remaining analyses within Cimicomorpha are restricted to the Miroidea. Tingidae have received some recent attention (58, 82) and the sister-group relationship of Vianaidinae with Cantacaderinae + Tinginae now seems well corroborated with the description of macropterous Vianaidinae (111). An overarching subfamily-level phylogeny for the largest family in Heteroptera, the Miridae or plant bugs, to test theories proposed by Schuh (106, 107) is still lacking. On a lower level, a wealth of revisionary work at the genus and tribal levels has been conducted in more recent years. Many of these revisions and monographs include cladistic analyses, such as, for example, the works of Stonedahl (129), Henry (64), Schuh (109), Cassis et al. (31), Weirauch (148), Forero (48), and Schwartz (119).

PENTATOMOMORPHA: STINK BUGS AND ALLIES

Pentatomomorpha comprise more than 14,000 species in 5 or 6 superfamilies (the Aradoidea, Coreoidea, Idiostoloidea, Lygaeoidea, Pentatomoidea, and Pyrrhocoroidea) and 40 families (66, 78, 114). With the exception of certain predatory and even hematophagous clades within Pentatomidae and Rhyparochromidae, members of this group are phytophagous and exploit resources from the roots to the seeds of their host plants. Schaefer (103) reviewed the systematic history of the group and discussed different superfamily-level arrangements. The mitochondrial genome analysis by Hua et al. (70) (see also Molecular Markers: Present and Future, below) considered the entire infraorder Pentatomomorpha and found the following relationships: Aradoidea + (Pentatomoidea + (Pyrrhocoroidea + (Lygaeoidea + Coreoidea))). However, the small number of taxa (16 species) and the omission of several critical taxa (i.e., Idiostolidae, Piesmatidae) render this hypothesis a relatively weak test of the suprafamilial classification of the infraorder. Henry's (66) morphology-based analysis of Pentatomomorpha found evidence for the monophyly of the six superfamilies named here. That study emphasized relationships within Lygaeoidea, found Lygaeidae to be paraphyletic, and accordingly raised these 11 clades to family rank (Artheneidae, Blissidae, Cryptorhamphidae, Cymidae, Geocoridae, Heterogastridae, Lygaeidae, Ninidae, Oxycarenidae, Pachygronthidae, and Rhyparochromidae). Li (78) used a small molecular dataset (16S rDNA; 29 taxa, 12 of them original) to test theories proposed by Henry and found the molecular dataset to provide congruent results, except that Rhyparochromidae and Lygaeidae were polyphyletic in their analyses.

Several additional analyses have focused on subordinate groups within Pentatomomorpha. Gapud (50) used morphological characters to investigate relationships within the Pentatomoidea and found Urostylididae to be the sister group to all remaining stink bugs, a result that was later corroborated by Grazia et al. (54). Cladistic analyses for three of the superfamilies, the Aradoidea, Coreoidea, and Pyrrhocoroidea, are missing to date. Especially in the case of Coreoidea, this gap results in the lack of a single accepted classification, with competing classifications differing widely in numbers and composition of subfamilies and tribes (80, 114). Schemata on relationships within Aradidae published by Vasarhelyi (141) and Grozeva & Kerzhner (57) were not based on rigorous cladistic procedures but nevertheless provide testable hypotheses on relationships within the flat bugs. We discuss the analyses of Trichophora (79) and Pentatomoidea (54, 70, 79) below.

On a lower systematic level, Pentatomomorpha has been the subject of cladistic tribal and generic analyses. The Aradidae, Coreidae, and Pyrrhocoridae have seen substantial revisionary activity (1, 24, 25, 62, 63, 140), but with the exception of Zrzavy & Nedved's (161, 162) studies on *Dysdercus*, these revisions usually do not include phylogenetic hypotheses. The situation is similar in Lygaeoidea, where most recent systematic work has consisted of taxonomic revisions (124, 125). Notable exceptions are the cladistic analyses of the genera of Berytidae (65) based on morphological characters and of the Australian *Allocasuarina*-feeding genus *Laryngodus* Herrich-Schaeffer based on morphological and molecular data (126). Thanks to the efforts of Jocelia Grazia and her students, cladistic analyses on Pentatomidae started to appear in the late 1990s. Among them are revisions and cladistic analyses of genera (19, 45), genus groups (53), and tribes (28).

COMPARATIVE MORPHOLOGY AND CHARACTER EXPLORATION

Comparative morphological and anatomical studies on Heteroptera have not yet made the transition to twenty-first century approaches that ideally combine large-scale studies and cutting-edge morphological methods as seen in some groups of insects and other arthropods (22, 154). Important comparative morphological studies on Heteroptera published in the mid- to late-twentieth century are summarized by Schuh & Slater (114). Some of the studies focused on character systems including the nervous system and mesodermal components of the genitalic system and required specialized preparation methods such as microdissections or histological sectioning. These comparative studies usually included small taxon samples and were often restricted to Geocorisae. Together with the lack of anatomical studies on outgroup representatives such as Coleorrhyncha, these prevented these character systems from being coded as part of comprehensive cladistic analyses. Notable exceptions are the thorough treatments by Cobben (34, 35), but also Margaret Parson's studies that focused on external and internal head and thoracic anatomy of Nepomorpha and Leptopodomorpha. The works of both authors were used by Wheeler et al. (153) and Mahner (88) in compiling the morphological matrices for their analyses and discussions. Cobben's work is especially important in that it spans a wide range of understudied character complexes such as eggs and immature stages, mouthparts, salivary glands, and male genitalia, but many of these characters have not yet been coded in a formal cladistic analysis. In more recent years, only a few publications have focused on comparative morphology of Heteroptera including all or most infraorders. Comparative studies at the level of Hexapoda or Pterygota have often incorporated only a few heteropteran taxa, usually Geocorisae or Gerromorpha, for example the study on circulatory organs by Pass et al. (93).

Surprisingly, this lack of comprehensive studies applies not only to internal anatomical structures, whose examination and documentation are time-consuming, but also to easily accessible body parts such as male and female external and ectodermal genitalic structures (**Figure 3*b***). Homologies are far from resolved even for parts of the male intromittent organ or the female styloids. This lack of well-documented empirical studies severely affects our ability to generate more robust cladistic hypotheses and prevents heteropterists from proposing and testing hypotheses on the evolution of these character systems.

A discussion of some recently published comparative morphological studies on the suborder provides glimpses into the potential of these approaches. Fischer et al. (47) studied the structure of the ommatidia using transmission electron microscopy in representatives of all heteropteran infraorders and Coleorrhyncha, mapped characters on an infraordinal-level hypothesis, and concluded that open rhabdoms are a synapomorphy of Heteroptera; the patterns in Gerromorpha and Leptopodomorpha are each apomorphic; and a unique V-pattern supports the sister-group relationship of Cimicomorpha + Pentatomomorpha. Building on Cobben's work, Brozek & Herczek (26) investigated locking mechanisms of the mandibular and maxillary stylets in representatives of 19 families in five infraorders, finding variation that may be of value for cladistic analyses at the superfamily level and above. Empirical studies on homologies of wing venation and axillary sclerites in Heteroptera are restricted to Geocorisae and Nepomorpha (21, 155), but the authors aimed at deriving some conclusions on patterns in the common ancestor of Heteroptera. Weirauch & Cassis (150) made an effort to include a small (i.e., 20 species) but representative sample of Heteroptera in all infraorders and other Hemiptera in their study of wing-coupling devices in Heteroptera. They concluded that frenae are plesiomorphic for Heteroptera and that a "druckknopf" (**Figure 3*c***) evolved convergently within Dipsocoromorpha and at the base of Panheteroptera. On an infraordinal level, Schuh et al. (118), in their analysis of cimicomorphan relationships, mapped data from Weirauch (147) and showed that hairy attachment structures known as fossula spongiosa evolved at least three times independently within Cimicomorpha (**Figure 3*e***).

A much larger body of work has been devoted to comparative morphological treatments within families. Keffer (73) studied and cladistically analyzed male genitalia of Nepidae with the resulting phylogeny differing from Mahner's (88) hypothesis in almost all sister-group relationships. Tingidae have received some attention with studies on little-studied structures such as the striking ornamentation patterns of immatures (59) and the ductus seminis (83). A series of papers on comparative morphology of glands (143, 145, 146), pretarsus (144), and attachment structures (147) laid much of the ground work for a cladistic

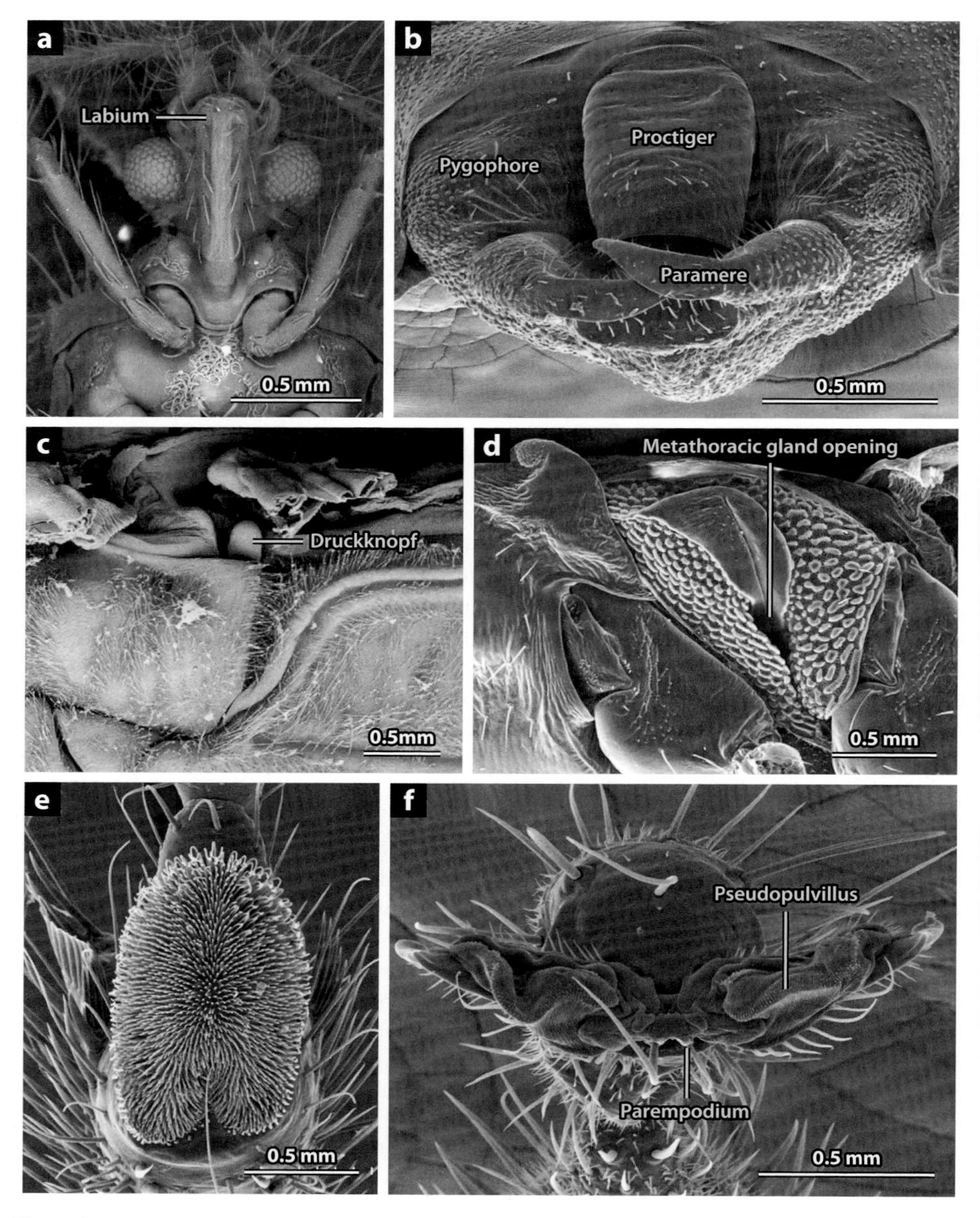

Figure 3

Examples of systematically relevant morphological character complexes in Heteroptera (*a*, *Ptilocnemus pallidus* [Reduviidae]; *b*, *Cantacader* sp. [Tingidae]; *c*, *Velitra* sp. [Reduviidae]; *d*, *Loricula* sp. [Microphysidae]; *e*, *Nabis americanus* [Nabidae]; *f*, *Felisacus* sp. [Miridae]). (*a*) Labium inserted anteriorly on the head. (*b*) Male genitalic structures. (*c*) "Druckknopf" wing coupling mechanism. (*d*) Opening and evaporatorium of the metathoracic gland. (*e*) Fossula spongiosa on the apex of the fore tibia. (*f*) Pretarsus and associated structures.

analysis of Reduviidae (149). The largest family within Cimicomorpha, the Miridae, has spurred only a few comprehensive treatments in recent years (75), but a wealth of character data on male and female genitalia and other structural systems (**Figure 3*f***) has developed through revisionary studies that may provide the basis for a morphological matrix including all subfamilies of Miridae. Finally, within Pentatomomorpha, studies have focused on dorsal abdominal scent glands, trichobothrial patterns, cephalic chaetotaxy, tibial combs, and spermatheca, among other character complexes in Cydnidae (84–87, 95), establishing the basis for future cladistic coding and hypothesis testing.

MOLECULAR MARKERS: PRESENT AND FUTURE

One of the fundamental changes in systematic practice during the past 25 years is the broader application of molecular approaches. The first molecular systematic study with focus on Heteroptera was by Wheeler et al. (153), who combined 31 morphological characters with partial 18S rDNA sequence data (670 bp) and proposed the phylogeny shown in **Figure 1**. After a gap of several years, Gerromorpha became the focus of several molecular systematic studies. Muraji & Tachikawa (91) analyzed 16S and 28S rDNA for 30 terminals of Gerroidea. Simultaneously, Jakob Damgaard, then in Nils Moeller Andersen's laboratory, spearheaded the generation and analysis of molecular data for Gerromorpha. These efforts resulted in a series of papers focusing on the genera *Aquarius* Schellenberg (41), *Halobates* Eschscholtz (39), and *Gerris* Fabricius (44); then moving into infra- and suprageneric relationships of *Aquarius*, *Gerris*, and *Limnoporus* Stål (42, 43); and finally conducting analyses at the infraordinal level (38, 40). The analyses by Damgaard and coworkers are based on mitochondrial cytochrome *c* oxidase I (COI) and II, ribosomal 16S rDNA and 28S rDNA, nuclear protein-coding elongation factor (EF) 1α, fixed alignment and parsimony methods, and dense taxon sampling.

Another group of Heteroptera started receiving considerable attention at about the beginning of the current century: the blood-feeding Triatominae (Reduviidae) that transmit Chagas disease in Central and South America. Early studies focused on relationships within the tribes that contain the most important Chagas vectors, Rhodniini (89) and Triatomini (51). Subsequent studies investigated the disputed monophyly of the subfamily based on relatively small samples of Triatomini, Rhodniini, and nontriatomine reduviid outgroups (72, 105, 123). Most of these analyses were based on limited amounts of sequence data. Weirauch & Munro (151) analyzed five species of Triatominae representing Triatomini and Rhodniini in their molecular subfamily-level analysis of Reduviidae that treated 89 taxa of Reduviidae and five outgroups for ~3,300 bp of 16S, 28S, and 18S rDNA and recovered a monophyletic Triatominae. No analysis to date has included representatives of the remaining three tribes of Triatominae or combined morphological and molecular data.

Beyond Reduviidae, Cimicomorpha as a whole have also been the focus of molecular studies. Tian et al. (138) analyzed ribosomal data (partial 16S, 28S, 18S rDNA) for 46 taxa and recovered a result largely incongruent with that of Schuh & Stys (116), with a paraphyletic Miroidea (Miridae and Tingidae) rather than Reduviidae as the sister group to the remaining Cimicomorpha. A combined morphological and molecular analysis was published a few months later (118), with 92 taxa, the same gene regions, and 73 morphological characters analyzed. Hypotheses in this analysis are largely congruent with those previously proposed by Schuh & Stys (116) as discussed above, notably the sister-group relationship of Reduvioidea with other Cimicomorpha and the monophyly of the Miroidea (less Thaumastocoridae).

Nepomorpha was treated in an analysis (61) combining 960 bp of 16S and 28S rDNA and 65 morphological characters that were derived mostly from Mahner (88). Parsimony and direct optimization analyses recovered Nepomorpha and included families as monophyletic, corroborated many but not all of Mahner's hypotheses, and found Naucoroidea (Aphelocheiridae, Potamocoridae, Naucoridae) to be paraphyletic. An analysis of mitochondrial genomes of nine Nepomorpha and five other Hemiptera came to different conclusions (71). Nepomorpha were not monophyletic in this analysis, Pleoidea, or the pygmy backswimmers, formed the sister group to Nepomorpha + Geocorisae, and the Naucoroidea (sensu Aphelocheiridae + Naucoridae) were monophyletic.

Several molecular or combined analyses have focused on Pentatomomorpha. Li et al. (79) (40 taxa; partial 18S rDNA and COI mtDNA; parsimony; likelihood; minimal evolution) found support for Pentatomomorpha, a sister-group relationship of Aradoidea and Trichophora, Pentatomoidea, and a clade comprising Lygaeoidea, Coreoidea, and Pyrrhocoroidea. Within the last clade, the superfamily-level groupings were recovered as nonmonophyletic in all (Pyrrhocoroidea) or some (Coreoidea, Lygaeoidea) analyses. A smaller study (26 taxa; partial 18S rDNA; MrBayes) that focused on Trichophora and coincided with the publication of Li et al. (79) supported the basal split between Pentatomoidea and Coreoidea, Pyrrhocoroidea, and Lygaeoidea, but added little resolution to the problems mentioned above (157). Hua et al. (70) analyzed complete or nearly complete mitochondrial genomes of 16 taxa of Geocorisae with emphasis on Pentatomomorpha and found the infraorder to be monophyletic, with Aradoidea as the sister taxon to the Trichophora. This analysis also recovered monophyletic Pentatomoidea (three taxa included), Pyrrhocoroidea (two taxa), Lygaeoidea (four taxa), and Coreoidea (three taxa). Finally, a combined morphological and molecular analysis of 84 terminals (52 terminals for molecular data; ~3,500 bp of 18S rDNA, 16S rDNA, 28S rDNA, and COI) recovered the monophyly of Pentatomoidea and many of the family-level taxa (i.e., Acanthosomatidae and Pentatomidae) but showed Cydnidae to be polyphyletic (54).

On a much larger scale, a genome sequencing project for *Rhodnius prolixus* (Reduviidae: Triatominae) is in progress and was in the draft assembly phase in 2010 (R.K. Wilson, unpublished data). In addition, EST (expressed sequence tag) data are being assembled for the bed bug *Cimex lectularius*. These genome and transcriptome data for relatively closely related organisms, together with other hemipteran genomes currently being processed, will provide a unique resource for clustering and rapid processing of future EST data for Heteroptera.

THE ROLE OF FOSSILS IN HETEROPTERAN SYSTEMATICS

Fossils allow dating by assigning minimal ages to nodes and provide often stunning perspectives on historical distributions and character evolution. Sherbakov & Popov (121) summarized information on fossil Heteroptera. As in other groups of insects, heteropteran fossils are often studied in isolation, without taking into account the systematics and classification of Recent taxa. However, there are notable exceptions to this practice. Anderson, in a series of papers on fossil Gerromorpha starting in the early 1980s and continuing until 2004, treated fossil taxa as part of cladistic analyses that were otherwise based on Recent taxa (2, 9, 10, 12, 13, 142). In the case of Mesozoic fossils such as those from Burmese amber, this approach allowed for dating basal splits in families such as Hydrometridae (13) and provided a unique perspective on the early evolution of this group. Cassis & Schuh (30) analyzed a morphological dataset of mostly extant terminals (43 taxa; 78 characters) to test theories on the placement of a Cretaceous fossil that had been described as a piesmatid, *Cretopiesma* Grimaldi and Engel, and found evidence to place it in the Aradoidea instead, leaving the oldest known fossil Piesmatidae in the Eocene, not the Cretaceous.

Amber fossils, despite their often young ages, provide surprising insights into extralimital historical distributions of entire lineages. The reduviid subfamily Centrocneminae is a small group with all extant members restricted

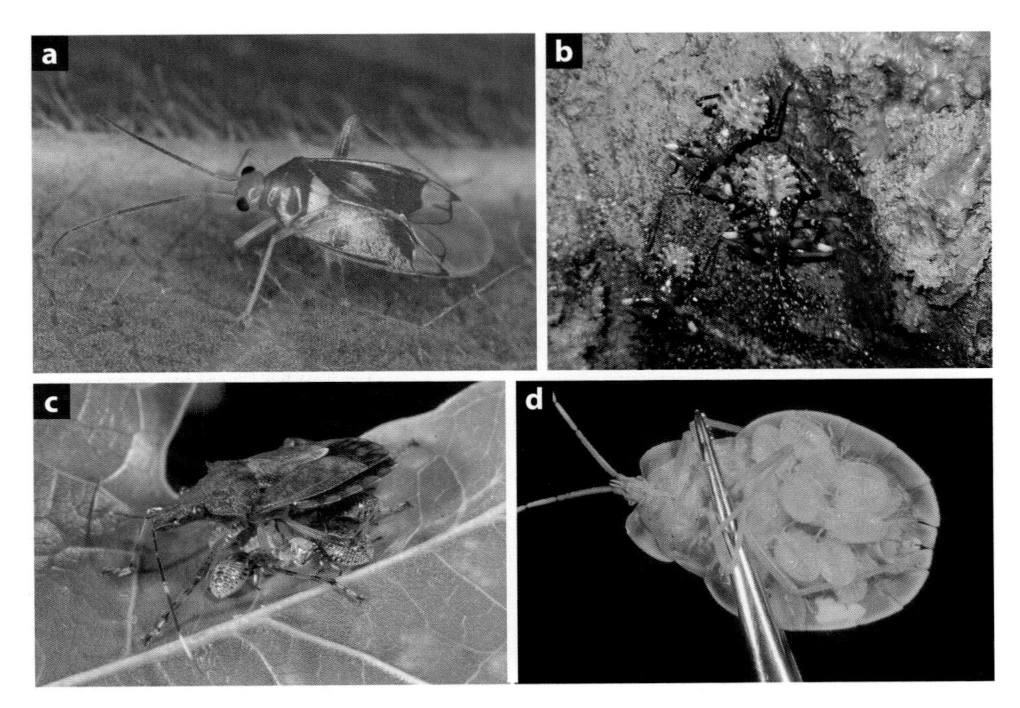

Figure 4

Life-history strategies of Heteroptera. (*a*) Orthotyline plant bug (Miridae) on its host plant (Photo: T. Yasunaga). (*b*) Immature resin bugs (Reduviidae) in ambush for meliponine bees. (*c*) Female *Bromocoris souefi* (Pentatomidae) guarding immatures (Photo: J. Wright, Queensland Museum). (*d*) *Peltocopta crassiventris* (Tessaratomidae) female carrying immatures (Photo: J. Wright, Queensland Museum).

to Southeast Asia and East Asia. Popov & Putshkov (97) described from Baltic amber a new genus that closely resembles some of the Recent genera, thus providing evidence that the distribution range of this subfamily extended much farther west only about 30 mya. Bechly & Wittmann (20) described a species of Xylastodorinae (Thaumastocoridae) from Baltic amber, a group that had heretofore been known only from the New World tropics, greatly extending the known range of the taxon.

CLASSIFICATIONS AND BEYOND: HYPOTHESIS TESTING, BIOGEOGRAPHY, AND EVOLUTIONARY SCENARIOS

Heteroptera show a wide array of ecological, behavioral, and morphological adaptations to a plethora of microhabitats and life-history strategies in virtually all ecosystems. It is therefore not surprising that an increasing number of studies are using phylogenetic hypotheses to test theories on historical biogeography and life-history traits such as host associations and mating strategies (**Figure 4**). However, the lack of detailed ecological, behavioral, and comparative studies, as well as cladistic analyses for many taxa within Heteroptera, still impedes broader application of hypothesis testing in an evolutionary context.

Biogeographical Studies

We here make reference only to historical biogeographic studies that use strictly cladistic approaches. Schuh & Stonedahl (115) spearheaded cladistic biogeography in Heteroptera with their analysis of the Indo-Pacific based on 10 clades of Miridae with results that emphasized the composite biotic nature of the fauna of New Guinea. Andersen (4) revisited the Indo-Pacific using cladistic data for 110 species of marine water striders, defined areas of endemism, conducted component analyses, and similarly concluded that the traditional subdivision of the area into Ethiopian, Oriental, and Australian regions oversimplified actual biogeographical patterns. Schuh (109), in an analysis of the mirid tribe Pilophorini, showed Southern Hemisphere patterns with only the most recent lineages in the Holarctic. Several analyses have focused on distribution patterns in the Neotropical region. Among them are studies on Pentatomidae (19, 28) and Peiratinae (36, 90) and Triatominae (72, 122) within Reduviidae.

Other biogeographic regions have received only little attention. Asquith & Lattin (17) analyzed the Holarctic subgenus *Limnonabis* (Nabidae) and concluded that current distribution patterns are best explained by a vicariance split between Europe and North America. Asquith (16) also investigated theories on the evolution of a diverse lineage of orsilline Lygaeidae on Hawaii. His cladistic interpretations were consistent with a previously proposed single New World origin of the Hawaiian clade. However, Asquith preferred a less parsimonious hypothesis based only on characters with little or no homoplasy that postulated multiple colonizations from the New World. Damgaard (37) focused on three genera of water striders, including the Holarctic *Limnoporus* and cosmopolitan *Aquarius*.

Host-Endosymbiont Relationships and Host-Plant Associations

Several studies have used cladistic methods to investigate endosymbiont-host associations within Heteroptera. Hosokawa et al. (69) found that the phylogeny of plataspid stink bugs in the genera *Brachyplatys*, *Coptosoma*, and *Megacopta* matched one of their obligate *Gammaproteobacteria* endosymbionts and concluded that this association had been formed by strict cospeciation. A similar study focusing on five genera of Acanthosomatidae and their symbiotic bacteria, also *Gammaproteobacteria*, came to a similar conclusion (76). These studies indicate that vertical symbiont transfer is widespread but that occasional horizontal transfer must occur to explain mismatches in the host and endosymbiont phylogenies.

A much larger body of work has been devoted to associations between plant-feeding Heteroptera and their host plants (**Figure 4*a***). Most of these studies have focused on the often host-specific Miridae but have also investigated other plant-associated groups such as rhyparochromid seed bugs (126). Because plant phylogenies with the detail or scope required for cospeciation studies are still largely unavailable, researchers have usually arrayed host data on their plant bug analyses to minimize the number of host shifts, a process referred to as optimization. This approach was spearheaded by Schuh (109), who found that an association with monocots is ancestral for Pilophorini. The study further found no clear evidence for cospeciation between plant bugs and their host plants, although genus-level clades are often restricted to closely related groups of plants. Studies focused on North American and Australian Phylinae and Orthotylinae allowed for tracing often unpredicted host group switches within the plant bug clades (48, 110, 148).

Evolution of Structural Traits and Ecological Adaptations in Gerromorpha

Andersen pioneered cladistic approaches to the study of wing polymorphism and adaptation in using Gerromorpha as a model. In a study published in 1993 (5), he mapped wing polymorphism data on cladograms of several genera of water striders and concluded that wing dimorphism and/or short wingedness is the ancestral state in two of these genera. Andersen (11) revisited dispersal dimorphism in Gerridae and found that the evolution of this phenomenon probably proceeded several times independently from wing dimorphism to obligatory long-winged or wingless forms. His analysis also corroborated the prediction that wing dimorphism with a high percentage of wingless forms is associated with temporary habitats, whereas stable habitats are inhabited by permanently dimorphic (with a high percentage of long-winged forms), seasonally dimorphic, or monomorphic long-winged forms. Andersen took these analyses even further (6) by investigating environmental zones, mode of locomotion, and morphological characters including leg and other thoracic structures associated with locomotion based on a cladistic analysis for the entire infraorder Gerromorpha. He concluded, among other things, that a hygropetric environment is plesiomorphic for Gerromorpha and that morphological modifications occurred after transitions to a given environment, thus qualifying as adaptations in a strict sense. The rigor of these arguments unfortunately remains an exception among studies making reference to adaptations in Heteroptera.

Evolution of Color Patterns and Mimicry

Given the array of striking examples of mimicry in Heteroptera, the lack of detailed analyses of these phenomena is somewhat surprising. Zrzavy & Nedved (161) analyzed the evolution of color patterns in cotton stainers in New World *Dysdercus* using a cladistic analysis of morphological characters based on van Doesburg's (140) revision of the group. They argued that a yellow coloration pattern with a dark median spot or stripe is plesiomorphic for the clade and that dark color patterns have evolved multiple times within the group. Later, Zrzavy & Nedved (162) explored the eight mimicry rings proposed by van Doesburg (140), combining their morphology-based cladistic analysis and distributional data, and found support for some, but not all, of these mimicry rings and argued for more in-depth studies.

Mating Systems and Parental Care

The Gerromorpha are a well-established system for studying the evolution of sexual conflict in animals (14, 15, 46, 52). Most studies have focused on microevolutionary processes within species, but Andersen (8) presented a historical and cladistic perspective on the evolution of sexual dimorphism and mating systems in Gerridae with a number of exciting conclusions. On the basis of his prior phylogenetic hypotheses, Andersen established that the last common ancestor of Gerridae was characterized by males slightly smaller than females, males with strong grasping legs to hold females, and genitalic grasping structures. More dramatic sexual size dimorphism evolved multiple times within the group, usually in conjunction with extended postcopulatory mate-guarding behaviors. In addition, he found that coevolution of male clasping and female anticlasping devices occurred in some, but not all, water striders. Finally, Andersen showed that scramble competition polygyny is most likely the ancestral mating system within water striders and that resource defense polygyny evolved at least four times independently. Other groups of Heteroptera have emerged as additional models for sexual conflict and sexual selection research, among them Coreidae (60) and phymatine Reduviidae, but the lack of phylogenetic hypotheses for the taxa in question precludes their interpretation in a broader evolutionary context.

A unique mating strategy, traumatic insemination, where the male pierces the female's integument to deposit sperm into the hemocoel, occurs in seven families of cimicomorphan Heteroptera. Based on relationships proposed by Schuh & Stys (116), traumatic insemination therefore arose at least three times independently within Cimicomorpha (137). This scenario was further corroborated by relationships recovered by Schuh et al. (118).

Tallamy & Schaefer (136) examined occurrences of maternal care within Hemiptera and Heteroptera and indicated that this strategy is present in at least some members of each of the four infraorders of Panheteroptera. Because maternal care occurs in some members of Reduviidae, the sister group to the remaining Cimicomorpha, and in what may be the most basal groups within the Pentatomoidea, the authors concluded that this strategy is likely plesiomorphic for both clades. This conclusion can be questioned in light of recently published cladistic analyses for Reduviidae (149) and Pentatomoidea (54) that show taxa displaying maternal care to be rather distantly related within these groups and thus maternal care is likely the result of convergence (**Figure 4*c,d***).

Evaluations of theories on the evolution of paternal care in belostomatid water bugs are more feasible based on currently available data. Mahner (88) proposed Lethocerinae to be the sister taxon to Horvathiniinae + Belostomatinae. Lethocerinae brood outside the water with males watering the eggs (emergent-brooding), males in Belostomatinae carry eggs on their backs (back-brooding) (127), but the reproductive behavior of species of the third subfamily, the Horvathiniinae, is unknown, thus precluding unambiguous conclusions on the evolution of the paternal strategies within the group. Information on the plastron in the three subfamilies indicates that the plastrons in Horvathiniinae and Lethocerinae are similar, which may hint at an emergent-brooding reproductive strategy in Horvathiniinae (94). The authors concluded that emergent-brooding is likely plesiomorphic within Belostomatidae, with back-brooding in Belostomatinae derived from that strategy.

CONCLUSIONS

The slow progress in comparative morphological heteropteran research comes into perspective in light of an opinion piece by Wheeler (152) that calls for the use of modern tools to drive systematic research from a morphological perspective. Only morphological research allows exciting insights into the complexity of character evolution and incorporation of fossil taxa into phylogenies. In Heteroptera, comparative morphological research can be taken to new, unexplored areas as demonstrated by the study of ommatidia rhabdom structure by Fischer et al. (47).

Molecular systematics in Heteroptera has made considerable progress since 1993, and multigene and combined morphological and molecular analyses have become the standard rather than an exception. However, this field is still lagging behind the achievements seen in some other groups of insects. Most analyses still rely on ribosomal data, with little effort to establish protein-coding nuclear markers or movement toward whole-genome analyses. Among ribosomal genes, 18S rDNA has been widely used in arthropods for recovering ancient nodes, but most heteropteran data currently available in GenBank pertain only to Geocorisae.

At the highest taxonomic levels, conflicting hypotheses on infraordinal relationships within Heteroptera affect our ability to understand the early evolutionary history of Heteroptera. The Geocorisae are supported by multiple analyses, but the Panheteroptera (Nepomorpha + Leptopodomorpha + Cimicomorpha + Pentatomomorpha) are supported by only one. Relationships of Enicocephalomorpha, Dipsocoromorpha, Gerromorpha, and Nepomorpha beg for comprehensive analytical work with larger samples of taxa and characters.

Within the infraorders, the Enicocephalomorpha and Dipsocoromorpha ask for attention. Since the mid-1980s, neither group has received sufficient taxonomic investigation. Relationships within Enicocephalomorpha could benefit greatly from additional comparative morphological studies on male genitalic structures in Heteroptera as well as Coleorrhyncha. A homology scheme for the extreme morphology of the male genitalia in Dipsocoromorpha will be important in order to use this character complex in the classification of the group. Both infraorders are sorely lacking in molecular data, an area of inquiry that might well help to understand relationships of groups that manifest so many autapomorphic characteristics.

The most critical open questions within Cimicomorpha in our view are the monophyly and systematic position of Thaumastocoridae, the monophyly and infrafamilial relationships of Nabidae, the position of Velocipedidae, the relationships within Cimicoidea, and a subfamily-level analysis of Miridae. Each of these taxonomic groupings should benefit from a broader sample of molecular data, including a broader sample of genes. Among the most pressing priorities within Pentatomomorpha in our view are tests of the superfamilial arrangements, but also cladistic analyses within several superfamily-level taxa (Aradoidea, Coreoidea, and Pyrrhocoroidea). Subfamily- and tribal-level analyses are also needed to test existing classifications for the Pentatomidae, Coreidae, and Rhyparochromidae.

The influence of cladistics on heteropteran classification as of 1986 was limited and certainly did not pervade the field. That situation has clearly changed, with workers on all continents having embraced the method for use with morphological and molecular data. One obvious change in many papers is the move from ground-plan constructs for terminal taxa to the use of exemplar taxa. While more explicit cladograms of relationships appear in published papers, authors also usually make their data available, producing a much higher degree of transparency of method. Nonetheless, cladistic approaches are not monolithic. In the molecular world, the choice of methods, and consequently the comparison of results, is a contentious issue. Even in parsimony-based studies, the choice of outgroups has drawn the results of some studies into question.

DISCLOSURE STATEMENT

The authors are not aware of any affiliations, memberships, funding, or financial holdings that might be perceived as affecting the objectivity of this review.

ACKNOWLEDGMENTS

We thank many of our colleagues for fruitful discussions during the last decade. We acknowledge G. Monteith and J. Wright (**Figure 4*c,d***) and T. Yasunaga (**Figure 4*a***) for allowing us to use their photo material and Steve Thurston for imaging true bugs for **Figure 2**. Lily Berniker, Dimitri Forero, Wei Song Hwang, and Guanyang Zhang critically read a prior version of this review. Financial support came from NSF DEB-0316495 (to R.T.S.) and NSF DEB-0933853 (to C.W.).

LITERATURE CITED

1. Ahmad I, Zaidi RH. 1989. A revision of the genus *Scantius* Stål Hemiptera Pyrrhocoridae from Indo-Pakistan subcontinent with description of three new species and their relationships. *Pakistan J. Zool.* 21:153–74
2. Andersen NM. 1981. A fossil water measurer (Insecta, Hemiptera, Hydrometridae) from the Paleocene/Eocene of Denmark and its phylogenetic relationships. *Bull. Geol. Soc. Denmark* 30:91–96
3. Andersen NM. 1982. The semiaquatic bugs (Hemiptera, Gerromorpha). Phylogeny, adaptations, biogeography and classification. *Entomonograph* 3:1–455
4. Andersen NM. 1991. Cladistic biogeography of marine water striders (Insecta, Hemiptera) in the Indo-Pacific. *Aust. Syst. Bot.* 4:151–63
5. Andersen NM. 1993. The evolution of wing polymorphism in water striders (Gerridae): a phylogenetic approach. *Oikos* 67:433–43
6. **Andersen NM. 1995. Cladistic inference and evolutionary scenarios: locomotory structure, function, and performance in water striders. *Cladistics* 11:279–95**
7. Andersen NM. 1995. Phylogeny and classification of aquatic bugs (Heteroptera, Nepomorpha). An essay review of Mahner's 'Systema Cryptoceratum Phylogeneticum'. *Entomol. Scand.* 26:159–66
8. Andersen NM. 1997. A phylogenetic analysis of the evolution of sexual dimorphism and mating systems in water striders (Hemiptera: Gerridae). *Biol. J. Linn. Soc.* 61:345–68
9. Andersen NM. 1998. Water striders from the Paleogene of Denmark with a review of the fossil record and evolution of semiaquatic bugs (Hemiptera, Gerromorpha). *K. Dan. Vidensk. Selskab Biol. Skr.* 50:1–157

6. A milestone study using cladistics to test evolutionary scenarios.

10. Andersen NM. 2000. Fossil water striders in the Oligocene/Miocene Dominican amber (Hemiptera: Gerromorpha). *Insect Syst. Evol.* 31:411–31
11. Andersen NM. 2000. The evolution of dispersal dimorphism and other life history traits in water striders (Hemiptera: Gerridae). *Entomol. Sci.* 3:187–99
12. Andersen NM. 2003. Early evolution of a unique structure: a fossil water measurer from Baltic amber (Hemiptera: Gerromorpha: Hydrometridae). *Insect Syst. Evol.* 34:415–26
13. Andersen NM, Grimaldi D. 2001. A fossil water measurer from the mid-Cretaceous Burmese amber (Hemiptera: Gerromorpha: Hydrometridae). *Insect Syst. Evol.* 32:381–92
14. Arnqvist G. 1988. Mate guarding and sperm displacement in the water strider *Gerris lateralis* Schumm. (Heteroptera: Gerridae). *Freshw. Biol.* 19:269–74
15. Arnqvist G, Jones TM, Elgar MA. 2006. Sex-role reversed nuptial feeding reduces male kleptoparasitism of females in Zeus bugs (Heteroptera; Veliidae). *Biol. Lett.* 2:491–93
16. Asquith A. 1994. An unparsimonious origin for the Hawaiian Metrargini (Heteroptera: Lygaeidae). *Ann. Entomol. Soc. Am.* 87:207–13
17. Asquith A, Lattin JD. 1990. *Nabicula (Limnonabis) propinqua* (Reuter) (Heteroptera: Nabidae): dimorphism, phylogenetic relationships and biogeography. *Tijdschr. Voor Entomol.* 133:3–16
18. Azar D, Fleck G, Nel A, Solignac M. 1999. A new enicocephalid bug, *Enicocephalinus acragrimaldii* gen. nov., sp. nov., from the Lower Cretaceous amber of Lebanon (Insecta, Heteroptera, Enicocephalidae). *Estud. Mus. Cienc. Nat. Alava* 14:217–30
19. Barcellos A, Grazia J. 2003. Cladistic analysis and biogeography of *Brachystethus* Laporte (Heteroptera, Pentatomidae, Edessinae). *Zootaxa* 256:1–14
20. Bechly G, Wittmann M. 2000. Two new tropical bugs (Insecta: Heteroptera: Thaumastocoridae—Xylastodorinae and Hypsipterygidae) from Baltic amber. *Stuttg. Beitr. Naturkunde Ser. B (Geol. Palaeontol.)* 289:1–11
21. Betts CR. 1986. The comparative morphology of the wings and axillae of selected Heteroptera. *J. Zool. Ser. B* 1:255–82
22. Beutel RG, Pohl H. 2006. Head structures of males of Strepsiptera (Hexapoda) with emphasis on basal splitting events within the order. *J. Morphol.* 267:536–54
23. Bourgoin T, Campbell BC. 2002. Inferring a phylogeny for Hemiptera: falling into the 'Autapomorphic Trap'. *Denisia* 4:67–82
24. Brailovsky H. 1992. The genus *Laminiceps* and description of three new species (Hemiptera, Heteroptera, Coreidae, Acanthocephalinae). *An. Inst. Biol. Univ. Nac. Auton. Mex. Ser. Zool.* 63:61–74
25. Brailovsky H. 2006. New genus and new species of Cloresmini (Insecta: Heteroptera: Coreidae) from Papua New Guinea. *Pan-Pac. Entomol.* 82:375–80
26. Brozek J, Herczek A. 2004. Internal structure of the mouthparts of true bugs (Hemiptera: Heteroptera). *Pol. Pismo Entomol.* 73:79–106
27. Campbell BC, Steffen-Campbell JD, Sorensen JT, Gill RJ. 1995. Paraphyly of Homoptera and Auchenorrhyncha inferred from 18S rDNA nucleotide sequences. *Syst. Entomol.* 20:175–94
28. Campos LA, Grazia J. 2006. Cladistic analysis and biogeography of Ochlerini (Heteroptera, Pentatomidae, Discocephalinae). *Iheringia Ser. Zool.* 96:147–63
29. Carver M, Gross GF, Woodward TE. 1991. Hemiptera (bugs, leafhoppers, cicadas, aphids, scale insects, etc.). In *The Insects of Australia*, ed. ID Naumann, PB Carne, pp. 429–509. Melbourne, Victoria: Melbourne Univ. Publ.
30. **Cassis G, Schuh RT. 2010. Systematic methods, fossils, and relationships within Heteroptera (Insecta). *Cladistics* 26:262–80**

30. Exemplary for its critique on the use of fossils in cladistics.

31. Cassis G, Schwartz MD, Moulds T. 2003. Systematics and new taxa of the *Vannius* complex (Hemiptera: Miridae: Cylapinae) from the Australian region. *Mem. Qld. Mus.* 49:123–51
32. Cassis G, Silveira R. 2001. A revision and phylogenetic analysis of the *Nerthra alaticollis* species-group (Heteroptera: Gelastocoridae: Nerthrinae). *J. N. Y. Entomol. Soc.* 109:1–46
33. Cassis G, Silveira R. 2002. A revision and phylogenetic analysis of the *Nerthra elongata* species-group (Heteroptera: Gelastocoridae: Nerthrinae). *J. N. Y. Entomol. Soc.* 110:143–81
34. Cobben RH. 1968. *Evolutionary Trends in Heteroptera. Part 1. Eggs, Architecture of the Shell, Gross Embryology and Eclosion.* Wageningen, The Nether.: Cent. Agric. Publ. Documentation. 475 pp.

35. Cobben RH. 1978. Evolutionary trends in Heteroptera. Part 2. Mouth part structures and feeding strategies. *Meded. Landbouwhogesch. Wagening.* 78:5–407
36. Coscaron MdC, Morrone JJ. 1997. Cladistics and biogeography of the assassin bug genus *Melanolestes* Stål (Heteroptera: Reduviidae). *Proc. Entomol. Soc. Wash.* 99:55–59
37. Damgaard J. 2006. Systematics, historical biogeography and ecological phylogenetics in a clade of water striders. *Denisia* 19:813–22
38. Damgaard J. 2008. Phylogeny of the semiaquatic bugs (Hemiptera-Heteroptera, Gerromorpha). *Insect Syst. Evol.* 39:431–60
39. Damgaard J, Andersen NM, Cheng L, Sperling FAH. 2000. Phylogeny of sea skaters, *Halobates* Eschscholtz (Hemiptera, Gerridae), based on mtDNA sequence and morphology. *Zool. J. Linn. Soc.* 130:511–26
40. **Damgaard J, Andersen NM, Meier R. 2005. Combining molecular and morphological analyses of water strider phylogeny (Hemiptera-Heteroptera, Gerromorpha): effects of alignment and taxon sampling. *Syst. Entomol.* 30:289–309**
41. Damgaard J, Andersen NM, Sperling FAH. 2000. Phylogeny of the water strider genus *Aquarius* Schellenberg (Heteroptera: Gerridae) based on nuclear and mitochondrial DNA sequences and morphology. *Insect Syst. Evol.* 31:71–90
42. Damgaard J, Cognato AI. 2003. Sources of character conflict in a clade of water striders (Heteroptera: Gerridae). *Cladistics* 19:512–26
43. Damgaard J, Cognato AI. 2006. Phylogeny and reclassification of species groups in *Aquarius* Schellenberg, *Limnoporus* Stål and *Gerris* Fabricius (Insecta: Hemiptera-Heteroptera, Gerridae). *Syst. Entomol.* 31:93–112
44. Damgaard J, Sperling FAH. 2001. Phylogeny of the water strider genus *Gerris* Fabricius (Heteroptera: Gerridae) based on COI mtDNA, EF-1α nuclear DNA and morphology. *Syst. Entomol.* 26:241–54
45. de Fortes NDF, Grazia J. 2005. Review and cladistic analysis of *Serdia* Stål (Heteroptera, Pentatomidae, Pentatomini). *Rev. Bras. Entomol.* 49:294–339
46. Fairbairn DJ. 1993. Costs of loading associated with mate-carrying in the water strider, *Aquarius remigis*. *Behav. Ecol.* 4:224–31
47. **Fischer C, Mahner M, Wachmann E. 2000. The rhabdom structure in the ommatidia of the Heteroptera (Insecta), and its phylogenetic significance. *Zoomorphology* 120:1–13**
48. **Forero D. 2008. Revision and phylogenetic analysis of the *Hadronema* group (Miridae: Orthotylinae: Orthotylini), with descriptions of new genera and new species, and comments on the neotropical genus *Tupimiris*. *Bull. Am. Mus. Nat. Hist.* 312:5–172**
49. Forero D. 2008. The systematics of the Hemiptera. *Rev. Colomb. Entomol.* 34:1–21
50. Gapud V. 1991. A generic revision of the subfamily Asopinae with consideration of its phylogenetic position in the family Pentatomidae and superfamily Pentatomoidea (Hemiptera-Heteroptera). *Philipp. Entomol.* 8:865–961
51. Garcia BA, Moriyama EN, Powell JR. 2001. Mitochondrial DNA sequences of triatomines (Hemiptera: Reduviidae): phylogenetic relationships. *J. Med. Entomol.* 38:675–83
52. Goodwyn PP, Fujisaki K. 2007. Sexual conflicts, loss of flight, and fitness gains in locomotion of polymorphic water striders. *Entomol. Exp. Appl.* 124:249–59
53. Grazia J. 1997. Cladistic analysis of the *Evoplitus* genus group of Pentatomini (Heteroptera: Pentatomidae). *J. Comp. Biol.* 2:43–48
54. **Grazia J, Schuh RT, Wheeler WC. 2008. Phylogenetic relationships of family groups in Pentatomoidea based on morphology and DNA sequences (Insecta: Heteroptera). *Cladistics* 24:932–76**
55. Grimaldi D, Michalski C, Schmidt K. 1993. Amber fossil Enicocephalidae (Heteroptera) from the Lower Cretaceous of Lebanon and Oligo-Miocene of the Dominican Republic, with biogeographic analysis of *Enicocephalus*. *Am. Mus. Nov.* 3071:1–30
56. Grozeva S, Nokkala S. 1996. Chromosomes and their meiotic behavior in two families of the primitive infraorder Dipsocoromorpha (Heteroptera). *Hereditas* 125:31–36
57. Grozeva SM, Kerzhner IM. 1992. On the phylogeny of aradid subfamilies (Heteroptera, Aradidae). *Acta Zool. Hung.* 38:199–205

40. Presents the first combined analysis of Gerromorpha.

47. Documents the value of exploratory morphological studies for systematics.

48. Tightly integrates a complex taxonomic monograph with cladistic approaches.

54. The first study to combine morphological and molecular data within Pentatomomorpha.

58. Guilbert E. 2001. Phylogeny and evolution of exaggerated traits among the Tingidae (Heteroptera, Cimicomorpha). *Zool. Scr.* 30:313–24
59. Guilbert E. 2005. Morphology and evolution of larval outgrowths of Tingidae (Insecta, Heteroptera), with description of new larvae. *Zoosystema* 27:95–113
60. Hardling R, Kaitala A. 2001. Conflict of interest between sexes over cooperation: a supergame on egg carrying and mating in a coreid bug. *Behav. Ecol.* 12:659–65
61. Hebsgaard MB, Andersen NM, Damgaard J. 2004. Phylogeny of the true water bugs (Nepomorpha: Hemiptera-Heteroptera) based on 16S and 28S rDNA and morphology. *Syst. Entomol.* 29:488–508
62. Heiss E. 1990. A review of the genus *Dysodius* Lepeletier and Serville 1828 with descriptions of two new species (Heteroptera, Aradidae). *An. Inst. Biol. Univ. Nac. Auton. Mex. Ser. Zool.* 61:279–96
63. Heiss E. 2008. A new genus and two species of apterous Mezirinae from Madagascar (Heteroptera, Aradidae). *Linz. Biol. Beitr.* 40:303–9
64. Henry TJ. 1991. Revision of *Keltonia* and the cotton fleahopper genus *Pseudatomoscelis*, with the description of a new genus and an analysis of their relationships (Heteroptera, Miridae, Phylinae). *J. N. Y. Entomol. Soc.* 99:351–404
65. Henry TJ. 1997. Cladistic analysis and revision of the stilt bug genera of the world (Heteroptera: Berytidae). *Contrib. Am. Entomol. Inst.* 30:1–100
66. **Henry TJ. 1997. Phylogenetic analysis of family groups within the infraorder Pentatomomorpha (Hemiptera: Heteroptera), with emphasis on the Lygaeoidea. *Ann. Entomol. Soc. Am.* 90:275–301**

66. Recognizes the Lygaeidae to be paraphyletic and implements a new classification based on monophyletic groups.

67. Hill L. 1990. A revision of Australian *Pachyplagia* Gross (Heteroptera: Schizopteridae). *Invertebr. Taxon.* 3:605–17
68. Hill L. 2004. *Kaimon* (Heteroptera: Schizopteridae), a new, speciose genus from Australia. *Mem. Qld. Mus.* 49:603–47
69. Hosokawa T, Kikuchi Y, Nikoh N, Shimada M, Fukatsu T. 2006. Strict host-symbiont cospeciation and reductive genome evolution in insect gut bacteria. *PLoS Biol.* 4:1841–51
70. **Hua J, Li M, Dong P, Cui Y, Xie Q, Bu W. 2008. Comparative and phylogenomic studies on the mitochondrial genomes of Pentatomomorpha (Insecta: Hemiptera: Heteroptera). *BMC Genomics* 9:610**

70. The first study in Heteroptera to use mitochondrial genomes for cladistic analysis.

71. Hua J, Li M, Dong P, Cui Y, Xie Q, Bu W. 2009. Phylogenetic analysis of the true water bugs (Insecta: Hemiptera: Heteroptera: Nepomorpha): evidence from mitochondrial genomes. *BMC Evol. Biol.* 9:11
72. Hypsa V, Tietz DF, Zrzavy J, Rego ROM, Galvao C, Jurberg J. 2002. Phylogeny and biogeography of Triatominae (Hemiptera: Reduviidae): molecular evidence of a New World origin of the Asiatic clade. *Mol. Phylogenet. Evol.* 23:447–57
73. Keffer SL. 2004. Morphology and evolution of waterscorpion male genitalia (Heteroptera: Nepidae). *Syst. Entomol.* 29:142–72
74. Kerzhner IM. 1981. Nasekomye khobotnye (Insecta: Rhynchota). In *Fauna SSSR*, Vol. 13(2)1–326. Leningrad: Nauka. 326 pp. In Russian
75. Kerzhner IM, Konstantinov FV. 1999. Structure of the aedeagus in Miridae (Heteroptera) and its bearing to suprageneric classification. *Acta Soc. Zool. Bohemoslov.* 63:117–37
76. Kikuchi Y, Hosokawa T, Nikoh N, Meng X-Y, Kamagata Y, Fukatsu T. 2009. Host-symbiont cospeciation and reductive genome evolution in gut symbiotic bacteria of acanthosomatid stinkbugs. *BMC Biol.* 7:2
77. Lee W, Kang J, Jung C, Hoelmer K, Lee S, Lee S. 2009. Complete mitochondrial genome of brown marmorated stink bug *Halyomorpha halys* (Hemiptera: Pentatomidae), and phylogenetic relationships of hemipteran suborders. *Mol. Cells* 28:155–65
78. Li H-M. 2007. Phylogenetic research on the Lygaeoidea (Hemiptera: Heteroptera) based on mt16S rDNA sequences. *Guangzhou Daxue Xuebao Ziran Kexue Ban* 6:30–34
79. Li H-M, Deng R-Q, Wang J-W, Chen Z-Y, Jia F-L, Wang X-Z. 2005. A preliminary phylogeny of the Pentatomomorpha (Hemiptera: Heteroptera) based on nuclear 18S rDNA and mitochondrial DNA sequences. *Mol. Phylogenet. Evol.* 37:313–26
80. Li X-Z. 1997. Cladistic analysis of the phylogenetic relationships among the tribal rank taxa of Coreidae (Hemiptera-Heteroptera: Coreoidea). *Acta Zootaxonomica Sin.* 22:60–68

81. Lindskog P, Polhemus JT. 1992. Taxonomy of *Saldula*: revised genus and species group definitions, and a new species of the pallipes group from Tunisia (Heteroptera: Saldidae). *Entomol. Scand.* 23:63–88
82. Lis B. 1999. Phylogeny and classification of Cantacaderini (=Cantacaderidae stat. nov.) (Hemiptera: Tingoidea). *Ann. Zool.* 49:157–96
83. Lis B. 2004. Comparative studies on the ductus seminis of aedeagus in Tingoidea (Hemiptera: Heteroptera). *Pol. Pismo Entomol.* 73:245–58
84. Lis JA, Hohol-Kilinkiewicz A. 2002. Abdominal trichobothrial pattern and its taxonomic and phylogenetic significance in Cydnidae (Hemiptera: Heteroptera). IV. Cydninae. *Pol. Pismo Entomol.* 71:133–43
85. Lis JA, Hohol-Kilinkiewicz A. 2002. Adult dorso-abdominal scent glands in the burrower bugs (Hemiptera: Heteroptera: Cydnidae). *Pol. Pismo Entomol.* 71:359–95
86. Lis JA, Pluot-Sigwalt D. 2002. Nymphal and adult cephalic chaetotaxy of the Cydnidae (Hemiptera: Heteroptera), and its adaptive, taxonomic and phylogenetic significance. *Eur. J. Entomol.* 99:99–109
87. Lis JA, Schaefer CW. 2005. Tibial combs in the Cydnidae (Hemiptera: Heteroptera) and their functional, taxonomic and phylogenetic significance. *J. Zool. Syst. Evol. Res.* 43:277–83
88. Mahner M. 1993. Systema Cryptoceratorum Phylogeneticum (Insecta, Heteroptera). *Zoologica* 48:1–302
89. Monteiro FA, Wesson DM, Dotson EM, Schofield CJ, Beard CB. 2000. Phylogeny and molecular taxonomy of the Rhodniini derived from mitochondrial and nuclear DNA sequences. *Am. J. Trop. Med. Hyg.* 62:460–65
90. Morrone JJ, Coscaron MDC. 1998. Cladistics and biogeography of the assassin bug genus *Rasahus* Amyot & Serville (Heteroptera: Reduviidae: Peiratinae). *Zool. Med.* 72:73–87
91. Muraji M, Tachikawa S. 2000. Phylogenetic analysis of water striders (Hemiptera: Gerroidea) based on partial sequences of mitochondrial and nuclear ribosomal RNA genes. *Entomol. Sci.* 3:615–26
92. Ouvrard D, Campbell BC, Bourgoin T, Chan KL. 2000. 18S rRNA secondary structure and phylogenetic position of Peloridiidae (Insecta, Hemiptera). *Mol. Phylogenet. Evol.* 16:403–17
93. Pass G, Gereben-Krenn B-A, Merl M, Plant J, Szucsich NU, Toegel M. 2006. Phylogenetic relationships of the orders of hexapoda: contributions from the circulatory organs for a morphological data matrix. *Arthropod Syst. Phylogeny* 64:165–203
94. Perez-Goodwyn PJ, Ohba S, Schnack JA. 2006. Chorion morphology of the eggs of *Lethocerus delpontei*, *Kirkaldyia deyrolli*, and *Horvathinia pelocoroides* (Heteroptera: Belostomatidae). *Russ. Entomol. J.* 15:151–56
95. Pluot-Sigwalt D, Lis JA. 2008. Morphology of the spermatheca in the Cydnidae (Hemiptera: Heteroptera): bearing of its diversity on classification and phylogeny. *Eur. J. Entomol.* 105:279–312
96. Pluot-Sigwalt D, Pericart J. 2003. The spermatheca of the Dipsocoridae with special reference to the strange "loculus capsulae" in *Harpago* species (Heteroptera, Dipsocoromorpha). *Ann. Soc. Entomol. Fr.* 39:129–38
97. Popov YA, Putshkov PV. 1998. *Redubinotus liedtkei* n. gen. n. sp.—the second Centrocnemina from the Baltic amber (Heteroptera: Reduviidae, Centrocnemidinae). *Ann. Upper Silesian Mus. Bytom Entomol.* 8–9:205–10
98. Popov YA, Shcherbakov DE. 1991. Mesozoic Peloridioidea and their ancestors (Insecta: Hemiptera, Coleorrhyncha). *Geol. Palaeontol.* 25:215–35
99. Popov YA, Shcherbakov DE. 1996. Origin and evolution of the Coleorrhyncha as shown by the fossil record. In *Studies on Hemipteran Phylogeny*, ed. CW Schaefer, pp. 9–30. Lanham, MD: Entomol. Soc. Am.
100. Pruthi HS. 1925. The morphology of the male genitalia in Rhynchota. *Trans. Entomol. Soc. Lond.* 1925:127–254
101. Redei D. 2007. A new species of the family Hypsipterygidae from Vietnam, with notes on the hypsipterygid fore wing venation (Heteroptera, Dipsocoromorpha). *Dtsch. Entomol. Z.* 54:43–50
102. Rieger C. 1976. Skeleton and musculature of the head and prothorax of *Ochterus marginatus* contribution towards clarification of the phylogenetic relationships of the Ochteridae Insecta Heteroptera. *Zoomorphologie* 82:109–91
103. Schaefer CW. 1993. The Pentatomomorpha (Hemiptera: Heteroptera): an annotated outline of its systematic history. *Eur. J. Entomol.* 90:105–22

104. Schlee D. 1969. Morphology and symbiosis its use in the determination of the relationships of Coleorrhyncha Insecta Hemiptera phylogenetic studies of Hemiptera. Part 4. Heteropteroidea Heteroptera and Coleorrhyncha as a monophyletic group. *Stuttg. Beitr. Naturkunde* 210:1–27
105. Schofield CJ, Galvao C. 2009. Classification, evolution, and species groups within the Triatominae. *Acta Trop.* 110:88–100
106. Schuh RT. 1975. The structure distribution and taxonomic importance of trichobothria in the Miridae Hemiptera Heteroptera. *Am. Mus. Nov.* 2585:1–26
107. Schuh RT. 1976. Pretarsal structure in the Miridae Hemiptera Heteroptera with a cladistic analysis of relationships within the family. *Am. Mus. Nov.* 2601:1–39
108. Schuh RT. 1986. The influence of cladistics on heteropteran classification. *Annu. Rev. Entomol.* 31:67–94
109. Schuh RT. 1991. Phylogenetic, host, and biogeographic analyses of the Pilophorini (Heteroptera Miridae Phylinae). *Cladistics* 7:157–90
110. Schuh RT. 2006. Revision, phylogenetic, biogeographic, and host analyses of the endemic Western North American *Phymatopsallus* group, with the description of 9 new genera and 15 new species (Insecta: Hemiptera: Miridae: Phylinae). *Bull. Am. Mus. Nat. Hist.* 301:1–115
111. Schuh RT, Cassis G, Guilbert E. 2006. Description of the first recent macropterous species of Vianaidinae (Heteroptera: Tingidae) with comments on the phylogenetic relationships of the family within the Cimicomorpha. *J. N. Y. Entomol. Soc.* 114:38–53
112. Schuh RT, Polhemus JT. 1980. Analysis of taxonomic congruence among morphological, ecological, and biogeographic data sets for the Leptopodomorpha (Hemiptera). *Syst. Zool.* 29:1–26
113. Schuh RT, Polhemus JT. 2009. Revision and analysis of *Pseudosaldula* Cobben (Insecta: Hemiptera: Saldidae): a group with a classic Andean distribution. *Bull. Am. Mus. Nat. Hist.* 323:1–102
114. Schuh RT, Slater JA. 1995. *True Bugs of the World (Hemiptera: Heteroptera): Classification and Natural History*. Ithaca, NY: Cornell Univ. Press. 336 pp.
115. Schuh RT, Stonedahl GM. 1986. Historical biogeography in the Indo-Pacific: a cladistic approach. *Cladistics* 2:337–55
116. Schuh RT, Stys P. 1991. Phylogenetic analysis of cimicomorphan family relationships (Heteroptera). *J. N. Y. Entomol. Soc.* 99:298–350
117. Schuh RT, Weirauch C, Henry TJ, Halbert SE. 2008. Curaliidae, a new family of Heteroptera (Insecta: Hemiptera) from the Eastern United States. *Ann. Entomol. Soc. Am.* 101:20–29
118. Schuh RT, Weirauch C, Wheeler WC. 2009. Phylogenetic relationships within the Cimicomorpha (Hemiptera: Heteroptera): a total-evidence analysis. *Syst. Entomol.* 34:15–48
119. Schwartz MD. 2008. Revision of the Stenodemini with a review of the included genera (Hemiptera: Heteroptera: Miridae: Mirinae). *Proc. Entomol. Soc. Wash.* 110:1111–201
120. Schwartz MD, Foottit RG. 1998. *Revision of the Nearctic Species of the Genus* Lygus *Hahn, with a Review of the Palaearctic Species (Heteroptera: Miridae). Memoirs on Entomology, International.* Vol. 10 Gainesville, FL: Assoc. Publ. 428 pp.
121. Scherbakov DE, Popov YA. 2002. Superorder Cimicidea Laicharting, 1781. Order Hemiptera Linne, 1758. The bugs, cicadas, plantlice, scale insects, etc. In *History of Insects*, ed. AP Rasnitsyn, DLJ Quicke, pp. 143–57. Dordrecht, The Nether.: Kluwer Academic. 517 pp.
122. Silva de Paula A, Diotaiuti L, Galvao C. 2007. Systematics and biogeography of Rhodniini (Heteroptera: Reduviidae: Triatominae) based on 16S mitochondrial rDNA sequences. *J. Biogeogr.* 34:699–712
123. Silva de Paula A, Diotaiuti L, Schofield CJ. 2005. Testing the sister-group relationship of the Rhodniini and Triatomini (Insecta: Hemiptera: Reduviidae: Triatominae). *Mol. Phylogenet. Evol.* 35:712–18
124. Slater A. 1992. A genus level revision of Western Hemisphere Lygaeinae (Heteroptera: Lygaeidae) with keys to species. *Univ. Kans. Sci. Bull.* 55:1–56
125. Slater JA. 1993. A new genus and two new species of Drymini from Africa (Hemiptera: Lygaeidae). *J. Afr. Zool.* 107:373–81
126. Slater JA, Schuh RT, Cassis G, Johnson CA, Pedraza-Penalosa P. 2009. Revision of *Laryngodus* Herrich-Schaeffer, an *Allocasuarina* feeder, with comments on its biology and the classification of the family (Heteroptera: Lygaeoidea: Rhyparochromidae). *Invertebr. Syst.* 23:111–33

127. Smith RL. 1997. Evolution of paternal care in the giant water bugs (Heteroptera: Belostomatidae). In *The Evolution of Social Behavior in Insects and Arachnids*, ed. JC Choe, BJ Crespi, pp. 116–49. Cambridge, UK: Cambridge Univ. Press
128. Sorensen JT, Campbell BC, Gill RJ, Steffen-Campbell JD. 1995. Non-monophyly of Auchenorrhyncha ("Homoptera"), based upon 18S rDNA phylogeny: eco-evolutionary and cladistic implications within pre-Heteropterodea Hemiptera (s.l.) and a proposal for new monophyletic suborders. *Pan-Pac. Entomol.* 71:31–60
129. Stonedahl GM. 1990. Revision and cladistic analysis of the Holarctic genus *Atractotomus* Fieber (Heteroptera, Miridae, Phylinae). *Bull. Am. Mus. Nat. Hist.* 198:1–88
130. Stys P. 1983. A new family of Heteroptera with dipsocoromorphan affinities from Papua New-Guinea. *Acta Entomol. Bohemoslov.* 80:256–92
131. Stys P. 1984. Phylogeny and classification of lower Heteroptera. *Int. Congr. Entomol. Proc.* 17:12
132. Stys P. 1984. Plural origin of the male intromittent organ in Heteroptera. *Int. Congr. Entomol. Proc.* 17:49
133. Stys P, Buning J, Bilinski SM. 1998. Organization of the tropharia in the telotrophic ovaries of the dipsocaormophan bugs *Cryptostemma alienum* Herrich-Schaeffer and *C. carpaticum* Josifov (Heteroptera: Dipsocoridae). *Int. J. Insect Morphol Embryol.* 27:129–33
134. Stys P, Kerzhner I. 1975. The rank and nomenclature of higher taxa in recent Heteroptera. *Acta Entomol. Bohemoslov.* 72:65–79
135. Sweet MH. 2006. Justification for the Aradimorpha as an infraorder of the suborder Heteroptera (Hemiptera, Prosorrhyncha) with special reference to the pregenital abdominal structure. *Denisia* 19:225–48
136. Tallamy DW, Schaefer C. 1997. Maternal care in the Hemiptera: ancestry, alternatives, and current adaptive value. In *The Evolution of Social Behavior in Insects and Arachnids*, ed. JC Choe, BJ Crespi, pp. 94–115. Cambridge, UK: Cambridge Univ. Press
137. Tatarnic NJ, Cassis G, Hochuli DF. 2006. Traumatic insemination in the plant bug genus *Coridromius* Signoret (Heteroptera: Miridae). *Biol. Lett.* 2:58–61
138. Tian Y, Zhu W, Li M, Xie Q, Bu W. 2008. Influence of data conflict and molecular phylogeny of major clades in Cimicomorphan true bugs (Insecta: Hemiptera: Heteroptera). *Mol. Phylogenet. Evol.* 47:581–97
139. Tinerella PP. 2008. Taxonomic revision and systematics of New Guinea and Oceania pygmy water boatmen (Hemiptera: Heteroptera: Corixoidea: Micronectidae). *Zootaxa* 1797:1–66
140. van Doesburg PH. 1968. A revision of the New World species of *Dysdercus* Guérin Méneville (Heteroptera, Pyrrhocoridae). *Zool. Verhand.* 97:373–424
141. Vasarhelyi T. 1987. On the relationships of the eight aradid subfamilies (Heteroptera). *Acta Zool. Acad. Sci. Hung.* 33:263–67
142. Wappler T, Andersen NM. 2004. Fossil water striders from the Middle Eocene fossil sites of Eckfeld and Messel, Germany (Hemiptera, Gerromorpha). *Palaeontol. Z.* 78:41–52
143. Weirauch C. 2003. Glandular areas associated with the male genitalia in *Triatoma rubrofasciata* (Triatominae, Reduviidae, Hemiptera) and other Reduviidae. *Mem. Inst. Oswaldo Cruz* 98:773–76
144. Weirauch C. 2005. Pretarsal structures in Reduviidae (Heteroptera, Insecta). *Acta Zool.* 86:91–110
145. Weirauch C. 2006. Dorsal abdominal glands in adult Reduviidae (Heteroptera, Cimicomorpha). *Dtsch. Entomol. Z.* 53:91–102
146. Weirauch C. 2006. Metathoracic glands and associated evaporatory structures in Reduvioidea (Heteroptera: Cimicomorpha), with observation on the mode of function of the metacoxal comb. *Eur. J. Entomol.* 103:97–108
147. Weirauch C. 2007. Hairy attachment structures in Reduviidae (Cimicomorpha, Heteroptera), with observations on the fossula spongiosa in some other Cimicomorpha. *Zool. Anz.* 246:155–75
148. Weirauch C. 2007. Revision and cladistic analysis of the *Polyozus* group of Australian Phylini (Heteroptera: Miridae: Phylinae). *Am. Mus. Nov.* 3590:1–60
149. Weirauch C. 2008. Cladistic analysis of Reduviidae (Heteroptera: Cimicomorpha) based on morphological characters. *Syst. Entomol.* 33:229–74
150. Weirauch C, Cassis G. 2009. Frena and druckknopf: a synopsis of two forewing-to-body coupling mechanisms in Heteropterodea (Hemiptera). *Insect Syst. Evol.* 40:229–52

149. Proposes and tests relationships within one of the largest families of true bugs.

151. Weirauch C, Munro JB. 2009. Molecular phylogeny of the assassin bugs (Hemiptera: Reduviidae), based on mitochondrial and nuclear ribosomal genes. *Mol. Phylogenet. Evol.* 53:287–99
152. Wheeler QD. 2008. Undisciplined thinking: morphology and Hennig's unfinished revolution. *Syst. Entomol.* 33:2–7
153. Wheeler WC, Schuh RT, Bang R. 1993. Cladistic relationships among higher groups of Heteroptera: congruence between morphological and molecular data sets. *Entomol. Scand.* 24:121–37

153. The first combined morphological and molecular analysis within Heteroptera.

154. Wirkner CS, Prendini L. 2007. Comparative morphology of the hemolymph vascular system in scorpions—a survey using corrosion casting, MicroCT, and 3D-reconstruction. *J. Morphol.* 268:401–13
155. Wootton RJ, Betts CR. 1986. Homology and function in the wings of Heteroptera. *Syst. Entomol.* 11:389–400
156. Wygodzinsky PW, Schmidt K. 1991. Revision of the New World Enicocephalomorpha (Heteroptera). *Bull. Am. Mus. Nat. Hist.* 200:1–265
157. Xie Q, Bu W, Zheng L. 2005. The Bayesian phylogenetic analysis of the 18S rRNA sequences from the main lineages of Trichophora (Insecta: Heteroptera: Pentatomomorpha). *Mol. Phylogenet. Evol.* 34:448–51
158. Xie Q, Tian Y, Zheng L, Bu W. 2008. 18S rRNA hyper-elongation and the phylogeny of Euhemiptera (Insecta: Hemiptera). *Mol. Phylogenet. Evol.* 47:463–71
159. Yang C-T. 2002. Preliminary thoughts on the phylogeny of Coleorrhyncha-Heteroptera (Hemiptera). *Formos. Entomol.* 22:297–305
160. Zrzavy J. 1992. Evolution of antennae and historical ecology of the hemipteran insects Paraneoptera. *Acta Entomol. Bohemoslov.* 89:77–86
161. Zrzavy J, Nedved O. 1997. Phylogeny of the New World *Dysdercus* (Insecta: Hemiptera: Pyrrhocoridae) and evolution of their colour patterns. *Cladistics* 13:109–23
162. Zrzavy J, Nedved O. 1999. Evolution of mimicry in the New World *Dysdercus* (Hemiptera: Pyrrhocoridae). *J. Evol. Biol.* 12:956–69

Cumulative Indexes

Contributing Authors, Volumes 47–56

D

E

F

G

H

I

J

K

L

M

N

O

P

R

S

T

Chapter Titles, Volumes 47–56

Behavior and Neuroscience

Biochemistry and Physiology

Biological Control

Bionomics (See also Ecology)

Ecology (See also Bionomics; Behavior and Neuroscience)

Forest Entomology

Medical and Veterinary Entomology

Miscellaneous

Morphology and Development

Vectors of Plant Pathogens